T0137330

Smart Innovation, Systems and Technologies

Volume 100

Series editors

Robert James Howlett, Bournemouth University and KES International,
Shoreham-by-sea, UK
e-mail: rjhowlett@kesinternational.org

Lakhmi C. Jain, University of Technology Sydney, Broadway, Australia;
University of Canberra, Canberra, Australia; KES International, UK
e-mail: jainlakhmi@gmail.com; jainlc2002@yahoo.co.uk

The Smart Innovation, Systems and Technologies book series encompasses the topics of knowledge, intelligence, innovation and sustainability. The aim of the series is to make available a platform for the publication of books on all aspects of single and multi-disciplinary research on these themes in order to make the latest results available in a readily-accessible form. Volumes on interdisciplinary research combining two or more of these areas is particularly sought.

The series covers systems and paradigms that employ knowledge and intelligence in a broad sense. Its scope is systems having embedded knowledge and intelligence, which may be applied to the solution of world problems in industry, the environment and the community. It also focusses on the knowledge-transfer methodologies and innovation strategies employed to make this happen effectively. The combination of intelligent systems tools and a broad range of applications introduces a need for a synergy of disciplines from science, technology, business and the humanities. The series will include conference proceedings, edited collections, monographs, handbooks, reference books, and other relevant types of book in areas of science and technology where smart systems and technologies can offer innovative solutions.

High quality content is an essential feature for all book proposals accepted for the series. It is expected that editors of all accepted volumes will ensure that contributions are subjected to an appropriate level of reviewing process and adhere to KES quality principles.

More information about this series at http://www.springer.com/series/8767

Francesco Calabrò · Lucia Della Spina
Carmelina Bevilacqua
Editors

New Metropolitan Perspectives

Local Knowledge and Innovation Dynamics Towards
Territory Attractiveness Through the Implementation
of Horizon/E2020/Agenda2030 – Volume 1

 Springer

Editors
Francesco Calabrò
Mediterranea University of Reggio Calabria
Reggio Calabria
Italy

Carmelina Bevilacqua
University of Reggio Calabria
Reggio Calabria
Italy

Lucia Della Spina
Mediterranea University of Reggio Calabria
Reggio Calabria
Italy

This Volume is part of a project that has received funding from the European Union's Horizon 2020 research and innovation programme under grant agreement N°645651

ISSN 2190-3018 ISSN 2190-3026 (electronic)
Smart Innovation, Systems and Technologies
ISBN 978-3-030-06362-7 ISBN 978-3-319-92099-3 (eBook)
https://doi.org/10.1007/978-3-319-92099-3

© Springer International Publishing AG, part of Springer Nature 2019
Softcover re-print of the Hardcover 1st edition 2019
This work is subject to copyright. All rights are reserved by the Publisher, whether the whole or part of the material is concerned, specifically the rights of translation, reprinting, reuse of illustrations, recitation, broadcasting, reproduction on microfilms or in any other physical way, and transmission or information storage and retrieval, electronic adaptation, computer software, or by similar or dissimilar methodology now known or hereafter developed.
The use of general descriptive names, registered names, trademarks, service marks, etc. in this publication does not imply, even in the absence of a specific statement, that such names are exempt from the relevant protective laws and regulations and therefore free for general use.
The publisher, the authors and the editors are safe to assume that the advice and information in this book are believed to be true and accurate at the date of publication. Neither the publisher nor the authors or the editors give a warranty, express or implied, with respect to the material contained herein or for any errors or omissions that may have been made. The publisher remains neutral with regard to jurisdictional claims in published maps and institutional affiliations.

Printed on acid-free paper

This Springer imprint is published by the registered company Springer International Publishing AG part of Springer Nature
The registered company address is: Gewerbestrasse 11, 6330 Cham, Switzerland

Preface

This volume contains the proceedings for the third International *"NEW METROPOLITAN PERSPECTIVES. Local Knowledge and Innovation dynamics towards territory attractiveness through the implementation of Horizon/Europe2020/Agenda2030"*, which took place on 22–25 May 2018 in Reggio Calabria, Italy.

The Symposium is jointly promoted by LaborEst (Evaluation and Economic Appraisal Lab) and CLUDs (Commercial Local Urban Districts Lab), Laboratories of the PAU Department, *Mediterranea* University of Reggio Calabria, Italy, in partnership with a qualified international network of academic institution and scientific societies.

The third edition of *"NEW METROPOLITAN PERSPECTIVES"* aims to deepen those factors which contribute to increase cities and territories attractiveness, both with theoretical studies and tangible applications.

It represents the conclusive event of the Multidisciplinary Approach to Plan Smart Specialisation Strategies for Local Economic Development (MAPS-LED) Research Project funded by the European Union's Horizon 2020 Research and Innovation Programme under the Marie Skłodowska Curie Actions—RISE 2014.

This edition of the Symposium is going to give a specific attention to those linkages between innovation dynamics and territories attractiveness, as will be better explained by our colleague Carmelina Bevilacqua.

In the last decades, metropolitan cities have been studied from different perspectives, according to diverse academic and scientific points of view, but under the common attitude towards their spatial dynamics.

Recent economic and political developments press the scientific community addressing two issues of current relevance:

- The spatial implications of the economic and demographic decline of large areas in Europe and Western Countries;
- The impact of ICT dissemination on urban/rural environment and, broadly, on the idea of society.

For decades, technical tools, especially in the field of urban planning, have been developed to allow urban and territorial transformations in a context characterized by expansive dynamics. Looking at the productive system and, within it, the job's organization as one of the discriminating elements of territorial transformation, the following question arises: what is the destiny of the industrial and post-industrial city, known as a place of concentration of workforce and market?

These considerations focus the attention of the academic community on the dimension of distant future. The spread of new communication technologies and new production systems is increasingly pushing everywhere towards the progressive "liquefaction" of the social structures, organizational models and systems that have been known so far as Bauman's intuition. Such a long horizon necessarily requires the renewal of a visionary, utopian vision that imagines society of the future through a dreamlike dimension of avant-garde. It becomes crucial to debate the running direction; the profound changes are going on in contemporary society and its impact on urban/rural environment of the future.

We might suggest the anthropic desertification is a phenomenon shared among the lagging Regions: increasingly, many people move from their hometowns to reach better places, such as the metropolitan areas, to improve their conditions of life. Such process inevitably contributes to the general poorness of those Regions, already weakened, by increasing such declining status even more.

One of the most important topics to be considered, which more than others characterizes all metropolitan regions, is surely their capacity to attract people, and consequently capitals. Indeed, territorial policies aim mostly to catch investments in order to enhance job creation and to positively influence socio-economic indicators. Nevertheless, attractiveness, as explained, is also about people: that is the indicator which can really synthesize a concept which includes both competitiveness and receiving capability.

If we go deeper, competitiveness means: research and innovation, public administration efficiency, skilled workforce, facilities, accessibility, credit access, international perspectives, energy cost consumption. As far as the receiving capability, it could be explained as carefulness for urban quality, housing policies, mobility, welfare, health care, security and, of course, job opportunities.

Particularly, the papers accepted, about 150, allowed us to develop six macro-topics, about *"Local Knowledge and Innovation dynamics towards territory attractiveness"* as follows:

1 Innovation dynamics, smart cities, ICT;
2 Urban regeneration, community-led practices and PPP;
3 Local development, inland and urban areas in territorial cohesion strategies;
4 Mobility, accessibility, infrastructures;
5 Heritage, landscape and identity; and
6 Risk management, environment, energy.

We are pleased that the International Symposium NMP, thanks to its interdisciplinary character, stimulates growing interests and approvals from the scientific community, at national and international levels.

We would like to take this opportunity to thank all who have contributed to the success of the third International Symposium "NEW METROPOLITAN PERSPECTIVES. *Local Knowledge and Innovation dynamics towards territory attractiveness through the implementation of Horizon/Europe2020*": authors, keynote speakers, session chairs, referees, the scientific committee and the scientific partners, the "Associazione ASTRI" for technical and organizational support activities, participants, student volunteers and those ones that with different roles have contributed to the dissemination and the success of the Symposium; particularly, the academic representatives of the University of Reggio Calabria: the Rector Prof. Pasquale Catanoso, the Vice Rector Prof. Marcello Zimbone, the responsible of internationalization Prof. Francesco Morabito, the Chief of PAU Department Prof. Francesca Martorano.

Thank you very much for your support.

Last but not least, we would like to thank Springer for the support in the conference proceedings publication.

Francesco Calabrò
Lucia Della Spina

Local Knowledge and Innovation Dynamics: The MAPS-LED Perspective

The third edition of the International Symposium "New Metropolitan Perspectives" aims at facing the challenges of Local Knowledge and Innovation dynamics towards territory attractiveness through the implementation of the Horizon/EU2020 Agenda. The Symposium is jointly promoted by the LaborEst and the CLUDsLab Laboratories of the PAU Department, Università Mediterranea of Reggio Calabria (IT), in partnership with a qualified international network of prestigious academic institutions and scientific associations. It represents the conclusive event of the Multidisciplinary Approach to Plan Smart Specialisation Strategies for Local Economic Development (MAPS-LED) Research Project funded by the European Union's Horizon 2020 Research and Innovation Programme, under the Marie Skłodowska Curie Actions—RISE 2014. The main aim of RISE Action is to favour the mobility of experienced and early-stage researchers between Europe, associated and third countries. The project empowers the strong international research network built up with the CLUDs Project (7FP) through the exchange of researchers, ideas and practices between EU and USA. To date, about 40 experienced and early-stage researchers benefited by the project mobility towards USA, at the Northeastern University of Boston and the San Diego State University. The researchers, coming from the Higher Education Institutions (HEIs) belonging to the MAPS-LED network, had the opportunity to increase their research, training and networking skills thanks to the high exposure to the international scientific community. The majority of the early-stage researchers belong to the International Doctorate Program in Urban Regeneration and Economic Development (URED), active since 2012 at the Università Mediterranea of Reggio Calabria (Project Coordinator). The Program is funded by the Calabria Region European Social Fund (ESF), making effective the operative linkage between Horizon 2020 (Research) and Cohesion Policy (ESIF).

The MAPS-LED Symposium represents an important event for disseminating research findings and for stimulating a fruitful debate among scientific and policy-makers' community.

The core of the research activities has earmarked for exploring how Smart Specialisation Strategies (S3) can be implemented by incorporating the place-based approach towards regenerating local economies.

The S3 has been designed in order to capture knowledge and innovation dynamics strictly connected with characteristics of context. According to the Maps-led perspective, the key concepts of S3 lie in the mutual correlation among entrepreneur, innovation and economic development. The entrepreneur is pushed by a local entrepreneurial culture activated by enhancing local knowledge. This process is called "entrepreneurial discovery" towards knowledge convergence and informational spillover for clustering phase, as precondition of competitive advantages.

Among the theoretical standpoints that explained how cluster policy and S3 share many similarities in their rationale, the research activities led to focus on the place-based approach as nexus in spurring the innovation process towards emphasizing the role of the city.

Thanks to the exchange scheme of the RISE programme, the MAPS-LED project has delivered a methodology to spatialize economic clusters in Boston and San Diego, as expression of how innovation is experimented in the modern economy and how the "place" works.

The "spatialization cluster methodology" has brought about a proxy for inno-vation concentration, by turning clusters in physical configurations at city level. This interpretation comes from the rationale grounded into cluster definition, val-idated by Porter with the model in which innovation, specialization and job creation are connected among those productive sectors related to shaping a cluster. The preliminary research findings pushed towards the explanation of how cluster per-formance factors can be combined with the context characteristics, by highlighting the spatial implications of knowledge dynamics. The case studies have been grouped into two frameworks of cluster rationale—Traded, to enhance competitive advantages, and Local, to reinforce comparative advantages. In synthesis, the first framework considers innovation as the main drive to define the relativeness of productive sectors to shape traded cluster, and the second ones bring into spe-cialization the main impulse in forming local cluster.

The spatially oriented methodology adopted for Traded clusters in the Boston area analysed the occurrence of "innovation spaces" in the places characterized by the presence of cluster, in order to identify specific urban areas (target areas) in which investigating the interaction of cluster (demand of innovation) with the urban fabric, its sociability and sustainability. The findings from "target areas" analysis allowed, on one hand, at identifying the link between city and S3 by introducing the innovation-driven urban policy as an important phase of the Entrepreneurial Discovery Process (EDP). On the other hand, gentrification and inequality issues resulted as the main negative effects in both cities, Boston and Cambridge, due to the evident increase, more than proportional, of the rent and property values.

The link between city and S3 is mainly stemmed by the emerging business environment or the atmosphere for innovation that acquires an important role in what Foray calls *structuring entrepreneurial knowledge*. Inside the "target areas", anchors institutions, public and private research centres, the entrepreneurs' com-munity and citizens concentrate their efforts supported by public policies (economic development and urban planning). The occurrence of such dynamic forces, able to trigger socio-economic and physical transformation, has brought to investigate how innovation policy can be harnessed in driving growth in specific localities. This aspect called for a better understanding and the exploration of innovation as a source for socio-economic and urban transformation, highlighting urban regeneration initiatives driven by the increasing demand for innovation[1].

The analysis of surrounding conditions has been considered important to give a practical explanation of how the EDP could be structured as policy action. The role of the city has emerged in spurring the innovation process and, in particular, how it can be the start point of the EDP, in terms of public policy action. The possible result of these research activities lies in finding a new concept of *urban dimension* within S3. The urban dimension inside the S3 implementation could be part of the EDP as engine of the quadruple helix model for knowledge dynamics. It is possible to group under the innovation-oriented urban policy's concept the increasing phenomena of innovation districts (in a broadly sense) to refine a different perspective of the role of the city in the creation of an innovation ecosystem. Another aspect emerged from the research activities in Boston is connected to how innovation has become a source of urban form and its transformation, pushing urban regeneration initiatives driven by the demand for innovation.

The spatially oriented methodology adopted for Traded clusters in the Boston area has been implemented also for the spatialization of Local clusters in San Diego. Here, the focus shifted from mapping innovation concentration towards mapping specialization in the innovative milieu perspective. Clusters and knowledge networking reveal how territorial milieu can influence the knowledge dynamics and how knowledge can be shared along the territorial milieu. The aim was to find a connection between urban and inland areas through the territorial milieu as an explanation of innovative milieu. Local Clusters have been examined through Dynamic Analysis, Innovation Ecosystems and their relationship with Community Plans and Zoning providing interesting insights into the activation of social innovation thanks to the interaction of three driving elements: knowledge, innovation and place. The different socio-economic and spatial configuration allowed to identify different development dynamics for local innovation ecosystems. In San Diego, harnessing innovation ecosystem is not limited only to local actors, even regulatory agencies and municipal or regional governments that create a dynamic, innovation-driven economy can be involved in the orchestration process.

In both cases (Boston and San Diego), innovation-oriented public policies pivot around the entrepreneurial spirit, in line with the desired entrepreneurial knowledge convergence of the S3 approach. The MAPS-LED project proposes the Entrepreneurial Discovery Process as a trigger for the coordination of the efforts at local level—public administrations, research institutions, entrepreneurs,

---

[1]The MAPS-LED has been appointed as "success story" in European Commission: New thinking to drive regional economic development. EU Cordis Research and Innovation success stories available at: http://ec.europa.eu/research/infocentre/article_en.cfm?id=/research/headlines/news/article_17_11_15-2_en.html?infocentre&item=Infocentre&artid=46436.

communities—in boosting the local knowledge convergence and generating the expected change.

The MAPS-LED project emphasized how the linkage between planning and innovation policy empowers EDP through bottom-up approaches. In other words, local communities and organizations are in the best position to know what can drive a city's regeneration and deliver economic change reinforcing the urban dimension of S3. The research activities highlighted how EDP could be the mean to design tailor-made policy acting on the fruitful relationship among knowledge, innovation and place. This process should be managed at local level and embedded in the urban development agenda due to its ability to activate urban regeneration mechanisms and expand innovation in distressed areas through public–private partnership and innovative financial instruments. In this sense, the MAPS-LED approach works as cross-cutting element in the understanding of knowledge dynamics, which are complex and difficult to trigger in specific places. The interaction of knowledge, innovation and place—and the related potential **output** indicators provided by the MAPS-LED project—attributes the local asset to the entrepreneurial discovery process activated by urban policy aiming at regenerating urban areas through innovation-led processes. In synthesis, the analysis of the local context shed the light on EDP as evidence-based and horizontal policy for S3 by considering two drivers: the urban regeneration mechanism joint with Knowledge-Based Urban Development to guide the identification of output indicators of EDP; the cluster life cycle analysis to guide the result indicators of the EDP.

Furthermore, the cluster spatialization methodology could help in finding out the regional areas of innovation towards focusing on public and private financial resources. The methodology developed could help in the understanding "where" entrepreneurial knowledge and forces are active and concentrated, lighting up the potential for the discovery phase. This is a cross-sectorial approach because the identification of potentials with respect to the local context allows to discover concentration of knowledge and feed innovation at local level. The identification of local potential areas of innovation, coherently with the principle of Smart Specialisation, can favour the discovery of new domains through an evidence-based territorial perspective rather than a mere analysis of regional economies. Further insights from these findings reveal the potential transformation of these urban areas of innovation in Economic Special Zones.

The results coming from the MAPS-LED project research activities stimulated the scientific debate around the key elements able to trigger the desired change through S3 as well as the understanding of its (current and potential) limits. The participation of international experts involved in the S3 design and the RIS3 implementation, as well as the academic contributions coming from different disciplines, highlighted the potentials of the multidisciplinary approach proposed by the project, allowing to boost up knowledge convergence in an a sectorial rationale. The Symposium represents the opportunity to stimulate the development of innovation-oriented models for the exploitation and valorization of local assets involving different disciplines and in a multilevel governance perspective.

The contribution offered through the Symposium, by either enriching the academic debate or providing evidence-based solutions for the implementation of economic development strategies, is attributed by wide scope and marked cross-cutting dimension. The multidisciplinary nature of the MAPS-LED project is reflected by the structure of the Symposium itself. Each session presents topics and arguments which are, to some extent, ingrained within overall framework of the MAPS-LED project, while they are expected to open up windows of opportunity for further studies and research. Consistently, the Symposium focuses on analysing, at different scales and under numerous perspectives, the strategies, objectives and impacts of local economic development and innovation processes, to achieve a smart sustainable and inclusive growth. In a sentence, the Symposium, and the contributions to its sessions, manifests the effort to re-proposing the multidisciplinary approach implemented within the MAPS-LED research project in a conference-based-dimension.

While the Symposium encompasses a number of sessions dealing with specific topics, it is reasonable framing them within the streamlined "smart, inclusive and sustainable" growth paradigm enacted by the EU 2020 strategy (EU, 2010). By following this logic, the Symposium kicks off by trying to overcome current limits and gaps in the implementation of plans and models (session TS01, TS13 and TS16), while it further develops by bringing to light the importance of the place-based approach to deliver successful urban regeneration processes (session TS02 and TS23). In this regard, the prominent role played by territorial peculiarities in affecting decision-making processes is taken into high consideration (session TS04). Drawing on the belief that the "place" matters, the Symposium devoted different sessions to the study of the territorial-specific developmental mechanisms aiming at identifying logical-operational tools that can interpret urban phenomena (i.e. urban safety with the session TS11), but also evolutionary and community involvement processes in coastal areas (session TS19). The dichotomy urban–rural is treated in more than one session, as it is further scrutinized under the lens of geospatial analysis and modelling tools in the way of identifying how landscapes transition from rural-to-urban (session TS25) as well as under the lens of inclusive knowledge and innovation networks (session TS08). Still on the territorial dimension, the sustainable-led and ecological approach is analysed (thanks to the contribution in session TS14, TS15 and TS17) as well as the cultural heritage territorial network valorization perspective (session TS18), as a mean to favour the discovery of territorial-specific developmental opportunities. The discovery of opportunities in general, and hidden economic potentials in particular, is also the central point of discussion in the session on urban and regional development (TS20). This session refers to innovation spaces as catalyst for the disclosure of latent economic strengths of territories, at different geographic scales. The importance of innovation spaces seems especially relevant in the context of the knowledge economy, where and when the re-combination of "pieces of knowledge" can drive towards unveiling novel products and processes novel products and processes. The challenges and potentials posed by innovative activities are also investigated under an economic perspective by catching up with the complexity of knowledge dynamics. Following their cross-cutting rationale, innovation and knowledge are also the focal point of the session TS05 focused on ICT and heritage for a sustainable development as well as for the territorial innovative networks for public services (session TS26).

This synthetic description of the sessions gives a clear idea of the complexity of the themes treated as well as their alignment with respect to the MAPS-LED project. Moreover, the participation to the Symposium of international experts as well as academics from different disciplines provides interesting insights for the RIS3 evaluation and monitoring processes for the post-2020 programming period. The multidisciplinary approach to plan Smart Specialisation Strategies proposed with the MAPS-LED project emerged as crucial to properly pursue the local economic development in the S3 perspective. Hence, the MAPS-LED project appears at forefront into this research domain.

Carmelina Bevilacqua

Organization

Programme Chairs

Carmelina Bevilacqua   Mediterranea University of Reggio Calabria
Francesco Calabrò   Mediterranea University of Reggio Calabria
Lucia Della Spina   Mediterranea University of Reggio Calabria

Scientific Committee

Stefano Aragona   INU—Istituto Nazionale di Urbanistica
Angela Barbanente   Politecnico di Bari
Filippo Bencardino   SGI—Società Geografica Italiana
Jozsef Benedek   RSA—Babes-Bolyai University, Romania
Christer Bengs   SLU/Uppsala Sweden and Aalto/Helsinki, Finland
Andrea Billi   Università La Sapienza di Roma
Adriano Bisello   EURAC Research
Nicola Boccella   Università La Sapienza di Roma, Presidente SISTur
Mario Bolognari   Università degli Studi di Messina
Kamila Borsekova   Matej Bel University, Slovakia
Nico Calavita   San Diego State University, USA
Roberto Camagni   Politecnico di Milano, Presidente Gremi
Farida Cherbi   Institut d'Architecture de Tizi Ouzou, Algeria
Maurizio Di Stefano   Icomos Italia
Yakup Egercioglu   Izmir Katip Celebi University, Turkey
Khalid El Harrouni   Ecole Nationale d'Architecture, Rabat, Morocco
Gabriella Esposito De Vita   CNR/IRISS Istituto di Ricerca su Innovazione e Servizi per lo Sviluppo

Rosa Anna Genovese	Università degli Studi di Napoli "Federico II"
Giuseppe Giordano	Università degli Studi di Messina
Olivia Kyriakidou	Athens University of Economics and Business, Greece
Ibrahim Maarouf	Alexandria University, Faculty of Engineering, Egypt
Lívia M. C. Madureira	Centro de Estudos Transdisciplinares para o Desenvolvimento—CETRAD, Portugal
Tomasz Malec	Istanbul Kemerburgaz University, Turkey
Ezio Micelli	IUAV Istituto Universitario di Architettura di Venezia
Nabil Mohäreb	Beirut Arab University, Tripoli, Lebanon
Mariangela Monaca	Università di Messina
Bruno Monardo	Università degli Studi di Roma "La Sapienza"
Pierluigi Morano	Politecnico di Bari
Fabio Naselli	IEREK International Experts for Research Enrichment and Knowledge Exchange
Peter Nijkamp	Vrije Universiteit Amsterdam
Davy Norris	Louisiana Tech University, USA
Leila Oubouzar	Institut d'Architecture de Tizi Ouzou, Algeria
Sokol Pacukaj	Aleksander Moisiu University, Albania
Aurelio Pérez Jiménez	University of Malaga, Spain
Keith Pezzoli	University of California, San Diego, USA
María José Piñera Mantiñán	University of Santiago de Compostela, Spain
Fabio Pollice	Università del Salento
Vincenzo Provenzano	Università di Palermo
Ahmed Y. Rashed	Founding Director Farouk ElBaz Centre for Sustainability and Future Studies
Riccardo Roscelli	Politecnico di Torino
Michelangelo Russo	SIU—Società Italiana degli Urbanisti
Alessandro Saggioro	Università La Sapienza di Roma
Helen Salavou	Athens University of Economics and Business, Greece
Paolo Salonia	CNR—Istituto per le tecnologie applicate ai beni culturali, Rome, Italy
Stefano Stanghellini	IUAV, Presidente SIEV
Luisa Sturiale	Università di Catania
Chiara O. Tommasi Moreschini	Università di Pisa
Claudia Trillo	University of Salford, UK

Internal Scientific Board

Giuseppe Barbaro   Mediterranea University of Reggio Calabria
Concetta Fallanca   Mediterranea University of Reggio Calabria
Giuseppe Fera   Mediterranea University of Reggio Calabria
Massimiliano Ferrara   Mediterranea University of Reggio Calabria
Giovanni Leonardi   Mediterranea University of Reggio Calabria
Francesca Martorano   Mediterranea University of Reggio Calabria
Domenico E. Massimo   Mediterranea University of Reggio Calabria
Carlo Morabito   Mediterranea University of Reggio Calabria
Gianfranco Neri   Mediterranea University of Reggio Calabria
Francesco S. Nesci   Mediterranea University of Reggio Calabria
Simonetta Valtieri   Mediterranea University of Reggio Calabria
Santo Marcello Zimbone   Mediterranea University of Reggio Calabria

Scientific Partnership

Regional Studies Association, Seaford, East Sussex, UK
A.I.S.Re—Associazione Italiana di Scienze Regionali, Milan, Italy
Eurac Research, Bozen, Italy
Icomos Italia, Rome, Italy
INU—Istituto Nazionale di Urbanistica, Rome, Italy
Società Italiana degli Urbanisti, Milan, Italy
Società Geografica Italiana, Rome, Italy
SIEV—Società Italiana di Estimo e Valutazione, Rome, Italy
Urban@it, Bologna, Italy
Federculture—Federazione Servizi Pubblici Cultura Turismo Sport Tempo Libero,
Rome, Italy
FICLU—Federazione Italiana dei Club e Centri per l'Unesco, Turin, Italy
Rete per la Parità—APS per la Parità uomo-donna secondo la Costituzione Italiana,
Rome, Italy

Organizing Committee

Associazione ASTRI—Associazione Scientifica Territorio e Ricerca Interdisciplinare
URBAN LAB S.r.l.

Contents

Contents xxiii

Innovation Dynamics, Smart Cities, ICT

Beyond Innovation Districts: The Case of Medellinnovation District

Arnault Morisson[(✉)] and Carmelina Bevilacqua

Mediterranea University of Reggio Calabria, 89100 Reggio Calabria, Italy
arnault.morisson@unirc.it

Abstract. Innovation districts are emerging as local economic development strategies in diverse cities around the world. They have however, been criticized for being non-participative top-down initiatives that encourage gentrification and economic polarization. Ruta N, a public organization, is leading the transformation of the innovation district of Medellin (Colombia), dubbed as Medellinnovation District. The paper investigates the programs that are being implemented in the Medellinnovation District in order to mitigate the negative externalities that such strategy can generate. The research methodology is based on a case study approach, using Medellinnovation District as a significant and high-impact case. The paper finds that the programs that Ruta N is implementing can be regrouped into two categories: attraction and absorption. The programs under attraction aim to attract knowledge companies and workers to the innovation district. The programs under absorption aim to activate the absorptive capacity of the residents living in the innovation district in order to make them full participants of the development of the innovation district.

Keywords: Innovation district · Medellin · Creative city · Public policy
Social inclusion · Innovation culture

1 Introduction

Innovation districts are being adopted in cities around the world as local economic development strategies. Barcelona (Spain), Boston (Massachusetts), Chattanooga (Tennessee), Detroit (Michigan), Medellin (Colombia), Montréal (Canada), Philadelphia (Pennsylvania), Rotterdam (the Netherlands), and San Diego (California) are some of the cities that are building their own version of an innovation district. The main objective of innovation districts is to accelerate the technological innovation process by clustering knowledge companies and workers [1, 2].

In 2009, the city of Medellin and the public-utility and telecommunications company EPM-UNE, launched the nonprofit organization Ruta N Medellin [3]. The organization has the mission to support companies, public institutions, and universities to foster technological innovations while at the same time transforming Medellin into a knowledge city [3]. In 2012, Ruta N unveiled its plan to create an innovation district around the Ruta N innovation center [4]. The innovation district of Medellin aims to redevelop 172 hectares of the northern part of the city [5]. Innovation districts and the innovation economy are criticized for contributing to gentrification and economic

© Springer International Publishing AG, part of Springer Nature 2019
F. Calabrò et al. (Eds.): ISHT 2018, SIST 100, pp. 3–11, 2019.
https://doi.org/10.1007/978-3-319-92099-3_1

polarization [6–9]. The development of innovation districts is also criticized for being non-participative and undemocratic where local governments are pushing a neoliberal agenda favoring the middle- and upper-class [10–12].

The paper investigates the programs that Ruta N implemented in order to mitigate the negative externalities of building its own version of an innovation district, the Medellinnovation District. The innovation district of Medellin is selected as a single significant and high-impact case study. The research conducted for this paper is based on three sources of data: semi-structured interviews, secondary data, and direct observation. The paper is highly relevant for urban policymakers who wish to mitigate the negative externalities from the creation of an innovation district in their cities. The paper finds that the programs that Ruta N is implementing can be regrouped into two categories: attraction and absorption. The programs under attraction aim to attract knowledge companies and workers to the innovation district. The programs under absorption aim to activate the absorptive capacity of the residents living in the innovation district in order to make them full participants of the development of the innovation district.

2   Innovation Districts, Gentrification, and Polarization

In the 1990s, capitalist countries started to undergo an economic transition towards post-Fordism or knowledge-based economies [13, 14]. In the knowledge economy, technological innovation is a precondition for high-standard of living and economic prosperity [15]. The academic literature provides, both across nations and over time, a solid theoretical background linking technological innovation to the progress of countries, regions, cities, and firms [16–21].

The OECD [22] argues that "innovation provides the foundation for new businesses, new jobs and productivity growth and is thus an important driver of economic growth and development" (p. 13). Technological innovation drives productivity growth as well as the quality and quantity of jobs, which are critical to improve standards of living [23]. Indeed, technological innovation is considered as a major force in economic growth and in a post-2008 era characterized by low economic growth, innovation is seen as a transformative force for developed and developing economies [22, 24].

The concept of an innovation district emerges in order for cities to harness the transformative power of technological innovations and to become knowledge cities. Cities are increasingly seen as the key administrative units to spur technological innovations [25, 26]. The concept of an innovation district is the policy-response to the increasingly spatial and urban dimensions of the knowledge-based economy, combining innovation theories with socio-economic trends that have emerged in the knowledge-based economy [2].

Innovation districts are criticized for being no less than gentrification programs [10, 11, 27]. Gentrification is a process that involves "the transition of inner-city neighborhoods from a status of relative poverty and limited property investment to a state of commodification and reinvestment" [28]. In gentrified neighborhoods, the population structure changes from being working-class to upper-middle-class and most often from being black to white [7, 29, 30]. In the United States in the 1980s, local and regional governments stepped out from implementing urban renewal programs, which led gentrification to strictly be the outcome of market forces [30]. In the United States, most innovation districts are pushed by real-estate development companies and as a result, lack some conceptual dimensions in order to be fully functioning innovation districts [2].

The innovation economy is criticized for polarizing the workforce and hollowing out the middle class [31–33]. Technological innovations are concentrated in large urban centers, which favors the emergence of "superstar cities" [6, 34–37]. Superstar cities concentrate economic wealth and are also highly unequal cities [8, 38]. In superstar cities, the middle- and upper-middle-class are being priced out due to rising rents and costs of living [39]. Superstar cities are thus becoming cities that only multi-millionaires and billionaires can afford to live in [7]. The innovation economy has fostered such economic polarization in some cities, like in San Francisco, that it has led to demonstrations against knowledge companies and knowledge workers [40].

3 Methodology

The research methodology is based on a single case-study approach, using primary and secondary data. The authors use the case study approach "out of the desire to understand complex social phenomena" [41]. The purpose of this case study is to uncover the programs that can be implemented to mitigate the creation of an innovation district. The case selected is the innovation district of Medellin that is being developed by the public organization, Ruta N Medellin. The paper investigates in a descriptive manner a contemporary phenomenon in which the researcher has no control on the actual phenomenon [42]. Moreover, the investigation of programs to mitigate the negative externalities of innovation districts has not been fully examined. A qualitative approach is as a result, the most appropriate method [42, 43].

The research conducted for this paper is based on three sources of data: semi-structured interviews, secondary data, and direct observation. The semi-structured interviews were conducted in Medellin with key informants who have extensive knowledge on the programs implemented by Ruta N Medellin. In total, eleven interviews were conducted. The persons interviewed were: employees at Ruta N Medellin, professors at EAFIT University and MIT, and employees at the Agency for Cooperation and Investment of Medellin and the Metropolitan Area (ACI). The stakeholders were selected according to their strong knowledge and diverse perspectives on the phenomena studied [44]. The interviews were conducted in order to uncover the strategies adopted to create an inclusive innovation district in Medellin. The secondary data that were used for the research are, but were not limited to: Ruta N's websites; Ruta N's planning documents; articles in news websites, newspapers, and magazines, such as *El Colombiano*, *El Tiempo*, and *Semana*; and Ruta N's annual reports. The direct observations involved non-participatory observations in the innovation district of Medellin. In total, the researcher conducted about 20 h of formal and informal observations in order to investigate the contacts between the newly arrived knowledge workers and the district's residents.

The case study aims to provide rich and deep understanding of the programs behind building inclusive innovation districts, and, as such, is a significant and high-impact case. The data are analyzed in an inductive manner in order to uncover patterns in the programs implemented [45]. From the pattern recognition, the authors decided to categorize the programs adopted into two groups: attraction and absorption. Validation is achieved through prolonged engagement, persistent observation, and triangulation in order to "assure that the right information and interpretations have been obtained" [46]. The rich description allows readers to make decisions regarding transferability [45].

4   Case Study - The Medellinnovation District

4.1   The Context

The City of Medellin is located in the Aburrá Valley in midst of the Andes in Colombia. Medellin is the second most populous city in Colombia with a population of 2,464,322 inhabitants as of 2015, after the capital city Bogotá [47]. The city of Medellin was the industrial capital in the 1970s of Colombia and one of the most important industrial powerhouses in Latin America [48]. In 1991, the city of Medellin became infamous for being the murder capital of the world during the heyday of the Medellin cartel [49]. In 2013, the Wall Street Journal and Citi announced Medellin to be the "most innovative city of the year" recognizing the city's social urbanism programs and innovative transportation system connecting the poorest neighborhoods with the city center [50].

In 2009, the city of Medellin and the public-utility and telecommunications company EPM-UNE launched Ruta N Medellin, a public organization that has the mission to transform the Medellin into a knowledge city [3]. In 2012, Ruta N unveiled its plan to create an innovation district in the northern part of the city [4]. The innovation district, dubbed as Medellinnovation District, aims to redevelop an area of 172 hectares with a population of 12,244 inhabitants as of 2015, comprising the districts of Chagualo, Jesús Nazareno, Sevilla, and San Pedro [5]. The innovation district is to be planned around Ruta N innovation center, a 33,140-square-meter three-building complex that houses Ruta N offices, EPM-UNE research laboratories, the ViveLab animation learning center, international companies, and international startups. The northern part of the city was selected due to its existing infrastructures, such as the University of Antioquia, the National University, the Hospital San Vicente de Paul, the Parque Explora, and the Botanical Garden of Medellin, two metro stations, the Ruta N innovation center, and the close proximity to the city center. Most importantly, the northern part of the city of Medellin has historically been an impoverished area, concentrating most of the city's poverty.

4.2   The Programs

In 2012, Ruta N established the innovation district division, which has the mission to supervise and to promote the development of the Medellinnovation District [4]. In 2012, experts from 22@ Barcelona, the first innovation district to have been planned by policymakers, came to Medellin as consultants to help structure the strategy of the Medellinnovation District. In 2013, urban planning professors from MIT, including Dennis Frenchman and Carlo Ratti, designed the masterplan for the Medellinnovation District. The following programs are current or former programs that were implemented by the Ruta N's innovation district division between 2012 and August 2017.

The soft landing program was established in 2012 in order to attract knowledge-intensive international startups to the Ruta N innovation center. In the Ruta N innovation center, there are three floors dedicated to the landing platforms comprising 660 workstations. The landing program aims to connect arriving international startups to the wider innovation ecosystem. Ruta N prioritizes international startups from three sectors, namely, ICT, health, and energy. International startups can rent a flexible number of workstations for a period of up to two years at a competitive price. After the period of two years, the knowledge-intensive startups can then rent spaces through Space N with Ruta N's partners in other parts of the city. From 2012 to 2017, the number of international companies participating in the landing programs increased from seven to more than 90. In total, since the inception of the landing program, 197 companies from 27 countries landed, generating more than 4,216 jobs. The landing program is managed by Ruta N innovation district team and the ACI. In 2014, Ruta N and the ACI visited 7 countries to promote Medellin as a hub for innovation and the landing program [51]. The city of Medellin grants, through the Agreement 67 of 2010, tax breaks such as the property taxes and industry and commerce taxes for companies of the following clusters that locate in the Medellinnovation District: ICT, energy, health, textile, construction, design, and tourism [4]. The tax breaks scheme aims to attract companies to the Medellinnovation District. MIT's and 22@ Barcelona's experts were involved in designing the innovation district's masterplan. The objective of the masterplan is to make the innovation district, an inclusive, sustainable, mixed-use, green, dense, compact, open, and walkable neighborhood. The overall strategy is to make the Medellinnovation District an attractive neighborhood with a high quality of life for the knowledge workers living and working there.

The planning of the Medellinnovation District included co-creation with its residents. Indeed, the Ruta N innovation district's team conducted interviews, observations, focus groups, innovation bazars, creative lunches, census, co-creation activities, and conferences with the inhabitants and associations in the Medellinnovation District. The co-creation process included four phases: approaching the community, co-creating with the community, communicating with the community, and including the community in the development of the innovation district. The DistritoLab program was conducted in 2014 and in 2015 in order to promote science, technology, and innovation to the students living in the Medellinnovation District. DistritoLab is a program that involves the participation of local high school students to find urban solutions and to create innovative prototypes for the innovation district. The students participating in the program learned to use 3D printing machines and laser cutters in order to create working prototypes for the innovation district. Ruta N innovation district's team also organizes events such as the innovation conferences in order to explain the innovation process, intellectual property, and competitive intelligence to the inhabitants of the innovation district. One other event was the innovation bazar where city's innovation institutions presented their conceptions of innovation in an interactive manner with the inhabitants of the Medellinnovation District. In 2017, Ruta N innovation district's team launched the open kitchen program in order to help existing restaurants and bars to adapt their offerings through capacity-building courses and coaching to the new demand that is generated by the newly arrived knowledge workers employed in the Medellinnovation District. The main objective of the open kitchen program is to make existing restaurants and bars full participants of the development of the innovation district. Similar programs will be organized for auto repair shops and landlords located in the Medellinnovation District. Ruta N innovation district's team is also developing a program to create a Living Lab in order for the local community to be more connected with the knowledge-intensive companies in the Medellinnovation District. The Living Lab will be a platform where local residents can test prototypes and new services.

5   Discussion

The programs developed and implemented by the innovation district division at Ruta N can be regrouped into two categories: attraction and absorption. Attraction refers to the programs that aim to attract companies, startups, and workers to invest, to live, and to work in the Medellinnovation District. Absorption refers to the programs that focus on the residents of the Medellinnovation District in order to make them full participants in the development of the innovation district.

The absorption programs such as DistritoLab, innovation events, Open Kitchen, the Living Lab, and the community co-creation aim to reduce the cognitive distance

Table 1. Summary of Programs with Programs' Key Findings and Goals.

Strategy	Program	Key Finding	Program's Goal
Attraction	Soft Landing	Incentives to attract companies to the district, such as low rent, knowledge spillovers, high amenities	Local economic development
	Tax-Breaks	Incentives to attract companies to the district	Local economic development
	Masterplan	Non-pecunary incentives to attract companies, such as high amenities and quality of life	Urban development
Absorption	Co-Creation	To familiarize local residents with changes brought by knowledge workers and companies	Participatory urbanism
	DistritoLab	To familiarize local residents with technologies used by knowledge workers and companies	Residentes' training
	Open Kitchen	To adapt local businesses to changing customers' preferences in the district	Residentes' business opportunities
	Living Lab	Formal and informal contacts between residents and knowledge-based companies	Residentes' inclusion
	Events	Formal and informal contacts between residents and knowledge-based companies	Residentes' inclusion

between the residents and the newly arrived knowledge workers. Indeed, the absorption programs aim to increase the absorptive capacity of the residents to the external knowledge that the knowledge workers are bringing with them to the Medellinnovation District. Nooteboom [52] shows that knowledge diffusion is constrained by the "cognitive distance" between actors. This cognitive distance should not be too wide nor too similar. Indeed, knowledge that is already known is just as useless as it is for knowledge that cannot be understood [53]. In reducing the cognitive distance between the residents and the incoming knowledge workers, Ruta N innovation district's team aims to build an inclusive innovation district that does not exclude residents from participating in the knowledge economy (Table 1).

6 Conclusions

The strategy adopted in the Medellinnovation District potentially offers a response on how to limit the negative externalities, such as gentrification, non-participation, and segregation, arising from the development of an innovation district. Organizations that implement innovation districts should design programs in order to reduce the cognitive distance, if such cognitive distance exists, between the residents and the incoming knowledge workers. In doing so, residents can become full participants in the development of the innovation district, and even accelerate the innovation process of the incoming companies through recombination of novel and diverse ideas that the residents can convey. The findings can be generalized to other innovation districts that have a large population and where there are concerns of possible gentrification. Although the strategy adopted in the Medellinnovation District is promising on a theoretical standpoint, it is difficult to assess its efficacy in the face of real-estate speculation and land pressures. Indeed, the Medellinnovation District has been slow to develop, and real estate developers have manifested little interest in investing in the district. Further research should look at strategies to mitigate negative externalities in innovation districts that have experienced strong real-estate speculations.

Acknowledgments. This article is part of the MAPS-LED research project, which has received funding from the European Union's Horizon 2020 research and innovation programme under the Marie Skłodowska-Curie grant agreement No 645651. The authors would like to thank the two anonymous referees as well as Ron Boschma and Pierre-Alexandre Balland for their comments on an earlier version of this article.

References

1. Katz, B., Wagner, J.: The Rise of Innovation Districts: A New Geography of Innovation in America, Brookings Institution, Washington, (2014)
2. Morisson, A.: Innovation Districts: An Investigation of the Replication of the 22@ Barcelona's Model in Boston, Master Dissertation. FGV-EAESP, São Paulo, (2014)
3. Ruta, N.: Informe de Gestión 2010. http://www.rutanmedellin.org/es/nosotros/ruta-n/informes-de-gestion. Accessed 21 Nov 2017

4. Ruta, N.: Informe de Gestión 2012. http://www.rutanmedellin.org/es/nosotros/ruta-n/informes-de-gestion. Accessed 21 Nov 2017
5. Ruta, N.: Tomo I Diagnóstico Documento Técnico de Soporte: Distrito Medellionnovation. In: Ruta, N. Medellin (2015)
6. Edlund, L., Machado, C., Sviatschi, M.M.: Bright Minds, Big Rent: Gentrification and the Rising Returns to Skill, National Bureau of Economic Research, Cambridge (2015)
7. Florida, R.: The New Urban Crisis: How Our Cities Are Increasing Inequality, Deepening Segregation, and Failing the Middle Class and What We Can Do About It. Basic Books, New York (2017)
8. Glaeser, E.L., Resseger, M., Tobio, K.: Inequality in cities. J. Reg. Sci. **49**(4), 617–646 (2009)
9. Stehlin, J.: The post-industrial "shop floor": emerging forms of gentrification in San Francisco's innovation economy. Antipode **48**(2), 474–493 (2016)
10. Moulaert, F.: Globalization and Integrated Area Development in European Cities. Oxford University Press, Oxford (2000)
11. Swyngedouw, E., Moulaert, F., Rodriguez, A.: Neoliberal urbanization in Europe: large-scale urban development projects and the new urban policy. Antipode **34**(3), 542–577 (2002)
12. Shin, H., Stevens, Q.: how culture and economy meet in South Korea: the politics of cultural economy in culture-led urban regeneration. Int. J. Urban Reg. Res. **37**(5), 1707–1723 (2013)
13. Amin, A.: Post-Fordism: A Reader. Blackwell, Oxford (1994)
14. Drucker, P.F.: The Discipline of Innovation. Harvard Bus. Rev. **76**(6), 149–157 (1998)
15. Aghion, P., Howitt, P.: A Model of Growth through Creative Destruction. National Bureau of Economic Research, Cambridge (1990)
16. OECD: The Knowledge-Based Economy, OECD Publishing, Paris (1996)
17. Fagerberg, J.: International competitiveness. Econ. J. **98**(391), 355–374 (1988)
18. Freeman, C.: Technical innovation, diffusion, and long cycles of economic development. In: Vasko, T. (ed.) The Long-Wave Debate, pp. 295–309. Springer, Berlin Heidelberg (1987)
19. Rosenberg, N.: Innovation and Economic Growth. OECD Publishing, Paris (2004)
20. Schumpeter, J.A.: The Theory of Economic Development: An Inquiry into Profits, Capital, Credit, Interest, and the Business Cycle. Transaction Publishers, New Jersey (1934)
21. Solow, R.M.: Technical change and the aggregate production function. Rev. Econ. Stat. **39**, 312–320 (1957)
22. OECD: The Innovation Imperative: Contributing to Productivity, Growth and Well-Being, OECD Publishing, Paris (2015)
23. OECD: Regions and Innovation Policy, OECD Reviews of Regional Innovation, OECD Publishing, Paris (2011)
24. Metcalfe, S., Ramlogan, R.: Innovation Systems and the Competitive Process in Developing Economies. Q. Rev. Econ. Finan. **48**(2), 433–446 (2008)
25. Castells, M.: The Rise of the Network Society. Blackwell Publishers, Oxford (1996)
26. Florida, R., Adler, P., Mellander, C.: The City as Innovation Machine. Reg. Stud. **51**(1), 86–96 (2017)
27. Smith, N.: New globalism, new urbanism: gentrification as global urban strategy. Antipode **34**(3), 427–450 (2002)
28. Ley, D.: Artists, aestheticisation and the field of gentrification. Urban Stud. **40**(12), 2527–2544 (2003)
29. Ehrenhalt, A.: The Great Inversion and the Future of the American City. Knopf, New York (2012)
30. Harvey, D.: Rebel Cities: From the Right to the City to the Urban Revolution. Verso Books, London (2012)

31. Acemoglu, D., Autor, D.: Skills, tasks and technologies: implications for employment and earnings. Handb. Labor Econ. **4**, 1043–1171 (2011)
32. Autor, D., Levy, F., Murnane, R.J.: The skill content of recent technological change: an empirical exploration. Q. J. Econ. **118**(4), 1279–1333 (2003)
33. Goos, M., Manning, A.: Lousy and lovely jobs: the rising polarization of work in Britain. Rev. Econ. Stat. **89**(1), 118–133 (2007)
34. Acs, Z.J., Audretsch, D.B.: Innovation in large and small firms: an empirical analysis. Am. Econ. Rev. **78**(4), 678–690 (1988)
35. Gyourko, J., Mayer, C., Sinai, T.: Superstar Cities. Am. Econ. J. Econ. Policy **5**(4), 167–199 (2013)
36. Jaffe, A.B., Trajtenberg, M., Henderson, R.: Geographic localization of knowledge spillovers as evidenced by patent citations. Q. J. Econ. **108**(3), 577–598 (1993)
37. Rodríguez-Pose, A., Crescenzi, R.: Mountains in a flat world: why proximity still matters for the location of economic activity. Camb. J. Reg. Econ. Soc. **1**(3), 371–388 (2008)
38. Aghion, P., Akcigit, U., Bergeaud, A., Blundell, R., Hémous, D.: Innovation and Top Income Inequality, National Bureau of Economic Research, Cambridge, (2015)
39. Priced Out of Paris. https://www.ft.com/content/a096d1d0-d2ec-11e2-aac2-00144feab7de. Accessed 21 Nov 2017
40. Oakland, The City that Told Google to get Lost. https://www.theguardian.com/technology/2014/feb/10/city-google-go-away-oakland-california. Accessed 21 Nov 2017
41. Yin, R.K.: Case Study Research: Design and Methods. Sage Publications, Thousand Oaks (1994)
42. Eisenhardt, K.M.: Building theories from case study research. Acad. Manage. Rev. **14**(4), 532–550 (1989)
43. Creswell, J.W.: Qualitative Inquiry and Research Design: Choosing Among Five Approaches. Sage Publications, Thousand Oaks (2013)
44. Eisenhardt, K.M., Graebner, M.E.: Theory building from cases: opportunities and challenges. Acad. Manag. J. **50**(1), 25–32 (2007)
45. Patton, M.Q.: Qualitative Evaluation and Research Methods. Sage Publications, Thousand Oaks (2015)
46. Stake, R.E.: Multiple Case Study Analysis. Guilford Press, New York (2013)
47. City of Medellin. https://www.medellin.gov.co/irj/portal/medellin?NavigationTarget=navurl://ecd9e39fad34752203a60e8a84a34ba1. Accessed 21 Nov 2017
48. Caballero, Argáez, C.: La Economía Colombiana del Siglo XX: Un Recorrido por la Historia y sus Protagonistas, Penguin Random House, Bogotá. (2016)
49. Maclean, K.: The 'Medellin Miracle': The Politics of Crisis, Elites and Coalitions, Development Leadership Program, University of Birmingham, Birmingham (2014)
50. The City of the Year. http://online.wsj.com/ad/cityoftheyear. Accessed 21 NOv 2017
51. Informe de Gestión 2014. http://www.rutanmedellin.org/es/nosotros/ruta-n/informes-de-gestion. Accessed 21 NOv 2017
52. Nooteboom, B.: Innovation, learning and industrial organisation. Camb. J. Econ. **23**(2), 127–150 (1999)
53. Boschma, R.: Proximity and Innovation: A Critical Assessment. Reg. Stud. **39**(1), 61–74 (2005)

Spatial Data Infrastructures in Santiago de Compostela

From the Heritage Information System to the City Council's Geoportal

Yamilé Pérez-Guilarte(✉) and Miguel Pazos Otón

University of Santiago de Compostela, Santiago de Compostela, Spain
yamile.perez@uscs.es, miguel.pazos.oton@usc.es

Abstract. Spatial Data Infrastructures constitute an organisational and technological bet on the part of Public Administrations to make available the geographic information they collect or generate to institutions and citizens. In Spain, various initiatives have been developed at state, regional, local and cross-border levels, which have been integrated into the Spatial Data Infrastructure of Spain. In the case of Santiago de Compostela, a World Heritage historic city, two experiences have taken place: the Heritage Information System (SIP, by its Spanish acronym) and recently the Geoportal of Santiago de Compostela City Council. This work aims to present the experience of both initiatives as tools for heritage management and public knowledge. An evaluation of the design and functionality of the SIP over the 7 years since its implementation has also been carried out. For this purpose, four indicators have been established: use of international and European standards, compatibility, proactivity, and synergies between documentation centers. The study found that despite the functionality that the SIP had in its creation, the lack of foresight to maintain its systematization and updating led to it being outdated. However, the Geoportal of Santiago de Compostela City Council was created as a new tool for disseminating and managing heritage and council services among institutions, professionals and citizens.

Keywords: Spatial data infrastructure · Cultural heritage · Management

1 Introduction

Considerable advances have been made globally in the development of Spatial Data Infrastructures (SDIs) in recent years. Many national, regional, and international programs and projects are working to improve access to available spatial data, promote its reuse, and ensure that additional investment in spatial information collection and management results in an ever-growing, readily available and useable pool of spatial information [1].

An SDI is an infrastructure, a set of agreements, rules and technologies that aim to enable the access and maintenance of spatial information in an interoperable way. The most visible component of an SDI can be a geoportal, a website with a cartographic

© Springer International Publishing AG, part of Springer Nature 2019
F. Calabrò et al. (Eds.): ISHT 2018, SIST 100, pp. 12–20, 2019.
https://doi.org/10.1007/978-3-319-92099-3_2

interface that allows access to spatial information of some kind, although not all geoportals are necessarily interfaces to access an SDI [2].

In Spain, SDIs have been developed at state, regional and local levels. They are integrated in the Spatial Data Infrastructure of Spain (IDEE, by its Spanish acronym) (www.idee.es) which follows the INSPIRE regulation. It belongs to the Superior Geographic Council and integrates the data, metadata, services and geographic information that are produced throughout the country. In this way, 50 state, 37 regional, 49 local and 3 cross-border SDIs have been implemented in areas such as hydrology, geology, archaeology, urban planning, transport, etc. Among the local initiatives cited in this portal the experiences carried out in the capital of the Autonomous Community of Galicia, Santiago de Compostela does not appear. However, in the last 7 years we can mention two examples of SDI implementation in the city: the Heritage Information System and the Geoportal of the City Council.

Santiago de Compostela, together with other urban spaces with a monumental and artistic legacy, belongs to the network of World Heritage Cities of Spain. In order to understand the reason for the implementation of a Spatial Data Infrastructure in Santiago de Compostela, it is necessary to take into account the networking which has been developing over several decades with cities, such as Alcalá de Henares, Segovia, Toledo, Cáceres or Córdoba (to take the most outstanding examples). This networking is advanced on a regular basis, sharing experiences and generating synergies in fields, such as urban planning, rehabilitation, mobility, urban landscape or security, to mention a few examples.

All of these cities share a similar problem, since they have to achieve the balance between the conservation of a centennial heritage with the adaptation to increasingly demanding urban tourism which demands contemporary quality services. In the daily management of historic centers, therefore, a property-by-property detailed analysis of all the elements of the urban plot is necessary.

As in the case of Santiago de Compostela, cities such as Cáceres or Córdoba are pioneers in the creation of geo-referenced and computerized databases that allow the consultation and management of information in real time. The case of Cáceres can be mentioned as a precedent for the one in Santiago de Compostela, and it stands out due to the cooperation between the academic field and the municipal administration.

Santiago de Compostela has been a World Heritage Site since 1985 due to its historical center, the main protagonist of the city. This space between the 60s and 80s of the twentieth century had declined in terms of its traditional tertiary functions and its demography (aging, feminization and loss of residents). However, the transfer of the political and administrative capital of Galicia to Santiago in 1980, as well as the declaration of World Heritage, led the local government to launch a series of measures aimed at the protection and recovery of the historical center. In this way, in 1994 the Consortium of Santiago began an ambitious project that harmonizes the necessary renovation of housing with the respect and preservation of all the elements that identify and characterize these architectures [3].

With the precise aim of creating an inventory geodocumental system of administration and management of heritage for the actions performed or likely to be carried out within the framework of the Special Plan for the Protection and Rehabilitation of the Historical City, the idea of the SIP arose [4]. This tool came into operation in 2010, however as will be discussed in this work, despite being a novel experience at the time of its creation, it has not been able to maintain its functionality and systematicity. On the other hand, the Geoportal of Santiago de Compostela City Council, recently launched in October 2017, is presented as a tool to make available to institutions, entrepreneurs and citizens all the existing geographic information about the city.

This work presents a balance of the experience over the last 7 years of the SDIs in Santiago de Compostela as tools for heritage management and general knowledge. An evaluation of the design and functionality of the Heritage Information System was carried out. As a result, the need for the information management to be adapted to a continuously changing digital sector and the necessity to count on permanent updating systems were proved.

2   Spatial Data Infrastructures for Cultural Heritage

The term Spatial Data Infrastructure was coined in 1993 by the U.S. National Research Council to denote a framework of technologies, policies, and institutional arrangements that together facilitate the creation, exchange, and use of geospatial data and related information resources across an information-sharing community [1]. The adoption of the IDE philosophy has essentially occurred thanks to the development of regulatory frameworks that have driven its implementation, and in some cases have made it mandatory. The Open Geospatial Consortium (OGC) is the international organization in which administrations, companies and academics meet to define and agree on these standards. There are OGC standards, or derivatives from them, for example, the INSPIRE standards, the Spatial Information in the European Community. It aims to provide the standards and policies that will enable the integration of these services in the European Union [5].

The use of digital spatial data in management, preserving and disseminating cultural heritage has been thoroughly addressed [5–14]. At present, a number of SDIs for managing and disseminating cultural heritage exist. One outstanding example is the Canmore system in Scotland, which catalogues entries for archaeological sites, buildings, industry and maritime heritage across the region (https://canmore.org.uk/). In addition, open source data management platforms for the heritage field exist. This is the case of Arches a geospatially-enabled software platform for cultural heritage inventory and management, (https://www.archesproject.org/) that is freely available to use and customizable without licensing costs for any organization worldwide.

Digitization has a substantial impact on heritage management since on-line availability of cultural data may enable managers in decision making and in preserving physical heritage. In historical towns there is a limitation regarding the accessibility to the abundant documentary information, either historical or contemporary. They are especially complex cases, not only because the amount and variety of information is inherently big (documents, texts, maps, etc.), but also because of the coexistence of elements in the same spatial location together with the human pressure [13].

The issues affecting the adequate inventorying of cultural resources around the world are a global problem [15, 16]. Several international initiatives have tackled this problem, allowing new approaches to be transmitted to experts and heritage organizations [8]. Some of the documents that provide an initial guide for the preparation of heritage information systems are: the ICOMOS Charter- Principles for the analysis, conservation and structural restoration of architectural heritage [17]; the Council of Europe's Guidance on inventory and documentation of the cultural heritage [18]; or the UK Historic Environment Data Standard [19].

According to Niccolucci [7], in order to develop appropriate digitization strategies some factors must be taken into account such as: using refined technologies derived from the nature of tangible cultural assets such as 3D scanning; supporting and disseminating the results of ongoing research and guidelines for its best use; providing feedback to programs and projects in this domain; involving all the relevant stakeholders; supporting the people who work in this interdisciplinary domain; and promoting synergies with other domains (e.g. tourism, planning and heritage management).

The digitization of cultural heritage, whilst initially framed by institutions, is now increasingly a collective process involving community access and the collective sharing of knowledge. Using web services, the so-called geoportals, as the delivery mechanism for spatial datasets has the significant advantage that they can be consumed by a range of devices and software applications providing a user experience appropriate to the consumer's requirement. Geoportals enable data browsing by theme, provider, temporal or spatial relevance, all of which is derived from the standardized metadata [5].

3 Methodology

For the study of the implementation and use of SDIs in Santiago de Compostela over the last 7 years, a detailed analysis of the web viewer of the Heritage Information System and the Geoportal of the Santiago de Compostela City Council was carried out. Likewise, the documentary fund of the Santiago Consortium was consulted and an exhaustive review of the international reference literature on SDIs was carried out. In this way, based on the main problems for documentation and information management

Table 1. Indicators for assessing the SIP.

Indicator	Description
International and European standards	For designing SDIs some standards should be consulted, such as: – Spatial Information in the European Community (INSPIRE) – Open Geoespatial Consortium (OGC)
Compatibility	Periodical backup and migration of digital data to new software and hardware is needed which means a long-term financial commitment
Proactivity	New records should be systematically collected and archived to ensure their availability for future reference and use
Synergies between documentation centers	Efforts should be made to link information centers electronically avoiding duplicity and lack of data

Source: Own elaboration from Letellier, Schmid and LeBlanc [20].

in heritage sites established by Letellier, Schmid and LeBlanc [20], four indicators were set (Table 1). These parameters were used to assess the Heritage Information System experience in Santiago de Composela and show both its potential and its weaknesses. As the Geoportal is such a recent initiative, it has only been mentioned briefly.

4   Spatial Data Infraestructures in Santiago de Compostela

The Santiago Consortium[1] in collaboration with the National Geographic Institute developed the Heritage Information System that was launched in 2010 (http://www. consorcio-santiago.org/es/sistema-de-informacion-patrimonial).

The Higher Council for Scientific Research, the Government of Galicia and the Santiago City Council collaborated on the project. It is a computer tool for the management and knowledge of archaeological, architectural and urban information of the Historical City of Santiago de Compostela. The idea of the SIP stems from the need to gather all the information generated over several years of interventions in the historical city that was dispersed among different institutions and agents. Therefore, the basic objectives of the Heritage Information System were [13]:

1. To support the technical staff of the Consorcio in the management of projects concerning heritage elements in the town.
2. To support research actions about the history of the town.
3. To disseminate heritage information to the public.
4. To support the decision-making processes by providing an assessment on the condition of the town and on priority areas for financing.

The SIP is a geodocumental system designed to unite the fields of document managers (CRM-Enterprise Content Management) and Geographic Information Systems (GIS), based on the paradigm of Spatial Data Infrastructures (SDI). The SIP respects the normative INSPIRE and the rules of the Open Geospatial Consortium (OGC). Its cartographic base is the *Cartociudad* project (http://www.cartociudad.es) which, from a general perspective, could be defined as an SDI that deals with the production of spatial data services, with state coverage: cadastral, postal, road, cartographic, geographical and statistical [4].

The SIP web portal is divided into 4 thematic blocks (Fig. 1). In the History section there are descriptions related to the origins of the city and its evolution from the Middle Ages to the present. The section of Buildings of Interest was designed to search for information on the most important buildings in the city, taking into account nine types, including civil architecture, cathedral, churches and chapels, among others. As for the Heritage Information, this allows access to specialized documentation, such as the Special Protection and Rehabilitation Plan, historical plans, documents from the University Archive and the Cathedral Archive, as well as other urban and archaeological information. Finally, the Geographic Information System is constituted by a set

---

[1] The Consortium of Santiago is an executive body of the Royal Patronage of the City. It was established in 1992 to realize the institutional cooperation between the Government of Spain, the Government of Galicia and the City of Santiago, under the Presidency of the King of Spain.

of cartographic information in raster and vectorial format, to which the documentary information is added, thus managing the geodocument section in a single interface.

Fig. 1. Screenshot showing the thematic sections of the SIP.

For its part, the newly launched Geoportal of Santiago de Compostela City Council (http://xeoportal.santiagodecompostela.gal) aims to disseminate among institutions, entrepreneurs and citizens all geographic information that is available in the city. In this way, the geoportal has a web viewer that provides information on transport, the school system, service networks, buildings of historical interest and contemporary architecture (Fig. 2) or public parks. It also has a download center for cartography that includes historical maps from 1783. Apart from being accessible in Spanish and Galician, the geoportal is available in English, so it is also an informative tool for visitors to the city.

Fig. 2. Screenshot showing buildings of interest in the Geoportal of Santiago de Compostela City Council.

5 Discussion and Conclusions

Having presented the two examples of Spatial Data Infrastructures implemented by public institutions in Santiago de Compostela, in this section we will discuss their functionality as tools for heritage management and dissemination. The Heritage Information System has a greater orientation as a management system. Making an analysis of the indicators established for its evaluation, it is first of all necessary to mention that it complies with the main regulations that govern SDIs, the international regulation of the OGC and the European INSPIRE. However, an exhaustive exploration of the web viewer of the Heritage Information System allowed us to verify that searches in some sections are no longer possible, as is the case of Buildings of Interest, where descriptions of buildings are no longer accessible. In terms of compatibility, it is appreciated that an update of the software and hardware had not been made, and therefore, the system was outdated.

In relation to the synergies between the documentation centers, the Urbe Program, included in the Patrimonial Information section, allows linking the SIP with the documents of the University Archive and the Cathedral Archive. However, greater integration with other asset management systems is necessary, such as with the City Council. The decoupling with the latter has impeded the proactivity of the system. Systematicity in the introduction of contents to the SIP has not been achieved, since it is not linked to the procedure of processing the files of the works carried out by the technicians of the municipality.

Obviously, the SIP was a useful tool for asset managers, managing to integrate Geographic Information Systems (Arcgis) with heritage management technologies (Livelink). However, the issues discussed and perhaps the lack of introduction of other elements such as three-dimensional, multi-temporal, collaborative or multimedia environments, contributed to it becoming obsolete, both for managers and for the general public. Instead, the city gained a new tool, the Geoportal of Santiago de Compostela City Council. It constitutes a more global tool than the SIP, since it aims to be a management system, but which is also very accessible to citizens. In this regard, the fact of its accessibility in English, an important aspect taking into account the vocation of Santiago de Compostela as a cultural and pilgrimage destination is remarkable.

Regarding the evaluation of the Heritage Information System of Santiago de Compostela, it must be mentioned that it was a very positive experience, as corroborated by the fact that it has been used uninterruptedly since its creation until 2017. In any case, the rapid evolution of the software and web applications has forced the implementation of a new generation Spatial Data Infrastructure, which has been launched recently. This study shows the fact that in a changing technological world, ensuring long-term financing is essential for the maintenance of up-to-date heritage management tools.

References

1. GSDI: Spatial Data Infrastructure Cookbook 2012 Update. http://gsdiassociation.org/index.php/publications/sdi-cookbooks.html. Accessed 20 Nov 2017
2. Parcero-Oubiña, C.: La arqueología y las infraestructuras de datos espaciales. In: Manual de tecnologías de la información geográfica aplicadas a la arqueología, pp. 341–359. Museo Arqueológico Regional, Madrid (2016)
3. Estévez, X.: Santiago de Compostela, conservación y transformación. Arbor **170**(671–672), 473–488 (2001)
4. López-Felpeto, M.A., Carrero, M.: Geographic Information Systems applied to urban archaeology: the case of the "SIP" of Santiago de Compostela. SÉMATA, Ciencias Sociais e Humanidades **27**, 349–363 (2015)
5. Corns, A., Shaw, R.: Cultural heritage Spatial Data Infrastructures (SDI) - unlocking the potential of our cultural landscape data. In: Rainer, R. (ed.) 30th EARSeL Symposium Remote Sensing for Science, Education, and Natural and Cultural Heritage, pp. 1–8. Curran Associates, Paris, (2010)
6. Pavlidis, G., Koutsoudis, A., Arnaoutoglou, F., Tsioukas, V., Chamzas, C.: Methods for 3D digitization of Cultural Heritage. J. Cult. Heritage **8**(1), 93–98 (2007)
7. Niccolucci, F.: European digitization policies: the cultural and political background. In: Niccolucci, F. (ed.) Digital Applications for Tangible Cultural Heritage Report on the State of the Union Policies, Practices and Developments in Europe, vol. 2, pp. 7–14. Archaeolingua, Budapest (2007)
8. Santana-Quintero, M., Addison, A.C.: Digital tools for heritage information management and protection: the need of training. In: Wyeld, T.G., Kenderdine, S., Docherty, M. (eds.) Virtual Systems and Multimedia. VSMM 2007. Lecture Notes in Computer Science, vol. 4820, pp. 35–46. Springer, Heidelberg (2008)
9. Myers, D., Dalgity, A., Avramides, I., Wuthrich, D.: Arches: an open source gis for the inventory and management of immovable cultural heritage. In: Ioannides, M., Fritsch, D., Leissner, J., Davies, R., Remondino, F., Caffo, R. (eds.) Progress in Cultural Heritage Preservation. EuroMed 2012. Lecture Notes in Computer Science, vol. 7616, pp. 817–824. Springer, Heidelberg (2012)
10. Nakamura, T.: Construction of GIS database on historic buildings in Buddhist temples and Shinto shrines in the historic center of Kyoto: an evaluation on distribution of cultural heritage in the historic city. AIJ J. Technol. Des. **18**(39), 765–770 (2012)
11. Hadjimitsis, D., Agapiou, A., Alexakis, D., Sarris, A.: Exploring natural and anthropogenic risk for cultural heritage in Cyprus using remote sensing and GIS. Int. J. Digital Earth **6**(2), 115–142 (2013)
12. Spano, A., Pellegrino, M.: Craft data mapping and spatial analysis for historical landscape modeling. J. Cult. Heritage **14**(3), S6–S13 (2013)
13. Parcero-Oubiña, C., Vivas White, P., Güimil Fariña, A., Blanco Rotea, R., Pavo López, M. F., Silgado Herrero, Á., Hernández Caballero, A., Granado García, C.: GIS-based tools for the management and dissemination of heritage information in historical towns: the case of Santiago de Compostela (Spain). Int. J. Heritage Digital Era **2**(4), 655–675 (2013)
14. Al-Kheder, S., Haddad, N., Fakhoury, L., Baqaen, S.: A GIS analysis of the impact of modern practices and policies on the urban heritage of Irbid, Jordan. Cities **26**(2), 81–92 (2009)
15. Dearstyne, B.W.: Managing Historical Records Programs: A Guide for Historical Agencies. Altamira Press, Walnut Creek (2000)

16. Garg, N.K., Gupta, A.K.: Database generation of heritage buildings and monuments: a strategy. Conserv. Cult. Property India **34**, 56–60 (2001)
17. ICOMOS: Charter- Principles for the analysis, conservation and structural restoration of architectural heritage (2003). https://www.icomos.org/fr/chartes-et-normes. Accessed 20 Nov 2017
18. Council of Europe: Guidance on Inventory and Documentation of the Cultural Heritage. Council of Europe Publishing, Strasbourg (2001)
19. FISH: MIDAS Heritage - The UK Historic Environment Data Standard, v1.1. English Heritage, Bristol (2012)
20. Letellier, R., Schmid, W., LeBlanc, F.: Recording, Documentation, and Information Management for the Conservation of Heritage Places: Guiding Principles. Getty Conservation Institute, Los Angeles (2007)

From Periphery to City, from City to Metropolitan Area: Growth of Urban Periphery, Strategies and Transformations

Alessandra Parise[✉]

Mediterranea University of Reggio Calabria, 89100 Reggio Calabria, Italy
alessandraparise.235@gmail.com

Abstract. The objective of this work is to analyze the structure, organization and attributions that characterize the Metropolitan Cities or Areas present in certain European Union countries. This is a complex path, developed taking into account the considerable differences between the historical experiences, the institutional structures and the economic-social conditions that have characterized the various forms in which the metropolitan experience was "built" in the various national contexts. In this research work we present the results of an analysis that took care of the city and of the relationships between the urban transformation paths and the processes of socio-economic development at the local level, the expansion of the same and the strategies of development. First of all we tried to analyze contemporary trends dominating the urban development and the real dynamics that underline the changes occurred in urban planning especially in recent years. Economic and social changes at the international level, have heavily influenced the evolution and unfolding of the urban dimension and this work, first proposes a theoretical and interpretative framework of urban reality that changes and subsequently it focuses on Europe and Italy, in order to investigate institutional aspects, normative indications, settlement models, public policies adopted and examples of realized practices, which have intervened on urban transformations and related processes of development. The aim is to offer a reasoned contribution to the possible and significant models of organization and governance of metropolitan areas in Europe, with ideas or solutions to outline our metropolitan model.

Keywords: Strategies · Metropolitan model · Grow · Smart City Better life

1 Basic Concepts for Urban Development

1.1 The Idea of City Sample

After events so decisive for civilization as concentration settlement and after several times events of planting and formation of cities, expansion and flowering, transplantation or decline until death, with or without resurrection, or persistent plurimillenary renewal on site and territorial restructuring, we must reach up to times extremely close together because the very idea of the city is represented in all its evidence and the functions of the settlements human beings on the territory appear in all their dynamic

© Springer International Publishing AG, part of Springer Nature 2019
F. Calabrò et al. (Eds.): ISHT 2018, SIST 100, pp. 21–29, 2019.
https://doi.org/10.1007/978-3-319-92099-3_3

complexity: in short, to understand, as Patrick Geddes taught towards the end of the last century, that a village, a city, a region is not just a "place in space", but a "drama in time", therefore in a process of dynamic development [1]. Essentially static and spatially delimited is the informative idea of the cities in the ancient world, from the palatial settlements to the "polis", which transpires from the descriptive fragments of historians, geographers and travelers, from urban regulations and archaeological evidence, as well as from the same Platonic and Aristotelian hypotheses of ideal formation and political regiment. Urban art developed in those centuries, which enriches the princely cities of new major architectural events; at the same time it begins to theorize on the "forma urbis" up to give life to a flowering of new urbanistic ideas that under the guise of "Ideal cities", are placed, as regards the existing ones, as many possible global alternatives; very often the cherished innovations are only formal, geometric and defensive, but in this inventive research new ideas break out both in the technical and in the of social organization field, paving the way for utopias.

It is no coincidence that it is precisely the Cerdà, which first employs systematically the statistical analysis in the preparatory studies for the Barcelona plan, to precede the printed illustration of the plan (1867) with an essay titled "Teoria general de la Urbanizaciòn" where the same word is used in the twofold meaning of concentration of urban population and physical expansion of the city [2]. This twofold meaning is currently in common use in Spanish and French, while in Italian and English two meanings are expressed in two distinct words: the demographic-social meaning respectively with the words "urbanism" and "urbanization", the physical one with "urban expansion" and with "city development" (as is the book by Geddes and Mawson, 1903), where "development" it is used in a broad sense of "development", or with "physical growth of towns". In Italian, the word "urbanization" is of very recent use and is used exclusively to indicate the process of transformation of use from agricultural land to urban land through design and implementation of works, installations, services and buildings at various destinations; and has its English correspondent in technical and legal meaning of 2development" (since 1947), with the warning that this word has meaning even more extensive, including even the simple transformations in the use of real estate, so that its translation into Italian maybe, according to the case, "urbanization" or "transformation of Use"; also the French word "urbanization" and the Spanish word "urbanizaciòn" used in the meaning physicists are also recently used in the more specific sense of urbanization and development [3].

2   The Matrix of the European City

2.1   European City and Metropolis: Birth of the Modern Concept

Talk about European city and then propose general statements that are valid for all cities it seems difficult. On the other hand, European cities today are just as profoundly transformed from the dynamics of globalization - economic, political and cultural - enough to make them similar to the cities of many other parts of the world.

In this context the category European city seems to have lost its meaning; yet when we look to the American city we discover diversities that strike us precisely because we compare them with them experiences and images of cities closer to us and family about three-quarters of the total population lives in urban areas. An in-depth analysis on the peculiarities of European cities and carried out by which identifies the typical characteristics of these metropolises [4, 5].

Vicari firstly affirms that the European city has a distinct morphological dimension: it is characterized for its compact and densely built form around a central area (historic center) where public buildings, churches, monuments, areas for trade and exchanges are concentrated. Starting from this center, the city develops along the radial lines, articulating in streets and squares. Despite this diversification, for the vast majority of European urban agglomerations the structuring of cities with reference to its historical center remains one common element and strong meaning. Bruzzo states that in different periods of time, urbanization processes had great importance throughout the European continent, so much so that Europe is currently the more urbanized continent, since it is not only that more ancient urbanization, but also that in which about three quarters of the total population lives in urban areas [6].

2.2 Metropolitan Model Governments

The identification and the delimitation of the "metropolitan areas" are, in fact, extremely issues complex. In general, used for this operation can be grouped into three large categories: (a) homogeneity criteria, on the basis of which entities or areas can be combined similar characteristics according to various parameters (demographic dimension, density, economic characteristics and sociological and so on); (b) criteria of interdependence, on the basis of which they can be grouped together entities or areas in which exchanges of people, goods or communication flows occur (commuting, areas of commercial gravitation, telephone exchanges or other); (c) morphological criteria, such as contiguity spatial or belonging to the same orographic or geographical configuration systems in a broad sense. In recent years, some international research institutions have carried out studies aimed at delimiting the metropolitan areas or similar territorial aggregates. In particular, in the research promoted by the European Spatial Planning Observation Network (Espon) are identified, for the European territory, the Morphological Urban Areas (MUA) and Functional Urban Areas (FUA) composed of those municipalities in which one significant share of the resident population moves for work purposes in the territory of a particular one Morphological Urban Area. The two units of analysis (MUA and FUA) are of equal importance for the classification of the urban and metropolitan phenomenon: if the MUAs allow to highlight the presence of contiguous densely populated territories that transcend the boundaries of individual municipal administrations, the FUA, on the other hand, allow the detection of the sphere of influence of the MUAs in terms of attracting the population with work capacity. In particular, the European urban system is characterized by the existence of a dense urban network, formed by urban regions that belong to large metropolitan cities [7]. These agglomerations are located at various levels of the urban hierarchy: so, at the highest level we find the so-called, "global" cities, characterized by the concentration of command and control structures of the system economic, industrial and financial world-wide, from the presence of infrastructures and infrastructures research training centers at a higher level. This is, in particular, London and Paris, where such concentration is much higher than in other European cities, followed by Milan, Berlin, Munich, Rome, Copenhagen, Barcelona, Amsterdam (Fig. 1).

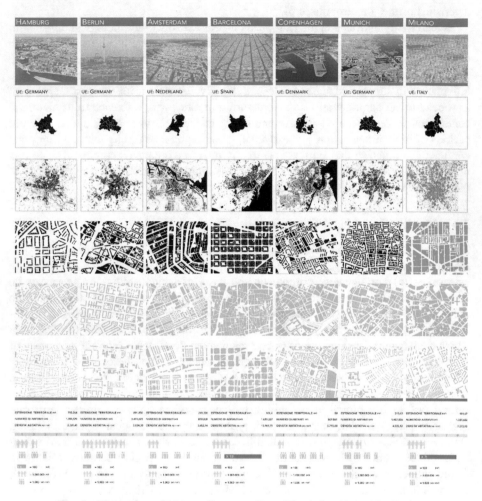

Fig. 1. Illustration of Metropolian area. Font: Thesis by Alessandra Parise.

3 From City to Metropolitan Area

3.1 The Concept of the Metropolitan Area

The "traditional" city was characterized by a substantial coincidence between the population that inhabited it and the one that worked there. During the day, the presence of people living outside the city was completely marginal. The metropolis manifests itself initially when this coincidence begins to fail. This happens at the time when cities, especially the larger ones, due to their economic strength linked to the industrialization and development of means of transport, they begin to exercise an influence on territorial areas so extensive as to generate the phenomenon of commuting, for which quotas important populations enter the city-metropolis daily to work there but live elsewhere [8]. In this sense, the transformation of large cities into metropolises began during the first decades of the twentieth century in the United States and then spread to Europe, reaching maturity in the decades immediately following the second post-war period. The metropolitan question is a very current topic. From this reason comes the idea of being able to put a comparison city and be able to extrapolate winning strategies. The concept of confrontation between Reggio Calabria and Oslo starts from this fundamental point: highlighting urban policies, connections, the importance of greenery and the functions that make Oslo a real Smart City (Fig. 2). Being able to draw inspiration from this small city and grasping the main functions to be able to implement it in Reggio Calabria is the goal of the work. The metropolitan question is a very current topic, for this reason we identify the macro areas to identify the two cities of comparison (Fig. 3).

Fig. 2. Various views of Oslo. Font: Sketch by Alessandra Parise.

The question, at the time, it was posed by the fact that the middle classes had moved their residence into counties out of big cities, continuing to use its services and infrastructures (from theaters to transport networks), which involved the not inconsiderable contradiction of large urban centers that they had to support high-level services relying on a low-income population while the wealthier classes were paying their taxes in small suburban centers; the need for a recognizable referent, a subject able to be interlocutor of the other great subjects that operate on the metropolitan area or anyway interfere.

Fig. 3. Metropolitan areas of the world. Font: Thesis by Alessandra Parise.

3.2 Strategic Planning of Cities

The metropolitan place of Oslo has become an increasingly complex and elusive place, whose definition it passes through the contribution of different disciplines and sometimes very sophisticated conceptualizations. This variety of contributions, in the absence of a process of convergence towards the construction of clear and above all useful paradigms, certainly does not benefit from the unambiguous definition of the metropolitan area, definition from which later elements could then emerge to attempt a delimitation. In the past, cities were geographically well defined, identifiable, self-contained entities to which they corresponded precise institutional levels [9]. The physical growth of the urban agglomeration is extended beyond the administrative boundaries, the population and economic activities have been redistributed throughout the territory interesting places around the central core. Over time the concept of urban system has changed and with it also the metropolitan area. The definition of the concept of "metropolitan area" can be refer to different conceptions that arise from as many different philosophies of approach to the problem but, for a correct delimitation of metropolitan area, it is not enough to refer to one or the other conception as it is necessary to search for a delimitation which takes into account at the same time all the approaches originating from synergistic integration of the conceptions of city:

- "political city" or government;
- "physical city", the city seen as a building continuum;
- "system city", the city seen as a system of production, distribution and continuum;
- "functional city", the city seen as a place of exchange e as a center of flows of goods, people and information.

Indeed, cities are the places where new and great inequalities become tangible social issues, environmental emergencies, the effects generated by migratory phenomena, the problems of social coexistence between groups belonging to different nations and/or ethnic groups, the themes of a quality of living in some more and more compromised cases [10, 11] (Fig. 4).

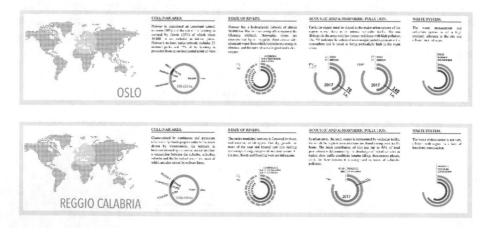

Fig. 4. Data and statistics of the environmental components of Oslo and Reggio Calabria. Font: Thesis by Alessandra Parise.

4 Grow of the City

4.1 The Model of the "Green City" for the Construction of the "Better City"

In urban planning, the term "intelligent city" refers to a set of city governance strategies aimed at optimizing and innovating service public materials of cities (called "Physical capital") with the human, intellectual and social capital of those who live there ("capital intellectual "and" social capital"), which, thanks to the widespread use of new technologies ICT (information and communication technologies) for communication, mobility, energy and environmental efficiency, determine urban performance and urban competitiveness of contemporary cities [12, 13]. Any definition of useful work for a Smart City needs to incorporate everyone these factors, to allow the development of good practices and relevant policy frameworks without losing the potential of scale. Despite this, the 'smart city' model seems to be what, according to an approach holistic and inclusive, it manages to contain all the meanings that the various concepts present in literature are able to express. The study is of particular interest because it identifies and compares the most important in being of 'Smart City', codifying a definition "Operative" of the term, valid for conducting a series of analyzes and collection of examples virtuous smart cities in Europe. Below is an excerpt from the article" Smart Cities in Europe "presented, in 2009, on the occasion of the 3rd Central European Conference on Regional Sciences: "We believe that a city is intelligent when it invests in human and social capital and traditional (transport) and modern (ICT) fuel communication infrastructure economic growth and high quality of life, with prudent management of natural resources, through participatory governance".

The awareness of the importance of safeguarding the urban environment, combined with the awareness that the quality of the urban environment has proved to be fundamental for quality of life and the well-being of citizens, has created a synergy between the issues of Smart City and that of sustainable urban development (Fig. 5).

According to a preliminary study, city of Oslo, these are the first strategic points for starting an urban change, improvement for the city of Reggio Calabria. The truth is that Reggio and Oslo have in common above all the naturalistic aspect.

Oslo has managed to extrapolate this factor and make it an attractor and a strong point (Fig. 6). Reggio Calabria could take an example, because it is characterized by a lot of environmental aspects that if they were maintained and valued they would be able to give a different face to the area. Not only attractors such as the urban system, connections, maintenance, the built-up to make Oslo one of the small but better cities, it is above all the enhancement of the green. Reggio Calabria, it is true, is lacking on the first aspects, but with the right strategies it would also be able to give a different face to the area. The truth is that Reggio and Oslo have in common above all the naturalistic aspect. Oslo has managed to extrapolate this factor and make it an attractor and a strong point. Reggio Calabria, it is true, is lacking on the first aspects, but with the right strategies it would also be able to give a different face to the area. It all depends on the quality of the environment that each of us wants to have. After all, we are the environment and we must improve more and more.

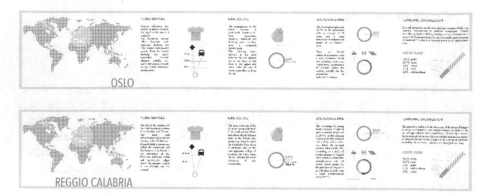

Fig. 5. Data and statistics of the urban developments of Oslo and Reggio Calabria. Font: Thesis by Alessandra Parise.

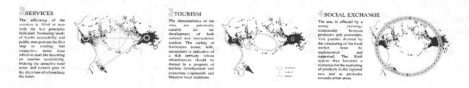

Fig. 6. Final strategies of Oslo. Font: Thesis by Alessandra Parise.

References

1. Alluli, M.: Città Metropolitane in Italia. La lunga attesa. Marsilio, Venezia, p. 16 (2014)
2. Atriparldi, V.: Il governo delle Aree Metropolitane. Esi, Napoli, p. 163 (1993)
3. Balboni, E.: L'Area Metropolitana tra funzioni e finzioni. Le Regioni **5**, 815 (1997)
4. Frasca, P.: Le Aree Metropolitane in Italia e nel mondo: quadro teorico e riflessi territoriali. Bollati Boringhieri, Torino, p. 15 (2009)
5. Vicari, H.: La città contemporanea. il Mulino, Bologna (2004)
6. Bruzzo, A.: Analisi del territorio, Letture sulla scienza economica regionale. Istituzione e Territorio dell'Università di Ferrara, Dipartimento di Economia (2006)
7. Carro Fernandez-Valmayor, J.L.: Una reflexion general sobre las area metropolitanas. Revista de Estudios de la Administraction Local y Autonomica **302**, 9 (2006)
8. Campilongo, G.: Aree Metropolitane, città metropolitane: individuazione dell'area metropolitana. Ricerca arpa, Lombardia (2011)
9. Crosetti, A.: Sul Governo delle Aree Metropolitane. Amministrare, p. 149 (1989)
10. Mantini, P.: La riqualificazione delle Aree Metropolitane: profile giuridici, in La riqualificazione delle aree metropolitane: quale futuro? In: Atti del XXVI Incontro di Studio, p. 23, Milano (1996)

11. Ruggiero, L.: La programmazione strategica e le politiche urbane. In: Zinna, S., Ruggiero, V. E., Grasso, A.A. (a cura di): Programmazione e linee strategiche per la progettazione del Masterplan di Catania, pp. 25–50. FrancoAngeli, Milano (2003)
12. Ruggiero, V., Scofani, L: Change in landscape and dynamics of cities, in Territorial cohesion of Europe and integrative planning. In: Relazione al 49th Congresso ERSA, 25th–29th August, Polonia (2009)
13. Tortorella, W., Alluli, M.: Le Città Metropolitane secondo la legge 135/2012. Amministrare **1**, 153 (2013)

Investigating Tourism Attractiveness in Inland Areas: Ecosystem Services, Open Data and Smart Specializations

Francesco Scorza$^{(\boxtimes)}$ (iD), Angela Pilogallo, and Giuseppe Las Casas (iD)

University of Basilicata, 85100 Potenza, Italy
Francesco.scorza@unibas.it

Abstract. From the beginning of the 21st century, following major European and global initiatives such as the Millennium Ecosystem Assessment (2005) [1, 2] and The Economics of Ecosystem and Biodiversity [3], the approach of Ecosystem Services could be considered an effective way to rebuild the traditional approach oriented to identify the impacts of territorial transformations in decision making processes. This research is oriented to contribute to the wider methodological framework of the Millennium Ecosystem Assessment [1]. Starting from this, the present work contributes to build interpretative models for the evaluation of a relevant part of the fourth class of ecosystem services: the territorial touristic attractiveness. The INVEST model, an open source toolkit, has been applied to assess the attractiveness of the Basilicata Region considering both natural and cultural heritage in order to highlight strengths and weaknesses of the investigated methodology, compared with Strategic development perspectives (also defined Smart Specialization Strategy).

Keywords: Ecosystem services · Millennium ecosystem assessment
Tourist attractiveness

1 Introduction

The 2005 Millennium Ecosystem Assessment (MA) could be considered one of the main effort to promote worldwide environmental assessments approaches sponsored by the United Nations. The most innovative contribution promoted by MA had been the unprecedented overview of the state of the world's natural environment founded on the basic idea that ecosystem value in decision making should be grounded on the idea of services provided to humans. That approach opened to a wide range of research linked to the demand by decision makers to hold new assessment tools in order to develop comprehensive scenario analysis. We prefer to agree on the implicit assumption that additional knowledge (deriving from such new interpretative assessment models) will reinforce the rational 'decision makers' in making 'better' decisions and policies [4–6].

This research aims to assess the territorial specialization level of the Basilicata region in terms of tourism attractiveness through the construction of a synthetic territorial index. The results can be used to support resource planning processes for the upgrading of the tourism sector. This improvement is in terms of "local specializations"

© Springer International Publishing AG, part of Springer Nature 2019
F. Calabrò et al. (Eds.): ISHT 2018, SIST 100, pp. 30–38, 2019.
https://doi.org/10.1007/978-3-319-92099-3_4

identified with the presence of natural, cultural and landscape resources (tourism attractors) and a supply system of proper specific services. The attention to territorial specialization belongs to the strategic setting of the 2014–2020 European Union Cohesion Policy [7–10], the Smart Specialization Strategy (cf. [11]) and the Regional Operational Programs according to the indications already formalized by Barca [12]. This work wants to be an attempt to build decision-making tools (DSS) (i.e. ToolKit [13–16]) that allow to apply the place-based approach [17] (or context-based approach according to Las Casas and Scorza definition [17, 18]) in an interpretative system of the territory elaborated at regional scale. The model of territorial interpretation built with this research is based on open data and open source tools, so it guarantees a high replicability of the evaluation procedures and the results extension to other contexts and case studies. The main result of this research is a synthetic territorial indicator of the tourism specialization level in the Basilicata region built using the spatial analysis tools and techniques. The analytical model used is included in the suite proposed by the Invest (Integrated Valuation of Ecosystem Services and Tradeoffs) software [19], an effective suite of tools supporting the development of Ecosystem Services approach in assessing territorial environmental features (among others [20–24]). After a brief description of the territorial context studied, highlighting the Basilicata's uniqueness in terms of tourism attractions, a suitable work was done to collect data and delivering elaborations in order to outline the comprehensive territorial index which, in this specific case, takes the significance of regional tourism attractiveness.

2   A Descriptive Overview of the Case Study and the Territorial Assessment Model

The case study focuses on Basilicata region, in particular on the relation between local resources and tourism development strategies. Basilicata territory is rich for natural habitats, cultural values [25] and traditions that make possible to indicate tourism development as the main key element for socio-economic progress of the entire region. Such values emerge also compared with current trends and research results concerning land take [26–29]. Tourism has gained a consistent weight in the economic and productive system of Basilicata thanks to significant public and private investment. This situation has led to a significant increase in the number of 'beds' and 'new accommodation facilities', with positive effects on the entire hospitality chains and a substantial increase of tourism demand, with the consequent strengthening of the tourist flows.

From 2016 emerges, with the role of major regional tourist destinations, the Metapontino area, focusing in seaside tourism, and Matera city, which is confirmed as the main cultural-historical attractive pole. Positive signals also come from other areas of the region where new tourist attractors are promoted, such as the Vulture area, the National Park of the Appennino Lucano, the city of Potenza - the main cultural services center - and, in general, all the coastal destinations.

The attraction system declines thematic areas (history, culture, landscape, etc.), each capable to generate significant added value depending on the communication modes, presentation and usage that will be realized in the short term: MATERA 2019 (see also [30, 31]).

Among the technological solutions available to produce territorial assessment of Ecosystem Services operating according with framework methodologies we used INVEST. The INVEST[1] model (Integrated Valuation of Ecosystem Services and Tradeoffs) was used as an instrument for assessing the tourist attraction of the Basilicata region. It is an open source software developed by the Stanford University Natural Capital (NCP) project and The University of Minnesota. This tool allows quantifying and mapping the consistency of ecosystem services and interpreting changes in ecosystems and relationships on human well-being. Invest includes eighteen distinct models and, in particular, for the purpose of the analysis carried out, the "Recreation and Tourism" package was applied. The software uses a graphical interface, elaborates a multiple linear regression model that includes geospatial functions on a complex system of input variables, which makes possible the interpretation of regional tourism specialization level. It is an input-output model based on the solution of a liner multiple regression according whit spatial functions applied on input variables[2].

Two territorial domains were considered in the analysis: natural heritage, cultural heritage.

The model was applied on each territorial domain and on the combination of the two domains. We show in the next figures the results obtained for each single domain and the result provided by the model for each variable (Fig. 1).

---

[1] INVEST- related documentation is available online at http://data.naturalcapitalproject.org/nightly-build/invest-users-guide/html/index.html.

[2] Based on these variables, the software processes a linear regression model. The regression equation used is the following:

$$y = \beta_0 + \beta_1 x1 + \beta_2 x2 + \ldots + \beta_i xi + e_i \quad i = 1, \ldots, N \tag{1}$$

where: βi are the linear regression coefficients; xi are the territorial components considered as predictive variables to input into the software; y matches with the expected value of the model, which in the specific case is the Basilicata region tourism specialization level.

Linear regression analysis is a technique that allows to analyse the linear relationship between a dependent variable (or response variable) and one or more independent variables (or predictors). It is an asymmetric methodology that is based on the hypothesis of the existence of a cause-effect relationship between several variables. The equation shown here contributes to the formation of a global index: the regional tourism attraction index (in this work). The template was applied on each domain and on proper combinations of domains to identify the most significant variable combination. The determination coefficient, better known as R2, is used as a measure of the good adaptation of the multiple linear regression model. This is a value between 0 and 1 and expresses the relationship between the variance explained by the model and the total variance. If the result is close to 1, it means that the predictors (input variables) are a good interpreter of the dependent variable value in the sample; if it is close to 0, they don't.

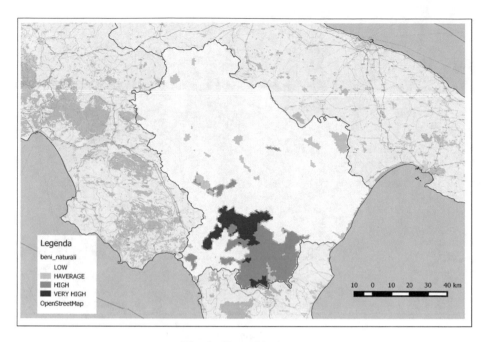

Fig. 1. Natural heritage.

The model allows to identify territorial specialization of the southwest of the region in terms of naturalistic-environmental attractors system supply [32]; it also highlights the role of other peculiar landscapes that characterize the area such as Monte Vulture, the graves area in Matera and part of the Appennino Lucano National Park. The contribution of coastal areas and that of many protected areas is also visible and it represent one of the most attractiveness of the Basilicata Region Tourism offer (Fig. 2).

Concerning the cultural heritage category, the result shows the distribution of the historical/cultural sites and the contribution that, in particular, the historical centers and the archaeological sites have on the regional tourism specialization. Matera city, where are concentrated specific development activities in the sector of cultural tourism connected with Matera 2019 EU Capital of Culture Program, is the main historical/cultural attraction pole in Basilicata while Potenza is the center of cultural services (Tables 1, 2 and 3).

The combination of "natural heritage" and "cultural heritage" domains provides a representation in which the natural components of the analysis prevailed on the others. The model allows to specify areas whit the integration of several tourism specialization sector of the entire regional territory (see Fig. 3).

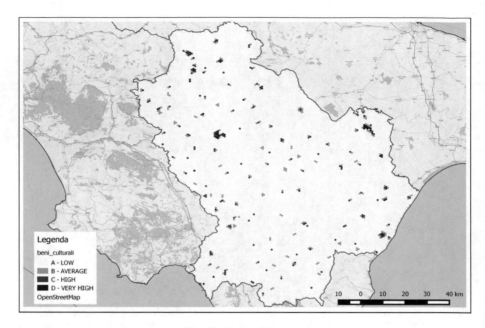

Fig. 2. Cultural heritage

Table 1. Territorial components and associated linear regression coefficients.

Territorial components	β_i
Intercept	+6.487 e−02
Oasis	+0.000 e+00
Areas_euap	+3.184 e−08
Coast_line	+2.074 e−03
Attractions	−3.536 e−17
Riverside	−9.983 e−06
Naturalistic trails	+5.741 e−05
Areas_protect	−2.724 e−09

Table 2. Territorial components and associated linear regression coefficients.

Territorial components	β_i
Intercept	+2.959 e−02
Cultural heritage	+0.000 e + 00
Archaeological sites	+2.384 e−06
Theatres	+3.031 e−04
Pro_Loco	+2.279 e−04
Museums	+1.875 e−04
Storic_centres	+2.529 e−06
Associations	+3.047 e−02

Table 3. Territorial components and associated linear regression coefficients.

Territorial components	β_i
Intercept	+2.747 e−02
Oasis	+7.957 e−15
Areas_euap	+4.033 e−08
Areas_protect	+8.858 e−10
Coast_line	+2.012 e−03
Attractions	+2.220 e−16
Riverside	−1.790 e−06
Cultural heritage	+0.000 e + 00
Archaeological sites	+2.373 e−06
Theatres	+3.036 e−04
Pro_Loco	+2.232 e−01
Museums	+1.872 e−04
Storic_centres	+2.532 e−06
Associations	+3.070 e−02

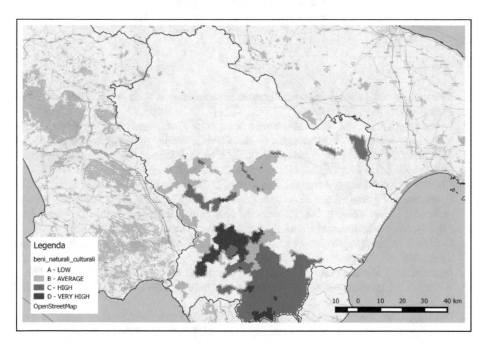

Fig. 3. Combination of "natural heritage" and "cultural heritage" domains.

3   Conclusions

This research proposes a global interpretation of the regional territory in terms of tourist attraction with the ambition that this information could support the governance processes of tourism development [33]. In other terms, it could be considered a kind of land suitability map for tourism development policies.

The research highlights incomplete information in terms of availability of spatial data and on the other hand the opportunity to exploit non-conventional data sources in order to improve the level of spatial knowledge as an input factor for INVEST. About the methodological profile, the INVEST model was useful for the geospatial features offered. In order to reach a more significant representation of the territorial tourism attractiveness levels, there are few possibilities to define a weighing system for input variables [24].

As a potential extension of the analysis, in the specific case of Basilicata region, it should be compared with the spatial distribution of natural risks, including land take, and spatial planning system [17, 34, 35] in order to provide a more comprehensive representation.

A potential extension of research is to integrate the main components of Ecosystem Services in order to achieve effective decision support tools enhancing sustainability assessment capacity of territorial transformations (see also [36–38]).

References

1. Millennium Ecosystem Assessment, Ecosystems and Human Well-being: Synthesis. Island Press, Washington, DC (2005)
2. Leemans, R., De Groot, R.S.: Millennium Ecosystem Assessment: Ecosystems and human well-being: a framework for assessment. Island Press, Washington, Covelo, London (2003)
3. Millenium assessment contribution) (2003). https://islandpress.org/book/ecosystems-and-human-well-being-0?prod_id=474. Accessed 12 Nov 2017
4. The Economics of Ecosystems and Biodiversity (TEEB) The Economics of Ecosys-tems and Biodiversity: Ecological and Economic Foundations. Pushpam Kumar. Earthscan, London and Washington (2010). http://www.teebweb.org/our-publications/teeb-study-reports/eco-logical-and-economic-foundations/. Accessed 12 Nov 2017
5. Owens, S.: Making a difference? some perspectives on environmental research and policy. Trans. Inst. Br. Geogr. New Ser. **30**, 287–292 (2005)
6. Sanderson, I.: Making sense of what works. Public Policy Public Adm. **17**, 61–75 (2002)
7. Weiss, C.: The many meanings of research utilization. Public Adm. Rev. **39**, 426–431 (1979)
8. DG Regio, The Programming Period 2014–2020 - Monitoring and Evaluation of European Cohesion Policy (European Regional Development Fund and Cohesion Fund) - Concepts and Recommendations, Guidance Document, Directorate-General for Regional Policy, European Commission, Brussels (2011)
9. EC, Green Paper on Territorial Cohesion Turning territorial diversity into strength Communication from the Commission to the Council, the European Parliament, the Committee of the Regions and the European Economic And Social Committee. European Commission COM (2008) 616 final, 6 October, Brussels (2008), http://ec.europa.eu/regional_policy/archive/consultation/terco/paper_terco_en.pdf. Accessed 12 Nov 2017

10. EC, Investing in Europe's future: Fifth report on economic, social and territorial cohesion. European Commission. Brussels (2010). http://ec.europa.eu/regional_policy/sources/docoffic/official/reports/cohsion5/pdf/5cr_part1_en.pdf. Accessed 12 Nov 2017

11. EC, EUROPE 2020 - a strategy for smart, sustainable and inclusive growth, Communication from the Commission, Brussels (2010a)

12. McCann, P., Ortega-Argilés, R.: Smart Specialization, Regional Growth and Applications to European Union Cohesion Policy. Reg. Stud. **49**(8), 1291–1302 (2015)

13. Barca, F.: An Agenda for a Reformed Cohesion Policy: A Place-Based Approach to Meeting European Union Challenges and Expectations (2009)

14. Las Casas, G., Scorza, F.: Sustainable Planning: A methodological toolkit. In: Gervasi, O., Murgante, B., Misra, S., Rocha, C.A.M.A., Torre, C., Taniar, D., Wang, S. (eds.) Computational Science and Its Applications - ICCSA 2016: 16th International Conference, Beijing, China, July 4–7, 2016, Proceedings, Part I, pp. 627–635. Springer, Cham (2016)

15. Las Casas, G., Scorza, F., Murgante, B.: Razionalità a-priori: una proposta verso una pianificazione antifragile. Italian J. Reg. Sci. in printing (2018)

16. Las Casas, G., Scorza, F., Murgante, B.: Conflics and sustainable planning: peculiar instances coming from val d'agri structural inter-municipal plan. In: Fistola, R., Papa, R. (eds.) Smart Planning: Sustainability and Mobility in the Age of Change. Springer (2018)

17. Las Casas, G., Scorza, F.: A renewed rational approach from liquid society towards anti-fragile planning. In: International Conference on Computational Science and Its Applications, pp. 517–526 (2017)

18. Las Casas, G., Scorza, F.: Un approccio "contex based" e "valutazione integrata" per il futuro della programmazione operativa regionale in Europa. In: Bramanti, A., Salone, C. (eds.) Lo sviluppo territoriale nell'economia della conoscenza: teorie, attori strategie, vol. 42, pp. 253–274. FrancoAngeli Editore (2009)

19. Las Casas, G., Lombardo, S., Murgante, B., Pontrandolfi, P., Scorza, F.: Open data for territorial specialization assessment territorial specialization in attracting local development funds: an assessment. procedure based on open data and open tools. Tema, J. Land Use, Mobil. Environ. (2014)

20. Natural Capital Project (NCP), InVEST User Guide (2015). http://data.naturalcapitalpro-ject.org/nightly-build/invest-users-guide/html/#. Accessed 12 Nov 2017

21. Antognelli, S., Vizzari, M.: Landscape liveability spatial assessment integrating ecosystem and urban services with their perceived importance by stakeholders. Ecol. Ind. **72**, 703–725 (2017)

22. Antognelli, S., Vizzari, M.: Ecosystem and urban services for landscape liveability: A model for quantification of stakeholders' perceived importance. Land Use Policy **50**, 277–292 (2016)

23. Vizzari, M., Modica, G.: Environmental effectiveness of swine sewage management: a multicriteria AHP-based model for a reliable quick assessment. Environ. Manage. **52**(4), 1023–1039 (2013)

24. Cannas, I., Zoppi, C.: Ecosystem services and the natura 2000 network: a study concerning a green infrastructure based on ecological corridors in the metropolitan city of Cagliari. In: Gervasi, O., Murgante, B., Misra, S., Borruso, G., Torre, C.M., Rocha, A.M.A.C., Cuzzocrea, A. (eds.) Computational Science and Its Applications, pp. 379–400. Springer, Cham (2017)

25. Scorza, F., Fortino, Y., Giuzio, B., Murgante, B., Las Casas, G.: Measuring territorial specialization in tourism sector: the basilicata region case study. LNCS (LNAI and LNBI), vol. 10409. LNCS, pp. 540–553 (2017)

26. Amato, F., Martellozzo, F., Nolè, G., Murgante, B.: Preserving cultural heritage by supporting landscape planning with quantitative predictions of soil consumption. J. Cult. Herit. **23**, 44–54 (2017)
27. Martellozzo, F., Amato, F., Murgante, B., Clarke, K.C.: Modelling the impact of urban growth on agriculture and natural land in Italy to 2030. Appl. Geogr. **91**, 156–167 (2018)
28. Modica, G., Laudari, L., Barreca, F., Fichera, C.R.: A GIS-MCDA based model for the suitability evaluation of traditional grape varieties: the case-study of "Mantonico" grape (Calabria, Italy). Int. J. Agric. Environ. Inf. Syst. **5**(3), 1–16 (2014)
29. Modica, G., Zoccali, P., Di Fazio, S.: The e-participation in tranquillity areas identification as a key factor for sustainable landscape planning. In: Murgante, B., Misra, S., Carlini, M., Torre, C.M., Nguyen, H.Q., Taniar, D., Gervasi, O. (Eds.): Computational Science and Its Applications - ICCSA 2013. LNCS, vol. 7973, pp. 550–565. Springer, Heidelberg (2013)
30. Modica, G., Pollino, M., Lanucara, S., La Porta, L., Pellicone, G., Di Fazio, S., Fichera, C. R.: Land Suitability Evaluation for Agro-forestry: Definition of a web-based multi-criteria spatial decision support system (MC-SDSS): preliminary results. In: Gervasi, O., Murgante, B., Misra, S., Rocha, A.C.A.M., Torre, M.C, Taniar, D., Wang, S. (Eds.): Computational Science and Its Applications - ICCSA 2016. LNCS, vol. 9788, pp. 399–413. Springer, Cham (2016)
31. Lucio, A., Gennaro, I.: Capitale Europea della Cultura 2019. Un'analisi delle candidature italiane, pp. 141–158 (2014)
32. Pontrandolfi, P., Scorza, F.: Sustainable urban regeneration policy making: inclusive participation practice. In: Gervasi, O., Murgante, B., Misra, S., Rocha, C.A.M.A., Torre M. C., Taniar, D., Wang, S. (eds.) Computational Science and Its Applications - ICCSA 2016: 16th International Conference, Beijing, China, July 4–7, Proceedings, Part III, pp. 552–560. Springer, Cham (2016)
33. PTR, (Piano Turistico Regionale). www.regione.basilicata.it/giunta/files/docs/DOCUMENT_FILE_523485.pdf. Accessed 12 Nov 2017
34. Amato, F., Martellozzo, F., Nolè, G., Murgante, B.: Preserving cultural heritage by supporting landscape planning with quantitative predictions of soil consumption. J. Cult. Herit. (2015)
35. Amato, F., Pontrandolfi, P., Murgante, B.: Supporting planning activities with the assessment and the prediction of urban sprawl using spatio-temporal analysis. Ecol. Inform. **30**, 365–378 (2015)
36. Amato, F., Maimone, B.A., Martellozzo, F., Nolè, G., Murgante, B.: The effects of urban policies on the development of urban areas. Sustainability **8**, 297 (2016)
37. Lai, S., Leone, F., Zoppi, C.: Land cover changes and environmental protection: a study based on transition matrices concerning Sardinia (Italy). Land Use Policy **67**, 126–150 (2017)
38. Lai, S., Lombardini, G.: Regional drivers of land take: a comparative analysis in two Italian regions. Land Use Policy **56**, 262–273 (2016)
39. Dvarioniene, J., Grecu, V., Lai, S., Scorza, F.: Four perspectives of applied sustainability: Research implications and possible integrations. LNCS (LNAI and LNBI), vol. 10409. LNCS, pp. 554–563 (2017)

Telemedicine and Impact of Changing Paradigm in Healthcare

Domenico Marino[1]([⊠]) [iD], Antonio Miceli[2],
Demetrio Naccari Carlizzi[3], Giuseppe Quattrone[4], Chiara Sancin[5],
and Maurizio Turchi[5]

[1] Mediterranea University of Reggio Calabria, 89100 Reggio Calabria, Italy
dmarino@unirc.it
[2] University of Messina, 98122 Messina, Italy
[3] P4C, Prepare 4 Change, Reggio Calabria, Italy
[4] GTechnology, 41123 Modena, Italy
[5] Dida, 98123 Messina, Italy
https://www.demetrionaccari.it/
articoli/categoria/Iniziative/sotto_categoria/Prepare
%204%20Change

Abstract. Health monitoring, crisis prevention and support for everyday activities represents an emerging field of application at a national level, with particular reference to fragile individuals, the elderly and people with chronic diseases.

An important aspect that should be explored by the end of this decade is how the technologies of artificial intelligence, as applied in the health context, might ultimately improve the quality of the current system and whether the work done as part of the efforts now being made is optimised and sufficient to achieve new objectives. In particular, the ability to process large quantities of data will act as a catalyst, triggering an extremely high number of benefits in the health and wellness sector in terms of prevention, diagnosis and individual treatment.

Keywords: E-Health · IA · Impact

1 Introduction

Health monitoring, crisis prevention and support for everyday activities represents an emerging field of application at a national level, with particular reference to fragile individuals, the elderly and people with chronic diseases. With this in mind, the prevention of functional decline and the treatment of physical frailty and cognitive weakness take on particular importance, as does the development of solutions for independent living, including by studying new diagnostic models and monitoring tools capable of providing for the clinical risk and at the same time reducing the related health and care costs.

An important aspect that should be explored by the end of this decade is how the technologies of artificial intelligence, as applied in the health context, might ultimately improve the quality of the current system and whether the work done as part of the

© Springer International Publishing AG, part of Springer Nature 2019
F. Calabrò et al. (Eds.): ISHT 2018, SIST 100, pp. 39–43, 2019.
https://doi.org/10.1007/978-3-319-92099-3_5

efforts now being made is optimised and sufficient to achieve new objectives. In particular, the ability to process large quantities of data will act as a catalyst, triggering an extremely high number of benefits in the health and wellness sector in terms of prevention, diagnosis and individual treatment.

2 The Emerging Chronic Pathologies

In the healthcare sector, the problem of chronic health conditions is increasingly a key priority. According to the WHO, the world's most widespread chronic diseases, with high mortality rates include cardiovascular diseases (causing 17.5 million deaths per year), cancer (7.6 million), chronic respiratory diseases (4.1 million) and diabetes (1.1 million)[1].

For instance, untreated hypertension, one of the key risk factors (atherosclerosis of the coronary arteries, the carotid artery, the renal arteries and the lower extremity arteries) could be controlled remotely in a relatively simple way.

According to WHO experts, in the year 2000 about 1 billion people were affected by hypertension worldwide (over 25% of the global population above 20 years of age).

According to estimates, in 2025 the number of people suffering from hypertension could reach 1.5 billion (almost 30% of the population).

Hypertension, for example, is the cause of about 14% of all deaths worldwide, and is also one of the most common causes of disability. It has been shown that the proportion of patients with hypertension increases in parallel with population age.

Type 2 diabetes mellitus especially represents worldwide an out-and-out pandemic.

According to the estimates of the International Diabetes Federation, 415 million people in the world suffer from diabetes, but the number is destined to rise to 642 million by 2040. In Italy, according to ISTAT [National Statistics Institute], around 3.5 million people suffer from diabetes, one million more than just 10 years ago. At the current rate of growth, by 2030, there will be more than 5 million Italians suffering from the disease.

The implementation of strategies designed to prevent diabetes through lifestyle intervention represents the only possible way of stemming this highly serious clinical, social and economic problem.

3 Telemedicine, Chronic Pathologies and Innovative Healthcare Strategies

Telemedicine can be a new frontier to follow patients, boosting their return to independence and facilitating their self-care capacity (allowing them to be self-sufficient in their homes and neighbourhoods). The basic concept is to create an integrated network between the hospital, community care services and the home, linking together the existing healthcare professional roles (GPs, paediatricians, nurses, podiatrists, medical

---

[1] http://www.who.int/chp/chronic_disease_report/contents/en/page2.

specialists) and perhaps other relevant professionals (such as pharmacists with appropriate expertise), who could handle certain tasks, such as scheduling medical appointments, preparing pharmaceutical formulation, educating and informing patients and care-givers [1, 2].

In the current social and economic context it will be essential to develop a healthcare model designed to ensure efficiency and continuity in healthcare delivery, to improve the quality of life of patients (not only elderly patients) optimising the existing resources and reducing costs. In this perspective, it is especially important to decentralise treatment, possibly by moving from the hospital setting to home-based care, while improving the quality and efficacy of professional healthcare delivery.

In summary the key drivers of the growing demand for telemedicine services are:

• the ageing population and the limited capacity to provide full-time care in care homes for the elderly;
• the government schemes to combat diseases and bad lifestyle choices, by means of large-scale prevention campaigns;
• the increase in the number of patients with chronic conditions who live at home;
• the reduction in the numbers of healthcare professionals and easier access to specialists;
• new treatment technologies.

In particular, the aims of telemedicine services are:

• respond in the early stages of disease with monitoring, to reduce the rate of disease progression and mortality;
• reduce the length of hospital stays;
• optimise decisions by means of remote consultation by specialists;
• reduce the cost of care for patients.

4 The Impact of Changing Paradigm in Healthcare

Achieving this paradigm shift in healthcare systems involves different focusing of care levels and healthcare provision channels. While the hospital would continue to play a key role in intensive care, community and home-based care would be the focus of the prevention effort and of monitoring to avert acute episodes.

This process, in which Italy is lagging behind other western European countries, requires increasingly holistic and personalised healthcare solutions. This approach focuses on systems for monitoring vital signs, on prevention and on patient well-being; it should be achieved by strategies ensuring quality and cost effectiveness.

The financial burden of healthcare delivery is on course to become unsustainable. On average, it absorbs 10.3% of the national GDP in Europe. In Italy, the figure is about 9.1%. This figure is set to rise in parallel with the steady increase in chronic diseases, which account for 75% of expenditure, also on account of the progressive ageing of the population. Projections show that the percentage of people over 65 will rise from 21% at present to 34% in 2051. Furthermore, the ratio of elderly people (over 65) to younger individuals (up to 64 years) is set to increase by more than 1/3 (nowadays there are 3 young people for every elderly person, in 2051 the ratio will be 1.9 younger persons per elderly person).

As concerns the commitments made with the Italian Digital Agenda for the eHealth service, between the present time and 2020 public funding of healthcare should rise from EUR 2 to EUR 7.8 billion, reaching in 2020 a total estimated commitment between EUR 9.5 and 10.2 billion. Part of these funds should be used to develop Telemedicine services.

Looking at the growth of demand for home care (both public and private), this coincides largely with the "growth in telemedicine services".

Starting with Hippocrates, basing medicine on the observation of events has long been the guiding epistemological criterion of the healthcare profession.

This criterion evolved as medicine progressed, resulting in the formula of *Evidence-Based Medicine (EMB),* which can be defined as "the process of systematic search, assessment and use of the results of contemporary research as a basis for clinical decisions" or also as "the use of mathematical estimates of risks, benefits and damage derived from high-quality studies conducted on population samples to support the clinical decision-making process in diagnosis or in the management of individual patients". The possibility of using big data and artificial intelligence has had a strong impact on this epistemological assumption of present-day clinical practice. The use of big data and artificial intelligence is ushering in a type of medicine based on elements that are not apparent to human doctors but can be extracted by using big data and deep learning techniques, due to the ability of computers to cover and process a far larger amount of information than a human being.

Ordinary tools of analysis, although not yet in use in certain areas of the country, are becoming obsolete. This requires rapid adjustments also to the regulatory framework. The Gelli law on medical liability is based on the by now outdated concept of Evidence-Based Medicine and fails entirely to consider the "non-evident evidence" of big data and artificial intelligence.

It is therefore necessary to update rapidly the legislation relating to evidence-based medicine to make it compatible with predictive medicine. This must be accompanied by reform and re-engineering of the health service programming cycle.

Today, by using big data and deep learning techniques we can deliver effective preventive medicine long before the onset of symptoms. For chronic and degenerative diseases, this provides a significant advantage. Instant access to the entire set of data makes it possible to plan evolution of the clinical presentation by means of algorithms supporting decision-making, improving the overall efficiency of the process. The overall process is constructivist and is aimed at delivering significant benefits in terms of patients' treatment and care.

The diagnostic and care model also based on the patient's personalised electronic medical record will respond to the demand for increasingly effective, efficient and high-quality diagnosis, prognosis and treatment services. A good trade-off between quality of service and implementation costs can be achieved via the application of innovative technologies, systems and procedures for management of the clinical process, based on an e-Health Service Management logic. Creation of the electronic medical record, constantly updated with data from remote monitoring will favour very early diagnosis of many diseases, the identification of risks and the remote delivery of treatment and care.

Health status monitoring, prevention of acute episodes and support in daily life are all emerging areas for e-health services, in particular for fragile and elderly individuals and people with chronic diseases.

This revolution can help to cut significantly the costs of healthcare, by reducing sharply the number of acute cases, preventing the development of many chronic diseases and delivering tele assistance and telemedicine [3].

To ensure that this revolution is achieved, the regulatory framework, health policies and clinical approaches to diseases must be adjusted as appropriate. A new, "4.0" generation of doctors must be trained, and much remains to be done in this area.

5   Conclusion

The creation of a telemedicine and remote monitoring system, will offer support for the long-term care of/provision for diabetic diseases by: (i) guaranteeing continuity of care at a hospital and regional level, (ii) integrating social and healthcare activities, (iii) favouring the continuation of such activities within the patient's own living environment for as long as possible and (iv) improving the patient's quality of life, in addition to providing improved support for diagnosis and treatment.

This approach is consistent with both international and national strategies for innovation, above all since it is developing new decision-making paradigms. As the result of the instantaneous access to the entire data set, provision can be made for the development of the health record through decision-making support algorithms, which make the entire process more efficient.

This revolution can help to cut significantly the costs of healthcare, by reducing sharply the number of acute cases, preventing the development of many chronic diseases and delivering tele assistance and telemedicine.

References

1. AGENAS, La mobilità sanitaria, 9° Supplemento al numero 29 di Monitor (2012)
2. Cislaghi, C., et al.: Per valutare l'intensità della mobilità ospedaliera non basta contare quanti escono da una regione per farsi ricoverare. Epid. Prev. **34**, 97–101 (2010)
3. Marino, D., Quattrone, G.: Mobilità sanitaria, prime valutazioni, forthcoming (2018)

SDI and Smart Technologies for the Dissemination of EO-Derived Information on a Rural District

Monica Pepe[1]([⊠]) [iD], Gabriele Candiani[1] [iD], Fabio Pavesi[1] [iD], Simone Lanucara[1] [iD], Tommaso Guarneri[2] [iD], and Daniele Caceffo[2]

[1] CNR-IREA Milano, 20133 Milan, Italy
pepe.m@irea.cnr.it
[2] Società Cattolica di Assicurazione Verona, 43762 Verona, Italy

Abstract. The Po Plain (Italy) is a complex mixture of urban and rural landscapes. Between Lombardy and Piedmont, the rural zone includes the largest rice crop area in Europe, accounting for 40% and 90% of the European and Italian rice production, respectively. The monitoring of this crop system is important by both environmental and economic points of view, because of its impacts on ecosystems, society and markets. In particular, the near real time (NRT) provision of information about crop status at farm scale is relevant for different players (i.e. farmers, consultants, policy makers, insurance companies). In this study, a Spatial Data Infrastructure (SDI), implemented for the provision and dissemination of NRT crop status information, is presented. These information, derived from very high resolution satellite imagery, can be helpful to support loss adjusters in their workflows. The SDI architecture is designed as a seamless solution from satellite data download to presentation of added value information, which are retrieved and displayed on a mobile device, directly in the field.

Keywords: Rural landscape · Spatial Data Infrastructures (SDIs)
Earth Observation

1 Introduction

The Po Plain in Northern Italy, is a continuous and complex mixture of urban and rural landscapes. The rural zone includes the largest rice crop area in Europe, accounting for 40% and 90% of the European and Italian rice production, respectively [1]. The monitoring of this crop system is important by both environmental and economic points of view, since it impacts on ecosystems, society and markets [2]. Rice in Europe is cultivated with modern techniques in intensive cropping system, rice farming is typically a highly mechanized monoculture, with a large use of agro-chemicals, which calls for up-to-date and distributed crop status information during on-going season, as well as for accurate information analysis concerning the processing product for managing the production and reduce its environmental impact.

© Springer International Publishing AG, part of Springer Nature 2019
F. Calabrò et al. (Eds.): ISHT 2018, SIST 100, pp. 44–50, 2019.
https://doi.org/10.1007/978-3-319-92099-3_6

In recent years the availability and forms of monitoring data increased. Earth Observation (EO) techniques reveal to be particularly efficient in crop status assessment [3]. Current systems exploits, together with EO data, also web-based and mobile-based software applications and technologies [4] to gather in-situ data streams with high spatio-temporal coverage. An example of coupling EO derived information modelling and Spatial Data Infrastructures (SDI) for providing rice services at different scales, is the ERMES FP7 project [5] in which the SDI and smart components are used to present information to decision makers in a way which is useful, understandable and exploitable. The near real time (NRT) provision of information about crop status at farm scale is relevant for different players (i.e. farmers, consultants, policy makers, insurance companies).

In this study, an SDI, implemented for the provision and dissemination of NRT crop status information, is shown. These information, derived from very high resolution satellite imagery acquired by RapidEye, can be helpful to support loss adjusters in their fieldwork. The objective is to create and disseminate added-value information regarding crop status to loss adjusters in a NRT and spatially contextual manner: a demonstration of how Information and Communications Technology (ICT) technologies could impact on the agriculture insurance sector, by technologically enabling novel workflows.

2 Overview of TELEMOD

2.1 TELEMOD Project

The main objective of the TELEMOD project is to provide crop monitoring tools supporting loss adjusters (indicated hereafter as "users") during their decision making processes. While the first important step is to create added-value updated information about crop status, the different crop conditions and the season characteristics, correspondingly the information dissemination step is very relevant. In fact, to properly support users sampling procedures and decision-making processes in the field, they need to be able to make sense of and properly handle the fine-grained products at parcel-level produced from EO data, to properly explore them through space and time. The dissemination of EO-derived maps is performed by a smart application rooted on a spatial data infrastructure, both designed for the purpose. Details of the TELEMOD system are reported in the following paragraphs.

2.2 TELEMOD System

The TELEMOD system is designed as a seamless solution from satellite data download to presentation of added value information, which are retrieved and displayed together with ancillary data (parcel boundaries and identification codes, municipal boundaries) on a mobile device, directly in the field. Several software tools and services have been developed within the system: automated EO data processing chains to produce NRT added-value products, supporting the users in their workflow; tools to automate the

NRT deployment of products according to a SDI; smart apps for the dissemination of products for their final exploitation.

An overview of the system architecture is depicted in Fig. 1, according to the three tier layered architecture based on the INSPIRE EU Directive specifications. The conceived SDI architecture has been implemented to allow sharing of geospatial data and to provide a user-friendly solution. We exploited the same three tier layered architecture in different researches [6, 7] successfully enabling the sharing of geospatial data from different domains. Service interoperability is partly accomplished by the adoption of Open Geospatial Consortium (OGC) standard specifications for service interfaces. Data are stored in different repositories: (i) the cloud solution of the EO data provider (BlackBridge) accessed via the EyeFind^{TM} APIs; (ii) the EO products are stored in a structured filesystem; (iii) ancillary data (information about parcels, municipalities, the index map of products) are stored in PostGIS and MongoDB databases. The service layer is composed by the EO data processing tools, together with the geospatial web-server (Geoserver) and a dedicated middleware for optimizing the product deployment. The presentation layer comprises two front-end applications: the Mobile version and the Web version of the TELEMOD-App, for the dissemination of EO product to users.

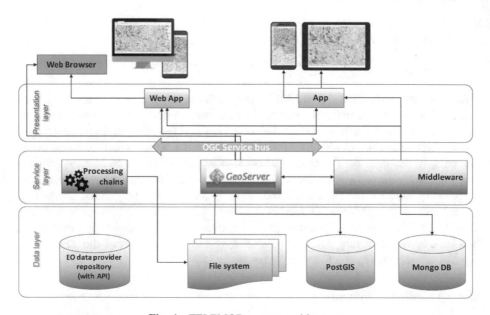

Fig. 1. TELEMOD system architecture.

The EO dataset used in TELEMOD is based the RapidEye system: a constellation of 5 satellites, equipped with sensors with 5 m spatial resolution and 5 spectral bands in the Visible and Near Infra-Red regions. Images were provided through the *Monitoring Program: Agriculture* developed by BlackBridge, which provide NRT access to a pool of RapidEye in-season imagery other the area of interest. It is worth to be noted that monitored area is covered by several image tiles (i.e. 42) acquired asynchronously, while the frequency of acquisition is not regular but depends on constraints (orbits, acquisition angle, cloud coverage). To timely produce the added-value products an *ad-hoc* automatic processing chains was developed to include all the steps involved in process: the download from image repository, the image pre-processing and the processing phase. During the download phase, an automatic script contacts the *Monitoring Program: Agriculture* server, searching for new images acquired in the last 7 days, and downloads them locally. After the download, a new script starts the following processing steps: each image is first radiometrically calibrated and atmospherically corrected to reflectance values. Then, a vegetation index is calculated for the whole image, whereas intra-field variability is computed for all the parcels within the study area (Fig. 2). All EO data analysis algorithms have been developed as modules to be encapsulated as reusable software libraries. All images and products (including intermediate ones) are stored on a server filesystem in which folders and data granules follow a given naming convention (containing, among others information, date and tile references).

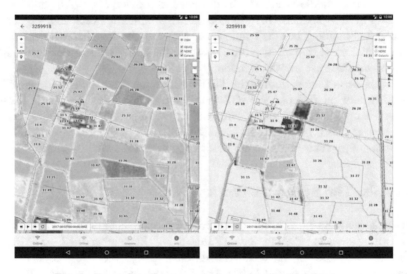

Fig. 2. Examples of map products as visualized by the App.

The maps of vegetation indexes and intra-field differences consists of time series data products, i.e. they are a series of raster files covering the same areas over a period of time. The data products, organized in an Image Mosaic data structure, are uploaded to GeoServer, by the Automatic Upload System (AUS), and then exposed as standard OGC Web Map Service (WMS) and Web Coverage Service (WCS). The Image Mosaic data structure is composed by a raster files folder structure, a data index file and a configuration file. The data index and the configuration files store the location of the data products in the file system, the geographical location of each single data product and the date it refers to following the ISO8601 standard. The data structure allows the creation of a time-mosaic from a number of raster contained in the file system folder structure. We used this model to gather the raster files into multi-dimensional rasters.

The AUS is envisaged for uploading the data products in Geoserver. It consists of two components, a data monitoring component, and an importing component. The data monitoring component examines the state changes in the Image Mosaic data structure. If a change occurs the data monitoring component updates the importing component. The importer exploits the GeoServer REST APIs to create/update the data index and associate the graphic style to the map products by means of OGC Style Layer Descriptor (SLD) standard. Thanks to the AUS when a new data is uploaded, or an existing data is deleted, the GeoServer repository and layer deployments are automatically updated, and then added-value products are available to users.

Geospatial Web Services allow client applications to discover, access, and visualize geospatial data. The types of services we employed are WMS and WCS. WMS is used to share data products as maps supporting the temporal dimension. The WMS deliver multi-temporal maps and relative legends in SLD standard to client applications for visualization. WCS is used to share data products as raster files, even in this case supporting the temporal dimension. We used the WCS to deliver the data products from GeoServer to the Middle-Ware. This is designed to provide the Mobile and Web App for: (i) tiles of EO products; (ii) pre-calculated pyramids for each tile, (iii) ancillary data, such as bounding box and time span for each tile together with municipality boundaries and place names. It allows also for uploading related files to the Mobile application for their use in offline mode.

The front-end of the system is the TELEMOD-App, designed and implemented to be installed both on Android and iOS mobile operating systems (OS). It allows for visualizing crop status and other information provided, both on-line and off-line. A web-app is also provided for making information reachable even using a web-browser. App functionalities are described in the following Sect. 3.

3   Results: The System at Work

The study area covers 26,250 km^2, quite the entire arable land in the Lombardy Region (Italy). During the acquisition period - 1st June to 30th September 2017 -, 517 RapidEye scenes were acquired on the area. The different image tiles were revisited with a frequency ranging from 6 to 17 days (10 days average). As described in previous sections the system automatically processes and deploys EO derived maps, while the TELEMOD-App allows visualization of them with ancillary data. Examples of such maps are shown in Fig. 2: on the left panel there is the vegetation index map, colors ranging from red (low) to green (high) represent vegetation vigor; on the right panel the map indicates how vigor varies within each parcel (overlaid on the map) – red/orange colors refer to conditions worse than the parcel average (represented in yellow), while greenish colors to a status over the average. The figure shows also the ancillary data provided (base maps, parcel boundaries and identifiers, municipalities). Parcel identifiers are important for users to properly report the damage.

When a registered user enters the App s/he is prompted to select a product tile by: (i) search: entering the placename (Fig. 3, left n.1); (ii) by clicking on the desired tile (Fig. 3, left n.7); as soon as the tile is selected, the most updated vegetation-index map for that tile is displayed (Fig. 3 right panel) with related identifier (n.1), legend (n.7), and date (n.4); the date is located on a slider which, whether the App is used in on-line mode, enables the navigation in time of all maps produced till then for the specific tile; the time sequence could be reproduced as a loop or by instants. A layer control tool (n.6) allows for switching the map between the two products (as in Fig. 2). Moreover, users can locate themselves in the field by clicking the placemark icon (n. 3). Zooming (n. 2) is provided by icons or pinch-in and pinch-out gestures. While cellular network is not always reachable in the field, the App could work in off-line mode (Fig. 3, left n.5), allowing for downloading the up-to-date map for a tile; the selection mode is the same described before. Another tab Fig. 2, left, n.6) provide the tools for managing the download sessions (list of downloadable files, presence of more recent maps on the server, local download/remove).

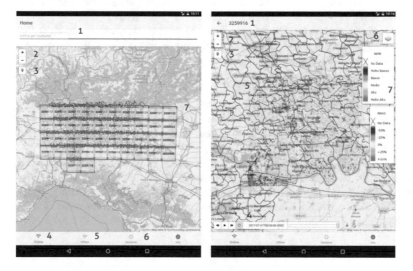

Fig. 3. The App at work: tile selection and visualization.

By using crop status maps, together with their position in the field, users make decisions about where to survey fields and to sample plants to account for the ranges of crop conditions in the farm (smart scouting). Moreover, the exploration of crop status maps through time helps the understanding of seasonal trends and recognition of anomalous conditions already present before the damaging event (heavy wind, hail-storms, etc.), for supporting the final estimations of yield losses.

4   Conclusions

We have introduced here the overall system architecture and the ecosystem of tools that shape the EO services for users developed in the realm of the TELEMOD project. We focus on the TELEMOD-App as it plays the role of a data dissemination and presentation tool that bridges the information about vegetation status, as recorded in space and time, to the loss adjusters to support them in damage assessment and estimations.

References

1. Global Rice Science Partnership, Rice around the world: Europe. http://ricepedia.org/rice-around-the-world/europe. Accessed 19 Feb 2018
2. Casas, G., Las, S.F.: Sustainable planning: a methodological toolkit. In: Gervasi, O., et al. (eds.) Computational Science and Its Applications-ICCSA 2016. Proceedings, Part I, pp. 627–635. Springer, Cham (2016)
3. Granell, C., Havlik, D., Schade, S., Sabeur, Z., et al.: Future Internet technologies for environmental applications. Env. Model Softw. **78**, 1–15 (2016)
4. Busetto, L., Casteleyn, S., Granell, C., Pepe, M., et al.: Downstream services for rice crop monitoring in Europe: from regional to local scale. IEEE JSTARS **99**, 1–19 (2017)
5. Granell, C., Miralles Tena, I., Rodríguez-Pupo, L.E., González-Pérez, A., et al.: Conceptual architecture and service-oriented implementation of a regional geoportal for rice monitoring. ISPRS Int. J. Geo-Inf. **6**(7), 191 (2017)
6. Lanucara, S., Oggioni, A., Modica, G., Carrara, P.: Interoperable sharing and visualization of geological data and instruments: a proof of concept. In: Gervasi, O., et al. (eds.) Computational Science and Its Applications, ICCSA 2017. Lecture Notes in Computer Science, vol. 10407. Springer, Cham (2017)
7. Modica, G., Pollino, M., Lanucara, S., La Porta, L., et al.: Land suitability evaluation for agro-forestry: definition of a web-based Multi-Criteria Spatial Decision Support System (MC-SDSS): preliminary results. In: Gervasi, O., et al. (eds.) Computational Science and Its Applications, ICCSA 2016. Lecture Notes in Computer Science, vol. 9788. Springer, Cham (2016)

Harmonization and Interoperable Sharing of Multi-temporal Geospatial Data of Rural Landscapes

Simone Lanucara[1]([⊠]) [iD], Salvatore Praticò[2] [iD],
and Giuseppe Modica[2] [iD]

[1] CNR, Istituto per il Rilevamento Elettromagnetico dell'Ambiente,
20133 Milan, Italy
`lanucara.s@irea.cnr.it`
[2] Mediterranea University of Reggio Calabria, 89100 Reggio Calabria, Italy

Abstract. Usually, rural landscape characterization is implemented through geomatics techniques and subsequent production and analysis of geospatial data. Thanks to internet diffusion, practitioners and researchers can share data in the World Wide Web. Data sharing process can improve participatory planning processes and allow an easy comparison between different landscape areas. Sharing can be done with varying degrees of interoperability and different software tools, proprietary or open source. A widespread way to share geospatial data and metadata is by Spatial Data Infrastructures (SDI) taking advantage on Open Geospatial Consortium (OGC) standards. Anyway, the sharing of data by OGC service lacks in data harmonization and in semantic enablement, making difficult compare, search and analyze data given by different sources. Different data schemas and linguistic barrier hinder the usefulness of data obtained from different sources. In this study we present a novel data workflow implemented for sharing in an interoperable, harmonized and semantically enriched way multi-temporal land cover datasets collected in a previous landscape characterization researches.

Keywords: Spatial Data Infrastructures (SDIs) · Inspire · Data harmonization
Sematic harmonization · Landscape

1 Introduction

Rural landscape characterisation and monitoring processes are usually implemented through geomatics techniques and subsequent production and analysis of geospatial data [1–5]. Nowadays, thanks to internet diffusion, practitioners and researchers can share data in the World Wide Web (WWW). Data sharing process can improve participatory planning processes [6], providing to stakeholders and decision-makers (DMs) information about past and ongoing landscape dynamics and allowing an easy comparison between different landscape areas. A useful way to share data and metadata entails the implementation of Spatial Data Infrastructures (SDI) [7, 8] that allow interoperability at different levels. The interoperability, indeed, can be subdivided in four main categories: systematic, syntactic, schematic and semantic. The semantic

© Springer International Publishing AG, part of Springer Nature 2019
F. Calabrò et al. (Eds.): ISHT 2018, SIST 100, pp. 51–59, 2019.
https://doi.org/10.1007/978-3-319-92099-3_7

interoperability turns data in knowledge into technological infrastructure [9]. The OGC services enable the interoperable distribution, at syntactic level, of maps and related legends, vector and raster data, graphic styles and metadata catalogs.The sharing of data and maps by OGC service lacks in data harmonisation and in semantic enablement [9], making difficult compare, search and analyse data given by different sources. In the last 10 years, the European Commission, with its Directive 2007/2/EC [10] "Infrastructure for Spatial Information in Europe" (INSPIRE), developed the interoperability guidelines and data specification to ease transnational access to different geospatial data as geological, risks, land use/land cover (LU/LC) data, etc. These guidelines concern metadata creation, data search services, data access (sharing; download; visualization) and vocabularies. Moreover, data specifications provide models and code lists for geospatial data sharing in order to allow their re-use in a multi-scale and multi-object environment [11].

In this paper, we describe the method used to share, in an interoperable, semantically enriched and harmonized way, multi-temporal LC data collected in previous landscape characterisation researches [1, 2, 12, 13].

2   Method

2.1   Study Area

Costa Viola study area is a narrow coastal strip falling in the Tyrrenian lowest part of Reggio Calabria Province (Calabria, Italy) (Fig. 1).

Fig. 1. The study area of 'Costa Viola' (Reggio Calabria Province, Italy).

The covered surface is above 24 km^2 while elevation ranges between 0–600 m a.s. l. The morphology of Costa Viola is characterized by very steep slopes, impressive cliffs and natural terraces. Because of this unfavourable morphology, in late

18th century, local people start to build terraces by means of dry-stone walls to allow agricultural practices in the area, mostly for vineyards and lesser to olive groves, citrus and other crops. The general abandonment of agricultural practices, mostly in those areas where it is no possible an efficient mechanisation, is the primary cause of the abandonment of agricultural terraces and the subsequent terraced landscape degradation with an increase of hydrogeological risk because of the landslides [14].

2.2 Base Data and Software Architecture

The base data collected and implemented in previous researches [1, 2, 12, 13] are seven LC dataset, one for each investigated year (1955, 1976, 1989, 1998, 2008, 2012, 2014). Each dataset was obtained by advanced geomatics techniques, stored in a PostgreSQL-PostGIS (http://postgis.net) spatial database.

To allow the access to geospatial data and to share them with stakeholders and DMs, we developed a SDI using Free and Open Source Software for Geospatial (FOSS4G). SDI can be defined as technological platform to manage and share geospatial data and metadata by means of appropriate standards. Software solutions to implement SDI architecture are many and widespread [15]. They can be proprietary as @ESRI [16] or FOSS4G, actually supported by huge availability of open source suits [17], or hybrid (a mixture among them).

The implemented FOSS4G SDI architecture has been conceived to allow sharing of geospatial data and to provide a user-friendly solution, characterized by accessibility and versatility. A multi-tier architecture (see Fig. 2) has been adopted, composed by four different layers implemented in Ubuntu-based virtual environment and managed by @Microsoft Server Datacenter. We exploited the same FOSS4G architecture in different researches [9, 11] successfully enabling the sharing of data from different domains.

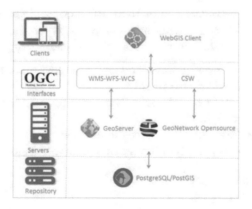

Fig. 2. Multi-tier SDI architecture

Each layer of the implemented SDI has a different function:

1. Repository, for data and metadata storage in a PostgreSQL geodatabase with PostGIS extension (http://postgis.net); Server, composed by GeoServer (http://geoserver.org) and GeoNetwork open source software (http://geonetwork-opensource.org), to data and metadata management and web sharing;
2. Standard OGC interface, to enable interoperable sharing of data and metadata;
3. Front-end client (WebGIS-client) developed in GeoExt (http://geoext.org) to enable a graphical user interface (GUI).

2.3   Data Harmonisation Process

As seen in Sect. 2.2, LC data were stored as tables in PostGIS geodatabase and shared through the SDI platform by means of Web Map Service (WMS) and Web Feature Service (WFS) OGC standards, achieving a syntactic interoperability. As expected, shared data are not easily reusable by other researchers and/or stakeholders. Linguistic barrier, different visual representations and different data schemas hinder the reuse of data [9]. To tackle this gap, the data have been converted following the INSPIRE Land Cover data specification (http://inspire.ec.europa.eu/documents/Data_Specifications/INSPIRE_DataSpecification_LC_v3.0.pdf) by exploiting Extract, Transform, Load (ETL) [18] procedures and achieving a European scale semantic harmonization. ETL procedures extract the geospatial data, transform the data model and allow users to retrieve the new data model in different formats.

To execute ETL procedure, we used the free and open source software Hale Studio (HS), already used in INSPIRE Data Harmonization Panel (http://www.dhpanel.eu/). HS allows users to import data and source schemas from different sources and formats and execute schema and data transformations by means of a GUI. Outputs and sources can be local files, web services or databases with Java Database Connectivity application programming interface.

To share data in accordance with INSPIRE specifications trough the SDI, we exploited the GeoServer app-schema module. We used the following workflow that describes and synthetizes the harmonization and LC data sharing procedures:

- App-schema module installation in the GeoServer environment of the SDI;
- Import source data and schema in HS;
- Import of INSPIRE LC schema, (http://inspire.ec.europa.eu/schemas/lcv/4.0/LandCoverVector.xsd), in HS;
- Schema mapping from different sources to outputs;
- Export of schema mapping configuration file in eXtensible Markup Language (XML) format;
- Import of XML file in GeoServer;
- WMS and WFS service sharing.

2.4   Land Cover (LC) Data Harmonization

LC data provide information about coverage of the Earth surface, including agricultural areas, woodlands, lakes, and natural and artificial areas. INSPIRE LC specifications provide LC data schemas and guidelines for web services implementation. These specifications provide two schemas, one for vector (LandCoverVector) and one for raster data (LandCoverRaster). Both provide the LC data observation date (such as aerial photo or satellite image date) that allow the representation of multi-temporal changes. To harmonize our data, stored on geospatial database tables, and to comply with the INSPIRE LC data specification, we mapped the source schema to the target schema, LandCoverVector. This schema contains four classes described with a Unified Modeling Language (UML): LandCoverDataset, LandCoverUnit, LandCoverObservation and LandCoverValue (Fig. 3). LandCoverVector schema model LC dataset (LandCoverDataset) as collections of LandCoverUnit. This last represent an area, defined by geometry, that can support more observations in different dates.

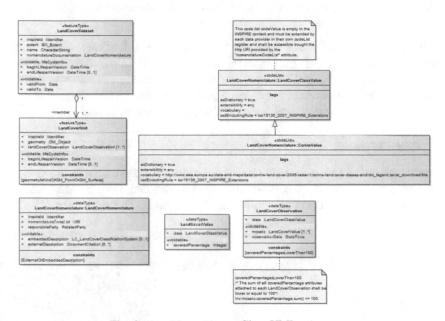

Fig. 3. LandCoverVector Class UML scheme.

The representation of multi-temporality is obtained by the element ObservationDate in the LandCoverObservation class. Also, to obtain a semantic enrichment, the schema provides the use of code list, classification code (e.g. CORINE Land Cover or other national or local codes) or multilingual vocabularies. As described in Sects. 2.2 and 2.4, starting data were different sets of LC vector data stored in different PostGIS tables, one for each observed year. The original schema is composed by: geometry ID, geometry, descriptive and coded LC for different hierarchical levels and presence or absence of dry-stone walls terraced areas (Fig. 4).

To harmonize all LC dataset have been mapped into a unique destination LandCoverVector scheme. In more details, the seven LC dataset have been combined into a single LandCoverDataset, where each dataset corresponds to one observation

CostaViola_LULC_1955	
gid	INTEGER
the_geom	geometry
uso_suolo	CHARACTER VARYING(50)
clc_code	INTEGER
terrazzo	CHARACTER VARYING(15)
clc_1_lev	CHARACTER VARYING(50)
clc_2_lev	CHARACTER VARYING(50)
clc_3_lev	CHARACTER VARYING(50)
clc_4_lev	CHARACTER VARYING(50)
clc_1_code	INTEGER
clc_2_code	INTEGER
clc_3_code	INTEGER
clc_4_code	INTEGER
area_km2	NUMERIC(6,2)

Fig. 4. Schematic representation of 1955 Land Use/Land Cover (LU/LC) attribute table.

(LandCoverObservation) stored for a specific year (ObservationDate). In addition, we implemented the European Environment Agency (EEA, http://dd.eionet.europa.eu/vocabulary/landcover/clc/view) vocabulary for LC description. Thanks to INSPIRE specifications, ETL procedure, schemes mapping and EEA vocabulary, a European level semantic harmonisation and interoperability has been achieved.

2.5   Client WebGIS

To view and query LC multi-temporal maps and information from a unique access point the implementation of a WebGIS client is needed [19]. In this work, we implemented a WebGIS client exploiting the GeoExt and D3js environment. GeoExt is

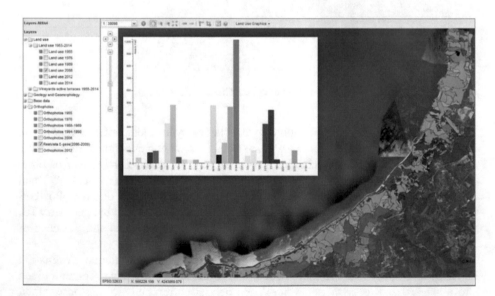

Fig. 5. WebGIS client showing an interactive graph on land cover classes surface distribution.

a powerful open source instrument that allows the combination of OpenLayers web-mapping libraries (https://openlayers.org/) with the user interface of Ext JS (https://www.sencha.com/products/extjs/). OpenLayers is an open source JavaScript library used to view maps on a web browser providing programming interfaces to access different geospatial information sources (e.g. OGC interfaces, satellite basemaps as Google Maps, Bing Maps, etc., and OpenStreetMap images). Ext JS is a JavaScript framework that allow web applications development for desktop, tablets and smartphones systems. D3js is a JavaScript library that develop web based dynamic interactive graphs. The implemented WebGIS client allows to view and query multi-temporal LC maps and analyze the LC changes by means of interactive graphs (Fig. 5).

3   Results and Discussions

The sharing of LC dynamics, and other space-based data and models, is crucial for the definition of adequate landscape-management strategies [2] specially in those areas where LC characteristics assume an important role in definition of cultural aspects, in tourism specialisation or in environmental degradation identifying [20–22]. It can improve participatory planning processes, providing stakeholders and DM the information about landscape dynamics; also it constitute a backbone in monitoring systems [23]. Geospatial data and LC information sharing at worldwide level faces different interoperability issues. To face this challenge, we developed and tested a method based on OGC WMS and WFS standards, INSPIRE LC specifications and EEA vocabulary. We implemented a FOSS4G SDI, harmonized multi-temporal LC data thanks to ETL procedures and EEA controlled vocabulary. The methodology allowed us to enable a transnational access to LC data in accordance with INSPIRE directive and guidelines. Moreover, we implemented a WebGIS client [24, 25] that allows browsing, zooming, and querying the harmonized LC data and show LC changes by web based dynamic interactive graphs.

References

1. Modica, G., Praticò, S., Pollino, M., Di Fazio, S.: Geomatics in analysing the evolution of agricultural terraced landscapes, Lecture Notes Computer Science (including Subseries Lecture Notes in Artifical Intelligence and Lecture Notes in Bioinformatics), vol. 8582, pp. 479–494 (2014)
2. Modica, G., Praticò, S., Di Fazio, S.: Abandonment of traditional terraced landscape: A change detection approach (a case study in Costa Viola, Calabria, Italy), L Degrad Dev, vol. 28 (2017)
3. Riguccio, L., Russo, P., Scandurra, G., Tomaselli, G.: Cultural landscape: stone towers on Mount Etna. Landsc. Res. **40**, 294–317 (2015)
4. Riguccio, L., Carullo, L., Russo, P., Tomaselli, G.: A landscape project for the coexistence of agriculture and nature: a proposal for the coastal area of a Natura 2000 site in Sicily (Italy), J. Agric. Eng. **47**(61) (2016)

5. Lasaponara, R., Murgante, B., Elfadaly, A., et al: Spatial open data for monitoring risks and preserving archaeological areas and landscape: Case studies at Kom el Shoqafa, Egypt and Shush, Iran. Sustain 9 (2017)

6. Torquati, B., Vizzari, M., Sportolaro, C.: Participatory GIS for integrating local and expert knowledge in landscape planning. In: Agricultural and Environmental Informatics, Governance and Management, pp. 378–396. IGI Global (2011)

7. Foley, R., Maynooth, N.: Spatial Data Infrastructure, pp. 507–511. Elsevier (2009)

8. Vizzari, M., Antognelli, S., Pauselli, M., et al.: Potential nitrogen load from crop-livestock systems. Int. J. Agric. Environ. Inf. Syst. 7, 21–40 (2016)

9. Lanucara, S., Oggioni, A., Modica, G., Carrara, P.: Interoperable sharing and visualization of geological data and instruments: a proof of concept. In: Computational Science and Its Applications - ICCSA 2017, pp. 584–599 (2017)

10. European Commission Directive 2007/2/EC of the European Parliament and of the Council of 14 March 2007 Establishing an Infrastructure for Spatial Information in the European Community (INSPIRE) (2007)

11. Modica, G., Pollino, M., Lanucara, S., et al: Land Suitability Evaluation for Agro-forestry: Definition of a Web-Based Multi-Criteria Spatial Decision Support System (MC-SDSS): Preliminary Results, pp. 399–413 (2016)

12. Di Fazio, S., Modica, G.: Le pietre sono parole: letture del paesaggio dei terrazzamenti agrari della Costa Viola. Iiriti Editore, Reggio Calabria (Italy) (2008)

13. Di Fazio, S.: I terrazzamenti viticoli della Costa Viola. Caratteri distintivi del paesaggio, trasformazioni in atto e gestione territoriale in un caso-studio in Calabria. In: Muri di sostegno a secco: aspetti agronomici, paesaggistici, costruttivi e di recupero. I Georgofil. Quaderni 2008-II, pp. 69–92. Edizioni Publistampa, Firenze (2008)

14. Lasanta, T., Arnaez, J., Oserin, M., Ortigosa, L.M.: Marginal lands and erosion in terraced fields in the Mediterranean mountains: a case study in the Camero Viejo (Northwestern Iberian System, Spain). Mt. Res. Dev. 21, 69–76 (2001)

15. Steiniger, S., Hunter, A.J.S.: Free and Open Source GIS Software for Building a Spatial Data Infrastructure, pp 247–261 (2012)

16. Maguire, D.J.: ArcGIS: general purpose GIS software system. In: Encyclopedia of GIS, pp. 25–31. Springer, Boston (2008)

17. Brovelli, M.A., Minghini, M., Moreno-Sanchez, R., Oliveira, R.: Free and open source software for geospatial applications (FOSS4G) to support Future Earth. Int. J. Digit. Earth 10, 386–404 (2017)

18. Vassiliadis, P.: A survey of extract - transform - load technology. Int. J. Data Warehous. Min. 5, 1–27 (2009)

19. Oliveira, A., Jesus, G., Gomes, J.L., et al.: An interactive WebGIS observatory platform for enhanced support of integrated coastal management. J. Coast Res. 70, 507–512 (2014)

20. Recanatesi, F.: Variations in land-use/land-cover changes (LULCCs) in a peri-urban Mediterranean nature reserve: the estate of Castelporziano (Central Italy). Rend. Lincei 26, 517–526 (2015)

21. Recanatesi, F., Clemente, M., Grigoriadis, E., et al.: A fifty-year sustainability assessment of Italian agro-forest districts. Sustain 8, 1–13 (2016)

22. Scorza, F., Fortino, Y., Giuzio, B., Murgante, B., Casas, G.L.: Measuring territorial specialization in tourism sector: the basilicata region case study. Lecture Notes in Computer Science (Including Subseries Lecture Notes in Artificial Intelligence and Lecture Notes in Bioinformatics), vol. 10409. LNCS, pp. 540–553 (2017)

23. Selicato, M., Torre, C.M., La Trofa, G: Prospect of integrate monitoring: a multidimensional approach. In: International Conference on Computational Science and Its Applications, pp. 144–156. Springer, Heidelberg (2012)

24. Pollino, M., Fattoruso, G., Della Rocca, A.B., et al: An open source GIS system for earthquake early warning and post-event emergency management. In: International Conference on Computational Science and Its Applications, pp. 376–391. Springer, Heidelberg (2011)
25. Pollino, M., Caiaffa, E., Carillo, A., et al: Wave energy potential in the Mediterranean Sea: design and development of DSS-WebGIS "Waves Energy". In: International Conference on Computational Science and Its Applications, pp. 495–510. Springer, Cham (2015)

Sentinel-2 Imagery for Mapping Cork Oak (*Quercus suber* L.) Distribution in Calabria (Italy): Capabilities and Quantitative Estimation

Giuseppe Modica[1](✉) [iD], Maurizio Pollino[2] [iD],
and Francesco Solano[1] [iD]

[1] Mediterranea University of Reggio Calabria, 89100 Reggio Calabria, Italy
giuseppe.modica@unirc.it
[2] ENEA - National Agency for New Technologies, Energy and Sustainable
Economic Development, Casaccia Research Centre, 00123 Rome, Italy

Abstract. The goal of this paper refers to the potential in using new Sentinel-2 (S-2) remote sensing imagery and *in situ* surveying for mapping Cork oak (*Quercus suber* L.) woodlands in Calabria Region (Southern Italy), comparing them to other satellite platforms such as Landsat 8 operational land imager (L8 OLI). Considering that S-2 spectral bands are particularly suitable for estimating different vegetation cover characteristics, we propose a methodology for mapping the actual consistence of this habitat, using the vegetation spectral reflectance to evaluate cork oak spectral response. A set of different S-2 and L8 OLI scenes where freely downloaded and pre-processed (topographic and atmospheric correction, band anomaly detection) in order to better investigate cork oak spectral signature. Normalized Difference Vegetation Index (NDVI) and ND Red Edge index where calculated to obtain a high spectral resolution vegetation mask. Digital Elevation Model (DEM), signature training sets, Ground Control Points (GCPs) and ancillary data where used to perform a supervised classification of both S-2 and L8 OLI images. Furthermore, an accuracy assessment was applied to the classified images in order to evaluate user's and producer's accuracies. S-2 provides a great opportunity for global vegetation monitoring due to its enhanced spatial, spectral and temporal characteristics compared with Landsat.

Keywords: Sentinel-2 · Landsat 8 OLI · Spectral signature · Cork oak woodlands

1 Introduction

Cork oak (*Quercus suber* L.) woodlands are one of the most representative forest ecosystems characterizing the Mediterranean environment and have a recognized natural value worthy of conservation at European level. Thus, methods and researches aiming at their mapping and monitoring at regional and national scale could give an operational aid to improve their knowledge. Moreover, can represent a significant base information to put into practice new conservation actions.

© Springer International Publishing AG, part of Springer Nature 2019
F. Calabrò et al. (Eds.): ISHT 2018, SIST 100, pp. 60–67, 2019.
https://doi.org/10.1007/978-3-319-92099-3_8

2 Materials and Methods

2.1 Study Area and Data Processing Workflow

The research was carried out in the Angitola (ANG) study area (Vibo Valentia province, Calabria, Italy), extended 49.2 km^2 (Fig. 1). Considering that Sentinel-2 (S-2) and Landsat 8 operational land imager (L8 OLI) spectral bands are particularly suitable for estimating different plant cover characteristics [1, 2], a methodology is proposed for mapping the actual consistency of this habitat using the spectral reflection of vegetation to evaluate cork oak spectral response and classify this type of vegetation (Fig. 2).

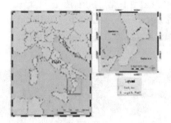

Fig. 1. Angitola (ANG) study area localization (Calabria Region, Italy).

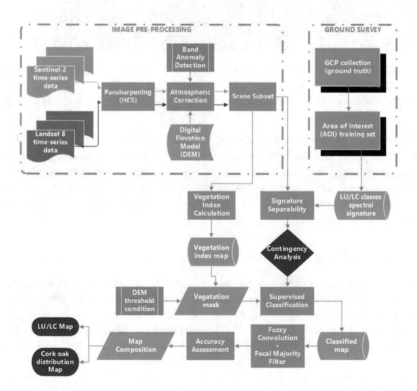

Fig. 2. Data processing workflow for cork oak detection.

2.2   Satellite Data and Images Pre-processing

For the investigated study area, cloud-free S-2 and L8 OLI were downloaded in the Universal Transverse Mercator (UTM) WGS84 map projection (Fig. 3). To improve the spatial resolution of the original L8 OLI bands, a merging algorithm was used between the 30 m spatial resolution (B1-B7) multispectral bands and the 15 m (B8) panchromatic band through the hyperspherical color space (HCS) algorithm [3]. For S-2 images, the same band-fusion procedure was performed between 20 m (B5-B6-B7) and 10 m (B2-B3-B4-B8) spectral bands. All pan-sharpened images were atmospherically and topographically corrected [4] to surface reflectance using all selected bands (Table 1) by means of the ATCOR3 module for Erdas Imagine® 2016. Atmospheric corrections (AC) implemented in ATCOR convert original digital numbers (DN) into at-sensor radiance (Lλ) using the calibration parameters, gain and offset values [5, 6], and then return surface reflectance values.

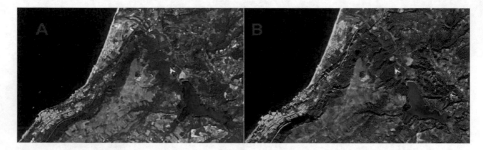

Fig. 3. ANG study area. (A) Landsat 8 OLI imagery acquired on 10 Aug 2017 showed in false IR colour band composite (RGB = 543), at 15 m resolution. (B) Sentinel 2A imagery acquired on 24 Feb 2017, showed in false IR colour band composite (RGB = 843), at 10 m resolution.

To better study the characteristics of smaller LU/LC classes, a subset of output images was performed to better locate the study area for both satellite platforms. Finally, an anomaly detection (AD) process was performed to identify pixels having a spectral signature that deviates considerably from many other pixel spectrum in the image (background spectrum). In this study, orthogonal subspace projection (OSP) was chosen as AD detection method [7, 8].

2.3   Data Processing and Classification

In order to better detect cork oak woodlands, the characterization and separability of cork oak spectral signatures was performed in both S-2 and L8 OLI imagery. To this end, a group of land uses was defined obtaining their spectral signatures starting form a set of 53 ground control points (GCP) - ground truths - collected by means of differential GNSS with accuracy ±0.5 m. On the other hand, these GCPs were utilized to obtain the spectral signatures of the defined LU/LC classes as follow: "Built-up areas"; "Crops"; "Coniferous woodlands"; "Cork Oak woodlands"; "Other Broadleaves woodlands"; "Inland waters"; "Marine waters". The spectral signatures from each class were collected using their area of interest (AOI) in order to perform the spectral separability of the cork oak woodlands, thus as training set to perform the image classification (Fig. 4). Euclidean distance (ED) algorithm was applied to measure spectral signature's separability and a contingency matrix (CM) to evaluate the consistency of the defined training sets. GCP coupled with other ancillary data have been selected as training samples to perform a supervised classification [9] by means of Maximum Likelihood Classification (MLC) algorithm (Table 2). For the extraction of the surface covered by vegetation, the calculation of vegetation indices was used. In particular, the Normalized Difference Vegetation Index (NDVI) [10] and the Normalized Difference Red Edge (NDRE) vegetation index [11] were used. These indices are combinations of reflectance measurements in the electromagnetic spectrum regions sensitive to the combined effects of chlorophyll concentration, surface and coverage by the canopy [12]. They measure quantity and quality of the photosynthetic material, essential for understanding the vigor state of vegetation [13]. The indices measure the reflection of the Red, NIR (in NDVI) and NIR, Red-edge (in NDRE) bands of electromagnetic spectrum regions in the reflection curve describing the transition from the absorption of chlorophyll to dispersion. The use of NIR measurements, with greater penetration depth through the canopy, allows to estimate the total amount of green material. Measurement in the Red-edge region allows these indices to be more susceptible to minor changes in vegetation health [14]. NDRE and NDVI where calculated respectively for the S-2 and L8 OLI images to obtain a high spectral resolution vegetation mask. Then a Digital Elevation Model (DEM) was used as threshold of altitudinal ecological limit to clip the vegetation mask where to run the classification (Fig. 2). In order to reduce the "salt and pepper" effects, a Fuzzy Convolution algorithm followed by a focal majority filter was applied (for both operations, a 3×3 pixels kernel matrix was adopted). Final cork oak woodlands distribution maps (Fig. 5), from both satellite platforms, have been developed with open source GIS software (QGIS Las Palmas 2.18).

Table 1. Synthesis of the spectral separability analysis performed for both satellite datasets for the ANG study area (down). "Crops" (Cr); "Coniferous woodlands" (Cw); "Cork Oak" (Co); "Other Broadleaves woodlands" (Bw). Higher values of Euclidean Distance (ED) and lower values of Divergence (D) mean the best spectral LU/LC class separability.

Satellite dataset	LU/LC class pairs	SENTINEL-2A (24 February 2017)	LANDSAT 8 (10 August 2014)
Euclidean Distance (ED)	Co-Cr	903	1510
	Co-Cw	581	625
	Co-Bw	1052	591
Divergence (D)	Co-Cr	83	586
	Co-Cw	137	224
	Co-Bw	162	140

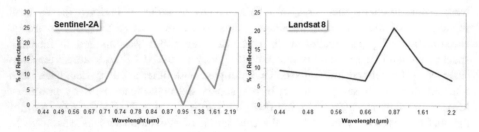

Fig. 4. Comparison of mean spectral profile of Cork oak woodlands over the ANG study area, for both S-2A and L8 OLI imagery.

3 Results and Discussion

3.1 Cork Oak Spectral Behavior and Separability

Cork oak spectral profiles analysis of both satellite imagery, as expected showed a higher definition of reflectance in the Red-NIR electromagnetic spectrum region in S-2 than in L8 OLI, in addition to the greater number of spectral bands (Shortwave infrared, 1.61–2.19 nm region) (Fig. 4). It means that, despite the bandwidth proximity of the two sensor, S-2 can provide more details on the phenology, and then on the reflectance values and on the general behavior of the cork oak's spectral signature than L8 OLI imagery, during the periods under investigation. In particular, the values of spectral separability (Table 1) obtained comparing Co with Cw and Bw are worth mentioning. Actually, with ED algorithm, even if slightly higher there is a better spectral separability with L8 OLI regarding Co-Cr and Co-Cw classes. Spectral discrimination between Co and Bw was very successful in S-2 imagery, with a separability value higher than more than twice between these two classes. Also with ED algorithm, the best spectral separability of Co was found with S-2 imagery.

3.2 Cork Oak Woodlands Mapping

The classified images obtained for ANG study area showed that the detected cork oak woodlands occupy the 4% of the scene for L8 OLI satellite image and the 8% of the examined scene for S-2 satellite image (Figs. 5 and 6). The surface occupied by the cork oak woodlands in the map derived from S-2 showed an increase of 4% compared to that derived from L8 OLI, probably at the expense of the other broadleaves woodlands classes. This result confirms the greater spectral separability obtained with S-2 and suggests a greater discriminative capacity between these two LU/LC classes, reflecting the marked difference between these two groups in the near infrared reflectance region. In addition, the Crops class showed an increase of 11% (from 49 to 60%) in S-2 generated map at the expense of the other broadleaves woodlands classes and Built-up class. These results can be explained by the greater potential for discrimination between vegetation elements that is improved in S-2 satellite platform due to the Red-edge bands and the greater spatial resolution [14].

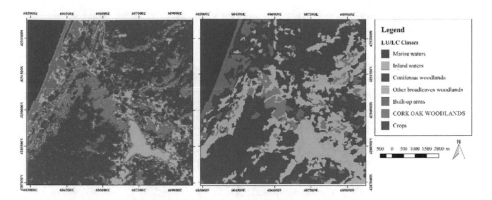

Fig. 5. Results obtained in detecting cork oak woodlands with S-2 (left) and L8 OLI (right) imagery in the ANG study area. All classified images were produced applying the maximum likelihood classification algorithm. LU/LC, land cover/land use.

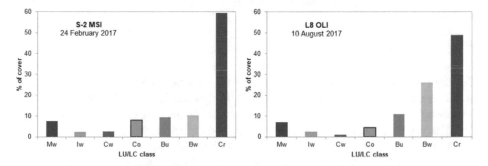

Fig. 6. Land distribution according to land cover/land use (LU/LC) classes referring to the image classification used for detecting cork oak woodlands. (Bu): Built-up areas; (Cr): Crops; (Cw) Coniferous woodlands; (Co)· Cork Oak woodlands; (Bw): Other Broadleaves woodlands; (Iw): Inland waters; (Mw) Marine waters.

3.3   Accuracy Assessment

In order to evaluate the classification accuracy, a confusion matrix was applied to the classified images [15]. Four significant indexes were derived from a contingency table and confusion metrics that express the accuracy in terms of omission/commission errors (Table 2). For two satellite imagery, S-2 classification was more accurate than L8 supervised classification. For the derived distribution map, the overall accuracy for S-2 and L8 was 90.6% and 63.3%, respectively. Regarding user's accuracy and producer's accuracy, the cork oak woodlands land cover class had consistently high values over S-2 classified map. As stated previously, cork oak has radically different spectral properties than the L8 OLI derived land cover class. This spectral difference allowed Co woodlands to be easily identified, thus resulting in higher accuracy values. The disparity in user's accuracy and producer's accuracy in L8 dataset indicates a tendency to underestimate the number of vegetation pixels.

Table 2. Accuracy assessment (%) at pixel level for MLC classification method and referring to the whole classified S-2A and L8 OLI imagery dataset.

Accuracy assessment (%)				
LU/LC	Satellite scene			
	Sentinel 2A MSI		Landsat 8 OLI	
	UA	PA	UA	PA
Mw	100	100	100	100
Iw	100	100	100	100
Cw	100	100	100	100
Co	83.33	100	25	100
Bw	50	100	46.15	66.67
Cr	100	83.33	28.57	33.33
Bu	100	100	100	100
Overall accuracy	**90.63**		**63.33**	
Kappa coefficient	0.8611		0.5346	

4   Conclusion

Results confirm the importance of the Red-edge bands on Sentinel-2 for vegetation applications, thanks to the combination with its high spatial resolution of 10 m and the possibility to use peculiar vegetation indices to analyze vegetation behavior [16]. Derived vegetation indices proved to be accurate and effective tools [12] in detecting and monitoring cork oak woodlands health and distribution. The performed classification of cork oak distribution suggested that, when the images are integrated with other ancillary data, it becomes possible to map species as well as to carry out other studies at regional scale with a high level of overall accuracy. As it emerges from other studies [17, 18] map and have [19] open data available with high accuracy, is crucial for further analysis of LU/LC changes and in a sustainable landscape planning in general.

References

1. Frampton, W.J., Dash, J., Watmough, G., Milton, E.J.: Evaluating the capabilities of Sentinel-2 for quantitative estimation of biophysical variables in vegetation. ISPRS J. Photogrammetry Remote Sens. **82**, 83–92 (2013)
2. Modica, G., Solano, F., Merlino, A., Di Fazio, S., Barreca, F., Laudari, L., Fichera, C.R.: Using Landsat 8 imagery in detecting cork oak (Quercus suber L.) woodlands: a case study in Calabria (Italy). J. Agric. Eng. **47**(205) (2016)
3. Padwick, C., Deskevich, M., Pacifici, F., Smallwood, S.: Worldview-2 Pan-Sharpening. In: ASPRS 2010 Annual Conference, San Diego, USA (2010)
4. Richter, R., Kellenberger, T., Kaufmann, H.: Comparison of topographic correction methods. Remote Sens. **1**, 184–196 (2009)

5. Balthazar, V., Vanacker, V., Lambin, E.F.: Evaluation and parameterization of ATCOR3 topographic correction method for forest cover mapping in mountain areas. Int. J. Appl. Earth Obs. Geoinf. **18**, 436–450 (2012)
6. Vanonckelen, S., Lhermitte, S., Balthazar, V., Van Rompaey, A.: Performance of atmospheric and topographic correction methods on Landsat imagery in mountain areas. Int. J. Remote Sens. **35**, 4952–4972 (2014)
7. Liu, Y., Gao, K., Wang, L., Zhuang, Y.: A hyperspectral anomaly detection algorithm based on orthogonal subspace projection. In: Sharma, G., Zhou, F., Liu, J. (eds.) 93012E (2014)
8. Wang, L., Li, Z., Sun, J.: Anomaly detection in hyperspectral imagery based on spectral gradient and LLE. Appl. Mech. Mater. **121–126**, 107–120 (2012)
9. Di Palma, F., Amato, F., Nolè, G., Martellozzo, F., Murgante, B.: A SMAP supervised classification of landsat images for urban sprawl evaluation. ISPRS Int. J. Geo-Inf. **5**(109) (2016)
10. Rouse, J.W., Haas, R.H., Schell, J.A., Deering, D.W.: Monitoring vegetation systems in the great plains with ERTS. In: Stanley, F.C., Mercanti, E,P., Becker, M.A. (eds.) Third Earth Resources Technology Satellite 1 Symposium. NASA, Washington, USA (1974)
11. Barnes, E., Clarke, T., Richards, S., Colaizzi, P., Haberland, J., Kostrzewski, M., Waller, P., Choi, C., Riley, E., Thompson, T., Lascano, R.J., Li, H., Moran, M.: Coincident detection of crop water stress, nitrogen status and canopy density using ground-based multispectral data. In: Fifth International Conference on Precision Agriculture (2000)
12. Benincasa, P., Antognelli, S., Brunetti, L., Fabbri, C.A., Natale, A., Sartoretti, V., Modeo, G., Guiducci, M., Tei, F., Vizzari, M.: Reliability of NDVI derived by high resolution satellite and UAV compared to in-field methods for the evaluation of early crop N status and grain yield in wheat. Exp. Agric. 1–19 (2017)
13. Vincini, M., Frazzi, E., D'Alessio, P.: A broad-band leaf chlorophyll vegetation index at the canopy scale. Precis. Agric. **9**, 303–319 (2008)
14. Pauly, K.: Applying Conventional Vegetation Vigor Indices To UAS-Derived Orthomosaics: Issues and Considerations. In: ISPA (2014)
15. Fichera, C.R., Modica, G., Pollino, M.: GIS and remote sensing to study urban-rural transformation during a fifty-year period. In: Murgante, B., Gervasi, O., Iglesias, A., Taniar, D., Apduhan, B.O. (eds.) Computational Science and Its Applications - ICCSA 2011, Part I. Lecture Notes in Computer Science, pp. 237–252. Springer, Heidelberg (2011)
16. Nolè, G., Lasaponara, R., Lanorte, A., Murgante, B.: Quantifying urban sprawl with spatial autocorrelation techniques using multi-temporal satellite data. Int. J. Agric. Environ. Inf. Syst. **5**, 19–37 (2014)
17. Recanatesi, F.: Variations in land-use/land-cover changes (LULCCs) in a peri-urban mediterranean nature reserve: the estate of Castelporziano (Central Italy). Rendiconti Lincei **26**, 517–526 (2015)
18. Recanatesi, F., Clemente, M., Grigoriadis, E., Ranalli, F., Zitti, M., Salvati, L.: A fifty-year sustainability assessment of Italian agro-forest districts. Sustainability **8**, 1–13 (2016)
19. Las Casas, G., Murgante, B., Scorza, F.: Regional local development strategies benefiting from open data and open tools and an outlook on the renewable energy sources contribution. In: Smart Energy in the Smart City, pp. 275–290. Springer (2016)

A Remote Sensing-Assisted Risk Rating Study to Monitor Pinewood Forest Decline: The Study Case of the Castelporziano State Nature Reserve (Rome)

Fabio Recanatesi[1]([⊠]), Chiara Giuliani[2], Carlo Maria Rossi[1], and Maria Nicolina Ripa[1]

[1] Tuscia University, 01100 Viterbo, Italy
fabio.rec@unitus.it
[2] Sapienza University, 00185 Rome, Italy

Abstract. The Multi-Spectral Instrument (MSI) aboard the ESA Sentinel-2 (S-2) allows satellite the Normalized Different Vegetation Index (NDVI) to be measured at much higher spatial resolution (10 m) than has been previously possible with space-borne sensors such as Medium Resolution Imaging Spectrometer aboard ENVISAT or Enhanced Thematic Mapper Plus aboard Landsat. Therefore, multi-spectral analysis of remote sensing data today represents an efficient tool for monitoring vegetation in a Mediterranean environment, where spatial resolution often represents a limiting factor due to high fragmentation and spatial distribution of forest stand.

The aim of this study has been to map the health conditions of the Castelporziano coastal pinewood forest (Roma). To this aim, we used a diachronic NDVI index, provided by ESA Sentinel-2 images and field observations, to monitor the health status in a historic pinewood forest that has recently been affected by a rapid diffusion of pests (*Tomicus destruens* Woll.). The monitoring performed allowed us to map the pinewood forest in risk classes and at the same time to provide data concerning the localization of areas showing a strong decline. Thus, we provide information useful for the correct management and planning of forestry thinning to preserve those areas of the pinewood forest not involved in the decline process.

Keywords: Remote sensing · NDVI · Sentinel-2 · Mediterranean forest

1 Introduction

Mediterranean landscapes are increasingly threatened by an intensification of forestry and agricultural activities, as well as urban development, tourism, and uncontrolled recreational usage. Indeed, the economic and demographic expansion which has occurred in recent decades, especially in coast urban and peri-urban areas, is altering the precarious balance achieved by environments characterized by high quality naturalistic values. Consequently, many protected areas located in such urban or peri-urban environments today are highly vulnerable towards landscaping changes that, in most cases, cause a decline of environmental quality as well cultural heritage. Forest health is

© Springer International Publishing AG, part of Springer Nature 2019
F. Calabrò et al. (Eds.): ISHT 2018, SIST 100, pp. 68–75, 2019.
https://doi.org/10.1007/978-3-319-92099-3_9

a key indicator of ecological conditions prevailing in any area, and in this specific context, forest areas perform an essential role in maintaining the ecological balance.

Among modern methods to monitor terrestrial ecosystems, remote sensing is of primary importance thanks to its capability of providing synoptic information over wide areas with high acquisition frequency [1–3]. For this reason scientists working in this field have developed vegetation indices (VI) for qualitatively and quantitatively evaluating vegetative cover using multi-spectral data. In fact, over forty vegetation indices have been developed during recent decades, amongst which the Normalized Difference Vegetation Index (NDVI) is the most widely used in monitoring the health conditions of forest surfaces [4].

The degree of vigor in forest vegetation cover can be classified according to its spectral response, which in the red (630–690 nm) is strongly correlated with chlorophyll concentration, while the spectral response in near infrared (760–900 nm) is correlated by the leaf area index and green vegetation density. Thanks to these properties, NDVI can be utilized as an indicator of possible vegetation stress, particularly that due to water shortage or pest diffusion [5].

In this study, we have developed a monitoring risk-rating system in a coastal protected area to map pinewood forest crown dieback due to increased populations of Tomicus (*Tomicus destruens* Woll.), which occurred suddenly in 2016. For this purpose, we applied a NDVI index using Sentinel-2 multispectral images in a Geographic Information Systems (GIS) environment, with spectral reflectance measurements acquired in the red (visible) and near-infrared regions, and field reliefs to correlate different degrees of pinewood crown dieback with respectively risk-rating classes.

The results of this monitoring allowed us to promptly locate the pinewood areas which have a high rate of decline, and consequently, to provide useful and precise information for the forestry interventions planning.

2 Materials and Methods

2.1 Study Area

The study area (41°42′50″N − 12°24′03″E) is the pinewood forest of Castelporziano, a State Nature Reserve located in a peri-urban area of the municipality of Rome (Fig. 1). Castelporziano has a total surface area of 6.000 ha and its land use is characterized mainly by forest and to a lesser extent by agricultural activities.

This territory is the last remnant of the ancient Mediterranean coastal forest, in which the predominant species are broadleaf oaks (4.000 ha) and pinewood (900 ha). In particular, the pinewood forest of Castelporziano, with numerous trees aged over one hundred years or more, is the last remaining example today of mature pinewood forest along the Tyrrhenian coast after the recent fire, which occurred in August 2017, causing huge damage to the neighboring ancient pinewood forest of Castelfusano. For these reasons the territory of Castelporziano can be considered as a unique environment in terms of natural and cultural values [6, 7]. Furthermore, it is listed in the Habitat Directive with two Sites of Community Importance (SCI) and the whole territory is classified as a Special Protection Area (SPA).

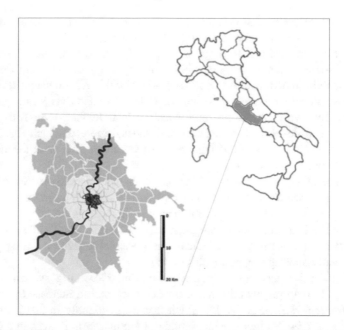

Fig. 1. Protected area of Castelporziano (yellow), Tiber river (blue), Municipality of Rome (dark gray), city of Rome (light gray and black).

In October 2016, the ongoing environmental monitoring program in the Castelporziano pinewood forest, carried out by remote sensing of multispectral Sentinel-2 images, allowed the detection of a diffuse crown dieback in several areas. This was due to a sudden infestation of *Tomicus destruens* Woll, resulting, in several areas, in a severe and rapid decline of the forest, as shown in Fig. 2.

2.2   Approach and Methods

A series of georeferenced Sentinel-2 images (06/21/2015; 06/08/2016; 06/06/2017), provided by the European Space Agency (E.S.A.), were analyzed to map the onset and recovery from pinewood forest decline.

The pinewood forest was assumed to be relatively healthy in 2015 and therefore the data relating to this period served as reference in this research. According to Richter's atmospheric correction method [8], the selected images were corrected with the Sentinel Application Platform (SNAP) program running in Sen2Core. This performs the atmospheric, terrain and cirrus correction of Top-of- Atmosphere Level 1C input data. Geometric distortion of the analyzed images was corrected with the rectified 10 m Digital Terrain Model (D.T.M.) provided by the Italian Ministry of Environment.

The forest inventory data were acquired from the Mediterranean Ecosystem Observatory Office of Castelporziano in a GIS resource. This data base contained spatial information of individual stands, topographic features, species composition and biophysical information such as tree age, basal area and site index for each forest stand.

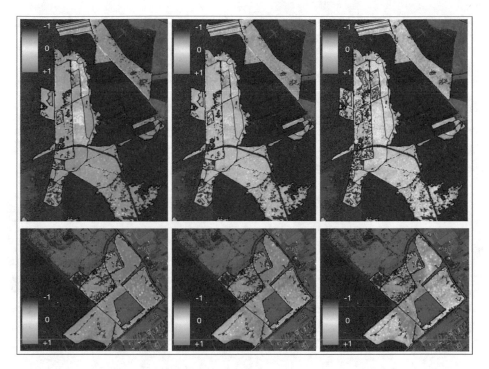

Fig. 2. Pinewood forest decline in Castelporziano detected by NDVI index, with reference to years: 2015, 2016 and 2017. (The green values [0; +1] indicate the different rates of forest health, while the red values [0; −1] indicate the different rates of forest decline).

In the present study all land use except pinewood forest were treated as non-forest and were excluded in the following analysis.

Using the NDVI image in 2015 as a reference, the 2016 and 2017 NDVI images were normalized by means of histogram match. The differential NDVI (ΔNDVI) could then be calculated as:

$$\Delta NDVI_{(t-t_{+1})} = \frac{NDVI_{(t+1)} - NDVI_{(t)}}{NDVI_{(t)}} \tag{1}$$

The histograms of the 2016 and 2017 NDVI images are characterized by an approximately standard Gaussian distribution, centered at approximately ΔNDVI = 0. In accordance with the interpretation used by Wang [9, 10], negative ΔNDVI in the left tail of the histogram reveals pinewood crown dieback and tree mortality attributed from loss of leaf moisture content. Positive ΔNDVI, in the right tail, of the histogram represents crown recovery of the pinewood. To create a map of the risk, a field survey was carried out to define the risk classes based on ΔNDVI percentage values. To this aim, according with the methodology proposed by Ogaya [11] a conspicuous number of pine trees with different ΔNDVI was examined so as to classify them in different thresholds of decline as reported in Table 1. In this phase we visually determined the percentage of dead vegetative apices for sampled trees and stand canopy density.

Table 1. Risk classes and relative ΔNDVI (2015–2017) percentage value.

Risk classes	ΔNDVI
Strong recovery	from +5 to +20%
Neutral recovery	from –5% to just less than +5%
Low decline	from –5% to just less than –15%
Medium decline	from –15% to just less than –25%
High decline	from –25% to just less than –45%
Dead plants	–100%

Once the risk classes had been defined and validated in the field, a preliminary analysis was conducted to correlate the risk classes to biometrics variables, such as volume and age derived by the Castelporziano Forest Management Plan. To do this we matched map risk class information with volume and age layers detected at the scale of forest parcel. No environmental variable, such as slope, aspect or elevation, was considered in this study because the territory analyzed is mainly flat, being located in a coastal area.

3   Results

In the observed period, 2015–2017, based on thresholding method reported in Table 1, in terms of pinewood decline four risk classes were identified based on the ΔNDVI values detected: low decline, medium decline, high decline and dead plants respectively with 10, 17, 21 and 5% of the whole surfaced analyzed, the remaining surface (47%) was classified as strong recovery or, weak recovery. In Fig. 3, the zoning of the pinewood forest in risk classes are reported, it can be noted that the spatial distribution of the class "dead plants" is distributed homogeneously throughout the whole pinewood forest. Concerning the spatial dimension of the single patches, they extend from 100 to 102.600 square meters, thus appearing very fragmented in the territory.

To determine/find out the effect of biological variables on the pinewood decline, we compared the occurrence rate for the class "high decline" (ΔNDVI from –25% to just less than –45%) with volume and age. For the latter, the correlation occurred just for that parcels in which the age was known. Nonetheless, from the results obtained, it emerges that a strong correlation exists between age and pinewood decline (R = 0.6), as shown in Fig. 4. On the contrary, only a weak correlation was observed between volume and pinewood decline (R = 0.3), as shown in Fig. 5. Regarding the role played by volume in influencing pinewood decline, it was observed that a significant increase in decline occurred starting from 350 m^3/ha.

Fig. 3. Castelporziano Nature Reserve: zoning of the pinewood according to the identified risk classes shown in Table 1.

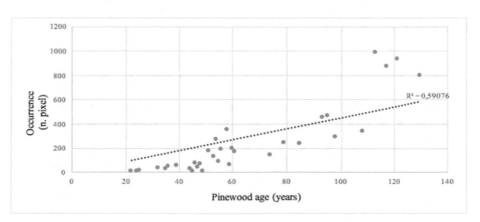

Fig. 4. Comparison between pinewood age (years) and occurrence of the risk class - high decline.

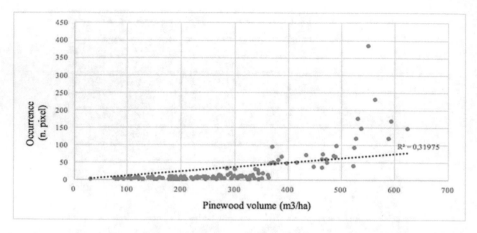

Fig. 5. Comparison between pinewood volume (m³/ha) and occurrence of the risk class - high decline.

4   Discussion and Conclusion

Sentinel-2 multi-spectral images used in the environmental monitoring of the protected area of Castelporziano, have been shown to be an accurate and efficient tool in monitoring the health conditions of vegetation, by application of an NDVI index.

The calibration of the NDVI values in risk classes allowed us to zone the whole territory in terms of pinewood crown decline rate and consequently to plan efficacious measures to mitigate the spread of *Tomicus* in the study area. In this way, from the first observation of *Tomicus* in the pinewood of Castelporziano, which occurred in September 2016, a program of forest harvesting was promptly designed to mitigate the diffusion of *Tomicus* in the rest of the pinewood. Thus, in just four months, almost 60 ha were subjected to forest thinning and, by 2018, another 100 ha is planned to be thinned to reduce pinewood vulnerability according to the detected risk classes.

References

1. Frampton, W.J., Dash, J., Watmough, G., Milton, E.J.: Evaluating the capabilities of Sentinel-2 for quantitative estimation of biophysical variables in vegetation. J. Photogrammetry Remote Sens. **82**, 83–92 (2013)
2. Sheeren, D., Fauvel, M., Josipović, V., Lopes, M., Planque, C., Willm, J., Dejoux, J.F.: Tree species classification in temperate forests using formosat-2 satellite image. Time Ser. **8**(9), 734 (2016)
3. Modica, G., Solano, F., Merlino, A., Di Fazio, S., Barreca, F., Laudari, L., Fichera, C. R.: Using landsat 8 imagery in detecting cork oak (Quercus suber L.) woodlands: a case study in Calabria (Italy). J. Agric. Eng. **47**(4), 205–215 (2016)
4. Bannari, A., Morin, D., Bonn, F.: A review of vegetation indices. Remote Sens. Rev. **13**, 95–120 (1995)

5. Maselli, F.: Monitoring forest conditions in a protected mediterranean coastal area by the analysis of multiyear NDVI data. Remote Sens. Environ. **89**, 423–433 (2004)
6. Recanatesi, F., Tolli, M., Ripa, M.N., Pelorosso, R., Gobattoni, F., Leone, A.: Detection of landscape patterns in airborne LIDAR data in the nature reserve of castelporziano (Rome). J. Agric. Eng. **XLIV**, 472–477 (2013)
7. Recanatesi, F.: Variations in land-use/land-cover changes (LULCCs) in a peri-urban Mediterranean nature reserve: the estate of Castelporziano (Central Italy), Rend. Fis. Acc. Lincei **26**, 517–526 (2014)
8. Richter, R., Wang, X., Bachmann, M., Schlaepfer, D.: Correction of cirrus effects in sentinel-2 type of imagery. Int. J. Remote Sens. **32**, 2931–2941 (2011)
9. Wang, C., Lu, Z., Haithcoat, T.L.: Detection forest dynamics responding to oack dieback in the Mark Twain National forest, Missouri. For. Ecol. Manage. **240**, 70–78 (2007)
10. Wang, C., He, H.S., Kabrick, J.M.: A remote sensing-assisted risk rating study to predict oak decline and recovery in the Missouri Ozark. GISci. Remote Sens. **45**(4), 406–425 (2008)
11. Ogaya, R., Barbeta, A., Başnou, C., Peñuelas, J.: Satellite data as indicators of tree biomass growth and forest dieback in a Mediterranean holm oak forest. Ann. For. Sci. **72**, 135–144 (2015)

Evaluations of Social Media Strategy for Green Urban Planning in Metropolitan Cities

Alessandro Scuderi$^{(\boxtimes)}$ and Luisa Sturiale

University of Catania, 95124 Catania, Italy
alessandro.scuderi@unict.it

Abstract. This study was intended to frame current changes of city and citizens in the new virtual contexts. This research analyzed different citizens need attitudes whether it occurs in a traditional or virtual context. Here we presented the first results, which may well explain the new strategies and modalities of city to join the new green urban planning and meet the new citizens' demand, by means of different tools and media. More specifically, the research carried out on the social media application for quality of metropolitan life. Information obtained has helped leading municipality towards a correct planning strategy in the different environments, traditional and virtual, also considering the impact of social media. In particular, the word-of-mouth mechanism among citizen seems to have gained a more and more relevant role in the information process of the virtual community, representing a strong generator of messages and experiences in the virtual and traditional environments.

Keywords: Social media · Citizen · e-Planning · e-Government
e-Democracy

1 Introduction

The use of Information and Communication Technologies (ICTs) in the public sector (e-government) has become a powerful strategy for administrative reform at all levels of government [1, 2]. According to Johannessen et al. [3], among the means of communication used by the urban stakeholders (politicians, administration and citizens) social media are placed after email and municipality web sites.

The target of this work is to understand the evolution of citizens behavior, in order to perceive the new attitudes that have evolved in the new context, especially with the advent of internet and the new communication modalities among citizens and between citizen and municipality, which in the last few years have started social networks. And, it aimed at defining the bases of such changes not only concerning citizens but also municipality that have been offering new activities to meet the new citizen demand and have to join the new virtual contexts. Several authors gave their contribute to the study of citizens behavior pointing out different key points for a theoretical definition in the urban planning management [4–6]. Seen the great number and variety of international approaches [7] concerning citizen behavior in the fruition process of the green urban areas, it was decided to follow the postmodern trend, that manages to catch the specific aspects of the society [8] of the new millennium [5, 9].

© Springer International Publishing AG, part of Springer Nature 2019
F. Calabrò et al. (Eds.): ISHT 2018, SIST 100, pp. 76–84, 2019.
https://doi.org/10.1007/978-3-319-92099-3_10

This approach also pointed out the reasons that have made consumers and citizens refer more and more to new information and communication technologies, ICTs, web and social networks. In particular, this study has approached deeply the attitudes and behaviors of citizens of metropolitan area of Catania about the fruition of the urban green areas, focusing on the influence of social networks on the decisions of fruition. Such information has turned to be interesting for policy and management of city to develop suitable strategies of e-democracy and e-government by the new tool of social. The exploratory research on which the article is based aims at evaluating the effect of the message in online and off-line communications in order to evaluate the reactions of the citizens about management policies of the urban green areas.

2 Materials and Methods

2.1 New Behavior of Web 4.0 Citizen

Web 4.0 and social media are progressively taking on a role of primary importance in the contemporary socio-economic context, contributing to change not only the processes and methods of communication of individuals, citizens and businesses, but also the organization and business management itself. Social media is at the heart of the Web 4.0 and consists of means of communication oriented to social interaction. The Italian cities, in recent years, have begun to enter the Web 4.0 environment, understanding its potential, using it for the story of their products and themselves, for the selection of staff or for the management and sharing of knowledge and information. The decision of the citizens to invest in social media is therefore transforming the way of constructing one's own image, of managing communication with the outside world and partly implying changes in the internal management of institutions [10, 11]. These tools are stimulating because they allow you to easily share information and materials and receive real-time comments and suggestions on what is shared. Among these Facebook, YouTube and Whatsapp, coherently with what happens in the rest of the world, are the most widespread social network in Italy. The other most popular social networks in Italy are Instagram, Twitter and LinkedIn (Messenger and Snapchat among young people).

In the public sphere, social networks in particular can support that process of innovation which, also through the use of new technologies, aims to make the action of the Municipalities more effective, efficient and participatory [12]. Due to their characteristics, social media seem to be able to contribute significantly to the development of e-Governance and e-Democracy, as tools for defining are based on dialogue and on the enhancement of the contribution of users-citizens or, more generally, of users-local stakeholders [13, 14]. In recent years, municipal administrations have been the subject of a vast reform and modernization movement, largely inspired by New Public Management [15], Public Governance, which emphasizes the systemic dimension and governance of the administrations, and to the Network Management, which highlights the importance of the network organization in the effective management of the public sector. The new technologies were then adopted with the aim of improving both the internal and external performance of the public sector, the quality of services, and the trust and collaboration with the administrations, both in the communication and involvement of the citizen and other local stakeholders [16]. Literature and practice agree that new technologies can play an enabler role in carrying out processes of change in the public sector [1, 6]. However, change in the public sector does not appear to be totally "technology driven", but can be influenced by other factors both endogenous, such as people and their skills, available resources, climate and organizational culture, exogenous, like the socio-economic conjuncture, the current regulatory system, the choices made within the public system [17]. E-Government can also be defined as the result of the interaction between new technologies, the public sector and individuals using the technologies themselves [9]. This is a complex phenomenon for which it is therefore difficult to offer a univocal and exhaustive definition, as it usually involves both technical, managerial, relational and institutional changes. In the logic of e-Government, the adoption of new technologies, however, requires a real revolution in the way of working within administrations, as it requires the use of tools, procedures and skills that were previously non-existent or poorly valued. According to Snellen and Van de Donk [14], there are three roles that ICTs can assume in e-Government: assisting the economy and efficiency of administrative activities, supporting the provision of public services and supporting democracy. In particular, the e-Government matrix considers these two dimensions: the level of sophistication in the adoption of ICTs for the provision of online services, and the type of stakeholders we turn to, we offer services and interactions are made. Crossing these two dimensions, the model identifies five stages in the process of adopting new technologies: (1) information (mono-directional from Public Administration to its stakeholders), (2) two-way communication, (3) transaction, (4) integration, (5) political participation. Going from one stage to the next, we tend to use more and more technologies and progressively more complex and moreover we are going to involve more users, be they citizens, businesses, public employees or other public administrations.

In this moment ICTs are more and more becoming a daily use feature and it's natural to involve them as one of the important tool to be used in public participation. Analysis of recent participation practices demonstrate that ICT-based tool can attract groups of citizens interested in participating in city planning strategies [10, 18, 19].

2.2 Methodology

The analysis carried out in 2017 in Catania metropolitan cities aimed at investigating the interaction among web users, measuring the intention of the citizen to tell others about their experience concerning a five green areas of urban planning, and the intention to use them [15, 17, 20]. In particular, in order to highlight the activity of citizen as source of messages that integrate or replace the emitting cities, it is important to differentiate the intention to talk about a personal experience with friends or other persons known, in a traditional offline environment, or with unknown people in virtual communities [21]. Besides, it was this study intention to verify how such will may vary according to the opinion of citizens to be referred to others, whether it was positive, negative or ordinary.

By means of an ad-hoc questionnaire, made up of 10 questions, 500 followers of the website of the city of Catania, interviewed by the Facebook account.

In particular, as far as the first phenomenon under examination is concerned, that we called *supply side*, in order to highlight the activity of citizens as source of messages that integrate or replace the emitting company, it is important to differentiate the intention to talk about a personal experience with friends or other persons known in a traditional offline environment, or with unknown people in virtual communities [22].

The second phenomenon under examination, called *demand side*, the study focused on the citizens as the receiver of a message and the one to decide in the fruition process, and on the intention to use the urban green areas that may be influenced by the suggestions and advice of other citizens and not directly by municipality. It included the possibility of such intention to use a area based on the comment, whether from a friend or unknown people, in a traditional/off-online context or a virtual/online one [22]. The other phase of the survey, however, aimed at identifying the structure of the fruition intention based on the influence of the two on/off cases. Its main aim, then, was that of establishing exactly how the suggestions and comments whether they derive from the off/online contexts and the reason to look for a specific green area, affected the final choice.

The research adopted a quantitative methodology through the experiment and the adoption of an inductive method, the systematic analysis, which experimented the phenomena under study including organization and interpretation. Questionnaire was distributed on line. The method was chosen out of a need to reveal a likely relation among the phenomena [23].

Such relation between one or more independent variables (x1, x2, ...xn) and one dependent variable (y) was random and the scientific method did only disturb without proving it. The limits of it, in fact, were the long time to obtain the requested results, the high costs to pay out the control units, and finally the complex management that led to results. Two of the three limits mentioned, time and costs, were reduced by mailing the questionnaire through Facebook.

Questions were randomized in a experimental design aiming at building up the questionnaire as much realistic as possible, showing simple scenarios and adopting widely used green area, both by frequent and non-frequent citizens, including them in a daily context of the style of interviewees.

The supply side was given three scenarios characterized by three experiences: positive, negative, neutral, for 5 green areas of Catania.

Each interviewee answered to questions for green area, while the value of the experience was randomized. On the demand side each interviewee answered to one question and the experimental design was built by crossing the three value conditions of the online experience, with the three offline ones (awful/ok/great) for one area. In order to understand the very efficacy of the experiment, as far as the demand is concerned, the "treated" interviewees, that is, those who were given questions with different randomized opinions, were studied together with a control group of "un-treated" ones, who were asked to tell their opinion about use green area without receiving any comments or suggestions from other citizens. The questionnaire aimed at evaluating from one side the will of people to tell others of the experience: to a friend for traditional purchasing or writing a online comment. It was then asked 65% of interviewees to indicate the level of probability, from 0 to 100%, to tell others about their experience concerning a product by telling a friend, while the other 35% by online posting. Randomized statements were given for the supply side, deriving from the union of the three values, positive, negative, neutral, and for the 5 green infrastructures area of category (Green Belt, Green Corridor, Green Network, Green Patch, Green Matrix, as required by the planning of the green areas of the city of Catania), 10 randomized statements were given to the demand side by joining the 3 level of values for green area from the offline environment (awful/ok/great) and from the online one, and one case where the interviewee did not received any kind of opinion.

This study utilized the statistics R package, to implement data analysis and verify the hypotheses. Estimation of parameters was carried out by minimizing the distances between data included in the model and those observed. Functions of estimation were different, LM, linear mode was used to study the relation between a dependent variable (y) and a series of independent ones (x1, x2, ...xn), in order to understand the impact of such variables on the subject under examination. Concerning the "supply side" the dependent variable y was the probability to tell others about a personal experience concerning a green area, both to friends of a offline context, and to online communities.

This model highlighted the logic scheme of variables that were related with the dependent variable USE-WOM prob, that is, the probability of telling others about the personal experience of the fruition. USE-WOM prob was function of a series of independent variables, such as, GREEN AREA CATEGORY in other words, the category of products that in this case were five, the experience value, POS (positive), NEG (negative), MED (neutral).

Here after is the base model with all the variables analyzed:

$$\text{SOCIAL probi} = \beta 0 + \beta 1\,\text{AREA} + \beta 2\,\text{POS} + \beta 3\,\text{NEG} + \beta 4\,\text{MED} + \beta 5\,\text{GENDER} + \beta 6\,\text{POST} - I_c$$
$$+ \beta 7\,\text{WHY} - I_c + \beta 8\,\text{AGE} - I_c + \beta 9\,\text{EDU} + \beta 10\,\text{PERS.f} + \varepsilon t$$

Where "SOCIAL probi" is the probability to report others about the experience of fruition, where i stands for individuals, while AREA stands for the typology of urban green area, POS for the positive experience, NEG for the negative experience, MED for the neutral experience, GENDER for the sex, POST for the online posting activity by interviewees, WHY for the motivation of the online posting activity, AGE for the age, EDU for the education, and pers.f the variable that collected the individual charac-teristics of each interviewee, and finally εt is the error of prediction/residue. Note that the first paragraph of a section or subsection is not indented.

3   Results

Models of linear regression were elaborated for the "Supply side" are 22, and 11 for the "Demand side". As far as the main results of the supply side are concerned, they are summarized in Tables 1 and 2.

In the second column of below tables are the coefficients of the different variables chosen to verify a possible influence of the same on the dependent variable, or the will to tell others about the personal experience of fruition, that measured in percentage terms varies from 0 to 100%. Such links were detected for the sample's will to spread

Table 1. Estimated structural parameters for the sample concerning the will to spread traditionally USE–WOM urban green areas (offline).

	Estimate	Std. Error	t value	Pr (>ltl)
(Intercept)	66.5297	7.4359	5.349	0,006135
med	-16.5341	3.4218	-5.251	0,039527
pos	21.5873	2.8742	6.492	0,023530
Green Belt	4.5291	3.1125	1.257	0,006983
Green Corridor	14.5318	3.4513	4.421	0,003312
Green Network	12.2351	3.6321	3.217	0,006217
Green Patch	12.3129	4.0125	3.874	0,000794
Green Matrix	12.0127	3.9253	2.527	0,000856

Table 2. Estimated structural parameters for the sample concerning the will to spread traditionally USE -EWOM urban green areas (online)

	Estimate	Std. Error	t value	Pr (>ltl)
(Intercept)	71.5326	6.1138	5.957	0,005689
med	-4.5126	2.7712	-2.112	0,06513
pos	26.7425	3.2587	5.387	0,09569
Green Belt	3.6713	2.0125	1.021	0,000945
Green Corridor	11.3551	4.5278	6.278	0,007758
Green Network	12.4578	3.2587	4.251	0,00741
Green Patch	15.258	4.7423	4.123	0,00521
Green Matrix	13.2374	3.1123	3.132	0,00235

the experience of use of the urban green areas in the community (off-line) (Table 1) and online (Table 2).

Specifically, there was a significant statistics link as far as the green urban area category is concerned and the relative will to tell others about the fruition experience. In the traditional channel (offline) all categories, which does not weight online, showed a significant and positive coefficient for the will to spread USE-EWOM relative to such category.

Considering the variables that referred to the value of the purchase experience, neg, med, and pos respectively, or to the relation that occurred between the typology of the citizens' experience for a specific green area and the intention to tell others, there was a significant relation for all typologies of value for the sample in both cases, traditional and virtual.

The positive experience generated more whom than a negative one, and the latter caused on its turn, a bigger will to transfer such experience than a ordinary one, for which, for both samples considered, we obtained all negative coefficients statistically significant. This means that, the presence of a med experience negatively affects the will to tell others information relative to a urban green area.

The same trend, where the will to tell others about the experience prevailed, was for the sample of the traditional environment.

Passing to consider the series of regression models concerning the experiment called "Demand side", the study focused on some models, that related the opinions deriving from the traditional offline environment and the online context on the possible impact they may have had on the intention to buy of citizens that received such information.

Without considering the specific results, this study allowed obtaining the following results for the urban green areas offered directly to citizens:

- the wish to tell others about the personal use experience varies whether this refers to a different level of past knowledge,
- the intention to refer others of one's own personal fruition experience varies whether the message is sent in a offline or online context,
- the will to tell others about one's own personal experience varies whether the message is from an offline or online context,
- the value of the fruition experience affects differently the will to refer others about one's own experience,
- the will to tell others about one's own personal experience varies whether it concerns area already known.

4   Conclusions

The interactions that can be activated in metropolitan cities with the use of social media in the more general framework of e-Government, while not directly addressing the issue, seems to suggest that social networks can improve the dialogue of Municipalities with citizenship and territory. ICTs tools allow creating a multi-layer stakeholder platform that is essential feature of planning strategies [19]. In general, Italian metropolitan cities can't fully speak of a massive evolution because the activities and services of e-Democracy and e-Governance are often still at an embryonic stage. On their whole, the metropolitan cities have late understood the opportunities offered the social media [24], in order to refer directly to final citizens who are here involved actively. It is necessary, however, to avoid failing, to point at the right target and build up specific supports then, such as events and advertisement, that involve municipality more and integrate their activities. The virtual environment has strongly influenced the fruitioning process' phases introducing innovative elements. Firstly, by guaranteeing a wide range of information on the web at sensibly reduced costs, which have allowed citizens to compare competitive green urban areas and know users' opinions. It is necessary for municipality, then, to start a suitable route to guarantee new forms of collaborations for citizens, with particular reference to the online environment, whose logic passes three levels: connection, conversation and building of new relationships [20]. The direct evaluation of citizens' behavior has allowed demonstrating its utility for green urban areas, to orient towards a strategy of communication in the different place, traditional and virtual, understanding firstly what's the impact of social media through citizens' word of mouth about the fruition decisions of the green areas referred by others, and also how such experience relative to the power of messages in the two considered environments, affects differently the fruition decisions.

References

1. Bonsón, E., Royo, S., Ratkai, M.: Citizen's engagement on local governments' Facebook sites. An empirical analysis: the impact of different media and content types in Western Europe. Gov. Inf. Q. **32**, 52–62 (2015)
2. Linders, D.: From e-Government to we-government: defining a typology for citizen e-Participation in the age of social media. Gov. Inf. Q. **29**, 446–454 (2012)
3. Johannesen, M.R., Flak, L.S., Sæbø, Ø.: Choosing the right medium for municipal e-participation based on stakeholder expectation. In: Electronic Participation, pp. 25–36. Springer, Heidelberg (2012)
4. Scuderi, A., Sturiale, L., Bellia, C., Foti, V.T., Timpanaro, G.: The redefinition of the role of agricultural areas in the city in relation to social, environmental and alimentary functions: the case of Catania. Rivista di Studi sulla Sostenibilità **2**, 1–10 (2017)
5. Sturiale, L., Trovato, M.R.: The smart management and e-cultural marketing of UNESCO heritage. Int. J. Sustain. Agricult. Manage. Inform. **2**(2/3/4), 155–173 (2016)
6. Wallin, S., Horelli, L., Saad-Sulonen, J.: Towards and ecology of digital tools as embedded in participatory e-planning. In: Wallin, S., Horelli, L., Saad-Sulonen, J. (eds.): Digital Tools in Participatory Planning, pp. 135–142. Aalto University, Centre of Urban and Regional Studies, Series C27 (2010)
7. Brown, J., Broderick, A.J., Lee, N.: Word of Mouth communication within online communities: conceptualizing the online social network. J. Interact. Market. **21**(3), 2–20 (2007)
8. Sturiale, L., Scuderi, A.: Information and Communication technology (ICT) and adjustment of the marketing strategy in the agrifood system in Italy. In: Salampais, M., Matopolous, A. (eds.) Proceedings of V International Conference of Information and Communication Technologies in Agriculture, Food and Environment (HAICTA), 8–11 September, Skiados, Greece (2011)
9. Welch, E.W., Hinnant, C.C., Moon, M.J.: Linking citizen satisfaction with e-Government and trust in government. J. Public Adm. Res. Theory **15**(3), 371–391 (2005)
10. Hiller, J.S., Bélanger, F.: Privacy Strategies for Electronic Government. The Pricewater-housecoopers endowments for the business of Government (2001). A Public Management for all Seasons? Public Adm. **69**(1), 3–19 (2001)
11. Klijn, E.H., Koppenjan, J.F.M.: Public management and policy networks: foundations of a network approach to governance. Public Manage. **2**(2), 135–158 (2000)
12. Pollitt, C., Van Thiel, S., Homburg, V. (eds.): New Public Management in Europe: Adaptation and Alternatives. Palgrave Macmillan, Houndmills (2007)
13. Kluemper, D., Rosen, P.: Future employment selection methods: evaluating social networking web sites. J. Manag. Psychol. **24**(6), 567–580 (2009)
14. Snellen, I.T.M., Van De Donk, W.B.H.J.: Public Administration in an Information Age: A Handbook. IOS Press, Amsterdam (1998)
15. Ferlie, E., Ashburner, L., Fitzgerald, L., Pettigrew, A.: The New Public Management in Action. Oxford University Press, Oxford (1996)
16. Torres, L., Pina, V., Acerete, B.: E-governance development in European Union cities: reshaping government's relationship with citizens. Governance **19**(2), 277–302 (2006)
17. Kickert, W.J.M., Klijn, E.H., Koppenjan, J.F.M.: Managing Complex Networks. Strategies for the Public Sector. Sage Publications, London (1997)
18. Saad-Sulonen, J., Horelli, L.: The value of community informatics to participatory urban planning and design: a case study in Helsinki. J. Commun. Inform. **6**(2), 1–16 (2010)

19. Stauskis, G.: Development of methods and practices of virtual reality as a tool for participatory urban planning: a case study of Vilnius City as an example for improving environmental, social and energy sustainability. Energy Sustain. Soc. **4**(7), 1–13 (2014)
20. Lipsman, A., Graham, M., Rich, M., Bruich, S.: The power of "like": how brands reach (and influence) fans through social-media marketing. J. Advertising Res. **52**, 40–52 (2012)
21. Kozinets, R.V.: The field behind the screen: using netnography for marketing research in online communities. J. Market. Res. **34**, 61–72 (2002)
22. Sandoval-Almazan, R., Gil-Garcia, J.R.: Are government internet portals evolving towards more interaction, participation, and collaboration? Revisiting the rhetoric of e-government among municipalities. Gov. Inf. Q. **29**, 72–81 (2012)
23. Gatautis, R., Kazakeviciute, A.: Consumer behavior in online social networks: review and future research directions. Econ. Manage. **17**, 1457–1463 (2012)
24. Sturiale, L., Scuderi, A.: Evaluation of social media actions for the agrifood system. Procedia Technol. **7**, 200–208 (2013)

Multi-stage Strategic Approach in Spatial Innovation: How Innovation District Matter?

Carmelina Bevilacqua, Luana Parisi$^{(\boxtimes)}$, and Laura Biancuzzo

Mediterranea University of Reggio Calabria, 89100 Reggio Calabria, Italy
cbevilac@unirc.com, {luana.parisi,
laura.biancuzzo}@unirc.it

Abstract. In the 21st century globalised economy, innovation is a crucial factor within strategies targeted at growing and sustaining competitiveness of regions and cities. Accordingly, the creation of knowledge process, along with sharing and commercialisation, became an effective response to the pressures generated by globalisation in order to increase the competitive advantage. The emerging trend of innovation-led urban planning initiatives provides strong evidence of how cities are implementing strategies to promote innovation mainstreaming. Hence, these innovation-oriented policies, which are targeted at reshaping cities, are currently translated in the creation of innovation districts. This paper aims at identifying the actors who foster the innovation process at urban level, and analysing their influence throughout the innovation district life cycle. Firstly, the authors assess the role played by public and private sector in the different stages of innovation district development, by adapting the Urban Land Institute conceptual framework in the Innovation Life Cycle District Assessment. Secondly, empirical research works are defined in order to test the ILCDA. The Boston Innovation District and the IDEA District are the two case studies under investigation, by pointing out the policies and planning initiatives undertaken in the Seaport area of Boston and in Downtown San Diego, respectively. Findings from this research highlight the level of public private partnership effectiveness in supporting the development of innovation districts. Useful lessons can be drawn in encouraging planners and policy-makers towards undertaking combined actions at the different stages of the development process.

Keywords: Innovation economy · Urban regeneration · Maps-Led

1 Introduction

Over the last decades, innovation has increased its importance within the pattern of economic growth, moving to the central stage of economists and policymakers concerning the factors that enable the process. According to the "innovation based growth theory", economic prosperity results from increase in knowledge, scientific and technological improvements, along with the development of an effective private-public partnership [1, 2]. Innovation is therefore considered as crucial factor of nations central strategies targeted at growing and sustaining competitiveness in the 21st century globalised economy [3]. As an extensive body of knowledge corroborates, cities and innovation are nowadays strongly linked and their tangible effort in providing a

© Springer International Publishing AG, part of Springer Nature 2019
F. Calabrò et al. (Eds.): ISHT 2018, SIST 100, pp. 85–94, 2019.
https://doi.org/10.1007/978-3-319-92099-3_11

favourable context for innovation to prosper, can be read in the emerging trend of innovation districts proliferating globally. Although several research studies attempted to scrutinise the dynamics that lie behind the creation of an innovation district, complying with the innovation economy forces [4–7], less emphasis has been placed on the key role played by the actors involved in the district development process. This paper puts the body of knowledge forward on the triggering actions, implemented by city governments and investors, influencing the innovation space patterns, through the application of a LCA methodology. The paper investigates the Boston Innovation District and the IDEA District case studies, by focusing on the implementation process of the innovation-led strategies undertaken in Boston and San Diego. Accordingly, the research is developed to identify the success factors to grow, develop, and sustain technology innovation ecosystems in cities in order to perform policy actions. This paper is organised in three parts. After a scientific background in describing the spatial dimension of innovation, the paper highlights the physical environment where the dynamic innovation ecosystem takes shape, i.e. the innovation district. An overview of the LCA implementation on the newly conceived urban model is then provided. In the second part, the methodology inherent with the breakdown of the innovation district evolutionary process is applied to the case studies analysis. In particular, the Innovation Life Cycle District Assessment (ILCDA) adapted by the Urban Land Institute conceptual framework, is defined to analyse multi-stage strategic approach undertaken in the BID and IDEA districts. Finally, findings and conclusions are discussed.

2   Innovation Economy and Spatial Patterns

2.1   The New Geography of Innovation

In recent times, the research on the spatial dimension of innovation provides controversial views, confirming the complexity of the phenomenon. On the one hand, an extensive body of knowledge corroborates the idea that innovation economy prefers regional systems as location for creating and spreading new knowledge [8–10]; on the other hand, the opinion that cities and innovation are strongly linked is becoming progressively popular [11], given the high concentration of innovation across and within cities and metro areas [12]. As a matter of fact, the urban environment proves evidence to encompass the suitable economic and cultural dynamics in order to generate radical innovations and boost the development of new industries [13]. Nevertheless, significant is the innovation economy potential in regenerating local economic areas and promoting local assets [14]. In this regard, cities are experiencing massive transformations by fostering "knowledge-intensiveness and technological advancement … in order to become competitive providers of first class living for highly skilled global work-force" [15]. All the above mentioned observations lead to the conclusion that a process of urbanisation of innovation is now occurring. A physical shift of innovative businesses from suburban corridors and science parks to inner-cities areas is taking place, prompted by companies' need to relocate in areas that ensure close connectivity among people and give direct access to markets and finance, in order to support the innovative entrepreneurial activities [16]. It follows that, policymakers are responsible for the institutional and regulatory framework in order to manage the re-urbanisation and influence the amount of innovative activity through the adoption of designated policies.

2.2 Innovation Districts

The tangible effort of cities at providing a favourable context for innovation to prosper, can be read in the emerging trend of innovation districts proliferating globally. Specifically, they are "geographic areas where leading-edge anchor institutions and companies cluster and connect with start-ups, business incubators, and accelerators. Compact, transit-accessible, and technically-wired, fostering open collaboration, and offering mixed-used housing, office, and retail" [17]. By bringing together in geographical proximity this unique combination of economic, physical, and networking assets, the idea generation is stimulated and the entrepreneurial activity facilitated [18]. Innovation district urban form and function cannot be defined a priory. However, according to their location and the type of businesses settled within their boundaries, they have been categorised into three models: (i) Anchor Plus Model; (ii) Re-imagined Urban Areas Model; (iii) Urbanised Science Parks Model [17]. From the above mentioned considerations, it stands to reason that innovation districts represent the physical environment where the innovation ecosystem takes shape. However, the relation between the two has a multi-dimensional and non-linear nature: on the one hand, innovation districts reflect the city's wider economic, social and political systems, and they cannot flourish without the innovative ecosystem in which they are embedded; on the other hand, innovation districts on their own do not generate any innovation ecosystem. This leads to the conclusion that "a city does not become an innovation hub simply by promoting the establishment of an innovation district … successful districts are driven by larger trends than site availability" [20].

2.3 Life Cycle Assessment

Innovation spaces are required as key component of new urban regeneration initiatives within cities wider strategy of urban growth, in enhancing competitiveness by nurturing and accelerating the innovation process, and in improving liveability by providing solutions for a more efficient land use [14]. However, achieving successful innovation districts requires tools to guide the actions implemented by city governments and investors at different points of their development. A valid response to this challenge is provided by the Life Cycle Assessment (LCA), a methodological framework traditionally focused on the improvement of goods and services [21]. Although it was born from the increasing environmental awareness of businesses towards achieving sustainability goals [22], the LCA provides a holistic perspective that is increasingly applied to policy-making issues analysis [23]. Accordingly, Belussi and Sedita [24] investigated the factors that influence the origin, development and maturity of industrial districts within their evolutionary processes. Specifically, "districts do indeed often follow an evolutionary path from in-fancy to a growth phase, followed in turn by maturity and subsequent stages of stagnation and decline or revitalization". This methodological framework allowed to examine the specific triggering factors at the basis of the district existence, and to describe the mechanisms that characterise their evolutionary path.

3   How Cities Build Their Innovation Economy Through a Multi-stage Strategic Approach: Evidence from Boston and San Diego

3.1   Methodological Framework

The study gives validation of the LCA analytical method in identifying the triggering actors, and relative actions, affecting the innovation districts' evolution at different stages of their life cycle. In so doing, the 'product oriented' conceptual framework built within the Urban Land Institute research on the innovation economy [20], was adapted to evaluate the wide-range of issues concerning the multi-stage strategic approach undertaken in two purposely selected case studies, namely the BID and IDEA districts.

Hence, the Innovation Life Cycle District Assessment (ILCDA) has been built (see Fig. 1), and the three stages of the above mentioned districts' development have been analysed, i.e. start-up, activation, and maturing. Firstly, the role played by the public and private sectors in nurturing the innovation ecosystem and in supporting specific locations as urban innovation districts has been investigated. Secondly, the efforts to foster the ecosystem conditions and to catalyse development in a specific location have been considered. Finally, the analysis of the strategies to sustain the environment for innovation as the district matures have been scrutinised. Thus, after a brief description of the two case studies, the actions implemented by the different actors involved at different stages will be examined.

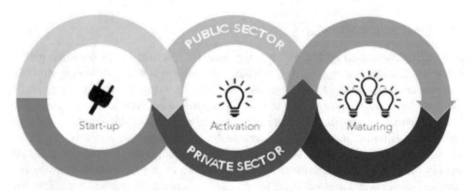

Fig. 1. ILCDA framework to analyse the innovation district multi-stage strategic approach. Source: Author's elaboration, 2017.

3.2   Case Studies Description and Results' Discussion

The Boston Innovation District provides an outstanding case study of thousand acres transformation into a centre of knowledge economy, fostering innovation, collaboration, and entrepreneurship [25]. In 2010, Major Menino declared his vision to redevelop the declined industrial area of the South Boston Waterfront through a District able to meet the needs of innovators, creating a job magnet, and an urban lab on the shore. The vision had a few main features: (i) the desire to cluster innovative entrepreneurs to increase proximity and density; (ii) the openness to industries of every kind; (iii) the adoption of a framework characterised by expedited decision making and planning flexibility, allowing the neighbourhood to develop organically and disperse innovation across the city [26].

On the other hand, Downtown San Diego stands for the innovation economy attempt to create a vibrant city centre driven by a "Design jobs cluster, nourished by Education, enriched by the Arts and focused on Innovation" [27]. Specifically, following the 1980s Centre City Development Corporation's general strategy targeted at revitalising the 'dormitory' character of Downtown San Diego, in 2010 two developers brought forward the **I.D.E.A. District**, in order to make the downtown attractive for the emerging workforce and bring jobs back to the city centre. The vision targeted 35 blocks located in the Upper East Village neighbourhood, where the presence of growing design businesses and educational institutions hungry for collaboration, as well as the availability of urban land all contributed to create a new design industry cluster framework.

Starting from the analysis of the *start-up stage*, it has to be acknowledged that coordinated actions between city government, landowners and developers are critical to transform the vision of an innovation district into reality. In the case of Boston, the public sector acted as the main operator, attracting private investments, creating jobs and providing the necessary services [28]. Indeed, the start-up stage was launched by the Mayor Menino, who entrusted the main public planning agency, namely, the BRA (Boston Redevelopment Authority) for the management of the District. The BRA, from the beginning, partially funded the project and, through public-private partnerships, helped to "ease the financial burden of the project on the City's budget" [29]. Several development tools have been promoted by public actors in this stage (e.g. variation of zoning regulations, and tax relief programs), attracting businesses in the area and increment tax revenues to fund specific public projects [28]. The city of Boston has led the project also in the *Activation stage*, being the host institution "instead of the host being a university or research firm" [30]. This brought to the identification between the District and the city and the adoption of a "hands-off approach" with some exceptions, such as the move of Vertex Pharmaceutical, facilitated by the Mayor. The Public actors further facilitated the development of physical and social infrastructures to build up a community [28], the move of educational institutions and the establishment of entertainment options. Public actors managed some of the innovation spaces also in the *Maturing stage* through the room rental model, that discounts off the fees for helping those organisations, mission-based programs, and start-ups that cannot afford the rent (personal communication, June 25, 2016). They further facilitated the establishment of shared workspaces, incubators and new residential options, including flexible housings. Boston has also worked to "institutionalise more dynamic processes of public planning and service delivery" [26]. As the District continues to be transformed by new mixed-use development projects, and retain workspace variety for different firm types and sizes, the perceptions of the area are starting to change. This is due to the disproportionate increase of both home values and average rents experienced over the last years. Thus, the district is attracting talents, but after they grow their business, they move out because economically they cannot sustain it in the long term (personal communication, June 22, 2016).

In the San Diego case there were unbalanced efforts, since the private sector made several of the most important enabling interventions. Indeed, the I.D.E.A. Partners have been the lead agents in the process of change, by replacing the public sector in understanding the city's competitive advantage and identifying the innovative industries to attract in order to create the critical mass capable of driving economic growth, together with the selection of the most appropriate location for the development. Furthermore, they started developing a shared vision by involving residents, local businesses and civic leaders in order to build consensus around the principles of the plan. Engaging the community of residents and innovators, through an effective outreach strategy, turned out to be crucial given the little support demonstrated from the outset by the local administration that failed in defining a tailored long-term strategy, as well as in simplifying the urban regulations to speed up the planning process. The private sector leadership was paramount also in the *Activation stage*, when, besides important catalytic investments, significant were the efforts to draw the attention of some anchor firms, universities and innovation hubs to settle in the district. Partnerships with other investors and developers have been established, leading to further development initiatives, such as the Makers Quarter, for fostering the work-live-play environment required by start-ups, tech companies, and talents. Moreover, the developers re-shaped the image of the neighbourhood as a more vibrant location through the strategic use of the tactical urbanism. The public sector role, also in this stage, can be defined somehow idle; the neighbourhood did not benefit from a centralised plan and the zoning requirements of the Community Plan 2006 remained unchanged. Thus, the city government did not undertake any effort to facilitate the mixed-use development and make the area more attractive to new businesses, since neither financial tools, nor system of development rights have been used in order to encourage strategic firms to relocate. The I.D.E.A. District is still at the very beginning of the *third stage* of its development, thus, the actions further implemented by the public and private sectors can be deduced by the current state of affairs. The public actors are starting to show their interest in the innovation processes going on in the downtown area and are facilitating the move of educational institutions, such as the UCSD (personal communication, May 16, 2017).

4 Conclusions

Given the shift of the geographical distribution of innovation from suburbs to urban areas, cities must constantly reinvent themselves in order to provide an environment that is conducive to innovation and remain competitive in the 21st century globalised economy. Figure 2 shows the synthesis of the results of the analysis of the main actions implemented by different actors at different stages of development.

As clearly visible, although on the one hand downtown San Diego naturally provides a compact urban structure, vital for productive collisions to take place between

STAGE	ACTIONS	BOSTON	SAN DIEGO
Start-up	City government as lead partner	●	○
	Public land ownership	●	○
	Public provision of the institutional framework	●	○
	Public regulatory framework	●	○
	Innovation Industries identification	○	●
	Zoning regulation change	●	○
	Tax incentives	●	●
Activation	Innovation-oriented urban policies	●	○
	Development of public Infrastructures	●	○
	Presence of a leading-edge anchor institution	●	○
	Strong set of private initiatives	○	●
	Tie between public leadership and private financing	●	○
Maturing	Public price controls	○	○
	Community involvement	●	●
	Provision of cultural projects and events	●	●
	Metropolitan connotation of the district	●	●
	Gentrification and homeless management	○	○

Fig. 2. Main actions of the public and private sectors in different stages of innovation district development. Source: Author's elaboration, 2017.

firms, people, capital, and ideas, on the other hand the city government has not been a leading partner in boosting the innovation ecosystem by providing the institutional and regulatory framework in order to manage the re-urbanisation and influence the innovative activities through the adoption of designated policies. The exception is represented by the UCSD Extension moving to downtown, where the city government put some regulations and guidelines given its ownership of a small property within the block subject of intervention. Thus, the landownership issue played a key role in discouraging any collaboration between actors: since the city owned a considerably small portion of downtown land, around the 20%, the prevailing private interests led the entire intervention of redevelopment. The innovation economy, indeed, took over from the manufacturing industry sector located in the downtown area, and the role of real estate turned out to be essential for its physical transformation. The reasons are well explained by the rate and capability of the real estate to adapt to the new innovation market requirements, determining a strong competitive advantage for the innovation system in which they operate [32]. Thus, the set of private initiatives currently happening, although operating independently, pursues the same objective to deliver an inspiring and accessible environment that attracts talents and fosters innovation. In addition, cultural projects and events turned out to be paramount in tailoring the district vision to the specific needs of the future users, creating value and sense of place within the community. The landownership has had an important role also for the spatial innovation. Indeed, the prevailing private ownership of the already consolidated urban fabric of Downtown San Diego does not allow to read a clear innovation spatial matrix, except for some regeneration interventions. On the other hand, the experience in Boston, linked to a public entrepreneurship approach, shows the results of a clear public intention to physically regenerate the area from the outset. Thus, the innovation spatial matrix is expressed through the creation of a dynamic living laboratory delivering built-environment goals, arising from scratch and targeting the community of innovators, and expressing place-based strategies. The public entrepreneurship approach adopted in Boston helped also to speed up the process and make it more efficient. It also gave the sparkle for establishing strong partnerships and absorbing the energy from the private sector and non-profit organisations. "The entire project relies heavily upon the principles of the shared economy and the connections between public leadership and private financing" [30]. The City changed the zoning regulation to accommodate R&D functions and offered tax incentives to attract businesses in the area [29]. As in the case of San Diego, the community involvement has been among the main goals, implemented through social and physical infrastructures [29]. Yet, the BID has become, over time, "less about start-ups and more about the expansion of Boston itself", so that "there's innovation going on there, but that's no longer the primary focus" [31]. Several amenities and facilities have been built in order to create a new neighbourhood to serve the whole city. Thus, in both cases, the Districts assumed a metropolitan connotation, becoming a pretext for physically regenerating the areas and expanding the cities. This is reflected by the rocketing real estate prices that created a tension between the economic growth driven by innovation and the hidden negative externalities generated by the Innovation District model. The dark side refers to the fact that the rewards tend to benefit only a few people, widening the gentrification gap and worsening, possibly, also the homeless issue. The "business model" of Innovation Districts, becoming unaffordable to the most and favouring the concentration of poverty in a few areas, adds to the effects of the existing high poverty rate in the two cities in question that rank already among the first places nationally.

Overall, it emerged that the Innovation District model cannot be started in a vacuum, since, being a place-based tool, it got attached to the rest of the ecosystem. In view of the two analysed case studies, it emerges the necessity to have a leading-edge anchor institution helping a critical mass of innovators and companies to take shape. The lack of innovation-oriented economic urban policies and economic development measures to foster the ecosystem preconditions and control the city's urban regeneration has proved crucial to the attraction and retention of anchor institutions and the development of human capital. Indeed, the high rents and the lack of any tax incentives are the main factors that can discourage companies from locating in these districts. Thus, the case studies provide clear evidence that the multi-stage strategic approach, implemented by concerted actions of public and private sectors, is crucial to create and nourish a successful innovative environment. The support of the public actors is fundamental for the coordination of the initiatives and the public benefits provision, for avoiding the unintended consequences linked to the phenomenon of aggregation of talents, such as the rocketing real estate prices and the consequent gentrification, that could benefit mainly middle and upper class people. Public initiatives, including zoning and investments, are fundamental also for supporting diversity, necessary for triggering innovation. The easiness of the bureaucratic processes can help to employ less public resources, encouraging also the public actors to be creative "in aligning stakeholder interests to move the project forward" [29].

References

1. Porter, M.E.: The Competitive Advantage of Nations. Free Press, New York (1990)
2. Baily, M., Katz, B., West, D.: Building a long-term strategy for growth through innovation. Metropolitan Policy Program at Brookings (2011)
3. West, D.: Technology and the innovation economy. Center for Technology Innovation at Brookings (2011)
4. Muro, M., Katz, B.: The new 'Cluster moment': how regional innovation clusters can foster the next economy. Metropolitan Policy Program at Brookings (2010)
5. Delgado, M., Porter, M.E., Stern, S.: Cluster and entrepreneurship. J. Econ. Geogr. **10**(4), 495–518 (2010)
6. Feldman, M.: The Geography of Innovation. Kluwer Academic Publishers, Dordrecht (1994)
7. Feldman, M.: The character of innovative places: entrepreneurial strategy, economic development, and prosperity. Small Bus. Econ. **43**(1), 9–20 (2014)
8. Asheim, B., Gertler, M.: The geography of innovation: regional innovation system. In: The Oxford Handbook of Innovation (2006)
9. Cortright, J.: Making sense of clusters, regional competitiveness and economic development. Metropolitan Policy Program at Brookings (2006)
10. Sallet, J., Paisley, E., Masterman, J.: The geography of innovation: the federal government and the growth of regional innovation clusters. Science Progress (2009)
11. Shearmur, R.: Are cities the font of innovation? A critical review of literature on cities and innovation. Cities **29**, S9–S18 (2012)
12. Florida, R., Adler, P., Mellander, C.: The city as innovation machine. Reg. Stud. **51**, 1–11 (2017)
13. Montgomery, J.: The New Wealth of Cities: City Dynamics and the Fifth Wave. Ashgate, London (2007)
14. MAPS-LED: S3 Cluster Policy & Spatial Planning. Knowledge Dynamics, Spatial Dimension and Entrepreneurial Discovery Process. MAPSLED Project (Multidisciplinary Approach to Plan Smart Specialization Strategies for Local Economic Development). Horizon 2020 - MarieSlowdoskwa Curie Actions - RISE P.R.645651 (2017)
15. Inkinen, T.: Reflections on the innovative city: examining three innovative locations in a knowledge bases framework. J. Open Innov. Technol. Market Complex. **1**, 8 (2015)
16. Mulas, V., Minges, M., Applebaum, H.: Boosting tech innovation ecosystems in cities: a framework for growth and sustainability of urban tech innovation ecosystem. The World Bank (2015)
17. Katz, B., Wagner J.: The rise of innovation districts, a new geography of innovation in America. Brookings Institution (2014)
18. Giuffrida, G., Clark, J., Cross, S.: Putting innovation in place, Georgia Tech's Innovation Neighbourhood of 'Tech Square' (2015)
19. Clark, G., Moonen, T., Peek G.: Building the innovation economy. City-level strategies for planning, placemaking and promotion. Urban Land Institute (2016)
20. Rebitzer, G., Ekvall, T., Frischknecht, R., Hunkeler, D., Norris, G., Rydberg, T., Schmidt, W., Suh, S., Weidema, B.P., Pennington, D.W.: Life cycle assessment: framework, goal and scope definition, inventory analysis, and applications. Environ. Int. **30**, 701–720 (2004)
21. Curran, M.: Life cycle assessment, Kirk-Othmer encyclopedia of chemical technology, pp. 1–28 (2016)
22. Guinee, J., Heijungs, R., Huppes, G., Zamagni, A., Masoni, P., Buonamici, R., Ekvall, T., Ry-dberg, T.: Life cycle assessment: past,present, and future. Environ. Sci. Technol. **45**(1), 90–96 (2011)

23. Belussi, F., Sedita, S.: Life cycle vs multiple path dependency in industrial districts. Eur. Plan. Stud. **17**(4), 505–528 (2009)
24. Menino, T.: Panel comments, Florence (2010)
25. Cohen, A.: The development of Boston's innovation district: a case study of cross-sector collaboration and public entrepreneurship (2014)
26. IDEA District vision. http://www.ideadistrictsd.com/vision-document-available-now/. Accessed 9 Oct 2017
27. Britto, N.: The evolving roles of local government: insights from Boston's Innovation District (2015)
28. National League of Cities (NLC): Boston Innovation District (2016). https://www.nlc.org/resource/boston-innovation-district. Accessed
29. Wagner, J., Davies, S., Sorring, N., Vey, J.: Advancing a new wave of urban competitiveness: the role of Mayors in the rise of innovation districts (2017)
30. Clark, G., Moonen, T.: Technology, real estate, and the innovation economy. Urban Land Institute (2015)
31. Mohl, B.: The next Kendall Square? Harvard tries its hand at innovation on the Boston side of the Charles. Common Wealth (2016)

Arco Latino: A Model of European Resilience

Massimo Corsico[1] and Elisabetta M. Venco[2(✉)]

[1] J&A Consultants, 27110 Pavia, Italy
[2] University of Pavia, 27100 Pavia, Italy
elisabettamaria.venco@unipv.it

Abstract. The cities of the Mediterranean basin are treasure chests of stories, memories and symbols: they are the essential nodes of communication and migratory currents between populations. Western culture took birth from the traditional city, the centre point of all the contaminations and all trade networks. Ports and cities are historically strongly linked, although some markedly different relationships exist between them. The strength of these links depends on local and global circumstances, and peculiar local challenges.

Authors focus on the Mediterranean dimension of European coastal cities, and in particular on those of the so-called Arco Latino, as representatives of an exported and exportable model of city, with typical recognizable characteristics that are capable to define the identity of places. Arco Latino is a complex system in which socio-cultural, economic and ecological environments are dynamically interrelated.

In particular, the paper aims to describe the main resilient features and peculiarities of "port cities" and "cities with port" in the Arco Latino area, resilience being meant as the ability to absorb, adapt to and/or rapidly recover from internal and/or external stresses due to the continuous change in citizens' needs and/or from potential disruptive events.

Keywords: Arco Latino · Mediterranean city-port's resilience
Cultural landscape's resilience

1 Introduction

During this period of constant change and uncertainty in the European situation, debate on culture is always stimulating and there is a need to seek out the root elements to ensure greater and lasting cohesion among the member states. At the end of the 6th century San Colombano, abbot of the Bobbio Monastery (located along the *Via del Sale*, Salt Route, between Liguria - Mediterranean coast - and *Pianura Padana*, Po Plain) expressed the concept of a single Europe in the Latin expression *totius Europae* [1]. The search for a culture of territory and landscape as a stronghold of European culture remains compelling.

It is obvious that one of the greatest fortunes in the development of the European mainland has been its vast access to the Mediterranean Sea, with a coastline extending from Gibraltar to Istanbul. In particular, Arco Latino (in English: Latin Arc, in French: Arc Latin, in Catalan: Arc Llatí, in Italian/Portuguese: Arco Latino) was, and still is, fundamentally [2] a very extended geographical area with more than seventy million

© Springer International Publishing AG, part of Springer Nature 2019
F. Calabrò et al. (Eds.): ISHT 2018, SIST 100, pp. 95–104, 2019.
https://doi.org/10.1007/978-3-319-92099-3_12

inhabitants scattered over various territories, coastal regions, islands, peninsulas and border territories. The area can be defined as the Euro-territory of the northwest coast of the Mediterranean basin, which extends from Sicily, through the Italian peninsula, southern France and the Iberian Peninsula as far as the Strait of Gibraltar and the Portuguese Algarve. It has an arch shape and corresponds to the heart of Latin Europe (Fig. 1). A series of common cultural, historical, socio-economic, geo-climatic and environmental characteristics defines this macro-region and gives it its own definite identity in the European context [3].

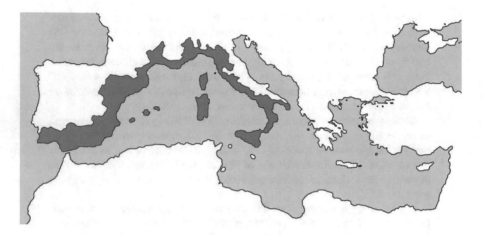

Fig. 1. Arco Latino. Source: authors.

The cities of the Mediterranean coastal system are the essential nodes in the communication and migratory currents between populations. Here, with ports to act as natural interfaces, gateways and connection devices between land and sea, many cities started as trading posts. Ports enabled small towns to become cities, and enhanced urban development thanks to trade. The old city maps show the strong relationship between port and urban development: many scholars have [4, 5] highlighted the importance of port-cities in the birth and growth of the global market economy.

Nowadays, ports are still closely connected to the city: even if they have disappeared or become detached from the city, they continue to influence it, in terms of urban structure, capacity for economic and trade development and any kind of change in social, economic, industrial environment.

Every coastal community, every coastal city, every port city and port area is unique and has its own development plan: in Italy, the first attempt to overcome the separation that had formed between ports and cities was made when law 84/1994 [6] and later Legislative Decree 169/2016 were issued [7]. Nevertheless, there are no truly integrated urban-port plans and this is one of the problems that slows down the new economic take-off of ports as driving forces for the development of cities [8]. Moreover, climate change is putting a strain on the survival of ports, coastal cities and shorelines.

Authors highlight and describe the elements which, notwithstanding the complex present scenario, help and develop resilience in the coastal cities of the Arco Latino, also taking into consideration the role of all the actors involved (institutions, government, stakeholders, companies and citizens).

Is the key to resilience the sea itself, as the basis for the shape and development of urban settlements? Or is it the dynamism of the communities that inhabit these places? Or the geographical peculiarities? Or the community's cultural background?

2 "Port Cities" and "Cities with Port" in the Mediterranean Basin

In order to understand the mutual importance of ports and cities, it is necessary to define the relationship between them in the Mediterranean basin and, in particular, in the Arco Latino.

Because of the complexity of the systems involved, there is no single model or explanation: there are cities that have grown and developed around a port, there are cities that live with one in a symbiotic relationship; there are cities like Genoa, Barcelona, and Marseille that live with it in a relationship of interdependence and there are cities that ignore their port and push it to the edges, like Naples. Despite these differences, each city was and still is linked to its port, to the sea and to the nearby mainland areas [9].

In the past, the urban identity and its shape included the port. This organic relationship is clear in city plans, drawings and classic treatises (i.e. Vitruvius, Leon Battista Alberti, Antonio da Sangallo, Francesco di Giorgio Martini), in which ports are considered as a public buildings and are assimilated into the overall design of the city. During the 16th century the geometrical figure of the fortified city incorporated the port: it was part of a defensive device, a space for mooring and protecting ships, that subordinated the commercial port facilities to the city walls. During this period, the development of the Atlantic routes and the colonial expansion of the great European nations rapidly brought to the fore the strategic advantages of the Atlantic ports.

Between the 17th century and the beginning of the 19th, investments were poured into the integration of the needs of ports and cities', and the most important architects worked on port areas and structures: amongst others, in Italy Bramante, Michelangelo, Sangallo, Leonardo, Buontalenti, Sanmicheli, Juvarra, and Vanvitelli [10].

To better understand this complex scenario, it is important to make a distinction between "port cities" (or "city-port") and "port areas" (or "cities with port").

In "port cities", ports were built out of necessity and they are the foundation and the *raison d'être*, central dynamic force and organising principle of the city [11].

They are core places in terms of economic strength, incomes and employment, fundamental nodes for import and export trade, the location for many enterprises and services, and attractive areas for tourism and cultural exchanges. As historical places of creativity and innovation in economy, culture, and society with a specific architectural landscape, port cities have always been cosmopolitan places with a strong social impacts and influences on the life of local communities and tourists. However, they are also places of conflict: for example, there are negative environmental impacts due to the high level of energy consumption, air and water pollution, natural resource consumption, land use, natural habitat and biodiversity degradation.

In "cities with port", ports were created according to the nature of places and are a mediation or a completion of facilities. They represent the physical and relational connection between city and sea. Here, the port, as driving force of economic wealth and development, can influence the entire urban settlement and the surrounding area – from local to regional, national and trans-national – thanks to the economic importance of the port itself. The presence of a particular landscape is a key characteristic of port areas and such landscape configurations are recognized by the European Union as an important economic resource [12].

3   Resilience in Urban and Port Areas

In coastal zones, seaports and their intermodal connectors are key element that support and enhance the global supply chain, regional economic activity, local transportation system services, community jobs, and cultural and natural tourism. The rhythm of social, cultural and economic changes due to natural, technological and global causes (i.e. Bauman, Sennett) nowadays ensures that modifications occur more rapidly than before. Different types of events, natural or man-made, continually generate multiple direct and indirect effects capable of producing temporary or permanent damage to the natural environment, urban settlements, human beings, the economy, and society. During and after an unexpected situation, cities, port areas, communities, the waterfront facilities and all the actors involved need a remarkable capacity to transform and adapt themselves to the new condition. This process is described by the organic-adaptive-evolutionist theory (e.g. Geddes) and it involves the concept of resilience and the ability of a system to find a new equilibrium after a shock [13]. It is also important to underline that the location and the nature of a port make it susceptible to both natural and human-made disasters: in particular, the geographical location (on the seashore and usually adjacent to rivers or streams as for almost all the cities of Arco Latino, e.g. the whole area from Provence to Tuscany) and their interdependencies (industrial and societal) with communities and municipalities [14].

Enhancing coastal resilience has become an important response to these events. In particular, to be resilient, port communities should be able to keep marine transportation moving, businesses open, and people working [15].

From materials science and engineering fields, the word resilience (from the Latin verb *resilio*, to bounce) refers to the physical property of a material to return to its original shape or position after a deformation (stress) that does not exceed its elastic limits. The term has been used in various disciplines, especially those engaging in ecological research [16]. In the literature, it is possible to distinguish three main approaches to the study of resilience (the engineering approach [17]; ecological approach [18]; adaptive, systemic and socio-ecological approach [19, 20]). According to the different meanings and contexts in which resilience develops, there are several definitions. Among these, it is useful to mention:

– The time, following a disturbance, to recover to some prior state or condition; or, in ecological or psychological terms, the amount of disruption (or stress) a system (or person) can absorb before it (or they) changes state. In particular, it is the ability to absorb, adapt to, and/or rapidly recover from a potentially disruptive event [21];
– Social resilience is the ability of communities to cope with external stresses and disturbances as a result of social, political and environmental change [...] ability of communities to withstand external shocks to their social infrastructure [22];
– Building resilience is about making people, communities and systems better prepared to withstand catastrophic events and able to bounce back more quickly and emerge stronger from these shocks and stresses' [23];
– Urban resilience is the capacity of individuals, communities, institutions, businesses, and systems within a city to survive, adapt, and grow no matter what kinds of chronic stresses and acute shocks they experience [24];
– Infrastructure resilience is the ability to reduce the magnitude, impact, or duration of a disruption [25];
– Disaster resilience is the ability to prepare and plan for, absorb, recover from, and more successfully adapt to adverse events [26];
– In port security, it is also the ability of a port to return to its normal mode of operation after a disruption caused by a natural or man-made attack [27].

Nowadays, resilience has a fundamental role in the development planning of urban, port areas and port cities. In particular, it is defined as the ability of a complex system (e.g. urban settlement, city port) to cope with external and internal stresses and continuous changes through adaptation and mutation strategies and to find a new equilibrium state (not necessarily equal to the original one). So, resilience can be read as a new way of understanding and managing urban-port planning. In the urban/port context, it is possible to identify the following types of resilience:

– Infrastructural resilience (buildings, infrastructure networks - mobility and technological services);
– Institutional resilience (governmental and non-governmental systems);
– Economic resilience (community's economic diversity and peculiarities);
– Social resilience (community's demographic and cultural characteristics).

The capacity of an urban area to be resilient depends on the organization and the relationships existing among the internal elements. A flexible system allows rapid recovery and the necessary swing of activities, and a flexible approach helps to reach resilience goals [16, 28]. Robustness is not enough, nowadays urban and port areas also need to be more flexible and engage community groups and other partners to work together [29]. In order to be resilient urban-port systems must:

– Pursue a balanced and sustainable development model based on integration of social, environmental, and economic aspects;
– Preserve and enhance local resources (natural and cultural);
– Reduce environmental impacts due to human phenomena (industrial systems, fuel pollution, etc.);
– Encourage social participation both in planning and management phases;
– Enhance social cohesion, information and education;

– Engage in rapid energy transition and reduction of emissions;
– Find new technologies, ways of construction, and processes of development to enhance effective climate adaptation.

The definition of resilience in the present paper is closest to the adaptive, systemic and socio-ecological approach: i.e. the capacity of a system to anticipate unexpected event in order to minimize potential negative impacts; the focus is on adaptive system capabilities and learning mechanisms [19, 20].

4 Resilient Peculiarities of Arco Latino

Over the centuries, water resources have played an important role in many parts of the world for the establishment and expansion of urban settlements and the definition of their identities [30]. Water is a very important planning element, used to improve the existing environment with aesthetic (visual, audial, tactual and psychological) and functional (climatic comfort, noise control, circulation and recreational) effects. This is the reason why, starting in Neolithic times, and throughout the Ancient and Classic ages the coastline population contributed enormously to the enhancement of continental development and civilization (i.e. the Ligurians, later included in the Roman Empire). During the Neolithic era, camps started to become villages and, then, villages become cities. In many Mediterranean cities, a public space constructed during the Middle Ages soon became the centre for meeting, trading and exchange activities that culminated in the flourishing economies of the mercantile and renaissance periods. Over the centuries, society has considered the sea as a fundamental resource and often the only possibility for sustenance, creating a symbiotic relationship with it.

So, it is clear the founding matrix of port-cities: it has become a model of basic-simple-exported and exportable city (strong parallels can be found between European cities of the eastern part of Mediterranean basin and extra-European cities such as: Muscat in Oman, Macao in the Peoples' Republic of China, the Genoese fort of Tabarka in Tunisia and of Sudak in Crimea) especially in areas where the economic, trading and cultural exchanges were really strong and persistent.

In the Mediterranean coastal area, all these cities have heterogeneous but easily recognizable urban features, inherited from a rich history: churches, bell towers, cemetery, military fortresses, places of exchange (piazzas) enclosed by buildings, possible also by the sea itself, alleyways, gardens and public spaces that dissolve into private spaces through the aesthetic perception of the 'familiar' [2]. Finally, of all the cities belonging to the European coastline of Mediterranean Sea, those in Arco Latino (and in particular the Savona - Imperia - Sanremo - Ventimiglia - Menton – Nice area) present a very high building density in coastal urbanizations (as infinite city): already in 1961, a report of the weekly journal *L'Europeo* denounced the uninterrupted urbanization of *Riviera di Ponente* (the western part of Liguria), where out of a total of 124 kilometres, only 35 were uninhabited: "the coastline of *Riviera dei Fiori* (Flower's Riviera) is a wall of houses for almost 90 kilometres, 72% of the overall landscape". Therefore, it appears as an almost continuous urbanization from west to east, adapting itself flexibly to fill the available spaces. In the same way, the society is a dynamic and fluid system that can easily follow the complex relational changes of the backcountry (at local, regional and national scales): a few square meters have the difficult and fundamental task of being the 'cell membrane' of the entire continental system.

Since the 1970s, the relationship between ports and cities has changed a lot, primarily because of new shipping technology that has increased the demand for vast back up land and for better access to inland transport networks. Today, most of the old ports of Arco Latino have lost their prominence: some have resigned themselves to being only marinas ("cities with port", like Ventimiglia, Sanremo and, Imperia), some are instead "city-port" (Fig. 2). One of the latter is Savona, a paradigmatic Mediterranean city, which bears the marks of its many transformations: after the Second World War, it boosted its industrial and maritime activities, while from the 1990s, because of the changed economic situation and deindustrialization, it reinvented itself, starting a further transformation that it is still undergoing: it is a place that, despite continual change, is able to maintain the highly recognizable features that represent and define its identity over time.

Fig. 2. "Port cities" (with the triangle-shape symbol) in Arco Latino. Source: authors.

Indeed, during the modern era, the Mediterranean city as a whole has followed a process of stratification while the places of the post-modern Mediterranean city created new equilibriums. The changes in mainland needs (economic, social, demand for raw materials and so on) have always set the pace for changes in the main activities of the coastal cities. As already mentioned, over the centuries "city-ports" have developed a succession of activities related to trade, industrialization/deindustrialization (i.e. Rosignano Solvay in Tuscany [31]) and tourism (on a local, regional and global scale). All the main "port-cities" have adapted their main characteristics but at the same time maintained fixed and recognizable elements (bearers of identity) by carrying out a continuous, flexible and organic transformation.

The great strength of Arco Latino is its centuries-long history, that plays out in the physical and functional/relational characteristics established within the city system; in the ability to overcome geographical barriers through a language common to the territories despite regional nuances; in the cultural elements that bear identity; in the geography; in the natural resources (above all the sea itself, but also the peculiarities of the coast and backcountry, the climate, the vegetation etc.); in the desire and in the capacities for Auto- and Ethero-organization of the communities (clear from the enormous difficulty of building and developing cities and ports on land, and on the use of the sea not only as an economic resource but also as physical support for development); and in the ability of these communities to create strong bonds and relations both with the mainland and with the other coastal cities of Arco Latino.

In addition, the marine protected areas for fish and wildlife repopulation are a clear example of adaptation of the local communities: from a fishing industry that is no longer economically profitable has arisen a conservation operation that makes use of the same devices and workers for the care and protection of marine parks. This has generated new, increasingly popular forms of tourism (i.e. Pelagos Sanctuary [32]) (Table 1).

Table 1. Main resilient peculiarities of Arco Latino coastline cities. Source: authors.

Inner physical/functional/relational capacities of adaptation	
Recognizable urban features (identity)	Multicultural society
Communities' capacities of Auto/Ethero organization	Strong bonds with mainland and other coastal cities
Centuries-long history	Shared language
Cultural elements, traditions, and cultural historic landscape	Presence of protected natural areas
Natural and climatic resources	Mediterranean Sea

These examples could form the foundation of a new, holistic vision of Europe in which nature abolishes the simple cartographic border. It is therefore necessary to enhance these attitudes, following criteria of efficiency and usefulness, and also of aesthetics and integration with the existing urban fabric.

5 Discussion and Conclusion

Over recent decades, the large ports have transformed rapidly, adapting themselves to the new requirements of maritime transportations. Nowadays, port areas are autonomous and distinctly separate from the adjunct urban centres, unlike in the cities of the past. For this reason, it is almost always more appropriate to speak about "cities with ports" than about "city-ports". While in the past urban identity was completely integrated with that of the port, the relationship between port and city now appears extraordinarily intricate and discontinuous. City and port areas have taken on distinct, opposing identities, with dynamic, complex relationships between them. The characteristics of "port cities" remain in some major cities and in some smaller realities of Arco Latino: here, ports continue to undergo changes especially in their main functions/purposes (livelihood, trade, tourism, defence) but they are an integral part of and a driving force behind their community's dynamism and life, and form dense networks of connections with various degrees of flexibility and transformability.

In line with EU strategy, the importance of the role of Mediterranean basin and its ports has been revived and they are now key places where economic strength, competitiveness, human capital and global appeal, population and migration processes are increasingly concentrated. The main challenge is to define an overall re-organization of maritime and port systems (as connection infrastructures and services nodes) and the associated coastal cities: port areas are increasingly becoming new spaces where an integrated smart development process, supported by an overall urban planning, contributes to urban ecological resilience, mostly thanks to their own peculiar resilient characteristics (see Table 1).

Moreover, Arco Latino municipalities and regions have a strong interest in looking to the Mediterranean basin not only for economic and logistic activities, or merely for tourism but also for the rationalization-distribution of the traffic of internal goods, for better development of infrastructural nodes with high economic-administrative-organizational efficiency (i.e. Ligurian ports), for better access to all territories (including islands) and for the development of economic, social and political relationships with the southern Mediterranean area.

In addition, today, Arco Latino is also facing the new challenge of migration: an example of proper inclusion is given by Nice, a city which is operating good practice such as the reconversion of public spaces into meeting and aggregation places, and the creation of cultural hubs. Here, the idea and the necessity to put in the underground some roads and the creation of Promenade du Paillon re-connects the city's archaeology with its social fabric. Once again, the ability to imagine new spaces and to renew the city itself is a manifestation of the potential for resilience in the coastline cities of Arco Latino.

This paper presents a general overview of the Arco Latino, its coastal cities and their main features as related to resilience issues. It is intended as starting point for further in-depth researches.

References

1. Zanuzzi, R.: San Colombano d'Irlanda Abate d'Europa, Ed. Pontegobbo, Piacenza (2000)
2. Corsico, M.: Forma maris antique - imago urbis. In: De Lotto, R., Zhuang, Y. (eds.) Urban Design. Maggioli Editore, Santarcangelo di Romagna (2013)
3. Arco Latino Homepage. http://en.arcolatino.org/. Accessed 03 May 2017
4. Ducruet, C., Lee, S.W.: Frontline soldiers of globalisation: port-city evolution and regional competition. GeoJournal 67(2), 107–122 (2006)
5. OECD, Port-Cities Programme (2010). http://www.oecd.org/governance/regional-policy/oecdport-citiesprogramme.htm. Accessed 03 May 2017
6. National Law 84/1994: Riordino della legislazione in materia portuale
7. Legislative Decree 169/2016: Nuovo Statuto sull'autorità portuale
8. Cannatella, D., et al.: I porti come generatori di resilienza nelle nuove città metropolitane costiere italiane. Urbanistica Informazioni 257, 20–24 (2014)
9. Capasso, M.: The ports and the cities Naples, Genoa, Trieste, Marseilles and San Francisco. Mediterranea Adriatic Sea (2005)

10. Pavia, R.: City Ports. http://retedigital.com/wp-content/themes/rete/pdfs/portus/Portus_15/I_porti_delle_città.pdf. Accessed 03 May 2017
11. Reeves, P., et al.: Studying the Asian Port City. In: Broeze, F. (ed.) Brides of the Sea: Port Cities of Asia from the Sixteenth to Twentieth centuries. University of New South Wales Press, Kensington (1989)
12. Council of Europe. European Landscape Convention. Florence, Italy (2000)
13. De Lotto, R.: Flexibility principles for contemporary cities. In: Shiling, Z., Bugatti, A. (eds.) Changing Shanghai - from Expo's after use to new green towns, pp. 73–78. Officina Edizioni, Roma (2011)
14. Wakeman, T., et al.: Port Resilience: overcoming threats to maritime infrastructure and operations from climate change. University Transportation Research Center, City College of New York (2015)
15. Scott, H., et al.: Climate change adaptation guidelines for ports, Enhancing the resilience of seaports to a changing climate report series. National Climate Change Adaptation Research Facility, Gold Coast (2013)
16. Venco, E.M.: La pianificazione preventiva per la riduzione del rischio: definizione di scenari preventivi nel contesto della città flessibile e resiliente. Maggioli Editore, Sant'Arcangelo di Romagna (2017)
17. Odum, P.E.: The strategy of ecosystem development. Science 164(3877), 262–270 (1969)
18. Holling, C.S.: Resilience and stability of ecological systems. Annu. Rev. Ecol. Syst. 4, 1–23 (1973)
19. Holling, C.S.: Understanding the Complexity of Economic. Ecol. Soc. Syst. Ecosyst. 4, 390–405 (2001)
20. Folke, C., et al.: Resilience thinking: Integrating resilience, adaptability and transformability. Ecol. Soc. 15(20) (2010)
21. NRC-National Research Council: Disaster Resilience: A National Imperative, National Academies (244) (2012)
22. Adger, W.N.: Social and ecological resilience: are they related? Prog. Hum. Geogr. 24(3), 347–364 (2000)
23. Rockfeller Foundation Homepage. https://www.rockefellerfoundation.org/our-work/topics/resilience/. Accessed 03 May 2017
24. Resilient Cities Homepage. www.100resilientcities.org/. Accessed 03 May 2017
25. Olsen, J.R.: Adapting Infrastructure and Civil Engineering Practice to a Changing Climate. ASCE. Committee on Adaptation to a Changing Climate, American Society of Civil Engineers 93 (2015)
26. Cutter, S.L., et al.: Disaster resilience: a national imperative. Environ. Sci. Policy Sustain. Develop. 55(2), 25–29 (2013)
27. Mansouri, M., et al.: A Policy making Framework for Resilient Port Infrastructure Systems. Mar. Policy 34, 1125–1134 (2010)
28. Jha, A.K., et al. (eds.): Building Urban Resilience: Principles, Tools and Practice. Directions in Development. Washington, DC: World Bank (2013)
29. Rotterdam Resilience strategy. Ready for the 21st century. Ed. Veenman+, Rotterdam
30. Pekin, U.: Urban Waterfronts Regeneration: a Model of Porsuk Stream in Eskişehir. In: 6TH INTERNATIONAL SYMPOSIUM AGRO ENVIRONMENT "NATURAL RESOURCES CONVERSATION, Use & Sustainability", pp. 410–413 (2008)
31. Caso Rosignano Solvay. https://www.docsity.com/it/caso-rosignano-solvay/2149766/. Accessed 03 May 2017
32. Pelagos Sanctuary Homepage. http://www.sanctuaire-pelagos.org. Accessed 03 May 2017

The Role of Spatial Models in Tourism Planning

Svjetlana Mise[(✉)]

Mediterranea University of Reggio Calabria, 89100 Reggio Calabria, Italy
svjetlana.mise@unirc.it

Abstract. Tourism is becoming one of the world's largest growing industries with a continuous growth of around 4% per year for seven years straight according to UNWTO. Consequently, interest in utilizing the potential of tourism to boost economic development has grown dramatically. And although tourism's beneficial impact on the fostering country's economic growth and the developmental force tourism generates is undeniable, non-strategic development of tourism industry can result in negative outcomes. The extractive character of the tourist industry poses a challenge to urban policies which have to avert negative outcomes, control the expansion of the tourist industry while responding to a particular set of problems which vary depending on the given political, economic, socio-cultural, environmental context. The article utilizes literature analysis in order to illustrate the role of spatial models, as an extension of those policies, in mediating the tourist-host conflict.

Keywords: Tourism industry · Policies · Morphology · Models

1 Introduction

Tourism has been the subject of academic studies for decades which resulted in a wide scholarship covering a variety of aspects. Early studies focused mainly on the economic impacts of tourism. Gradually the focus shifted to socio-cultural aspects and offered new insights into the impact tourism has on a community. The rising awareness of environmental and sustainability issues veered the discussion into new directions offering alternative forms of tourism development with an aim to minimize its exploitative character. This paper gives an overview of the accumulated body of knowledge on the impact of tourism activities on the hosting community. Thus, reviling a niche within spatial tourism development: this scholarship has failed to account for the key causal role spatial models play in current crises of sustainability and spatial democracy.

© Springer International Publishing AG, part of Springer Nature 2019
F. Calabrò et al. (Eds.): ISHT 2018, SIST 100, pp. 105–112, 2019.
https://doi.org/10.1007/978-3-319-92099-3_13

2   Key Aspects of Contemporary of Tourism

2.1   The Development Motor

Over the past few decades, tourism has experienced continues growth and an increasing diversification [1]. These dynamics have positioned the touristic industry as one of the fastest growing economic sectors in the world and into a key driver of socio-economic progress. The literature on spatial development resonates with this notion placing significant focus on development spillovers from globalization dynamics in developing areas, peripheral regions and small islands [2]. And although literature supporting the notion of tourism as a positive economic force are legion [3–5], the development of tourism industry can result in uneven growth and development patterns causing dispersal of domicile population and subsequently affect endogenous social, economic and cultural development [6]. The introduction of tourism activities affects all aspects of a community [7] in both positive and negative ways [3]. In their work Mathieson and Wall [8] synthesized much of the research on the impact of tourism and put forward an overview which shifted the focus away from the hitherto emphasized economic perspective.

2.2   The Impact of Tourism Activities on the Hosting Community

The four categories on which tourism impact studies focus are economic, social, cultural and environmental [8]. A variety of academic fields recognizes tourism impact as a crucial element in tourism development and destination management [3, 8]. Following up on this notion, tourism planners typically consider the nature of the impact and how it can be managed in order to boost optimal outcomes [9]. Much of the research that presents positive impacts of tourism are based on objective indicators such as income per capita, crime rates, and pollution [10]. Recent progress in this field has made significant advancement in understanding the links between residents' perception of tourism effects and their satisfaction with life in a destination area. The findings show an uneven rate of satisfaction with life varying over time which may suggest that tourism impact changes over time based on the development stage of the tourist area [11]. This discrepancy called for a more in-depth approach and ultimately placed the community in the center of the debate.

3   A Place-Based Approach

As Richards and Hall [7] elaborated: "[…] although the concept of community has shifted in meaning and application in the tourism field over the years, the recent rediscovery of the 'local' and the growing importance of identity have placed 'community' at the forefront of discussions about tourism development."

3.1 Community-Centered Approach to Tourism Development

Richards and Hall refer to Murphy's [12] classic review of community tourism, who stressed that it was necessary for each community to connect tourism development to local needs. This was often not the case, as Urry [13] will come to point out, in many cases the process of creating tourist infrastructure takes precedence over the needs of the local population. As a result of this basic argument, subsequent studies have gradually expanded the notion of community-based tourism to incorporate a wide range of issues, such as local participation, democracy, and ecological aspects. In these discussions on tourism community development four key areas can be discerned [7, 14, 15]: community participation, sustainability, community economic development, and heritage.

Community Participation

In a balanced dynamic partnership and collaborative approach toward tourism planning the emphasis should be placed on planning with as wide a group of stakeholders as possible, thereby making an attempt to accommodate the public interest rather than planning for a narrow set private interests under a corporatist perspective [16]. Important points of discussion are the definition of community participation, the level of community participation in tourism development initiatives and how it has been and should be implemented [17, 18]. Tosun identified four important constraints in fully achieving community participation, which he followed by developing the framework "Stages in the emergence of participatory tourism approach in the developing world" in a later work [19]. Mitchell and Reid [20] developed the "Integration tourism framework" in order to investigate how public participation and related external and internal factors possibly influence or determine planning processes for a certain tourism project.

Social Sustainability

Community participation is often linked to the concept of sustainability, as this is often recognized as an ex-ante condition for tourism to develop in a sustainable way. Mowforth and Munt's [21] criteria for sustainability in tourism include the criterion "local participatory" as community empowerment is an important concept within the sustainability and participation debate [22]. As Hall [16] stated: "A sustainable approach to tourism would state that all stakeholders are relevant because of the contribution they bring to the creation of social capital." However, a working model for benefiting community economic development still has to be developed, while a set of paradoxes and contradictions between sustainable development and tourism development remain unresolved [23].

Community Economic Development

The first and second areas within the literature on tourism community development ("community participation" and "sustainability") are linked to the third category, as "tourism continues to be driven by levels of government rather than community interests" [14]. Governments, especially in developing countries, have seen the economic potential of tourism as a way to create jobs, reduce debts, boost economic and regional development, and are encouraged to do so by international organizations such as the United Nations and the World Bank [24].

Cultural Heritage

The issue of cultural heritage plays an important role in the literature on tourism community development. How cultural heritage in a community can be and is being commoditized to meet the need of the tourists [14] is dissected. Notions such as staged authenticity or communicative staging are an important part of the debate [21, 25].

3.2   Managing the Turmoil – Issues and Implications

Today tourism continues to grow at around 4% per year, for seven years straight seeing international tourist arrivals reach 1,235 million in 2016 [1]. And while this growth has generated a considerable income for the hosting countries, it has also caused a string of rebellions in destination areas. In cities such as Venice, Barcelona and Dubrovnik citizens have organized a series of protests and actions to draw attention to the unconstrained expansion of the tourist industry which is taking its toll on the quality of citizen's life in destination areas. This growing reluctance in hosting communities, has contributed to the year "2017" being addressed by UNWTO as the "International Year of Sustainable Tourism for Development" with significant focus placed on the community and environmental aspects while the concept of "sustainable development" notably achieved a widespread acceptance as a desirable objective for tourism development policies and practice. Furthermore, it is becoming clear that the traditional strategy in tourism development by reducing barriers and simulating market interest [26] will not necessarily produce the most appropriate or sustainable solution [27, 28]. A variety of forms for intervention are necessary to protect the environmental and cultural assets on which tourism is based upon. In that sense, policies play a great role in mitigating the exploitative character of the tourism industry. And while the outcome for these interventions depends on the dominant type of tourism as well as the administrative and political context, there is an increasing recognition of the importance of integrated approach in all levels of planning which should be comprehensive [29]. Tourism, as a global phenomenon, utilizes local aspects and should be observed as an integrated part of local processes resulting in both desirable and less desirable outcomes-which are often a direct result of the physical manifestation of tourism in space [30].

4   Spatial Models in Tourism Planning

In order to communicate problems and concepts, an adequate *vocabularium* was established. Tourism infrastructure is thus referred to as a model, "[…] an abstract, generalized, ideal and simplified construct that serves to reduce the complexity of the real world in the interest of explanation by highlighting the fundamental elements or characteristic of an actual situation or process" [31]. The main concern in destination planning is to identify a vision of spatial development, and consequently by method of design put forward a preferred model of land use [32] which will in return maximize revenue.

4.1   The Challenge of Integrated Spatial Tourism Planning

The process of spatial planning is carried out at the local or regional level, unlike market-oriented tourism which is carried out on regional or higher levels [32], rendering these processes divergent. Consequently, spatial and urban planners attempt to assess projects on an *ad hoc* basis using planning documents which have been elaborated in a non-corresponding manner with respect to tourism development issues [33]. Unfortunately, there is a discrepancy in between the advancements being made in regards to methodological processes of tourism planning [28, 29, 34] and spatial concepts, models and corresponding theories from which planners can draw. This problem can in return aggravate the difficulty of integrating destination management into the urban planning framework and subsequently the spatial model [32].

4.2   Development of Spatial Models in Tourism

During the last decade, a number of models have emerged as references to planners in order to facilitate the processes of destination planning and design. Planning tools utilized in spatial planning of tourism models can generally be divided into three categories [32]. The first group is focused on the nature of planning processes and is closely related to the field of decision-theory and policy analysis [35]. In tourism literature, several planning models exist which make an attempt at following a rational comprehensive paradigm [27, 28, 34]. The second group is the functional tools. This group consists of a broad range of theories, models, and concepts, which illustrate the causation behind specific settlement patterns and their functional aspects. They are principally derived from systems theory. They are also divided into those which make an attempt at being holistic or simply present one aspect of a larger system [36]. In tourism literature, these tools encompass the center periphery [37], the analysis of travel behavior patterns [38] and morphogenic studies of destination regions [39, 40]. Normative tools are the third group and they deal with the linkages between human values and settlement forms, by dealing with the links between architecture, urban design, landscape, and society. Some of the examples of normative tools in planning literature include "Good City Form" [41] and "A Pattern Language" [42]. However, examples of normative tools in tourism are scarce. There are a few notable exemptions, including the "model of attractions" [43] and the "integrated model" [32].

   Nevertheless, the problems of tourism destination management do not simply stem from a lack of planning tools and models as such, but the considerable fragmentation of spatial tourism models which are developed independently of one another, with little or no effort to build on previous efforts [44] and with little regard towards the impact the models have on their immediate context: the community.

4.3   The Role of Spatial Models in Supporting Spatial Justice — A Case Study

Another issue coming recently into focus is the rise of the privatization of public space [45]. As tourism becomes a privatized product it has caused a chain reaction in what Sorking [46] calls the end of public space in the city. This is an important aspect in the debate on the question of spatial justice spanning many areas of social science embedded firmly in urban studies, and yet still is rarely discussed in tourism and leisure with a few exceptions [47, 48]. Even though conservation and urban policies seek out strategies to lessen the pressure on communities and avert the overuse of built and natural heritage, there are tourism-related problems which cannot be shaped by public policies but rather urban policies and more directly spatial planning. A notable attempt in that sense has been made in a recent study of causes for conflicts in tourist-historic cities through a morphological analysis Bálint Kádár [49]. By analyzing urban morphology in two historical city centers of Prague and Vienna, Kadar [49] concluded that, unlike Prague, which is denser, Vienna had a more dispersed spatial configuration allowing tourists various options while moving from one attraction to another, lessening the congestion. With the method of analyzing syntactic space systems introduced by Hillier [50], he illustrates that Vienna is more attractive to tourists as it offers more choices of exploration, and therefore more freedom, an essential value in leisure activities.

5 Conclusion

New emerging studies [49], as a result of the rebellion taking place in hosting communities, epitomize the new paradigm which provides an alternative framework to rethink the role of spatial models in tourism planning on the basis of the principles of social justice, equity, spatial democracy, and sustainability. They shed new light on the question of how touristic infrastructure mediates the host-tourist conflicts by imposing a specific order with the physical form and the role spatial models play in skewing the outcomes of public policies and urban policies. In other words, they shift the emphasis in favour of certain policy perspectives and thus reveal the bias of the public agency. Spatial logic would seem to hold insight into why policy outcomes are enabled or restricted to perform efficiently and with respect to conflicting groups of actors. We must take the performative characteristics of space into account in order to optimize the spatial dimension of the urban form and maximize the positive outcomes of tourism with respect to the needs of the community.

References

1. World Tourism Organization (UNWTO): Tourism Highlights (2017)
2. Read, R.: The implications of increasing globalization and regionalism for the economic growth of small Island States. World Dev. **32**, 365–378 (2004)
3. Kim, K., Uysal, M., Sirgy, M.: How does tourism in a community impact the quality of life of community residents? Tour. Manag. **36**, 527–540 (2013)
4. Richards, G.: The market for cultural attractions. In: Richards, G. (ed.) Cultural Attractions and European Tourism, pp. 31–53. CABI, Wallingford (2001)
5. Schubert, S., Brida, J., Risso, W.: The impacts of international tourism demand on economic growth of small economies dependent on tourism. Tour. Manag. **32**, 377–385 (2011)
6. Hadjimichalis, C.: Uneven Development and Regionalism. Croom Helm, London (1987)

7. Richards, G., Hall, D.: Tourism and Sustainable Community Development. Routledge Advances in Tourism, vol. 7. Taylor & Francis Group, London (2000)
8. Wall, G., Mathieson, A.: Tourism. Pearson Prentice Hall, New York (2006)
9. Murphy, P.: Perceptions and attitudes of decisionmaking groups in tourism centers. J. Travel Res. **21**, 8–12 (1983)
10. Crotts, J., Holland, S.: objective indicators of the impact of rural tourism development in the state of Florida. J. Sustainable Tour. **1**, 112–120 (1993)
11. Butler, R.: The concept of a tourist area cycle of evolution: implications for management of resources. Can. Geogr./Le Géographe canadien **24**, 5–12 (1980)
12. Murphy, P.: Tourism: A Community Approach. Methuen, New York (1985)
13. Urry, J.: The Tourist Gaze: Leisure and Travel in Contemporary Societies. Sage, London (1990)
14. Joppe, M.: Sustainable community tourism development revisited. Tour. Manag. **17**, 475–479 (1996)
15. León, M.: Women and empowerment: community participation in tourism in Ecuador. Msc thesis, Department of Environmental Sciences, Social-spatial Analysis. Wageningen University, Wageningen (2006)
16. Hall, C.: Tourism Planning: Policies, Processes and Relationships. Prentice Hall, Harlow (2000)
17. Pretty, J., Hine, R.: Participatory Appraisal for Community Assessment: Principles and Methods. University of Essex, Colchester (1999)
18. Tosun, C.: Limits to community participation in the tourism development process in developing countries. Tour. Manag. **21**, 613–633 (2000)
19. Tosun, C.: Stages in the emergence of a participatory tourism development approach in the developing world. Geoforum **36**, 333–352 (2005)
20. Mitchell, R., Reid, D.: Community integration: island tourism in Peru. Ann. Tour. Res. **28**, 113–139 (2001)
21. Mowforth, M., Munt, I.: Tourism and Sustainability: New Tourism in the Third World. Routledge, London and New York (2003)
22. Scheyvens, R.: Ecotourism and the empowerment of local communities. Tour. Manag. **20**, 245–249 (1999)
23. Sharpley, R.: Tourism and sustainable development: exploring the theoretical divide. J. Sustainable Tour. **8**, 1–19 (2000)
24. Perrottet, J., Garcia, A.: Pacific Possible Background Paper, No. 4. World Bank, Washington, DC (2016)
25. Chhabra, D., Healy, R., Sills, E.: Staged authenticity and heritage tourism. Ann. Tour. Res. **30**, 702–719 (2003)
26. Getz, D.: Tourism planning and research: traditions, models and futures. In: Strategic Planning for Tourism: An Australian Travel Research Workshop. 5–6 November 1987. Conference Papers and Workshop Notes, pp. 2–43. Lord Forrest Hotel, Bunbury (1987)
27. Inskeep, E.: Environmental planning for tourism. Ann. Tour. Res. **14**, 118–135 (1987)
28. Inskeep, E.: Tourism planning: an emerging specialization. J. Am. Plann. Assoc. **54**, 360–372 (1988)
29. Getz, D.: Models in tourism planning: towards the integration of theory and practice. Tour. Manag. **7**, 21–32 (1986)
30. Konsolas, N., Yacharatos, G.: Regionalisation of tourism activity in greece: problems and policies. In: Briassoulis, H., Straaten, J. (eds.) Tourism and the Environment: Regional, Economic. Cultural and Policy Issues. Kluwer Academic Publishers, Dordrecht (2000)
31. Pacione, M.: Models of urban land use structure in cities of the developed world. Geography **86**(2), 97–119 (2001)

32. Dredge, D.: Destination place planning and design. Ann. Tour. Res. **26**, 772–791 (1999)
33. Dredge, D., Moore, S.: Planning for the integration of tourism and town planning. J. Tour. Stud. **3**, 8–21 (1991)
34. Baud-Bovy, M., Lawson, F.: Tourism and Recreation Development. Architectural Press, London (1977)
35. Campbell, S., Fainstein, S.S.: Introduction: the structure and debates of planning theory. In: Campbell, S., Feinstein, S.S. (eds.) Readings in Planning Theory, pp. 1–4. Blackwell Publishers, Cambridge (1996)
36. MacLoughlin, J.: Urban and Regional Planning. Faber and Faber, London (1973)
37. Britton, S.G.: A conceptual model of tourism in a peripheral economy. In: Pearce D.G., (ed.) Tourism in the South Pacific: The Contribution of Research to Development and Planning. NZ MAB Report, 6. NZ National Commission for UNESCO, pp. 1–12. University of Canterbury, Christchurch (1980)
38. Lundgren, J.: The tourist frontier of Nouveau Quebec: functions and regional linkages. Tour. Rev. **37**, 10–16 (1982)
39. Smith, R.: Beach resort evolution. Ann. Tour. Res. **19**, 304–322 (1992)
40. Stansfield, C., Rickert, J.: The recreational business district. J. Leisure Res. **2**, 213–225 (1970)
41. Lynch, K.: Good City Form. The MIT Press, Cambridge (1981)
42. Alexander, C., Ishikawa, S., Silverstein, M.: A Pattern Language: Towns, Buildings, Construction. Oxford University Press, New York (1975)
43. Gunn, C.: Vacationscape. Board of Regents of the University of Texas System, Austin (1972)
44. Pearce, D.: Tourism Today: A Geographical Analysis. Addison Wesley Longman, Harlow (1995)
45. Davis, M.: Ecology of Fear: Los Angeles and the Imagination of Disaster. University of California, Berkeley (1999)
46. Sorkin, M.: Variations on a Theme Park: The New American City and the End of Public Space. Hill and Wang, New York (1991)
47. Hall, C., Page, S.: The Geography of Tourism and Recreation: Environment, Place and Space. Routledge, London (2006)
48. Page, S., Connell, J.: Leisure. Pearson Education Limited, Harlow (2010)
49. Kádár, B.: A morphological approach in defining the causes of tourist-local conflicts in tourist-historic cities. In: Resourceful Cities. International RC21 Conference, Berlin (2013)
50. Hillier, B.: Space is the Machine. Cambridge University Press, Cambridge (1996)

Scenarios for a Sustainable Valorisation of Cultural Landscape as Driver of Local Development

Lucia Della Spina(✉) iD

Mediterranea University, 89100 Reggio Calabria, Italy
lucia.dellaspina@unirc.it

Abstract. This study explores the potential of the multiple values and considerable resources that characterize a territory related to the Tyrrhenian cultural landscape in Southern Italy. The approach presented here is an experimentation that employs evaluative experiences tested in similar contexts. The aim of this study is to outline incremental and adaptive decision-making processes that focus on the identification of values and needs and are supported by a bottom-up decision-making process. Evaluation is defined as a multidimensional, dynamic, incremental and cyclical learning process, in which integrated assessment techniques are combined with public participation techniques in order to outline shared and transparent intervention scenarios.

Keywords: Cultural landscape · Adaptive decision-making process
Multi-group analysis · Multi-Criteria analysis · Social Multi-Criteria Evaluation
NAIADE method

1 Introduction

In the last thirty years the notion of landscape has evolved: not only does it refer to landscapes characterized by the presence of natural elements, but it includes cultural landscapes as well, which are defined as the result of the combined work of nature and man. Cultural landscapes thus are the expression of the identity of local communities in relationship with the places where they have settled [1].

In particular, the growing interest in landscape as a relevant perspective in the processes of sustainable development is promoted globally by many competent institutions through important regional policies and directives. The new opportunities presented by a landscape approach for the European continent identify landscape as a new paradigm for a development model aimed at the harmonious integration of social, economic and environmental factors in space and time [2].

The Italian territory is largely made up of small cities: 70.4% of Italian municipalities have less than 5,000 inhabitants [3]. These municipalities are often characterized by extreme geographic marginality, low population growth rates and high rates of old age and immigration. Moreover, these areas present negative structural conditions such as lack of services, unemployment, depletion of productivity and the inability to attract new business and promote their identity [4]. Such conditions cause

© Springer International Publishing AG, part of Springer Nature 2019
F. Calabrò et al. (Eds.): ISHT 2018, SIST 100, pp. 113–122, 2019.
https://doi.org/10.1007/978-3-319-92099-3_14

profound socio-economic disadvantage, for which these territories are defined as inland marginal areas [5]. These areas have escaped modernization but are gifted with a rich and authentic heritage made of monuments, traditions and unique landscapes [6], and they have recently become relevant in government policies that acknowledge their considerable potential for development. The Italian cultural landscape of inland areas is the result of the combination of tangible and intangible values, of a complex network of relations between the social system and the eco-system [7].

In order to identify a sustainable valorisation and development strategy for the Tyrrhenian cultural landscape, a multi-methodological evaluation process was organized to support the development of alternative intervention strategies. To this end, the study has selected a significant area of the Metropolitan City of Reggio Calabria, in southern Italy (see Fig. 1), characterized by conflicting values and resources. The area under examination covers a surface of 1,111.76 Km², for a total of 43 municipalities. Though it has many km of coasts, this area can be classified as an inner area [8].

The 43 municipalities are very different from one another because of their position and territorial extension and their cultural and economic resources. The territory is characterized not only by coastal municipalities for a very small extent, but also by

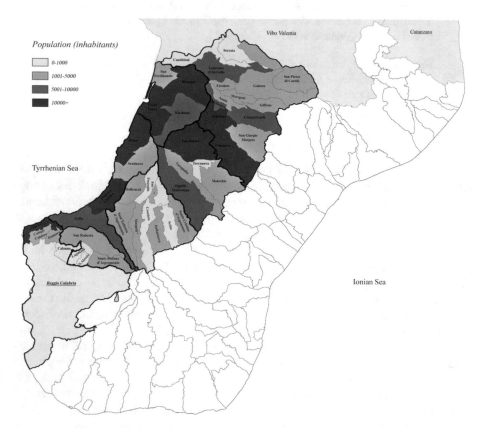

Fig. 1. The Tyrrhenian cultural landscape of the Metropolitan City of Reggio Calabria (Italy).

larger hill towns; some municipalities also feature both coastal and rural landscape within the same municipal territory (Fig. 1). The basic features to be considered for the pursuit of a sustainable development of the landscape system of the Tyrrhenian cultural landscape are: the valuable landscape heritage, characterized by rich biodiversity and numerous cultural values not adequately enhanced, which are not well connected with each other; the fragility of the heterogeneous territorial system; the hydrogeological risk; the abandonment of terrace cultivation, a fundamental component of the cultural landscape system; the fragmentary nature and the parcellation of agricultural terraces; the heterogeneity of local activities and productions; seasonal tourism; the numerous Community funds often spent without an integrated design of the projects [7–9].

Furthermore, among the main considerations that emerged from the study of planning tools currently underway, there was a lack of an overall and unitary vision of the whole area. This study proceeded to identify the existing lines of development that emerge from the current planning in order to understand the evolution of the landscape system. In this way, we could propose project scenarios that aimed at comparing the current state of affairs with a more beneficial vision and at defining site management plans.

From this perspective, sustainable strategies for the valorisation and transformation of the Tyrrhenian cultural landscape have been identified and evaluated three specific scenarios through a structured evaluation process that used the interaction between the different areas of expertise as a basis on which to build bottom-up transformation choices.

2 Methodology

The aim of the research project was to identify a decision-making process that could involve the various stakeholders, identify the relationships that linked them and build shared and sustainable valorisation strategies. The study of dynamics, processes and potentials was essential to understand the complex decisional issue and to build possible strategies for future transformations from bottom up, with a focus on the peculiarities of the cultural landscape under consideration.

Within its context, which is characterized by strong contradictions and conflicts, an integrated and adaptive evaluation approach is effective for dealing with and managing problems which arise from the presence of multiple interests and points of view.

These can become the starting point of a dialogue process in the community [10, 11]. The methodology consists of three main phases (see Fig. 2).

The first phase, "Knowledge of Cultural Landscape (CL)" consisted in the participant observation of the context through the processing of hard and soft data. In the first phase the implementation of Hard System Analysis instruments [12] allowed to elaborate specific indicators with reference to specific thematic areas such as society, quality of life, economy and production, tourism, transport, infrastructure, landscape, cultural heritage, and services [8].

On the other hand, the application of Soft System Analysis tools [13] produced subjective perceptions of CL, such as the out-comes of in-depth interviews with selected categories of stakeholders. At the same time, the selection of soft data was

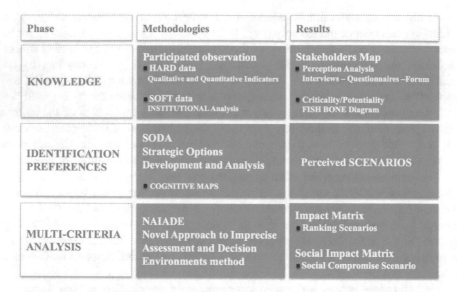

Phase	Methodologies	Results
KNOWLEDGE	**Participated observation** ■ HARD data Qualitative and Quantitative Indicators ■ SOFT data INSTITUTIONAL Analysis	**Stakeholders Map** ■ Perception Analysis Interviews – Questionnaires –Forum ■ Criticality/Potentiality FISH BONE Diagram
IDENTIFICATION PREFERENCES	**SODA** **Strategic Options** **Development and Analysis** ■ COGNITIVE MAPS	**Perceived SCENARIOS**
MULTI-CRITERIA ANALYSIS	**NAIADE** **Novel Approach to Imprecise** **Assessment and Decision** **Environments method**	**Impact Matrix** ■ Ranking Scenarios **Social Impact Matrix** ■ Social Compromise Scenario

Fig. 2. Methodology

structured from an institutional analysis [14, 15] that has produced a map which identified the various stakeholders and leading and important figures in the local culture. The stakeholders were divided into ten main groups: institutions, consortia, associations/groups of active citizenship, social workers, traders, entrepreneurs and four groups of inhabitants divided by age groups: ages 5–19, 20–35, 36–70, 71+.

From the analysis of qualitative information (critical and potential issues, actions, future visions, obstacles, actors, and environmental limits) contained in the verbal protocol of the interviews, it was possible to develop cognitive maps for the different categories of stakeholders (institutions, hotel managers, owners of restaurants and traders, experts, associations, farmers, and inhabitants). Appropriate focus groups were also organized for some of the most relevant topics. The results obtained were decodified and interpreted to allow the identification of resources, potential and significant critical issues for CL. For this purpose, we used the Ishikawa diagram [16], a cause-effect diagram also known as the 'Fish Bone diagram', which allowed the study to identify the most likely causes of a given problem, summarize the results obtained and highlight the importance assumed with respect to the context.

The second phase, "Identification of Preferences and Construction of the Perceived Scenarios", consisted in explaining the most significant questions for the stakeholders that could outline the components of the possible alternative scenarios for the valorisation and transformation of CL. Through the application of the Strategic Options Development and Analysis (SODA) methodology [17, 18], it was possible to identify the questions that the community deemed relevant and to analyse both the qualitative and quantitative components [10].

After having decoded the interviews and the results which emerged from the focus groups, the related cognitive maps have been drew. Thanks to the analyses carried out it was possible to draw up a strategic map of perceptions, which helped us in identifying the main goal of the development strategy for the Tyrrhenian cultural landscape. In order to create an autonomous tourism-based work system and to stop the impoverishment and depopulation process we defined three consistent visions of the future: "Renewal of the Tyrrhenian Cultural Landscape for Tourism"; "Modern and Cultural Tyrrhenian Cultural Landscape"; "Sustainable Tyrrhenian Cultural Landscape".

The aim of the last phase was to identify the preferable perceived scenario through the evaluation of the scenarios which emerged as most significant in the previous phase.

Consistently with the Social Multi-Criteria Evaluation (SMCE) approach and by applying the Novel Approach to Imprecise Assessment and Decision Environments method (NAIADE) [18, 19] it was possible to structure a multi-criteria decision model and multi-group in the fuzzy field.

The NAIADE method is a discrete multi-criteria method whose impact matrix may include either crisp, stochastic or fuzzy measurements of the performance of an alternative with respect to a judgment criterion, thus it is very flexible for real-world applications [18, 19]. NAIADE also performs an equity and conflict analysis in order to identify those alternatives which could reach a certain degree of consensus or would provide a higher degree of equity among different interest groups. It is a very flexible method for decision problems where fuzzy uncertainty or indeterminacy is recognised.

Criteria / Alternatives	Turistically renewed	Modern and cultural	Sustainable
Strengthening of public transport	More or Less Bad	Very Bad	More or Less Bad
Creation of cultural structures	More or Less Bad	Very Bad	Bad
Organization of events throughout all the year	Very Bad	More or Less Bad	Extremely Bad
Increase and improvement of professional and entrepreneurial training	Bad	Extremely Bad	Very Good
Creation of a reception center and tourist information	Moderate	Very Bad	Bad
Implementation of a Tourism Marketing Plan	Very Bad	Bad	More or Less Good
Upgrading and adaptation of existing hotel facilities	Moderate	Very Bad	Very Bad
Promotion of new accommodation activities	More or Less Good	Bad	Bad
Promotion of new tourist itineraries	Good	More or Less Bad	More or Less Bad
Creation of quality certification marks of typical local products	More or Less Good	More or Less Bad	More or Less Bad
Promotion and dissemination of local products	More or Less Bad	More or Less Bad	Extremely Bad
Dissemination of organic farming practices	Very Good	More or Less Bad	Moderate
Increase in the number of farms	More or Less Bad	Very Bad	Good
Promotion and dissemination of handicraft products in the area	More or Less Bad	Very Bad	Very Good
Recovery of the system of terracing and irrigation channels	More or Less Good	Extremely Bad	Very Good
Restoration and enhancement of the pathway system	More or Less Bad	Extremely Bad	Moderate
Recovery / transformation of the existing building stock	Bad	More or Less Bad	Moderate
Use of energy sources from renewable resources	Bad	Very Bad	Extremely Bad
Ordinary mountain maintenance	More or Less Bad	More or Less Good	Very Bad
Development of tourism within the coast	Very Bad	More or Less Good	Very Bad
Improvement of accessibility and usability of the entire area	Moderate	More or Less Good	Moderate
Protection and conservation of ecosystems and habitats	Extremely Bad	Good	More or Less Good

Fig. 3. The multi-group analysis: the impact matrix

For the case study of the Tyrrhenian cultural landscape, NAIADE method elaborated an impact matrix (criteria/alternative) (Fig. 3) and an equity matrix (groups/alternative) (Fig. 4) and provided a ranking of scenarios considered as alternatives according to the stakeholders' preferences, and indications of the distance of the positions of the interest groups and possibilities of convergence of interests and/or of coalition formation. The equity analysis was performed by the completion of an equity matrix (Fig. 5) where a similarity matrix was calculated. It shed light upon the level of decision conflicts among the different interest groups and highlighted the possible formation of coalitions through a dendrogram of coalitions, and by showing the impact of each alternative according to the perception of the different stakeholders. In this way, NAIADE provided the following information: distance indicators between the interests of the different groups of stakeholders, as an indication of the coalition formation possibility, or interest convergence; rankings of alternatives for every coalition, in accordance with the impacts over the groups of stakeholders, or the social compromise solution. Every group then gave an evaluation of each scenario according to a semantic scale (perfect, very good, good, more or less good, moderate, more or less bad, bad, very bad, extremely bad), in order to identify the most sustainable scenario. From the equity matrix, a similarity matrix was computed and gave an index which expressed the similarity of judgement over the alternatives for each pair of interest group.

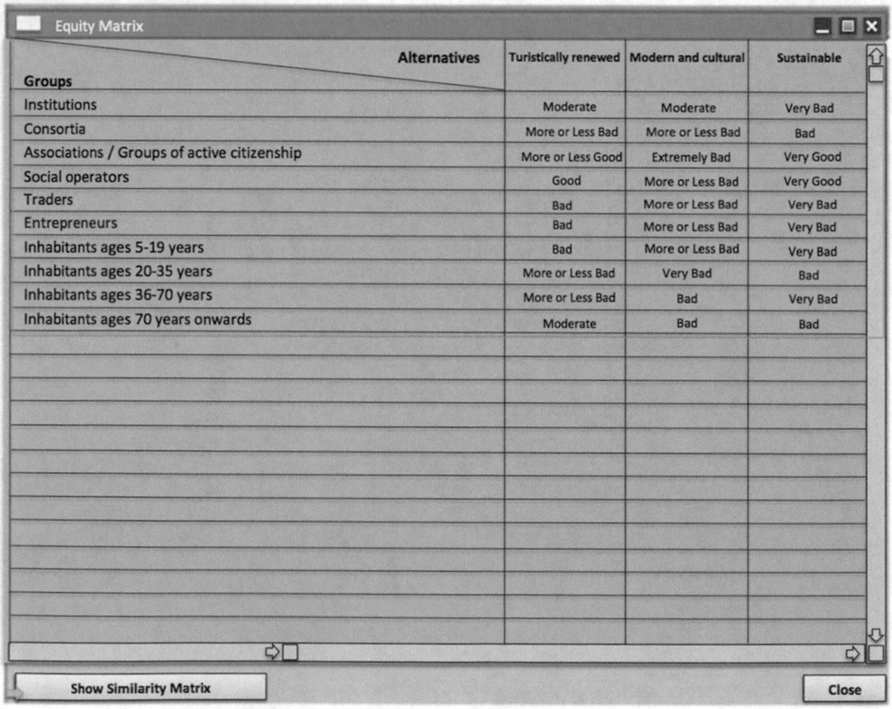

Groups \ Alternatives	Turistically renewed	Modern and cultural	Sustainable
Institutions	Moderate	Moderate	Very Bad
Consortia	More or Less Bad	More or Less Bad	Bad
Associations / Groups of active citizenship	More or Less Good	Extremely Bad	Very Good
Social operators	Good	More or Less Bad	Very Good
Traders	Bad	More or Less Bad	Very Bad
Entrepreneurs	Bad	More or Less Bad	Very Bad
Inhabitants ages 5-19 years	Bad	More or Less Bad	Very Bad
Inhabitants ages 20-35 years	More or Less Bad	Very Bad	Bad
Inhabitants ages 36-70 years	More or Less Bad	Bad	Very Bad
Inhabitants ages 70 years onwards	Moderate	Bad	Bad

Show Similarity Matrix   Close

Fig. 4. The multi-group analysis: the equity matrix

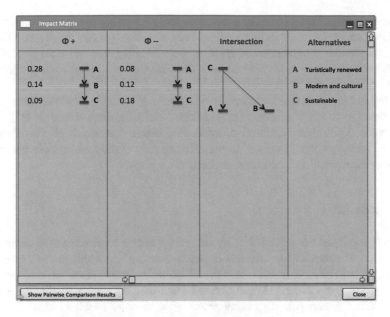

Fig. 5. Results of the equity matrix

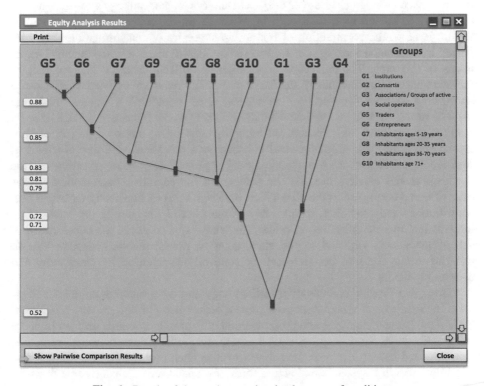

Fig. 6. Result of the equity matrix: dendrogram of coalitions

The outcome of the NAIADE method is an order of preference between scenario alternatives and a dendrogram of coalitions (Fig. 6) that expresses the degree of conflict among stakeholder groups and the level of consensus for each scenario [18–20]. This led to the definition of a defensible and fair decision that reduce the level of conflict and reach a certain degree of consensus.

At the end of the two assessments, two different preferable scenarios emerged. The first alternative in order of preference obtained from the impact matrix (Fig. 5) was "Sustainable Tyrrhenian Cultural Landscape", while "Renewal of the Tyrrhenian Cultural Landscape for Tourism" is the alternative that reduces conflict and reaches consensus throughout the various stakeholders (Fig. 6).

3 Conclusions

This study explored the potential of an integrated approach aimed at the development of strategies to promote the area under discussion, and focused on the peculiarities, values and considerable resources that characterize the cultural landscape of the Tyrrhenian coast of Calabria. The structured multi-methodological evaluation approach is an experimentation part of a broader research process, which employs the contribution and evaluation experiences tested in other contexts [21, 22] and is aimed at outlining incremental and adaptive decision-making processes oriented to the elaboration of shared design choices [22, 23]. Said choices include strategies and actions that contribute to the creation of a pool of expertise on the richest and most complex territory and to the construction of bottom-up strategies of transformation [24], supported by a decision-making process focused on identifying assets and attentive to the needs of this area. The combined application of different methods and techniques, included those from disciplinary fields not necessarily related to evaluation, allowed to approach a complex decision problem characterized by multiple variables and a high level of uncertainty, in an incremental and cyclical evaluation process characterized by continuous feedback and constant interactions, essential to outline a project of conscious and shared transformation and enhancement, consistent with the principles of sustainable development. A methodological approach developed according to the proposed model requires that the construction of the cognitive framework develops over time and accompanies the elaboration of design choices, constantly nurturing new contributions and providing, in turn, new stimuli useful for guiding the selection of information, the identification of values, the analysis of conflicts, the construction of shared preferences oriented to the elaboration of transformation scenarios able to respond to the needs of decision-making contexts characterized by complexity and uncertainty [25].

Through a flexible and adaptive methodology and by combining complex assessment techniques and stakeholder engagement techniques [14, 24], it was possible to build development strategies and promote good governance processes that can improve local deliberative democracy by activating effective collaboration between promoters, operators and users [26]. With the support of integrated evaluation approaches it was possible to begin a systemic and active form of preparation for change, as a basis for shared actions in a long-term vision aimed at developing and effectively constructing public decisions.

References

1. Rössler, M.: World heritage cultural landscapes. In: The George Wright Forum, vol. 17, no. 1, pp. 27–34. George Wright Society (2000)
2. Brandt, J., Tress, B., Tress, G.: Multifunctional landscapes. In: Interdisciplinary Approaches to Landscape Research and Management. Centre for Landscape Research, Roskilde, Denmark (2000)
3. Tortorella, W., Marinuzzi, G. (eds.): Atlante dei piccoli comuni. Centro Documentazione e Studi Comuni Italiani ANCI-IFEL con ANCI, Rome (2013)
4. AmbienteItalia: Indicatori Comuni Europei, verso un profilo di sostenibilità locale. Ancora Arti Grafiche, Milano, Italy (2003)
5. Legambiente: Ecosistema urbano XIX edizione. http://www.legambiente.it. Accessed 11 Jan 2018
6. Bassanelli, M.: Borghi sostenibili. La valle di Zeri http://www.lablog.org.uk. Accessed 11 Jan 2018
7. Cassalia, G., Tramontana, C., Lorè, I., Zavaglia, C.: Statistiche culturali - Il censimento del patrimonio culturale nell'area Tirrenica della Provincia di Reggio Calabria. LaborEst **13**, 12–18 (2017)
8. Cassalia, G., Lorè, I., Tramontana, C., Zavaglia, C.: L'analisi socio-economica a supporto dei processi decisionali: il caso dell'area tirrenica della Città Metropolitana di Reggio Calabria. LaborEst **14**, 26–33 (2017)
9. Calabrò, F., Cassalia, G.: Territorial cohesion: evaluating the urban-rural linkage through the lens of public investments. In: Bisello, A., Vettorato, D., Laconte, P., Costa, S. (eds.) Smart and Sustainable Planning for Cities and Regions, Results of SSPCR 2017. Green Energy and Technology. Springer, Cham (2018). https://doi.org/10.1007/978-3-319-75774-2_39
10. Fusco Girard, L., Cerreta, M., De Toro, P.: Valutazioni integrate: da "processo di apprendimento" a "gestione della conoscenza". Valori e valutazioni **4**(5), 101–115 (2010)
11. Calabrò, F.: Local communities and management of cultural heritage of the inner areas. An application of break-even analysis. In: Gervasi, O., et al. (eds.) Computational Science and Its Applications – ICCSA 2017, ICCSA 2017. Lecture Notes in Computer Science, vol. 10406. Springer, Cham (2017). https://doi.org/10.1007/978-3-319-62398-6_37
12. Breiling, M.: Systems analysis and landscape planning. In: European Landscape Theory Course, Module Two: Issues in Contemporary Landscape Theory, Erasmus Landscape Studies Network ICP UK 2028/02, Swedish University of Agricultural Sciences. Department of Landscape Planning, Alnarp, Sweden (1995)
13. Rosenhead, J., Mingers, J.: Rational Analysis for a Problematic World Revisited. Wiley, Chichester (2001)
14. De Marchi, B., Funtowicz, S.O., Lo Cascio, S., Munda, G.: Combining participative and institutional approaches with multicriteria evaluation: an empirical study for water issues in Troina, Sicily. Ecol. Econ. **34**(2), 267–282 (2000)
15. Funtowicz, S.O., Martinez-Alier, J., Munda, G., Ravetz, J.: Multicriteria-based environmental policy. In: Abaza, H., Baranzini, A. (eds.) Implementing Sustainable Development, pp. 53–77. UNEP/Edward Elgar, Cheltenham (2002)
16. Wittwer, J.W.: Fishbone diagram/cause and effect diagram in Excel. From Vertex42.com. http://www.vertex42.com/ExcelTemplates/fishbone-diagram.html. Accessed 11 Jan 2018

17. Eden, C., Simpson, P.: SODA and cognitive mapping in practice. In: Rational Analysis for a Problematic World, pp. 43–70 (1989)
18. Munda, G.: Multicriteria evaluation in a fuzzy environment: theory and applications in ecological economics. Springer, Heidelberg (2012)
19. Munda, G.: Social multi-criteria evaluation: Methodological foundations and operational consequences. Eur. J. Oper. Res. **158**(3), 662–677 (2004)
20. Joint Research Centre of the European Commission. NAIADE: Manual and Tutorial (1996)
21. Della Spina, L.: Integrated evaluation and multi-methodological approaches for the enhancement of the cultural landscape. In: International Conference on Computational Science and Its Applications, pp. 478–493. Springer, Cham (2017). https://doi.org/10.1007/978-3-319-62392-4_35
22. Della Spina, L.: The Integrated evaluation as a driving tool for cultural-heritage enhancement strategies. In: Bisello, A., Vettorato, D., Laconte, P., Costa, S. (eds.) Smart and Sustainable Planning for Cities and Regions. Results of SSPCR 2017. Green Energy and Technology. Springer, Cham (2018). https://doi.org/10.1007/978-3-319-75774-2_40
23. Guarini, M.R.: Self-renovation in Rome: Ex Ante, in Itinere and Ex Post Evaluation. In: Gervasi, O., Apduhan, B.O., Taniar, D., Torre, C.M., Wang, S., Misra, S., Murgante, B., Stankova E., Rocha, A.M.A.C. (eds.): Computational Science and its Applications - ICCSA 2016, PT IV. Lecture Notes in Computer Science, vol. 9789, pp. 204–218. Springer, Cham (2016). https://doi.org/10.1007/978-3-319-42089-9_15
24. Concilio, G.: Bricolaging knowledge and practices in spatial strategy-making. In: Making Strategies in Spatial Planning: Knowledge and Values, pp. 281–301 (2010)
25. Bencardino, M., Nesticò, A.: Demographic changes and real estate values. A quantitative model for analyzing the urban-rural linkages. Sustainability **9**, 536 (2017). MDPI AG, Basel, Switzerland (2017). https://doi.org/10.3390/su9040536
26. Morano, P., Tajani, F.: Saving soil and financial feasibility: a model to support the public-private partnerships in the regeneration of abandoned areas. Land Use Pol. **73**, 40–48 (2018)

The Impact of Users' Lifestyle in Zero-Energy and Emission Buildings: An Application of Cost-Benefit Analysis

Cristina Becchio⬤, Martina Bertoncini, Adele Boggio,
Marta Bottero⬤, Stefano Paolo Corgnati⬤,
and Federico Dell'Anna$^{(\boxtimes)}$⬤

Politecnico di Torino, 10129 Turin, Italy
federico.dellanna@polito.it

Abstract. The increase in energy exploitation and air pollution have forced the European Union to deal with energy saving and CO_2 emissions reduction. With specific reference to buildings, important targets by 2050 have been established, implementing 2020 energy saving goals, and the concept of NZEB (nearly zero energy building) has emerged. In this context, clear understanding of buildings energy performances achieves primary importance and the inclusion of new variables in the analysis such as the user behavior can offer some insights into the solution of the discrepancy between predicted and real energy performances. In the research, an apartment block settled in Turin (Northern Italy) is selected as case study, energy retrofit measures formulated and evaluated including different users' lifestyles by means of Cost-Benefit Analysis.

Keywords: Benefit/cost ratio · Users' lifestyle · Economic/energy performance
NZEB · Decision-Support System

1 Introduction

Nowadays building sector plays a significant role in energy and carbon emissions reduction, as it has a significant impact on the environment contributing to 40% of the total energy consumption and 36% of total CO_2 emissions [1]. The recast of the Energy Performance of Building Directive recast (EPBD recast, 2012) introduced the concept of nearly zero energy building (NZEB) as a building with very high-energy performances able to cover the residual energy demand with technologies for the exploitation of renewable energy sources installed on-site or nearby [2]. The main focus is on new buildings, that should be NZEB by the end of 2020. Furthermore, new European targets have shifted the focus on CO_2 emissions reduction, setting reduction goals for 2050 [3]. Aiming to integrate two high efficiency building models (in terms of energy and emissions), the concept of zero-energy and -emission building (ZEEB) rises, in line with the national minimum energy performance requirements [4, 5]. Dealing with ZEEB definition, the precise understanding of building energy consumption data has primary importance. Building energy use is affected by many factors not always taken into account when calculating its performance. These factors can influence a lot the

© Springer International Publishing AG, part of Springer Nature 2019
F. Calabrò et al. (Eds.): ISHT 2018, SIST 100, pp. 123–131, 2019.
https://doi.org/10.1007/978-3-319-92099-3_15

simulated energy use and lead to a significant discrepancy between the designed and real energy consumption in buildings [6]. The reasons for this divergence should be mainly found in the role of users' behavior [7, 8]. With the introduction of users' lifestyle in the calculation, it is necessary to employ new economic evaluation methods because the Cost-optimal methodology introduced by the EPBD is not sufficient. Considering only costs is not appropriate to carry out a complete evaluation, stressing the need to introduce benefits as economic profits depending on users' lifestyle. In this scenario, Cost-Benefit Analysis (CBA) [9] seems to be more appropriate. With this awareness, an existing residential building in North Italy was selected as case study with the aim of demonstrating previous considerations. The following paragraphs will explain the used methodology and will present the case study analysis and results, till the identification of the economically efficient solution for the building energy retrofit, taking into account users' lifestyle.

2 Methodological Framework

2.1 Cost-Optimal Methodology

The cost-effectiveness of each scenario was firstly measured with the cost-optimal approach, focusing on the energetic and economic parameters. The Global Cost (1) was calculated following the methodology suggested by the EPBD with the formula:

$$C_g(\tau, r) = I_0 + \sum_{i=1}^{\tau} \frac{FC(i)}{(1+r)^i} = I_0 + \sum_{i=1}^{\tau} FC(i) * R_d(i) \tag{1}$$

where the initial investment cost (I_0) is summed with the annual cash-flow ($FC(i)$) referred to year (i), represented by maintenance, energy, greenhouse gas emissions and replacement costs, multiplied by the discount factor ($R_d(i)$) ((r) is the discount rate). The residual value at the end of the time horizon (τ) enters in the cash-flow ($FC(i)$) and has to be subtracted to previous costs. Combining energy performances and Global Costs, the Cost curve was drawn and the retrofit scenario representing the Cost-optimal level identified. Proving that the resulting cost-optimal scenario complies with ZEEB definition, it entered in the following analysis.

2.2 Cost-Benefit Analysis Approach

A further step was the analysis of the impact of users' lifestyles on building energy performances of the resulting ZEEB scenario, carried on with DesignBuilder software (http://www.designbuilder.co.uk). Different users' lifestyles were defined varying different parameters depending on occupants' habits and their interaction with the building. The subsequent economic analysis was performed using CBA, as an analytical tool aimed at defining and evaluating costs and benefits associated to alternative projects investments, judging their economic advantages and disadvantages, and contributing to the selection of the convenient investment based on clear criteria. This analysis was based on the identification of the monetary amount of costs and benefits

Table 1. Description of considered benefits

Benefits	Description	Reference
Energy saving	Direct benefit resulting from increased energy efficiency. It positively affects both environmental and economic domains	[8] [9] [10] [11]
GHG emissions' reduction	Retrofit interventions can allow the reduction of the amount of energy consumption and relative carbon emissions. The criterion is considered in terms of avoided damage costs of the impacts of climate change	EPBD recast, 2012 [9] [10] [12]
Indoor thermal comfort	Improving building energy efficiency could determine an increase of indoor thermal comfort and an improvement of physical conditions and air quality, raising occupants' satisfaction	[9] [11] [13] [14]
Water saving	Users behavior can also determine the amount of consumed water. A sustainable lifestyle can lead to save water, influencing both environment and incurred costs	[9] [11]

recognized by the society to an investment [9]. In this research, costs were the one included in the cost-optimal analysis and benefits are explained in Table 1. Monetized costs and benefits were discounted to be comparable one each other. Then, the evaluation of the different quantities was done as difference between benefits and costs in order to quantify the net total benefit.

Once displayed the cash-flow calculation, the discounted Benefit-Cost Ratio (B/C) was calculated as the ratio between the discounted value of the benefits (B) sum and the discounted value of the costs (C) sum, as shown in (2). This index should be greater than one to let judging the investment positive.

$$B/C = \frac{\sum \frac{B}{(1+r)^t}}{\sum \frac{C}{(1+r)^t}} \qquad (2)$$

The scenario corresponding to the highest (B/C) ratio represents the economically efficient situation. The results stability at the variation of input parameters is proved by sensitivity analyses.

3 Case Study

3.1 The Reference Building

In order to test the methodology, an existing building settled in Turin (Northern Italy) was chosen as Reference Building (RB). It is an apartment block built between 1991 and 2005 and it is composed of six heated floors divided into 34 apartments and one unheated floor represented by the basement. The RB represents a typical Italian residential building built without considering aspects related to energy saving and characterized by lower quantity of insulation; moreover, in the considered RB the system, represented by two centralized standard gas-fired boilers (installed before 1996), that supply space heating and domestic hot water production, have very low efficiency. Results proceeded by DesignBuilder simulation demonstrate the possibility of reducing building's energy consumption with retrofit measures. The RB primary energy consumption is equal to 116.85 kWh/m^{2}y (considering space heating and cooling, ventilation and DHW-Domestic Hot Water consumptions and the emissions are 40.53 kgCO$_2$/m^{2}y, taking into account all building energy uses (space heating and cooling, ventilation, DHW consumption, lighting and equipment). These values are very high and validate the choice of this case study to experiment the methodology described above.

3.2   The Energy Retrofit Actions

Retrofit scenarios, characterized by different energy efficiency measures (EMMs) applied to building envelope and systems were defined and combined with the installation of technologies for the exploitation of renewable energy sources (RESs). Different levels of building envelope retrofit, from the least performing solutions to the most performing ones, were studied in compliance with Italian standards requirements, as shown in Fig. 1. A further step for improving building energy efficiency was the combination of envelope measures with three systems configurations. In addition, the potential energy saving in installing a controlled mechanical ventilation system (CMV) was tested in packages with the highest energy performances, allowing the decreasing of heating consumptions thanks to the heat recovery and a better control of indoor thermal comfort.

		System					
		A- Gas-fired condensing boiler | Multisplit | RES		B- District heating | Chiller | Radiant floors | RES		C - Heat pump | Radiant floors | RES	
Envelope	Level 1	P1A	-	P1B	-	P1C	-
	Level 2	P2A	-	P2B	-	P2C	-
	Level 3	P3A	P3A CMV	P3B	P3B CMV	P3C	P3C CMV
	Level 4	P4A	-	P4B	-	P4C	-
			CMV		CMV		CMV

Controlled mechanical ventilation system

Fig. 1. Matrix of alternative retrofit scenarios. Level 1 and 2 refer to values in force till 2015 and 2021 referred to the Decreto Interministeriale 26th June 2015. Level 3 refers to subsidized level of the Allegato energetico ambientale number 2010-08963/38. Level 4 concerns thermal transmittance of the opaque envelope referred to Level 3, glazing referred to Level 2. RESs are in compliance with the Decreto Legge 28/2011 and the Allegato Energetico Ambientale.

Finally, depending on systems configurations, different areas of solar panels were installed on the building, in line with the Italian standard in force. Using cost-optimal methodology, packages' energy performance and global cost were combined (Fig. 2). The cost-optimal level was identified in P2C solution, characterized by the second level of insulation for the envelope, the installation of an air-to-water heat pump combined with radiant floors and RES integration.

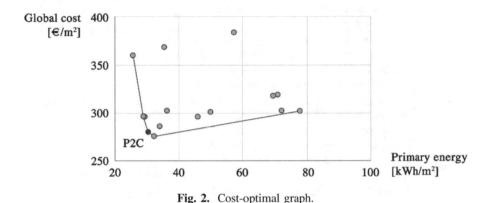

Fig. 2. Cost-optimal graph.

3.3 Users Lifestyle Impact on Building Energy Performances

Achieving ZEEB targets allow to reach an unequivocal value describing building energy efficiency, but the inclusion of users' actions in the evaluation is necessary to estimate building's consumptions. Users' behavior and interaction with the building can be described analyzing different variables, such as heating and cooling temperature set-points, ventilation (the number of fresh air rates from outdoor), lighting and equipment (regularity in the use of lights and electric appliances) [15]. The combination of different values of these variables, depending on users' habits allowed us to determine different lifestyles, such as an energy conscious lifestyle (low consumer – LC), a standard lifestyle (average consumer – AC) and an energy wasteful lifestyle (high consumer - HC). The heating/cooling set-points and the ventilation rates refer to comfort categories described in EN15251. With the aim of investigating the household composition, two additional situations referred to real habits of some types of families were analyzed; the behavior of a young couple (A) and that of an elderly couple (B), as described in Table 2. It has to be considered that these last two profiles come from recent studies on the behavior of the users' lifestyles that reflect the socio-demographic characteristics of the inhabitants [15]. The energy performances of the RB and of ZEEB retrofit scenario were assessed with dynamic simulations, including the aforementioned users' characterizations.

Table 2. User lifestyles analyzed

Lifestyle	Description	Primary energy consumption	CO_2 emissions
LC	User actions are characterized by low thermal, low electricity and low water consumptions	82.6 kWh/m^2y	14.9 kg/m^2y
AC	User is less concerned about energy saving and his actions are the conventional ones described in the Standards	100.28 kWh/m^2y	18.10 kg/m^2y
HC	User does not care about energy saving due to high expectations for indoor environmental quality. High thermal energy, high electricity and high water consumptions are considered	153.13 kWh/m^2y	27.64 kg/m^2y
A	A young couple has a large number of electric devices leading to a high electricity consumption. At the same time, spending few hours at home, the energy consumption for space heating and cooling is quite low	119.67 kWh/m^2y	21.60 kg/m^2y
B	An elderly couple usually has a high-energy use for space heating and cooling, spending most of daily time at home, but a low electricity consumption, given by the little number of used devices	105.93 kWh/m^2y	19.12 kg/m^2y

3.4   Cost-Benefit Analysis to Evaluate Users' Lifestyle

In order to evaluate the best configuration considering the energy retrofit intervention and the impact of users' lifestyle, the CBA was set up. RB and ZEEB retrofit models, combined with users' lifestyles, defined alternative scenarios (namely RB LC - RB HC – RB A – RB B – ZEEB LC – ZEEB AC – ZEEB HC – ZEEB A – ZEEB B) that were compared with the base one (RB AC). Once costs and benefits were identified, the following steps consisted in their quantification and monetization (Table 3). Considering a calculation period of 30 years and a discount rate equal to 2% [9], all annual costs and benefits were discounted and summed up separately. In this way, different B/C ratios were calculated.

Table 3. Identification and monetization of costs and benefits.

Costs	Description	Estimation procedure
Investment cost	Initial amount of money spent for the retrofit interventions	EEMs investment costs with reference to the Piedmont Region price-list [€]
Maintenance cost	Annual cost to maintain the initial performances of building system	Cost calculated as a percentage of the investment cost [€] (EN 15459/2007)
Replacement cost	Cost to replace building components at the end of their service life	Investment cost of components replaced at the end of their working life [€] (EN 15459/2007)

(continued)

Table 3. (*continued*)

Costs	Description	Estimation procedure
Energy cost	Annual cost of energy consumed for all domestic uses	Energy cost [€/kWh] × Energy used [kWh]
Residual value	Value of building's components at the end of the calculation period	Depreciation of the investment costs at the end of the calculation period [€]
Benefits	Description	Estimation procedure
Energy saving	Direct benefit resulting from reduced energy consumption	Energy cost [€/kWh] × Energy saved [kWh]
GHG emissions' reduction	Increased energy efficiency let the reduction of GHGs and pollutants	GHG emissions' cost [€/kgCO$_2$eq] × GHG emissions saved [kgCO$_2$eq]
Indoor thermal comfort	Direct benefit in terms of higher indoor temperatures and indoor air quality	Hourly cost of work [€/h] × Numb. of avoided discomfort hours [h][a]
Water saving	Amount of water saved thanks to sustainable user behavior	Water cost [€/m^3] × Saved water [m^3]

[a]Indoor thermal comfort was monetized associating it to users' productivity at work. This methodology is generally used for offices and similar destinations [18].

4 Discussion of Results

The results of CBA show the prevalence of the elderly couple lifestyle combined with ZEEB retrofit (ZEEB B in Table 4). Even if costs for ZEEBs scenarios are higher due to the investment cost for the energy retrofit [16], benefits are more significant and determine positions in the ranking. The improvement of indoor thermal comfort, in addition to energy saving and CO$_2$ emissions reduction, positively affects B/C ratio for ZEEB B scenario. Moreover, energy cost of ZEEB B scenario is one of the lowest due to the low electric consumption and it is the only component of cost decisive to determine results.

Table 4. B/C ratio results.

Solution	ZEEB LC	ZEEB AC	ZEEB HC	ZEEB A	ZEEB B	RB LC	RB HC	RB A	RB B
B/C ratio	0.81	1.56	1.42	0.47	1.95	0.43	0.73	0.04	1.2

5 Conclusions

The research aim was to demonstrate the importance of the experimentation of new economic evaluation methods of building energy performances. To fill the discrepancy between predicted and real energy consumption it was verified the necessity of the inclusion of users' behavior, as the major source of uncertainty especially in high-performing buildings. The application of the CBA instead of the cost-optimal methodology allows the evaluation of benefits influenced by occupants' lifestyle. Results obtained reveal ZEEB scenarios as preferred, confirming cost-optimal results and the elderly couple lifestyle combined with ZEEB scenario (ZEEB B) as the economically efficient solution thanks to the amount of benefits relapsing on users once invested in the retrofit.

The variable that most influence results stability is related to the monetization of thermal indoor comfort, motivating the need of a more accurate estimate including stated and revealed preference methods that permit to measure users' willingness to pay to reach certain comfort conditions [17]. In order to further improve the analysis, it is necessary to increase the amount of benefits included, evaluating also those ones not immediately quantifiable but important from users' point of view. It would be useful to investigate the impacts of energy efficiency related to the reduction of health effects caused by unhealthy conditions in the building, due to stale air or the presence of condensation and fogging. Finally, further investigation could explore the fact that energy-efficient buildings typically have longer life cycles, lower maintenance costs and lower operational costs which could be resulted in an increase in resale value.

References

1. BPIE: Smart buildings in a decarbonised energy system. Buildings Performance Institute Europe (2016)
2. EBPD 2010: Directive 2010/31/EU. Directive of the European Parliament and of the council of 19 May 2010 on the Energy Performance of Building (2010)
3. European Parliament: Communication from the commission to the European parliament, the Council, the European economic and social Committee and the Committee of the regions. Roadmap for moving to a competitive low-carbon economy in 2050, EU (2011)
4. BPIE: Nearly Zero Energy Buildings definitions across Europe. Buildings Performance Institute Europe (2015)
5. Torcellini, P., Pless, P., Deru, M.: Zero Energy Buildings: A Critical Look at the Definition. National Renewable Energy Laboratory (2006)
6. IEA: Total Energy Use in Buildings: Analysis and Evaluation Methods (Annex 53). Energy in Buildings and Communities Programme EBC (2016)
7. Fabi, V., Andersen, R., Corgnati, S.P., Olesen, B.W., Filippi, M.: Description of occupant behaviour in building energy simulation: state of art and concepts for improvements. In: Proceedings of Building Simulation, 12th Conference of International Building Performance Simulation Association, Sydney (2011)
8. Janda, K.B.: Buildings don't use energy: people do. Architectural Sci. Rev. **54**, 15–22 (2011)
9. European Commission: Guide to Cost-Benefit Analysis of Investment Projects. Economic appraisal tool for Cohesion Policy 2014–2020. EU (2015)
10. IEA: Capturing the Multiple Benefits of Energy Efficiency. International Energy Agency, Paris (2014)
11. Urge-Vorsatz, D., Herrero, S.T., Dubash, N.K., Lecocq, F.: Measuring the co-benefits of climate change mitigation. Annu. Rev. Environ. Resour. **39**, 549–582 (2014)
12. Copenhagen Economics: Multiple benefits of investing in energy efficient renovation of buildings, CE (2012)

13. Clinch, J.P., Healy, D.H.: Cost-benefit analysis of domestic energy efficiency. Energy Policy **29**(2), 113–124 (2001)
14. Fang, X., Bianchi, M.V., Christensen, C.: Monetization of thermal comfort in residential buildings. In: ACEEE Summer Study on Energy Efficiency in Buildings. ACEEE (2012)
15. Barthelmes, V., Becchio, C., Corgnati, S.P.: Occupant behaviour lifestyles in a residential nearly zero energy building: effect on energy use and thermal comfort. Sci. Technol. Built Environ. **22**, 960–975 (2016)
16. Barthelmes, V.M., Becchio, C., Bottero, M., Corgnati, S.P.: Cost-optimal analysis for the definition of energy design strategies: The case of a nearly-Zero energy building. Valori e Valutazioni **16**, 61–76 (2016)
17. Buso, T., Dell'Anna, F., Becchio, C., Bottero, M., Corgnati, S.: Of comfort and cost: Examining indoor comfort conditions and guests' valuations in Italian hotel rooms. Energy Res. Soc. Sci. **32**, 94–111 (2017)
18. Wargocki, P., Seppanen, O., Andersson, J., Boerstra, A., Clements-Croome, D., Fitzner, K., Olaf Hanssen, S.: Clima interno e produttivita negli edifici. Translator: Schiavon, S., Collana Tecnica AICARR, Milano (2008)

EUSALP, a Model Region for Smart Energy Transition: Setting the Baseline

Silvia Tomasi[1], Giulia Garegnani[1], Chiara Scaramuzzino[1],
Wolfram Sparber[1], Daniele Vettorato[1], Maren Meyer[2], Ulrich Santa[2],
and Adriano Bisello[1(✉)]

[1] Institute for Renewable Energy, Eurac Research,
Viale Druso 1, 39100 Bolzano, Italy
adriano.bisello@eurac.edu
[2] Agency for Energy South Tyrol - CasaClima,
Via A. Volta 13/A, 39100 Bolzano, Italy

Abstract. Nowadays energy transition is a recurring topic, which describes the process of an energy system moving from fossil-based sources towards renewables. The transition can unfold at different levels, from the single initiative of a local community to a complex cross-border agreement. The latter type is well represented by EUSALP, the European macro-regional strategy for the Alpine region. One of its aims is to transform its territory into a model region for energy efficiency and renewable energy. To support a well-informed decision making process, this study provides the first insight about the status quo of energy balances in EUSALP, at local as well as aggregated level. Moreover, it offers an overview on the various energy targets defined by the territorial units that constitute the EUSALP region. Data has been retrieved via a bottom-up quality-oriented process consisting of (i) a survey targeted at responsible person in local energy departments; and (ii) data control and harmonization. We found that the EUSALP region is actually a model region only in clean power production, whereas starting point as well as energy targets of territories are highly heterogeneous. We also identified the need of more harmonized data collection methodologies. We conclude that this bottom-up process can support and legitimate policy makers in cross-border cooperation activities under a smart macro-regional energy strategy, which pursues an increment in energy savings, renewable energy production and a broad engagement of relevant stakeholders.

Keywords: Renewable energy · Energy efficiency · EUSALP area
Energy planning · Smart energy transition · EU macro strategies
Bottom-up data collection

1 Introduction

Macro regional strategies provide a framework for cooperation, coordination and consultation between and within various states and regions. They depict an opportunity for greater regional cohesion and coordinated implementation of European sectoral policies among territories confronted with common challenges and opportunities. The EUSALP, the European Strategy for the Alpine Region, is the fourth macro

© Springer International Publishing AG, part of Springer Nature 2019
F. Calabrò et al. (Eds.): ISHT 2018, SIST 100, pp. 132–141, 2019.
https://doi.org/10.1007/978-3-319-92099-3_16

regional strategy of the European Union, following those for the Baltic Sea Region [1], for the Danube Region [2] and for the Adriatic and Ionian Region [3]. The EUSALP concerns 7 Countries, of which 5 EU Member States (Austria, France, Germany, Italy and Slovenia) and 2 non-EU countries (Liechtenstein and Switzerland); overall fifty territorial entities from national to subnational level. Its main objective is to cope with shared challenges, endorsing trans-border cooperation among its territories [4]. The EUSALP Action Plan aims to translate the identified common challenges and potentials into concrete actions. It is built upon three thematic policy areas (economic growth and innovation, mobility and connectivity, environment and energy) and one cross-cutting policy area (governance) [5]. The Action Plan focuses on nine actions, to be implemented by as many Action Groups (AG). Under the 3rd objective "for a more inclusive environmental framework for all and renewable and reliable energy solutions for the future", AG9 has the mission to *"make the EUSALP territory a model region for energy efficiency and renewable energy"* [5].

EUSALP territorial entities significantly differ by extension, population and GDP. In general, the EUSALP area is economically strong and densely inhabited: it covers nearly 10% of the EU surface (about 470.000 km^2), encompasses 16% of the population, and generates 20% of the GDP (including non-EU countries Liechtenstein and Switzerland).

Up to now, a few studies have been carried out in order to assess energy consumption in some of the EUSALP territories [6, 7], but detailed energy consumption, and energy targets assessment of the whole macro regional EUSALP area has remained unstudied. Similarly, the narrative about the renewable energy potential of this territory has traditionally focused only on the core alpine area, as the *"water tower"* or the *"the green battery"* of Europe [5, 8]. Hence, the goal of this study is to summarize latest findings on energy-related issues in the EUSALP, beyond preconceived notions. Thus, it depicts for the first time the EUSALP energy balance, and sheds light on detailed consumption by energy sources. Moreover, it collects and compares the local energy strategies and targets, defined by regions and states within the EUSALP area, to establish a common baseline on which to build trans-border cooperation for a harmonized low-carbon energy transition.

2   Methodology

2.1   The EUSALP Energy Survey 2017

EUSALP AG9 took up its work in 2016 with the mission *"to make the territory a model region for energy efficiency and renewable energy"*. In order to gain an overview over the "state of the art" of energy policy goals and energy consumption and production in the EUSALP, AG9 commissioned the "EUSALP Energy Survey 2017". The Institute for Renewable Energy of Eurac Research (EURAC) conducted the collection and analysis of the relevant data. The Survey consisted of twenty nine open questions organized in 7 sections. Starting from March 2016, the English version of the Survey was made accessible online, by using the tool "Survey Monkey", and an invitation to fill it in was sent to all EUSALP territories. Please refer to the on-line published version of the Report for the details of the data collection process [9]. The choice to collect data by means of a survey was motivated by (i) the absence of public available and updated energy data at local level in some EUSALP territories, and (ii) the need to collect qualitative information from informed stakeholders.

2.2   Structure of the Survey

The survey is designed to collect energy production and consumption data in the EUSALP territories, and their energy strategies. The main items of the survey are questions about contact person information (Q1 and Q2), general data of the territory (Q3), energy data (Q4–Q9), remaining potentials of renewable energy sources (RES) (Q10–Q17), energy strategies (Q18–Q25), governance (Q26–Q28), and topics of regional interest as feedback for EUSALP (Q29). Tables 1 and 2 report the details of

Table 1. EUSALP survey 2017 (Section 3 - Energy Data)

Energy Data				
	Q5 Please indicate the energy related data of your Region/Canton/Province/State (GJ per annum).		Q5-a Primary energy consumption Q5-b Final energy consumption Q5-c Primary energy production Q5-d Final energy production Q5-e Final energy consumption from RES Q5-f Final energy production from RES	
Energy consumption	**Q6** Please indicate the final energy consumption by sector (in GJ per annum).	Q6-a Residential Q6-b Industrial Q6-c Transportation Q6-d Public Administration Q6-e Agriculture	**Q7** Please indicate the final energy consumption by energy source (in GJ per annum).	Q7-a Electricity Q7-b Coal Q7-c Natural gas Q7-d Liquid gas Q7-e Mineral oil Q7-f Solar thermal energy Q7-g Geothermal energy and ambient heat Q7-h Biofuels and biomass Q7-i Waste
Energy production	**Q8** Please indicate the electricity production (final energy) by energy source (in GJ per annum).	Q8-a Coal Q8-b Natural gas Q8-c Nuclear Q8-d Mineral oil Q8-e Wind power Q8-f Hydro power Q8-g PV Q8-h Cogeneration Q8-I Waste	**Q9** Please indicate the heat production (final energy) by energy source (in GJ per annum).	Q9-a Coal Q9-b Gas Q9-c Liquid gas Q9-d Electricity (direct) Q9-e Mineral oil Q9-f Solar thermal energy Q9-g Geothermal energy and ambient heat Q9-h Biofuels and biomass Q9-i Waste

the questions regarding energy data, and medium and long-term targets of energy strategy, since these are the topics of this study. As the first table shows some of the questions of this section are related, and they offer the possibility to double-check the data, or conversely pose doubts if divergences arise.

Table 2 shows the symmetrical structure of questions concerning medium and long-term targets. In fact, both ask about targeted percentages of energy savings, though not specifying if in primary or final units, then targeted share of RES in final consumption, electricity production and heat production.

Table 2. EUSALP survey 2017 (Section 5 - Energy Strategies)

Energy Strategies			
Q21 Medium Term Targets Of The Energy Strategy. Please indicate the medium term targets (by 2020) defined by the energy strategy.	Q21-a Baseline year Q21-b Target year Q21-c Energy savings in % Q21-d Share of RES in final energy consumption in % Q21-e Share of RES in electricity production in % Q21-f Share of RES in heat production in %	**Q22** Long Term Targets Of The Energy Strategy. Please indicate the long term targets (e.g. by 2050) defined by the energy strategy.	Q22-a Baseline year Q22-b Target year Q22-c Energy savings in % Q22-d Share of RES in final energy consumption in % Q22-e Share of RES in electricity production in % Q22-f Share of RES in heat production in %

2.3   Energy Data Analysis Rationale

To synthetize collected data, we categorized the energy consumption into electricity consumption, thermal consumption, and consumption by the transport sector, according to the European Renewable Energy Directive [10] and other works on energy flows (see for example [11]). Later on, we differentiated renewable energy sources from fossil sources, and estimated their share in final energy consumption. Finally, we assessed the electricity exported or imported. This categorization shapes the energy balance in the EUSALP area.

To cope with some accounting differences in provided data we did the following assumptions: (i) heat consumption is equal to the difference between final energy consumption and electricity consumption plus the consumption of the transport sector. Where consumption data by sectors are missing, we assume the heat consumption and production to be equal; (ii) to calculate the electricity export, yearly inland electricity need is assumed to be satisfied firstly by energy production from the local RES, then from fossil fuel plants or nuclear plants. The remaining amount of energy is considered as exported; (iii) transport is always considered fossil based. It is worth mentioning, that in this study Switzerland has been considered as a whole [12].

Provided data has been carefully revised, in order to harmonize measurement units, to discover and remove evident errors or inconsistent figures. This process required a strong involvement of responsible persons, to double check the data in an iterative way.

2.4   Energy Strategies Analysis Rationale

To analyze the local energy strategies, we considered firstly the targeted shares of RES (overall share, share in electricity and in heat) in the medium (by 2020, specified when different) and long-term (by 2050, specified when different). Then, we compared them with the medium and long-term targets of the EU28 [13].

Finally, we focused on the energy saving targets, making the same distinction between medium- and long-term ones, and comparing them with the EU28 energy savings targets [13, 14]. For energy strategies too, Switzerland has been considered at federal level, therefore energy targets are national [15, 16].

3   Results

3.1   Energy Consumption and Share of Renewable Energy Sources

Energy data provided by respondents cover near 71% of the overall calculation of energy consumption, 78% of the overall calculation of energy production from renewable sources, and ranges from year 2008 to 2015. Missing data has been retrieved from official data sources.

Based on these elaborations, it is possible to sum up the main energy figures in the EUSALP area, showed in Fig. 1.

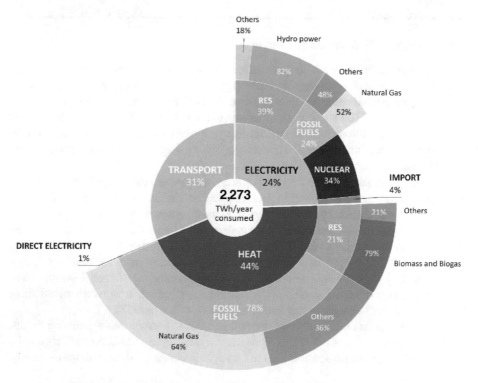

Fig. 1. EUSALP energy consumption by sectors and energy source

Yearly energy consumption in the EUSALP area is approximately 2,270 TWh, respectively related to heating needs (44%), transport (31%) and electricity (24%). This means a per capita energy consumption of about 28 MWh.

EUSALP figures are compared to the EU28 ones in Table 3; transport consumption in the EUSALP area seems to be slightly lower, whereas electricity consumption results higher than in the EU28, and thermal consumption similar. The share of RES in the local power generation is quite high (39%), especially due to a strong hydropower production in the core Alpine area. As expected, the share of RES in electricity production in EUSALP is higher than in the EU28, as highlighted in Fig. 2. On the other hand, nuclear plants, located in France and German EUSALP regions, as well as in Slovenia and Switzerland, cover the 34% of electricity demand, resulting higher than the European average. The remaining 4% of the EUSALP electricity demand is covered by imported power.

Table 3. Energy consumption figures, comparison between EUSALP and the EU28

Energy consumption in the EUSALP area			EUSALP	EU28
ELECTRICITY			24%	22%
	RES		39%	28.8%
		Hydropower	82%	11%
	Nuclear		34%	27%
HEAT			44%	45%
	RES		21%	18.6%
		Biomass/biogas	79%	N.A.
TRANSPORT			31%	33%
PER CAPITA			28 MWh	25 MWh

Concerning the thermal consumption, 79% is satisfied by non-renewable sources, mostly natural gas. Remaining 21% from RES, where biomass and biogas (no further distinguished by the Survey) constitute by far the most relevant source of clean carriers.

Figure 2 shows how these average EUSALP values are built up on a great variety of local shares, far from a normal distribution, and graphically relates these to the EU 28.

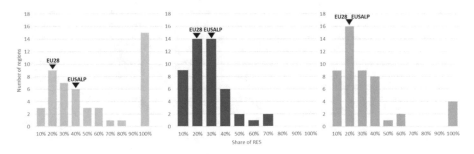

Fig. 2. Frequency distribution of the share of RES in (a) electricity, (b) thermal consumption, and in (c) final energy consumption in the EUSALP

Overall, the share of RES in electricity figure for EUSALP as a whole (40%) widely overcomes the value for the EU28 (29%), but it does not stand out in thermal consumption, where only two percentage points part EU28 (19%) and EUSALP (21%).

It emerges how some virtuous regions exceed their overall energy demand with local energy production from renewable energy sources, while almost two thirds of EUSALP territories cover less than the 30% with their RES production. As expected, electricity consumption in the EUSALP area is cleaner than thermal consumption, and the share of renewable electricity shows a wide heterogeneity. It is higher in the Alpine core area, where hydropower production is significant and lower in the marginal regions. On the other hand, electricity consumption is higher in the more industrialized and densely populated regions where the share is less than 50%.

3.2   Energy Strategies

About 70% of EUSALP regions answered the section of the survey related to energy strategies. Hence, Fig. 3 gives only a partial insight on this topic, albeit the most complete one in the state of art.

Figure 3 gives at a glance the status quo in this field, comparing medium and long-term targets of EUSALP regions, which aim to enhance the share of RES in the energy mix, with the European targets in the same field. A great variety of targets emerges for medium as well as for long-term targets. The comparison of energy saving targets, among the EUSALP territories and the EU28, results even more challenging, due to the uncertainties of baseline years and baseline consumptions.

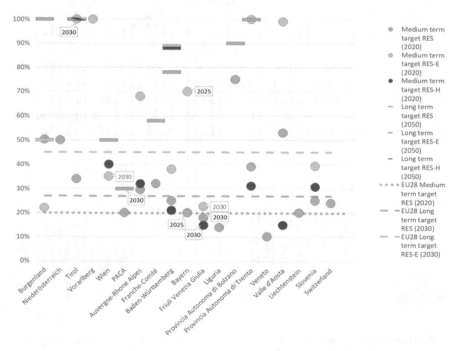

Fig. 3. Future targets of RES penetration in the EUSALP area

4   Discussion and Conclusions

This study provides important insights into the current energy mix and trends of the EUSALP area, a relatively new and therefore less investigated macro-regional context, which embraces the regions and states located in the Alps and the pre-Alpine territories.

Energy data and aggregated results come for the first time from a bottom-up approach, thanks to an energy survey conducted in 2016-2017 among the fifty EUSALP territorial entities. This snapshot fills an important gap in the background knowledge needed to develop a sound scientific based decision-making process. The study offers, at the best of our knowledge, the first global information on the energy balance of the EUSALP area. Interestingly, the EUSALP area has a per capita energy consumption slightly above the EU28 average, and its energy consumption from nuclear plants is higher than the European rate. Moreover, even if both the share of RES in thermal and electricity consumption are higher in the EUSALP area than in the EU28, despite the large availability of natural resources in the alpine regions, only 19% of energy in the overall balance comes from RES, a share slightly above to the EU28 (17%). Thus, becoming a "model region" for a smart energy transition is an ambitious goal to pursue in the coming years, and the journey has only just begun.

Moreover, we also provide an overview of the diversified set of local energy targets, both of renewable energy sources penetration and energy saving in the medium and long term. The heterogeneity of the EUSALP territories reflects the diversity of their energy targets. Local strategies are often hard to compare, due to multiple factors, first of all the starting point of each territory in terms of current share of RES, as well as differences in accounting methodologies. Comparison with European targets shows that the local long-term targets of the EUSALP regions are more ambitious, even if related to a more distant target year. On the other hand, for the medium as well as for the long term the achievement of thermal goals seems to be more challenging. An enhanced collaboration among the EUSALP territories, aiming to harmonize their energy targets and to coordinate their energy strategies, could lend a relevant support to policy makers in defining and implementing long-term suited energy targets and strategies for the Alpine territory. This cohesion will be possible only thanks to the broader awareness of the involvement in the EUSALP territories, maybe grouped into more homogeneous clusters, sharing socio-economic and energy conditions, and thus facing similar barriers and drivers to the energy transition [17]. To foster the commitment and effectiveness, strategies and action plans towards the smart energy transition should even more stress the possible multiple benefits [18, 19] and should rely on trustable assessment of RES potential [6].

To conclude, further research should focus on standardization of energy data accounting methodologies, taking into account seasonal fluctuations in energy production and consumption, and on the assessment of trustable energy savings data and targets. The resulting information should be available to all stakeholders, especially to policy makers, to facilitate the definition of shared energy targets, as part of an ad hoc EUSALP energy strategy.

Acknowledgements. This study has been realized in the frame of the Alpine Space project "AlpGov", co-financed by the European Union via Interreg Alpine Space. The EUSALP Energy survey design was developed by Ulrich Santa, Maren Meyer, Ulrich Klammsteiner, and was further elaborated with members of the EUSALP Action Group 9, sub-group 1, concerned with regional and national energy strategies. The authors are extremely grateful to the Agency for Energy South Tyrol - CasaClima, and all members of EUSALP Action Group 9 and contact persons involved in filling the EUSALP Energy Survey 2017 for the time and effort dedicated. The authors would also like to acknowledge the colleagues of the Urban and Regional Energy Systems group of the Institute for Renewable Energy at Eurac Research, who supported and contributed with their expert knowledge to this research.

References

1. European Commission: Communication the European Union Strategy for the Baltic Sea Region (2012)
2. European Commission: Communication European Strategy for the Danube Region (2010)
3. European Commission: Communication from the commission to the European parliament, the council, the European economic and social committee and the committee of the regions concerning the European Union Strategy for the Adriatic and Ionian Region, p. 2004 (2016)
4. Regio, D.G.: An eu strategy for the alpine region (eusalp) core document (2015)
5. European Commission: EUSALP Action Plan (2015)
6. Grilli, G., De Meo, I., Garegnani, G., Paletto, A.: A multi-criteria framework to assess the sustainability of renewable energy development in the Alps. J. Environ. Plan. Manag. **60**(7), 1276–1295 (2017)
7. Hecher, M., Vilsmaier, U., Akhavan, R., Binder, C.R.: An integrative analysis of energy transitions in energy regions: a case study of ökoEnergieland in Austria. Ecol. Econ. **121**, 40–53 (2016)
8. Alpine Convention: Alpine Convention - 4th International Conference Water in the Alps Sustainable Hydropower - Strategies for the Alpine Region (2012). http://www.alpconv.org/en/organization/groups/WGWater/waterinthealps/pages/default.aspx?AspxAutoDetectCookieSupport=1. Accessed 6 Dec 2017
9. Bisello, A., Tomasi, S., Garegnani, G., Scaramuzzino, C., Segata, A., Vettorato, D., Sparber, W.: EUSALP Energy Survey 2017 Report (2017)
10. European Commission: Directive 2009/28/ec of the European parliament and of the council of 23 April 2009 on the promotion of the use of energy from renewable sources and amending and subsequently repealing Directives 2001/77/EC and 2003/30/EC, pp. 16–62 (2009)
11. Mathiesen, B.V., Lund, H., Connolly, D., Wenzel, H., Østergaard, P.A., Möller, B., Nielsen, S., Ridjan, I., Karnøe, P., Sperling, K., Hvelplund, F.K.: Smart energy systems for coherent 100% renewable energy and transport solutions. Appl. Energy **145**, 139–154 (2015)
12. Bundesamt für Energie, Schweizerische Gesamtenergiestatistik 2015 Statistique globale suisse de l' énergie 2015, Bundesamt für Energ., p. 64 (2015)
13. European Commission: Energy Roadmap 2050 (2011)
14. European Commission: Clean energy for all Europeans, Communication from the Commission to the European Parliament, the Council, the European Economic and Social Committee and the Committee of the Regions, vol. COM(2016), 860 final (2016)
15. Bundesamt für Energie (BFE) Eidgenössisches Departement für and E. und K. (UVEK) Umwelt, Verkehr, Aktionsplan, Erneuerbare Energien (2008)
16. Bundesamt für Energie (BFE), Energiestrategie 2050 (2017). http://www.bfe.admin.ch/energiestrategie2050/index.html?lang=en. Accessed 6 Dec 2017

17. Mosannenzadeh, F., Bisello, A., Diamantini, C., Stellin, G., Vettorato, D.: A case-based learning methodology to predict barriers to implementation of smart and sustainable urban energy projects. Cities **60**, 28–36 (2017)
18. Bisello, A., Grilli, G., Balest, J., Stellin, G., Ciolli, M.: Co-benefits of smart and sustainable energy district projects: an overview of economic assessment methodologies. In: Smart and Sustainable Planning for Cities and Regions, pp. 127–164 (2017)
19. Nippa, M., Meschke, S.: Germany's 'Energiewende' as a role model for reaching sustainability of national energy systems? History, challenges, and success factors. In: Handbook of Clean Energy Systems, pp. 1–23 (2015)

Geographically Weighted Regression for the Post Carbon City and Real Estate Market Analysis: A Case Study

Domenico Enrico Massimo[1], Vincenzo Del Giudice[2],
Pierfrancesco De Paola[2(✉)], Fabiana Forte[3], Mariangela Musolino[1],
and Alessandro Malerba[1]

[1] Mediterranea University of Reggio Calabria, 89124 Reggio Calabria, Italy
[2] Federico II University of Naples, 80125 Naples, Italy
pierfrancesco.depaola@unina.it
[3] Luigi Vanvitelli University of Campania, 81031 Aversa, Italy

Abstract. Geographically Weighted Regression is a statistical technique for real estate market analysis, particularly adequate in order to identify homogeneous areas and to define the marginal contribution that the geographical location gives to the market value of the properties. In this paper a GWR has been applied, in order to verify the robustness of the real estate sample, this for the subsequent individuation of progressive real estate sub-samples in able to detect and to identify possible potential market premium in real estate exchange and rent markets for green buildings. The model has been built on a large real estate dataset, related to the trades of residential real estate units in the city of Reggio Calabria (Calabria region, Southern Italy).

Keywords: Post carbon city · Green buildings
Geographically Weighted Regression (GWR) · Spatial Statistics
Real estate market analysis

1 Introduction

Economic activities and scientific research have always been a pressing need to analyze the data by evolving the general or so-called "global" regression models (such as the classic Multiple Regression Analysis), which although they contain the zonal variables in parametric form, to more detailed models that distribute the original point-discrete values in a continuous mode in the space [1–3].

From these models belongs Geographically Weighted Regression (GWR). This method is a spatial analysis technique mainly intended to indicate where non-stationarity is taking place on the map, that is where locally weighted regression coefficients move away from their global values.

The authors contributed equally to the study.

© Springer International Publishing AG, part of Springer Nature 2019
F. Calabrò et al. (Eds.): ISHT 2018, SIST 100, pp. 142–149, 2019.
https://doi.org/10.1007/978-3-319-92099-3_17

In this study a GWR has been applied, in order to verify the robustness of the real estate sample considered, this for the subsequent individuation of progressive real estate sub-samples in able to detect and to identify possible potential market premium in real estate exchange and rent markets for green buildings [21–28].

The GWR is a "local" form of linear regression used to model spatially varying relationships.

One of the comparisons between Multiple Regression Analysis (MRA) and is inferred from the GWR [1] (Table 1):

Table 1. Comparison between MRA and GWR.

Global classic regression (MRA)	Local regression (GWR)
Research of all the data of the study	Breakdown of the overall statistical local
Not mappable	Mappable
Spaceless	Spatial
Emphasis on similarities in space	Emphasis on the differences in space
Finding regular laws	Search exceptions spatial
Dialoguing with little space systems	Maximally dialoguing with space systems

The classical analysis parameters models of the data represent an average of phenomena, then a summary of the diversity existing in breakdown territorial or "local" with respect to the reference area or to the "global".

The parameters of the classical models of analysis of the data represent an average of phenomena [4–28, 30, 31] has particularly focused on the need for analytical models "local" supplementary to those "global", developing previous research, and defining as equivalent to microscopes in the analysis of phenomena.

Without them reality description would be uniformly monochromatic. Considering a classical regression "global" regression "local" extends the traditional structure of the MRA for entering the variable "ZONE" local rather than global parameters. In fact, the model passes from the global:

$$y_i = \beta_0 + \sum_k \beta_k \cdot x_{ik} + \varepsilon_i, \tag{1}$$

to the local (GWR):

$$y_i = \beta_0(u_i, v_i) + \sum_k \beta_k(u_i, v_i) \cdot x_{ik} + \varepsilon_i. \tag{2}$$

In GWR: (u_i, v_i) are the coordinates in the point i in space; $\beta_k (u_i, v_i)$ is the continuous function $\beta_k (u, v)$ in point i. This leads to a continuous surface of parameterized values, with random measurements denoting the variability in surface space and consequently of the values analyzed. As a first conclusion, the GWR take place from the existence of spatial variations of values and provides the tools to measure them [3].

Not only might the variables in the model exhibit spatial dependence (that is, nearby locations will have similar values) but also the model's residuals might exhibit spatial dependence. The latter characteristic can be observed if the residuals from the basic regression are plotted on a map where commonly the residuals in neighboring spatial units will have a similar magnitude and sign.

The locations at which parameters are estimated are generally the locations at which the observations in the dataset have been collected. This allows predictions to be made for the dependent variable and residuals to be computed. These are necessary in determining the goodness of fit of the model and we shall discuss this below. The locations at which parameters are estimated can also be non-sample points in the study area – perhaps the mesh points of a regular grid, or the locations of observations in a validation dataset which has the same dependent and independent variables as the calibration dataset.

2   GWR Applied to the Case Study

Thanks to the realization of a dedicated real estate Geodatabase and consequent geolocation of the data sample it was possible to perform a GWR using the Geographically Weighted Regression tool contained in the tools box <Spatial Statistics Tools> in ArcGis ® [29] (Fig. 1).

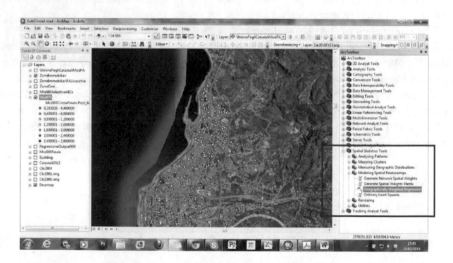

Fig. 1. Capture from the real estate Geodatabase.

The Kernel type used is the ADAPTIVE, i.e. the spatial context (the Gaussian kernel) is a function of a specified number of neighbors. Where feature distribution is dense, the spatial context is smaller; where feature distribution is sparse, the spatial context is larger.

It was choosen the AICc bandwoth method, i.e. the extent of the kernel is determined using the Akaike Information Criterion (AICc) (Fig. 2).

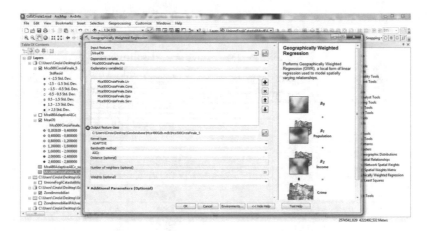

Fig. 2. GWR input data.

A summary of the variables and a descriptive statistics of the real estate sample follows:

- The dependent variable, or explained, used in the model is the <elicited> Market Price of the property (PRZ) expressed in thousands of Euros.
- Net Area (SUP): cardinal progressive variable, referring to the net area.
- Floor level (LIV): discrete variable, understood as the storey above ground level, i.e. the level of strong foundations, where the housing unit is located; floor above ground level (LIV) of the apartments range is between a minimum of 1 to a maximum of 7.
- Maintenance (CONS): ordinal progressive variable, for the case study, maintenance status was considered as a progressive variable with a range of 1 (= insufficient) to 4 (= excellent).
- Construction Age (AGE): progressive cardinal variable, evaluated considering the construction age of the building thanks to a perfect knowledge about the year of construction of the building and by reference to the date shown on the certificate of fitness as human shelter.
- Zone (ZONA): ordinal progressive variable, agreed as the area in which the property is located, including those in which, from the analysis carried out, the spatial reference area was divided, i.e. the municipality of Reggio Calabria. The Zone range goes from 1 (= periphery) to (6 = center).
- Bathrooms (SERV): ordinal variable, agreed as the number of bathrooms in the property. The feature services (SERV) comprises apartments with one service up to a maximum of 4.

- Date of purchase (SALE): progressive cardinal variable, evaluated considering the year in which the sale took place by reference to the date shown on the purchase agreement and counting the years since the sale.

The data described so far are summarized in the following table (Table 2):

Table 2. Descriptive statistics.

Variable	Range	Type of variable: Regressive/Progressive	Mean	Std. deviation	Number of observations
prz	€		122,21	54,42	470
sup	m²	Progr	108,25	29,53	470
liv	1–7	Progr	3,07	1,47	470
cons	4–1	Progr	2,98	,85	470
sale	# years	Regr	11,00	6,55	470
zona	1–6	Regr	2,97	1,56	470
serv	from 1 up	Progr	1,61	,521	470
age	# years	Regr	28,52	18,65	470

3 Output of GWR Model

The GWR tool produces a variety of different outputs such as: Output feature class and a table with the tool execution summary report diagnostic values. The name of this table is automatically generated using the output feature class name and "_supp" suffix. The Output feature class is automatically added to the table of contents with a hot/cold rendering scheme applied to model residuals (Table 3).

Table 3. GWR output for the data sample.

ID	Var. name	Variable	Def
1	Neighbors	78	
2	Residual squares	318.193,54	
3	Effective number	120,00	
4	Sigma	29,73	
5	AICc	4.717,07	
6	R2	0,778	
7	R2 adjusted	0,697	
8	Dependent field	0	Prz
9	Explanatory field	1	Liv
10	Explanatory field	2	Cons
11	Explanatory field	3	Data
12	Explanatory field	4	Epo
13	Explanatory field	5	Serv

As a minimum, GWR produces parameter estimates and their associated standard errors at the regression points. If the regression points are the same as the sample points then GWR produces predictions for the dependent variable (fitted values), residuals and standardized residuals.

Some implementations output local R^2 values and influence statistics based on the hat matrix.

If the regression points are not the same as the sample points, and there are no independent variables available for the regression points, then little else besides parameter estimates and standard errors will be available – fitted values, residuals, and a hat matrix will not be available. If independent variables are available, then fitted values will be available. If there is also a dependent variable present as well, then the whole range of outputs can be created.

4 Conclusion

Spatial heterogeneity is a relevant factor for complex process of real estate appraisal. The development of statistical techniques, such as GWR, is constantly evolving in mass appraisal of real estate properties and in the analysis of real estate markets.

A Geographically Weighted Regression analysis has been implemented for the appraisal data set using an ADAPTIVE Kernel type and an AICc bandwoth method.

The application of the GWR model to the data sample in the Case Study confirmed the validity of the Multiple Linear Regression Analysis for both the estimated values and regarding the goodness of the model. This result may aid for the verification of progressive real estate sub-samples in able to detect and to identify possible potential market premium in real estate exchange and rent markets for green buildings [21–28].

The experimentation also highlights the benefits achieved with the use of the model and the consistency of the results in real estate market conditions. The model also allows an optimal adaptability of real estate market to territory transformation and its related dynamics.

References

1. Fotheringham, A.S.: Spatial structure and distance-decay parameters. Ann. Assoc. Am. Geogr. **71**(3), 425–436 (1981)
2. Fotheringham, A.S.: Geographically weighted regression and multicollinearity: dispelling the myth. J. Geogr. Syst. **18**(4), 303–329 (2016)
3. Manganelli, B., Pontrandolfi, P., Azzato, A., Murgante, B.: Using geographically weighted regression for housing market segmentation. Int. J. Bus. Intell. Data Min. **9**(2), 161–177 (2014)
4. Del Giudice, V., De Paola, P.: Spatial analysis of residential real estate rental market. In: d'Amato, M., Kauko, T. (eds.) Advances in Automated Valuation Modeling. Studies in System, Decision and Control Series, vol. 86, pp. 9455–9459. Springer (2017)
5. Del Giudice, V., De Paola, P., Forte, F.: The appraisal of office towers in bilateral monopoly's market: evidence from application of Newton's physical laws to the Directional Centre of Naples. Int. J. Appl. Eng. Res. **11**(18), 9455–9459 (2016)

6. Del Giudice, V., Manganelli, B., De Paola, P.: Depreciation methods for firm's assets. In: ICCSA 2016, Part III. Lecture Notes in Computer Science, vol. 9788, pp. 214–227. Springer (2016)
7. Manganelli, B., Del Giudice, V., De Paola, P.: Linear programming in a multi-criteria model for real estate appraisal. In: ICCSA 2016, Part I. Lecture Notes in Computer Science, vol. 9786, pp. 182–192. Springer (2016)
8. Del Giudice, V., De Paola, P., Manganelli, B.: Spline smoothing for estimating hedonic housing price models. In: ICCSA 2015, Part III. Lecture Notes in Computer Science, vol. 9157, pp. 210–219. Springer (2015)
9. Del Giudice, V., De Paola, P.: Geoadditive models for property market. Appl. Mech. Mater. Trans. Tech. Publ. **584–586**, 2505–2509 (2014)
10. Del Giudice, V., De Paola, P., Manganelli, B., Forte, F.: The monetary valuation of environmental externalities through the analysis of real estate prices. Sustainability **9**(2), 229 (2017)
11. Del Giudice, V., De Paola, P., Cantisani, G.B.: Rough Set Theory for real estate appraisals: an application to Directional District of Naples. Buildings **7**(1), 12 (2017)
12. Del Giudice, V., De Paola, P., Cantisani, G.B.: Valuation of real estate investments through Fuzzy Logic. Buildings **7**(1), 26 (2017)
13. Del Giudice, V., De Paola, P., Forte, F.: Using genetic algorithms for real estate appraisal. Buildings **7**(2), 31 (2017)
14. Del Giudice, V., Manganelli, B., De Paola, P.: Hedonic analysis of housing sales prices with semiparametric methods. Int. J. Agric. Env. Inf. Syst. **8**(2), 65–77 (2017)
15. Del Giudice, V., De Paola, P., Forte, F., Manganelli, B.: Real estate appraisals with bayesian approach and Markov Chain Hybrid Monte Carlo Method: an application to a central urban area of Naples. Sustainability **9**, 2138 (2017)
16. Morano, P., Locurcio, M., Tajani, F., Guarini, M.R.: Fuzzy logic and coherence control in multi-criteria evaluation of urban redevelopment projects. Int. J. Bus. Intell. Data Min. **10**(1), 73–93 (2015)
17. Guarini, M.R., D'Addabbo, N., Morano, P., Tajani, F.: Multi-criteria analysis in compound decision processes: the AHP and the architectural competition for the chamber of deputies in Rome (Italy). Buildings **7**(2), 38 (2017)
18. Antoniucci, V., Marella, G.: Immigrants and the city: the relevance of immigration on housing price gradient. Buildings **7**, 91 (2017)
19. Antoniucci, V., Marella, G.: Small town resilience: housing market crisis and urban density in Italy. Land Use Policy **59**, 580–588 (2016)
20. Saaty, T.L., De Paola, P.: Rethinking design and urban planning for the cities of the future. Buildings **7**, 76 (2017)
21. Del Giudice, V., Massimo, D.E., De Paola, P., Forte, F., Musolino, M., Malerba, A.: Post carbon city and real estate market: testing the dataset of Reggio Calabria market using spline smoothing semiparametric method. In: Calabrò, F., Della Spina, L., Bevilacqua, C. (eds.) Local Knowledge and Innovation Dynamics Towards Territory Attractiveness Through the Implementation of Horizon/E2020/Agenda2030. Springer, Berlin (2018). ISBN 978-3-319-92098-6
22. De Paola, P., Del Giudice, V., Massimo, D.E., Forte, F., Musolino, M., Malerba, A.: Isovalore maps for the spatial analysis of real estate market: a case study for a central urban area of Reggio Calabria, Italy. In: Calabrò, F., Della Spina, L., Bevilacqua, C. (eds.) Local Knowledge and Innovation Dynamics Towards Territory Attractiveness Through the Implementation of Horizon/E2020/Agenda2030. Springer, Berlin (2018). ISBN 978-3-319-92098-6

23. Massimo, D.E.: Green building: characteristics, energy implications and environmental impacts. Case study in Reggio Calabria, Italy. In: Coleman-Sanders, M. (ed.) Green Building and Phase Change Materials: Characteristics, Energy Implications and Environmental Impacts, pp. 71–101. Nova Science Publishers (2015)
24. Massimo, D.E.: Valuation of urban sustainability and building energy efficiency: a case study. Int. J. Sustain. Dev. **12**(2–4), 223–247 (2009)
25. Massimo, D.E.: Emerging Issues in Real Estate Appraisal: Market Premium for Building Sustainability. Aestimum, pp. 653–673 (2013). ISSN 1724-2118. http://www.fupress.net/index.php/ceset/article/view/13171. Accessed 21 Sept 2017
26. Massimo, D.E., Musolino, M., Barbalace, A.: Stima degli effetti di localizzazioni universitarie sui prezzi immobiliari. In: Marone, E. (ed.) La valutazione degli investimenti pubblici per le politiche strutturali. Firenze University Press, Firenze (2011)
27. Massimo, D.E.: Stima del green premium per la sostenibilità architettonica mediante Market Comparison Approach, Valori e Valutazioni (2010)
28. Spampinato, G., Massimo, D.E., Musarella, C.M., De Paola, P., Malerba, A., Musolino, M.: Carbon sequestration by Cork Oak forests and raw material to built up post carbon city. In: Calabrò, F., Della Spina, L., Bevilacqua, C. (eds.) Local Knowledge and Innovation Dynamics Towards Territory Attractiveness Through the Implementation of Horizon/E2020/Agenda2030. Springer, Berlin (2018). ISBN 978-3-319-92098-6
29. Malerba, A., Massimo, D.E., Musolino, M., De Paola, P., Nicoletti, F.: Post Carbon City: building valuation and energy performance simulation programs. In: Calabrò, F., Della Spina, L., Bevilacqua, C.: Local Knowledge and Innovation Dynamics Towards Territory Attractiveness Through the Implementation of Horizon/E2020/Agenda2030. Springer, Berlin (2018). ISBN 978-3-319-92098-6
30. ESRI 2011: ArcGIS Desktop: Release 10. Environmental Systems Research Institute, Redlands, CA (2011)
31. De Ruggiero, M., Forestiero, G., Manganelli, B., Salvo, F.: Buildings energy performance in a market comparison approach. Buildings **7**, 16 (2017)
32. Ciuna, M., Milazzo, L., Salvo, F.: A mass appraisal model based on market segment parameters. Buildings **7**, 34 (2017)

The Challenge of Augmented City: New Opportunities of Common Spaces

Pierfrancesco Celani$^{(\boxtimes)}$, Roberta Falcone, and Erminia d'Alessandro

Università della Calabria Arcavacata di Rende, Rende 87036, Italy
pierfrancesco.celani@unical.it

Abstract. Infrastructures represent the element able to reunite territories, giving back their identity. In this paper, we propose the results of the research RES NOVAE (Reti, Edifici, Strade, Nuovi Obiettivi Virtuosi per l'Ambiente e l'Energia), made by University of Calabria and other public/private partners with the aim of defining an upgrade of the meaning of infrastructures. The project describes the realization of Cosenza Augmented city, integrating the energy smart grid, focus of the technological research, with a more complex social infrastructure. In this new model of city, citizens are the main users of urban intelligence. The social infrastructure is built on points and connections that weave projects and cities. Some points are physical, others are virtual. These new elements will allow the transformation from a classic Smart City to a model based on human dimension: Open Source city.

Keywords: Augmented city · Network · Common space

1 New Challenges for Cities

1.1 From Contemporary to Smart

Nowadays, smart city is the new paradigm of the city that is moving towards the future, in terms of environmental, social and economic sustainability. It is the city that becomes smart by using constantly evolving technologies, providing high value services. But we need to keep in mind that cities are places where people live, work, spend their free time and where they create relationships and opportunities. Historically, cities are the place where people can cooperate, live, be able to realize their projects and express their attitudes. Therefore, a city becomes smart only when the transformation starts from the community that animates it (smart people).

Community is the crucial node for the city and its future because true changes can happen only if they start from people. This is why the matrix of development of smart cities is a cultural and ethical matrix; technology is only a "tool" for better sharing resources, increasing energy efficiency and environmental quality, raising the cultural level, creating communities around common objectives [1]. It is not possible to neglect the role that education plays in the cultural growth of communities, that, today more than ever, have to share and express a conscious opinion on the smart governance guidelines. In fact, satisfaction of each individual living in a welcoming environment and its creative expression have to be the first concerns of smart cities' administrators.

© Springer International Publishing AG, part of Springer Nature 2019
F. Calabrò et al. (Eds.): ISHT 2018, SIST 100, pp. 150–156, 2019.
https://doi.org/10.1007/978-3-319-92099-3_18

A Smart city is a place where a new humanism of technology is realized, it is a vital organism in which each part is related to the others, that does not consume resources that should be saved for future generations, that does not claim to dominate the environment but that uses technology to increase energy efficiency, to raise cultural level and to share resources [2].

1.2 Urban Infrastructural System: Physical and Intangible Networks

Smart city is based on physical and immaterial networks that, by interweaving and overlapping, ensure the survival of the urban system. Material networks, as infrastructures, are those that characterize the development of cities: mobility networks, energy networks, lifelines etc. [3].

These networks, that have historically articulated and structured urban systems, are connected to others that are now indispensable for the city: the socio-anthropic component. New technologies allow the creation of new networks that can be used by people to share information, participate in specific issues, express consent etc. There is a clear distinction between the different types of networks: infrastructural networks belong to the physical subsystem of the city and define the circulation that distributes the flows and ensures the survival of the socio-anthropic component and of the entire city. Then there are immaterial networks: relationships between citizens, stakeholders and city users, relationships between social groups, innovative relationships created according to the social medias that dematerialize and virtualize behavior and functions of the city [4].

The goal of our generation should be finding the best interconnection and integration between these networks aiming to generate a structure for the urban system that changes and renews itself incessantly to ensure the city a sustainable evolution. In its physical form, the metropolis of tomorrow may not be too different from what we see today but it will increasingly become an extraordinary theater for new relational dynamics. We will not live the environment of the future as a new *urbs* (the physical space of the city) but, above all, as *civitas*, a supportive and participatory community of a common future. A community that, thanks to the networks and its interconnections, will be open and inclusive [5].

2 From Smart City to Augmented City

The classical vision of Smart City is usually based on an approach that prioritizes the influence that technology has in the management of the city. This can be considered a significant advantage when the whole technological infrastructure is mature. A fully developed Smart City scheme applies a similar logic to all the functional elements of the city (transportation networks, energy distribution, waste management, air and water quality monitoring), allowing an integrated control of the urban system. Moreover, the information provided by sensors can be used on mobile devices (especially when the data is open access) allowing the user's personalization of the city's services [6]. Smart city is not only this; there is an increasing attention to everything that is related to the fulfillment of the real needs of the citizens, that have to actively participate as urban actors, leading the administration from being a provider to become an enabler (facilitator and development promoter) [7]. Before being a set of technological items, smart city, in fact, tries to respond to emerging social needs on a urban scale and it is an expression of a new generation of innovative policies that already invest different levels of local government; we are talking about a city that becomes more competitive on a global level when culture, knowledge sharing, learning skills and innovation will play a strategic role [8].

Recent experiences have brought out a new vision that upgrades the original concept of smart city in a more humanistic perspective, gained thanks to the adoption of participatory and citizen-centered approaches in the co-design and development of services. In fact, it is becoming clear that the smartness of sensors, apparatuses and infrastructures sometimes leaves citizens out of the process.

This is the starting point for a new paradigm for the city: Augmented City; a city where citizens and communities are the main actors of urban intelligence. They do not have to use technologies that have been selected only by local administrations; rather, they are encouraged to create and co-design their services using the available technologies. An Augmented City adopts services that arise from the real needs of people in the city. Augmented City is a spatial/cultural/social/economic device designed to improve contemporary urban life, both individual and collective, informal and institutional, generating well-being and happiness [9]. Augmented city is a place capable of improving the lives of those who live it every day, and those who are only passing by, establishing a new way of perceiving spaces and creating new citizenship models in which every user becomes active and aware participant in the processes that determines life in the city.

It is important to underline that the creation of an Augmented City can be carried out by using simple technologies and does not always require sophisticated and complex infrastructures. This is essentially relevant to the scalability of the solution; simple and creative solutions can rise from local communities and allow big cities to extend their strategies or small cities to integrate new ones. This is an important advantage for city administrations that can exploit the potential to enable the creation of intelligent services without significant investments. The main requirement to foster innovation and competitiveness is to give importance to human dimension, assuming the active role of leadership networks as social capital, able to select and guide technological hardware, to structure and interpret learning, favoring processes of innovation really effective because citizen-driven, as well as human-centered [8]. It is therefore necessary to recreate those social mechanisms that belong to the urban tradition, with trusting urban communities, high quality of life, sense of security and belonging. Methods and tools such as Urban Living Lab, Design Thinking and Social Gaming supported by online platforms foster relationships among residents, pushing them to work in partnership to create, innovate and implement local services based on real needs [7].

3   RES NOVAE: The Platform for Cosenza Augmented City

In 2012, the Ministry of Education and Research (MIUR) through the National Operational Program Research and Comprehension (PON R & C) 2007-2013 published a call for proposals on the topic £Smart Cities and Communities and Social Innovation" which financed projects and proposals in the following topics:

- Integrated Action for the Information Society;
- Integrated Action for Sustainable Development.

On the second area and in particular on Renewable energy and smart grid intervention line, the RES NOVAE (Reti, Edifici, Strade - Nuovi Obiettivi Virtuosi per l'Ambiente e l'Energia)[1] project was presented and financed. The project supports the realization of pilot demonstrators in the city of Bari and Cosenza, aiming at the creation of efficient and sustainable environment for citizens. The added value of the RES NOVAE project, compared to other numerous smart projects presented, strongly unbalanced on the smart grids and ICT aspects, lies in the creation of a platform, a social infrastructure intended as a network of nodes (innovative public spaces) and connections (new technologies), which creates the necessary link between the experiments carried out in the project and the city. The nodes are both part physical places, and part virtual places, in which the community learns and creates. The main nodes of this network are: Urban Lab CreaCosenza, Smart Street and 3D City Model.

3.1   Urban Lab "CreaCosenza": A Living Lab for Cosenza

UrbanLab CreaCosenza aims to involve the city and its citizens in the daily use of energy technologies, showing how all the technologies and communication and control systems developed during the research activity can be integrated with each other and can interact with other projects (ongoing or planned) of intelligent and sustainable use of the city. The Urban Lab allows people to know more about the RES NOVAE project, demonstrating and disseminating research results, with methods, techniques and languages that simplify communication, attract users' interest and encourage energy saving and sustainable use of resources. It is a living lab, a physical place that can be clearly identified in the urban context which, thanks to technology, overcomes its physical limits, opens up to the city and interacts with it, welcoming participation, collaboration and sharing. The living lab of Cosenza wants to be the dynamic tool of the active participation of the community and all others involved in the creation of the smartness of the city. With this choice, we wanted to affirm the overcoming of the traditional smart grids approach and the establishment of networks of subjects (public and private). We also want to emphasize the direct involvement of the population, which is stimulated to become a virtuous protagonist in the strategic sectors of urban life [10].

---

[1] Partners of the project RES NOVAE are: Enel Distribuzione, IBM Italia, General Electric Transportation System, Università della Calabria, Politecnico di Bari, Consiglio Nazionale delle Ricerche, Enea e Datamanagement PA.

The main aim pursued by the project, that represents the gap between more conventional smart city approach and ours, concerns the possibility of creating social cohesion and stimulating processes of social awareness and self-organization. The Urban Lab "CreaCosenza" was born for spreading awareness on strategic issues of urban life, but at the same time has the task of facilitating active participation of all possible stakeholders in user-driven innovation processes. This aspect of the project, initially marginal (relegated to the mere fulfillment of the obligations of dissemination and communication), during implementation has assumed a bigger role, determining concrete effects also in terms of regeneration of the existing city, with spatial and architectural quality interventions.

This second goal, a direct consequence of the first one, become clearer with the transformation of the building hosting the Urban Lab. The new use has reinserted the building in the city life, triggering a further process of physical requalification of the square behind it, downgraded for years to illegal parking and finally given back to the city as a public space.

3.2 The Smart Street Corso Mazzini

Another physical node of the Augmented City social infrastructure is the Smart Street, a smart urban location equipped with sensors able to interact with each other, the surrounding environment and the citizen through a cloud computing platform.

Along the Smart Street Corso Mazzini are located a series of sensors/actuators (smart objects) able to connect and cooperate with each other processing input from the surrounding environment.

The system of Smart Street is based on the Rainbow platform, which consists in a cloud data store, a series of servers and the sensor network. The data collected by the smart objects are sent to the servers that process them and transmit them to the cloud in real time, generating an extensive data repository (big data) accessible through public web platforms. The information collected (and returned) by the smart objects are many (and many others can be added at any time, in an extremely simple way) and can concern: air quality, temperature, humidity, noise, traffic, just to cite some examples. They can also provide information on different types of points of interest, for example on the statues of the Bilotti Open Air Museum (MAB).

3.3 The 3D City Model of Cosenza

The third node, the 3D city model of the city of Cosenza, is a virtual place, a three-dimensional and georeferenced model of the city, which through augmented experiences and virtual reality will allow anyone to interact with the city, consulting databases (acquired by sensors or available online), and integrating the system with new information.

In the RES NOVAE project, the creation of the 3d model represents the tool that makes innovative methodologies of participatory urban planning possible (Fig. 1).

The 3D City Model is, therefore, the tool for the observation of the complex levels of information available today for a heterogeneous and dynamic object such as the urban environment; starting from this, it is possible to add the data coming from sensor

Fig. 1. Application of augmented reality with 3D City Model. Source: ETT S.p.A., http://www. ettsolutions.com

systems (Smart Object) and informative databases available, as well as the information that citizens give, observing the reality directly or interacting with the model itself [11]. The integrated management of heterogeneous information at the urban level will allow the creation and dissemination of applications, which could benefit from a shared database containing detailed information on buildings and other entities relevant to urban planning [12].

4 Augmented City as an Open Source City

The augmented city is built through the assiduous experimentation of decoding processes and interpretative cooperation; the open source city is the city of the narrators, who experiment new geographies through hidden, forgotten and abandoned plots, it is the city of discovery through impetuous practices, born from the citizens [13].

This is why Open source culture is important. Open source changes the way of doing urban planning, making it an informal modality progressively improvable, so citizens are allowed to interact and propose continuous changes in the structure of their city through informal hacking actions and modification of the identity code of the urban system. It is necessary to create a civic platform that facilitates participatory and collaborative processes, activating a new system of network thinking based on a widespread social creativity. The open and transparent city is the one that frees the information and resources, focusing on their characters of common good [14].

Open knowledge is the prerequisite for collective intelligence, that makes it possible to realize the main advantage of openness: increasing the ability to control, explore and combine different databases and develop new products and services. An open source city will not be the one that adds technology and efficiency to its traditional organism, but it will have to contribute to the increasing rate of collective intelligence, sustaining virtuous behavior from below and giving visibility to the individual and collective advantages of a new way of thinking and planning [15].

References

1. D'Alessandro, E., Celani, P.: UrbanLab CreaCosenza. Un living lab per smart city di Cosenza. Urbanistica Informazioni **263**(VII), 14–18 (2015)
2. Riva Sanseverino, E.: L'umanesimo della tecnologia nelle comunità intelligenti. In Città intelligenti per comunità intelligenti. Smart city tra tecnologia, cultura, cittadinanza e partecipazione, pp. X–XI. Energia Media srl, Milano (2015). https://issuu.com/energiamedia/docs/citt___intelligenti_per_comunit___i?e=14848929/15268709 Accessed 12 Sept 2017
3. Fistola, R.: La "non-city" e il disegno delle reti urbane. Urbanistica Informazioni **263**(IX), 14–18 (2015)
4. Fistola, R., La Rocca, R.A.: The virtualization of urban functions. NETCOM Netw. Commun. Stud. **15**(1–2), 39–48 (2001)
5. Ratti, C.: Un po' meno urbs un po' più civitas: una dimensione sociale e socievole, La Lettura, Supplemento culturale del Corriere della Sera, vol. 27 (2016)
6. Oliveira, A., Campolargo, M.: From smart cities to human smart cities. In: Proceedings of the 48th Annual Hawaii International Conference on System Sciences, HICSS 2015, pp. 2336–2344 (2015)
7. Pultrone, G.: Partecipazione e governance per smart cities più umane. Tema, J. Land Use Mobility Environ. **7**(2), 159–172 (2014)
8. Campbell, T.: Beyond Smart Cities. How Cities Network, Learn and Innovate. Earthscan, New York (2012)
9. Carta, M.: The Augmented City: A Paradigm Shift. LISt Lab, Trento (2017)
10. Zupi, M.: L'esperienza del PON RES NOVAE a Cosenza: dalla smart city alla città resiliente. In: Arena, M. (ed.) atti del Convegno, Città metropolitane e resilienti. Messina progetta il futuro, 22 giugno 2015, Messina (2015)
11. Borga, G.: City Sensing. Approcci, metodi e tecnologie innovative per la Città Intelligente, Franco Angeli, Milano (2013)
12. Agugiaro, G.: I modelli digitali 3D di città come hub informativo per simulazioni, energetiche a scala urbana. In: XVIII Conferenza Nazionale ASITA, Firenze, 14–16 Ottobre 2014. http://atti.asita.it/ASITA2014/Pdf/019.pdf Accessed 12 Sept 2017
13. Vitellio, I.: La Città Open Source, Urbanistica Dossier Online 6, 11–15 (2014). http://www.urbanisticainformazioni.it/IMG/pdf/ud006.pdf. Accessed 12 Sept 2017
14. Infante, C., Massaro, S.: Performing Media per l'Urban Experience. La via ludico-partecipativa alla cittadinanza educative, Urbanistica Dossier Online 6, 20–24 (2014). http://www.urbanisticainformazioni.it/IMG/pdf/ud006.pdf. Accessed 12 Sept 2017
15. Carta, M.: Reimagining Urbanism. Creative, Smart and Green Cities for the Changing Times. LISt Lab, Trento (2013)

Multiple-Benefits from Buildings' Refurbishment: Evidence from Smart City Projects in Europe

Stefano Zambotti[✉], Simon Pezzutto, and Adriano Bisello

Institute for Renewable Energy, EURAC Research, 39100 Bolzano, Italy
stefano.zambotti@eurac.edu

Abstract. Given the necessity of strengthening the transition towards a smarter, more sustainable low-carbon future, Smart Cities are considered a powerful tool. However, Smart City projects involving the refurbishment of existing buildings carry key barriers to implementation. The most prominent ones are: (i) a wide time discrepancy between appreciable environmental and economic benefits and immediate costs of action and (ii) economic benefits that might not accrue to who bears the cost of the intervention. This research provides a clue to solving this impasse based on the concept of multiple-benefits evaluation stemming from a shift in perspective from mitigation costs to development opportunities. We considered the costs of interventions on the European building stock under the Smart City projects to assess the multiple-benefits delivered to society. Starting from the monetary aspects of single projects, we identified multipliers to assess three different types of multiple-benefits: (i) Energy savings; (ii) Health and well-being; and iii.) Employment. Our findings indicate that in a time span of 14 years (2005–2018), an amount of about 260 million Euros invested in such projects lead to: (i) an accumulated saving potential of approximately 40 kilotons of oil equivalent, corresponding to 465 GWh; (ii) a reduction in air pollution corresponding to a value of 3 million Euros in avoided costs; and (iii) the creation of around 1,000 jobs with an average duration of 5 years. Considering that most of such investments occurred during the latest economic recession, the impact of the aforementioned multi-benefits appears to be not negligible.

Keywords: Multiple-benefits · Smart city projects · Deep energy retrofits
European building stock

1 Introduction

The European Union (EU) is facing unprecedented challenges related to climate, energy, social and economic aspects, and has therefore set specific goals for 2020, 2030 and 2050 [1]. Cities are recognized as pivotal players for the development of a smart, sustainable, and low-carbon economy [2]. One side of the coin identifies cities as key elements of social and economic innovation, as milieus where consumers, workers, and businesses are concentrated, delivering about 67% of EU's gross domestic product (GDP) [3]. The other side points out how poverty, segregation, energy consumption, and pollutant emissions often manifest themselves there [4].

© Springer International Publishing AG, part of Springer Nature 2019
F. Calabrò et al. (Eds.): ISHT 2018, SIST 100, pp. 157–164, 2019.
https://doi.org/10.1007/978-3-319-92099-3_19

There is necessity for a rapid transition toward more efficient and sustainable urban settlements [5]. Smart Cities (SC), when properly implemented, represent the current way of creating more livable areas, which are both sustainable and energy efficient [6, 7].

Energy systems are a major domain of intervention of SC projects. The goal of SC projects is a transition towards self-sufficient, sustainable, and resilient energy systems. Moreover, these interventions aim at the optimization of the integration of energy conservation, energy efficiency, and local energy sources. Integration is mainly achieved exploiting information and communication technologies [6, 8].

A major area of intervention of SC projects concerns the refurbishment of existing buildings [9]. Unfortunately, there exist two main key barriers to the refurbishment of buildings, they are i.) the wide time discrepancy between the environmental and economic benefits resulting from the refurbishment and the immediate cost of actions [10]; and also ii.) the split incentive, which is the situation where the owner of the building pays for the refurbishment but the tenant is the person that accrues the benefits [11].

Trying to solve this impasse, we analyze the concept of multiple-benefits, to shift the perspective from mitigation costs to development opportunities [12] and to highlight other socio-economic advantages [13]. Multiple-benefits indicate the broad range of positive spillovers, energy as well as non-energy related, without prioritizing them. Moreover, multiple-benefits express a holistic balance among the various aims addressed by a project [14, 15].

2   Materials and Methods

In this research, we considered the costs of the interventions carried out for refurbishments of the European building stock under SC projects. To roughly estimate the share of SC projects' total investments concerning interventions on buildings we had to consider the different sectors that each project addressed. What we found is that SC projects have three main sectors of interest. These are mobility, energy networks and infrastructure, and buildings. Once we identified the number of sectors that each project addressed (1, 2 or 3), we divided the total amount allocated for that specific project by that number. This way, we were able to roughly estimate the share of the total amount of euros spent on buildings.

Our attention focused on projects started in 2005 or afterwards, projects that have already been concluded or whose year of completion starts in 2018. We tried to assess a set of multiple-benefits that these interventions deliver to society at large. In the present work we analyzed available documentation, provided especially by the European Commission (EC), on SC projects financed within the Sixth and Seventh Framework Programme (FP6 - 2002 until 2006 and FP7 - 2007 until 2013) "CONCERTO initiative" [16] as well as the FP7 "Smart Cities & Communities" activities [17].

Starting from the share of funds spent for refurbishment activities on existing buildings, multipliers (e.g. number of jobs created by million Euros invested) estimated using the ratios provided by scientific sources (in particular [18] and [19]) were applied to generate assumptions for three different types of multiple-benefits assessed: (i) Energy savings, (ii) Health and well-being, and (iii) Employment.

The rebound effect, that is the potential cancelling out of benefits stemming from increased efficiency due to changings in people's behavior, was not taken in consideration in this study.

(i) Energy savings

Energy savings through reduced energy consumption is a direct consequence stemming from increased energy efficiency. In the following, energy savings have been quantified by units of kilotons of oil equivalent (ktoe) [19]. Energy savings are the source of three primary benefits, namely cost savings, climate change mitigation and energy security. It is generally shown that interventions aimed at reducing energy consumption have significant cost-effective potential, even without including climate change mitigation or energy security improvements in the calculation [20]. The calculation method and the proportions used to estimate the energy savings were retrieved from a Fraunhofer's report [21]. For this estimation Fraunhofer considered the specific building stock of all EU Member States considering age, climatic zones, buildings' energetic standards and countries' energy demands. Information on material cost, labor cost, costs for different sorts of refurbishment was also considered. Energy savings have important consequences in terms of CO_2 emissions reduction, contributing significantly to the fight against global warming [22]. Furthermore, energy savings have the potential to improve energy security at national and EU level [19, 23].

(ii) Health and well-being

A more indirect benefit occurs through health benefits. Most energy renovation measures will improve the indoor climate, and by doing so health benefits can be obtained through fewer diseases, reduced mortality, improved worker productivity, and improved overall quality of life. While most of these benefits accrue to society in general, public budgets may also be improved through fewer hospital expenses and fewer sick days. Health benefits also occur as power and heat production from power plants, district heating (DH) plants and local heating is reduced. Power and heat generated in these facilities give rise to air pollution and particularly to dangerous chemical compounds such as nitrogen oxide (NOx), sulphur dioxide (SO_2), small particle matters and carbon dioxide (CO_2). By reducing energy consumption, air pollution can be reduced. The health impact of air pollution from different inputs is well defined [24]. In the following, we therefore calculate the economic value of reduced air pollution stemming from the refurbishment of buildings under the considered SC projects using the ratios provided by the Copenhagen Economics [19].

(iii) Employment

The construction sector is considered a significant source of low- and high-skilled jobs [25]. Given the current slow recovery from the global economic downturn, investments concerning buildings' refurbishment can increase economic activity, and improve public budgets by reducing unemployment costs and increasing tax revenue from the raised economic activity. The direct impact on the local labor market is mainly related to the implementation phase, and it widely varies due to some characteristics.

One is the size of the intervention: how many buildings are refurbished or how large is the demonstration site. The second concerns the different approaches of physical intervention on buildings [18]. In the following, we calculate the number of jobs stemming from the refurbishment of buildings under the considered SC projects using the ratios provided by The Energy Efficiency Industrial Forum [26].

3   Results

Nearly one-third of the identified SC projects result to have carried out buildings' refurbishment activities. Considering only the interventions on buildings, the CON-CERTO projects account for approximately 65% of the investments while the remaining 35% concerns the FP7 Smart Cities & Communities projects. These activities of energy renovations of buildings are performed in nearly one-hundred locations (i.e. cities, provinces, regions). Specific areas have been home to several SC projects - e.g. Amsterdam for nearly 30 times. Figure 1 shows these locations across Europe. The 51% of the capital invested in these projects has been provided by the EU and the remaining part by private investors.

Fig. 1. SC Projects locations in Europe in which building's refurbishment took place (indicated by dots on Europe's map) [16, 17].

Figure 2 shows an estimation of the amount of euros invested per year in Europe to carry out refurbishment interventions on buildings under SC projects from 2005 until 2018 (2018 corresponds to the last year FP7 spendigs are provided for SC activities on buildings' refurbishment).

In total, we estimated that more than 260 mil. € have been invested for buildings' refurbishments under SC projects, within the time period indicated above. A peak can be seen in 2009 with around 32 mil. € and the lowest value is reported for 2012 with

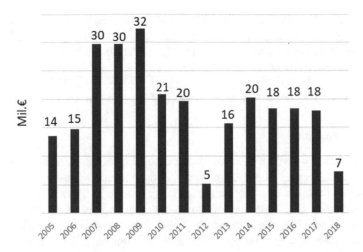

Fig. 2. Amount of mil. € invested per year (2005–2018) to carry out SC projects in Europe (values indicated in this figure and next Figs. 3, 4 and 5 have been rounded) [16, 17].

5 mil. €. The year 2012 separates the two main clusters of investments (2005–2011, 2012–2018), the one related to the "CONCERTO initiative" and the one concerning the "Smart Cities & Communities" projects. The average amount spent per year is about 19 mil. €. Starting from the expenditures per year at European level, the following three sections measure the multiple-benefits resulting from buildings' refurbishments, with regard to energy savings (Sect. 3.1), health benefits (Sect. 3.2), as well as employment (Sect. 3.3).

3.1 Energy Savings

Figure 3 displays the amount of ktoe saved by SC refurbishment activities on buildings. The peak of about 40 ktoe (equivalent to 465 GWh) saved in 2018 is the result of

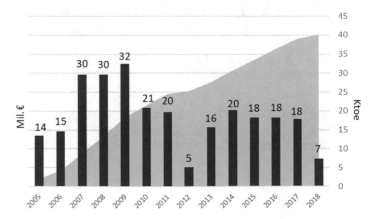

Fig. 3. Energy savings in ktoe reached from 2005 until 2018 (orange) based on SC expenditures for buildings' refurbishment in the same period (blue) [16, 17, 19].

energy saving potentials accumulated until date. Taking advantage of the calculations carried out by Fraunhofer [21] we can assume that the monetary value of such accumulated energy saving potential resulting from refurbishment activities is about 40 mil. €. against total investments of about 264 mil. €.

The increase of ktoe saved per year grows rapidly from 2005 onwards with a complication in 2012 where the least amount of funding provided is shown.

3.2   Health Benefits

Figure 4 visualizes the value of reduced air pollution in mil. €, generated by buildings' refurbishment performed under SC projects. A drop in energy consumption generally corresponds to a drop in air pollution since energy production from conventional power plants is reduced. Our estimation follows the approach used by the Copenhagen Economics [19] in which the reduced air pollution is considered as avoided costs from other investment measures that the EU would have otherwise needed to undertake to reach its goals of pollution reduction. The peak of more than 3 mil. € reached in 2018 is the result of the spending accumulated until date.

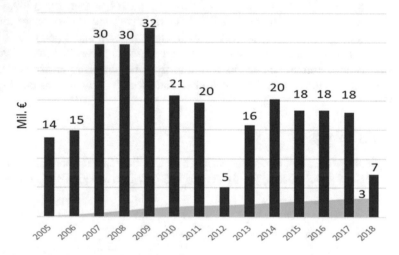

Fig. 4. Value of reduced air pollution in mil. € achieved from 2005 until 2018 (orange) based on SC expenditures for buildings' refurbishment in the same period (blue) [16, 17, 19].

As it was already the case for Fig. 3, also in Fig. 4 the increase for the value of reduced air pollution per year grows from 2005 onwards with a complication in 2012, were the least amount of funding provided is shown. Afterwards it rises continuously until 2018.

3.3   Employment

Using the multiplier found in the literature [18, 27], that is 19 jobs with an annual duration every million spent in energy efficient buildings' refurbishments, we estimated that the SC projects that we took in consideration generated about 33 jobs each, with a duration equal to the length of the projects that in most cases last 5 years. In total, around 1000 such jobs have been created under the CONCERTO and SSC projects that we considered, about 634 associated with the former and 367 with the latter.

4   Conclusions

Considering that these multiple-benefits occur mainly during times characterized by the deepest economic recession since the 1930s in Europe [28], the impact of refurbishments of buildings under SC projects at European level appears to be relevant. To further increase stakeholders' acceptance and political commitments towards SC projects, it is crucial to make the public, as well as policy makers more aware about the multiple benefits arising from such interventions. Moreover, linking economic values to the multiple benefits can positively change the overall economic figure of these interventions. We consider our results as a useful starting point for future research.

However, we also believe them not to be very accurate as they are the outcome of rough estimations based on large aggregate values.

The scope of our research was limited to the assessment of only few multiple-benefits that stem from the refurbishment of the European Building Stock. However, the literature on multiple-benefits [13, 29] present a vast set of positive spillovers arising from the transition towards more sustainable energy systems, such as (i) enhanced energy access and affordability; (ii) provision of ecosystem services; (iii) improved energy security; and (iv) other macroeconomic benefits that we didn't consider. We believe there is a gap to be filled within this area of research so as to shift the perspective from mitigation costs to the very attractive development opportunities stemming from the refurbishment of buildings.

References

1. EC. https://ec.europa.eu/clima/policies/strategies_en. Accessed 28 Nov 2017
2. EC. http://eur-lex.europa.eu/legal-content/EN/ALL/?uri=celex:52014DC0015. Accessed 28 Nov 2017
3. Ricci, L., Macchi, S.: Città e cooperazione allo sviluppo: permanenze e novità delle politiche UE per il post 2015. In: urban@it Centro nazionale di studi per le politiche Urbane, Rapporto Sulle Città 2015. Metropoli Attraverso La Crisi (2015)
4. EC. http://eur-lex.europa.eu/legal-content/EN/TXT/?uri=celex:52014DC0490. Accessed 29 Nov 2017
5. Droege, P.: Urban Energy Transition from Fossil Fuels to Renewable Power, 1st edn. Elsevier Science, Newcastle (2008)
6. Mosannenzadeh, F., Bisello, A., Vaccaro, R., D'Alonzo, V., Hunter, G.W., Vettorato, D.: Smart energy city development: a story told by urban planners. Cities **64**, 54–65 (2017)

7. Cugurullo, F.: Exposing smart cities and eco-cities: Frankenstein urbanism and the sustainability challenges of the experimental city. Environ. Plann. A **50**, 73–92 (2017)
8. Pezzutto, S., Fazeli, R., De Felice, M.: Smart city projects implementation in Europe: assessment of barriers and drivers. Int. J. Contemp. Energy **2**(2), 46–55 (2016)
9. FP7 SINFONIA. http://www.sinfonia-smartcities.eu/en/knowledge-center/d21–swot-analysis-report-of-the-refined-conceptbaseline. Accessed 29 Nov 2017
10. Mayrhofer, J.P., Gupta, J.: The science and politics of co-benefits in climate policy. Environ. Sci. Policy **57**, 22–30 (2016)
11. Economidou, M.: Overcoming the split incentive barrier in the building sector. Workshop Summary (2014)
12. Davis, D., et al.: Ancillary benefits and costs of greenhouse gas mitigation: an overview. In: OECD, Proceedings of an IPCC Co-Sponsored Workshop, pp. 273–274. OECD Publishing (2016)
13. Bisello, A., Grilli, G., Balest, J., Stellin, G., Ciolli, M.: Co-benefits of smart and sustainable energy district projects: an overview on economic assessment methodologies. In: Green Energy and Technology, pp. 127–164 (2017)
14. BERKELEY LAB. https://pubarchive.lbl.gov/islandora/object/ir%3A158591. Accessed 30 Nov 2017
15. IEA, Capturing the Multiple Benefits of Energy Efficiency: A Guide to Quantifying the Value Added. OECD/IEA, Paris (2014)
16. EC. https://www.concertoplus.eu/. Accessed 08 Dec 2017
17. EC. https://ec.europa.eu/inea/en/horizon-2020/smart-cities-communities. Accessed 30 Nov 2017
18. EEIF. https://euroace.org/wp-content/uploads/2016/10/2012-How-Many-Jobs.pdf. Accessed 04 Dec 2017
19. Copenhagen Economics. https://www.copenhageneconomics.com/publications/publication/multiple-benefits-of-investing-in-energy-efficient-renovation-of-buildings. Accessed 05 Dec 2017
20. Sorrell, S., et al.: The Economics of Energy Efficiency. Edward Elgar, Northampton (2004)
21. Fraunhofer ISI et al.: Study on The Energy Savings Potential In EU Member States, Candidate Countries and EEA Countries (2009)
22. EC, Energy efficiency in public and residential buildings Final Report Work Package 8 (2015)
23. EU Energy Security Strategy.: Communication From The Commission To The European Parliament And The Council (2014)
24. Becchio, C., et al.: Evaluating health benefits of urban energy retrofitting: an application for the city of Turin. In: Green Energy and Technology, forthcoming (2018)
25. Assessing the Employment and Social Impact of Energy Efficiency Final report vol. 1, Main report. Cambridge Econometrics (2015)
26. Janssen, R.: How Many Jobs? A Survey of the Employment Effects of Investment in Energy Efficiency of Buildings. The Energy Efficiency Industrial Forum (2012)
27. Casey, J.B.: Energy Efficiency Job Creation: Real World Experiences (2012)
28. EC. http://ec.europa.eu/economy_finance/publications/pages/publication15887_en.pdf. Accessed 07 Dec 2017
29. Vorsatz, Ü., et al.: Measuring the Co-Benefits of Climate Change Mitigation (2014)

The Portuguese Coastal Way and Maritime Heritage

An Outstanding Debt with the New Technologies

Lucrezia Lopez(✉)⬤, María de los Ángeles Piñero Antelo⬤,
and Inês Gusman⬤

University of Santiago de Compostela, 15782 Santiago de Compostela, Spain
lucrezia.lopez@usc.es

Abstract. This diagnostic study explores the use of Information and Communications Technologies (ICTs) to highlight the visibility of the maritime heritage as a complementary product of The Way of St. James. In order to achieve this objective, a case study was carried out on two stages of The Portuguese Coastal Way, Baiona and Combarro, in the province of Pontevedra (Galicia, Spain). Three types of sources were used to analyse the suitability of the ICTs and their content to promote the maritime cultural heritage. The identified problems were synthesized into three diagnostics. The results show that despite the progress made by the responsible institutions towards using ICTs to promote heritage, there are dysfunctions that hinder the potential of these tools for economic and cultural valorisation of Galician maritime cultural heritage.

Keywords: Maritime cultural heritage · ICT · The Coastal Way of St. James

1 Introduction

Information and Communications Technologies (ICTs) are becoming useful ways to study, manage and preserve cultural assets. Considering this, the present paper introduces a preliminary diagnostic study of their use for promoting maritime cultural heritage. At this first stage, the study aims to provide an insight into their potential and usefulness as drivers for local sustainable heritage management. This study focuses its interest on the ICTs of The Portuguese Coastal Way of St. James[1]. The specific objectives are: (1) To analyse the suitability of the ITCs type and content used to promote the maritime cultural heritage of Baiona and Combarro (the two case studies); (2) To investigate their visibility on The Portuguese Coastal Way; (3) To evaluate the opportunities for using ICTs to promote their maritime heritage. It might interest stakeholders and heritage managers, urging them to reflect upon the effective use of

---

[1] The abbreviation "The Way" will be used henceforth to refer to The Way of Saint James that is one of the most important Christian pilgrimages, with Santiago de Compostela (Galicia, Spain) as the final destination. In this study, The Way is considered as a tourist product [1]. Due to its complexity, it focuses its attention on the relation between the territory, its cultural asset and technological potentialities. Although relevant, tourists and pilgrims' practices and their perceptions and engagement with the sites, are not object of the investigation.

© Springer International Publishing AG, part of Springer Nature 2019
F. Calabrò et al. (Eds.): ISHT 2018, SIST 100, pp. 165–172, 2019.
https://doi.org/10.1007/978-3-319-92099-3_20

contemporary technologies. The next section gives a short presentation of the background literature about Maritime Cultural Heritage and the potential of ICTs in the improvement of the cultural and economic value of these elements. The villages of Baiona and Combarro were selected as case studies because they are stages of the spiritual variant of The Portuguese Coastal Way and 2018 has been declared the Year of the Portuguese Route. The content and the functions of ICTs sources related with their maritime heritage are examined through a content analysis methodology. The last section summarises the study's contribution and suggests future directions.

2   Background Literature

Over centuries coastal communities have developed strategies, knowledge, traditions, beliefs and professional skills that were related to exploitation, trading and trafficking in marine resources [2]. For this reason, cultural assets that witness the relationship between people and sea are agglomerated in the category of Maritime Cultural Heritage. This definition refers to all the cultural materials (in the water and on the nearby land) and intangible assets that are an expression of a water-based culture (saltwater and fresh water) that has anthropological, archaeological, historical, architectural, artistic, scientific or literary values or interests, among others [3]. Communities which are dependent on fisheries seem to have recognised the historical and cultural value of the maritime heritage they inherited, especially as the cornerstone of a new tourism. The fishing economic industries therefore provide coastal towns and communities with an identity as well as an economy [4, 5]. In these coastal areas, the maritime cultural heritage is one of the forces on which it is necessary to base the local sustainable development strategies [6, 7].

As a matter of fact, the 2020 Galician Tourism Strategy[2] [8] emphasises the need to coordinate the actions of the administration and the tourist sector with those initiated by the Fisheries Local Action Groups (FLAGs) in coastal areas in order to develop maritime tourism in fishing towns and their historic centres. It mentions innovation and new technologies as key factors in increasing tourism in the coming years. This is due to the fact that, nowadays, ICTs are important communication tools that make it possible to disseminate information quickly and easily. Institutions and/or organisations dedicated to heritage increasingly depend on their intelligent use. This is especially true in the case of the heritage-related economic activities, like the tourist industry. Therefore, due to the competition between tourist destinations, the selection and the choice of a particular destination depend on the information that is available to potential tourists [9]. Data from a study carried out by Google showed that around 77% of travellers use a smartphone in their spare time to get ideas for future travels [10]. According to Garau [11], devices like smartphones and tablets can contribute to cultural and urban

---

[2] The 2020 Galician Tourism Strategy centres around preparations for the 2021 Holy Year (a Jubilee Year that is celebrated in Santiago de Compostela, whenever July 25th, the day of Santiago the Elder, falls on a Sunday). This document emphasises the challenge of creating awareness of the unique heritage features of Galicia's coastal areas. It involves promoting maritime tourism as a way of connecting tourists with maritime culture, by implementing projects that bring together the tangible and intangible heritage and gastronomy of small ports and fishing villages.

development. Also, the digitalised promotion of cultural assets and the spread of web connections have reduced access costs and overcome geographical and time constraints and allow the product differentiation among the international competition.

3   An Introduction to the Area of Study

The area studied combines two territorial and characteristic factors of Galicia: the coast and The Way. The Portuguese Coastal Route is the part of The Way that goes through Portugal to enter Galicia and has a spiritual variant where lie Baiona and Combarro (Fig. 1). The case studies have historic centres of greatly valued heritage, protected by regional legislation. Having identified the poor knowledge of their maritime heritage as a weakness of these areas, provisions were made in the local development strategies. These included platforms and applications that are seeking alternative ways to diversify the economy by promoting the maritime heritage. This action follows the 2015–2021 Strategic Plan for The Way of St. James [12]. It aims to protect and preserve the identity of The Way. It sets out guidelines for promoting its cultural and natural heritage. With a medium-term vision, it is designed to enhance social and territorial cohesion and to continue building the universal identity of Galicia and Europe.

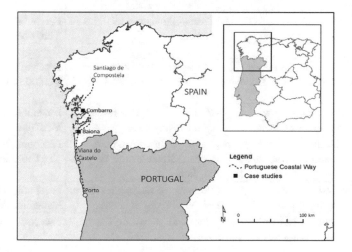

Fig. 1. The Portuguese Coastal Route and its spiritual variant. Source: Prepared by the authors.

3.1   Sources and Methodology

This study uses a combined examination of three different sources of web information: websites, social media and mobile applications (apps). The criterion for selection was their institutional nature[3]. Websites are important travel planning tools and resources

---

[3] Sources developed or promoted by national, regional and local government institutions or by representative organisations and associations.

and they are capable of attracting more potential visitors [13, 14]. APPS complete this planning process by helping users while they are doing the visit. After the trip, social media allow visitors to keep in contact with the destination visited [15]. A content analysis was applied as a technique for gathering and analysing the content of text based on categories [16]. They are: sources' main purpose and target users, ITCs' type and main functions, presence of information about case studies and users' interactivity. This study examines 10 webpages and their social media profiles and 3 apps. Google was used as search engine. Given the comparative nature of this study, the results are presented using a simple grid system (Tables 1 and 2).

Table 1. Sample of content analysed: Websites and related social media.

Name | URL | Abbreviation	Main purpose	Social media
Mar Galaica Turismo Mariñeiro | www.margalaica.net | MGTM	Boost the facilities of Mar Galaica (the seacoast of Galicia)	Facebook, Twitter, YouTube
Turismo de Baiona - www.turismodebaiona.com | TB	Present the tourist facilities in Baiona	Facebook, Twitter and YouTube
Rías Baixas.info - www.riasbaixas.info | RB	Promote the tourist resources of the "Rías Baixas"	Not activated
Turgalicia - www.turismo.gal |TGAL	Present the tourist facilities of Galicia	Facebook, Twitter, YouTube, Instagram, Google+
Terras de Pontevedra, Xunta de Galicia - www.terrasdepontevedra.og | TP	Promote the tourism resources of the Pontevedra area	Facebook, Twitter, YouTube, Instagram.
Turismo Rías Baixas - www.caminoporlacosta.es | TRB	Promote the tourist resources of the "Rías Baixas"	Facebook, Twitter, YouTube
Descubre el Camino Portugués por la Costa - www.caminoporlacosta.es | DCPC	Publicise the Portuguese Coastal Route of the Way of St. James	Facebook, Twitter, YouTube, Instagram
Camino Xacobeo - www.caminodesantiago.gal |CX	Promote the tourism resources of the Way of St. James	Facebook, Twitter, Instagram, YouTube
Camino de Santiago. El camino de las Estrellas - www.caminosantiago.org |CS	Help with preparing for the Way	Facebook and Twitter
Variante Espiritual.com - www.varianteespiritual.com | VE	Present the spiritual variant of the Coastal Route	Facebook, Twitter, Google+

Table 2. Sample of content analysed: Apps.

Name | Abbreviation	APP Store	Main purpose
Camino Santiago en Galicia | CSG	Google Play Store and iTunes	Provide information on all the routes of the Way of St. James from where it enters Galicia
Baiona Guía Oficial | BGO	Google Play Store and iTunes	Present the tourist facilities in Baiona
Turismo de Poio | TP	Google Play Store and iTunes	Present the tourist facilities in Poio. With special reference to Combarro

4   Results

The results are presented below according to the specific objectives listed in the introduction:

1. To analyse the suitability of the ITCs type and content used to promote the maritime cultural heritage of Baiona and Combarro (the two case studies).

The survey on the websites analysed shows that most of the technological resources were not created by the institutions themselves but were based on others already available on the Internet, such as Google Maps geolocation maps. This might indicate a low involvement of the institutions. The multimedia resources which are common to all the websites are videos and photo galleries. In some cases, there are also electronic leaflets. There is also an audio-guide and QR codes for the heritage attractions in Combarro. Indeed, these are very simple multimedia resources, which just in one case are accompanied by a very important resource: trip planning, where the tourists' needs can be customised according to their preferences. This is the case of the CX website. VE website makes it possible to organise a sea route. Lastly, the TGAL website offers a section devoted to 360° visual experiences.

In regard to the behaviour on the social media of the analysed sources, it was possible to identify a significant presence on Facebook, Twitter and YouTube. Although other media are used, these are the most common and dynamic. It could be seen that the social media are used by institutions as communication tools to supplement their websites. Only websites like the RB, CX and TGAL post original publications on their social media that are appropriate to the type of communication on these platforms, although the resources used on the three social media are very similar. Also, clear differences can be detected in their popularity. The institution with the best positioning on the social media is CX, with 89,427 followers on Facebook. It is also noticeable that the social media webpages with more technological resources and dynamism have more user interaction.

Three institutional apps were analysed: CSG, which grew out of the Smart Tourism Plan of Regional Government; the BOG, implemented by the Villas Marineras Association, and, lastly, TP, developed by the municipality to which Combarro belongs. The first one offers the possibility of accessing the heritage databases, which allows users to find up-to-date information on heritage, although with no specific category of maritime heritage. However, there are no multimedia resources and, in general, the heritage attractions of the stages are not the main concern. The second one provides information on points of interest in the town, with an audio guide, geolocation system and information on hostel accommodation. The same is true of the TP, which provides basic tourist information, organised like a traditional paper guidebook, and with no possibility of user interaction. The BOG and TP Apps have an attractive user interface in terms of graphic design, but their functionality, to be understood as the number of tasks that can be performed, is quite limited.

2. To investigate their visibility on the Portuguese Coastal Way.

The presentation of the content on Baiona and Combarro on the Turgalicia website is accompanied by an interactive map (Google Maps) and a photo gallery; in the case of Baiona, The Portuguese Way is mentioned. On the CX website, the Baiona-Vigo stage of The Portuguese Coastal Way is described using a geolocation map and TB website also makes a reference of this route. Although there is some heritage promotion on the social media of all these websites, there was a different type of communication and different levels of involvement with maritime cultural heritage. The most important sites in this regard are the TB, VM, DCPC, VE and the RB. On these social media pages is possible to identify a clear intention to communicate a tourism product related with maritime cultural heritage, especially in Baiona and Combarro, whether separate from or associated with the Portuguese Coastal Way. Posting promotional and historical videos to publicise the maritime cultural heritage of the case studies is a tool with great potential and a high number of views and shares, and these kinds of resources have been used by TB, RB and TGAL.

Only two of the institutional apps refer to the route studied here. Firstly, there is the CSG, with information focusing on The Way with routes, scheduling and experiences, and pilgrims can share their experiences. The interactive capability for users is high, because it allows them to create and share information and to collect text and multimedia materials and store them in a journal, in order to publish an e-book, using the available templates, and share it with other users. Secondly, the TP app has a section on the part of The Portuguese Coastal Way and Combarro. Even so, the information is limited to a static map showing this variant of The Way, which makes it of little use. However, this same application includes recommendations for visits and it is in this section that the cultural heritage of Combarro becomes a point of interest.

3. To evaluate the opportunities for using ICTs to promote their maritime heritage.

The most eye-catching multimedia resources from the visual point of view are omitted, such as 360° virtual visits, which might anticipate the real visit. There is no use of interactive 3-D reconstructions and displays, through which it would be possible to see how the heritage features really look. There are no interactive maps, so a wider use of them needs to be made. Nowadays, a wide variety of marketing tools exists on the social, however, only TGAL frequently uses some marketing resources provided by Facebook to increase the visibility of sites on the web, such as ads or the creation of promotional campaigns, events and groups. Tools related with entertainment (e.g. social media game-based marketing) and social commerce were not identified. The apps studied underuse the opportunity to promote maritime heritage attractions. They include heritage features using catalogues of cultural assets but a significant number of those relating to maritime heritage tend to be missing. The app CSG includes augmented reality tools so that users can search for nearby resources along their route using a geo-positioning system. In apps BGO and TP, less attention is paid to tangible and intangible heritage assets, which are mentioned in some cases but are not the focus of the application.

5   Conclusion

The strategic prioritisation of using ICTs to enhance heritage through new multimedia resources improves promotion of Galician historical areas, such as those along The Way. In addition, the ICTs and their speed and immediacy when generating information are advantages for the public bodies charged with protecting and promoting cultural heritage. A rich and varied tangible and intangible heritage exists, which forms part of the cultural background of coastal towns.

The present analysis argues that promoting the maritime heritage of the coast, specifically the two case studies of Baiona and Combarro, is consistent with a growing tourist product like The Portuguese Coastal Way, which even has the ability to bring value added to the users' tourist experience along The Way. New technologies have a great potential for highlighting the complementarity between these elements and bringing visibility to the valuable heritage of Galicia's maritime past. The maritime heritage of Baiona and Combarro is not considered a territorial resource, and even less within The Portuguese Coastal Way. This route is underestimated within the opportunities offered by the ITCs to exploit the maritime heritage of Baiona and Combarro. Thus, territorial innovation of The Portuguese Coastal Way might go through the combined valorisation of ITCs and maritime cultural asset of attractive factors. Consequently, it is essential to reinforce a discourse that links The Portuguese Coastal Way with maritime heritage values. The present analysis warns about weaknesses both from the point of view of the message and of the technological tools used to promote this heritage segment. The criterion of selection of official sources and, thus, official actors has pointed out a low level of involvement of stakeholders and even less involvement of territorial agents to promote their resource. This resulting gap requires substantial investment, in terms of both effort and resources, to bring about ITCs-based territorial promotion. To this purpose, territorial synergies and intensification of innovative dynamics can enhance new perspectives and fortify territorial attractiveness. For instance, at the moment, the technologies that users have access are intended for a wider public. As a matter of fact, their effectiveness might be improved by adopting age as a criterion for elaboration and presentation of the web content (17% of pilgrims walking The Portuguese Way were over 60 years old in 2016). A further improving strategy might be the incorporation of volunteered generated geographic content, which would stress public participation. For all this, it is necessary to reposition the information drawn up by the institutions on ITCs resources and user interaction. These can be key factors in jointly presenting and processing the heritage of a rich territory like the one of The Way.

References

1. Santos, X.: El Camino de Santiago: Turistas y Peregrinos hacia Compostela. Cuadernos de Turismo **18**, 135–150 (2006)
2. Howard, P., Pinder, D.: Cultural heritage and sustainability in the coastal zone: experiences in south west England. J. Cult. Heritage **4**(1), 57–68 (2003)
3. Baron, A.T.O.: Constructing the Notion of the Maritime Cultural Heritage in the Colombian Territory: Tools for the Protection and Conservation of Fresh and Salt Aquatic Surroundings. Division for Ocean Affairs and the Law of the Sea, Office of Legal Affairs, United Nations (2008)
4. Brookfield, K., Gray, T., Hatchard, J.: The concept of fisheries-dependent communities. a comparative analysis of four UK case studies: Shetland, Peterhead, North Shields and Lowestoft. Fish. Res. **72**, 55–69 (2005)
5. Nadel-Klein, J.: Granny baited the lines: perpetual crisis and the changing role of women in Scottish fishing communities. Women's Stud. Int. Forum **23**(3), 363–372 (2000)
6. Santana, A.: Patrimonio cultural y turismo: reflexiones y dudas de un anfitrión (2005). https://www.naya.org.ar/congreso/ponencia3-10.htm. Accessed 04 Sep 2012
7. Carbonell, E.: Maritime heritage and fishing in Catalonia. Coll. Antropol. **38**(1), 289–296 (2014)
8. Xunta de Galicia, Turismo de Galicia, Clúster Turismo de Galicia (eds.): Estrategia del Turismo de Galicia 2020 (2017). https://www.turismo.gal/docs/mdaw/mjk2/~edisp/turga296028.pdf?langId=es_ES. Accessed 06 Nov 2017
9. Fodness, D., Murray, B.: Tourist information search. Ann. Tourism Res. **24**(3), 503–523 (1997)
10. The 2014 Traveler's Road to Decision. https://storage.googleapis.com/think/docs/2014-travelers-road-to-decision_research_studies.pdf. Accessed 19 Oct 2017
11. Garau, C.: From territory to smartphone: smart fruition of cultural heritage for dynamic tourism development. Plann. Pract. Res. **29**(3), 238–255 (2014)
12. Xunta de Galicia: Plan Director e Plan Estratéxico do Camiño de Santiago 2015–2021. http://www.caminodesantiago.gal/documents/17639/293816/Plan%20director%20estratexico%20Cami%C3%B1o%20de%20Santiago.pdf?version=1.0. Accessed 23 Nov 2017
13. Palmer, A.: The internet challenge for destination marketing organizations. In: Morgan, N., Pritchard, A., Pride, R. (eds.) Destination Branding: Creating The Unique Destination Proposition, 2nd ed, pp. 128–140. Elsevier, Oxford (2004)
14. Law, R., Qi, S., Buhalis, D.: Progress in tourism management: a review of website evaluation in tourism research. Tourism Manage. **31**(3), 297–313 (2010)
15. Pepe, M.S., Biunique, R.: Using social media as historical marketing tool for heritage sites in eastern New York state. J. Appl. Bus. Res. (JABR) **33**(1), 123–134 (2016)
16. Neuman, W.L.: Social research methods: quantitative and qualitative approaches, 6th edn. Allyn & Bacon, Boston (2006)

Geospatial Analysis to Assess Natural Park Biomass Resources for Energy Uses in the Context of the Rome Metropolitan Area

Francesco Solano[1]([✉]) [iD], Nicola Colonna[2], Massimiliano Marani[2],
and Maurizio Pollino[2] [iD]

[1] Mediterranean University of Reggio Calabria, 89100 Reggio Calabria, Italy
francesco.solano@unirc.it
[2] ENEA - National Agency for New Technologies, Energy and Sustainable
Economic Development, Casaccia Research Centre, 00123 Rome, Italy

Abstract. The Metropolitan city of Roma Capitale (Italy) represents a vast area, which purposes are not only institutional but also specific functions, such as the promotion and coordination of economic and social development. The park areas included in the Metropolitan area are able to provide ecosystem services and resources such as agricultural and forest products. The rational exploitation of biomass resources produced around the Metropolitan area can be an opportunity to replace fossil fuels, make the city more climate friendly and, at the same time, to relaunch the sustainable management of forest that are often abandoned and prone to degradation risk. The goal of this paper is to investigate and update the actual distribution of the main forest types of the Bracciano-Martignano Regional Natural Park, through GIS and Remote Sensing techniques, in order to assess the biomass potential present in the forest areas. Results confirmed the importance of Sentinel-2 satellite data for vegetation applications, allowing to map species and surfaces as well as to carry out other studies at regional scale with a high overall accuracy. The forest types distribution analysis performed inside the park showed that there are about 20,000 t of woody biomass per year available, indicating that rationale forest management can be strategic to deal both with forest degradation and city energy supply.

Keywords: Sentinel-2 (S-2) · Remote sensing · GIS · Biomass
Energy · Forestry spatial analysis · Rome Metropolitan Area

1 Introduction

The Metropolitan city of Roma Capitale was established in 2015, according to the Italian law n. 56/2014. It represents a large area, including 121 Municipalities, which purposes are not only related to the institutional role, but are also aiming at specific functions, including the promotion and coordination of economic and social development, ensuring support for economic and research activities that are innovative and consistent with the vocation of the Metropolitan area. The area of the Metropolitan City is able to provide ecosystem services (air, clean water) and also resources (agricultural

© Springer International Publishing AG, part of Springer Nature 2019
F. Calabrò et al. (Eds.): ISHT 2018, SIST 100, pp. 173–181, 2019.
https://doi.org/10.1007/978-3-319-92099-3_21

products, fish, wood and firewood), but it is also affected by intense touristic pressure. Moreover, such area also have the potential of offering wood-based products for energy, which are present in large quantities but underutilized, while the city is a major energy consumer that has a significant emissive impact. The rational use of the biomass resource can be an opportunity to partially replace fossil fuels and make the city more climate friendly than post-Kyoto commitments and at the same time provide an opportunity for the economic exploitation of forest that are often abandoned and subject to risks of degradation (fires, landslides). Given that within the Metropolitan area are included many green areas to be used in a sustainable way, we set ourselves the goal of estimating the potential contribution of biomass present in the territory to meet a fraction of the consumption of fuels of the neighbouring areas, through GIS-based analyses [1] that uses cartographic and satellite data to calculate the forestry biomass potential. The challenge is how to properly manage local forestry biomass resources in the context of a natural park area suitable for biodiversity conservation as well as tourism development without compromising ecosystem services but rather improving them.

2   Materials and Methods

2.1   Study Area and Processing Workflow

The research was carried out in The Regional Natural Park of Bracciano-Martignano, in the context of the Rome Metropolitan Area (Fig. 1).

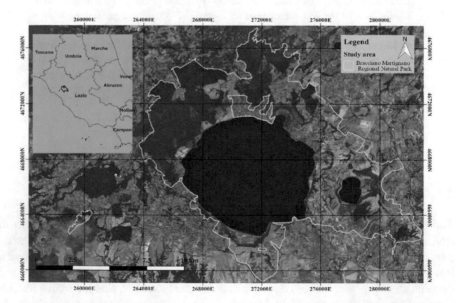

Fig. 1. Study area localisation. Sentinel-2 image (acquired on August 29, 2017), showed in false infrared colour band composite, (RGB = 432). Yellow lines show the park boundaries.

The area extends for 16,682 ha through the towns of Anguillara Sabazia, Bassano Romano, Bracciano, Campagnano di Roma, Manziana, Monterosi, Oriolo Romano, Rome (XV Municipality), Sutri and Trevignano Romano, falling in the vast area of the Sabatini Mountains (Italy). The area under analysis is characterized by an heterogeneous landscape, with the presence of Bracciano Lake, several agricultural activities, built-up areas up to natural resources such as beech woods, chestnut woods and deciduous oaks in the mountain stands.

Referring to the goal of this research, a specific methodology is proposed for mapping the current consistency of forest resources and estimate the biomass potential present in these habitat (Fig. 2).

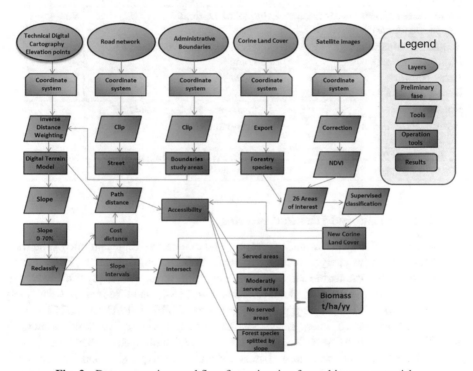

Fig. 2. Data processing workflow for estimating forest biomass potential.

2.2 Data Availability and Retrieval

In order to carry out the geospatial analyses, a data availability survey was firstly made. The following Table 1 contains a comprehensive list of data and information exploited. In particular, some data have been properly customised (e.g., GIS import procedures, selection by attributes or by area, spatial overlay, points interpolation, etc.) in order to tailor them to the peculiarity and to the particular territorial situation under study.

Table 1. Data availability and source.

Data	Source	Customisation
Corine Land Cover level 4 (shape file format)	https://dati.lazio.it/	Yes
Lazio Digital Cartography, 1:5,000 shape file format	https://dati.lazio.it/	No
Elevation points (5 × 5 m) .dxf file format	https://dati.lazio.it/	Yes
DTM and Slope (5 × 5 m) Derived from elevation points (raster file format)	https://dati.lazio.it/	Yes
Boundaries of Monumental areas (shape file format)	https://dati.lazio.it/	Yes
Sentinel-2 satellite image (raster file format)	https://scihub.copernicus.eu/dhus/#/home	Yes
Road Network (shape file format)	https://dati.lazio.it/	Yes
Road Network	https://www.openstreetmap.org/	Yes
Admin. Boundaries (shape file format)	https://dati.lazio.it/	Yes
Orthophotos (RGB) Via WMS service	http://www.pcn.minambiente.it/mattm/servizio-wms/	No

2.3   Satellite Data and Images Processing

Remote sensing data, opportunely processed and elaborated, represent one of the most important sources to analyse land cover characteristics and dynamics [2], giving an effective effort to sustainable landscape planning and management [3]. To investigate and study the area of interest, cloud-free Sentinel 2 (S-2) satellite images were freely downloaded with UTM-33 map projection, datum WGS84 (EPSG code 32633).

In order to improve image spatial resolution of the original S-2 bands a merging algorithm was used between the 20 m (B5-B6-B7) multi-spectral bands and 10 m (B2-B3-B4-B8) spatial resolution bands through the hyperspherical colour space (HCS) algorithm [4]. All the pan-sharpened images were atmospherically and topographically corrected [5] to surface reflectance using all selected bands by means of the ATCOR3 module for Erdas Imagine® 2016. With the aim to get a detailed update of Corine Land Cover (CLC) map and to better detect/describe forest resources, the characterization and separability of the spectral signatures of the different forest type was performed in S-2 imagery. To this end, a group of land uses was defined obtaining their spectral signatures starting form a set of 26 areas of interests (AOI) - collected by means of original CLC classes, thus as training set to perform the image classification. For the extraction of the surface covered by vegetation, the calculation of vegetation indices was used. In particular, the Normalized Difference Vegetation Index (NDVI) [6] was used to obtain a high spectral resolution vegetation mask where to perform the classification. Euclidean distance (ED) algorithm was applied to measure spectral signature's separability and a contingency matrix (CM) to evaluate the consistency of the defined training sets. The AOI data have been selected as training samples to perform a supervised classification by means of Maximum Likelihood Classification (MLC) algorithm. The result was an improved land cover (ILC) map, more suitable for the purposes of the study and focused on forest areas.

2.4   Geomorphological Analysis

In this phase, the depiction of topography as a Digital Terrain Model (DTM) was fundamental for describing geomorphological features and characteristics of the study area [7]. To this end, was exploited the Technical Digital Cartography, at 1:5,000 scale, provided by the Lazio Region (http://dati.lazio.it/en/), which contains a specific layer about elevation points (in .dxf format) describing the topography of all the Region. By using the inverse distance weighting (IDW) method to interpolate such points, a DTM raster surface was generated, with a 5×5 m of spatial resolution. Then, a slope map (in percentage) from the above mentioned DTM was derived. The slope layer was sliced in 4 different classes (20% equal intervals) [8]. The next step, in GIS environment, was to overlay the slope layer and the ILC map, in order to further classify the forest land cover classes also according the local terrain slope.

2.5   Viability Analysis and Accessibility Mapping

The map of the areas served by roads (i.e., accessibility map) for the purposes of silvicultural activities was produced in the GIS environment on the basis of Map Algebra operations conducted in other Italian experiences [9, 10] and taking into account the Euclidean distance of each area considered in relation to the slope of the mountainsides. The method for determining the forest area served by roads refers to the forest accessibility evaluation method proposed by Hippoliti [11], which is based on the determination of the time that an operator can spend to reach on foot the area of use from the nearest vehicle accessible road (suitable for forestry machinery operations). The parameters constituting the assessment of accessibility to the area are: (i) the distance from the area to the road and (ii) the difference in elevation of the area with respect to the starting point on the road. The area of interest was then indexed according to the travel time [11] and in relation to the ground identity and a coast raster map was obtained. The distance from the road (real distance) and the difference in elevation are calculated by means of the above mentioned DTM (5×5 m). Through a GIS function for calculating the cumulated distances (Path Distance) a new dataset was created representing the degree of accessibility of each point of the map. The classification of the area of interest in relation to accessibility for silvicultural purposes was then carried out according to three classes: (1) served areas; (2) moderately served areas; (3) poorly served areas.

2.6   Forest Biomass Estimation

The last step of the approach pursued was to take into account the most frequent forest types, including chestnut, oak, beech woodlands and artificial pine forests. For the estimation of the obtainable forest biomass, data of Wood Commission, per hectare, were used, provided by experts with a deep and specific knowledge of the territory coupled with literature research datasets. This datum was then referred to the forest area, obtained by the updated CLC forest type that led to a wood supply expressed in $t \cdot ha^{-1}$.

3   Results

3.1   Forest Land Cover Mapping

The forest cover is mainly characterized by oak [12], chestnut, beech and artificial pine forests. Beech forests are considered an UNESCO heritage site and sites of community interest (SIC). The forest species that occupy the largest surface are chestnut with 1,992 ha and oak with 2,087 ha total (Fig. 3).

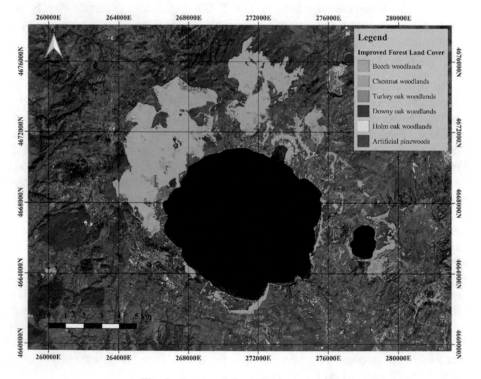

Fig. 3. Improved forest land cover map.

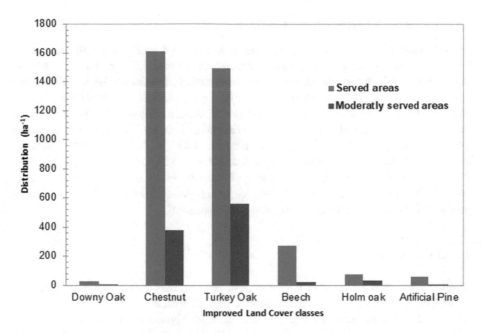

Fig. 4. Distribution of the main forest type according to the accessibility for silvicultural purposes, based on the derived improved land cover (ILC) classes.

Table 2. Improved Land Cover (ILC) distribution of the main forest type in the study area.

Level of accessibility	Land cover	Area [ha]	Usable area (4% of total surface) [ha]	Biomass [t]
Served areas	Oak	1,521	61	6,710
	Chestnut	1,614	65	9,100
Moderately served areas	Oak	566	23	2,530
	Chestnut	378	15	2,100
Total		**4,079**	**164**	**20,440**

Moreover, it is possible to observe that most of the territory has an excellent degree of accessibility (Fig. 4). Accessibility cartography has been classified into two levels: served areas and moderately served areas. On a total surface of 4,533 ha, about 3,547 ha have been classified as served areas and 1,006 ha as moderately served areas (Table 2).

3.2 Viability and Accessibility Assessment

The most of the territory under study is provided with a good road network. This network is characterized by useful roads/tracks, both for access to the forest by forest operators and for logging with forestry tractors.

3.3   Biomass Characterization and Distribution

After estimating the surface of accessible areas and moderately accessible areas, the forest biomass obtained on the whole surface was estimated. As a starting point, all forest species present in the territory were evaluated, to the define the so called gross biomass potential, then moved on to identify the species actually usable for energy purposes (potentially available), considering oak and chestnut woods. For oak forests, an average yield of about 100 m^3 ha^{-1} is estimated at 110 t ha^{-1}, while for the chestnut woods an average of 140 m^3 ha^{-1} is estimated at 140 t ha^{-1}. In accordance with the forest law, to manage the forest heritage sustainably, we hypothesized that only 4% of the total area of study can be ceded yearly. From the results obtained it is possible to estimate that the area of interest is able to provide about 20,440 t of forest biomass yearly without affecting local forest production potential.

4   Conclusions

Results confirm the importance of the red-edge bands on Sentinel-2 imagery for vegetation applications, thanking the combination of with its high spatial resolution of 10 m and the possibility to use peculiar spectral signature to analyse vegetation behaviour. The performed images classification of forest distribution suggested that, when these images integrate with other ancillary data, it becomes possible to map species as well as to carry out other studies at regional scale with a high overall accuracy [13]. In the ILC map, the forest species that occupy the largest surface are chestnut with 1,992 ha and oaks with 2,087 ha in total. Accessibility analysis, focused on forest areas, defined according to road availability and slopes showed that the overall amount of forest can be easily utilized for biomass harvesting for energy purpose because the overall resources fall in area classified as served ones. According to a large number of experiences [14, 15] a good accessibility is a precondition for the technical and economic exploitation of forest biomass resources but is not enough. The next steps will be to analyse woody biomass logistics and local heat demand to properly fit the energy demand with the resource availability and promote the biomass use as a flywheel for local development.

References

1. Modica, G., Laudari, L., Barreca, F., Fichera, C.R.: A GIS-MCDA based model for the suitability evaluation of traditional grape varieties. Int. J. Agric. Environ. Inf. Syst. **5**, 1–16 (2014)
2. Nolè, G., Lasaponara, R., Lanorte, A., Murgante, B.: Quantifying urban sprawl with spatial autocorrelation techniques using multi-temporal satellite data. Int. J. Agric. Environ. Inf. Syst. **5**, 19–37 (2014)
3. Di Palma, F., Amato, F., Nolè, G., Martellozzo, F., Murgante, B.: A SMAP supervised classification of landsat images for urban sprawl evaluation. ISPRS Int. J. Geo Inf. **5**, 109 (2016)

4. Padwick, C., Scientist, P., Deskevich, M., Pacifici, F., Smallwood, S.: WorldView-2 pan-sharpening. Asprs **48**, 26–30 (2010)
5. Pflug, B., Makarau, A., Richter, R.: Processing Sentinel-2 data with ATCOR. 18, 15488 (2016)
6. Rouse, J.W., Haas, R.H., Schell, J.A., Deering, D.W.: Monitoring vegetation systems in the great plains with erts. In: Freden, S.C., Mercanti, E.P., Becker, M.A. (eds.) Third Earth Resources Technology Satellite 1 Symposium. NASA, Washington, DC (1974)
7. Modica, G., Pollino, M., Lanucara, S., La Porta, L., Pellicone, G., Di Fazio, S., Fichera, C.R.: Land suitability evaluation for agro-forestry: definition of a web-based multi-criteria spatial decision support system (MC-SDSS): preliminary results. In: Lecture Notes in Computer Science, pp. 399–413. Springer, Cham (2016)
8. Hippoliti, G., Piegai, F.: Tecniche e sistemi di lavoro per la raccolta del legno. Compagnia delle Foreste, Arezzo (2000)
9. Chirici, G., Marchi, E., Rossi, V., Scotti, R.: Analisi e valorizzazione della viabilità forestale tramite G.I.S.: la foresta di Badia Prataglia (AR), L'Italia For e Mont, pp. 460–481 (2003)
10. Bulfoni, D., De Vetta, R., Magrini, A.: Pianificazione della viabilità forestale. Impiego di dati lidar nella Foresta di Fusine (UD). Sherwood (2010)
11. Hippoliti, G.: Sulla determinazione delle caratteristiche della rete viabile forestale. L'Italia Forestale e Montana (1976)
12. Modica, G., Solano, F., Merlino, A., Di Fazio, S., Barreca, F., Laudari, L., Fichera, C.R.: Using Landsat 8 imagery in detecting cork oak (Quercus suber L.) woodlands: a case study in Calabria (Italy). J. Agric. Eng. **47**, 205 (2016)
13. Di Fazio, S., Modica, G., Zoccali, P.: Evolution trends of land use/land cover in a mediterranean forest landscape in Italy. In: ICCSA 2011. Lecture Notes in Computer Science, vol. 6782, pp. 284–299. Springer, Heidelberg (2011)
14. Colonna, N., Del Ciello, R., Petti, R.: Biomasse agroforestali: valutare il potenziale a scala regionale. Ambiente Risorse e Salute **127**, 20–24 (2010)
15. Lupia, F., Pulicani, P., Colonna, N.: Un modello di processamento per lo sfruttamento ottimale delle biomasse legnose. Atti IX Conferenza Utenti ESRI, Roma (2006)

Promoting Cultural Resources Integration Using GIS. The Case Study of Pozzuoli

Carlo Gerundo$^{(\boxtimes)}$ and Guilherme Nicolau Adad

University of Naples Federico II, Naples, Italy
`carlo.gerundo@unina.it`

Abstract. In the Phlegraean Fields, in the Metropolitan City of Naples, the interactions between local geological phenomena and the urban development that has taken place in the city of Pozzuoli since the Greek occupation, have inevitably had an impact on the way of life of the local population and on the use of the city's built heritage over time.

In order to set a cultural strategy able to valorize the exceptional worth of the city and to understand its dynamics over time, it is necessary to create a narrative that guides visitors and locals to understand these interactions and the integrity of the landscape.

The paper describes a methodology carried out to collect, analyze and synthesize information in order to identify the cultural resources, to understand them within their context and to create links between these elements. Firstly, it is crucial to constitute a cultural inventory based on several resources coming from conventional and non-conventional data, such as geo-tagged social media data. The result is a database that contains information pertaining to all forms of cultural resources, uploaded in a graphic map generated using GIS software.

The aim of the project is to make freely available supporting material designed for public authority, to be used by the municipality to promote Pozzuoli's cultural, economic and touristic potential.

Keywords: Cultural planning · Cultural mapping · GIS

1 Introduction

1.1 A Different Perspective on Cultural Resources: From Cultural Planning to Cultural Strategy

For most of the history of urban planning, scholars and planners have mainly been concerned with land use regulation, without focusing their attention on the immaterial aspects such as social services, education, arts and culture. In the same way, the much more long-lived world of the arts had never wanted to be delineated in a planning framework. Starting from the '70 s, however, artists and art lovers have begun to rethink their role within the community and, analogously, planners have reconsidered the role of culture in the activation of economic and touristic development processes and in the revitalization of urban areas and historic centers [1]. This effort of reconciliation has generated the concept of cultural planning, understood as the management

© Springer International Publishing AG, part of Springer Nature 2019
F. Calabrò et al. (Eds.): ISHT 2018, SIST 100, pp. 182–188, 2019.
https://doi.org/10.1007/978-3-319-92099-3_22

of cultural resources for the sustainable development of the territory, formulated in the United States in the late '70s, and resumed a decade later in Britain and Australia [2].

According to the definition given by Colin Mercer, cultural planning is the planning process of the strategic and integrated use of cultural resources for urban and community development. While the term 'strategic' implies that this process has to be part of a wider strategy of territorial development, the word 'integrated' highlights the need to establish connections with physical environment planning and with socio-economic development targets [3, 4].

However, it is acknowledged that the notion of cultural planning is based on the concept of 'cultural resources'. Even if it is not necessary to dissert in detail about what is meant by 'culture', it is useful to quote the first organic definition offered by the English anthropologist Edward Burnett Tylor in 1871. According to Tylor, culture, or civilization, if understood in its broad ethnographic sense, represents "that complex whole which includes knowledge, beliefs, arts, morals, law, customs, and any other capabilities and habits acquired by [a human] as a member of society" [5]. This definition combines a dual vision of culture, understood both anthropologically, as ways of living, values and beliefs, and humanistically as any kind of performance [2].

The British theoretician Raymond Williams, about a century later, traced the definition of culture back to three great categories [6]:

- the *ideal* culture, understood as a process of civilization, of human perfection, in terms of certain absolute or universal values;
- the *documentary* culture that is as a "the body of intellectual and imaginative work, in which, in a detailed way, human thought and experience are variously recorded", which could synthetically be approximated with the concept of 'arts';
- the *social* culture, in which culture is a description of a particular way of life, which expresses certain meanings and values not only in art and learning but also in institutions and ordinary behavior.

This last definition goes beyond the concept of culture as a simple descriptive assortment of customs and traditions of societies, as claimed by some anthropological theories, and espouses the concept of culture as sum of the interrelation of all social practices [7].

Although there are some conceptual differences about 'cultural resources' definition, there is an indissoluble link between the territory and these resources. It's evident that cultural resources characterize the territory itself (landscape, city, architecture, monuments, works of art) and urban space, at the same time, contains them or plays the role of the scenography when the some intangible cultural resources (traditions, uses, activities, social relations, performances) arise.

Therefore, it is clear that cultural planning in no way is a process having dangerous references to the Stalinist regime, but rather a culturally sensitive approach to spatial planning.

The activation of urban strategies and policies that pursue a harmonious systematization of cultural resources can contribute to enhance the sense of community and promote of urban regeneration. However, in order to ensure that these efforts could be effective, it is appropriate to bring them back within the broader and more structured urban planning and cultural planning processes. In other words, it is necessary to shift from the concept of cultural resources planning for a territory, to the concept of cultural strategies applied to the plan processes, from cultural planning to cultural strategy.

The drafting of a cultural strategy requires a preliminary and capillary cognitive phase aimed at recognizing the cultural resources on the territory, known as cultural mapping. As far as cultural mapping is concerned, there are many definitions in the scientific literature that, however, converge in identifying the cultural mapping as a tool to identify, catalogue, classify and analyze cultural resources of a territory, conducted, using geographical information systems (GIS). It is achieved through a wide consultation process of local community, and is used for promotion and valorization strategies, planning processes or other initiatives aimed at social, economic and cultural development [8–10]. The involvement of the population is an indispensable aspect of cultural mapping in order to recognize the intangible cultural heritage which, as specified in the *UNESCO Convention for Safeguarding of the Intangible Cultural Heritage*, is not only made of traditions and customs, but also of contemporary elements (performing arts, events, social relationships, local crafts). The intangible cultural heritage can be considered as such "when it is recognized by the community, by groups or individuals who create, preserve and transmit it" [11].

The drafting of the actual strategy represents a subsequent phase. This strategy should be able to strongly express local identities since it could enhance the potential of a place for attraction, which communicates cultural values also to externa actors [12]. In the cultural strategy it will be essential to outline the overall vision, roles, partnerships, and actions to strengthen the management of cultural resources, establish a governance model and evaluate progresses achieved. This last point is linked to the third and last phase concerning monitoring mechanisms to maintain the cultural strategy on the path traced, through the selection of performance indicators able to identify significant changes. Progresses monitoring can also be used to help keep the community informed about the results achieved by the cultural strategy.

1.2   Campi Flegrei, a Cultural Landscape Lacking Cultural Resources Integration

Urban settlements are the result of humans interacting with their environment, and these interactions produce the exceptional value of a particular urban agglomeration. The Phlegrean Fields (Campi Flegrei), in the Metropolitan City of Naples, is a unique landscape due to the presence of a volcanic system that has influenced the circulation of people and the settlement patterns of the constant urban development that has taken place in this site since the foundation of the Greek colony of Cuma (VIII cent. b.C.).

Its geological peculiarity is not only proved by the irregular topography but also by the effects produced on urban environment by the bradyseism, a telluric phenomenon consisting in the gradual uplift or descent of part of the Earth's surface caused by the inflation and deflation of the volcanic caldera. Over the centuries, history and nature have left traces that have resisted the transformation and the development of society despite the way they have been perceived and narrated is changed over time.

The picturesque features of Campi Flegrei have always inspired travelers, painters and writers from Italy and abroad from ancient times to the present day. Today, the information brought by many of these traces of the past is not adequately safeguarded and valorized. Actually, Campi Flegrei currently lacks the integration of different elements that compose its natural and cultural landscape, as well as a connection to bind its cultural resources. This shortage makes difficult to do an overview of the importance of the heritage and to understand what makes it important.

Nevertheless, the safeguarding and valorization of landscape should not 'hibernate' externality but rather make room for local communities, retaining their modes of expression and finding adequate ways of perpetuating them.

2 Targets

This paper describes the attempt to valorize the exceptional worth of the territory and to understand its dynamics over time through the implementation of a method able to guide both local governments and private initiatives to understand the integrity of this cultural landscape, to link the geological characteristics of the site (the volcanic system and the bradyseism phenomenon) to the cultural heritage, and, eventually, to produce policies oriented towards its sustainable development.

The aim of the research described in the present paper is to propose a method that uses conventional and non-conventional data, as geotagged social media data (GSMD) to collect, measure and interpret information concerning the cultural resources of a territory and to be used as a participative support for the debate and the decisions on the cultural and territorial planning discussion.

3 Methodologies and Techniques

In order to accomplish these targets, we propose a hybrid methodology based on two different data source: conventional data and "Big Data".

The first one, conventional data, consists in investigation of iconography, photography, cartography, archives, bibliography, databases and statistics data as indirect resources.

The second one is based on new non-conventional sources of geotagged information, as derived from social media like Instagram, Flickr and Twitter, known as "Big Data". In the age of social media, people use to share their experiences and emotions on the web, especially when they are surrounded by elements having such historical and natural relevance, as it happens in Campi Flegrei. Therefore, this data source, known as geotagged social media data (GSMD), give possibility to analyze and collect precious information to interpret spatial and temporal characteristics, as flux and frequency of people in a determinate site; and cognitive features, as perception and appropriation of a place [13]. GSMD provides accurate data and contains user information that makes possible to build character profiles for demographic analyses.

Later, this information were added into a Geographic information system (GIS) software aiming at interpreting and analyzing them in order to understand the actual dynamics of the territory and to address specifics issues regarding its governance. It can be applied in a large-scale case study and englobe all the territory of the Campi Flegrei through quantitative indicators to determinate, for example, the flux and the frequency of the tourism in the region, or it can be applied in minor-scale to analyze a lower quantity of data to read the actual perception of the space, by the cognitive interpretation of the content of each user entry. The second approach can also serves as an element of comparison between historical documents, as iconography and photography, to understand how a given place has developed across time.

4   Case Study

This method was applied to Pozzuoli, the larger and most populated Municipality of Campi Flegrei (43 km^2, 82.000 inhabitants) in order to understand and characterize the dynamics of an urban cultural landscapes and to promote its development, either in the touristic and cultural fields.

An interactive cultural mapping project was developed for the city of Pozzuoli, as part of a more extensive project of cultural planning, resorting to ESRI Story Map tool [14].

This work is the result of a process of collecting, analyzing and synthesizing information in order to identify the cultural resources, to understand them within their context and to create links between these elements.

Firstly, a cultural inventory based on several resources such as cartography, iconography, filmography and literature was set up [15–18]. The result of this phase was the creation of a database that contains information pertaining to all forms of cultural resources, such as archeological sites and monuments localization, points of view of paintings over the centuries, movies scenes filmed in Pozzuoli (Figs. 1 and 2).

Fig. 1. Cultural resource map, archeological sites.

Later, this information were interpreted over three different levels (object, context and connection) and were added to a graphic map generated using ESRI Story Map.

Fig. 2. Cultural resource map, geotagged paintings point of view.

Then, GSMD were imported in the map, filtering the contents through the hashtag #pozzuoli, aiming at displaying the most popular places in Pozzuoli and check if they match with the cultural resources previously identified (Fig. 3).

Fig. 3. Cultural resource map, geotagged social media data.

5 Conclusion

The aim of the project was to make freely available supporting material designed for the public, to be used by the municipality to set up a strategies and policies to promote Campi Flegrei cultural, economic and touristic potential. Moreover, we wanted to create a model that can be followed and applied in the future by other sites that intend their inscription as a Cultural Landscape on the World Heritage List of UNESCO.

References

1. Jones, B.: Current directions in cultural planning. Landsc. Urban Plan. **26**, 89–97 (1993)
2. Amari, M.: Progettazione Culturale. Metodologia e strumenti di cultural planning. Franco Angeli, Milano (2006)
3. Mercer, C.: What is cultural planning? In: Community Arts Network National Conference, Sydney, Australia (1991)
4. Mercer, C.: By accident or design. Can culture be planned? In: Matarasso, F., Halls, S. (eds.) The Art of Regeneration: Nottingham 1996: Conference Papers. Comedia, UK (1996)
5. Tylor, E.B.: Primitive Culture: Researches into the Development of Mythology, Philosophy, Religion, Language, Art and Custom. John Murray, London
6. Williams, R.: The Long Revolution. Chatto & Windus, London (1961)
7. Hall, S.: Il soggetto e la differenza. Per un'archeologia degli studi culturali e postcoloniali. Meltemi Editore, Roma (2006)
8. Duxbury, N., Garrett-Petts, W.F., MacLennan, D.: Cultural mapping as cultural inquiry: introduction to an emerging field of practice. In: Duxbury, N., Garrett-Petts, W.F., MacLennan, D. (eds.) Cultural Mapping as Cultural Inquiry. Routledge, New York and London (2015)
9. Pillai, J.: Cultural Mapping: A Guide to Understanding Place, Community and Continuity. Strategic Information & Research Development Centre, Petaling Jaya, Malesia (2013)
10. Rashid, M.S.A.: Understanding the past for a sustainable future: cultural mapping of malay heritage. Procedia Soc. Behav. Sci. **170**, 10–17 (2015)
11. UNESCO: The Convention for Safeguarding of the Intangible Cultural Heritage (2003)
12. Petroncelli, E., Stanganelli, M.: Place values and change. In: Collins, T., Kindermann, G., Newman, C., Cronin, N. (eds.) Landscape Values, Place and Praxis. Centre for Landscape Studies National University of Ireland, Galway (2016)
13. Chua, A., Servillo, L., Marcheggiani, E., Vande Moere, A.: Mapping cilento: using geotagged social media data to characterize tourist flows in southern Italy. Tour. Manag. **57**, 295–310 (2016)
14. Pozzuoli, un paesaggio Culturale. https://www.arcgis.com/apps/MapJournal/index.html?appid=c05bcaf745f642a791c88515c0d2aaeb. Accessed 15 Dec 2017
15. Raso, C.: Golfo di Napoli - Guida Letteraria: Da Cuma a Sorrento in 43 itinerari. Franco di Mauro Editore, Sorrento (2007)
16. Di Liello, S.: Il paesaggio dei Campi Flegrei: realtà e metafora. Electa, Napoli (2005)
17. Alisio, G.: Campi Flegrei. Franco di Mauro Editore, Sorrento (1995)
18. Amalfitano, P., Camodeca, G., Medri, M.: I Campi Flegrei: un itinerario archeologico. Marsilio Editori, Venezia (1990)

Urban Regeneration, Community-led Practices and PPP

Investigating Local Economic Trends for Shaping Supportive Tools to Manage Economic Development: San Diego as a Case Study

Carmelina Bevilacqua⬤, Giuseppe Umberto Cantafio⬤,
Luana Parisi⬤, and Giuseppe Pronesti$^{(\boxtimes)}$⬤

Mediterranea University of Reggio Calabria, 89100 Reggio Calabria, Italy
{cbevilac, giuseppe.pronesti}@unirc.it

Abstract. During the last decades, the urge to support Regions' competitive positioning within the knowledge-based global economy has been perceived as a priority. Given this emerging need, the concept of regional clusters has become increasingly central to support the decision-making process and guide economic development strategies. This paper aims at contributing to the debate on the role of clusters as the key engine of regions economic development, with a specific focus on local clusters. This argument is discussed through actualizing the study on the case of the San Diego Metropolitan Statistical Area (SD MSA), California. Firstly, an insightful review of relevant works on the topic of clusters is offered. Secondly, the attention shifts on the Case of the SD MSA, which is investigated through a twofold methodological approach, validated in compliance to the MAPS LED project. Accordingly, a sample of highly performing local cluster is selected and studied both quantitatively (using indicators) and spatially (through the GIS mapping). Ultimately, the study offers a comprehensive picture about San Diego local economy by coupling quantitative and spatial analysis. In addition, the study offers insights to policy makers on the potentials of clusters spatialisation as a supportive tool for effective decision-making.

Keywords: Local clusters · Policy-making · San Diego · MAPS-LED

1 Introduction

In recent years, supporting the creation of localized economic advantages has been perceived as a central challenge from regions seeking to compete in the fast-changing global economy. In view of this emerging need, the concept of cluster has become increasingly popular by attracting growing consideration from both academics and practitioners. Accordingly, clusters are currently held in high consideration both as unit of analysis for regional economies and as crucial drivers of decision-making. The allure of the latter academic concept has lead through an extensive production of cluster-based scholarly works, which offer a wide array of definition and categorisations [1]. Despite the plethora of academic contributions on the topic, still the most

© Springer International Publishing AG, part of Springer Nature 2019
F. Calabrò et al. (Eds.): ISHT 2018, SIST 100, pp. 191–200, 2019.
https://doi.org/10.1007/978-3-319-92099-3_23

relevant theorisation about clusters is attributable to Michael Porter [2]. The latter author has conducted extensive studies on clusters by also devising a crucial taxonomy, which subdivides clusters into two categories based on their geographical scope. In this sense, Porter suggests that clusters can be typologically classified as Local and Traded [2, 3]. Accordingly, Local clusters are seen as an inherent element of every region' economic structure as they serve the local market. As opposite, traded clusters occur just in some regions where special competitive advantages are available and serve the global market. The two typologies of clusters differ by nature; however, they play a complementary role in supporting the strengthening of regional economies. Local clusters provide the necessary economic structure to foster the prosperity of traded clusters, which in turn enhance the overall performance of regional economies. By drawing on this backdrop, the present study aims at offering a comprehensive snapshot of the local clusters' panorama in the San Diego area towards pointing out the most relevant features of the local economy. Consistently, the study identifies first, and later spatializes a set of six best performing Local Clusters located within the San Diego area. The focus on the San Diego MSA is justified by the fact that its regional economy is widely recognized as one of the fastest growing and diverse of the nation as a whole. Specifically, the choice to study its local economic pattern is supported by the fact that Local Clusters drive San Diego's local prosperity, covering approximately two-thirds of the region's employment. Indeed, the economic pattern of San Diego is predominantly characterized by a 'small and medium size' business pattern of economic activities serving local market, hence locally producing, and distributing goods and services [5]. In fact, San Diego is not exploiting its international exporting potential, showing a good performance just in the case of the export of components to Mexico [5].

This approach to the study allows the authors to interpret current and potential development trajectories of local economies. The present work targets a wide audience of academics and practitioner by presenting the application of an innovative methodology for the analysis of local clusters and disclosing its potentials to guide decision-making process.

2 Literature Review

The academic and political consideration of clusters is constantly increasing since at least two decades. However, the economic arguments underlying the idea of firms' co-location, have distant roots. The very birth of such long-running cluster concept dates back to the work of the economist Alfred Marshall [6] issued at end of the 19th century. Since then many other scholars have contributed towards expanding knowledge about clusters. Notwithstanding the relevance of such plethora of scholarly works, the very popularisation of the concept of cluster is attributed to Michael Porter [7, 8]. The latter scholar defined clusters as "a geographically proximate group of interconnected companies and associated institutions in a particular field, linked by commonalities and complementarities." According to this definition, geographic proximity can be considered as the core element enabling both interactions within, and development of clusters. Besides defining the cluster phenomenon, Porter also devised a four-factors-based model, the so-called diamond model, to explain clusters success. The model included the following elements: (i) factor conditions, (ii) demand conditions, (iii) firm strategy/rivalry, and (iv) Beyond the diamond model and various.

Beyond the diamond model and various other reflections, Porter recently provided a classification of clusters based on their economic scope [2, 3]. The scholar named above has devised a two-category-based framework, which distinguishes traded from local clusters. These two typologies play different, but complementary roles towards supporting the strengthening of regional economies. On the one hand, traded clusters highly concentrate in few regions, which features some specific competitive advantage. As their name suggests, they tend to export products and services, and consequently are significantly exposed to cross- and extra-regional competition. The occurrence of traded clusters in certain territories usually signals the opportunity for the region to achieve high levels of overall economic performances. On the other hand, local clusters are almost equally occurring in every region, since the location choice of the firms composing clusters does not depend on the availability of territorial competitive advantages. Local clusters are strongly tied to the region they are located in, hence are not exposed to cross- and extra-regional competition. This synthetic characterisation of traded and local clusters suggests some considerations. First, traded and local cluster manifest divergent market targets. The former cluster type competes on, and serves for the global market, in contrast, the latter one works within the boundaries of the local market. Second, traded and local clusters have different features in terms of job creation and innovative potential. Traded clusters typically register higher wages, and levels of innovation (in term of patented products or services), while local clusters feature higher employment (which is usually proportional to the region's population size). Despite the clear-cut discrepancies in scope and scale, traded and local cluster still expect to operate jointly. Specifically, local clusters provide the necessary economic infrastructures to foster the prosperity of traded clusters, which in turn enhance regional economic performances. Despite the importance of the dichotomy local vs. traded, it seems that in recent years some part of the academic world has focused more towards the study of local clusters. Such tendency is witnessed by a flourishing production of local-cluster-based scholarly articles [9–12]. Recently Demarchi, Dimaria and Gereffi analysed the co-evolutionary trajectory of local clusters and global value chain towards envisaging the possibility for territories containing local clusters to prosper in the new global scenario. The same argument is also supported by Nadvi and Halder who investigated two case studies in Pakistan and Germany [13]. The growing importance of local cluster calls for consideration from academics to better comprehend the dynamics through which these local forms of economic agglomeration can contribute to the enhancement of local economies.

3 Methodology

In accordance to the claims presented in the introductory paragraph, the essential aim of this work is to snapshotting the local clusters' panorama in the SD MSA towards pointing out the most relevant features of the SD local economy. In other words, the study seeks to disentangle and describe the role played by local clusters, representing the regional high-specialisation pattern, towards shaping the actual, and influencing the potential structure of San Diego local economy. By drawing on this research objective, the study implements a two-stage methodology by taking cues from the MAPS-LED project.[1] Firstly, the essential attributes of a sample of local clusters are pointed out, by drawing on the quantitative examination of some key indicators. Secondly, the spatialisation of the selected local clusters is accomplished towards highlighting cluster-specific morphological features.

Initially, a set of indicators is selected to streamline the structure of each sampled cluster. These indicators have been collected for the length of time 1998–2014, using the U.S. Cluster Mapping Project as the main data source [14]. The indicators can be classified into three categories representing the following features of clusters:

- Composition, which is indicated by the number of establishments per year;
- Economic structure, which is signalled by Location Quotient, National Employment Share, Employment, and Wage;
- Innovation Ecosystem, which is quantitatively described by Poverty rate, job creation, patent count, patent count growth, Venture Capital per $10,000 GDP, Total receiving high school diploma or more, Total with some college or associate degree, Total completing a bachelor's degree or more, and Cluster strength).

Then, the study delineates cluster-specific morphology, representative of the regional high specialisation pattern, both at the County and the urban level. In view of this objective, it has been set up a relationship between NAICS codes, strictly connected to the classification of Clusters operated by Delgado et al. [15] and Land Use data, which reflect the Regional Growth Forecast of the SANDAG Public Agency [16]. The rationale of the spatialisation process is rooted in the acknowledgment that each Land Use code can be combined with the economic activities that are classified through NAICS codes, allowing to create a morphology of subclusters and, in turn, clusters.

The general methodology mentioned above [17] is applied to the case of the San Diego MSA, by sampling the regional local clusters and analysing them through quantitative indicators and spatial investigation. Given the exploratory nature of the research, the sample size determination is based on the criterion of high-specialisation, considering that the specialisation pattern is measured by the value of the Location Quotient (LQ) of a cluster and that a LQ greater than 1 indicates a concentration of the cluster higher than average. As a result, the list of case studies - content of the present investigation - includes the six most performing local clusters in terms of high-specialisation, thus, with a LQ greater than 1, specifically: Local Personal Services (Non-Medical); Local Household Goods and Services; Local Hospitality Establishments; Local Commercial Services; Local Real Estate, Construction, and Development; Local Industrial Products and Services.

Shape files and Metadata have been gathered from the main US authorities' web sites. The process of data gathering has represented one of the main challenges over the workflow of the study, as it has required an intensive online research of reputable

---

[1] http://www.cluds-7fp.unirc.it/maps-led.html.

sources. The following Table 1 shows the metadata used throughout the implementation of the study:

Table 1. Metadata about the San Diego Metropolitan Statistical Area.

Data Name	Authority	Data Typology	Years
San Diego Metropolitan Statistical Area Delineation	U.S. Census Bureau	Metadata	2009
U.S. Nation Boundaries	Census.Gov	Shapefile	2014
American Continent Boundaries	Arcgis.com	Shapefile	2012
State with County Boundary	SANDAG	Shapefile/Metadata	2014
Municipal Boundaries	SANDAG	Shapefile/Metadata	–
Census	SANDAG	Shapefile/Metadata	2010
Zoning Base	SANDAG	Shapefile/Metadata	2015
Land Use	SANDAG	Shapefile/Metadata	2008
Air Runways	SANDAG	Shapefile/Metadata	–
County - Medical Services	SANDAG	Shapefile/Metadata	–
Zip Codes	SANDAG	Shapefile/Metadata	2015
Colleges	SANDAG	Shapefile/Metadata	–
Hospitals	SANDAG	Shapefile/Metadata	–
Transportation	SANDAG	Shapefile/Metadata	–
Jurisdiction	SANDAG	Shapefile/Metadata	–

As we will see in the next section, this research activity, conducted at the parcel level, will allow to give a physical meaning to the abstract units of the industrial sectors and to observe their level of concentration in specific areas of the San Diego MSA.

4   Discussion and Results

Consistently with the methodology described in the previous paragraph the sample of six clusters has been firstly investigated through a quantitative lens towards analysing the indicators accounting for clusters': (i) composition, (ii) economic structure, and (iii) innovation ecosystem.

The study of the clusters' composition revealed a general increase in the number of firms belonging to the six local clusters selected in the San Diego MSA, during the time frame 1998–2014. The highest increase in number of firms was registered for the Real Estate Local Cluster, with an increase of firms of 100%. The other components of the sample showed lower increases in the indicator number of firms (e.g. the 20% of increase in the case of the establishments of the Local Commercial Services Cluster). The investigation on clusters' economic structure suggests a general tendency of the best performing clusters towards improving their structure. In this sense, the percentage change in employment (for the entire sample of clusters) from 1998 to 2014, increased by 30%, with a certain loss of jobs in the years 2007–2011, probably due to the economic downturn that hit the global economy. Also, the percentage change in Wages by County from 1998 to 2014 increased, without any diminution, neither in the economic crisis period. The percentage change in Patents by County from 1998 to 2014 showed some up and downs until 2008, when it started increasing until 2014. In 2014 an increase of 60% of patents has been appreciated in comparison with the levels of 2008. The National Employment Share by County considering its percentage change in the selected length of time, presented an increase until 2007, when it was registered a peak. From 2007 to 2010 a decrease in the specialisation has been recorded, then the indicator started back to increase until 2014. The location quotient showed the same trend as the national employment share, as they both represent the specialisation of local clusters.

The study of Clusters' innovation ecosystem targets that, as we have seen, have been divided in the two categories of Performance indicators and Business environment indicators, have been collected for the period 2008–2013. Over the selected period, the poverty rate increased for all the strongest local clusters, showing a rising gap in society in terms of equality. The job creation has been found negative for the year 2008 in most of the local clusters, while it has been registered generally positive in the years 2011–2013. It seems that every selected cluster manifest, to a certain extent, a tendency towards innovation. This is witnessed by the increasing number of patents counted in the years 2008, 2011, and 2013. Concerning the Business Environment indicators, the venture capital investment is very small especially for local industrial products and clusters seems to have a similar behaviour. This first stage of the study has disclosed some quantitative features of the selected clusters; however it falls somewhat short of providing links to connect local economic phenomena with a detailed spatial dimension. In this sense, the analysis of indicators needs to complement with the spatialisation of clusters. The spatialisation of the six selected clusters within the boundaries of San Diego MSA has been carried out by considering 5-digit zip codes as the main territorial unit. Accordingly, the authors have collected and processed the entire set of 185 zip codes composing the San Diego MSA, namely from the zip code 91901 to 92199.

The mapping enables an in-depth understanding of the relevance of the urban scale in the way of disclosing concentration and localisation of the sampled clusters. Such urban-level analysis displays clear-cut differentiations in the spatial configuration of the selected clusters. The spatial pattern and economic features allow the authors outlined two macro categories.

The first category, namely "supporting clusters", includes Local Hospitality (in light blue), Local Personal Services (in orange), and Local Industrial Services (in purple). These clusters feature a scattered spatial configuration, which is likely to depend on the typology of services that such clusters are expected to provide. The clusters mentioned above are mainly located in proximity to the most important touristic sites: on the coastal area, in the downtown area, La Jolla area, and in the inland areas.

The second category, namely "specific clusters", includes Real Estate (in yellow), commercial services (in red) and Household Goods (in green). These clusters manifest a concentrated morphology, and feature high specialisation (e.g. the location quotient of the real estate cluster is 1.3%). Considering the Real Estate cluster, it is possible to observe that it is concentrated around University City area, because of the presence of high skilled labour pool specialized in the design and construction of high-tech laboratories for the facilities and firms located around the UCSD campus. The other two clusters, namely the Local Commercial Services (in red) and Local Household Goods (in green) are localized in specific areas. The former is located in the downtown area, where there is more density and therefore more presence of customers and tourists, while the latter is located close to the Mexican border. This cluster has been found to be located at the southern limit of the city of San Diego, bordering with Mexico, where it reveals a greater concentration of establishments.

5 Conclusions and Further Studies

The present study investigates the case of local clusters in the San Diego County, California. Drawing insights from previous research activities carried out in compliance to MAPS-LED project, the aim has been to define the local clusters' panorama in the SD area towards pointing out the most relevant features of its local economy. Specifically, the study has disclosed the attributes of the regional high-specialization pattern in the way of understanding its influence on the County's economic structure. The paper suggests a novel methodology for mapping local cluster based on quantitative indicators, drawing insights from the idea developed by the Land Development Code Commerce City of Colorado [18].

The implemented methodological approach has aimed at displaying where clusters are physically localized within an urban territorial scale.

The core of the clusters spatialisation process has been represented by the connection between NAICS codes and the Land Use categories of the San Diego County. Such process allowed to build up a descriptive picture of clusters at the Local level, giving a physical meaning to the abstract units of the industrial sectors and observing their level of concentration in specific areas of the examined County.

The mapping activity allowed to understand where clusters are more concentrated, and also to observe their level of concentration. As shown in Fig. 1, although clusters are concentrated in the inner city area, they look geographically dispersed. Moreover, it emerged the multi-scalarity of the geographic concentration of cluster, so that each territorial unit, at different scales, provides an independent structure able to give important information and describe some significant aspects of that particular territorial dimension.

The cluster occurrence characterizes each area also in terms of innovation and specialisation, as indicated by the increasing number of patents and Location Quotient, and deducted by the strict connection between clusters, innovation and specialisation theorised by Michael Porter. Additionally, several innovation initiatives are active in the San Diego area, with a crucial role in spurring Local Economic Development, increasing jobs, attracting new businesses, and empowering local communities. The analysed clusters further show a high propensity to physical transformations driven by urban planning tools, that help also to augment innovation. Moreover, unlike the other regions of the U.S., San Diego presents a richness in terms of clusters' variety and local employment, that should be leveraged further for expanding the region's ability to

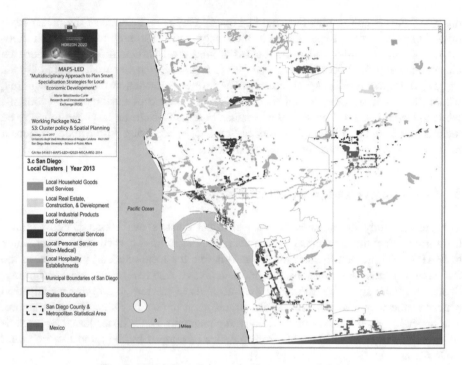

Fig. 1. San Diego County, local cluster spatialisation.

attract investments and reinforce the economic structure through the consolidation of professional networks.

The work is a lens that has the potential to give to the policy-makers a real picture of the existing economic trends at the urban level, identifying the structure of specific economies, their specialisation and the role played by clusters. Thus, if paired with the analysis of social indicators and sided by policy initiatives for spatial planning, the visualisation methodology can help as a supportive tool in taking place-based decisions for the future pockets of development, within the Smart Growth Framework.

This analytic instrument can reveal what are the features that influence regional growth, showing what are the factors that characterize a region by making it unique. Moreover, considering that the cluster activation, with its geographical concentration, is highly related to innovation, "the cluster approach is part of a set of innovation system approaches" [19]. Accordingly, the mapping methodology can be a tool first for regional policy makers to design and implement the Smart Specialisation Strategy framework. Furthermore, in view of the fact that clusters do not rely on administrative boundaries, this work can be extremely useful for entrepreneurs and its lens can help entrepreneurs and regions to see each other, defining new opportunities for the growth of new businesses, the implementation of new joint ventures and investments, and the creation of new jobs.

The economic structure of clusters based on Porter's methodology can bring also some challenges linked to the fact that each specific place has its own clustering model, thus, by "standardising" the process the risk is to miss the dynamics related to the aggregation process of some specific places. Moreover, the statistical mapping is not sufficient to identify clusters' methodologies, because of the existence of inconsistencies and gaps in the databases. This can make difficult the interpretation but also the comparison of data. The necessity to overcome these limitations consider a further step that validates the statistical findings and includes qualitative information and analysis that will be combined with the quantitative methodology.

Further studies will establish the role of Community Planning Areas in supporting the growth of local clusters. This will be done on the basis of the land use categories defined by the Plans, the morphology of clusters and the significance of their concentration. This correlation will include the investigation of the occurrence of localized competitive advantage linked to territorial-specific economic attributes.

Acknowledgement. This work is part of the MAPS-LED research project, which has received funding from the European Union's Horizon 2020 research and innovation programme under the Marie Skłodowska-Curie grant agreement No. 645651.

References

1. Cruz, S.C.S., Teixeira, A.A.C.: The evolution of the cluster literature: shedding light on the regional studies-regional science debate. Reg. Stud. **44**(9), 1263–1288 (2010)
2. Porter, M.E.: The economic performance of regions. Reg. Stud. **37**(6–7), 549–578 (2003)
3. Delgado, M., Porter, M.E., Stern, S.: Clusters, convergence, and economic performance. Res. Policy **43**(10), 1785–1799 (2014)
4. Bevilacqua, C., Spisto, A., Cappellano, F.: The Role of Public Authorities in supporting Regional Innovation Ecosystems: the cases of San Diego and Boston regions (USA). In: International Research Conference 2017: Shaping Tomorrow's Built Environment Conference Proceedings, Salford, UK, pp. 1005–1018 (2017)
5. Porter, M. E.: Clusters of Innovative Initiative (2001)
6. Marshall, A.: Principles of Economics. Macmillan and Co., London (1890)
7. Porter, M.E.: Clusters and the new economics of competition. Harvard Bus. Rev. **76**(6), 77–90 (1998)
8. Porter, M.E.: Porter on the competitive advantage of clusters. Strategic Direction, 21 (1999)
9. Mytelka, L., Farinelli, F.: Local Clusters, Innovation Systems and Sustained Competitiveness. In: UNU/INTECH Discussion Papers, pp. 7–37 (2000)
10. Porter, M.E.: Economic development: local clusters in a global economy. Econ. Dev. Q. **14**(1), 15–34 (2000)
11. Brenner, T.: Local Industrial Clusters: Existence, Emergence, and Evolution. Routledge, London (2004)
12. Brenner, T.: Identification of local industrial clusters in Germany. Reg. Stud. **40**(9), 991–1004 (2006)
13. Nadvi, K., Halder, G.: Local clusters in global value chains: exploring dynamic linkages between Germany and Pakistan. Entrepreneurship Reg. Dev. **17**(5), 339–363 (2005)
14. U.S. Cluster Mapping. Cluster Mapping Methodology. http://www.clustermapping.us/content/cluster-mapping-methodology. Accessed 10 May 2017
15. Delgado, M., Bryden, R., Zyontz, S.: Categorisation of traded and local industries in the US economy. Mimeo (2014)

16. SANDAG: San Diego Forward. The Regional Plan (2015)
17. Bevilacqua, C., et al.: MAPS-LED Multidisciplinary Approach to Plan Smart Specialisation Strategies for Local Economic Development (2015)
18. Land Development Code Commerce City, Colorado, pp. 1–111. Article V. Uses and Accessory Structures (2015)
19. Terstriep, J.: Cluster Mapping - Analysis Grid. Institute for Work and Technology (2008)

Integrated System of Training and Orientation: Towards a Measurement of Outcomes

Domenico Marino[1]($\boxtimes$) (iD), Antonio Miceli[2], and Pietro Stilo[1]

[1] Mediterranea University of Reggio Calabria, 89100 Reggio Calabria, Italy
dmarino@unirc.it
[2] University dm of Messina, 98122 Messina, Italy

Abstract. The need for training and guidance to be provided synergistically as part of the same path is well accepted.

Identification of the level of learning and commitment, individual aptitude and aspirations cannot be left only to the choices of individuals and their parents. In personal, family and social terms, the investment, not only economic-financial but also in terms of cultural and professional growth and social advancement is too high and strategically important to be implemented without the support necessary to reduce the risk of failure or wasted effort (dropping out, delays etc.).

Keywords: Education · Evaluation

1 Introduction

The need for training and guidance to be provided synergistically as part of the same path is well accepted.

Identification of the level of learning and commitment, individual aptitude and aspirations cannot be left only to the choices of individuals and their parents.

In personal, family and social terms, the investment, not only economic-financial but also in terms of cultural and professional growth and social advancement is too high and strategically important to be implemented without the support necessary to reduce the risk of failure or wasted effort (dropping out, delays etc.). This does not affect the choice of a precise political-social model and the State has a duty to ensure social mobility for deserving individuals as well as equal opportunities (also in a gradual manner, if necessary) across the national territory.

Likewise, urgent new investments in schools, universities and research are essential for the country in the current adverse economic climate. These should always be seen as high-return investments and not costs. In 2009, the Bank of Italy commissioned an important, detailed study on this topic.

"In the long term, the higher public expenditure required to fund an increase in education levels would be more than offset, especially in the South, by the increase in tax revenue, the taxation system being unchanged, and by the lower costs generated by the increase in employment rates."

© Springer International Publishing AG, part of Springer Nature 2019
F. Calabrò et al. (Eds.): ISHT 2018, SIST 100, pp. 201–205, 2019.
https://doi.org/10.1007/978-3-319-92099-3_24

Therefore, a comprehensive scheme must be put in place to connect education and training with civil society.

2 Objectives

We intend to highlight the links and the move from secondary school to university education, also considering certain aspects of post-university education.

To this end, the following aims should be considered:

- Increase the number of university graduates, aligning it with that of more advanced European States;
- Enable deserving students to access advanced education. In this regard, it is essential to make investments in the "Right to Education";Those students who are "falling behind" with their university studies should receive guidance to help them choose alternative education or employment pathways more in line with their aptitude and experience;
- Respond positively to the strong demand coming from young people to "renegotiate" the conditions of their future, recognising their right to work, to social advancement based on merit, to make free and informed choices about their future;
- Improve the reputational capital of the education system and its effectiveness in preparing young people for employment.

3 Tools Needed and How to Use Them

To achieve these objectives it is necessary to:

- Consider education and guidance as a single system along the whole educational supply chain;
- Draw up a single, multi-annual plan covering: objectives, personnel, facilities, services and financial resources for the development and operation of the whole system and its subsystems;
- Provide for operational flexibility of peripheral structures, based on their greater autonomy and self-programming capacity. This would be a participatory model, in which the teachers of the various educational agencies, students and families discuss the issues of the different educational levels with stakeholders and local authorities;
- Choose a model based on merit and personal aptitude in which - at the points of transition between the different subsystems - assessment is mainly systemic and only occasionally based on interviews or tests. Hence, assessments based on the "continuity and consistency" of the previous educational path and on respect of planned time-frames (especially for working students);
- Prepare ranking lists (adjustable over time) for exercise of the students' admission rights. This right is especially important for courses with a pre-set or planned number of admissions;

- Change the current curricula for the various types of degree and establish 4 macro-areas:

  (A) Life Sciences
  (B) Technology
  (C) Social-economic and legal
  (D) Literary

- Draw up a new guidance and educational model for students lagging behind their university schedule. Establish clear deadlines for students currently lagging behind and ensure they can either complete their education (even by means of individual agreements) or transition to work;
- Create a systematic link between education, research, and the labour market, at both central and local level. This systemic and synergistic link must perforce involve all education and training institutions;
- Pay special attention to models and tools for assessing aptitude, educational levels and expectations. These models, apart from being defined and tested at central level, must ensure the highest possible level of objectivity and transparency. In the initial phase, they should be binding only for individual universities. In the event of transfer, a committee could be set up to assess the comparability of assessments and hence admission to the ranking lists of the "receiving" university.
- Design a dynamic and open paradigm for transfers between different degree courses. This should include courses with a cap on admissions and ensure regular monitoring and updating of ranking lists. This is to allow the exit of those who failed to meet the required standards and the entry of students who have achieved outstanding results in less in-demand degree courses [1].

4 Programming and Updating Model at Peripheral and Regional Level

As stated in the introduction, it is necessary to coordinate and plan at national and regional level the various components of the education system (Primary, Secondary, University and Postgraduate). This coordination must rest on the autonomous pro-gramming capacity of each school, cluster of schools in a single municipality or district and the schools of a whole region. This is because the constantly changing social and economic environment requires constant updates that cannot be left to the central government alone.

One allied benefit of local coordination would be to limit national and regional red tape and ensure that the evolution of the organisation of educational processes is not entrusted entirely to rigid and complex legislative procedures [2, 3].

Another important task of the "Peripheral Programming and Updating Model" would be to periodically review the currency and relevance of the educational contents to be delivered by teachers and guidance counsellors (in the transition between lower and upper secondary school and between upper secondary school and university). Review at secondary school level would be ensured by coordination of the various local school managers and authorities (Guidance Council), while review in the transition to the university would be carried out by creating a joint school-university council. In this case too, the stakeholders' opinion should be considered, while not being binding. This approach would remove the risk of a self-referential approach and would harmonise the boundaries between the different educational subsystems. The link to be achieved between education, the community and the territory is equally important. The aim is also to monitor constantly the effectiveness of the educational supply chain in enabling young people to enter the labour market and social processes.

5   Conclusions

For this reason, too we propose a general education-guidance-selection model, participated and informed, in which the different stakeholders shape and control the entire educational supply chain, from primary and secondary education to university and postgraduate education.

Thus, this approach offers an overall model for participated guidance, in which the teachers at the different educational levels operate within a system approach and interact dynamically and constantly with stakeholders along the entire educational path. An education and guidance system designed around students, to support their right to employment, to have their needs met, to shape their own destinies. This approach would empower stakeholders (students, families, representatives of businesses and industry) to control actively a system in which educational pathways are in line with social and economic changes, avoiding rigidities stemming from the defence of privileges and unfair benefits. A single education and guidance system supporting students in making their educational choices, helping them to correct those choices if necessary and supporting them in their personal, cultural and civic growth and in accessing the labour market. A system able to plan its activities autonomously and to monitor that each phase in the education pathway is as close as possible to the general public interest.

Accordingly, when students are assessed for admission to courses subject to quota restrictions - by reference to structural and operational capacity and to regional, national and European demand - it is necessary to assess student capacity, aptitude, knowledge and consistency of learning effort. Young people must be aware that their education, testing and guidance process is essential for their future transitions and choices. This awareness can be boosted by a system of incentives and consequences based on their learning effort and test results. This would take the form of systematic recording of learning credits and/or debts accompanying individual selection tests.

Accepting the principles of this preliminary and indicative model could be a first step for reflecting on the objectives of a process to rationalise schools and, especially, universities, in terms of scientific-didactic and professional contents.

References

1. Cingano, F., Cipollone, P.: I Rendimenti dell'Istruzione. Questioni di Economia e Finanze **53**, 15 (2017)
2. Organisation for Economic Cooperation and Development: The PISA 2003, Assessment Framework. Mathematics, Reading, Science and Problem Solving Knowledge and Skills. OECD, Paris (2003)
3. Trapasso, R., Staats, B.: Policy complementarities and OECD national skills policies, working paper. OECD (2017)

Post Carbon City and Real Estate Market: Testing the Dataset of Reggio Calabria Market Using Spline Smoothing Semiparametric Method

Vincenzo Del Giudice[1], Domenico Enrico Massimo[2],
Pierfrancesco De Paola[1(✉)], Fabiana Forte[3], Mariangela Musolino[2],
and Alessandro Malerba[2]

[1] Federico II University of Naples, 80125 Naples, Italy
pierfrancesco.depaola@unina.it
[2] Mediterranea University of Reggio Calabria, 89124 Reggio Calabria, Italy
[3] Luigi Vanvitelli University of Campania, 81031 Aversa, Italy

Abstract. In this paper a hedonic price function built through a semiparametric additive model is tried out for the real estate market analysis of the central area of Reggio Calabria. The semiparametric model uses Penalized Spline functions and aims to achieve an improvement in the prediction of the market prices of housing properties in the central area of Reggio Calabria. More in particular, the final objective of the research is to detect and to identify possible potential market premium in real estate exchange and rent markets for green buildings. This is the first preliminary phase for the unavoidable verification of the robustness of the real estate sample, or for the subsequent individuation of progressive real estate sub-samples.

Keywords: Post carbon city · Green buildings
Penalized spline semiparametric method · Semiparametric regression
Real estate market analysis

1 Introduction

The evolution of real estate markets is influenced by quantitative and qualitative characteristics, as well as by differentiation and the change in the mode of appreciation of the real estate goods. These aspects suggest the development of new and advanced models for hedonic analysis of property prices, able to recognize the different forms of appreciation, based on survey and statistical analysis of market data [1–29].

Final and telescopic objective of the paper is to detect and to identify possible potential market premium in real estate exchange and rent markets for green buildings [23–28]. This is the first preliminary phase for the unavoidable verification of

---

The authors contributed equally to the study.

© Springer International Publishing AG, part of Springer Nature 2019
F. Calabrò et al. (Eds.): ISHT 2018, SIST 100, pp. 206–214, 2019.
https://doi.org/10.1007/978-3-319-92099-3_25

robustness of the real estate sample, for the subsequent individuation of progressive real estate sub-samples.

In international literature many studies have applied some special non-parametric or semiparametric additive regressions for the formulation of hedonic price models for the analysis of housing market. Mainly, these studies make use to Generalized Additive Models, among the most common non-parametric multivariate regression techniques, and the "backfitting algorithm" [3] that represents main method for resolution of additive models in base to available statistics data [4, 5].

An alternative approach with limited computational difficulties in estimating the individual functions that define an additive model, consists to place and match to each of these functions some specific *smoothing spline* function.

Currently the use of *smoothing spline* functions interest many scientific fields, like chemistry, natural and physical sciences, medicine, economy (limitedly to production costs only).

Semi-parametric models applied to real estate appraisals are currently subject of specialized literature and, particularly, it concerns choice and processing of property prices and real estate features [13, 14, 16–24].

2 Model Specification

The relationship between selling price and explanatory variables is examined with a semi-parametric additive model, characterized by the combination of a generalized additive model, which expresses the relationship between the non-linear response and the explanatory variables, and a linear mixed effects model, which expresses the spatial correlation of observed values:

$$\text{PRICE} = \beta_0 + \beta_1 \text{LIV} + \beta_2 \text{MAIN} + \beta_3 \text{BATH} + \beta_4 \text{ZONE} \\ + f_1(\text{AREA}) + f_2(\text{EPOCA}) + f_3(\text{DAT}) \tag{1}$$

More precisely, in the expression (1) the additive component, the mixed effects and the erratic component (ε), are independent. Furthermore, in order to obtain a function estimated using the procedures relating to models mixed effects, it is considered a version of low rank both for the additive component both for the mixed effects [1].

The proposed semiparametric model can then be briefly defined by the following general formula:

$$y = X\beta + Zu + \varepsilon \tag{2}$$

where:

$$y = (y_1, \ldots, y_N)^T \tag{3}$$

$$X = [1 \ xi \ ]1 \le i \le N \tag{4}$$

Z contains $T \leq N$ truncated power basis functions of *p-degree* for the approximation of nonlinear structure in *f* functions:

$$Z = \begin{bmatrix} (x_1 - \kappa_1)^p_+ & \cdots & (x_1 - \kappa_k)^p_+ \\ \cdots & \cdots & \cdots \\ (x_n - \kappa_1)^p_+ & \cdots & (x_n - \kappa_k)^p_+ \end{bmatrix} \tag{5}$$

And alternatively, in reduced form:

$$Z = \left[(x_i - \kappa_k)^p_+ \right._{1 \leq k \leq K} \right]_{1 \leq i \leq n} \tag{6}$$

where u = (u1,..., uk)T is the vector of random effects with:

$$E(u) = 0, \ Cov(u) = \sigma_u^2 I, \ Cov(\varepsilon) = \sigma_\varepsilon^2 I, \tag{7}$$

considering the coefficients (uk) of knots (κk) as random effects independent of ε term [1].

Note that the formulation (2) is a particular case of linear mixed-effects model of Gaussian type. For non-linear components of the model are used penalized spline functions qualified by the following general expression:

$$f(x) = \alpha_0 + \alpha_1 x + \ldots + \alpha_p x^p + \sum_{k=1}^{K} \alpha_{pk}(x - \kappa_k)^p_+ \tag{8}$$

in which the base of the generic function (3) is represented by the following terms:

$$1, x, \ldots, x^p, (x - \kappa_1)^p_+, \ldots, (x - \kappa_k)^p_+ \tag{9}$$

Where the generic function $(x - \kappa_k)^p_+$ has $(p - 1)$ continuous derivatives.

For p > 0 the expression that is used to determine the fitted values is as follows:

$$\hat{y} = X(X^T X + \lambda^{2p} D)^{-1} X^T y \tag{10}$$

Where:

$$X = \begin{bmatrix} 1 & x_1 & .. & x_1^p & (x_1 - \kappa_1)^p_+ & \cdots & (x_1 - \kappa_k)^p_+ \\ .. & .. & .. & \cdots & \cdots & & \cdots \\ .. & .. & .. & \cdots & \cdots & & \cdots \\ .. & .. & .. & \cdots & \cdots & & \cdots \\ 1 & x_p & .. & x_n^p & (x_n - \kappa_1)^p_+ & \cdots & (x_1 - \kappa_k)^p_+ \end{bmatrix}$$

$$D = diag\left(0_{p+1}, 1_K\right) \tag{11}$$

Simplifying, the relation (4) becomes:

$$\hat{y} = S_\lambda \cdot y \tag{12}$$

The smoother matrix is defined as follows:

$$S_\lambda = X(X^T X + \lambda^{2p} D)^{-1} X^T \tag{13}$$

The λ term is usually referred to as smoothing parameter.

The smoothing parameter intervenes in the determination of the degrees of freedom for nonlinear component of the model and allows also to control the trade-off between fitting model to the observed values (smoothing parameter near to zero value) and the smoothness of the same (high values of smoothing parameters).

The selection of the smoothing parameter, for a spline function of p-degree, occurs by the Restricted Maximum Likelihood condition.

3 The Real Estate Market Analysis of the Central Area of Reggio Calabria

The market price analysis of the property carried out with the use of an additive semi-parametric model provides for the adoption of statistical tools (significance test, measures of residues, etc.) able to select both the sample data and the endogenous variables [9–28]; these tools also allow to verify the reliability and the quality of results.

The algebraic structure of proposed model has been specified on the basis of real estate data of the sample, as well with the help of statistical and empirical- argumentative tests, by implementing the following semiparametric additive model:

$$\begin{aligned} \text{PRICE} = {} & \beta_0 + \beta_1 \text{LIV} + \beta_2 \text{MAIN} + \beta_3 \text{BATH} + \beta_4 \text{ZONE} \\ & + f_1(\text{AREA}) + f_2(\text{EPOCA}) + f_3(\text{DAT}) \end{aligned} \tag{14}$$

The data sample refers to a defined real estate market segment of Reggio Calabria (Calabria region, Italy) and, specifically, no. 490 sales of residential property units located in an urban central area during twenty five years (Tables 1 and 2).

The sampled properties have the same build type and quality (residential units located in used multi-storey buildings), and they are included in a homogeneous central area in terms of qualification and distribution of main services.

In the absence of multicollinearity phenomena, given the low correlation between the explanatory variables, the main verification indexes of the model are shown for completeness in tables and graphs.

The amounts related to the standard error (€ 15,313.85) and absolute percentage error (11,40%) appear congruent, because the forecast values obtained using the proposed model show a trend compliant to observed data, also even residue analysis shows no abnormalities (Fig. 1).

Table 1. Variable description.

Variable	Description
Real estate price (PRICE)	Expressed in thousands of Euros
Property's age (EPOCA)	Expressed retrospectively in no. of months
Sale date (DAT)	Expressed retrospectively in no. of months
Internal area (AREA)	Expressed in sqm
Number of services (BATH)	No. of services in residential unit
Positional variable (ZONE)	Expressed with a score scale (from 1 to 6, passing from more central areas to peripheral areas)
Maintenance (MAIN)	Expressed with a score scale (1 for bad conditions, 2 for mediocre conditions, 3 for good conditions, 4 for optimal maintenance state)
Floor level (LIV)	No. of floor level of residential unit

Table 2. Statistical description of variables.

Variable	Std. Dev.	Median	Mean	Min.	Max.
PRICE	55,006	115,000	121,088	13,000	400,000
EPOCA	19	25	29	2	102
DAT	7	11	11	0	35
AREA	30	105	108	39	287
BATH	1	2	2	1	4
ZONE	2	5	4	1	6
MAIN	1	3	3	1	4

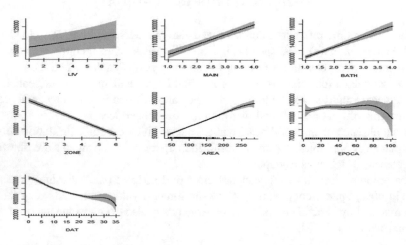

Fig. 1. Effects of nonlinear components on selling prices with representation of 95% confidence level.

From the statistical point of view, significant is the determination index, equal to 0,969 (corrected index equal to 0,968), as well as the F test is significant for a 95% confidence level.

The fixed effects of model's linear component that result statistically significant coincide with all variables.

With regard to the nonlinear part of model, there are no significant abnormalities encountered in the values assumed by smoothing parameters (spar) or freedom degrees (df), (see Table 3).

Table 3. Main results of semiparametric model and estimation of fixed effects for linear and nonlinear components of the model.

No. of variables				7	
Mean for PRZ variable (€)				121,088	
Standard error (€)				15,313.85	
Percentage error (SE/Mean for PRZ variable) (%)				12,65	
Determination index (R^2)				0,969	
Corrected determination index (R^2)				0,968	
Absolute percentage error (%)				11,40	
F-test				1,130.30 (> F (α, n, m-n-1))	
Variable	*coef*	*SE*	*Ratio*	*P-value*	
Intercept	− 9,732.0	61,040.0	− 0.1594	0.8734	
LIV	888.1	541.7	1.6390	0.1013	
MAIN	13,130.0	1,044.0	12.58	0.0000	
BATH	24,120.0	1,934.0	12.47	0.0000	
ZONE	− 15,400.0	543.4	− 28.33	0.0000	
Variable	*df*	*spar*	*knots*		
f(AREA)	5.000	299.0	27		
f(EPOCA)	5.479	35.97	18		
f(DAT)	5.874	10.22	7		

In the model's linear component, the variables' coefficients directly express the implicit marginal prices; for the nonlinear component, marginal prices for each variable are obtained by processing and examination of estimated functions.

For brevity of discussion, for each nonlinear variable of model, the marginal prices are not shown, being a primary objective of this paper the experimentation of proposed model and to verify the reliability of the real estate sample.

In conclusion, this work leads to results which, for their consistency with buying and selling prices detected, can be considered representative of the validity of the methodology used. The tool used for analyze the real estate data is the R-project software.

4 Conclusions

The results obtained with the application of the proposed model are excellent, and they suggest that semiparametric models can be successfully used for the prediction of residential property selling prices.

In the study case the error committed in the prediction of selling prices is lower about 3.26% respect to conventional multiple regression models, showing very high use's potential. This result may aid to detect and to identify, as further future research developments, possible potential market premium in real estate exchange and rent markets for green buildings, as well as progressive real estate sub-samples to analyze.

More generally, in line with the experimental results obtained in this paper, the semiparametric models can lead to improved estimated between 10 and 20% in the prediction of housing market prices, compared to conventional multiparametric techniques. The objectives pursued with the theoretical model proposed are many and varied, such as the study of the various segments of the local real estate markets, or even the prediction and interpretation of phenomena related to the genesis of the income housing, with particular reference to the problems of transformation of urban areas concerned from projects or plans of action and in order to optimize the user choices of goods and resources such as energy.

References

1. Ruppert, D., Wand, M.P., Carroll, R.J.: Semiparametris Regressions. Cambridge University Press, Cambridge (2003)
2. Hastie, T., Tibshirani, R.: Generalized Additive Models. Chapman & Hall, New York (1990)
3. Bao, H., Wan, A.: On the use of spline smoothing in estimating hedonic housing price models: empirical evidence using Hong Kong data. Real Estate Econ. 32(3), 487–507 (2004)
4. Opsomer, J.D., Claeskens, G., Ranalli, M.G., Kauermann, G., Breidt, F.J.: Nonparametric small area estimation using penalized spline regression. J. Roy. Stat. Soc. Ser. B 70(1), 265–286 (2008)
5. Montanari, G.E., Ranalli, M.G.: A mixed model-assisted regression estimator that uses variables employed at the design stage. Stat. Methods Appl. 15(2), 139–149 (2006)
6. Bin, O.: A prediction comparison of housing sales prices by parametric versus semiparametric regressions. J. Hous. Econ. 13(1), 68–84 (2004)
7. Clapp, J.: A semiparametric method for estimating local house price indices. Real Estate Econ. 32(1), 127–160 (2004)
8. Wand, M.P.: Smoothing and mixed models. Comput. Stat. 18, 223–249 (2003)
9. Morano, P., Locurcio, M., Tajani, F., Guarini, M.R.: Fuzzy logic and coherence control in multi-criteria evaluation of urban redevelopment projects. Int. J. Bus. Intell. Data Min. 10(1), 73–93 (2015). Inderscience Enterprises Ltd.
10. Del Giudice, V., De Paola, P.: Spatial analysis of residential real estate rental market. In: D'Amato, M., Kauko, T. (eds.) Advances in Automated Valuation Modeling. Studies in System, Decision and Control, vol. 86. Springer, Cham (2017)
11. Del Giudice, V., De Paola, P., Forte, F.: The appraisal of office towers in bilateral monopoly's market: evidence from application of Newton's physical laws to the Directional Centre of Naples. Int. J. Appl. Eng. Res. 11(18), 9455–9459 (2016)

12. Saaty, T.L., De Paola, P.: Rethinking design and urban planning for the cities of the future. Buildings **7**, 76 (2017)
13. Del Giudice, V., Manganelli, B., De Paola, P.: Depreciation methods for firm's assets. In: ICCSA, Part III, Lecture Notes in Computer Science, vol. 9788, pp. 214–227. Springer, Cham (2016)
14. Manganelli, B., Del Giudice, V., De Paola, P.: Linear programming in a multi-criteria model for real estate appraisal. In: ICCSA, Part I, Lecture Notes in Computer Science, vol. 9786, pp. 182–192. Springer, Heidelberg (2016)
15. Del Giudice, V., De Paola, P., Manganelli, B.: Spline smoothing for estimating hedonic housing price models. In: ICCSA, Part III, Lecture Notes in Computer Science, vol. 9157, pp. 210–219. Springer, Cham (2015)
16. Ciuna, M., Milazzo, L., Salvo, F.: A mass appraisal model based on market segment parameters. Buildings **7**, 34 (2017)
17. Del Giudice, V., De Paola, P., Manganelli, B., Forte, F.: The monetary valuation of environmental externalities through the analysis of real estate prices. Sustain. MDPI Switz. **9** (2), 229 (2017)
18. Del Giudice, V., De Paola, P., Cantisani, G.B.: Rough set theory for real estate appraisals: an application to Directional District of Naples. Build. MDPI Switz. **7**(1), 12 (2017)
19. Del Giudice, V., De Paola, P., Cantisani, G.B.: Valuation of real estate investments through fuzzy logic. Build. MDPI **7**(1), 26 (2017)
20. Del Giudice, V., De Paola, P., Forte, F.: Using genetic algorithms for real estate appraisal. Build. MDPI **7**(2), 31 (2017)
21. Del Giudice, V., Manganelli, B., De Paola, P.: Hedonic analysis of housing sales prices with semiparametric methods. Int. J. Agric. Environ. Inf. Syst. **8**(2), 65–77 (2017). IGI Global Publishing
22. Del Giudice, V., De Paola, P., Forte, F., Manganelli, B.: Real estate appraisals with Bayesian Approach and Markov Chain Hybrid Monte Carlo Method: an application to a Central Urban Area of Naples. Sustainability **9**, 2138 (2017)
23. Massimo, D.E.: Stima del green premium per la sostenibilità architettonica mediante Market Comparison Approach. Valori e Valutazioni (2010)
24. Massimo, D.E.: Green building: characteristics, energy implications and environmental impacts. case study in Reggio Calabria, Italy. In: Mildred, C.-S. (ed.) Green Building and Phase Change Materials: Characteristics, Energy Implications and Environmental Impacts, vol. 01, pp. 71–101. Nova Science Publishers (2015)
25. Massimo, D.E. Emerging issues in real estate appraisal: market premium for building sustainability, Aestimum, pp. 653–673 (2013)
26. Massimo D.E., Del Giudice, V., De Paola, P., Forte, F., Musolino, M., Malerba, A.: Geographically weighted regression for the post carbon city and real estate market analysis: a case study. In: Calabrò, F., Della Spina, L., Bevilacqua, C. (eds.) Local Knowledge and Innovation Dynamics Towards Territory Attractiveness Through the Implementation of Horizon/E2020/Agenda2030. Springer. Berlin (2018). ISBN 978-3-319-92098-6
27. De Paola, P., Del Giudice, V., Massimo D.E., Forte, F., Musolino, M., Malerba, A.: Isovalore maps for the spatial analysis of real estate market: a case study for a Central Urban Area of Reggio Calabria, Italy. In: Calabrò, F., Della Spina, L., Bevilacqua, C. (eds.) Local Knowledge and Innovation Dynamics Towards Territory Attractiveness Through the Implementation of Horizon/E2020/Agenda2030. Springer, Berlin (2018). ISBN 978-3-319-92098-6

28. Spampinato, G., Massimo D.E., Musarella, C., M., De Paola, P., Malerba, A., Musolino, M.: Carbon sequestration by cork oak forests and raw material to built up post carbon city. In: Calabrò, F., Della Spina, L., Bevilacqua, C. (eds.) Local Knowledge and Innovation Dynamics Towards Territory Attractiveness Through the Implementation of Horizon/E2020/Agenda2030. Springer, Berlin (2018). ISBN 978-3-319-92098-6

29. Malerba, A., Massimo D.E., Musolino, M., De Paola, P., Nicoletti, F.: Post carbon city: building valuation and energy performance simulation programs. In: Calabrò, F., Della Spina, L., Bevilacqua, C. (eds.) Local Knowledge and Innovation Dynamics Towards Territory Attractiveness Through the Implementation of Horizon/E2020/Agenda2030. Springer, Berlin (2018). ISBN 978-3-319-92098-6

The Life Cycle of Clusters: A New Perspective on the Implementation of S3

Giuseppe Pronesti(✉) ⓘ and Carmelina Bevilacqua ⓘ

Mediterranea University of Reggio Calabria, 89100 Reggio Calabria, Italy
giuseppe.pronesti@unirc.it

Abstract. In recent years the urge to sharpen strategic public actions, in the way of boosting regional economic performances, has become an imperative. Accordingly, the concept of Smart Specialization Strategy (S3) has attracted growing consideration, by bringing to light an innovative, place-based policy framework for regional economic development. Although S3 policies has been widely examined by many scholars, nonetheless some implementation gaps remain under addressed. The key issue where the debate is still open relates to the operationalization of the Entrepreneurial Discovery Process (EDP). The EDP is a crucial stage in S3 policy design, since it drives to identify priorities by focusing on exploration and experimentation of new opportunities to transfer them in a clustering phase. Considering this backdrop, the paper seeks to contribute in bridging the S3 implementation gaps by investigating the potentials of the Cluster Life Cycle (CLC) analysis to guide the operationalization of the EDP. The paper presents and discusses a conceptual model towards highlighting if, and how stage-specific features of clusters (in terms of dynamism, cooperation among firms, diversity of knowledge and actors, and spatial significance) provide potential input in the operationalization of EDP, to enhance S3 implementation. Ultimately, the authors find that EDP implementation could significantly benefit the framework conditions of dynamism, cooperation, variety provided by some specific stages of the CLC.

Keywords: Cluster · Life cycle · Smart specialization
Entrepreneurial discovery

1 Emerging Challenges and Opportunities in the Implementation of S3

During the last twenty years the urge to sharpen strategic public actions, in the way of boosting regional economic performances, has become an imperative. In view of this emerging need, the concept of Smart Specialization Strategy (S3) has attracted growing consideration, by bringing to light an innovative, place-based policy framework for regional economic development. Accordingly, S3-oriented policies are meant to provide place-specific strategies, which aim at unleashing territorial economic potentials. Therefore, S3 policy construct focuses on the "vertical" identification of regional economic priorities aiming at: (i) producing smart, sustainable, and inclusive growth, (ii) promoting research potential, and (iii) maximizing the usage of innovations [1].

© Springer International Publishing AG, part of Springer Nature 2019
F. Calabrò et al. (Eds.): ISHT 2018, SIST 100, pp. 215–225, 2019.
https://doi.org/10.1007/978-3-319-92099-3_26

Such a policy concept, which has been suddenly endorsed within the formal discourse on the EU 2020 innovation plan, roots in five policy principles, namely [1, 2]: (i) granularity, (ii) experimentation, (iii) inclusiveness, (iv) cyclic nature, and (v) entrepreneurial discovery. While all the principle mentioned above contribute remarkably towards defining the uniqueness of S3 policies, the entrepreneurial discovery holds a particular importance. Indeed, the entrepreneurial discovery process (EDP) is seen by Foray as both the conceptual and operational cornerstone of S3, enabling the identification of hidden economic potentials of territories. The EDP serves the collection of dispersed regional knowledge and supports the exploration and disclosure of novel domains of opportunity, towards paving the way for achieving innovation at the regional level. In this sense, the scope of EDP is very broad, since it does not refer to the identification of technological innovations per se, rather it aims at experimenting alternative ways to deploy innovative ideas. To sum up, it is possible to identify three stylized element which stay at the core of the EDP, namely: (i) integration of knowledge, (ii) engagement of stakeholders, and (iii) exploration of new economic domains. Ultimately, the logic is to combine, within a unique regional bundle, the entire set of science-, technology-, engineering-, and market-related knowledge available at the regional level by engaging multiple actors in the way of experimenting novel avenues of economic specialization and setting regional priorities.

Although the nature of S3 policies has been widely investigated by many scholars, nonetheless the practical avenues of S3's operationalization feature some underdressed issues. Recently, Capello and Kroll [3] have finely identified a set of emerging implementation gaps, and bottlenecks, referring to the lack of: (i) favourable framework condition for practice-based innovation at regional level, and (ii) social and political readiness to confront the paradigmatic shift embedded in the S3-oriented policy. In addition, the latter authors point out the scarce capability of certain regions (lagging regions) to engage actively in processes of regional entrepreneurial discovery. This element seems to be particularly problematic by hampering the effectiveness of S3 priority-setting process. The implementation difficulties pertaining to EDP do not surprise. Foray firstly, observed that the identification of entrepreneurial discoveries *"[is] not [an] easy empirical investigation"* [4]. Still on the EDP implementation difficulties, Gheorghiu, Andreescu, and Curaj [5], lamented the absence of a *"functional blueprint for the entrepreneurial discovery process"* (p. 2). On an analogous line of thoughts, Santini et al. [6], highlighted that entrepreneurial discovery gaps have contributed in the way of ratifying the failure of the EU research system. These evidences call for consideration from scholars and practitioners to disclose new opportunities and potential solutions to facilitate the operationalization of S3, and particularly of EDP. Notably, one of the major opportunities to further the effective implementation of S3, refers to the exploitation of EU regions' knowledge and experiences with cluster and cluster policies [1, 7, 8]. While much has been said on the role of cluster and cluster policies in supporting the implementation of S3, there are still very few pieces of work adequately considering the potential input of the cluster life cycle analysis. However, considering that clusters dynamics and spatial configurations change over time, it is expectable that the effectiveness of different policy measures vary over the life cycle of clusters. The latter idea seems to apply, to some extent, also to S3 and particularly EDP.

Considering this backdrop, this paper aims at contributing to the debate on the role of clusters in the arduous implementation of S3, towards investigating the potentials of the CLC analysis to guide the operationalization of S3, and in particular EDP. The work is structured as follows. Firstly, the study of the literature on the CLC allows the authors to understand which are the leading indicators accounting for the evolution of clusters. Secondly, the indicators drawn from the literature study, are used to build a conceptual model representing the evolution of clusters. Thirdly, the discussion on the model highlights how stage-specific features of clusters (in term of innovative dynamism, cooperation among firms, diversity of knowledge and actors, and spatial significance) can potentially input the operationalization of S3 and in particular EDP.

2 CLC Inputs on the Implementation of S3

By drawing insights from the literature, this article describes the CLC, according to a three-stage taxonomy, including the phases of (i) emergence, (ii) development and (iii) maturity of clusters.

Clusters usually emerges because of the trigger-effect of an exogenous economic shock [9]. The exogenous shock induces the take-off of the clustering process, and consequently drives a limited number of small-sized companies to co-locate in certain geographical areas [10, 11]. Such early agglomeration phenomenon manifest through a scattered spatial configuration and lacks consistency because the locational benefits are not evident yet [9]. At this stage, the flow of knowledge and information between cluster insiders is mainly involuntary and informal as it does rely nor on structured networks neither on consolidated partnerships. Despite the lack of sharpened inter-firm organisational forms, nonetheless, a stock of heterogeneous knowledge circulates among insider businesses [11]. This explorative stage of the CLC is also characterised by significant Venture Capital (VC) and Research and Development (R&D) investments. To summarise, the emergence is a very early, upstream, and explorative phase of the CLC and it is featured by a marked tendency of firms towards innovativeness. The role of start-ups, as well as the values of creativity, and willingness to risk added by entrepreneurs, are crucial to further the prosperity of clusters. The benefits deriving from network activities and knowledge spillovers are somehow available, and the stock of accessible knowledge is highly heterogeneous.

Developing clusters expand through a substantial proliferation of the companies entering the market, and a significant increase in employment. The locational benefits become high towards encouraging co-location. Accordingly, the profitability of insider businesses rises, reaching its peak. In this phase, the agglomeration economies, theorised by Marshall, are the key engine enabling the endogenous growth of the cluster [9]. Consequently, many positive externalities take place, including (i) specialised labour pooling; (ii) interactions among stakeholder, and (iii) knowledge spillovers. In addition to the Marshallian externalities, another factor contributing to the cluster prosperity is the medium/high level of heterogeneity of available knowledge within the clusters' environment [11]. Tersely, the success of clusters at this stage seems boldly rooted in regional self-reinforcing processes (such as networking activities, interactions, and cooperation) occurring among local firms and institutions. The number of

Start-ups and entrepreneurs is still relevant but no longer crucial to further the cluster's development. The R&D and VC investments remain significant as well as the level of heterogeneity of accessible knowledge.

Mature clusters reach a stable configuration, towards focusing on specific business segments, consolidating networks' structure, and acquiring cooperative routines [11, 12]. This state of quasi-equilibrium of clusters features a severe decrease in frequency and number of entries, which in turn makes the clusters' growth rate dropping down. At this point, although locational benefits and self-reinforcing effects are still somehow accessible, however they tend inevitably to dissolve [11]. Moreover, mature clusters usually features high specialisation (if not over-specialisation), as a consequence both the variety of economic activities and the heterogeneity of available knowledge narrow down [11]. Concisely, mature clusters reach the maximum size, have a well-shaped network structure, and a precisely-defined core business. In this context, the entry of Start-ups in the clusters becomes irrelevant, R&D and VC investments decrease, and the knowledge accessible becomes homogeneous.

2.1 A Conceptual Model to Disclose the Potential of CLC Analysis Towards Inputting EDP Operationalization

Drawing on the study of literature it is reasonable to claim that clusters evolution can be explained, to some extent, by variations in certain cluster's dimensions, namely: (i) dynamism, indicated through R&D investment, VC investment, new firms (start-ups) birth rate; cooperation, indicated by intensity of network activities; variety, signalled by heterogeneity of available knowledge; spatial significance, indicated through specialization and agglomeration. The dimensions of clusters' evolution as well as the indicators are framed in Table 1, which also defines the literature sources the indicators have been retrieved from.

Table 1. The selected indicators.

Dimension	Indicator	Literature source
Dynamism	**R&D investment**	Keeble and Wilkinson, 1999 | Brenner, 2000 | Bergman, 2008 | Brenner and Schlump, 2011
	VC Investment	Brenner, 2000 | Braunerhjelm, 2000 | Chatterji, Glaeser and Kerr, 2014 |
	Start-up birth rate	Fornahl and Menzel, 2003 | Mario A Maggioni, 2004 | Menzel and Fornahl, 2009 | Brenner and Schlump, 2011| Suire and Vicente, 2014
Cooperation	**Intensity of network activities**	Brenner, 2000 | Bergman, 2008 | Brenner and Schlump, 2011
Variety	**Heterogeneity of available knowledge**	Menzel and Fornahl, 2006 | Menzel and Fornahl, 2009 | D. H. Shin and Hassink, 2011| Biggiero et al., 2016
Spatial significance	**Specialization**	Maggioni, 2002 | Brenner and Schlump, 2011 | Handayani et al., 2012
	Agglomeration	Maggioni, 2002 | Maggioni, 2004

These indicators are deployed to build a conceptual model presenting the potential input of CLC in the way of operationalizing S3 and specifically EDP. Such research objective is pursued by:

1. scoring qualitatively the strength of each indicator at each stage of the CLC (Table 2). The scores are assigned by the authors on the base of deductions and insights drawn from the literature and previous works [13]. For the scoring, the authors used a scale based on five degrees of intensity: low, medium/low, medium, medium/high, and high.
2. framing the indicators within a conceptual model (Fig. 1) towards comprehending the potential input that stage-specific features of clusters (in term of innovative dynamism, cooperation among firms, diversity of knowledge and actors, and spatial significance) can provide towards the operationalization of S3, and in particular EDP.

According to Table 2, indicators belonging to the same dimension vary in equal manner during the evolution of clusters. Consistently the conceptual model which follows (Fig. 1) presents directly the four families of indicators, rather than the single indicators, and opens to a discussion.

Table 2. Indicators score.

	Dynamism			Cooperation	Variety	Spatial significance	
	R&D	VC	Start-ups birth rate	Intensity of network activities	Heterogeneity of available knowledge	Specialization	Agglomeration
Emergence	high	high	high	medium	high	low	low
Development	mid/high	mid/high	mid/high	mid/high	mid/high	mid	mid/high
Maturity	low	low	low	mid/high	low	high	high

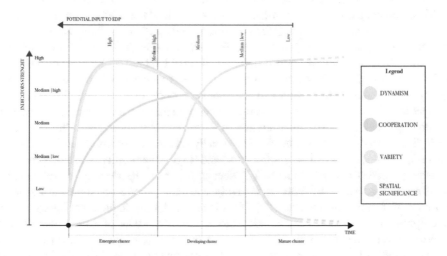

Fig. 1. Conceptual model of CLC analysis potential input to EDP.

The discussion firstly, provides an interpretation of the indicators and their variation in strength over the CLC stages, and secondly, highlights the relevance of clusters stage-specific features in terms of dynamism, cooperation, variety, and spatial significance towards inputting EDP.

R&D investment is considered both as an indicator of clusters innovativeness [14], and a determinant of entrepreneurship [15]. Indeed, R&D propels the generation and diffusion of knowledge, which in turn furthers the vibrancy of entrepreneurial environment and favors inventions. The benefits of R&D seem to be particularly important during the early life course of firms [16], as new-born firms (such as start-ups) tend to exploit investment on R&D more efficiently than the old ones. Consequently, clusters featuring high number of start-ups, attract and call for R&D investment, which in turn generate remarkable innovative outputs. Given these considerations, it seems logically more convenient to operate R&D investments during the initial phases of the CLC, namely emergence and development. These two initial stages appear to be more suitable for entrepreneurial discoveries, because of their high start-ups' birth rate, and flourishing innovative environment. Conversely, clusters in their maturity rely on aged firms, which operate according to consolidated, if not stagnating, industrial practices. Henceforth, envisioning that EDP is meant to "(…) logically identify (…) the domains where new R&D and innovation projects will (…) create future domestic capability" [17], it is reasonable to conclude that emergent and developing clusters offer optimal context conditions for EDP implementation. The same conclusion is also in the case of Venture Capital (VC) investments. VC investments refer to "a form of equity financing particularly important for young companies with innovation and growth potential but untested business models and no track record" [18]. This funding system is seen both as a marker of clusters' innovative potential, as well as an essential factor nourishing clusters' entrepreneurial environment [19]. Indeed, VC is especially advocated in, and appealed by, highly pioneering territorial contexts [20]. Such setting coincides with those of emerging and developing clusters. Therefore, VC investments, by focusing especially on the explorative stages of the CLC, trigger potential innovations which could be intercepted in the way of EDP. Along with R&D and VC, start-ups birth rate is the third indicator accounting for dynamism and goodness of the entrepreneurial environment within clusters. Start-ups include all newly born firms that are up to two years old [18]. Such "young" and usually small-sized businesses, because of their very explorative, and potentially innovative nature, are crucial endogenous drivers of territorial development. High values of the indicator start-ups birth rate also mean that entrepreneurial actors (the bearers of entrepreneurial knowledge) are particularly active. To sum up, there is a positive correlation at the territorial level between high values of the indicator start-ups birth rate (which usually attributes emerging and developing clusters), high density of entrepreneurs and high availability of entrepreneurial knowledge. Given that EDP has a "(…)special focus on the regional entrepreneurial environment, assessing whether it is lively and can generate a significant flow of experiments, innovation ideas (…)" [20], it is reasonable to deduce that emerging and developing clusters could provide valuable inputs in the way of entrepreneurial discoveries. As already stressed, high values of the indicator start-ups birth rate are a marked feature of clusters' emergence and development stage. Instead, the entry of start-ups, and their importance in the functioning of the cluster, drastically decreases during maturity. This theoretical evidence suggests that EDP can be effectively supported by the bold entrepreneurial, innovation-oriented, cross-sectoral environment manifested at the two initial stages of clusters' evolution.

Networks activities refer to actions generating or nourishing organisational forms of economic activities [21]. The intensity of network activities provides a measure of knowledge exchange and firms connectedness, within certain geographic boundaries (which are mutable and permeable). Empirical studies demonstrate that increases in network activities are positively correlated with the rise of firms' innovativeness [21]. The same studies also prove that the willingness to engage in knowledge-based networks has a negative correlation with firms' size. These evidences, suggest that network activities are more intense in the presence of new-born, small-sized firms (such as start-ups). The latter orientate towards more flexible, sometimes informal, forms of networks. On the contrary, big firms rely on routine-based, formally-regulated networks. These differences in the structure of, and willingness to engage in networks make small firms' more innovative, more adaptable and less sector-specific than big ones. However, networks and spillovers co-evolve with clusters. As previously highlighted, networks are mostly informal, and spillovers often happen involuntarily during clusters' emergence. This is due both to the scattered configuration of the spatial agglomeration of firms and to the explorative nature of the businesses entering the market (mainly start-ups). When clusters move on to the development stage, networks get gradually more structured and spillovers more formal. This condition evolves further on during the maturity stage, when networks become rigid and spillovers significantly decrease. Given these considerations, it is reasonable to affirm that EDP should focus on emergent and eventually developing clusters, which are featured by the "relational density" postulated by Foray. Indeed, the significant density of start-ups and entrepreneurs, the marked attitude of firms towards innovative activities and knowledge sharing, make emergent and developing clusters an exceptional source of various entrepreneurial and economic knowledge.

The heterogeneity of knowledge [11], indicates the variety of the available knowledge-stock inside clusters. Considering that entrepreneurs are the knowledge bearers, the variety of accessible knowledge also indicate, to some extent, the assortment of entrepreneurial actors. The more such assortment is various, the more clusters manifest a marked attitude towards adjusting to changing conditions [11]. It has been said that the heterogeneity of knowledge and actors evolve over the CLC. Specifically, while the initial phases of the CLC feature high and medium heterogeneity of accessible knowledge, during maturity, this variety tends to attenuate toward homogenization. This shift, from heterogeneous to homogeneous knowledge, is due both to a decrease in the number of diverse entrepreneurial actors entering the clusters and to an increase in specialisation. Considering that EDP calls for a diversity of economic actors and knowledge, the best match in the way of EDP operationalization seems to be manifested by the features of emerging and developing clusters.

Specialisation refers to the share of regional employment in a sector, relative to the national context. This indicator is widely endorsed in literature as a marker of spatial concentration of industries [9]. The discourse on specialisation presents a split-screen view. On the one hand, low specialization: (i) prevents clustered firms from exploiting the full potential of competitive advantages and (ii) allows clustered firms to benefit a vibrant, cross-sectoral and diversified entrepreneurial environment (typical attribute of emergent and developing clusters). On the other hand, high specialisation leads clustered firms to exploit competitive advantages fully, while eventually leading to stagnation and lock-in (a common attribute of mature clusters). Tersely, high specialisation can lead towards flattening clusters' economic vibrancy and innovativeness as well as losing the positive effects of the variety externalities theorised by Jacob. Once again, the best fitting ecosystem for EDP is expectedly the one provided by emergent and developing clusters. Indeed, considering that EDP pertains to the detection of potential domains for future regional specialisation, targeting already specialised clusters would mean pointing out traditional industrial sectors instead of S3-type domains. Another indicator accounting for the spatial configuration of clusters is the agglomeration. The latter indicates the number of firms concentrating in some geographical regions [9]. This indicator's value increases as clusters get holder, till reaching its peak during the maturity stage. At this point, the mass of economic activities located in a specific geographic area reaches its maximum. As a consequence, the attractiveness of such areas starts decreasing due to a scarce availability of locational benefits [9]. Conversely, in cases when spatial agglomeration presents a configuration not saturated yet, businesses from outside are encouraged to locate inside clusters because of potentially high locational benefits. These considerations reveal that the locational attractiveness should be found in clusters that have not reached the spatial agglomeration peak yet, namely: emerging and developing clusters.

3 Conclusion

This article presented a theoretical discussion on the role of clusters and cluster policies to support S3, and specifically EDP implementation. Although a significant body of scientific literature confirms that EU experience with clusters and cluster policies is a crucial element towards supporting the implementation of S3, nonetheless many operational gaps keep standing out. One of the most problematic factors pertains to the operationalization of the EDP. Consistently the authors intended to test whether the CLC analysis could eventually guide the discovery of regional economic potentials.

Considering the discussion presented in the previous section of this work it is reasonable to claim that the EDP implementation could significantly benefit the framework conditions for innovation, relational density, and diversity of knowledge and actors provided by some specific stages of the CLC. In detail, the authors find that emerging and developing clusters can provide a number of significant inputs towards implementing EDP (Fig. 1): (i) the significant strength (medium and high) of innovative dynamism (in terms of R&D and VC investment, and start-ups' birth rate) signal high quality framework conditions for innovation; (ii) the medium and high strength in intensity of network activities indicates a significant relational density among clusters insiders and a tendency towards innovative, cross-sectoral cooperation; (iii) the medium and high heterogeneity of available knowledge, which also indicate the variety in the assortment of economic actors, enables the opportunity to enlarge the regional knowledge-base, gathering economic and entrepreneurial knowledge; finally, (iv) the low/medium levels of firms' agglomeration and specialization suggest the existence of a territorially localized economic potential, which has not been fully exploited yet.

These findings call for consideration of policy-makers, to reflect more consciously on CLC analysis to overcome EDP implementation issues, and consequently get to a fully effective operationalization of S3.

By drawing on these findings it is possible to derive a couple of recommendations. First, the innovative potentials of regions need to be analysed under an evolutionary perspective by considering the study of the CLC. Disregarding the investigation of the CLC would produce an underestimation of regional economy's dynamics, and consequently a misinterpretation of the context. Given that S3 aims at enhancing the untapped potentials of regional economies, hence it cannot ignore the insights provided by such evolutionary-based, CLC-oriented context analysis. Second, different policies must be adopted in different regions on the base of the analysis of the CLC. As already stated in previous paragraphs, emerging, and developing clusters display suitable context conditions for the implementation of EDP, and S3, while mature clusters do not. Consistently, the exercise of S3 is more likely to produce a desirable restructuring of the economic systems in regional contexts dominated by emerging and developing clusters. Conversely, the application of S3 in regional contexts dominated by mature clusters can hardly produce the expected restoration of regional economies, which would benefit much more traditional industrial policies for cluster support. This differentiation, implies to the need for aligning public intervention with respect to the different phases of clusters' evolution.

However, the study presents some limitations pertaining to the conceptual model. Firstly, the lack of established conventions on indicators for the study of clusters makes the selection of the variables for the model, a relatively arbitrary process. Secondly, the CLC cannot be satisfyingly explicated through a single model. Given these considerations, some potentially influential factors are ignored, while the variables that are most frequently endorsed in the literature are included. Moreover, given that industries are not alike, and that different variables have different importance in the industries, it might be that the model does not represent the mechanism of some industrial sector. The same limit also applies when considering diverse territorial contexts. However, the theoretical literature provides evidence that a detailed modelling of all relevant processes might not be of crucial importance. From this, it is reasonable to conclude that while the model is not fully explanatory, it still reflects appropriately the potential contribution of the CLC and spatial analysis in the way of S3 and EDP.

Further studies are needed to empirically validate the model presented in this paper.

Acknowledgement. The MAPS-LED project has received funding from the European Union's Horizon 2020 research and innovation programme under the Marie Skłodowska-Curie grant agreement No. 645651.

References

1. Foray, D., Goenega, X.: The Goals of Smart Specialisation. Publications Office of the European Union, Luxembrug (2013)
2. Foray, D., et al.: Guide to Research and Innovation Strategies for Smart Specialisations (RIS 3). Publications Office of the European Union, Luxembrug (2012)
3. Capello, R., Kroll, H.: From theory to practice in smart specialization strategy: emerging limits and possible future trajectories. Eur. Plan. Stud. **4313**, 1–14 (2016)
4. Foray, D.: Smart specialisation: opportunities and challenges for regional innovation policy. Reg. Stud. **49**(3), 480–482 (2015)
5. Gheorghiu, R., Andreescu, L., Curaj, A.: A foresight toolkit for smart specialization and entrepreneurial discovery. Futures **80**(2015), 33–44 (2015)
6. Santini, C., et al.: Reducing the distance between thinkers and doers in the entrepreneurial discovery process: an exploratory study. J. Bus. Res. **69**(5), 1840–1844 (2016)
7. Ketels, C., et al.: The Role of Clusters in Smart Specialisation Strategies. Publications Office of the European Union, Luxembrug (2013)
8. Aranguren, M.J., Wilson, J.R.: What can experience with clusters teach us about fostering regional smart specialisation? pp. 1–22 (2013)
9. Maggioni, M.A.: The rise and fall of industrial clusters: technology and the life cycle of region. Documents de treball IEB (2004), http://dialnet.unirioja.es/servlet/articulo?codigo=1289144. Accessed 12 Sept 2017
10. Andersson, T., et al.: The Cluster Policies Whitebook. IKED, Malmö (2004)
11. Menzel, M.P., Fornahl, D.: Cluster life cycles-dimensions and rationales of cluster evolution. Ind. Corp. Change **19**(1), 205–238 (2010)
12. Brenner, T., Schlump, C.: Policy measures and their effects in the different phases of the cluster life cycle. Reg. Stud. **45**(10), 1363–1386 (2011)
13. Bevilacqua, C., Pronesti, G.: Clusters in designing S3-oriented policies. In: 13th International Postgraduate Conference 2017, pp. 1015–1027. The University of Salford, Manchester (2017)
14. Davis, C.H., et al.: What Indicators for Cluster Policies in the 21, Innovation, pp. 1–15, September 2006
15. OECD: OECD Factbook 2013: Science and Technology. OECD Publishing, Paris (2013)
16. Stam, E., Wennberg, K.: The roles of R&D in new firm growth. Small Bus. Econ. **33**(1), 77–89 (2009)
17. Foray, D., David, P.A., Hall, B.H.: Smart specialisation - from academic idea to political instrument, the surprising career of a concept and the difficulties involved in its implementation. MTEI-Working_Paper-2011-001 (2011), https://infoscience.epfl.ch/record/170252. Accessed 12 Sept 2017
18. OECD: Entrepreneurship at a Glance 2016. OECD Publishing, Paris (2016)
19. Breschi, S., Malerba, F.: Clusters, Networks, and Innovation. Oxford University Press, Oxford (2005)
20. Bevilacqua, C., Pizzimenti, P., Maione, C.: S3: cluster policy and spatial planning. Knowledge dynamics, spatial dimension and entrepreneurial discovery process. Second Scientific Report, MAPS-LED Project, Multidisciplinary Multidisciplinary Approach to Plan Smart specialisation strategies for Local Economic Development, Horizon 2020–Marie Swlodowska Curie Actions -RISE–2014-grant agreement No 645651 (2017)

21. OECD: Networks, Partnerships, Clusters and Intellectual Property Rights: Opportunities and Challenges for Innovative SMEs in a Global Economy, 2nd OECD Conference of Ministers Responsible for Small and Medium-sized Enterprises (SMEs), pp. 1–78. OECD Publishing, Paris (2004)

A Multi-level Integrated Approach to Designing Complex Urban Scenarios in Support of Strategic Planning and Urban Regeneration

Lucia Della Spina[(✉)]

Mediterranea University of Reggio Calabria, 89100 Reggio Calabria, Italy
lucia.dellaspina@unirc.it

Abstract. The purpose of this study is to assess the feasibility of an integrated multi-dimensional and multi-level approach that supports decision-makers in planning, designing and managing complex urban regeneration plans. The mixed methodological approach and the combined effect of various assessment tools provide rational arguments for using scarce public resources to determine intervention priorities and the most effective alternatives among different options. The definition of shared objectives and development scenarios can reduce problems and guarantee economic development over time. The integrated model applied to five areas of strategic value for urban regeneration of the metropolitan area of Reggio Calabria (Southern Italy) highlights the potential of this model. The methodological framework combines different tools: Stakeholder Analysis and Cognitive Maps to identify interests, objectives and values; and a Multi-Criteria Analysis to define the most effective and shared alternatives in order to determine the order of priority of public intervention.

Keywords: Urban regeneration · Strategic planning · Scenario Analysis
Stakeholder Analysis · Cognitive Maps · Multi-Criteria Analysis
Multi-Attribute Value Theory (MAVT)

1 Introduction

This research aims at investigating problems related to urban regeneration processes that require special programmes aimed at eliminating social decline, increasing the quality of life, supporting the development and valorisation of cultural resources, protecting the environment, fostering economic development and so on [1].

In this perspective, planning processes should be approached as decision-making processes, in which the assessment of urban transformation scenarios represents a complex decision problem which calls for the evaluation of different conflicting aspects, and for the definition of shared goals for the identification of possible solutions and their effects according to different development scenarios. Beginning with a problem structuring phase that analyses the strengths and weaknesses of the decisional context, as well as needs and preferences, the decision makers can define satisfactory goals and strategies that will be implemented and monitored [2]. During this series of

© Springer International Publishing AG, part of Springer Nature 2019
F. Calabrò et al. (Eds.): ISHT 2018, SIST 100, pp. 226–237, 2019.
https://doi.org/10.1007/978-3-319-92099-3_27

activities the evaluation provides a rational support for facing the complexity and makes it possible to verify the effectiveness and validity of the choices, thus increasing their transparency and improving the collective learning processes [3]. Furthermore, the growing attention to the concept of sustainable urban development underlined the importance of evaluation in assessing progress in the improvement of the quality of the environment, of social well-being and economic growth in the long-term. At the same time, given the need to assess the progress of cities towards sustainability, a value-focused approach has progressively emerged as a conceptual framework for integrating values into decision-making activities by making the direction of preferences explicit [4, 5].

The approach was conceived according to an incremental path in which, taking into account the evaluation context, the most appropriate analysis and evaluation techniques that could meet the needs of the decision-making process were selected.

More specifically, the case study of an urban regeneration programme of the metropolitan area of Reggio Calabria (Southern Italy), (see Fig. 2), aims on one side at catering to the economic and social needs of an area affected by unregulated development and the presence of many abandoned areas, both industrial and nonindustrial; and on the other side at assisting the public administration and decision makers in planning, designing and managing complex urban systems.

The application to the case study presented here shows how the integrated methodology is potentially useful not only to compare different functional solution related to a single action plan, but above all how it is possible to guarantee and justify, through rational, transparent and effective arguments, the use of scarce public resources to establish intervention priorities and more shared and effective solutions among different options.

This paper is divided into four sections. The introduction is followed by the second section, which outlines the methodological background and the evaluation techniques used. The third section presents the application of the evaluation approach to the case study and highlights the results obtained in reference to the initial settings. The last section summarizes the main conclusions of the work.

2 The Multi-methodological Framework

The proposed method (see Fig. 1) has been structured in order to allow the interaction between different techniques selected to outline a dynamic, flexible and adaptive decision support system, focused on the peculiarity of the context and oriented to the elaboration of strategies based on expert and common knowledge, and on shared and recognized values. The experimented approach was conceived according to an incremental path in which, for each of the decision-making phases, techniques consistent with the Systems Thinking Approach [6–9] applied to problem solving were selected. According to this approach, decision-making problems are considered as part of a global system that can influence habits, behaviours and routine, whose components can be better understood by taking into account mutual relationships, connections, and interactions among the elements that make up the whole system in a cyclic, rather than linear, process.

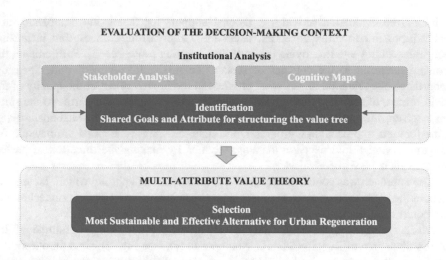

Fig. 1. The methodological approach.

The methodological approach followed in this study has been implemented in two phases. Through an 'Institutional Analysis' [10], the first phase, Evaluation of the Decision-Making context, allowed to identify the stakeholders involved in the problem context [11] and their relevant and shared goals and attributes by following 'Value Focused Thinking' concepts through a cognitive map [12, 13]. The second phase developed a Multi-Attribute Value Theory (MAVT) [4], a multi-criteria methodology increasingly used to support the decision-making process in the choice of the most sustainable, effective and shared alternative among multiple competing options. The evaluation of the various alternatives is based on the analysis of the problematic context, of the Stakeholder Analysis and of the related Cognitive Maps developed during the first phase.

2.1 Institutional Analysis

An Institutional Analysis [10, 14] of the decisional context enabled to outline a map that identified the different actors involved in the decision-making process. The structuring of a process of institutional analysis involves being able to identify the points of view, the perceptions and the hierarchy of interests of the different groups that make up the community.

The stakeholders were divided into three main categories: promoters, economic operators, and users [11]. Furthermore, for each category, three experts were chosen to represent the plurality of actors and visions derived from the analysis of the stakeholders.

The three experts have a scientific background in urban planning, architectural history and economic evaluation, and are specialised in urban regeneration. For example, the expert in economics was chosen because, thanks to his expertise, he could represent the interests and goals of public and private economic actors involved in the transformation.

The Institutional Analysis was carried out in two consecutive phases: in the first phase, the analysis of points of view and perceptions was structured by having the three experts called for individual, detailed and structured interviews; in the second phase, appropriate focus groups, where the three experts worked together.

The results of this analysis have been decoded and interpreted to identify some significant issues for the sustainable urban regeneration programme, with the aim of bringing out the perception of critical points and potentials, and identifying future scenarios of transformation and the related implementation strategies.

The decoding of the interviews and the results emerged from the focus groups led to the creation of the related strategic cognitive maps for each stakeholder category, as well as a general strategic map that presented the different points of view on the transformation objectives [12].

This has been an extremely important phase, since it led to the definition of problems and the construction of a shared tree of values that defines and structures the fundamental goals and their related attributes.

Cognitive mapping is a simple graphic tool that can be used to acquire and clarify the ideas and perceptions of the different actors [15, 16], so it is well suited to solving complex problems, when many aspects and dimensions of the problem are difficult to understand adequately or are just completely undetermined.

When there are several points of view and perspectives on a problem, cognitive mapping is a good way to gather them together and negotiate a new shared vision, which will allow all stakeholders to work in groups.

The main advantages of using cognitive mapping can therefore be summarized as follows: (i) it provides a well-organized systematization of available concepts and theories; (ii) it facilitates access to key concepts and theories to inexperienced actors as well; (iii) it can be very useful in organizing and structuring trees of value [15].

Furthermore, the use of cognitive maps to structure problems under the Multi-Criteria Decision Aiding (MCDA) has been discussed by several authors [17–20].

In this document, the cognitive mapping has been integrated with the Institutional Analysis of the interested parties to follow a more inclusive approach in the identification of the relevant objectives and criteria. These were chosen among the most relevant, to evaluate the intervention priorities and the best performing solutions among the different options.

The complementary relationship activated between the Systems Thinking Approach [6–9] techniques and the Multi-Criteria and multi-group evaluation methods allows to define a decision-making process that uses the peculiarities of each technique to improve knowledge of the context, explain the preferences of the different stakeholders, build shared visions of the future and identify preferred scenarios and actions.

3 Application

3.1 Transformation Scenarios for Urban Regeneration

In the application to the case study, the integrated approach has been applied to some public areas. The following areas play a strategic role for the Metropolitan Area of Reggio Calabria (Southern Italy) and are part of a broader programme of urban regeneration:

1. Isola del Gusto "Foro Boario"
2. Social housing, flat complexes, offices and shopping mall "Giardini"
3. Accommodation and multi-purpose services for the Hospital "Morelli"
4. Convention centre, hotel and flat complexes "Pentimele"
5. Sports and shopping mall and flat complexes "Italcitrus".

The main goal of the regeneration programme is to cater to the economic and social needs of a metropolitan city plagued by unregulated development and the presence of some abandoned areas, included industrial ones. Moreover, given the scarce availability of economic resources, it was fundamental to identify on the one hand a series of functions able to promote the economic development of the metropolitan area over time and on the other hand to identify and plan the priorities among five competing alternatives.

Most of these areas occupy a strategic position: some of them lie close to the main train station and the several executive offices of the municipality (named Ce.Dir). Despite their intrinsic value and the relevance of their position, they all present strong elements of urban decay (see Fig. 2).

In addition, in order to be applied to the case study, an assessment of feasibility and profitability of the investment projects was carried out before, and simultaneously with, the planning assessment in order to verify the activation of forms of public-private partnership and the affordability of the investments [21–23].

3.2 Multi Attribute Value Theory

The second phase of the process consisted in the development of a Multi-Criteria Analysis (MCA) [22, 24–26], a tool increasingly used to support the decision-making process for choosing between multiple competing options, in relation to multiple objective functions and attributes (criteria).

The MCA is especially useful as a tool for the assessment of sustainability in urban and territorial planning, when a complex and interconnected series of environmental, social and economic issues has to be taken into account, and where goals are often in competition and trade-off is inevitable [27]. Among the different methods of Multi-Criteria, a very important role is played by the Multi-Attribute Value Theory (MAVT) [4].

The MAVT methodology is a specific Multi-Criteria Analysis technique [28] that can be used to address problems that imply a finite and discrete set of alternative options that must be evaluated on the basis of conflicting objectives. The MAVT can manage qualitative and quantitative data, thus playing a crucial role in the field of

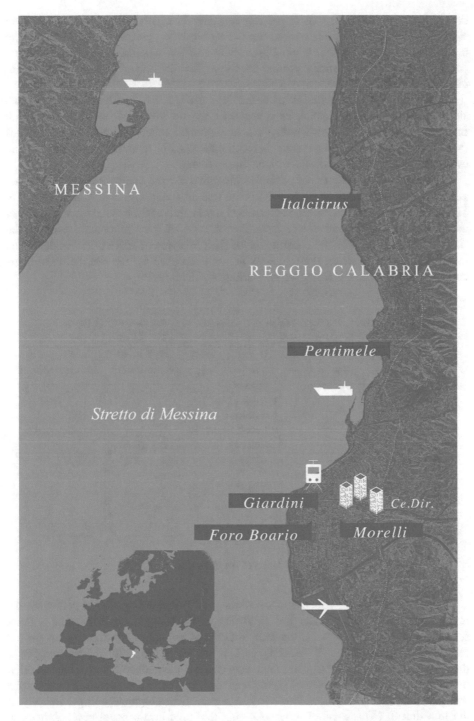

Fig. 2. The metropolitan area and the areas included in the urban regeneration programme.

decision-making problems of urban regeneration in which many aspects are often intangible. Moreover, in this context the decision-making process is often made more complicated by various and conflicting opinions of the stakeholders, which require a participatory decision-making process that includes different perspectives and facilitates discussion.

The MAVT process is illustrated in the literature [12, 19], and is structured in three steps. The first step is crucial, as it concerns the definition of the problem, which involves identifying and structuring the fundamental objectives and their attributes (which represent those measurable characteristics used to quantify the objectives and evaluate the performance of the various alternatives).

Figure 3 presents the set of attributes identified for the evaluation of alternatives, which were organized according to the value tree approach [12, 19]. It is necessary to highlight that such attributes are the technical translation of the objectives and needs of the stakeholders [28] and derive from the analysis of the problematic context, the stakeholder analysis, and the construction of their relative cognitive maps, as described in Sect. 2.1. In this sense, the attributes that will be used for the evaluation of alternatives represent the 'interests' of the various stakeholders.

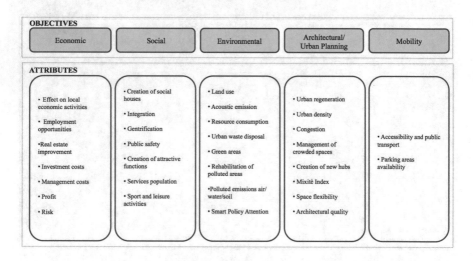

Fig. 3. Structure of the value tree.

The proposed model considers the full range of possible impacts. In particular, the assessment takes into account the following attributes: (1) environmental aspects concerning the effects of transformation in terms of pollution, consumption of natural resources and green areas; (2) social aspects, in relation to the multiple consequences of the intervention on the population, such as services for the inhabitants, public security, integration and social inclusion; (3) economic aspects such as job creation or synergies with local businesses, besides feasibility in terms of development and profitability of the investment; (4) aspects of urban planning that take into account accessibility and mobility elements, and aspects related to density, congestion of the area and integration between several functions; (5) aspects of mobility related to access to public transport and parking areas. These attributes were used to evaluate the performance of the various alternatives and identify the best compromise among the different competing options under consideration.

According to the MAVT methodology, the next step consisted in the elicitation of the value functions, which are a mathematical representation of human judgments. Every attribute is described by a value function [29, 30] that allows to grade the attributes with 0 (worst performance and low goal-achievement rate) or 1 (best performance and high goal-achievement rate). (see Fig. 4). This phase (second step) must be as impartial as possible and must preferably be expressed and constructed together with the experts.

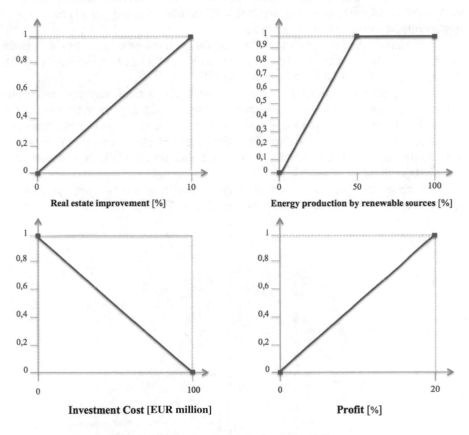

Fig. 4. Final value functions for some attributes.

For example, Fig. 4 describes the value functions result for some of the attributes considered in this study [31].

Once the alternatives have been evaluated from the point of view of their respective performances, it is necessary to define the level of importance and weight of the different attributes used in the decision-making process (third step).

For evaluating attribute weights, a focus group with experts in the fields of architecture, urban planning, sociology, environmental engineering, and economic evaluation was organized. During the focus group, the experts were asked to assign a predetermined amount of points (for example 100) to establish the relative importance of the attributes [6]. For the evaluation of the attributes, each expert was asked to answer a detailed questionnaire, while the experts answered together to the questionnaire on the evaluation of the objectives.

In the application, the set of weights established within the focus group is as follows: 43.4% of importance was assigned to the environmental aspects; 31.6% to social aspects; 14.3% to economic aspects; 6.8% to urban mobility aspects and 3.8% to architecture and urban planning aspects.

The results of the evaluation show that the best scenarios are the "Morelli" project (0.272 in the priority vector), followed by "Foro Boario" (0.217), "Italcitrus" (0.153), "Giardini" (0.126) and lastly "Pentimele" (0.119).

These results were further processed by developing a sensitivity analysis on the weights of the general objectives, in order to verify the stability and robustness of the model. In particular, an approach in which the weight of one objective at a time was increased to 40%, while the weights of the other four objectives were kept at 15% was used. In this final version, the final priorities of the alternatives have been recalculated considering the five scenarios.

The sensitivity analysis defined the final ranking with the two most performing options, namely "Foro Boario" and "Pentimele" (see Fig. 5).

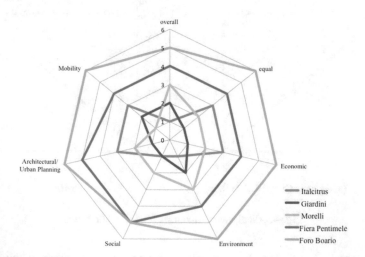

Fig. 5. Final results MAVT.

4 Conclusion

The study illustrates a multi-methodological evaluation model which allows to reduce uncertainties in the domain of decision-making problems related to urban regeneration processes. The method supports decision makers in managing heterogeneous qualitative and quantitative, monetary and non-monetary information, expressed on ordinal or cardinal scales, and thus gives a real contribution in the strategic decision-making phase, where detailed information on the performance of the projects is needed [32].

It is also interesting to note that the proposed approach is structured according to an iterative process, characterised by constant feedback. This greatly contributed to the definition of functional design scenarios during the evaluation phase.

Furthermore, the complementary relationship activated between the techniques of the Systems Thinking Approach [8, 9, 33] and the multi-criteria and multigroup evaluation methods allows to define a decision-making process that uses the specificities of each method to improve the knowledge of the context [34], explain the preferences of the various stakeholders, build shared visions of the future [35, 36] and identify scenarios and preferred actions considering both hard and soft data [37].

References

1. Roberts, P., Hugh S., Rachel G.: Urban Regeneration. Sage (2016)
2. Las Casas, G.B., Tilio, L., Tsoukiàs, A.: Public decision processes: the interaction space supporting planner's activity. In: Murgante, B., et al. (ed.) Computational Science and Its Applications - ICCSA 2012. Lecture Notes in Computer Science, vol. 7334. Springer, Heidelberg (2012)
3. Bentivegna, V.: Analisi multicriteri: applicazioni alternative. Genio rurale **1**, 68–79 (1995)
4. Keeney, R.L., Raiffa, H.: Decisions with Multiple Objectives: Preferences and Value Trade-Offs. Wiley, New York (1976)
5. Giuffrida, S.: The true value. On understanding something. In: Stanghellini, S., et al. (ed.) Appraisal: From Theory to Practice, pp. 1–14. Springer (2017). https://doi.org/10.1007/978-3-319-49676-4_1
6. Bánáthy, B.H.: Guided evolution of society: a systems view (Contemporary Systems Thinking). Springer, Berlin (2000)
7. Jackson, M.: Systems thinking: creating holisms for managers. Wiley, Chichester (2003)
8. Checkland, P., Poulter, J.: Learning for action. Wiley, Chichester (2006)
9. Ackoff, R.L.: Systems thinking for curious managers. Triarchy Press, Gillingham (2010)
10. Funtowicz, S.O., Martinez-Alier, J., Munda, G., Ravetz, J.: Multicriteria-based environmental policy. In: Abaza, H., Baranzini, A. (eds.) Implementing Sustainable Development, pp. 53–77. UNEP/Edward Elgar, Cheltenham (2002)
11. Dente, B.: Understanding Policy Decisions. Springer, Berlin (2014)
12. Eden, C.: Cognitive mapping: a review. Eur. J. Oper. Res. **36**, 1–13 (1988)
13. Keeney, R.L.: Value-Focused Thinking, a Path to Creative Decision Making. Harvard University Press, New York (1992)
14. De Marchi, B., Funtowicz, S.O., Lo Cascio, S., Munda, G.: Combining participative and institutional approaches with multi-criteria evaluation. An empirical study for water issue in troina, sicily. Ecol. Econ. **34**(2), 267–282 (2000)

15. Mendoza, G.A., Prabhu, R.: Evaluating multi-stakeholder perceptions of project impacts: a participatory value-based multi-criteria approach. Int. J. Sustain. Dev. World Ecol. **16**(3), 177–190 (2009)

16. Sheetz, S., Tegarden, K., Zigurs, I.: A group support system to cognitive mapping. J. Manage. Inf. Syst. **11**(1), 31–57 (1994)

17. Bana e Costa, C.A., Ensslin, L., Corrêa, E.C., Vansnick, J.: Decision support systems in action: integrated application in a multicriteria decision aid process. Eur. J. Oper. Res. **113**, 315–335 (1999)

18. Belton, V., Stewart, T.J.: Multiple Criteria Decision Analysis: An Integrated Approach. Kluwer Academic Publishers, Boston (2002)

19. Kpoumié, A., Damart, S., Tsoukiàs, A.: Integrating cognitive mapping analysis into multi-criteria decision aiding. Cahier du Lamsade 322 (2013). https://hal.archives-ouvertes. fr/hal-00875480. Accessed 14 Sept 2017

20. Stewart, T.J., Joubert, A., Janssen, R.: MCDA framework for fishing rights allocation in South Africa. Group Decis. Negot. **19**, 247–265 (2010)

21. Calabrò, F.: Local communities and management of the cultural heritage of the inner areas. An application of break-even analysis. In: Gervasi, O., et al. (ed.) Computational Science and Its Applications-ICCSA 2017. Lecture Notes in Computer Science, vol. 10406. Springer, Cham (2017). https://doi.org/10.1007/978-3-319-62398-6_37

22. Della Spina, L., Lorè, I., I write, R., Viglianisi, A.: An integrated assessment approach as a decision support system for urban planning and urban regeneration policies. Buildings **7**, 85 (2017). https://doi.org/10.3390/buildings7040085

23. Morano, P., Tajani, F.: The break-even analysis applied to urban renewal investments: a model to evaluate the share of social housing financially sustainable for private investors. Habitat Int. **59**, 10–20 (2017)

24. Nesticò, A., Sica, F.: The sustainability of urban renewal projects: a model for economic multi-criteria analysis. J. Property Investment Finance **35**(4), 397–409 (2017). https://doi. org/10.1108/JPIF-01-2017-0003

25. Figueira, J., Greco, S., Ehrgott, M.: Multiple Criteria Decision Analysis: State of the Art Survey. Springer, New York (2005)

26. Huang, I., Keisler, J., Linkov, I.: Multi-criteria decision analysis in environmental science: ten years of applications and trends. Sci. Total Environ. **409**, 3578–3594 (2011)

27. Belton, V., Stewart, T.J.: Multiple Criteria Decision Analysis: An Integrated Approach. Kluwer Academic Press, Boston (2002)

28. Munda, G.: Social multi-criteria evaluation: methodological foundations and operational consequences. Eur. J. Oper. Res. **158**(3), 662–677 (2004)

29. Beinat, E.: Value Functions for Environmental Management. Kluwer Academic Publishers, Dordrecht (1997)

30. Miller, D.: Project location analysis using the goals achievement method of evaluation. J. Am. Plan. Assoc. **46**, 195–208 (1980)

31. Canesi, R., Antoniucci, V., Marella, G.: Impact of socio-economic variables on property construction cost: evidence from Italy. Int. J. Appl. Bus. Econ. Res. **14**(13), 9407–9420 (2016)

32. Calabrò, F., Della Spina, L.: The public-private partnerships in buildings regeneration: a model appraisal of the benefits and for land value capture. Adv. Mater. Res. **931–932**, 555–559 (2014). https://doi.org/10.4028/www.scientific.net/AMR.931-932.555

33. Jackson, M.: Systems Thinking: Creating Holisms for Managers. Wiley, Chichester LNCS Homepage. http://www.springer.com/lncs. Accessed 14 Sept 2017

34. Calabrò, F., Cassalia, G.: Territorial cohesion: evaluating the Urban-Rural Linkage through the Lens of Public Investments. In: Bisello, A., Vettorato, D., Laconte, P., Costa, S., (eds.) Smart and Sustainable Planning for Cities and Regions. Results of SSPCR 2017, Green Energy and Technology. Springer (2018). https://doi.org/10.1007/978-3-319-75774-2_39
35. Mangialardo, A., Micelli, E.: Processi partecipati per la valorizzazione del patrimonio immobiliare pubblico: il ruolo del capitale sociale e delle politiche pubbliche. LaborEst **14**, 52–57 (2017)
36. Della Spina, L., I write, R., Ventura, C., Viglianisi, A.: Urban renewal: negotiation procedures and evaluation models. In: Gervasi, O., et al. (ed.) Computational Science and Its Applications - ICCSA 2015, Lecture Notes in Computer Science, vol. 9157. Springer, Cham (2015). https://doi.org/10.1007/978-3-319-21470-2_7
37. Carbonara, S.: The effect of infrastructural works on urban property values: the asse attrezzato in pescara, Italy. In: Murgante, B., Gervasi, O., Misra, S., Nedjah, N., Rocha, A. M.A.C., Taniar, D., Apduhan, B.O. (eds.) Computational Science and Its Applications - ICCSA 2012. Springer Verlang, Berlino (2012)

Urban Abusiveness, Planning and Redevelopment

Claudia de Biase, Salvatore Losco$^{(\boxtimes)}$, and Bianca Petrella

Luigi Vanvitelli University of Campania, 81100 Caserta, Italy
salvatore.losco@unicampania.it

Abstract. The present paper proposes a reading of urban abusiveness that has affected Italian cities in recent decades, distinguishing it from the unauthorized building. It addresses the phenomenon within the broader issues of physical planning of the territory, in the awareness of the strong environmental impact it determines. Urban abusiveness has a strong influence on the land's layout and on the consumption, use and protection of the soil, and heavily shapes the redevelopment of many southern Italian cities. The paper is organized in three parts: the first one will focus on recognition of urban abusiveness, that is, with urban/territorial effects, distinguishing it from the phenomenon of unauthorized building; the second one will propose a procedure, based on the overlay-mapping technique, capable of representing the urban-scale phenomenon and the third one proposes guidelines for the drafting of the urban plan (or other territory planning tool), which focuses on the urban regeneration of the settlements affected by this phenomenon.

Keywords: Urban/building abusiveness · Urban redevelopment
New standards

1 For the Identification of Urban Abusiveness

In 1985 the law No. 47 (Rules regarding controls of planning-building activity, sanctions, recovery and amnesty of illegal works) was enacted. It opened the way for administrative regularisation, or legitimacy, of most of building works that had transgressed the land and town planning laws. The Law No. 724 of 1994, the Law No. 326 of 2003 and the relevant regional provisions followed the first legislative measure. Excluding the first one, the subsequent two laws were public finance rules, containing special articles dedicated to the "ordinary" regularisation, and the "extraordinary" amnesty for infringements of building regulations, in their different legal meaning [1].

Within this contribution, the result of common elaboration of the authors, personal contributions can be identified as specified below: *For the identification of urban abusiveness* (Bianca Petrella), *An analysis procedure* (Claudia de Biase), *Redevelopment plan criteria of illegal built settlements* and *Conclusions* (Salvatore Losco).

© Springer International Publishing AG, part of Springer Nature 2019
F. Calabrò et al. (Eds.): ISHT 2018, SIST 100, pp. 238–248, 2019.
https://doi.org/10.1007/978-3-319-92099-3_28

Although the types of illegal constructions provided by the laws are only seven, the offenses can be of various nature and have different degrees of impact that constitute a varied casuistry. Here we neglect the ethical component of the legislative measures. The aim here is to clarify the difference between illegal building and urban abusiveness; that is between a crime that has no significant consequences on the urban and territorial system, and vice versa, a work that, by increasing the pressure, modifies the state of the urban and territorial system, risking breaking the equilibrium [2].

According to the definition in Italian regulations [3], the urban pressure is "the requirement of territorial provision for a particular property or settlement in relation to its size and land use. Changes in the urban pressure include the increase or reduction of such requirements resulting from the implementation of urban planning or construction changes or land use modifications". For territorial facilities have to be intended "Infrastructures, services, equipment, public spaces and any other work of urbanization and for sustainability (environmental, landscape, socio-economic and territorial) envisaged by law or plan". Services, equipment or public spaces, in residential and agricultural areas, vary with the number of inhabitants; while in the production, business and directorial areas, vary with the area of production and the gross floor area extent; instead, primary infrastructure varies with the quantity and spatial layout of building works.

Except the type of abuse n. 7 (extraordinary maintenance which cannot be estimated in terms of area or volume) the compensation to be paid must be calculated in relation to the area and the volume extent of the illegal constructions. This, in itself, does not mean that all misuses change the urban pressure already calculated in the urban plan, as it is necessary to distinguish between the works constructed in the absence of authorization, but respecting the predicted indicators and parameters, and the works carried out in contravention of the statutory requirements and current urban plans. Even in these cases it must be distinguished where failure to comply with the parameters leads to a change in the urban pressure; in fact, a construction work may not respect the indices which regulate the form of use and at the same time respect the intensity of land use index, without affecting the area carrying capacity.

Such reasoning can also be deployed for the illicit change of land use because it is possible to distinguish if the different land use requires primary and secondary infrastructures different from the original one, for types and quantities [4].

Ultimately, an infringement of local building regulations turns into urban abusiveness when:

- in the residential and agricultural areas the volume and the floor space can be changed and therefore the theoretical number of the inhabitants varies;
- in the production zones, the surface designated to production is varied and the gross floor surface varies accordingly.

In the first case, the changed number of inhabitants must achieve a corresponding variation of the utilities and public service of the relevant Zoning Classification: A (historical), B (high building density) C (low building density) and E (agricultural); together with the surfaces of general public facilities, that is ZTO F (territorial equipment). At the same time, it must be verified that technical structures are able to absorb the new demand for consumption. Also for the production zones, there is the need for a new quantification of urban services, assuming a different number of employees, energy needs, waste generation, etc.

A considerable amount of new volumes and land covered with building requires the revision of the territorial equipment necessary for the social and functional suitability of the different zones and the totality of the settlement. We must also observe if they are concentrated in one or more areas or if they are diffused in a more or less capillary way, since the way to intervene for the recovery and the renewal of the rehabilitated building stock will depend on the different location.

It should be recalled that Article 29 of Law No. 47/1985 (partly amended by Law No. 326/2003) requires that the Regions comply with the variants for the urbanization of abusive settlements. These must determine, inter alia, criteria for the perimeter redrawing and, in the event of a failure of the Regions, the law provides that in any case, special variants will be allowed.

At present, only a few Regions (seven) have legislated on the mapping of abusive settlements [5]. The criteria they dictate for delimitations, albeit with different terminology (cores, agglomerations, aggregates, …), all call for a continuity of substantially sized building, although, in some cases, the regional regulations provide that also for sparse illegal buildings the same intervention tools should be used.

Whatever the purpose, the boundaries of parts of the territory (as well as the setting of dimensional threshold limits) always results in a mismatch between rules and privileges of position among the inhabitants falling within the delimitation and those who are excluded; problematic that becomes more delicate for those who are close to the borderline. Is it possible to make a delimitation with objective and fair methodology? Probably not, although attempts to delineate the demarcation lines are already present in some cases, such as ecological networks, zoning of parks and, more generally, environmental issues where awareness of the complexity of the interaction of open and dynamic systems is more rooted [6].

Fuzzy logic, applied to territorial zoning, could become a useful tool for defining scalar rules in the bands across the boundaries of separation. This can also be valid to determine the perimeter of the areas of building development of illegal settlements. In these cases, the disparity is stronger because it involves persons who acted in accordance with the rules and persons who instead acted illegally. The application of the logic to the recognition of "territorial families" has already been experimented with a fair success [7]. The next step about to be made is to use the logic rule structure that allows the linguistic description of the system and admits the belonging of an element to more than one aggregate, to define implementation technical rules for town plans that can measures the rules between adjacent areas and thus equalize the rights and duties of land use.

Another issue that the research team is working on is the definitions of methodologies and techniques for redevelopment of the system of settlement in cases where the illegal building spread across the territory are many. This issue affects also the unresolved changes in the pressure determined by the increase, this time legitimate, of volumes and surfaces granted, in derogation of town planning instruments, from the so-called "Piano Casa" launched by individual regions under national laws issued since 2009.

2 An Analysis Procedure

The proposed methodology for analyzing the phenomenon can be a valid support for municipal administrations in identifying, quantifying and precisely locating buildings made illicitly. The procedure represents a useful technical tool from the point of view of construction, for a chronological quantification of the phenomenon, and from the point of view of urban planning, constitutes a support for the correct determination of perimeter of abusive settlements, prescribed by law, being still an open question. As already mentioned, Law No. 47/85, art. 29, delegated the Regions which, by their own law, should have been regulating [5] the criteria of the above-mentioned perimeter determination. As Paolo Urbani and Stefano Civitarese clarify, "the perimeter criterion has been interpreted differently by the regional legislator, based on the continuity of the settlement, or on the basis of the minimum number of buildings built, extending the perimeter also to legally built buildings but integral to the settlement" [8].

The proposed methodology is based essentially on the Overlay Mapping, which consists of overlapping thematic maps, including zoning. It facilitates the immediate visualization of individual abuses, zoning and areas with a high concentration of settlement, providing, as well, an indication of possible criteria for identifying abusive settlements.

The procedure begins with the collection of all the remission practices present in the municipal administration, divided into practices that are issued, suspended or rejected. The location on cartographic survey of all the questions, thus divided, is essential to identify which type of plan must be used; in fact, it must be remembered that the intervention to be foreseen varies according to the settlement morphology: in the case in which there is a concentration of conditional amnesty (or to obtain it in the future) buildings, the Plan Review can be used or the Redevelopment Plan. If, instead, the buildings are without the conditional amnesty, there are two options: either the demolition or the acquisition to the municipal real estate. Finally, if there are buildings for which the conditional amnesty is still suspended, it is not yet clear how to operate. The first thematic map, therefore, returns a picture of this situation in the investigated municipality and constitutes an initial reasoning supporting the possible interventions. The second step is the drafting of a further thematic map that localizes the volumes with the conditional amnesty by differentiating them in total or partial abuse and according to use, residential or non-residential. This second step returns a clear picture of the concentration and/or diffusion of the illegal types of building. Therefore, allows to recognize the presence of a prevalent type and in which parts of the territory it is located. This second step is also preparatory to the subsequent planning and directs towards the most suitable urban plan: Plan Review or the Redevelopment Plan.

In order to be able to direct planning choices it is essential to know also the zoning (otherwise the legal status of land use), or to know in which zoning classification of the urban plan it will be necessary to operate. Therefore the third step is the overlay mapping between the current zoning and the location of the types of abuse. It is clear that for each amnesty a specific thematic map has been drawn up, relating to the validity of urban plan on the date of the amnesties (1985, 1994 and 2003). When, especially in the first year (1985), there was not yet an existing urban plan (absence of a legal status of the land use), we can just use the perimeter of the inhabited center. This additional thematic map clarifies the relationship between type, amount of abuse and zoning of the urban plan and provides other indications for the correct choice of what kind of plan must be used for redevelopment purposes.

In order to check whether the phenomenon of abuse continues, another overlaying map of the volumes has been created - from before 1985 until today - showing the volumes which had or not a building permit, including the new housing volumes envisaged by the so-called "Piano-Casa". This makes visible all the building abuses perpetrated in time, both those of the past that did not require the amnesty, and those built after the last chance to use them (after 2003).

The procedure described has been applied in three municipalities of the Caserta Province with high rate of abusiveness: Casal di Principe, San Cipriano d'Aversa and Casapesenna [9].

Comparing the data of the three municipalities it is clear that the phenomenon of abusiveness has had a different trend over the years and, above all, in relation to the possibilities offered by the three national laws on urban amnesties.

- In relation to Law No. 47/85, the phenomenon presents itself as an abusive settlement around the historic areas of the city, filling - so to speak - those urban spaces still intertwined.
- With regard to Law No. 724/94 the axes of unauthorized buildings are decentralized: the "residual" space is populated, also preferring areas where facilities and structures are not guaranteed.
- Finally with regard to Law No. 326/03, it is noted that phenomenon is spread among sparse episodes. Such constructions are, from a numerical point of view, so few that they do not significantly affect the phenomenon in question as well as the territorial and urban planning. They have been built with the randomness of the location of the property areas.

The various cartographic overlays, together with the acquired data cross-referencing, produced as final result the elaboration of thematic maps that allow to compare and analyze the results obtained, making the understanding of the phenomenon, at an urban level and its social implications, more immediate. The results from the application of the procedure can be seen, as anticipated, as a support to the reading of the phenomenon, to its quantification and location, but, above all, as a starting point for defining criteria for the delimiting of the phenomenon of abuse with urban value in the specific territory. From the analysis of the land of the three municipalities, in fact, there arises a serious situation characterized by degraded areas, lack of urban standards, but above all by a considerable abusive building stock, such to attribute to the three municipalities the sad record in the sector. Moreover, if we analyze the municipalities in an overall picture focusing exclusively on the abuses falling within the so-called "type 1", it is clear that these properties are distributed throughout the territory: fall in both the current B and C zones, but a strong concentration there is also in A zone and there is a further number of abuses, mainly scattered in the current E zone. The problem of perimeter determination criteria emerges with even greater evidence from the reading of the three cases; it is clear, in fact, that if from the overlay arises a concentrated urban abuse, the same determination is - for the planner - easy and immediate. In the case of a scattered urban abuse, however, the problem of defining perimeter criteria remains open.

How to proceed in this situation? From this brief reflection, it is obvious that the Urban Plan, which operates in an illegal territory, will have to behave differently depending on the greater or lesser concentration of the works, the location and the spread of the abuses. The more the abuse is spread and does not assume connotations of "settlement", the harder will be the perimeter delimitation. Therefore, the importance of this phase emerges - and above all the definition of clear and objective criteria - which is preparatory to the drafting of the illegal settlements Redevelopment Plans [10].

3   Redevelopment Plan Criteria of Illegal Built Settlements

The mode, all about building, represents one of the most controversial topic with which illegal built settlements has been dealt with during the last thirty years in Italy.

The extension of the phenomenon, especially in southern Italy, requires a cultural and scientific-technical debate aimed towards a better understanding the illegal built city process.

Because these illegal built settlements condition the liveability and the environmental quality of extensive areas of Italy, it is necessary to find specific urbanistic techniques to regenerate this illegal building stock, regardless of the adjustments due to the amnesty laws.

It is therefore useful to identify criteria and tools for the redevelopment and/or regeneration and/or upgrading of these settlements that involve all the scales of the project in a transversal way.

The systematic consumption of agricultural areas produced so important environmental damage to induce the recognition of pre-eminent public interest and to recognize the urgency for an extensive redevelopment of the areas characterised by widespread and concentrated illegal built settlements.

This urban development typology, dense or sprawl [11], has produced, among other things, deficiencies or completely lacking of infrastructures and public facilities, has led to the un-sustainable increase of urban loads on the legal pre-existing settlement system and to an in-tolerable imbalance on its morphological, functional and management also. In this way the damage produced is neither eliminated nor compensated by a building-oriented solution only but through an urban vision that adopts a mitigating approach [12]. The latter was clearly stated in the recourse to redevelopment (Article 29 of Law No. 47/85), which entrusted the regulatory powers of the regions with the task of regulating the formation, adoption and approval of municipal urban plan review, aimed at the redevelopment of illegal built settlements, within a framework of economic and social convenience. In 2003 the Law No. 326 has integrated this article also allowing private initiative proposals for urban plan review.

The regions issue laws at different times, some simply incorporating the national standard, others detailing the application aspects and introducing specific implementation plans [5] too. In particular, the Campania Region, with the Regional Law No. 16/04 (art. 23), has only partially regulated the provisions of 47/85, entrusting the Municipal Urban Plan - PUC (Piano Urbanistico Comunale) - with the perimeter definition of the existing illegal built settlements subject to amnesty. Of this perimeter definition, however, the objectives [2] are regulated, replicating those already reported in 47/85, but the delimitation criteria are not. The PUC "defines the modalities of the urban and building redevelopment of illegal built settlements, the obligatory upgrading interventions and the procedures, also compulsory, for the realization of the same, also through the formation oft he building sectors. The PUC may make the implementation of the urban and building redevelopment interventions of the illegal built settlements conditional on the drafting of special implementation plan (Piano Urbanistico Attuativo) - PUA, called Settlement Redevelopment Plans – PRIA (Piani di Recupero degli Insediamenti Abusivi), whose training process follows the regulation of the implementing regulation provided for in article 43 bis". In the regional law, therefore, the plans to face the question of the illegal built settlements are the municipal urban plan review to the new municipal urban plan or, in the implementation phase, the PRIA; but the technical implementation regulation that should have indicated the criteria for the perimeter de-limitation of illegal built settlements (Article 23 paragraph 3, Regional Law No. 16/04) is not regulated neither by Regulation no. 5/2011 (Burc No. 53, 8 August 2011) to the Regional Law n. 16/04, nor from the operational manual of the Regulation (published on the website of the region in 2012).

The only technical pertinent solution, (when the urban the illegal built settlement constitutes and embodies itself the rule of the expansion and/or dominant transformation of the territory) is the municipal plan at general urban scale and/or implementation plan at sub-urban one. There is no doubt this illegal, dense and/or sprawl building stock, entails conditioning for the elaboration of the new PUC, in order to redevelope those city parts and insert them as an integrated element of the city to which it belongs; to this end it is useful to identify some first guidelines for the elaboration of these plans. They can be distinguished by scale of intervention: the municipal one for the drafting of the PUC or its general or partial review and the sub-municipal one for the implementation plan (sub-funds and PRIA). The present contribution introduces first criteria useful to redevelop dense illegal built settlements at the implementation urban scale while it sends back to a further study for the sprawl [13] illegal built settlements.

Both for the Urban Municipal Planning and for Implementation Plans (Plan Review and PRIA) environmental sustainability, together with the social and economic components, will have to constitute the macro-category to which refer for the redevelopment of the illegal assets, while the criteria for the choices of these plans must consist of the following elements:

- the realization or integration of infrastructures and local urban equipment in order to guarantee the liveability of the area (see Fig. 1);
- the realization of services and equipment able to connect recent, marginal and sometimes distant expansions with the main urban body to ensure integration with the consolidated city;
- urban renewal aimed at mitigating the effects of the dangerous event on the population, on building and infrastructural assets and, therefore, on the entire settlement system;

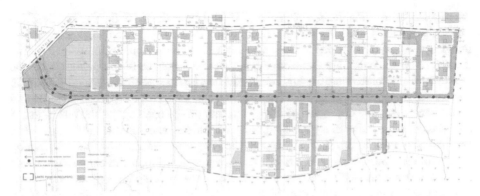

Fig. 1. PRIA - Casamarciano (Na) infrastructures plan and local public equipment zoning.

Fig. 2. PRIA - Casamarciano (Na) redevelopment plan.

- horizontal and vertical alignment with respect to the urban profiles established by the implementation plan to ensure the morphological quality of the urban environment (see Fig. 2);
- structural adjustment in compliance with the regulations for buildings in the seismic zone in order to reduce the vulnerability of buildings;
- the technological and functional adaptation of the entire building and, if necessary, of further interventions on those parts of it that require compliance with specific regulations to improve the performance of the building stock;
- the possibility of changing the land use and, above all, to volumetric rewards in order to achieve the necessary densification and to encourage the involvement of private individuals (owners, companies or companies of urban transformation) in the implementation to ensure the economic sustainability of the interventions.

The addresses of the PUA must therefore be traced both in the disciplinary tradition of the urban design [5] and in the innovation underway of the renewal and updating of the regulations towards the eco-planning [14]. The urban design defines in concrete and precise ways the technical, functional and formal characteristics of the open space, integrates the different open spaces and sorts them into sequences and paths. The themes of urban design are therefore the design and sequence of the public space, the treatment of the green and the environmental connections, the attribution of meaning and role to the open space between the buildings, the design of the pedestrian paths and road connections, the definition of the private and public open space, the resolution of the relationship with the ground floors of public and private buildings. The eco-planning aims to develop a favorable bio-integration of the environment built by man with the natural one, trying to conserve, restore and repair the compromised and stressed ecosystems, facilitating at the same time the development of the built environment within ecologically acceptable borders. The eco-planning is able to realize, through design, a single dynamic living system that is both interactive and functional and requires the bio-integration of four infrastructures: the green infrastructure, the blue infrastructure, the grey infrastructure and the red infrastructure. The green infrastructure: the eco-infrastructure about nature; the blue infrastructure: the water eco-infrastructure, that is natural drainage and water conservation systems and hydrological management in general; the grey infrastructure: the engineering infrastructure, that is roads, sewers, drain pipes, etc., as support systems for sustainable urban development; the red infrastructure: the human infrastructure, that is the built environment, including human activities and economic and legislative social systems.

The synthesis of these two approaches will be able to effectively guide the projects for the upgrading and redevelopment of illegal built settlements.

There remain many open questions, of a general nature, related to the redevelopment of these settlements, which will have to be deepened and addressed, for a complete understanding and control of the phenomenon, among the main ones we can mention: the assessment of environmental damage, the legal problem of double compliance, the misalignment of the town planning procedures with the real estate registry office, the parallelism without crossing between urban law and property right, the disconnection with the civil engineering authorization procedures, the issues related to the monitoring of the phenomenon related to technological innovation (webgis, satellite images).

4   Conclusions

The illegal built settlements have played an important role in the development and transformation of the southern Italian city with lack consideration in the Municipal Urban Plans - PUC. The progressive resort to local development policies, as an alternative to the extraordinary intervention of the State, can offer useful opportunities to integrate city development local cultures with the control of the spatial transformations envisaged by the urban planning in force. The traditional implementation recovery plan - PdR has lost the exclusive primacy among the tools to redevelop degraded urban contexts in favor of integrated intervention programs; to the advantage of the rapid approval of the PdR are opposed the disadvantages of its rigidity in the relationship between state of places and interventions in the individual Minimum Units of Intervention (UMI), its substantial finalization only to residential construction, funding difficulties for equipment and services and the exclusion of infrastructures. The greater adaptability of complex programs includes, in addition to the possibility of extending to heterogeneous urban areas, a variety of interventions including infrastructures and urban furniture.

The proposed criteria for the redevelopment combine and update the best disciplinary tradition to the new questions of the territory/environment and represent the references to correctly plan these settlements.

Some reflections, proposed to the debate, may prove useful to the implementation of new urban planning tools, aimed to urban regeneration and based on eco-planning, on the participation of the recipients and on the public-private partnership. This imposes, however, the search for economic advantages, within a framework of certainty of the rules, without which the commitment of private individuals is inadmissible.

It must, therefore, be concluded that the relevance and interest of the redevelopment of illegal settlements is related both to the vastness of the phenomenon and to the employment and development prospects that a general redevelopment work will be able to offer.

References

1. Italia, V., Bassani, M.: Sanatoria e Condono edilizio. Giuffrè editore, Roma (1985)
2. Perrone, C., Zetti, J. (eds.): Il valore della terra. Teoria e applicazioni per il dimensionamento della pianificazione territoriale, pp. 12–14. Franco Angeli, Milano (2010)
3. Quadro delle definizioni uniformi, Allegato A, G. U. della Repubblica italiana, Serie generale -n. 268, Presidenza del Consiglio dei Ministri-Conferenza Unificata, Intesa 20/X/2016
4. Sentenza C. Cass. pen. 05/10/2011, n. 36104, Aggravio del carico urbanistico: condizioni e rilevabilità (2011)
5. de Biase, C.: Le politiche per l'abuso edilizio-urbanistico. In: de Biase, C., Losco, S., Macchia, L. (a cura di): Abusivismo urbanistico e sostenibilità ambientale. Edizioni Le Penseur, Brienza (Pz) (2017)
6. Dittrich, T.M.: Buffer Zone Form, Function, and Design: A Review. Cornell University, NY (2004)
7. Arefieva, N., Terleeva, V., Badenko, V.: GIS-based fuzzy method for urban planning. Proc. Eng. **117**, 39–44 (2015)
8. Urbani, P., Civitarese, S.: Diritto urbanistico: Organizzazione e rapporti, sesta edizione, p. 169. Giappichelli, Torino (2017)
9. For the Municipality of Casal di Principe, the aerial photogrammetries of 1985, 1994 and 2003 and the 2016 image of Google Earth were used; as the Urban Planning Tools, the 1970 Manufacturing Program and the P.R.G. of 2003. For the Municipality of San Cipriano d'Aversa the orthophotos of the National Geoportal of 1988, the aerial photogram- metries of 1997 and 2003 and the image of 2016 of Google Earth; as far as the plans are concerned, only the P.R.G. of 2004. As regards, finally, the Municipality of Casapesenna was used the aerial photogrammetries of 1974, 2001 and 2005 and the image of 2016 of Google Earth. The Plan available is P.R.G. 2000. While, for all three municipalities, the practices relating to the amnesties of 1985, 1994 and 2003 have been used

10. Arosio, M.: Il recupero urbanistico degli insediamenti abusivi del capo III della legge n 47 del 28 febbraio 1985, Riv.giur.ed. 2, 249 (1985)
11. Ingersoll, R.: Sprawltown. Meltemi, Roma (2003)
12. Colombo, L.: Abusivismo e pianificazione consensuale. Urbanistica Informazioni **188**, 57–59 (2003)
13. Colombo, L., Cerreta, M., Palomba, G.I.: Urban Sprawl in Italy. Urban and suburban densification and the peri-urban border. In: Proceedings of the 5th Annual International Conference on Architecture and Civil Engineering, ACE 2017, Singapore, 8–9 May 2017 (2017)
14. Yeang, K.: Ecomasterplanning. Willey, Chichester (2009)

The Context of Urban Renewals as a 'Super-Wicked' Problem

Isabella M. Lami^(✉) (iD)

Politecnico di Torino, 10125 Turin, Italy
isabella.lami@polito.it

Abstract. Urban renewals (URs) were identified as a "wicked problem situations" in the original article which introduced the concept. Recently, transforming urban historic areas has become increasingly complex because of a number of reasons (including the property fragmentation, the values' creation mechanisms, the regulatory framework, the multitude of public and private stakeholders with divergent perspectives and values). In this paper it is argued that problem situations related to URs have become 'super-wicked', borrowing the term from the literature on climate change and introducing it urban planning and evaluation realm. In this sense, it is possible to distinguish a series of issues that can grouped in three main features: (i) the measurable dimension of the problem, mainly related to the physical and spatial dimension of that peculiar economic good represented by the urban tissue; (ii) the specificity of the decisional processes in this realm; (iii) the normative dimension.

Keywords: Super-wicked problem · Urban renewals

1 Introduction

In the '70s, "urban renewals" (URs) - the transformations of urban historic areas - was mentioned by [1] as an example of a 'wicked' problem situation. The term 'wicked' refers to a complex and uncertain problem situation for which there is neither a clear formulation and definition, nor a simple and straightforward solution. Differently from a chess problem or a mathematical equation, wicked problem situations are ones for which there is no clear stopping rule: working more on them might well bring forth a suitable solution. There is no single right answer and each attempt to resolve the problem situation can matter because it deeply affects society (e.g. the realisation of a new transport infrastructure can substantially modify the landscape of a region, its ecosystem and its real estate market) [1–3]. Recently, transforming urban historic areas has become increasingly 'wicked', especially because of a number of reasons related to: the property fragmentation, the values' creation mechanisms, the regulatory framework, the multitude of public and private stakeholders with divergent (also conflicting) perspectives and values which are involved in defining problems and solutions. In this paper it is argued that problem situations related to URs have become 'super-wicked', borrowing the term from the literature on climate change and introducing it urban planning and evaluation realm.

© Springer International Publishing AG, part of Springer Nature 2019
F. Calabrò et al. (Eds.): ISHT 2018, SIST 100, pp. 249–255, 2019.
https://doi.org/10.1007/978-3-319-92099-3_29

2   Urban Renewal as a "Super-Wicked Problem"

2.1   Process and Outcomes of Urban Transformations

Between the '60 s and the late '80 s, urban transformations were carried out following a specific plan, that is urban plans were determining and driving the process and type of urban transformation. Such plans were created and executed without closely verifying a fit between forecasts (made by planners) and the demand on the property market [4]. Despite decision makers' indifference concerning the implementation of the plans, urban transformations were successful, especially because the population was growing, the cities were expanding, the economy was booming, and the Public Administrations were experiencing a phase of financial well-being [5]. In the '90s, however, Public Administrations started facing a financial crisis causing a decrease in public incentives to invest in urban transformations, accompanied by both a fall in developers' motivation to undertake real estate transactions and decision makers' power to control and guide urban transformations. This crisis strengthened the role of private developers played in urban transformations and led to the emergence of public-private partnerships [6]. One of the consequences has been the gradual increase in players in the decision-making arena. In the past, the stakeholders interested in the transformation of urban areas were essentially two, the owner of the area and the user, often represented by one and the same person [7]. Today, urban transformations involve a multitude of public and private stakeholders that include not only those affected by and affecting the state of the area (e.g. users, owners and managers), but also those who practically realize the transformation (e.g. owners, decision makers, investors and actuators), as well as intermediate actors (e.g. real estate agents, actors looking for areas and buildings to transform, and funders). These stakeholders typically define, due to their different (also conflicting) interests and goals, urban problems and solutions from different perspectives and values [4, 8], thus triggering conflict that may slow down and hamper decision making and affect the quality of the decisions made [9, 10]. Hence, despite the presence of a clear plan, the social interactions and interdependencies between these stakeholders are the major determinant of the process and outcome of urban transformations. These outcomes are, however, unpredictable products of chance, coincidence and decision-making power (of developers and investors), because of the underlying conflicts in the debate [11]. Such debates are especially evident in the transformation of urban historic areas, or 'urban renewals', which currently represent the majority of urban transformations in Europe.

Transforming urban historic areas presents peculiar characteristics: (i) the asset's ownership is shared by a vast number of different public and private stakeholders that need to agree on and pursue a common goal of transformation. Generally, it is higher than in the case of non-historic areas, due to the characteristics of the urban fabric itself, with previously small housing units; to which are added the recent splitting operations for real estate speculation. The former industrial areas, large for the type of production they hosted, belong mostly to a single entity (natural or legal person). The former industrial areas, large for the type of production they hosted, belong mostly to a single entity (natural or legal person), allowing - from a point of view of the ownership structure - a simpler start of the urban renewals; (ii) the divergence of stakeholders' perspectives and values regarding the transformation is typically stronger than in the case of non-historic areas (historic areas are of higher value due to their central location, better accessibility and supply of services). This is particularly evident in protect areas that have "special architectural or historic interest, the character or appearance of which is desirable to preserve or enhance" (Civic Amenities Act, 1967, Section 1); (iii) established and completed buildings with a specific identity and value are transformed to include new uses and functions that might be contested by different stakeholders (due to their different interests and goals); (iv) the characteristics of a specific historic city (e.g. as underlined by [12], the medieval fabrics of Toledo; Copenhagen and Barcelona display a labyrinth complexity; and the shape and network structure of Paris Haussmannian blocks articulate density and versatility) determine which successive uses and functions can be enabled by gradually modifying form and structure; finally (v) there has been recently an increase in social conflict characterised by protests against URs that tend to generate and intensify inequalities amongst citizens [13–16]. If URs present a series of peculiarities, it seems important to redefine the problem: what are we talking about when we approach a UR?

2.2   A Super Wicked Problem

URs were identified as a wicked problem situations in the original article which introduced the concept [1]. If the wicked problem "defies resolution because of the enormous interdependencies, uncertainties, circularities, and conflicting stakeholders implicated by any effort to develop a solution" [17], recently some scholars presented the term of "super wicked problem" (SWP) referring to the climate change, because of its even further exacerbating features [17, 18].

According to [17], the three main features of climate change (Table 1, left columns) are, in extreme synthesis, related to:

1. the quantitative and in somehow measurable aspects of the problem (the greenhouse gases, the stock/flow nature of physical and chemical processes underlying it with its enormous temporal dimension; the lack of homogeneity in the geographical distribution of the effects);
2. the limits of the human mind (human beings think mostly in physiological time; there is a human tendency to judge the likelihood of an occurrence based on the relative ability to imagine its happening and climate change is an unimaginable problem; the complexity of the causal chains of human actions makes their consequences seem very far from each others);
3. the absence of an adequate institutional framework able to "address a problem of climate change's tremendous spatial and temporal scope".

Drawing on climate change literature [17], it is possible to re-define URs as SWPs. It is possible to borrow these concepts from the context of climate change to that of the URs, distinguishing a series of issues that can grouped in three main features (Table 1, right columns): (i) the measurable dimension of the problem, mainly related to the physical and spatial dimension of that peculiar economic good represented by the urban tissue; (ii) the specificity of the decisional processes in this realm; (iii) the normative dimension. These three main features can better specified, as follow.

Table 1. Features of climate change as super-wicked problem situations (SWP) versus urban renewals.

SWP CLIMATE CHANGE (Lazarus, 2009)		SWP URBAN RENEWALS	
Features	Characters	Features	Characters
THE SCIENCE OF CLIMATE CHANGE	Greenhouse effect	*THE MEASURABLE DIMENSION*	Ownership is highly fragmented
	Stock/Flow Nature of Atmospheric Chemistry		Urban Stock/Flows of urban fabric (values' creation mechanisms)
	Spatial Dimension of Climate Change: Global Cause vs. Global Effect		Spatial distribution of values
HUMAN NATURE AND COGNITIVE PSYCHOLOGY	Myopia and Climate Change's Temporal Dimension	*THE DECISIONAL PROCESSES*	Urban renewal's temporal dimension
	The Availability Heuristic, Space, and Complexity		The expansion of the decisional network
	Representativeness Heuristic and Climate Change Cause and Effect		Cause and effect of urban renewals
THE NATURE OF U.S. LAWMAKING INSTITUTIONS	The Challenges of Environmental Lawmaking in General	*NORMATIVE DIMENSION*	Regulatory framework
	The Making of Climate Change Law in Particular		Inclusive processes and administrative procedures

2.3   Measurable Dimension

Ownership is Highly Fragmented. URs are typically based on quite large conversion areas, whose ownership is highly fragmented. The process of "grouping" small areas belonging to different entities into a single project is the core challenge of contemporary urbanism. Decision makers and planners face difficulties in finding ways to enable URs, and introduce mechanisms to enhance the implementation of plans (that ensure the profitability and attractiveness of operations) [19];

Urban Stock/Flows. The city is a "collective" asset, created and defined by both public and private investments and decisions. From all this derives an important consequence: the economic value of its single parts is not determined by the single action, but by the collective action, external to the single actor, by the fact that synergies and externalities occur crossed with all the decisions - localization, investment, management - which take place or occurred in the physical surroundings of the place where the individual decision was made [20]. Moreover, when urban heritage is unused or underused, it can be considered as a sleeping fixed capital to be activated [21]. The economic challenge of URs - adjusting form and function in a long-term effort [12] or adopting the adaptive re-use, which can be carried out by adapting the content to the container rather than the convers and involves maximum conservation and minimal transformation [22, 23] - is to create a profitable relationship between stock and flows;

Spatial distribution of values. Benefits and costs of URs are often unevenly distributed (e.g. inclusion of undesirable facilities; new accessibility may change the use and perception of an area), triggering conflict amongst stakeholders [15]. If the benefits are widespread and the costs are concentrated on a small community that is forced to bear the costs of an intervention that benefits others, it is quite natural that conflict arises [16].

2.4   Human Nature and Decisional Processes

URs' Temporal Dimension. The transformation of the city is the product of an on-going social process that is enabled (guidance, support) and constrained (limitations, prohibitions) by a plan. At the same time, urban transformation operations are fragmented driving processes, with different time perspectives for operators. Each approach needs a specific metric to make the results achieved objectively measurable and comparable. The challenge is to coordinate different operations, which have different potential in terms of time/profitability/values;

The Expansion of the Decisional Network. Multiple stakeholders with divergent values, interests, goals and perspectives on the transformation [9] constitute new patterns of public decision-making. In a situation of high institutional and social fragmentation, the power of *veto* is in fact multiplied. It does not refer only to the traditionally strong interests, but also to the traditionally weak interests (as long as there is a concentration). Groups that are not involved in the decision process have the possibility to stop the choices made by others, or at least to delay them [10]. Moreover, the web connection among citizens enabled by ICT, potentially expands the decision making arena out of proportion [24];

Cause and Effect. URs are regulated by plans and projects, which can ensure direct effects (e.g. the realisation of a square or a shopping mall) but not desired and claimed side effects (e.g. the square will be popular, the shopping mall will be profitable and lead to the economic revitalization of the surrounding area). The effect of a project cannot be determined linearly from the origin: the final state achieved does not always correspond to the expected one [25–27]. Furthermore, the effect of URs can be much more than the physical transformation of the urban fabric: its media coverage, plans, standards, other projects, etc. The achievement of the latter is hardly negotiable and measurable [4].

2.5   Normative Dimension

Planning Instruments. Soil has potentially many possible uses, but it is intelligence, enterprise and human work - especially in conditions where incentives exist to do so and mechanisms that allow to arrive and exchange missing knowledge - to generate a creative plurality of the latter [28]. The State is the only entity authorized to use force - to impose rules, levy taxes, etc. - in a certain territory; has, in other words, the "monopoly of the use of coercion" [29]. Despite the variety of arguments for and against planning in the modern urban context, there is an implicit consensus about the need for public sector planning to perform four social functions: promoting common/collective interests of the community; considering external effects; improving the information base for decision making; and considering distributional effects [30]. The choice of instruments to enable these functions is, however, still controversial;

Inclusive Processes and Administrative Procedures. Over the past two decades participatory practices concerning URs have multiplied. The challenge consists in balancing the relationship between the inclusive processes and administrative procedures: the former can be part of the normative and in other cases they are held entirely *extra legem* (are not prohibited, but not even explicitly provided) [31].

3   Conclusions

The paper introduces the concept of the "super wicked problem" to the urban realm, arguing that transforming urban historic areas has become increasingly complex and defining the specific features of the issue. The problem setting is important in order to find new and effective ways to tackle it, because the reality of Italian cities shows how traditional forms of intervention in the urban tissue (in financing operations, in designing and building methods, in urban planning) are no longer working. There is considerable potential for further research. The following step, already on-going, is to study the conditions for activating urban regeneration operations with low financial capital intensity and high cultural and human capital intensity through a new culture of design and regulation.

References

1. Rittel, H.W.J., Webber, M.M.: Dilemmas in a general theory of planning. Policy Sci. **4**, 155–169 (1973)
2. Buchanan, R.: Wicked problems in design thinking. Des. Issues **8**(2), 5–21 (1992)
3. Weber, E.P., Khademian, A.M.: Wicked problems knowledge challenges, and collaborative capacity builders in network settings. Public Adm. Rev. **68**(2), 334–349 (2008)
4. Healey, P.: Urban Complexity and Spatial Strategies. Towards a relational planning for our times. Routledge, London and New York (2007)
5. Secchi, B.: Il racconto urbanistico. Einaudi, Torino (1984)
6. Ruegg, J.: Formes du PPP. In: Ruegg, J., Decoutère, S., Mettan, N. (eds.) Le partenariat public-privé. Presses Polytechniques et universitaires Romandes, Lausanne, pp. 79–98 (1994)
7. Faludi, A. (ed.): A Reader in Planning Theory. Urban and Regional Planning Series 5. Pergamon Press, New York (1973)
8. Fainstein, S.S., DeFilippis, J. (eds.): Readings in Planning Theory, 4th edn. Wiley-Blackwell, Hoboken (2015)
9. Dente, B.: Understanding policy decisions. PoliMI SpringerBriefs, Milano (2014)
10. Lami, I.M.: Evaluation tools to support decision-making process related to european corridors. In: Lami, I.M. (ed.) Analytical Decision Making Methods for Evaluating Sustainable Transport in European Corridors, pp. 85–102. Springer, Cham (2014)

11. Palermo, P.C.: I limiti del possibile - Governo del territorio e qualità dello sviluppo. Donzelli Editore, Roma (2009)
12. Salat, S. (ed.): Cities and Forms on Sustainable Urbanism. Hermann, Paris (2011)
13. Lefebvre, H.: Le droit à la ville. Anthropos, Paris (1968)
14. Lefebvre, H.: Writings on cities. Blackwell, Oxford (1996)
15. Harvey, D.: The right to the city. New Left Rev. **53**, 23–40 (2008)
16. Bobbio, L.: Conflitti territoriali: sei interpretazioni. TeMA **4**, 79–88 (2011)
17. Lazarus, R.J.: Super wicked problems and climate change: restraining the present to liberate the future. Cornell Law Rev. **94**(5), 1153–1234 (2009)
18. Levin, K., Cashore, B., Bernstein, S., Auld, G.: Overcoming the tragedy of super wicked problems: constraining our future selves to ameliorate global climate change. Policy Sci. **45**, 123–152 (2012)
19. Micelli, E.: Cinque problemi intorno a perequazione, diritti edificatori e piani urbanistici. Scienze Regionali **2**, 9–27 (2014)
20. Camagni, R.: Verso una riforma della governance territoriale (2011). http://storicamente.org/quadterr2/camagni.htm. Accessed 23 Dec 2017
21. Camagni, R.: Le ragioni della coesione territoriale: contenuti e possibili strategie di policy. Scienze Regionali **2**, 97–112 (2004)
22. Németh, J., Langhorst, J.: Rethinking urban transformation: Temporary uses for vacant land. Cities **40**, 143–150 (2014)
23. Robiglio M.: The adaptative reuse toolkit, Urban and Regional Policy Paper 38, GMF (2016)
24. Castells, M.: The Rise of the Network Society: The Information Age, Economy, Society, and Culture, vol. I, 2nd edn. Wiley- Blackwell, Hoboken (2009)
25. Armando, A., Durbiano, G.: Teoria del progetto architettonico. Dai disegni agli effetti. Carocci, Roma (2017)
26. Todella, E., Lami, I.M., Armando, A.: Experimental Use of Strategic Choice Approach (SCA) by Individuals as an Architectural Design Tool. Group Decision Negotiation (2018)
27. Tavella, E., Lami, I.: Negotiating perspectives and values through soft OR in the context of urban renewal. J. Oper. Res. Soc. (2018). https://doi.org/10.1080/01605682.2018.1427433
28. Moroni, S.: Suolo. In: Somaini, E. (ed.) I beni comuni oltre i luoghi comuni. IBLLibri, Torino (2015)
29. Moroni, S.: Fondamenti: principi e palette. In: Ponti, M., Moroni, S., Ramella, F. (eds.) L'arbitrio del Principe. IBLLibri, Torino (2015)
30. Klosterman, R.: Arguments for and against planning. In: Campbell, S., Fainstein, S. (eds.) Readings in Planning Theory, 2nd edn. Blackwell, Malden (2003)
31. Bobbio, L. (ed.): A più voci. Amministrazioni pubbliche, imprese, associazioni e cittadini nei processi decisionali inclusivi. Edizioni Scientifiche Italiane, Napoli (2004)

The Role of Physical Aspects in the City Plan Rules Definition

Roberto De Lotto[✉] and Cecilia Morelli di Popolo

Università degli Studi di Pavia, 27100 Pavia, Italy
{roberto.delotto, cecilia.morellidipopolo}@unipv.it

Abstract. Definition of city as a complex system opens up to a series of problems that involve the legitimacy of the design of the technocratic plan. The plan nowadays cannot be considered as a black-box tool, but it might be considered as an open process, able to adapt to the needs of citizens and to the socio-economic and environmental contexts. As many scholars underlined, the plan is sequence of phases that must be programmed and that considers the project as a possible scenario, not a definitive one. In a complex and flexible city, the planner (here considered both as the urban studies expert and the political decision maker) has the role to create the conditions for the development of the city, and for creating or maintaining the possibilities of evolution of the citizens who live in a certain territory. In this sense, the strategic aspects of the city plan and its programmatic role, have a relation more with the rules system than with the design one; but it is also very clear that the human space in urban context is made of physically well defined elements. Basing on the most commonly cited example of anti-planning city organization, the MVRDV project Oosterwold, authors underline the importance of the physical and geographical components inside the rules of a plan, nevertheless recognizing the difficulty to establish detailed boundaries to a complex and flexible city.

Keywords: Flexible and complex city · Role of rules · Physical aspects

1 Introduction

Considering the contemporary debate about the theory of complex systems applied to the city, the main aim of authors is to underline the relevance of territorial aspects even in structural and decision processes [1]. With territorial aspects, authors refer to physical and geo-graphical elements that, in the historical tradition were synthesized by the latin word "urbs": they are the material and spatially definable components of the city.

The uncertainty of the future on the one hand, and the natural tendency of scholars to reach universally valid rules on the other hand, risk to point out the problems at a very general scale and define solutions that may be not suitable for specific contexts and times [2–6]. The predictive capacity, that should have been the goal of urban planning, founded its legitimacy in particular in technical planning tools that, nowadays (after the season of ICT based urban models), show their weakness. In the same time, the speed of changing of the typologies of requests from citizens, does not correspond

© Springer International Publishing AG, part of Springer Nature 2019
F. Calabrò et al. (Eds.): ISHT 2018, SIST 100, pp. 256–263, 2019.
https://doi.org/10.1007/978-3-319-92099-3_30

to the immediate satisfaction by the local administrations. Moreover, if speed is a problem, economic uncertainty and the inability to predict market trends do not give the possibility to hypothesize credible and effective city plans [7].

"How do we decide what is of lasting value in ourselves in a society which is impatient, which focuses on the immediate moment? How can long-term goals be pursued in an economy devoted to the short term? How can mutual loyalties and commitments be sustained in institutions which are constantly breaking apart or continually being redesigned? These are the questions about character posed by the new, flexible capitalism" [8].

About flexibility, complexity and uncertainty we can refer to a definition by Portugali [9], that define city as a "dually complex systems", because composed by two fundamental elements: the artificial component (system of well-defined elements till the smallest details, even if complex, from bridge to bolt) and the urban agents (those who relate with the artificial elements and transform them into functional elements, for life and their survival).

So, we can say that the cities are modeled by the actions of urban agents on the artificial components and on the mutual connections among all components or, as defined by Bertuglia and Staricco [10], by the system's organization (see Fig. 1). For a complex system, the organization is a fundamental characteristic: the ability to act on its own organization is one of the fundamental properties of a system, and can be expressed as an evolution of interaction. As a consequence, organization becomes a constituent property of a system [11, 12].

COMPLEX CITY SYSTEM
(Portugali)

Artificial components
composed by simple elements
(from smallest to biggest)

'urban agents'
elements that interact with artificial components
and transform the system

Where:
Organization: systems' characteristics
Relation (tipology of) defines the structure organization

Fig. 1. Complex city system scheme.

Defining the city as a complex system means affirming that the city is linked to a set composed by elements in relation to each other (system). As it is well known, it is not possible to manage and control the processes inside a system with deterministic instruments; considering the city as a complex system [10] the development of the city is not linearly predictable, basing on the knowledge of initial conditions (dynamically complex system). The non-linearity of interactions among urban actors and the plurality of urban systems description effectively connect the definition of city with definition of complex system.

1.1 Hyper-determined Planning vs no-Planning

If urban planning is the science that studies the development of territorial and urban systems, it is easy to understand that the hyper-determination of urban planning does not coincide with current criticalities previously cited.

A deterministic system like the one used until modernism to role the planning cities, has already demonstrated its limits. These kind of plans has always been ineffective and inefficient; the rigidity of the technical tools and of the structure of decision making give little scope to unpredictable changes. To this basic elements, we must necessarily add the role of technological innovations and globalization, able to distribute knowledge, but also to simplify and conform, activating phenomena that can not be controlled a priori.

To work into such a complex conditions, researchers need firstly to redefine the role of the actors that act inside the process: expectations, rules, training.

Then, researchers need to change the approach to "planning" itself, with particular reference to the planners' role.

2 Actors of Complex System

We can consider three principal actors (that are a part of 'urban agents') of the development urban process:

- Citizens: intended as a group of private individuals;
- Public administration: intended as official bureaus hierarchically organized;
- Planners: urban science experts.

In what way they interact inside the urban process? What are the action they complete inside the urban process?

From the classical modernist urban system point of view (characterized by a deterministic process), the action of 'urban agents' is simple to be read:

- The citizens' action inside the planning process are related to the directly living space: house, workspace, free time spaces and connective systems (i.e. road system). The active action in the process is limited to the daily sphere. As a single individual, the citizen interacts partially with the other daily spheres, unless different types of processes are activated (i.e. collaboration, informal associations, urban conflicts, …);
- The action of the public administration have a top-down decision-making model: according to the decisions of the different administrations, the programs' development of the city are modified in a more or less incisive way, acting on the citizens without direct interaction;

- The main action of the planners is to translate public administration strategies on 'paper' by acting on the urban structure at all levels of detail (from large to small scale).

What happens if the system is seen as a complex system (more flexible system) [13]? How the various roles change considering also the invasion of the last technological and IOT applications?

- Citizen: they are the main actors, the ideas' actuators and those who have the power to modify space usage with specific tactics. Some examples can be all of self-organization processes born inside informal urban contexts starting from the more different situations. About the social media instruments, a nice example is the relatively new app "Nextdoor" that connect neighborhoods inhabitants in order to solve simple problems (basic goods, kids' necessities, security, etc.);
- Public administration: the introduction of the subsidiarity principle changed the role of the classical hierarchical structure and, also, the role of the whole public administration in relation with citizens. Rules are no more furnished directly from the higher level to the lower one, but they are shared and hopefully participative. In example, following the so called "negative law" [14] the action of urban agents is determined by shared rules that indicate what "not to do"; these rules give freedom of action to citizens and the role of the public administration might become superfluous;
- The role of planners changes considerably They become those who build the rules that can give the citizen the ability to act. The citizens pass from the rules to action. The city planners do not indicate the actions but the boundaries inside which the citizens can act.

3 Geography in Complex System

If in the deterministic system, the relation between scales is clearly and well defined in Rogers' statement "from the spoon to the town" (Rogers, 1952 in Charter of Athens); in the complex system the descendent scale from macro to micro becomes more complicated and difficult to be defined.

The inductive method, explained through specific episodes or through actions principally related to generic public space, is aimed not to consider the specific local characteristics such as physical and geographical ones in which actions take place.

A basic logical connection, may be synthesized by the couple of questions: "where, what?" or "what, where?" [15], that mean: define a space to find the correct actions, or on the other hand: define an action to find the correct space to put it into reality. As long as discussed by know, the actions depend on the rules; so, does a rule have a physical scale? Which is the scale of a specific action that derives from the application of a rule?

Another interesting question may be: does an action need a specific space? Or: is the action itself that define the space? Following the anthropo-genetic approach of Choay [16, 17] the human actions on the territory, and the conse-quent modifications of the territory are integral parts of the human being itself.

Moreover, considering the city plan as the result of the job of public administration and planners, does the rule system need a specific geographical and physical definition? Does the rule system born from the geographical and physical context?

Not all of these questions may have an exhaustive answer, but it may occur a certain dichotomy between rule nature and physical dimension.

4   Geography in Oosterwold Project

Freeland by MVRDV (2012) in Oosterwold, Almere (NL) is experimentation that represents the most radical application (in a very large scale considering the wideness is 4.300 ha) of the Organic Development Strategy [18]; the development is based on cohabitation rules. The aim of the project is to generate a city without any specific design detail but considering the actions of the individuals as engine of the development itself. What are fundamental are the ethical and responsible behaviors that citizens must have.

The Oosterwold rules are listed above [19]:

1. Everyone is equally welcome to initiate a project. All sorts of collaborations are possible too. Whoever comes to Oosterwold is obliged to invest for the long run.
2. An initiator can select a plot by themselves. Size, shape, and location are almost completely open. There are no minimum or maximum plot sizes.
3. The same 'generic' rules apply to all plots, no matter size or shape:
   (1) In principle, it is obligatory to provide 'development', paved surface, water, public green, and agriculture or horticulture. 'Development' is the conglomerate of housing, work, and amenities. The plot cannot be divided.
   (2) A ratio applies to the different functions that have to be accommodated.
4. This ratio is not absolute. Depending on its location a specific 'character' can be achieved by including more of one of the ingredients. The urban plan discerns between landscape plots, agricultural plots, core plots, and work plots.
5. Local infrastructure will be developed by the initiators themselves. Main infrastructure will be provided by local government. All infrastructure will be public in character, as far as interconnectedness is concerned. In every plot, the edges stay free of development to guarantee continuity and 'wadability'.
6. To prevent complete build up of a plot and guarantee Oosterwold's green character, the FAR is 0.5. For 'core plots' FAR is 1.0
7. More than two-thirds of Oosterwold's surface will be green. it will consist of a mix of private and public green (agricultural land is considered as private), though all green is privately owned and maintained.
8. Every plot will be self-sustaining to the max. Fresh water management, wastewater management, and energy are responsibilities of the initiator/plot owner. collaboration and cooperation with neighbors is unavoidable. connection to the electricity grid is considered a backup option.
9. Most plots will be financially self-sustaining, meaning that no subsidies are available, apart from plots with special requirements.

10. Public investment follows private investment. This reverses the current order of things. Public spending comes are earning money instead of the other way around.

Starting from these rules, we can understand the importance of the equilibrium between the two components: general rules and geographical components. Starting from point 1 and 2, it is necessary to define the "surrounding conditions" from which all the project takes moves. In this case the surrounding conditions are not made of physical aspects, but of the action of the first actors; but this action determines a physical bounder for all the other actors (no matter which is the specific shape of the bounder).

Point 3 synthesizes the "functional mix" with a simple ratio among functions, specifying in point 4 that the ratio may be variable. In point 3 and 4 the comprehensive idea of flexibility is totally expressed.

As indicated in the points 5 or 6, the role of the public administration is fundamental to generate important elements, such the main infrastructures, but also the role of the planner that is able to evaluate some aspect, like the 'core plots'. The total autonomy or self-organization, can generate a total chaos that needs a lot of time to find a new equilibrium (see also AnarCity [20]). Differently a balance between top-down and bottom-up processes, the so-called ethero-organization [21] can really help the development of the city following all the 'urban agents' desires and behaviors, in respect of the environment and artificial elements.

Point 6 translates the index that is usually related to the difference of real estate value depending on the relevance of the location (so called differential rent) and on the value of the function. The rules do not specify the locations but let to the public administration to define the 'core plots', giving to the public power the ability to design the density distribution.

Point 7 is a sort of environmental performance index that deeply involves the physical shape of the city.

Points from 8 to 10 have an economic character.

Looking at satellite maps, the actual realization of Oosterwold is still very small and it is proceeding by single plots located at a large distance one to the other. A small road and a basic system of wind turbines are the only visible infrastructures [22].

So, by now, it seems that Freeland rules are following a pure nomocratic approach; maybe it is because the conflicts among the different developed plots are safeguarded by a sufficient spatial distance and, at the moment, the density is too low to represent a minimal neighborhood unit. So, all the social interactions are simple and far to be complex.

5   Conclusions

As far as someone moves away from the physical and geographical dimension, this one remains indispensable. In Freeland the physical aspects prevail on the geographical ones, and all of them depend on the ethical behaviors.

Considering the general theme of complexity, in which way can planners help the dually complex system to develop under uncertain condition? The construction of scenario with solid tools, can help planners [23]. But the solid tools may not be structured with a technocratic point of view. The variability must be taken into account and the solidity of the tool does not depend on the permanence of the specific object (or space) but on its usability by citizens.

For some scholars, the role of the urban planner is intended to be no longer the designer for the 'urbs' but a 'lawnmaker': starting from a series of considerations about the environment surrounding of a given territory, planners can create a dataset useful to establish and build rules, which will have the task of 'guiding' the (Choay's) 'civitas' in shaping and designing their own spaces.

The first task that a town planner has to face, is to identify the objectives of a given intervention, which must take origin by the 'civitas'. With a look to flexibility, also the law structure must be under constant modification. Flexible city planning, considered as an answer to the cities' complexity, will mainly focus on the ability that 'urbs' will have to adapt to the new demands of the city system.

As declared by Viny Maas in "The shape of the law", 'It's legal, but is it legitimate?' and 'It's legitimate, but is it legal?' These are questions not widely discussed in fields like architecture and urban design.

Once the relevance of rules is defined, to have a comprehensive point of view about the city plan, it is necessary to conceive the possible physical and geographical consequences of the set of rules, in particular with reference to actual problems that cities have to face, such as (at least in Europe): soil consumption, densification, carrying capacity of infrastructural settlement depending on land exploitation, quantity and quality of urban spaces.

Yet the architectural scale has to deal with rules and regulations, and the physical and the geographical elements must maintain their relevant role, "Because in the end law is too important to be left to lawyers" [24].

References

1. Rabino, G.A.: Processi decision e Territorio nella simulazione multi-agente. Esculapio, Bologna (2005)
2. Bauman, Z.: Globalizzazione e glocalizzazione. Armando Editore, Roma (2005)
3. Bauman, Z.: La società dell'incertezza. Il Mulino, Bologna (1999)
4. Sassen, S.: Making the global economy run: the role of national states and private agents. Int. Soc. Sci. J. **51**, 409–416 (1999)
5. Cesareo, V.: Globalizzazione e contesti locali: Una ricerca sulla realtà italiana. FrancoAngeli Edizioni, Roma (2001)
6. Horgan, J.: From complexity to perplexity. Sci. Am. **272**(6), 104–109 (1995)
7. Batty, M.: Cities as Complex Systems: Scaling, Interaction, Networks, Dynamics and Urban Morphologies. Encyclopedia of Complexity and Systems Science. Springer, New York (2011)
8. Sennett, R.: The Corrosion of a Character: the personal Consequences of Work in the New Capitalism, p. 10. W.W. Norton & Company, New York (2000)
9. Portugali, J.: What makes cities complex? http://www.spatialcomplexity.info. Accessed 22 Oct 2013

10. Bertuglia, C.S., Staricco, L.: Complessità, autoorganizzazione, città. FrancoAngeli Edizioni, Roma (2000)
11. Ruelle, D.: Chance and Chaos. Princeton University Press, Princeton (1991)
12. Gargiulo, C., Papa, R.: Caos e caos: la città come fenomeno complesso. Per il XXI secolo: una enciclopedia e un progetto. Università degli Studi di Napoli Federico II
13. De Lotto, R.: Il progetto urbanistico nella città flessibile. In: De Lotto, R., Di Tolle, M. (eds.) Elementi di progettazione urbanistica. Rigenerazione urbana nella città contemporanea. Maggioli Editore, Santarcangelo di Romagna (2013)
14. Moroni, S.: Planning, liberty and the rule of law. Plann. Theory 6(2), 146–163 (2007)
15. De Lotto, R., Cattaneo, T., Venco, E.M.: Functional reuse and intensification of rural-urban context: rural architectural urbanism. Int. J. Agric. Environ. Inf. Syst. 7(1), 1–27 (2016)
16. Choay, F.: Del destino della città. Alinea, Milano (2001)
17. Choay, F.: La città: Utopie e realtà. Einaudi Editore, Torino (1973)
18. Rauws, W., De Roo, G.: Adaptive planning: generating conditions for urban adaptability, lessons from dutch organic development strategies. Environ. Plann. B Plann. Des. 43(6), 1052–1074 (2016)
19. MVRDV, Freeland project. https://www.mvrdv.nl/projects/freeland. Accessed 10 Dec 2017
20. The Why Factory lab. AnarCity Project. http://thewhyfactory.com/output/anarcity/. Accessed 10 Dec 2017
21. Morelli di Popolo, C.: La città flessibile. Le dimensioni della flessibilità nella città contemporanea e futura. Ph. D. thesis in Civil and Architectural Engineering, XXVI cycle (2014)
22. Moroni, S.: Rethinking the theory and practice of land use regulation: towards nomocracy. Plann. Theory 9(2), 137–155 (2010)
23. Blecic, I., Cecchini, A.: Verso una pianificazione antifragile: Come pensare al futuro senza prevederlo. FrancoAngeli Edizioni, Roma (2016)
24. Maas, W.: It's legal, but is it legitimate? Shape Law 38, 106–112 (2014)

Informal Settlements: The Potential of Regularization for Sustainable Planning. The Case of Giugliano, in the Metropolitan City of Naples

Claudia de Biase[1], Fabiana Forte[1], and Pierfrancesco De Paola[2(✉)]

[1] Campania University, 81100 Caserta, Italy
[2] Federico II University, 80125 Napoli, Italy
pierfrancesco.depaola@unina.it

Abstract. The present study tries to analyze the phenomenon of *informal settlements* in a specific territory. The choice is due to the fact that the above phenomenon has a strong territorial impact, as well as a strong influence on the use and consumption of the land. In Italy the problem of informal settlements is so old and consistent that three building amnesties have not solved it. The aim of the study is to analyze the phenomenon under both the perspectives: urban planning and local finance. A specific methodology, based on the overlapping of different maps, is implemented here. This allows figuring out the phenomenon of illegal settlements in the municipality of Giugliano, in the Metropolitan city of Naples. Giugliano in Campania is the third municipality in Campania Region for the number of inhabitants. The rapid increasing of population has caused an unplanned development, especially in the suburbs and along the coast, two areas that are frequently characterized by the scarcity of infrastructures and public facilities. By analyzing the municipal budgets of the last years, and, in particular, by analyzing the revenues deriving from the retrospective building permits, it is possible to evaluate the activity of the administrations in the sustainable urban management of the informal settlements. By simulating the regularization for some well identified informal settlements (in Giugliano the district of Casacelle with 21.000 inhabitants and almost all illegal) it is possible to demonstrate the effectiveness and the efficiency of the 'sanatorium' instrument, highlighting its potential from both the local finance and the sustainable planning perspective.

Keywords: Informal settlements · Sustainable planning · Local finance
Regularization · Evaluation

The paragraph 2 was drafted by Claudia De Biase; the rest of the article was drafted in equal part by the three authors.

© Springer International Publishing AG, part of Springer Nature 2019
F. Calabrò et al. (Eds.): ISHT 2018, SIST 100, pp. 264–271, 2019.
https://doi.org/10.1007/978-3-319-92099-3_31

1 Informal Settlements in the Perspectives of Urban Sustainability

The 2017 Ischia earthquake, in the Metropolitan city of Naples (Campania Region), has highlighted the phenomenon of the informal settlements that are very widespread all over the Italian territory and particularly in the South. According to Legambiente [1], the island of Ischia is a "perfect example of the defacement of the coast", positioning itself among the first five coastal cities marked by informal settlements; and, according to BES Report [2], Campania Region results one of the most affected by this phenomenon together with Calabria Region [3].

The issue of informal settlements has long been investigated through dedicated academic researches [4–6] strictly connected with the problematic of the territorial context where the University of Campania "Luigi Vanvitelli" is localized (between the metropolitan city of Naples and the province of Caserta).

These researches, aiming to the recognition and the interpretation of the phenomenon, have been carried out in several municipalities where a sustainable planning has to face with the *physical sustainability* or the "sustainability of built environment". This kind of sustainability concerns «the capacity of an intervention to enhance the livability of buildings and urban infrastructures for 'all' city dwellers and the efficiency of the built environment to support the local economy» [7]. In this perspective, the informal settlements are part of the built environment; they mobilize investments that remain outside of the formal economy and investment cycles, thus constituting a particular form of 'valuable capital assets'

Diffusely, in the province of Naples and Caserta, the built environment results by unplanned urban development, with the existence of a "real estate" of informal buildings and transactions [8, 9]. Very often the responsibilities cannot be attributed only to the transgressors who commit the abuse, but also to the public administrations which should control and manage the territory (the "political" dimension of urban sustainability). Campania Region still suffers a lack of interventions discipline: most of the municipalities are un-provided of Master Plan; the remaining ones have old planning tools (as the Programma of Fabbricazione, suppressed on 1982).

More than ten years ago was approved the Regional Urban Planning Law (Law n.16, 2004) that prescribed, for all the municipalities, the approval of a new and updated Master Plan. More than 90% of the 551 municipalities of Campania Region did not achieve the established aim and so their plans are still missing [10]. In this scenario, also the efficacy of the "Recovery Plan of Unauthorized Buildings"- introduced by LUR 16/2004 and allowing to the regularization process of unauthorized buildings- has been experimented very rarely.

Among the several approaches (repressive, mitigative and comprehensive [11]) that can be set out by Italian regulations in order to address informal settlements, the *comprehensive approach* - which provides the regulation on the basis of a pecuniary sanction to obtain a "retrospective building permit"- should be mostly implemented in territories, as that of Naples and Caserta. There, the phenomenon involves entire peri-urban areas, outside the boundaries of municipalities, where the residential component prevails and is characterized by a low level of physical quality, scarcity of standards and public facilities. If the *repressive approach* is economically un-sustainable (in the actual conjuncture municipal budgets are frequently in financial difficulties), the *mitigative approach* is still far from implementation. Therefore the possibility of the sanctions or 'sanatorium' for the regularization of a consistent part of the built environment (where it is possible, in accordance to the Presidential Decree No. 380/2001), could represents a significant item of revenue and capital expenditure or "investment" in financing the *public city* (in terms of urban facilities and infrastructures, urbanizations, etc.), being the revenues constrained funds.

Specifically, in case of interventions realized without the building permit, or in dissimilarity with it, or even in absence of the statement of the statement of the start of works, whoever is responsible for the illegal works or the actual owner of the building can obtain the permit "in sanatoria" or the "retrospective building permit". This can happen if the interventions are in compliance with the planning and buildings rules in force both at the time of the realization of the building and at the time of presenting the application. However, in order to get the "retrospective building permit" an amount equal to the double of "contribution of construction" (art. 16, Presidential Decree No. 380/2001) must be paid. The contribution is commeasured to the incidence of urbanization charges and construction costs, determined by the Region periodically [6].

In this perspective the article presents an explorative work carried out in the Municipality of Giugliano in Campania (in the metropolitan city of Naples), where the phenomenon of informal settlements has been analyzed with reference to a specific district, as in Sect. 2. Starting from the framework of the consistency of illegal buildings in the entire district, the Sect. 3 verifies the potential of the regularization.

2 The Case Study

Giugliano in Campania is located in the Northwest of the province of Naples, and covers an area of 94.4 km^2. Approximately 125,000 inhabitants live there. Infact, it is the third largest municipality in the province for its demographic dimension. Its coastline is long 3 sq km: it goes from Marina di Varcaturo to Licola. Another feature of the area is the presence of a salty lake born in the crater of a volcano, the Lago Patria. The city of Giugliano, within three decades, sees double its population from 50,000 to 125,000 residents of 1985 in 2015.

This rapid population growth has led to rapid and unplanned urbanization of four-surrounding area, in particular: Lago Patria, Varcaturo, Licola and Casacelle. All these areas miss services and infrastructure and over them it is possible to find countless unauthorized settlements. In this article we will focus on Casacelle, one of the above areas that until early 1900 did not exist.

The manufacturing program of 1975, as can be seen from Fig. 2, stipulated that part of the neighborhood will come in zone C, i.e. those parts the territory intended for new settlement complexes, while the remainder falls within the area intended for agricultural use (Fig. 1).

The planning scheme of the 1984 (see Fig. 3) locates – confirming the signs of manufacturing program – locates in the Casacelle area dedicated zones for the new residential and office development and identifies six homogeneous areas.

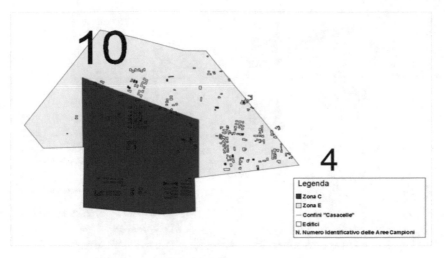

Fig. 1. Casacelle in the Manufacturing program of 1975.

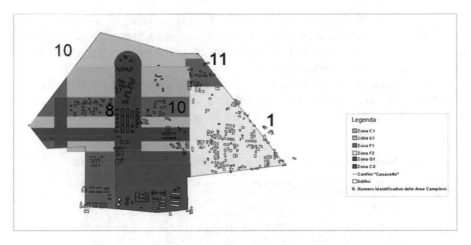

Fig. 2. Casacelle in the Planning Scheme of 1984.

- Area C1; residential urban expansion arca.
- Area E1; normal agricultural area.
- Area F1; area for standards and equipment, including higher education and social services-municipal- and extra-municipal scale.
- Area F2; urban park.
- Area G1; directional area of new urban expansions.
- Area CX; area of implementation of PEEP (Economic and popular housing plan) and PdL (allotments plan) approved under manufacturing program[1].

---

[1] Implementation laws with integration N°120 del 6/7/1984.

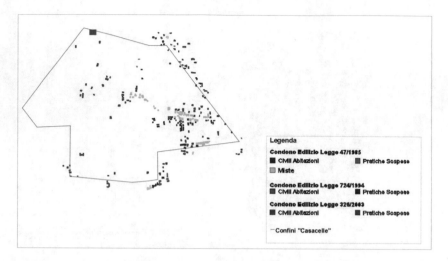

Fig. 3. Amnesties in Casacelle.

Using the method of overlay mapping [12–14], referred to 1975, the year of manufacturing program, and to 1985, the year of planning scheme, it is possible to note the increase in volumes edified in the agricultural area. And the building, often made from illicitly, did not stop over the years.

As is possible to see in the Table 1, for three amnesties (1985–1994–2003) were presented nicely 1035 applications for remission. Out of these, only 374 were approved and rejected, the remaining 13 questions are lying.

Table 1. Data of amnesties under L. 47/85, L. 724/94 and 326/03.

Laws	Practices presented	Permits issued	Rejected practices
L. 47/85	233	92 (39,5%)	
L. 724/94	469	171 (36,5%)	6 (1,27%)
L. 326/03	333	111 (33,3%)	7 (2,1%)
TOTALE	1035	374 (36,1%)	13 (1,25%)

These data have been located and on the map and from it is clear the spread of abusive public service areas intended for by the town planning scheme (see Fig. 4).

In this framework, to understand the overall consistency of the abuse, occurs to consider also all the volumes which are still without habilitation. It can therefore be concluded that in the District of Casacelle, among the 97 volumes emerged, 21 are those without any legitimate habilitation (see Fig. 5).

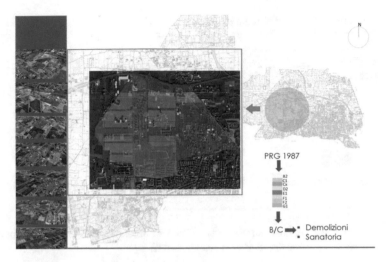

PRG 1987

B2
C1
C4
D2
E1
F1
F2
G1

B/C ➡ • Demolizioni
       • Sanatoria

Fig. 4. Overlay mapping in Casacelle.

3 The Potential of Regulation

It's well known that local finance plays a crucial role in the implementation of sustainable planning; the importance of budgeting has grown significantly as local administrations face increasingly difficult expenditure and revenue decisions in an context of increasing demand of services and infrastructure. In accordance to the DPR n. 380 del 2001(art.36), which assumes the regularization of some residential buildings in the Casacelle District (as in Fig. 5) through the "retrospective building permit", which could be the economic effects from the local finance perspective?

Analyzing the revenues registered in the Municipal Budget under Title IV, Category 5 ("Disposal of fixed assets, capital transfers from other governments and private entities") in the years 2011–2015, in regard with the "building permit" and the "retrospective building permit", it is possible to register a decrease (see Table 2).

On the side of the revenues deriving from the building permits, it is worth highlighting that the decrease, over the last five years, could be a good indicator of the reduction of the soil consumption. On the side of the revenues deriving from the "retrospective building permit"- which drastically reduced over the last two years- their consistency (from 2011 to 2014) demonstrates the effort in the managing the unauthorized buildings through the regularization by the Giugliano Local Administration.

This kind of revenues, in a municipality characterized by a high concentration of informal settlements (as shown in the Sect. 2) and a scarce equipment of public facilities and services, could represent a meaningful opportunity to collect resources for the pursuit of a sustainable planning. The assessment of the retrospective building permits for 8 residential buildings in Casacelle confirms this assumption, as in Table 3.

In fact, the evaluation of the amount of the construction contribution (composed by the amount commensurate to the construction costs and the amount commensurate to the urbanization fees) for all the buildings, as shown in Table 3, highlights the entity of

Table 2. Municipal Budget of Giugliano in Campania. Title IV, Category 5 (years 2015–2011).

Revenues	2015	2014	2013	2012	2011
Building permits (B.P.)	€ 163.303	€ 342.946	€ 932.979	€ 1.109.430	€ 1.288.265
Retrospective (R.B.P.)	€ 811.569	€ 812.046	€ 2.080.479	€ 2.779.490	€ 3.235.539
Total	€ 974.872	€ 1.154.992	€ 3.013.458	€ 3.888.920	€ 4.523.804

Table 3. Construction contribution for the regularization of 8 illegal buildings in Casacelle District

Buildings no.	Total area sm	Amount commensurate with construction cost (c.c)	Amount commensurate with urbanization fees (u.f.)	Total contribution (c.c. + u. f) × 2
1	106	€ 667	€ 14.804	€ 30.942
2	579	€ 3.390	€ 40.779	€ 88.338
3	119	€ 730	€ 20.711	€ 42.882
4	119	€ 730	€ 20.711	€ 42.882
5	468	€ 3.102	€ 32.008	€ 70.220
6	743	€ 4.533	€ 61.173	€ 131.412
7	302	€ 282	€ 28.284	€ 57.132
8	749	€ 4.189	€ 65.950	€ 140.278
			Total revenues	**€ 604.087**

the revenues. The Municipalities of Giugliano could earn about 600.000 euro which, compared with the total number of illegal buildings localized in the Casacelle District (97 illegal buildings, as in the Sect. 2), represents an appreciable source of revenue in the municipal budget.

4   Conclusions

The existence of informal settlements and the need for their regularization is a relevant issuer and an important task for the Campania Region, as in the recent draft law "Simplification measures and support guidelines for the municipalities in the government of territory" (Campania Regional Council, 9 June 2017).

The case presented in the article highlights how the regularization should aim not only to the urban regularization but also to the physical, economic and social integration of informal settlements and their residents into the existing urban tissue.

This approach is consistent both with the "Vienna Declaration on Informal Settlements in South Eastern Europe" endorsed in 2004 and with the most recent Habitat III (UN, October 2016).

References

1. Legambiente, Mare Monstrum, Osservatorio Ambiente e Legalità (2017)
2. ISTAT-CNEL, BES Report (2016)
3. Della Spina, L., Calabrò, F., Calavita, N.: Transfer of Development Rights as Incentives for Regeneration of Illegal Settlements, New Metropolitan Perspectives - The Integrated Approach of Urban Sustainable Development, vol. 11. Trans Tech Publications, Zurich (CHE) (2014)
4. De Biase, C., Forte, F.: Unauthorised building and financial recovery of urban areas: evidences from Caserta Area, Euromed (2013)
5. Forte, F.: Illegal buildings and local finance in new metropolitan perspectives. In: Bevilacqua, C., Calabrò, F., Della Spina, L. (eds.) New Metropolitan Perspectives. Advanced Engineering Forum, vol. 11 (2014)
6. Forte, F.: The management of informal settlements for urban sustainability: experiences from the Campania Region (Italy). In: Brebbia, C. (ed.): The Sustainable City X, pp. 153–164, Wessex Institute (2015)
7. Allen, A.: Sustainable cities or sustainable urbanization. UCL's J. Sustain. Cities. http://www.ucl.ac.uk/sustainable-cities/. Accessed 12 Sept 2017
8. Del Giudice, V., De Paola, P., Forte, F.: Using genetic algorithms for real estate appraisal, buildings. MDPI 7(2), 31 (2017)
9. Del Giudice, V., De Paola, P., Manganelli, B., Forte, F.: The monetary valuation of environmental externalities through the analysis of real estate prices, sustainability. MDPI Switz. 9(2), 229 (2017)
10. D'Angelo, G.: Urbanistica e territorio vince l'anarchia, Il Mattino, 05 January 2014
11. de Biase, C.: Le politiche per l'abuso edilizio-urbanistico. In: de Biase, C., Losco, S., Macchia, L. (a cura di): Abusivismo urbanistico e sostenibilità ambientale, Edizioni Le Penseur, Brienza (2017)
12. Petrella, B., de Biase, C.: Unauthorized building and land use: cases studies. Urbanistica Informazioni 257, 31–35 (2014)
13. Petrella, B., de Biase, C.: A tipycal italian phenomenon: the unauthorized building. In: Rakesh, K.: Proceeding of Second International Conference in Advances in Civil and Structural, pp. 70–73. Institute of Research Engineers and Doctors, NY (2014)
14. de Biase, C.: A methodology able to investigate the phenomenon of unauthorized building: the case of Giugliano in Campania. In: Gambardella, C.: Atti del XIII Forum Le vie dei Mercanti: Heritage and technology, pp. 1027–1036. La Scuola di Pitagora, Napoli (2015)

Urban Planning and Innovation: The Strength Role of the Urban Transformation Demand. The Case of Kendall Square in Cambridge

Carmelina Bevilacqua® and Pasquale Pizzimenti(✉)®

Mediterranea University of Reggio Calabria, 89124 Reggio Calabria, Italy
pasquale.pizzimenti@unirc.it

Abstract. The paper discusses two interrelate aspects that have been emerging in the current phase of Smart Specialisation Strategies (S3) implementation, that is the concept of dynamic location advantages (cluster) and the change in social demand for urban transformation (urbanization). Both concepts contribute to redefine the role of the city in the innovation policy, athwart renovating the tools of urban policy and planning, underlined also by the Urban Agenda 2030. However, it is also widely recognized from combining Schumpeter (1934) and Jacobs (1969) that the concept of dynamic location advantages finds at city level the conditions to launch real change in regenerating local economic areas and valorize local assets. From these considerations, it follows that it is crucial to investigate how cluster-oriented policies and urban policy and planning are related in transforming cities. The aim of the paper is to figure out how the connection of urban policy with place-based innovation approach allows at reaching the knowledge convergence to activate informational spill-overs through zoning and urban planning tools. The paper examines the case of the Kendall Square area in Cambridge (MAPS-LED project -Horizon2020), which is analyzed through the lens of urban planning and zoning adopted for the area. The development of the Kendall Square area is characterized by a mixed-use approach, and innovation spaces are included as a zoning requirement for the foreseen development of the area. Conclusions highlight how urban planning and zoning are pushing factors in supporting the innovation-oriented demand of socio-economic and physical transformation.

Keywords: Urban planning and innovation · Zoning and social demand
S3 and urbanization

1 Introduction

By 2050 the majority of the population, economic activities as well social and cultural interactions will concentrate in cities [1]. The challenges generated by rapid urbanization processes request an urban paradigm shift to reach a sustainable and inclusive development in the mid and long-term. The UN Agenda for Sustainable development 2030 [2] recognized "that sustainable urban development and management are crucial" for people's quality of life and "local authorities and communities should "renew and plan" cities fostering community cohesion, personal security and stimulate innovation

© Springer International Publishing AG, part of Springer Nature 2019
F. Calabrò et al. (Eds.): ISHT 2018, SIST 100, pp. 272–281, 2019.
https://doi.org/10.1007/978-3-319-92099-3_32

and employment". Particularly, the Goal No. 11 on sustainable cities and communities, represented the basis for the development of the UN Urban Agenda [1]. It represents a paradigm shift for cities underlying a new "recognition of the correlation between good urbanization and development" [1]. The implementation focuses on five main pillars: national urban policies, urban legislation and regulations, urban planning and design, local economy and municipal finance, and local implementation [1]. Therefore, urbanization represents an opportunity from which take advantage in the perspective of sustainable and inclusive growth [1]. If we take into account European – or widely western countries' – cities, the urbanization concept is often associated with urban areas' physical transformation and more precisely with the urban regeneration concept. Thanks to its comprehensive and integrated approach [3], urban regeneration has the potential to tackle the current urban challenges underlined by the UN New Urban Agenda. In Europe, the debate on sustainable urban development and urban regeneration as key elements in defining an EU urban policy [3], developed around the need to reach a smart, inclusive and sustainable growth (EU 2020 Strategy) consolidating an EU Urban Agenda for European cities and urban areas [3]. In 2016, with the Pact of Amsterdam [4], EU Member States has adopted the European Urban Agenda in order to disclose the potential of European cities and urban areas through a better Regulation, Funding, and Knowledge, in accordance with the Europe 2020 strategy and the Smart Specialisation Strategy. Particularly: "The EU and its Member States will seek to boost the potential of cities as hubs for sustainable and inclusive growth and innovation... In line with the UN's New Urban Agenda, they will promote sustainable land use planning, equitable management of land markets, sustainable urban mobility and smart, safe cities that make use of opportunities from digitalisation and technologies" [5]. The paper aims at figuring out how the connection of urban transformation with place-based innovation approach allows at reaching the knowledge convergence to activate informational spill-overs. The next section highlights the relevance of cities as the catalyst for innovation in the globalized economy triggered by creative, innovation and entrepreneurial activities. The central section examines the case of Kendall Square through the lens of zoning and urban planning tools adopted for the area. The choice of this case derives from the research activities conducted within the MAPS-LED Research Project (Horizon2020). The case analysis revealed how Kendall Square area presents a concentration of a high number of companies, start-ups - the highest density of start-up in the country [6] - incubators, research laboratories and facilities, and the presence of anchor institutions. Conclusions highlight how urban planning and zoning are pushing factors in supporting the innovation-oriented demand of socio-economic and physical transformation towards knowledge concentration. The risk of knowledge fragmentation or dispersion, which characterizes European cities, needs a policy action in order to favor Knowledge concentration and trigger a smart, inclusive and sustainable development.

2 Cities as Catalysts for Innovation: Towards Knowledge Concentration

Cities are increasingly becoming strategic hubs in the globalized economy triggered by creative, innovation and entrepreneurial activities [7]. As a result, cities and urban areas are becoming the dominant engines of economic growth in the current knowledge-driven innovation economy [8]. In EU, the need of an integrated and multilevel approach in urban policy stemmed from Lisbon strategy [9] and created the condition to reinforce the link between urban policy and regional innovation system through the Smart Specialisation Strategies (S3) approach. Particularly, the interaction among innovation, clusters, knowledge dynamics, and spaces provide interesting insights on the functional connection between urban policy and S3 through the concept of innovation-driven urban policy [10]. Clusters constitute "the breeding ground for innovation" [11] and cities can be considered as nodes of an international complex network [12], the center of economic activity, and the focal point of innovation [13]. The cluster provides a conceptual framework to describe and analyze important aspects of modern economies [14]. Its role does not lie in defining a specific area, but in characterizing the specific geographic area in terms of innovation, specialization, and capacity to activate competitive and comparative advantages [15]. In these areas, the social demand for innovation has become a source of urban form and its transformation. Hence, in order to better understand the connection of urban policy with place-based innovation approach it is relevant to take into account that a particular connection occurs between (cluster) policies in terms of factors related to the clusters' governance systems [10] and (spatial/urban) planning in terms of factors suitable to be mapped in physical terms (proximity and accessibility, spatial patterns etc.). In combining the contribution of Schumpeter on innovation [16] and those of Jacobs [17], also recalled recently by Florida et al. [8], it is arguable that this connection (cluster policies and spatial planning) starts at city level, where this it finds the "good atmosphere to nourish and feed knowledge dynamics and innovation in regenerating local economic areas and valorizing local assets". From these considerations, it follows that it is crucial to investigate how cluster-oriented policies and urban policy and planning are related in transforming cities.

3 Innovation and Zoning: A Focus on Kendall Square in Cambridge

Starting from the spatial configuration of clusters (based on Porter's definition) at city level developed for the MAPS-LED Project [18], we moved to the interpretation of the role played by those spaces (innovation spaces) expression of knowledge dynamics' source. The Greater Boston Area, as well many other urban regions in the US, present many interesting cases about the concentration of innovation and knowledge in the urban environment as a trigger for economic development. The cluster spatialization methodology developed for the MAPS-LED Project [18] and analysis of cluster-oriented initiatives as well the urban regeneration ones, revealed how the city of Cambridge

(US) offers interesting hints in providing (urban) innovation-oriented policy examples for boosting concentration of innovation, entrepreneurship, and creativity in reaching the knowledge convergence to activate informational spill-overs. Accordingly, the City of Cambridge performs two strongest Clusters: Education and Knowledge Creation and Business Services. The urban configuration proposed is a combination of the economic aggregation of Cluster with the City land use categories [18]. The city of Cambridge (MA) offers interesting insights on the process of knowledge convergence through the concentration of creativity, innovation and entrepreneurship [10]. The reason of their strength is mostly due to the presence of Research Institutions (Harvard, MIT) and a high number of related activities, remarking a high-density level of relationships among public, private sectors, cluster organizations, innovation stakeholders (such as start-ups, small-medium enterprises) and community [18]. During the last three decades, Kendall Square experienced a shift from a former industrial district to one of the world's leading centers for biotech research and innovation [19]. The development of the area started during the 1960s with the Kendall Square Urban Renewal Plan, which main aim was to locate the NASA's Electronic Research Centre [19]. During the 1970s, the willing of NASA to locate its facilities in other cities pushed the Cambridge Redevelopment Authority [20], owner of 70% of the land, to transfer the project for the construction of the Department of Transportation. This change signed the shift from a post-world war renewal phase to a significant physical transformation [20] of the area, which started being recognized as an opportunity for economic development through the exploitation of locational advantages [20]. During the last two decades, Kendal Square evolved into a livable mixed-use district thanks also to the attention of the City and of the Department of Community Development, which focused on crucial aspects such as housing afford-ability, open spaces and transportation accessibility.

3.1 The Vision for Kendall Square: The K2C2 Planning Study and Zoning Insights

The increased interest in the development of the Kendall Square pushed the City of Cambridge to coordinate a planning study for the area [19]. The vision of the study aims to inspire the Cambridge's sustainable and globally significant innovation community [21]. The K2C2 planning study focuses on four main pillars: Nurture Kendall's innovation culture; Create Great Places; Promote Environmental Sustainability; Mix Living, Working and Playing (see Table 1). The vision presents the characteristics of an urban regeneration scheme (social, economic and physical dimension) tailored on the local context characterized by an "innovation culture", which need to be nourished. In order to realize this vision zoning recommendation are needed. The vision aims to maintain the existing characteristics of a world center for biotech, entrepreneurship, high tech, and the knowledge economy, paying attention to livability aspects, housing and retail [19]. The approach is characterized by the willingness to increase density encouraging the development of housing, incubators spaces, open spaces and other amenities [19]. In order to achieve the goal of "Nurture Kendall's innovation culture" the strategy includes two actions: (a) Recognize that all aspects of the vision for Kendall Square need to work together if the innovation culture is to realize its full potential; (b) Retain and expand incubator spaces for entrepreneurs. The first recognizes the need to increase the livability of the area - identified in the past as office district - with a low presence of small and medium businesses. These actions focus on the livability aspects as an important element in thriving innovation in Kendall Square together with businesses' support, incubators spaces and entrepreneurs. Kendall Square is acknowledged as attractive for multi-national corporations and for the local community of entrepreneurs. In this perspective, the role played by anchors institutions is crucial. In fact, the MIT is one of the privileged actors involved in the transformation process of the area. The City of Cambridge's aims to strengthen the connection with MIT in attracting innovative businesses, together with the involvement of other developers in participating in the efforts for a vibrant environment [19].

Table 1. Kendall square plan: summary of zoning & urban design recommendations.

Pillar	Actions
Nurture Kendall's innovation culture	- Expand opportunities for Kendall Square knowledge economy to continue to grow; - Foster a strong connection between the MIT campus and the rest of Kendall Square. Enable MIT to develop in a manner consistent with its academic and research mission, so that it continues to be a magnet attracting innovative businesses to the area - Support a vibrant environment for creative interaction; - Three themes (below) working together supporting the central theme of nurturing Kendall's innovation culture
Create great places	- Support open space and recreation needs of a growing neighbourhood; - Create lively, walkable streets; - Expand opportunities for Kendall's diverse community to interact; - Development and public place improvements must happen in tandem
Promote environmental sustainability	- Expand convenient, affordable transportation and access choices - Enhance streets as public places - Create a healthier natural environment - Reduce resource consumption, waste emissions - Leverage the environmental and economic benefits of compact development
Mix living, working and playing	- Leverage community and innovation benefits of mixed-use environment - Focus intensity around transit - Minimize development pressures on traditional neighbourhoods - Continue to support city and state economic development

Besides the traditional zoning categories, the City of Cambridge Zoning Ordinance [22] includes a set (19) of special "districts" categories with tailored regulations for specific areas of the city. The Kendall Square area is included in two of these specific districts: The Cambridge Centre Mixed Use Development District (MXD) and the

Planned Unit Development Districts (PUD). The first was created to guide development in the Kendall Square Urban Renewal Area and requires a balanced mix of uses (light industrial, office, retail, institutional and residential uses) with the requirement of extensive public open spaces. The PUD district is a special category district allowing a more intense and diverse mixed-use development and permits the coordination of public and private development to implement the urban design for these areas [23]. Furthermore, part of the MIT campus area within Kendall Square is identified as Institutional Overlay District, with more flexible institutional use regulations than in areas where non-institutional uses predominate [22]. The so-called Planned Unit Development Special District (PUD), indeed, defines zoning rules for Kendall Square [22]. In this case, the designated PUD Kendall Square District (PUD-KS) has the purpose "to provide for the creation of a mixed-use district of high quality general and technical office and retail activity, with a significant component of residential use" [22]. Particularly, the PUD-5 (see Fig. 1) is recognized by the Zoning Ordinance as "a world-renowned center of innovation and a vibrant neighborhood" which should become a "mixed-use district of high quality general and technical office and laboratory uses with significant retail activity proximate to the MBTA station" [22]. Within the PUD 5-KS the development proposals containing new Office Uses, must include Innovation Office Spaces as zoning requirement: the 5% of the Gross Floor Area approved for Office Uses by the development plan [22]. Furthermore, the zoning ordinance provides also the characteristics of innovation spaces [22]:

- "Durations of lease agreements (or other similar occupancy agreements) with individual business entities shall be for periods of approximately one (1) month;
- No single business entity may occupy more than 2,000 sq. ft or 10% of the entire Innovation Office Space required to be provided in the PUD-5 District, whichever is greater;

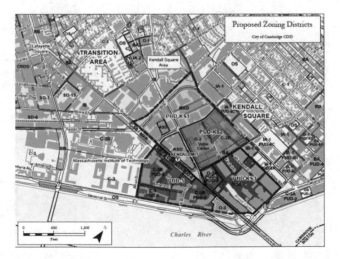

Fig. 1. Kendall Square Area Zoning - PUD KS Proposed Zoning District (Kendall Square Final Report 2013. City of Cambridge Community Development Department (2013).

- The average size of separately contracted private suites may not exceed 200 sq. ft of GFA;
- Innovation Office Space shall include shared resources (i.e., co-working areas, conference space, office equipment, supplies and kitchens) available to all tenants and must occupy at least 50% of the Innovation Office Space;
- Individual entities occupying Innovation Office Space may include small business incubators, small research laboratories, office space for investors and entrepreneurs, facilities for teaching and for theoretical, basic and applied research, product development and testing and prototype fabrication or production of experimental products".

In this case, the zoning recommendations [23] provide a specific innovation-oriented development trajectory for the physical development of Kendall Square. The requirement of innovation spaces is directly connected with the general vision of the city to empower the globally and knowledge-oriented features of the area. The link between innovation-oriented policies and urban transformation in the City of Cambridge has favored the concentration of knowledge-related activities in Kendall Square. Here, the presence of anchor institutions, private companies, a proactive entrepreneurs' community is stimulating the socio-economic and physical transformation. The City of Cambridge supports this attractive process with economic development measures. The Economic Development Division of the City of Cambridge set out specific measures in order to attract and support businesses in the highly innovative sectors such as Life Science and Technology. The awareness of the increasing role played by these sectors

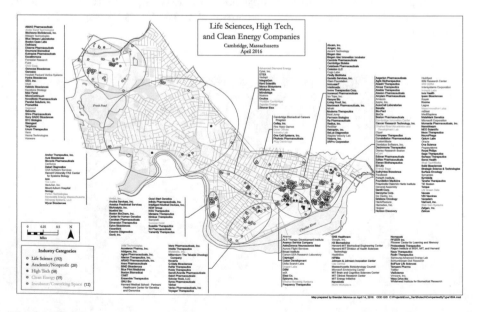

Fig. 2. Life Science, High Tech and Clean Energy Companies, City of Cambridge (2017). (http://www.cabridgema.gov/∼/mdia/Files/CDD/Maps/techcompanies/cddmap_tech_companies_2017 05.pdf?la=en).

in the Cambridge economy, pushed the administration to provide assistance and services in order to attract and create companies, start-ups and industries. The area shows a high concentration of start-ups and companies for the Life Science and Clean Tech sectors located within or in close proximity to, the Kendall Square Area. Here are located approximately 70 Life Science related companies, 13 Academic/no profit institutions, 31 High Tech related companies/start-ups, 7 High Tech related companies/start-ups, and 6 Incubators/co-working Spaces.

4 Conclusions

The recent international official documents on cities highlighted [1, 2] how urbanization should be seen as an opportunity for a more sustainable and inclusive development model characterized by innovation, entrepreneurship and knowledge dynamics. The case of Kendall Square shows how knowledge dynamics and innovation-oriented policies are contributing to the physical transformation of the area. The connection of urban and innovation-oriented policy in the City of Cambridge reveals a sort of innovation production oriented approach, in order to boost competitiveness and attract exogenous resources. Furthermore, the analysis of urban planning tools, together with economic development measures, shows up two integrated outcomes. On one side, the public actors support the implementation of innovation oriented policy, through economic development measures, for the retention and attraction of businesses (see Fig. 2). On the other side, the public actors enhance the above process through zoning and urban planning tools by adding innovation spaces requirement within mixed-use districts. In this case, innovation spaces represent the key triggering element in the socio-economic and physical transformation process in place in the area.

The presence of innovation spaces expands the opportunities for the knowledge economy facilitating knowledge sharing and transfer processes, the interaction among local communities, the promotion of more livable areas. The case of Kendall Square highlight how zoning and urban planning tools are contributing to support local innovation ecosystems acting as "big push" for innovation by setting new tools, or the adaptation of the existing ones, in response to the new and sophisticated socio-economic and physical demand of transformation. However, the innovation-oriented approach implemented in the Kendall Square case needs to be investigated in further studies in order to better understand the possible side effects/impacts especially in terms of gentrification phenomena, socially inclusive practices and urban governance. Cities, then, become crucial in the application of the desired bottom-up approach in S3 implementation, which needs innovation-driven urban regeneration interventions in order to calibrate the discrepancies in the demand/supply of services for innovation.

Acknowledgement. This work is part of the MAPS-LED research project, which has received funding from the European Union's Horizon 2020 research and innovation programme under the Marie Skłodowska-Curie grant agreement No. 645651.

References

1. United Nations: The New Urban Agenda, Habitat III, United Nations Conference on Housing and Sustainable Urban Development (2016)
2. United Nations: Transforming Our World: The 2030 Agenda For Sustainable Development, United Nations, United Nations Sustainable Development Summit (2015)
3. Couch, C., Sykes, O., Borstinghaus, W.: Thirty years of urban regeneration in Britain, Germany and France: the importance of context and path dependency. Prog. Plann. **75**, 1–52 (2011)
4. European Union: Establishing the Urban Agenda for the EU "Pact of Amsterdam" Agreed at the Informal Meeting of EU Ministers Responsible for Urban Matters on 30 May 2016 in Amsterdam, The Netherlands (2016)
5. European Commission: The New European Consensus On Development 'Our World, Our Dignity, Our Future, Joint Statement by the Council and the Representatives of the Governments of the Member States Meeting Within the Council, The European Parliament and the European Commission, European Commission International Cooperation and Development Building Partnerships for Change in Developing Countries (2017)
6. MIT New: Kendall Square: A global center for innovation grows alongside MIT. Once lined with old factories and abandoned buildings, Kendall Square is now a global center for innovation. Liz Karagianis | MIT Spectrum, 7 May 2015. http://news.mit.edu/2015/kendall-square-global-center-innovation-grows-alongside-mit-0507. Accessed 16 Nov 2017
7. Karlson, C., et al.: Knowledge, Innovation and Space, CESIS Electronic Working Paper Series Paper 367, CESIS (2015)
8. Florida, R., Adler, P., Mellander, C.: The city as innovation machine. Reg. Stud. **51**(1), 86–96 (2017)
9. Parysek, J.: Urban policy in the context of contemporary urbanization processes and development issues of polish cities. J. Urban Reg. Anal. **2**(2), 33–44 (2000)
10. Bevilacqua, C., Pizzimenti, P.: Urban innovation-oriented policies and knowledge dynamics: insights from Boston and Cambridge, US. In: Talia, M.: Un Nuovo Ciclo Della Pianificazione Urbanistica Tra Tattica E Strategia - A New Cycle Of Urban Planning Between Tactics And Strategies 2016, pp. 11–19. Planum, Roma-Milano (2016)
11. Ketels, C., Lindqvist, G., Sölvell, Ö.: Strengthening Clusters and Competitiveness in Europe. The Role of Cluster Organizations, The Cluster Observatory, Stockholm School of Economics, Center for Strategy and Competitiveness (2012)
12. Simmie, J.: Critical surveys edited by Stephen Roper innovation and space: a critical review of the literature. Reg. Stud. **6**(39), 789–804 (2005)
13. Tong Soo, K.: Innovation across cities, Economics Working Paper Series 2015/027. The Department of Economics, Lancaster University Management School, UK (2015)
14. European Commission: The Role of Clusters in Smart Specialisation Strategies. Directorate General for Research and Innovation (2013)
15. Porter, M.: Cluster and the New Economics of Competition. Harvard Business Review, Boston, pp. 77–90 (1998)
16. Schumpeter, J.A.: The Theory of Economic Development: An Inquiry into Profits, Capital, Credit, Interest, and the Business Cycle. Harvard University Press, Cambridge (1934)
17. Jacobs, J.: The Economy of Cities. Vintage, New York (1969)
18. Bevilacqua, C., Pizzimenti, P., Maione, C.: S3: Cluster Policy and Spatial Planning. knowledge dynamics, spatial dimension and entrepreneurial discovery. Second Scientific Report, MAPS-LED Project, Multidisciplinary Multidisciplinary Approach to Plan Smart

specialisation strategies for Local Economic Development, Horizon 2020 - Marie Swlodowska Curie Actions - RISE - 2014 - grant agreement No. 645651 (2017)

19. City of Cambridge, K2C2 Planning Study. Kendall Square Final Report 2013. City of Cambridge Community Development Department (2013)
20. Cambridge Redevelopment Authority: Kendall Square Urban Renewal Plan. http://www.cambridgeredevelopment.org/kendall-square-1. Accessed 16 Nov 2017
21. City of Cambridge, K2C2 Kendall Square Plan: Summary of Zoning & Urban Design Recommendation (2012)
22. City of Cambridge: Zoning Ordinance. http://www.cambridgema.gov/CDD/zoninganddevelopment/Zoning. Accessed 16 Nov 2017
23. City of Cambridge, the Zoning Guide Cambridge a User Guide to the City of Cambridge Zoning Ordinance 2nd edn. - Updated Fall, 2004. https://www.cambridgema.gov/~/media/Files/CDD/ZoningDevel/zoningguide/zguide.pdf. Accessed 16 Nov 2017

New Urban Agenda and Open Challenges for Urban and Regional Planning

Giuseppe Las Casas⬤, Francesco Scorza⁽✉⁾⬤, and Beniamino Murgante⬤

University of Basilicata, 85100 Potenza, Italy
francescoscorza@gmail.com, Beniamino.murgante@unibas.it

Abstract. Starting from the Ivan Blečić and Arnaldo Cecchini book "Verso una pianificazione antifragile" [1], this paper will identify main arguments that: (i) help to deal with the conflicts of a complex society that weakens the connections between pieces of society; (ii) recognize in Z. Bauman thought the elements of concern that characterize the liquidity of our society and its negative connection with urban and regional planning; (iii) highlight in "anti-fragile planning" an innovation instance for the discipline promoting new approaches that starting from the reduction of territorial vulnerability (resistent), are able to promote the regeneration of utility functions (resilient) by involving local communities in a collective form of creativity strategic development form. In one word: anti-fragile.

Keywords: Urbanism · Strategic planning · Resilience · Anti-fragility

1 A Reference Framework for the Proposed Perspectives

Both researchers and practitioners are well aware that urban and regional planning needs innovation. Prospects of change are provided by a prolific work on drawing up of fundamentals, guidelines, best-practices, defined at international level that contributes to relaunch and to legitimize urban and regional planning as an instrument to overcome contemporary challenges related to the management of natural risks, climate change, urban growth and the resulting abandonment of rural areas.

Efforts to express the "Right to Plan", as stated within the latest international UN HABITAT references [2–4], through planning instruments and procedures geared to guarantee an a-priori rationality represent a request of the discipline looking for renewed approaches.

Our proposal is facing the difficulty of referring to this request of a renewed approach to plan rationality that acknowledges following basic principles:

- equity;
- efficiency;
- conservation of non-renewable resources (trans-generational value).

Limits spotted in terms both of methodology and of operational implementation, briefly explained in this work, are due to the persistence of a dense network of conflicts

© Springer International Publishing AG, part of Springer Nature 2019
F. Calabrò et al. (Eds.): ISHT 2018, SIST 100, pp. 282–288, 2019.
https://doi.org/10.1007/978-3-319-92099-3_33

between groups and individuals who, following Simon [5, 6], limits rationality strength.

Our focus considered as methodological reference the Kantian a-priori rationality as reported by Karl Popper in well-known Conjectures and Refutations [7] and later projected by Faludi [7, 8] in the public decisions sphere.

Furthermore, we tried to follow Blečić and Cecchini's [1] reasoning about the ambition to an anti-fragile plan process that could represent a response to the liquid society highlighted by Bauman [10, 11]. It should be noted, indeed: if liquidness is convenient because a strategic dimension of choices together with long terms effects prediction is dropped, resilience and anti-fragility bring back up the burden of modelling to determine and monitor processes.

Therefore, we confirm straight away the modernity of regional analysis models that, in the most operational forms, become a part of the toolkit of planners invited to confront with complexity of territorial requests made by contemporary society. Within it, in fact, conflicts among social groups, operating on the context to which the plan is called to ensure representativeness in decision-making and inclusion, arise in many different shapes and severity depending on place to place.

Thus the regional model, whose sensitivity is aimed to include and systematize complexity of context request, is an operative instrument for planner who, encouraged by an increasing data availability for spatial phenomenon analysis and not, as well by a spread and considerable computational capability of information technologies (open technologies too), has the opportunity to support and inform decision process by promoting quantitative approaches that increase public authority accountability and encourage inclusion and participation by knowledge transfer.

2 Resilience and Anti-fragility

Anti-fragile planning refers to an ambition: going over the idea of resilience to appeal for forms of creativity able to make enduring systems which man-made territory consist of. According to Zygmunt Bauman's writings [10] comes out a getting liquid society in which bonds among individuals and groups become weaker always more. On the other hand, anti-fragile planning by Blečić and Cecchini [1] suggests the research of shared values and scenarios, toward which system develops its own regenerative and creative capabilities.

Quoting Rabino [12] "I think that someone else might take exception to the thesis that, referring to the context of "knowledge society", ever wider spaces are arising for modeling practices in policies and plans (even if not explicitly expressed), exactly as an effect of the need to understand (and to some extent to condition) a planning which is increasingly losing the characters of an understandable ideologically oriented action, to assume characters of a complex self-organized social mechanism, aimed to a totality of extremely varied and often intrinsically conflicting purposes, no longer appropriately transmitted through plans and policies by administrative machineries typical of traditional forms of representative democracy" it should be underlined that the motion to structure appropriate instruments of knowledge management to support a suitable level of sharing scenarios within the planning process represents once again a disciplinary request in respect of which the commitment in refining instruments and models is needed.

Looking at major topics that have internationally influenced our disciplinary debate, two different references come out: "resilience", New Urban Agenda. It should be noted that these concepts get their main place (in terms of formalization and institutional sharing) within the acts of the United States while they appear segmented into a multitude of sectorial approaches that sometimes alter their original meanings.

On one side it should be observed that the emergence of a sharing tension between scientific community and institutions in proposing "resilience complying" governance tools has contributed to reaffirm disciplinary renewal request able to boost the strategic role of plan and related instruments within medium and long term visions adapting current tools in the direction recommended by technical and political working group as from Disaster Risks Reduction (DRR) and Disaster Risks Management (DRM).

Examining "Sendai Framework for Disaster Risk Reduction 2015–2030" (SFDRR) [13, 14] appears an overall strategy aimed to guide and foster an inclusive approach as an essential instrument for the fulfillment of the four operative priorities proposed:

P1: To understand calamities risk;

P2: To strengthen the risk governance in order to manage it;

P3: To invest in calamities risk reduction in order to increase resilience;

P4: To strengthen "preparedness" to disasters for an efficient reaction and a better rebuilding ("Build Back Better") for the restoration and for urban and regional rehabilitation.

On the other side, the New Urban Agenda of the United Nations [2–4] fixes attention to anthropic and natural risks reduction and management, safeguard of ecosystemic resources and "civic engagement" promotion as involvement and inclusion into a system of regional governance able to reintroduce the role of regional planning within an integrated system of tools and resources, within the pivotal points of a sharing vision.

It clearly refers to plan as a rational instrument in which approaches for sustainable, people-centered, inclusive and gender-based urban and regional development should be implemented by carrying out policies, strategies and at every level capacity building based on essential drivers including:

- urban policies development and accomplishment at an appropriate level;
- reinforcement of urban governance;
- actions to relaunch long term integrated planning and design at urban and regional directed to optimize urban model spatial dimension, guaranteeing all the benefits from the urbanization process;
- support effective, innovative and sustainable funding frameworks and tools able to strengthen municipal financial resources and related fiscal system to create, support and share the added value produced by a sustainable urban development.

The UN's New Urban Agenda [3], at odds with this discipline systematic vulnerability, is organized in 15 categories ("pillars") that establish priorities in respect of which should be developed operative solutions that, all together, determine a methodological framework for a renewal of city and regional planning instruments.

NUA's 15 pillars define an index of priorities to be developed by plan. It concerns issues about systematic knowledge of actual regional requests which are involved in Basilicata region as well in Indonesia, although with different intensity/priority, defining a common basis of sensitivity and contents to be included within planning practice. These enhanced levels of systematic knowledge can be founded on models able to simplify debate between contexts and governance actions and to operationalize empowerment process descending from lessons learned and successes evaluation.

It looks like that both NUA and SFDRR lay the foundations for an effective renewal of an even more uncertain disciplinary where every researcher or professional offers his own perspective. Frequent shortage of significantly incisive answers leads often to a flexibility investigation which actually carries to "day by day" decision without ever introducing a vision for the future which citizens, operators and investors can deal with.

Among these contrasts, the most troubling is the transposition of the concept of Bauman's liquid society in the approach or in trends of a discipline that in the name of a flexibility, considered fundamental compared to turbulent modifications of society and that doesn't give credit to the plan in governing the stages of a reformation process in which costs and benefits for all the stakeholders have been evaluated and an agreement has been reached.

3 Conflicts and Uncertainty

Umberto Eco [15] says "…Is there a way to survive liquidity? Yes, there is and it is exactly to realize that we live in a liquid society that requires new instruments to be understood and perhaps overcome. However, the trouble is that politics and intelligentsia have not yet understood the real magnitude of the phenomenon. At present, Bauman is still a "vox clamantis in desert".

We focus on the part of conflicts as a liquefaction factor which considers the agreements' feebleness between the components of society among which we arrange regional planning choices (see also [16, 17]).

The experience of the studies for drafting "Val d'Agri Inter-Municipal Structural Plan" (Las Casas and Scorza [18, 19]) gave us the opportunity to confirm how the overlie of conflicting situations - referred to the most classic of problems: the conflict between nature conservation and industrial activities characterized by considerable emissions (due to oil drilling) and dramatic socio-economic repercussions – allows to describe a level of complexity in which intra-groups, among groups and, finally, among individuals conflicts overlap.

In this conflict that lies the complexity that, according to Simon [6] limits the possibility of knowledge and places consequent limits on rationality.

Conflicts examined in the case of Val d'Agri can be identified on two levels: those emerging between "encodable" groups - otherwise ones attributable to categories of decision-makers that interact for the formation of plan choices; and conflicts within those macro groups that are signs of a disaggregated social structure (or "becoming liquid") unless appropriate instruments of knowledge about sensitive issues: pollution, health risks, loss of identity resources, a persistent lack of opportunities for residents (in terms of work, services, infrastructure).

Having repeatedly affirmed [19] that new instruments mentioned by Eco [15] are those of a renewed approach to the rationality of the plan, in order to reduce impact of the lack of knowledge that determines limits of rationality according to Simon, the contribution that has been developed within the formation process of that plan has focused on two operational levels: anti-fragile strategies for comparison with communities; methodologies for monitoring the effects of transformations.

This approach [16, 17, 20] is based on the identification of three principles, intended as a solid anchorage point of an identification process of a problems hierarchical structure connected with a causal link (assumptions) to which the formulation of an objectives hierarchical structure (program structure) follows.

Since 1987, Faludi [9] indicated the definition of objectives as a crucial point of a strategic planning. They were identified by removing problems. On the other hand, the problem was defined as everything put between objectives that are being reached. The result was the classic unconditional loop of the "egg and hen".

Our position leads us to affirm that a regional problem exists if (at least one of them) principles of equity, efficiency and conservation of irreproducible resources is denied by the current territorial arrangements and by those on which the plan action is directed.

Acceptance of the three principles is taken for universal because they correspond to the essence of social contract that would ensure the coexistence of members of a community and, in fact, on them rest the constitutions of many countries.

It is clear that liquefaction of Bauman corresponds precisely to the tampering with these principles in favor of the search for the prevalence of one individual over another.

However, we are aware that acceptance of the three principles determines in all cases the need to manage competing principles, such as efficiency versus equity and ending up creating a further conflict between ethics and individuals or groups interests.

Conflicts reveal themselves as a consequence of a lack of knowledge resulting from substantive and procedural limitations related to the management of the planning process and from what Simon [6] proposed:

1. attentional limits: we cannot follow more than one event at the same time;
2. limits of working memory: we can think consciously only about a limited amount of information;
3. long-term memory limits: it is difficult to record results of our reasoning;
4. limits in the coherence of knowledge: it is impossible to compare all our beliefs so that we can make them congruent.

To these we would add a consideration that, in a different form, we encounter in Simon's: "those behaviors of the human soul dictated by breakthrough passions that lead to hide the truth of some facts that can make one part prevail over the other."

Since the 1960s, following the paradigm of comprehensive rational planning, this need to fully understand, predict and control the complex mechanism of urban and regional development has been confronted with a quantitative and modelling approach based on the dynamic relations between supply and demand, accessibility and land-use.

In order to better understand the references toward we are looking at, we concern to two important surveys [21–23] in which Giovanni Rabino's contribution is mentioned among the others, which has produced important progresses on these issues both in theoretical and application aspects since the early 1980s [24].

The above-mentioned contributions by Wegener and Harris largely prove to be very topical because they show us a great capacity to document (Wegener) and to argue (Harris) that we certainly did not find among the detractors of the modeling approach that, since the 1970s, have produced criticisms that were mostly not very constructive (among others: [25]).

4 Research Perspectives

Bauman (2006) says: "A society can be defined as a "modern liquid" if situations in which people act change before their ways of acting are consolidated into habits and procedures. The liquid nature of life and society feed and reinforce each other. Liquid life, like the liquid-modern society, is not able to maintain its own shape or keep on track for a long time".

It seems to be consequent to transpose the reflection on this second aspect to the identification of a "fragile planning" in which uncertainties that characterize the vision of the future, are exaggerated by the refusal to accept that the pact stipulated through the Plan cannot be continually questioned.

This fragility of the plan as a social pact cannot make us give up cultivating the commitment to develop and apply analytical regional analysis models with the awareness that they alone do not exhaust our research for a renewed rationality of the plan "as a process".

Rather, it seems necessary to ask ourselves the questions reaffirmed by NUA in terms of plan empowerment and accountability of decisions which impose requirements about measurability, monitoring and evaluation of results, and therefore the decision-making process is placed within those requests of a-priori rationality adhering to the scientific paradigm by Popper [7].

References

1. Blečić, I., Cecchini, A.: Verso una pianificazione antifragile. F. Angeli (2016)
2. UN HABITAT, International Guidelines on Urban and Territorial Planning, UN-Habitat (2015)
3. UN HABITAT, New Urban Agenda, UN-Habitat (2016)
4. UN HABITAT, Action Framework for Implementation of the New Urban Agenda UN UN-Habitat (2017)
5. Simon, H.A.: Theories of bounded rationality. Decis. Organ. 1(1), 161–176 (1972)
6. Simon, H.A.: Models of Bounded Rationality: Empirically Grounded Economic Reason, vol. 3. MIT Press, Cambridge (1982)
7. Popper, K.: Congetture e confutazioni, ed. it. 1969/72. Il Mulino (1969)
8. Faludi, A.: Critical Rationalism and Planning Methodology. Research in Planning and Design, vol. 14. Pion Ltd., London (1986)

9. Faludi, A.: A Decision-Centered View of Environmental Planning, Pergamon Pr. agosto (1987)
10. Bauman, Z.: Liquid Modernity. Wiley, Oxford (2013)
11. Bauman, Z.: Vita Liquida. Laterza, Bari (2006)
12. Rabino, G.: Modellistica: mission accomplished. In: EyesReg, vol. 1, no. 1 - Maggio (2011)
13. UN, Sendai Framework for Disaster Risk Reduction 2015–2030, (SFDRR) (2015)
14. UN, Hyogo Framework for Action (HFA) (2005)
15. Eco, U.: La società liquida. Con questa idea Bauman illustra l'assenza di qualunque riferimento "solido" per l'uomo di oggi. Con conseguenze ancora tutte da capire, Repubblica, L'Espresso (2015)
16. Las Casas, G., Scorza, F., Murgante, B.: Razionalità a-priori: una proposta verso una pianificazione antifragile. Ital. J. Reg. Sci. (2018). In printing
17. Las Casas, G., Scorza, F., Murgante, B.: Conflicts and sustainable planning: peculiar instances coming from val d'agri structural inter-municipal plan. In: Fistola, R., Papa, R. (eds.) Smart Planning: Sustainability and Mobility in the Age of Change. Springer (2018)
18. Las Casas, G., Scorza, F.: Sustainable planning: a methodological toolkit. In: Gervasi, O., Murgante, B., Misra, S., Rocha, C.A.M.A., Torre, C., Taniar, D., Wang, S. (eds.) Computational Science and Its Applications - ICCSA 2016: 16th International Conference, Beijing, China, 4–7 July 2016, Proceedings, Part I, pp. 627–635. Springer, Cham (2016)
19. Las Casas, G.B., Scorza, F.: I conflitti fra lo sviluppo economico e l'ambiente: strumenti di controllo. In: Atti della XIX Conferenza nazionale SIU, Cambiamenti, Responsabilità e strumenti per l'urbanistica a servizio del paese. Catania 16–18 giugno 2016, Planum Publisher, Roma-Milano (2017)
20. Las Casas, G.B.: Governo del territorio: innovare la ricerca per innovare l'esercizio professionale. In: Francini, M. (ed.) Modelli di sviluppo di paesaggi rurali di pregio ambientale. Franco Angeli (2011)
21. Wegener, M.: Operational urban models state of the art. J. Am. Plann. Assoc. 60(1), 17–29 (1994)
22. Harris, B.: The real issues concerning lee's "Requiem". J. Am. Plann. Assoc. 60(1), 31–34 (1994)
23. Oryani, K., Harris, B.: Review of land use models: theory and application. In: Sixth TRB Conference on the Application of Transportation Planning Methods (1997)
24. Lombardo, S., Rabino, G.A.: Nonlinear dynamic models for spatial interaction: the results of some empirical experiments. Papers Reg. Sci. Assoc. 55(1), 83–101 (1984)
25. Lee Douglass, B.: Requiem for large-scale models. J. Am. Inst. Planners 39(3), 163–178 (1973)

New Value from Stalled Real Estate Investments. Empirical Evidences from Some Italian Experiences

Agostino Valier[(⊠)]

University of Padua, 35122 Padua, Italy
agostino.valier@phd.unipd.it

Abstract. The recent economic global crisis has raised the default rate of mortgage loans. Real estate assets -buildings or areas- set as collateral for mortgages encounter numerous difficulties during the liquidation phase. They can remain in disuse without meeting any possible purchaser or they can be sold at an auctioned price much lower than the book value. However, some developers may see investment opportunities in these areas and undertake processes of value extraction. These subjects undertake processes of value extraction from these areas. This research aims to investigate whether the strategies and experiences of urban transformation can find valid strategies for the valorisation of these assets. Information derived from transformation works on unfinished buildings have been used as data set. For the purposes of the investigation, the stalled real estate investments were the closest representative comparable to the characteristics of real estate collaterals in non performing loans. The result is the elaboration of a grounded theory that theorizes three intervention guidelines for these assets, not based only on market indicators but also exploiting the potential of urban planning tools such as the transfer of development rights.

Keywords: Real estate NPLs · Distressed assets · Urban transformation
Value creation · Transfer of development rights

1 Introduction

The recent economic-financial crisis has left on the ground numerous tangible signs of the crisis: uninhabited houses, unfinished buildings, abandoned warehouses. It is not possible to investigate the causes of these abandonment phenomena entirely in the real estate market trends. To contribute to the phenomenon were also social phenomena, settlement dynamics, changes in production processes and other factors of various kinds [1]. However, focusing on the economic nature of these areas, it is not possible to ignore the role that credit. A large part of these real estate assets has been purchased or constructed using credit and setting the asset as collateral in the event of mortgage insolvency. The outbreak of the crisis in 2008 then increased, year after year, the rate of default in mortgages granted, forcing lenders to retaliate on collaterals. In the liquidation phase of the assets banks can meet many difficulties due to the state of crisis in which the real estate market currently pays off. Auctioning rarely succeeds in converting good into money, sometimes it does so at very low values. The consequence is

© Springer International Publishing AG, part of Springer Nature 2019
F. Calabrò et al. (Eds.): ISHT 2018, SIST 100, pp. 289–296, 2019.
https://doi.org/10.1007/978-3-319-92099-3_34

that these real estate, finished or not, often no longer meet end-user and remain in a state of abandonment, real open wounds in our territory [2, 3].

These financial issues have serious repercussions on the physical space. These assets can be interesting investment opportunities for subjects who can imagine intended uses different from the initial ones. The purpose of this research is to investigate the transformation processes, in order to identify which strategies and techniques can also be applied to these areas for an effective creation of value. It is useful to state now that this research does not aim at a comprehensive treatment on the extraction of value from the properties used as collateral in bad debts. It aims to investigate a precise strategy (the transformation for the purpose of repositioning the asset on the market) knowing that there are other methods of creating value on similar areas. For example, especially for public properties, there are valuation processes not directly aimed at the straight liquidation of the asset but which use socially innovative methodologies as "bottom-up" paths for regenerating spaces [4, 5].

2 The Value of Real Estate Stalled Investments: Assessment and Value Creation Issues

The real estate collateral is therefore the immovable property placed as a mortgage guarantee of the credit at the time of registration of the loan. Following the non-payment of 3 or more agreed repayment installments, the credit is defined as Non-performing loans, better known as Npl. At the first signs of insolvency by the creditor, the bank tries to agree with the client new financial instruments for the repayment of the credit disbursed, reshaping times and methods of return. The liquidation of collateral takes place only when there is no longer any credible hope of repayment of the loan. The phenomenon is growing strongly: in Italy in 2016 the percentage of Npl on total loans was 21.1%, while in 2008 it was 4.9%. Among the loans with no more reasonable hope of being recovered those secured (i.e. with real guarantees) are 48% of the total (data from the Bank of Italy). The European Central Bank, in its "Guidance to banks on non-performing loans" (2017) identifies three phases of the life cycle of NPLs. The third phase, the last and the most serious, consists in the enforcement of guarantees that can occur through judicial or extrajudicial proceedings. The asset is therefore acquired by the creditor at its updated market value, or auctioned for sale to third parties.

However, the auction mechanism always shows a gap between the forced sale value and the market value, which generally stands at 35% [6, 7]. Furthermore, together with the general decline in prices on the market, the fair value always deviates from the book value on which the loan was disbursed. This phenomenon is even more evident in contexts where the demand is poor [8, 9]. The auction sale, except in a few and circumstantial cases in which the asset is appreciated by the market, is not a useful strategy for the banks. The gap between the realizable value and the book value creates holes in the financial statements of the banks. Finally, many buildings and areas do not meet any buyer, so thus remain abandoned and without any reasonable hope of recovery. Some more farsighted investors and even some banks - the Real estate owned companies, also known as Re.o.co - start to exploit the potential value of these areas. They engage in processes alternative to liquidation, such as their long-term management or transformation. This last strategy can meet notable points of contact with the wide architectural and urban research carried out on the transformation processes [10, 11].

3   The Methodology and the Empirical Research

As regards to the collection of the data sets useful for this research, it was very difficult to not work directly with the real estate used to guarantee non-performing loans. It is worked to research on the numerous transformation processes that took place on the Italian territory, restricting the field of analysis to those that could be more likely to be comparable with collaterals of bad debts. The "funnel selection" led to choose the cases of transformation operated in areas where there were unfinished buildings. There are two components that have led to the assimilation of unfinished projects to valid surrogate data for the entire set of real estate guarantees in non-performing loans: the high residual value and the unavoidability of a massive transformation intervention. The grounded theory deduced therefore developed with the study of the 28 cases examined, collected in number useful for the satisfactory achievement of the theoretical data saturation.

Proceeding with the research, the treatment of the volume in the transformation process emerged as the core category, a direct expression of the strategies to be applied in relation to the extraction of value from the area. So many cases of transformation have been studied on the stasis of previous ongoing projects that were then interrupted. They are sorted according to their impact on the built volume. This paper reports only the three most significant cases, one for each category: rarefaction, conservation and densification. The process of creating economic value, the urban and legislative instruments and finally the more strictly architectural ones were analyzed for each transformation.

For the category of rarefaction processes, a significant example was found in the municipality of Milan with the former hotel Monluè, an abandoned concrete frame on the south-eastern outskirts of Milan. It had always been object of attention by the city administration; the hypotheses presented were numerous, but never one of them had ever seen an effective realization. The solution to this problem occured only in 2008 when a private real estate company agrees to sell the area to the municipality demolishing the unfinished building and renouncing the rights that allowed it to further build the area. In return, it obtains the change of intended use (from public offices to private offices) of the complex of Piazza Freud, better known as Garibaldi towers. With the same agreement both the needs of the private (an important real estate investment) and those of the public were met: a degraded area was in fact healed, expanding the green area of the municipal territory; today the area is rented to private subjects for agricultural purposes. All this without adding a single cubic meter more, but subtracting.

An example of conservation strategy is represented by a complex site in Villorba (near Treviso, in the north-east of Italy) which consisted of an unfinished concrete frame of a project never completed. The property, requested by the municipal administration, starts work to create a functional mix - hotel, convention center, restaurant, commercial spaces plus some residences – adapting it in the existing structure. In the case of Villorba, the Municipality has made a variation on both the urban planning regulations (with a variation to the General Regulatory Plan) and the building regulations (Technical Regulations) to allow the promoter to carry out a project which then found a excellent response in terms of the market. In return, he obtained the payment of the due charges, which were transformed into the realization of some public works near there.

The most significant case of urban densification in this research is the Porta Nuova area in Milan where, without the massive infrastructural intervention, the massive private urban development investments in the area would not have taken place. Specifically for this case, the large abandoned structure of Via De Castillia never reached the completion because of legal proceedings. Attempts to auction have never been successful. It was precisely the contemporary relaunch of Porta Nuova to change the intentions of the artifact and to convince the property to transform it into its headquarters, transforming the initial project then obsolete on the functional level and also on the formal-aesthetic one. Even proceeds to densify the area with the construction of a new skyscraper currently under construction.

4   Three Ways to Create New Value in Stalled Real Estate Investments Recovery

Gathered all the data from all the examined cases, especially from the three main ones described above, the research proceeded interpreting the main empirical dynamics.

All the rarefaction processes derive from the reduction of expectations (downsizing) initially envisaged for the area. These interventions constitute a category wider than only demolition; it's area rarefaction even in all those cases of renunciation of the construction of the already approved volume, reclassifying the planned land use from building to agricultural. However, when operating on areas where unfinished pre-existing buildings already stood, this strategy is implemented in dismantling and abatement measures.

From the economic point of view, the will to renounce the extraction of value from the area appears when the initial project is set aside, just as possible reconfigurations of the initial hypotheses are excluded. All this implies an awareness on the part of the developer that there are no more market conditions for the area to produce income, whatever its functional destiny. It must be reasonably considered that, in almost all cases, the fate of such goods will be that of abandonment. There is no interest - at least from a financial point of view - on the part of an investor to carry out demolition of the existing plant: the agricultural land market in fact does not settle on such high values as to be able to compensate for onerous expenses such as the demolition.

On the other hand, the public administration is very interested in resolution strategies for similar abandoned areas, which generate only negative externalities on the context. But how can the public encourage similar operations? Or through financial incentives - not advisable, given the current state of municipal budgets - or remunerating the private promoter with the urban planning "currency". Metaphors aside, release of building rights to be exercised on another area in exchange for the demolition of the building. In addition, the agreement will often include the sale to the public of the area in question [12, 13]. Clearly, since the agreements for the transfer of building rights can only take place within the same municipal territory, these will be effective only in medium-large municipalities. In fact, small municipalities rarely contain abandoned areas and at the same time areas of great attractiveness for new building investments.

Different is the case of those who do not renounce to extract value from the area, but decide to intervene on site with the tools of architectural transformation. It is a question of transforming existing buildings, keeping the volumes initially planned; the change to be made will not concern the volumetric quantities but the functions that can be expressed by the good. The original functional program does not appear to be more convenient, assuming that the promoter does not consider it more profitable; if not, the developer would not have hesitated to continue the interrupted construction site. Initial intentions are not always the result of an error in planning or a lack of competence in operating with investments in real estate: often the braking condition is the radical change in the economic scenario of recent years that has affected all sectors of the economy. Sudden system changes difficultly can be incorporated into the world of building production, which by its nature needs longer times that expand further in the event of unforeseen events.

Re-functionalisation - a practice widely dealt with by the world of architecture - imposes a reflection also in terms of urban planning instruments. In fact, a change of functional destination for the area is often necessary; the public administration is called upon to evaluate these requests for changes in the plan, considering the impact that the transformations would have on the urban balance of the territory. With the instrument of the agreement, the public will then proceed to participate in the surplus value that its choices have allowed to the private individual. Thus will be realized in the payment of the charges due, or in the realization of public works [14–17].

In a process of re-functionalisation, it's fundamental a promoter who knows how to work with the terms of the financial economy and at the same time knows the tools and possibilities of the architectural project: only in this way he can imagine a different destiny for the area, a new vocation that is viable and profitable at the same time. In order to choose an effective second life, the developer must have a good ability to read market trends and, more generally, to know what functions the context is actually asking. It would be limiting to exercise these analytical skills only by drawing on the real estate market trends, it is essential the dialogue with the productive and social realities of the territory.

Concept of densification of the areas conceptually starts from the same principle that guides the transformation interventions described above, that is the will to extract value from the area intervening on it. What differentiates them is the amount of investment that the property is about to face, in view of a return that can repay the costs. Densification means an intensification of what has been already built, a tangible increase in the physical volumes of the building. These processes appear as the only viable way for a new urban development, in fact they go to strengthen already existing areas instead of consuming empty land with new expansions. The promoter generally declares himself available to bear important expenses; both for the greater quantity to produce and to build, and because often similar interventions translate into typological solutions (tower buildings) and higher-cost technological solutions [18].

However, it remains to be clarified how an area that has been abandoned - for which it is therefore assumed that there is little interest on the part of the market - proves to be an excellent location for profitable investments, even with an increase in areas.

Obviously it has been a change in the context, a radical change in the positional qualities; but similar changes do not happen autonomously, they are always the result of huge investments.

5 Conclusions

This research tries now to draw some conclusions from the empirical experiences reported. Upstream of any kind of intervention, a selection to operate between the areas remains essential, without falling into the rhetoric and slogans of reuse. The positioning of the areas is fundamental, the positional incidence is still confirmed as one of the main factors that guide the choice of the project [19].

For many small centers of our country there will therefore not be any likely strategy because there is no market demand that can accept the good and therefore bear the cost of its transformation. Not even agreements for demolition will be easy to implement: it's difficult to find within the same municipal territory high-income areas and other ones without any interested buyer, unless it is a fairly large urban context.

Then there are numerous medium-income contexts, in which the market demand is active only on some segments, while in others it is close to zero. Here it is advantageous to operate with refunctionalization strategies in order to intercept those active parts of the demand. Usually no significant increases in volume are expected, as these interventions are limited to the area and do not generate a surplus value such as to change the context, even only at the infrastructural level. So it is almost never possible to make significant increases in surface area, which would overturn the urban balance of the place.

Finally, there are few isolated cases where rent is very high and where the pure transformation of the existing would not prove beneficial; in fact, the market requires large volumes that are much higher than the existing ones. Often these interventions will result in some densification models such as tower buildings.

As seen, hundreds of billions of euros are crystallized in this kind of areas: it is therefore essential to intervene in order to contribute to the recovery of the economy, and the project must play a key role in this contribution. The awareness that so far the real estate has seen an overproduction must not however result in the opposite extreme of not intervening anymore. There is a great need for building interventions, but they must be mainly aimed at transforming what is already there. It is not just a second life for the decommissioned, it is often a matter of imagining a real first life for these areas.

What remains indispensable is the right concertation between public and private, with agreements that allow both parties to achieve their goals. It's innovative the use in this perspective of the urban planning instruments, that were born to regulate a phase of expansion of the cities but that can also be useful in the recovery of brownfield sites.

The reshaping of agreements basing them on the gains generated by reconversion rather than those produced by new constructions, as well as being a wide field of investigation for future research, offers unprecedented opportunities for operators involved in the extraction of value from similar areas.

These operators are therefore starting to undertake some transformations of distressed assets, having grasped the great potential inherent in them. It is essential, however, that they possess a solid culture of the project, which should not be exhausted only in reading the market trends. It must, however, be wide-ranging, knowing how to capture even unprecedented possibilities such as those represented by town planning instruments.

References

1. Fregolent, L., Savino, M.: Città e politiche in tempo di crisi, 1st edn. Franco Angeli, Milano (2014)
2. Cui, L., Walsh, R.: Foreclosure, vacancy and crime. J. Urban Econ. **87**, 72–84 (2015)
3. Zhang, L., Leonard, T.: Neighborhood impact of foreclosure: a quantile regression approach. Reg. Sci. Urban Econ. **48**, 133–143 (2014)
4. Mangialardo, A., Micelli, E.: Simulation models to evaluate the value creation of the grass-roots participation in the enhancement of public real-estate assets with evidence from Italy. Buildings **7**(4) (2017)
5. Mangialardo, A., Micelli, E.: Urban reuse and public real estate: the valorisation of the bottom up heritage. Territorio **79**, 109–117 (2016)
6. Canesi, R., D'Alpaos, C., Marella, G.: Forced sale values vs market values in Italy. J. Real Estate Lit. **24**(2), 377–401 (2016)
7. Canesi, R., D'Alpaos, C., Marella, G.: Guarantees and collaterals value in NPLs. In: 2nd International Symposium New Metropolitan Perspectives, Reggio Calabria (2016)
8. Chow, Y.L., Hafalir, I., Yavas, A.: Auction versus negotiated sale: evidence from real estate sales. Real Estate Econ. **43**(2), 432–470 (2015)
9. Donner, H., Song, H., Wilhelmsson, M.: Forced sales and their impact on real estate prices. J. Hous. Econ. **34**, 60–68 (2016)
10. Scardovi, C.: Holistic Active Management of Non-Performing-Loans, 1st edn. Springer, Cham (2016)
11. Rottke, N., Gentgen, J.: Workout management of non-performing-loans. J. Property Investment Finan. **26**(1), 59–79 (2008)
12. Antoniucci, V., Micelli, E.: Il segno meno: la ristrutturazione di progetti di trasformazione urbana e accordi pubblico-privato ai tempi della crisi. In: 16th Società Italiana degli urbanisti national conference, Urbanistica per una diversa crescita, Napoli (2013)
13. Lanzani, A., Merlini, C., Zanfi, F.: Quando un nuovo ciclo di vita non si dà. Fenomenologia dello spazio abbandonato e prospettive per il progetto urbanistico oltre il paradigma del riuso, Archivio di studi urbani e regionali 109 (2014)
14. Micelli, E.: Development rights markets to manage urban plans in Italy. Urban Stud. **39**(1), 141–154 (2002)
15. Micelli, E.: La gestione dei piani urbanistici: perequazione, accordi, incentivi. Marsilio, Venezia (2011)
16. Micelli, E.: Five issues concerning urban plans and the transfer of development rights. Scienze regionali **13**(2), 9–27 (2014)

17. Micelli, E.: I diritti edificatori per il governo del territorio: strumento generalizzato o tecnica di nicchia? Scienze Regionali **15**(1), 123–130 (2016)
18. Antoniucci, V., Marella, G.: Torri incompiute: i costi di produzione della rigenerazione urbana in contesti ad alta densità. Scienze Regionali **13** (2014)
19. Bullen, P., Love, P.: The rethoric of adaptive reuse or reality of demolition: views from the field. Cities **27**(4), 215–224 (2010)
20. Manganelli, B.: Real Estate Investing: Market Analysis, Valuation Techniques, and Risk Management. Springer, Cham (2014). https://doi.org/10.1007/978-3-319-06397-3

The French Way to Urban Regeneration. Tangible and Intangible Assets in the Grands Projets de Ville

Anna Laura Palazzo[✉] [iD]

'Roma Tre' University of Rome, 00184 Rome, Italy
annalaura.palazzo@uniroma3.it

Abstract. The *Politique de la Ville*, grounded on an idea of equality that has not withdrawn despite political changes and economic downturns, was launched in the late 1970s, aiming at reducing territorial inequalities within disadvantaged neighborhoods dating back to the second post-war period.

In the *Communauté urbaine* of Lyon recently established as a *Métropole* and provided with a Strategic Plan (SCOT), several challenging generations of PdV have been set up addressing the traditional domains of social housing and urban environment and supporting widespread access to education and cultural facilities. In the deprived neighborhood of La Duchère, high standard urban renewal was deemed able to break down the invisible barriers of the social stigma by attracting new people and activities and promoting social inclusion.

Beyond the questionable and controversial displacement of the previous inhabitants, the long-standing mobilization of the PdV in La Duchère features the ideal fieldwork to assess whether and to what extent alongside current regeneration tools intangible assets have been releasing benefits to the community.

Keywords: Urban regeneration · Social inclusion · Cultural sustainability
Politique de la Ville · Lyon

1 Introduction

In France, where individual freedom and citizenship had resulted from the lesson of the French Revolution, over the second half of the Twentieth Century a broader meaning of equality would be reshaping the Welfare State and social spending: *From each according to his ability to pay, to each according to his needs*, paraphrasing a renowned slogan by Karl Marx.

The French political and legal discourse on the 'Right to the City' [1], in expanding the social guarantees originally entrusted to the State and subsequently to local authorities, launched the *Politique de la Ville* (PdV) in the late 1970s aiming at reducing territorial inequalities within disadvantaged neighborhoods (*Quartiers en crise*).

Social housing has long been a major goal in the French policy agenda, to the point that, according to Eurostat (2011), 89% of this building stock is owned by public or private organizations (*Habitations à Loyer Modéré*, HLM), directly or indirectly

© Springer International Publishing AG, part of Springer Nature 2019
F. Calabrò et al. (Eds.): ISHT 2018, SIST 100, pp. 297–304, 2019.
https://doi.org/10.1007/978-3-319-92099-3_35

subsidized by public funding. Since 1977, new social rented housing has been financed by loans with a reduced interest rate allocated by the *Caisse de Dépôts et Consignations* [2, 3].

As for education, from the 1980s onwards, the French system envisaged the need to support within the PdV so-called 'priority education target areas' where schools and colleges would be provided with additional resources and greater autonomy to deal with school truancy and social order. Whereas the saying originally was: si on travaille bien à l'école, on trouve un boulot, further cultural issues have gradually been embedded in the fight against discrimination launched by the Agence nationale pour la cohésion sociale et l'égalité des chances (ACSE), that complements the Agence nationale de rénovation urbaine (ANRU). The idea behind this approach is that intangible assets such as widespread access to educational facilities, cultural and art practices, events and festivals, keep the pace with social inclusion and economic integration [4].

The sphere of influence of the PdV would eventually encompass the inter-municipal level by means of by-law requirements in order to raise to 20% the share of social housing citywide (*Loi d'orientation pour la Ville*, LOV, 1991). Meanwhile, various laws addressed the unruly sprawl at the agglomeration level (notably the *Loi Solidarité et Renouvellement urbain*, SRU, 2000) offering incentives to urban planning through the so-called *Schémas de Cohérence territoriale* (SCOTs), strategic plans accommodating housing, infrastructure and environmental provisions for a 20 years period [5, 6]. In large urban areas, the SCOTs are generally committed to address social inclusion both by a dilution of social housing estates - *Grands Ensembles* - even through radical demolitions and reconstructions and by infill practices of social dwellings in downtown areas.

After briefly discussing the main goals and achievements within the PdV over time, this paper tackles the specific case study of Lyon La Duchère, a social housing estate built up on a hill some distance away from the City Center with heavy prefabrication techniques. La Duchère (12,000 inhabitants and 5,500 dwellings in 1975) has been experiencing for some years now a huge regeneration program using tangible and intangible assets to achieve a challenging living environment. The radical reshaping of the district layout coupled with an impressive displacement program for about 60% of the households settled in the buildings to be demolished, in compliance to the *Charte du relogement* (2006). Conversely, a brand new building stock to be sold on the free market was expected to break down the barriers of social stigma by attracting young middle-class households [7, 8]. For those who would remain in La Duchère, culture in general terms, and cultural facilities, have been targeted as a strong asset for social inclusion and awareness raising in individuals and in the community at large.

2 The Main Stages of the Politique de la Ville

In the summer 1981, the districts of Vénissieux, Villeurbanne and Vaulx-en-Velin belonging to the *Grand Lyon*, a territorial collectivity made up of some 60 municipalities, came to the fore due to a series of riots that shook the young *Politique de la Ville*. The hunger strike of 12 young residents in the Monmousseau des Minguettes district (Vénissieux) brought about a protest march for equality, started in Marseille with 30 demonstrators and ended in Paris with more than 100,000 people [9].

The social unrest due to the oil crisis was then exacerbated by the resurgence of unemployment, urban decay and racism towards immigrants concentrated in the *Cités*, the suburban housing estates.

Ever since, an array of policy reforms requiring strong cooperation between the State, municipalities, ONGs and local associative networks have sought to address community development issues, notably integration of young people into the labor market, along with traditional rehabilitation measures. Cross-cutting strategies had long been discussed within parliamentary reports to address social distress and exclusion, raise participation of city users in the functioning of public services, provide procedures and actions facilitating the voice of the inhabitants (*prise de parole*).

Therefore, several challenging generations of the PdV have been set up tackling the specific domain of housing and urban environment and more general issues in employment, health, law and order, security, education and culture [10].

In the mid 1990s, the PdV launched two new instruments.

The first one was devoted to the implementation of so-called *Contrats de Ville* (CdV), cooperation agreements between the State, local governments and other partners such as housing associations, where goals and commitments previously defined would stay in force over a limited time-period.

The second one, as a part of the *Pacte de relance pour la Ville*, was meant to select so-called *Zones urbaines sensibles* (ZUS) according to specific indicators such as the share of unemployment, school dropouts and average income. Among the two different categories of ZUS then envisioned - the *Zones de revitalisation urbaine* (ZRU) and the *Zones franches urbaines* (ZFU) - the latter were deemed able to stimulate economic development by a strong commitment in targeting these areas for new jobs and activities, notably small firms. Within the overall goal of creating in the ZFUs 100,000 new jobs over a 5-year period, along with renovating housing, restructuring business districts and improving public services, compelling requirements were set up to hire at least one third of the new recruitments from inside the target areas, and five-year tax exemptions were allocated.

Actually, the areas under contract generally overlapped with disadvantaged neighborhoods, and were unlikely to change in the short run. It is precisely for this reason that in 1999 50 *Grands Projets de Ville* (GPV) were established countrywide encompassing the previous ZUSs. The idea was to achieve a mass effect able to improve living conditions by an integrated set of relevant and place-specific physical, social and economic measures.

At the same time, a participatory process would allow residents to say over the priority action programs affecting their everyday life and to share place-specific development paths to be implemented and assessed with the institutional partners of the *Contrats de Ville*.

At the turn of the century, the most important innovation in the PdV was the *Programme national de rénovation urbaine* (PNRU) entrusted to the homonymous *Agence nationale* (ANRU) in charge with the funding and coordination of urban renewal projects in most vulnerable neighborhoods in order to achieve 200,000 demolitions, 200,000 new social housing, 200,000 heavy renovations by 2012.

The program aimed at rehabilitating some 530 neighborhoods affecting almost 4 million inhabitants through an investment of 40 billion euros. According to the *Loi d'orientation et de programmation pour la Ville et la rénovation urbaine* (2003), the ANRU acts as a single contact point gathering the State, the *Caisse de Dépôts et Consignations* and other public and corporate bodies, draining resources from the European Regional Development Fund [11, 12].

In 2007, after a new riots outbreak in several *Quartiers en crise*, the former city contracts were replaced by simpler and more readable contracts for social cohesion (CUCs), on the base of multi-year action plans and five key priorities: jobs and economic development; housing and the environment; education and equal opportunities; citizenship and crime prevention; healthcare. These contracts between the State and the municipalities were signed for a renewable three-year period as opposed to the six-year term of the CdVs.

In 2014, according to the *Loi de programmation pour la ville et la cohésion urbaine*, a new priority geography depending on the income per inhabitant was drawn by the *Commissariat général à l'égalité des territoires* (CGET) and shared with the local authorities.

The new contracts affecting some 1,500 *Quartiers prioritaires de la Ville* claimed for a wider spatial framework at the metropolitan scale. As such, they would collect in a single document the commitments taken by all partners, among which the State and its peripheral institutions, inter-municipalities and cities, the departments and regions, as well as others institutional actors and representatives from the civil society.

In 2016, these contracts entered their operational phase complemented by action plans for prevention of radicalization, funded at the level of 43 Meuros over three years.

At times, these measures would affect the very same *Quartiers en crise* tackled in previous generations of the PdV by elected representatives and city officials, entailing four levels of intervention: agglomeration, municipality, neighborhood, proximity. In such cases, thanks to long-standing cooperation attitude, the *travail social* entrusted to a *Chef de projet* featuring skills as a mediator, a facilitator and a manager, allows for *Projets de Territoire* locally grounded and supported by the citizens [13].

3   The Regeneration Process in Lyon – La Duchère

Lyon, located in Rhône Alps, belongs to the second region in terms of economy and wealth distribution in France. The *Grand Lyon*, established in 1969 as a *Communauté Urbaine* and in 2015 as a *Métropole*, consists of 59 municipalities with over 1 million three hundred thousand inhabitants and is featured as a strong authority responsible for the PdV and economic development.

The *Communauté* experienced urban disadvantage and social segregation in several phases, requiring increasing commitment by the PdV, notably in the *Grands Ensembles* dating back to the 1950s and 1960s [14].

Among them, La Duchère, occupying 120 ha of public land of which 30% devoted to open space, was built up to host mainly in social rental housing the many workers previously located in the unhealthy neighborhoods of Vaise. In 1968, the population had already grown to 20,000 inhabitants, and a half of them were *Pieds noirs*, people repatriated from Algeria.

Despite the excellent location on a hill close to the City Center and its facilities (the town hall, the *Maison départementale du Rhône*, a mall, high schools, a library and sports equipment), the neighborhood was confronted to deprivation and social exclusion. Building decay, poor access to home ownership and/or to rental market, dull homogeneity in housing typologies, no suitable dwellings for the elderly and students ranked among the main problems.

According to the 1989 census, due to ageing and decrease in households size, La Duchère counted 12,880 inhabitants that is nearly 40% less than its original population. The unemployment rate hit 21% and children less than 14 years old accounted for 23% of the total population. In addition, 27% had no degree at all and 80% had not reached the bachelor level.

La Duchère was thus ranked as a *Quartier prioritaire* and as such included in a *Contrat de Ville* in the 1994–1999 programming period, requiring heavy rehabilitation with funding specifically devoted to open space and new facilities, such as the Job Center.

The program *80 mesures pour La Duchère* signed by the Mayor of Lyon, Raymond Barre (1998), mostly aiming at strengthening the security in the area, enhanced cooperation in the field of social and economic development.

The initial idea focused on the rehabilitation of the public space without demolitions. Several circumstances, such as the obsolescence of the housing stock, would be taken into account over time, prompting decision-makers to a radical reshaping of the neighborhood with radical demolitions and reconstructions of new dwellings available on the free market.

Since some major problems such as unemployment rates and urban blight remained unaddressed, further policy measures were set up by the *Conseil général de Développement du Grand Lyon*, established in 2001. A thematic focus on solidarity and social cohesion was then set up and a cautious gentrification was scheduled in order to improve the living environment while promoting social mix, attracting new activities and decreasing the previous ratio of 80% to 20% among social housing and other tenures to a ratio of 55% to 45% (*Grand Projet de Ville Lyon Duchère*).

In the common perception, La Duchère stands out as a 'showcase' for urban renewal policy promoted by the ANRU, even though the initiative was prevailingly funded by the *Grand Lyon* [15, 16]. As a matter of fact, the overall regeneration program, to be obtained by a radical reshaping of the whole neighborhood and creation of new urban facilities, would be ruled by a body expressly put in place, the *Mission Grand Lyon*, a team directly in touch with the Mayor. In order to coordinate different contractors (the City of Lyon, the *Communauté urbaine*, the State, local authorities, public lessors…), the Mission encompassed the public developer as well - the *Société d'Equipement du Rhône et de Lyon* (SERL) - in charge of public spaces and roads, the marketing of constructible land and the development agreements in the target area. A specific tool, the *Zone à aménagement concerté* (ZAC), frames the intervention procedures and rules within a given time span (Convention signed on March 29, 2004).

More than a hundred partners have since been rallied around the program in its physical, social, economic and educational dimensions.

In the time period 2003–2017, the demolition of around 1,300 dwellings in the *barres* (multi-storey building several tens meters long) and reconstruction in-situ of as many apartments, with a larger proportion of private housing, would take place. The built environment would not exceed 7 levels and be provided with entrances directly from the streets, by gardens in the courtyards and underground parking (Fig. 1).

Fig. 1. The reshaping of La Duchère. The renewal program took place with layouts carefully designed for accommodating different building typologies.

Between 2003 and 2012, the relocation of more than half of the previous inhabitants of the *barres* (602 households) to the same neighborhood (47%) and to other *Arrondissements* (10%) was arranged. The *Office Public de l'Habitat du Rhône* (OPAC), as a major lessor, relocated 273 households. Such a questionable and controversial initiative mobilized human resources in the field of the *travail social* in a dialogue with the residents disoriented and worried about their destiny, even though, among a sample of displaced people, 80% would declare their satisfaction with new housing conditions.

In turn, the Cité, funded with 700 Meuro, was provided with a powerful neighborhood center and a square named after the Abbé Pierre, who had lived in Lyon, an Athletic stadium and other facilities. Around it, the Parc du Vallon, realized between 2011 and 2014, embeds playgrounds, a public garden, a glade and an undergrowth, includes a retention basin to manage storm water runoff and prevent flooding, and uncovers the Gorges Creek buried in the 1960s.

Along with education and training strategies for young people and adults, new dynamics and activities have been triggered by improvements in accessibility.

Beyond the *Duchérois* as tenants or owners, the program envisioned first-time homeowners and young workers (under 35 years) as ideal newcomers. In fact, according to the *Loi engagement national pour le logement* (ENL, 2006), households not exceeding certain income limits could benefit from tax relief for VAT.

Such provisions of mixed typologies available for sale or rent at competitive prices complying with the *Programme local de l'Habitat du Grand Lyon* (2007) are currently embedded within the planning routines carried out by the SCOT (2010). They perform a first step towards the transition of the PdV from a sort of special legal order to 'ordinary law' jurisdiction and from a protected economy to the market one.

Between 1999 and 2006, in Lyon the unemployment rate in the four ZUS areas (Vaulx-en-Velin, La Duchère, Rillieux and Vénissieux) decreased from 21.5% to 13.6%. In La Duchère, in the same period the growth in the number of enterprises was about 20%. Eventually, the inscription of the ZUS as *Territoires entrepreneurs* contributed to an increase in the number of activities: in 2009, new companies added up to 3,310, 60% more than in 2000. Construction and business services, finance and real estate represented 58% of this growth.

As for La Duchère, it experienced the establishment of 236 new firms in the district and in Greenopolis business park below the Parc du Vallon, resulting in the recruitment of 505 people, among which 131 from areas targeted PdV [17].

4 Conclusions. Lights and Shadows of the PdV

All over the country, the PdV has helped establish a different working style, prompting new attitudes towards governance and decision-making schemes expected to combine tangible and intangible assets.

The case of La Duchère conveys some more specific reflections. Whatever the appreciation of the *Grand Projet de Ville*, the huge displacement of households was perceived as a 'top-down' initiative permeated by ideology likely to disconcert people, notably the elderly, while weakening previous social links, whereas the restyling of the suburb was deemed poorly rooted in the local context and met strong opposition.

Despite these major criticisms, the building stock released in the first phase of the program was entirely sold out. The *désenclavement* of the *Cité* was also addressed by the new setting of the *Parc du Vallon*, which dialogues with the City Center, and by the provision of better connections with Lyon and other municipalities in the *Communauté urbaine*.

Far less evident are the outcomes of the intangible *désenclavement*, that is the participation enlargement within social life by all means: education and culture above all.

From the very beginning, accompaniment measures have been set up within a place-specific *Projet éducatif* for children and young people, thanks to an appropriate tariff policy supporting a broader access to culture (live-long learning and post-graduate training for adults), or in a prevention approach 'city-life-holidays' for the youth.

Still, such a strong commitment on educational activities, prevention and safety was not rewarded by a successful coordination degree among the many institutional tools, while associations and professionals of the *travail social* found it difficult to adjust their daily routines in the fieldwork within the PdV.

In conclusion, the French political and legal discourse and its theoretical achievements turn out to lie far above the practical fulfilments dealing with stories, experiences, behavioural patterns locally rooted.

The sphere of the 'Right to the City' has been incorporating a concept of public service, meant as an activity of general interest, under the control of public authorities, secured by a public or private body entrusted with the powers necessary to perform its mission and obligations. The 'Right to Housing', as a subset of the 'Right to the City', persists as a steady horizon: in this respect, the determination in marking the transition from equality to territorial equity leaves no doubts. The current challenge addressed by the PdV in terms of social mix and inclusion deals with regional planning tools entrusted to allocate spatial justice, from which the democratic game attains new legitimacy. Therefore, despite the general crisis of the Welfare State in itself, the *Politique de la Ville* is likely to continue its path.

References

1. Lefebvre, H.: Le Droit à la Ville. Anthropos, Paris (1968)
2. Hall, P., Hickman, P.: Neighborhood renewal and urban policy: a comparison of new approaches in England and France. Reg. Stud. **36**(6), 691–696 (2002)
3. Couch, Ch., Sykes, O., Borstinghaus, W.: Thirty years of urban regeneration in Britain, Germany and France: the importance of context and path dependency. Prog. Plann. **75**, 1–52 (2011)
4. Oblet, T.: La ville solidaire au pouvoir des mairies. In: Paugam, P. (ed.) Repenser la solidarité, pp. 653–669. Presses Universitaires de France, Paris (2007)
5. Jouve, B.: Planification territoriale, dynamique métropolitaine et innovation institutionnelle: la Région Urbaine de Lyon. Politiques et management public **16**(1), 61–82 (1998)
6. Guigou, J.L.: Aménager la France de 2020: Mettre les territoires en mouvement. DATAR, Paris (2000)
7. Grand Lyon: Programme Local de l'Habitat, Tome 1, Diagnostic et Programme d'Action, Secteur Centre (2007)
8. Grand Lyon: Programme Local de l'Habitat, Tome 3, Diagnostic et Programme d'Action, Secteur Centre (2007)
9. Querrien, A., Lassave, P. (eds.): Les Annales de la Recherche urbaine, 88, pp. 3–5 (2000)
10. Chaline, C.: Les Politiques de la Ville. Presses Universitaires Françaises, Paris (2014)
11. Bonneville, M.: Les ambiguités du renouvellement urbain en France. Les Annales de la Recherche urbaine **97**, 7–16 (2004)
12. Gilbert, P.: Social stakes of Urban renewal: recent French housing policy. Build. Res. Inf. **37**(5–6), 638–648 (2009)
13. Palazzo, A.L.: Politique de la Ville et Projets de Territoire. Urbanistica **149**, 74–77 (2012)
14. Carpenter, J., Verhage, R.: Lyon city profile. Cities **38**, 57–68 (2014)
15. Verhage, R.: Le renouvellement urbain à La Duchère. In: Boino, P. (ed.) Lyon. La production de la Ville, pp. 194–217. Parenthèses, Marseille (2009)
16. Stouten, P., Rosenboom, H.: Urban regeneration in Lyon: connectivity and social exclusion. Euro. Spat. Res. Policy **1**, 97–117 (2013)
17. Lyonduchere.org. http://www.gpvlyonduchere.org/. Accessed 10 Dec 2017

An Innovative Interpretation of the DCFA Evaluation Criteria in the Public-Private Partnership for the Enhancement of the Public Property Assets

Francesco Tajani[1], Pierluigi Morano[1($\boxtimes$)], Felicia Di Liddo[2],
and Marco Locurcio[2]

[1] Polytechnic University of Bari, 70125 Bari, Italy
pierluigi.morano@poliba.it
[2] Sapienza University of Rome, 00197 Rome, Italy

Abstract. With reference to the public-private partnership procedures for the enhancement of the public property assets, in this paper an innovative methodology for assessing the financial conveniences of the parties involved (private investor and Public Administration) is proposed. The developed method borrows the most widely-used evaluation criteria for the verification of the investment financial sustainability, and through basic logical assumptions, it allows to define combinations of the financial performance indicators easily interpretable by the parties involved and to be used in the negotiation phases. The aim is to provide a rapid tool for the verification of the investment financial viability, through an original interpretation of the classic DCFA evaluation criteria, that could be more relevant to the typology of public-private partnership agreements for the territorial regeneration.

Keywords: Public-Private partnerships · Discounted Cash Flow Analysis
Payback period · Evaluation criteria · Financial sustainability

1 Introduction

In recent years, the enhancement of the public property assets has been the subject of an extensive political, social and cultural debate about the modalities of the property management, the new uses and the territorial effects related to the investment decisions [1, 2]. The property enhancement, as the maximization of the efficiency in the use of the property, requires a precise plan for the property transformation, in terms of physical recovery and functional reconversion, which, by respecting the identity of the property asset, the urban context and the urban planning regulations, provides a compromise solution between the needs of the real estate market, the cultural and social vocation of the property to be enhanced and the demands of the local community [3].

In the current economic situation, the scarcity of public financial resources requires the involvement of the private operators for the enhancement of the public property assets [4]. The bank credit crunch, the crisis in the real estate market, the difficulty of placing complex property assets on the reference market, characterized by very large

© Springer International Publishing AG, part of Springer Nature 2019
F. Calabrò et al. (Eds.): ISHT 2018, SIST 100, pp. 305–313, 2019.
https://doi.org/10.1007/978-3-319-92099-3_36

sizes, and the danger of a "bad" sale of the public property assets, have led many European countries to seek alternative solutions to the transfer of the property.

In Italy, regulated by Art. 3-bis of D.L. No. 351/2001, as amended and integrated by Law No. 228/2012, the "enhancement concession" is among the public-private partnership (PPP) procedures, characterized by the involvement of the private operators in public initiatives [5]. In particular, with regard to the public buildings, the enhancement concession provides for private investors to be entitled to use the public buildings for a specified concession period, in exchange for their requalification, functional reconversion and ordinary and extraordinary maintenance. The private investors, as "managers" (and not "owners") of the public property asset, recognize a share of the revenues to the Public Administration, in terms of financial burden (as lump sum or leases) and/or of public works to be realized for the local community. Upon expiration of the concession period, the public entity falls within the availability of the property, and the acquisition of any transformation, improvement, addition and accession realized by the private entrepreneur.

From a financial point of view, the involvement of a private investor in a PPP procedure takes place if the convenience of the investment is satisfied, i.e. the ability of the investment to compensate for the initial monetary outlay and to generate a profit capable of remunerating the (market, financial, etc.) investment risk. The *investment (worth) value* [6] of the private entrepreneur will depend on the specific characteristics of the operator - the risk appetite, the expected return on investment, the "waiting time" aimed at the recovery of the invested capital - and on the burdens required by the Public Administration, in monetary terms and/or in terms of public works.

2   Aim

The financial sustainability of an investment is generally assessed by developing a *Discounted Cash Flow Analysis* (DCFA), which includes: (i) the cost and revenue valuation, related to the investment realization and management, (ii) the calculation of the cash flows generated during the analysis period; (iii) the determination of the performance indicators that allow to verify the feasibility of the initiative [7]. The *Net Present Value (NPV)*, the *Internal Rate of Return (IRR)*, and the *Discounted Payback Period (PbP)* are among the most widely used evaluation criteria. Specifically, a higher zero value of the *NPV* immediately confirms the investment convenience, whereas the *IRR* and the *PbP* are to be compared with "threshold" values. The determination of these acceptability thresholds requires, on the one hand, a careful market survey, aimed at identifying the values of the performance indicators generally referred to by investors that operate in the same sector; on the other hand, it is conditioned by the specific characteristics of the investor [8].

With reference to PPP initiatives, this research proposes an original methodology for assessing the financial conveniences of the parties involved (private investor and Public Administration). Starting from the mathematical expressions for the calculation of the *NPV* and the *PbP*, and on the basis of the assumptions regarding the distribution of the investment cash flows, the elaborated model allows to determine a combination of the main evaluation criteria (*NPV*, *IRR* and *PbP*) that guarantees the mutual financial conveniences. The aim is to provide an evaluation tool for the "quick" verification of the financial conveniences of the parties involved, through a "new" interpretation of the classic DCFA performance indicators, that could be more relevant to the typology of public-private partnership agreements for the enhancement of the public property assets.

The research is divided into three sections. In the first section the proposed methodology is outlined: the basic assumptions are introduced, the mathematical expressions for the calculation of the *NPV* and the *PbP* are recalled, the model equation for the assessment of the "new" DCFA evaluation criteria is presented; in the second section, the model is applied to a real case study, that is the requalification of the building that houses the Italian Securities Institute and the Italian Exchange Office, located in Rome (Italy): after having schematized the values of the economic parameters for the DCFA development, the items to be used for the model implementation are extrapolated and the results obtained are illustrated; in the third section, the conclusions of the work are finally discussed.

3 Model

Remembering that the *PbP* gives the number of years it takes to break even from undertaking the initial expenditure, in the development of the DCFA, as the discounted rate r increases, the *NPV* decreases and the *PbP* grows, that is consistent with the increase of the investment risk [9–12].

In the context of PPP initiatives for the use-conversion of the public property assets, the implementation of the DCFA has the advantage of explicating the financial conveniences of the parties involved (private investor and Public Administration) through an appropriate interpretation of the evaluation criteria: determined the transformation costs, the management costs and the revenues of the initiative, and set the actualization rate of the cash flows equal to the periodic expected yield of the private investor (r_{min}), the *NPV* represents the maximum amount that the Public Administration may require to the private investor in monetary terms and/or in terms of public works of equivalent value to be realized for the local community [13, 14].

In fact, if the financial threshold for the private investor is satisfied, the *NPV* will be positive, and it will represent an extra-profit for the private entrepreneur over the expected minimum yield from the initiative. Therefore, the Public Administration can formulate its requests on this extra-profit amount, guaranteeing the financial sustainability of the initiative for the private investor; on the other hand, monetary burdens higher than the calculated *NPV* do not ensure the financial convenience of the initiative for the private entrepreneur. Figure 1 shows the trends of the *NPV* and the *PbP* as the discount rate r (i.e. the minimum yield of the PPP initiative) increases.

The proposed methodology provides two basic assumptions: (i) the investment cost (and any financial charges for the debt capital) is concentrated at the time of the valuation $(t = 0)$; (ii) the investment cash flows that occur after the *PbP* - as the difference between the revenues generated by the initiative and the management costs - are periodically constant.

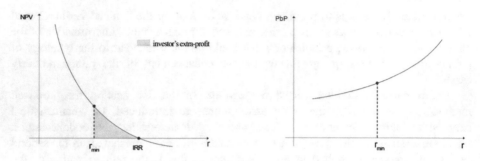

Fig. 1. Trends of the *NPV* and the *PbP* for the discounted rate *r* variations.

These two assumptions allow to simplify the mathematical expressions for the calculation of the *NPV* and the *PbP* through the Eqs. (1) and (2). Table 1 summarizes the economic terms involved in the two equations.

Table 1. Parameters and evaluation criteria of the DCFA in the defined model.

NPV	Net present value of the investment
r	Discounted rate
PbP	Discounted Payback Period (<T)
F_t	Cash flow of the investment in the period *t*
F_0	Realization cost of the investment (including any financial charges for the debt capital)
T	Analysis period of the investment

$$\sum_{t=1}^{T} \frac{F_t}{(1+r)^t} - F_0 = NPV \qquad (1)$$

$$\sum_{t=1}^{PbP} \frac{F_t}{(1+r)^t} - F_0 = 0 \qquad (2)$$

At this point, by subtracting Eq. (2) to Eq. (1):

$$\sum_{t=PbP+1}^{T} \frac{F_t}{(1+r)^t} = NPV \qquad (3)$$

For the constancy of the cash flows hypothesized through the assumption (ii), Eq. (3) is equivalent to Eq. (4):

$$F_t \cdot \frac{(1+r)^{(T-PbP)} - 1}{r \cdot (1+r)^{(T-PbP)}} \cdot \frac{1}{(1+r)^{PbP}} = NPV \tag{4}$$

Within the PPP procedures, the model formulated in Eq. (4) can be used in three different modalities: case (A), set the NPV, i.e. the financial burden required by the Public Administration to the private entrepreneur, it is possible to determine the combinations $[r_{min} - PbP]$ of financial viability for the private investor; case (B), set the investment yield expected by the private investor ($r = r_{min}$), the model allows to determine the $[PbP - NPV]$ combinations; case (C), fixed the time period within which the private entrepreneur intends to recover the invested capital (PbP), the model returns the combinations of $[r_{min} - NPV]$.

4 Case Study

The case study concerns a hypothesis for the enhancement concession of a property located in Rome (Italy), owned by the Bank of Italy, currently occupied by the Italian Securities Institute and the Italian Exchange Office.

It is assumed that a call for tenders is provided, aimed at the selection of a private operator for the redevelopment and the management of the property. In particular, the works on the property are finalized to the normative, functional and energy adjustment of the building complex that, since its realization in the early 1950s, has never been affected by an organic restructuring project.

In order to test the validity of the proposed model, the hypothesis assumed is that the Bank of Italy opted for a PPP procedure, according to which the works of modernization of the building are supported by a private investor, in exchange for the temporary use concession of the property for thirty years. Taking into account the current intended use of the building (large-sizes offices), the revenues for the private investor are related to the possibility of leasing the property during the concession period.

Table 2 summarizes the economic data for the evaluation of the financial viability of the initiative through a classic DCFA. It has been specifically assumed that 40% of the investment cost is loaned by a bank, at a rate of 6%, to be repaid for ten years, resulting in the financial charges reported in Table 2.

Table 2. Economic data for the development of the DCFA.

total investment cost	30,000,000 €
total financial charges	4,304,155 €
annual management cost	490,487 €
annual revenues	4,943,805 €
loan amortization period	10 years
concession period	30 years

The development of the DCFA has returned an *IRR* of the investment equal to 12.16%.

Figures 2 and 3 show the trends of the *NPV* and the *PbP* of the investment for increasing values of the discount rate *r*, starting from a rate of 6% and up to the value of the *IRR* that has been found.

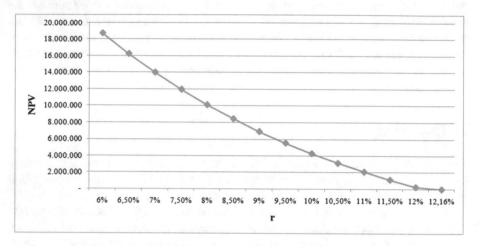

Fig. 2. Trend of the *NPV* of the investment for the discounted rate *r* variations.

It should be noted that for the values of the minimum yield expected by the private investor lower than the *IRR* (r_{min} < *IRR*), the remuneration to which the public operator - in the present case, the Bank of Italy - may target varies from a maximum value of 18,737,700 € (r_{min} = 6%) and a minimum value of 266,850 € (r_{min} = 12%). At the same time, the *PbP* of the private investor varies between a minimum value of 14,622 years (r_{min} = 6%) and a maximum value of 28,278 years (r_{min} = 12%).

In Table 2, the economic parameters necessary for the application of the model formalized in Eq. (4) are indicated through a light blue color. The results obtained are summarized in Figs. 4, 5 and 6.

Figure 4 describes the case (A) of the proposed model: assuming that the remuneration required by the public entity (i.e. the *NPV*) is known and is equal to 10,000,000 €, the possible combinations of the evaluation criteria [r_{min} − *PbP*] for the private investor are determined.

The developed methodology generates an inverse relationship [*r* − *PbP*] with respect to the one obtained from the usual application of Eq. (2) and reported in Fig. 3: this contingency is correlated to the invariance hypothesis in the proposed model of the *NPV* at the variation of the employed discount rate.

Figure 5 shows the case (B) of the model: assuming that the private investor has specified the minimum expected return on investment (r_{miz} = 7.50%), the resulting combinations [*PbP* − *NPV*] are obtained. The graph describes the empirical evidence

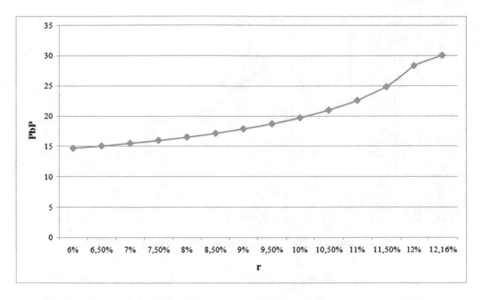

Fig. 3. Trends of the *PbP* of the investment for the discounted rate *r* variations.

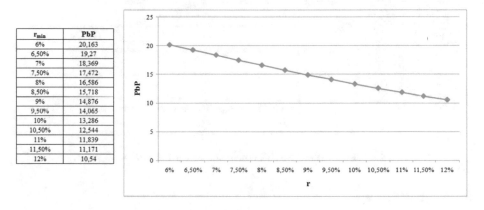

r_{min}	PbP
6%	20,163
6,50%	19,27
7%	18,369
7,50%	17,472
8%	16,586
8,50%	15,718
9%	14,876
9,50%	14,065
10%	13,286
10,50%	12,544
11%	11,839
11,50%	11,171
12%	10,54

Fig. 4. Case A: $[r_{min} - PbP]$ combinations for the $NPV = 10,000,000$ €.

that, as the time span required for the recovery of the invested capital increases, the economic compensation that the public subject may require to private investor decreases.

Finally, Fig. 6 shows the case (C) of the model: fixed the *PbP* equal to ten years, that is at least equal to the time span required for the repayment of the bank loan obtained, the implementation of Eq. (4) returns the displayed combinations $[r_{min} - NPV]$. The graph in Fig. 6 shows the inverse functional relationship between the two variables.

PbP	NPV
10	22.027.472
11	20.017.497
12	18.147.753
13	16.408.456
14	14.790.506
15	13.285.435
16	11.885.370
17	10.582.984
18	9.371.461
19	8.244.464
20	7.196.094
21	6.220.867
22	5.313.678
23	4.469.782
24	3.684.762
25	2.954.511
26	2.275.208
27	1.643.298
28	1.055.475
29	508.663
30	-

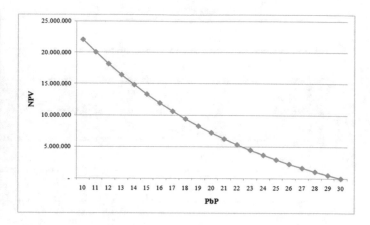

Fig. 5. Case B: [$NPV - PbP$] combinations for the r_{min} = 7.50%.

r_{min}	NPV
6%	28.522.364
6,50%	26.140.291
7%	23.983.166
7,50%	22.027.472
8%	20.252.364
8,50%	18.639.345
9%	17.171.979
9,50%	15.835.649
10%	14.617.337
10,50%	13.505.440
11%	12.489.601
11,50%	11.560.565
12%	10.710.056

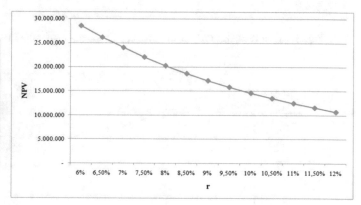

Fig. 6. Case C: [$r_{min} - NPV$] combinations for the PbP = 10 years.

5   Conclusions

With reference to the PPP procedures for the enhancement of the public property assets, in this research a "quick" methodology for the evaluation of the financial conveniences of the parties involved (public and private subjects) has been developed. The proposed method borrows the most widely used evaluation criteria for the verification of the financial sustainability of the investment - *IRR*, *NPV* and *PbP* - and, by through appropriate hypotheses, provides the combinations of the performance indicators, easily interpretable by the parties involved and to be used in the negotiation phases. In particular, the model does not consider the investment costs, and it only depends on the cash flows in the phase of normal operation of the investment. This simplification implies that the model constitutes a "first" performance indicator, to be used in the early evaluation phases of the financial feasibility of the investment, but to be complemented with the results of a classic DCFA. The results obtained from the application of the model to the case study for the three different cases considered highlight the potentialities of a simple-to-use model and easy to be repeated, that should be implemented as a further verification of the evaluation criteria generated by more complex but less controllable financial analyzes.

References

1. Las Casas, G., Scorza, F.: Sustainable planning: a methodological toolkit. In: Apduhan, B. O., et al. (eds.) ICCSA 2016. LNCS, vol. 9786, pp. 627–635. Springer, Heidelberg (2016)
2. Del Giudice, V., De Paola, P.: Geoadditive models for property market. In: CEABM 2014, Applied Mechanics and Materials, vols. 584–586, pp. 2505–2509. Trans Tech Publications, China (2014)
3. Pontrandolfi, P., Scorza, F.: Making urban regeneration feasible: tools and procedures to integrate urban agenda and UE cohesion regional programs. In: Murgante, B., et al. (eds.) ICCSA 2017. LNCS, vol. 10409, pp. 564–572. Springer, Heidelberg (2017)
4. Della Spina, L., Scrivo, R., Ventura, C., Viglianisi A.: Urban renewal: negotiation procedures and evaluation models. In: Gervasi, O., et al. (eds.) Computational Science and Its Applications. ICCSA 2015. Lecture Notes in Computer Science, vol. 9157. Springer, Cham (2015)
5. Della Spina, L.: Integrated evaluation and multi-methodological approaches for the enhancement of the cultural landscape. In: Gervasi, O., et al. (eds.) Computational Science and Its Applications. ICCSA 2017. Lecture Notes in Computer Science, vol. 10404. Springer, Cham (2017)
6. RICS, RICS Valuation - Global Standards 2017, London (UK) (2017)
7. Morano, P., Tajani, F., Locurcio, M.: GIS application and econometric analysis for the verification of the financial feasibility of roof-top wind turbines in the city of Bari (Italy). Renew. Sustain. Energy Rev. **70**, 999–1010 (2017)
8. Del Giudice, V., Passeri, A., Torrieri, F., De Paola, P.: Risk analysis within feasibility studies: an application to cost-benefit analysis for the construction of a new road. In: Liu, H. W., et al. (eds.) ICAEMAS 2014. Applied Mechanics and Materials, vols. 651–653, pp. 1249–1254. Trans Tech Publications, China (2014)
9. Manganelli, B., Morano, P., Tajani, F.: Risk assessment in estimating the capitalization rate. WSEAS Trans. Bus. Econ. **11**, 199–208 (2014)
10. Berto, R., Stival, C.A., Rosato, P.: Enhancing the environmental performance of industrial settlements: an economic evaluation of extensive green roof competitiveness. Build. Environ. **127**, 58–68 (2018)
11. Guarini, M.R.: Self-renovation in Rome: ex ante, in itinere and ex post evaluation. In: Murgante, B. et al. (eds.) ICCSA 2017. LNCS, vol. 9789, pp. 204–218. Springer, Heidelberg (2017)
12. Torrieri, F., Grigato, V., Oppio, A.: A multi methodological model for supporting the economic feasibility analysis for the renovation of the Valsesia railway system. Techne **11**, 135–142 (2016)
13. De Ruggiero, M., Forestiero, G., Manganelli, B., Salvo, F.: Buildings energy performance in a market comparison approach. Buildings **7**(1), 16 (2017)
14. Salvo, F., Ciuna, M., De Ruggiero, M., Marchianò, S.: Economic valuation of ground mounted photovoltaic systems. Buildings **7**(2), 54 (2017)

Innovative Milieu in Southern California: The Case of the San Diego Craft Breweries

Francesco Cappellano$^{(\boxtimes)}$ and Alfonso Spisto

Mediterranea University of Reggio Calabria, 89100 Reggio Calabria, Italy
francesco.cappellano@unirc.it

Abstract. Within the ethos of Smart Specialisation Strategies (S3), the EU regions seek to "discover specificities" in their territories to be harnessed in order to gain significant competitive advantages. This paper offers an illustration of a niche market which grew up in the North American context turning into a remarkable economic asset. First, we assess the consistency of the craft breweries instance with the "innovative milieu" concept. From this perspective, three spatial-based economic approaches, such as clusters and regional innovation ecosystems, are compared and debated. Second, we shed light on the tools implemented by the public to boost this industry sector. The methodology follows a qualitative approach, harnessing both secondary and primary data, specifically interviews with key informants. The results shed light on a multi-faceted business-friendly ecosystem which encompasses a quadruple helix array of actors. Nonetheless, the attitude of the public sector has proven keen to identify and tackle needs and hindrances.

Keywords: Craft breweries · Innovative milieu · Smart specialisation
Specialization strategies

1 Introduction

EU Regions seek to build their development strategies on *"existing dominant technological and skills profile and capabilities, but then diversify around this core base"* [1] in compliance with the S3 ethos. Regarding skills and capabilities, the innovative milieu refers to *"a cognitive set on which the functioning of this localized production system depends"* [2]. The concept surfaced during the 1980s in the field of regional sciences thanks to the GREMI group [3–5]. It came back into academic rhetoric because: (i) it rewards relationships between production systems and the territories where these systems were located [6–11]; (ii) it explains the success of certain regions where the presence of district economies and the synergistic relationships among the local actors boosted the innovation processes [12]; (iii) it is advanced under an endogenous development approach which lies at the heart of the S3 concept. In this view, the need to both identify and harness innovative milieus within the RIS3 design process looks promising.

This paper seeks to give an illustration of the craft breweries in Southern California in order to evaluate: (i) its consistency with the innovative milieu concept; and (ii) assess the policy choices by the local authorities meant to support the milieu.

© Springer International Publishing AG, part of Springer Nature 2019
F. Calabrò et al. (Eds.): ISHT 2018, SIST 100, pp. 314–321, 2019.
https://doi.org/10.1007/978-3-319-92099-3_37

Accordingly, the first section addresses the concept of innovative milieu, breaking it down into its main characteristics. The second section encompasses the reasons behind the success at the US level, whereas the third section focuses on the case of the San Diego brewing industry.

2 Conceptual Framework

In order to build up a framework to investigate the case study, we try to disentangle the different features scholars have associated with the concept over time. From a semantic standpoint, the French word *milieu* refers to the *context* or the *environment*. This "context" refers to "the complex network of mainly informal social relationships in a limited geographical area" [4]. Alike the social capital concept, the milieu's socially embedded relationships between organizations determine a successful regional development in terms of growing agglomeration of innovative firms [13]. Therefore, those ties are key to generating "externalities that are specific to innovation, and through the convergence of such learning processes, towards increasingly effective forms of joint resource management" [5]. From this perspective, the innovative milieu represents a coordinated and functioning institutional environment constituted by universities, research laboratories, public institutions and firms.

Finally, milieus are considered innovative when: (1) there are effective actors' relationships within the regional system; (2) social contacts enhance learning processes among the actors; (3) there is an image and sense of belonging [13]. With respect to the latter, great emphasis has been awarded by Camagni (1991) to "a specific external 'image' and a specific internal 'representation' and sense of belonging, which enhance the local innovative capability through synergetic and collective learning processes" [4].

Despite the different perspectives and approaches, the concepts that emerged can easily meet some of the characteristics of other spatial-based economic approaches, namely clusters and regional innovation ecosystems. In order to clarify the differences and go beyond the academic lingo, we proceed with a comparison between all three concepts that have been the focus of the MAPS-LED research project: Territorial Milieu, Clusters and Regional Innovation Ecosystems.

Unlikely with the other two concepts, the innovative milieu features the utmost important characteristics for the following analysis: (1) horizontal relationships between actors; (2) common behaviour intended as rationality, time frames and objectives underpinning the milieu; (3) its identity as a localized production system and the sense of belonging to a territory - and what the latter represents: "[...] space meant as just geographical distance is substituted with territory – relational space -, defined as the context where common cognitive models operate and tacit knowledge is created and transmitted" [14].

Our fieldwork, desk study and interviews allowed us to identify the San Diego craft beer industry as the one with the abovementioned behavioural patterns that match the identity features the most. Therefore, the next section provides an analysis of the craft beer industry to support the argument in choosing this sector as a case study.

3 Discussion

3.1 The American Craft Beer Industry: Identity, Place and Collaboration

The sense of identity in the American craft beer industry seems to emerge from two different but connected standpoints: from an organizational perspective – specifically, how they define themselves, the rules they must follow and, thus, how they differ from large beer producers – and from a more sociological perspective – identity as a way to reconnect to a place within a globalized society.

According to the American Brewers Association, an American brewery must meet the following specific criteria in order to be considered as craft [15]:

- Small: Annual production of 6 million barrels of beer or less. Beer production is attributed to the rules of alternating proprietorships.
- Independent: Less than 25% of the craft brewery is owned or controlled (or equivalent economic interest) by an alcoholic beverage industry member who is not themselves a craft brewer;
- Traditional: A brewer that has a majority of its total beverage alcohol volume in beers whose flavour derives from traditional or innovative brewing ingredients and their fermentation. Flavoured malt beverages (FMBs) are not considered beers.

The three features work as hints for the "the external image" [4] and the connection the industry builds with the places they are located. The identity - also due to socio-cultural features developed in a geographic area [16] - determine the milieu's characteristics. However, contrasting Governa (1998) who considers identity by "looking at the past" [17], Schnell and Reese (2014) believe the identity of the craft breweries can be seen as an active and conscious process. Thus, the creation of an identity leads to "neolocalism, considered as the conscious attempt of individuals and groups to establish, rebuild, and cultivate local ties, local identities, and increasingly, local economies" [18].

Neolocalism processes occurring through microbreweries have been remarked upon by Flack (1997), who believed craft breweries, heavily rooted in their local communities, represent a way to resist the homogenizing force of globalization [19].

At first glance, craft breweries contribute to this phenomenon by using imagery, stories and features associated with a specific place as a way to promote their beers. Pike argues that "brands and branding embody an inherent spatiality" [20]. However, someone may think that the contribution of local brewers to foster identity and sense of place is just a branding and marketing strategy. Indeed, along with such thoughts, local brewers also adopt an active and conscious civic engagement with the local communities and NGOs.

Eberts (2014) identifies three main broader categories of activities that breweries undertake to connect with their communities: the brewing of special events beers, tourism, and community economic development [21]. While the first two categories are easy to understand, the community economic development practices include: Training Programs to empower women in Craft brewing [22]; donations to local charities and NGOs; clean up initiatives within the neighbourhoods in which they are located [22, 23]; scholarship opportunities to attend craft brewing courses [24]; providing, free of charge, spent grain to local famers to use as feed for livestock; commitment to local sourcing

[25]; community engagement with local charities and non-profits [26]; advocacy to raise money for good causes such as literacy, local food, conservation, and food banks [27]; and Public Art [27].

This "internal representation" [4] craft brewers have built is not just formal. Rather it translates into the creation of a big collaborative environment among all the participants. In fact, craft brewers help each other in several ways: by lending hops, advice in beer production, time and machinery and sometimes even man-power.

Our interviewees all report this type of behaviour when it comes to the production or even the start-up phase of the business: "[…] For breweries to share ingredients, to be collaborative, to work together to support non-profit organizations or just to support the industry is very common… The industry is very focused on the success of the industry itself. The success of one is success for all" as assessed by a craft brewery manager interviewed.

3.2   Favourable Conditions for the Success at National Scale

Several factors at different levels contributed to the craft beer industry success in US and specifically in the San Diego region. Nationwide, in 1965, the first US micro-brewery was built in San Francisco, CA despite the homebrewing was still prohibited by that time. During the first craft brewing wave, brewers were completely inexperienced, and most of the time, their businesses were far from successful or even viable. Brewing was considered a hobby for the majority of them, so they were considered pioneers "not only for micro-brewing but the entire DIY movement" [28]. Since the very beginning, the craft beer movement featured a collaborative environment where brewers were teaching each other and helping each other regardless of any economic competition. By doing so, they pushed the demand for a unique product which was completely different from industrial beer produced by large companies.

Additionally, a friendly political environment triggered the movement. In 1976, the US Senate enacted the small beer differential which allowed a tax deduction for those business who brewed less than two million barrels per year. For their first 60.000 barrels, those breweries receive $2 per barrel reduction in the $9-per barrel federal excise tax (ibid.). Later in 1979, home-brewing activity was legalized despite several people who used to brew clandestinely in their backyards. Moreover, the craft breweries were facilitated by another important change in tax policy. In 1991, the tax fees for industrially brewed beer were doubled (18$ per barrel), but no change in taxation hit the craft beers. Nonetheless, taxation played a significant role in supporting the upheaval of the beer industry in the US. Craft breweries grew from 1 in 1985 to nearly 1500 in 2000 while non-craft regional/national beer industries faced a stark decline.

Besides tax policy, cultural factors contributed to the boost of the sector. Consumers started to look for more sophisticated tastes, rejecting the industrial mass-produced goods. Additionally, the idea of naturally crafted beers using only natural preservatives matched a strong connection to lifestyle choices, along with awareness of taste and health culture, namely the foodie movement which surfaced during the late 1990s. Moreover, the link between the brewery and the local clientele has been entrenching over time, building a sense of place as discussed earlier.

California, specifically the San Diego region experienced a boom for craft breweries in the last decade. The region particularly featured a strong history in craft brewing even in the 19th century. In fact, the San Diego Brewing Company, established in 1896, was the first pre-prohibition brewery in San Diego. Recently, San Diego breweries gained much popularity due to their goods' quality, as certified by the 18 medals awarded during the last Great American Beer Festival in 2016. The steady pattern of breweries' birth rate highlights the strength of the industry, which generated $851 million in 2015 and employed 4,512 workers in San Diego alone [29]. In retrospect, the reasons behind the success cannot exclude the choices made by the local public authorities which favoured nourishing the sector, as explained in the following.

3.3   The Successful Factors for the Milieu in the San Diego Region

The carried-out analysis indicated that the City of San Diego acknowledges the sector as part of the "emerging cluster" [30] featuring a multiplier effect estimated, 4.7, higher than comparable sectors (e.g. food manufacturing). Moreover, the sector is held at the core of the Economic Development Strategy (ibid.) as part of the endeavour to enlarge the local economic base. Consistently, the Economic department of the city government enacted an array of interventions to support the sector. In fact, different hindrances seem to constrain the full nourishing of the sector, including: access to capital, Land/space/available real estate, distribution, and water/wastewater management [29].

Regarding access to capital, the Economic Department provided two important tools: according to the interviewed official, the department adopted the "New Tax Market Credit" to provide grants to let breweries expand or purchase additional equipment, and so creating small jobs in the city. Additionally, a *"bridge loan"* was implemented and supported *"businesses unable to get the bank buy-in. The idea is that you get that loan and when you start to be profitable you begin to refinance it"* as assessed by a city official interviewee.

Since the region used to face some serious drought problems, the water supply represents a hazardous constraint for all water-intensive business, amongst: life science and craft beer industries in the San Diego region. In this respect, the City adopted funds from defunct state tax credit program to purchase underutilized sewer capacity and selling it at a discounted rate to the business looking at expanding their facilities.

The interventions also succeeded in tackling other types of deterrents, such as the procedures for obtaining permits needed for opening a brewery. A brewery owner interviewed highlighted that several local governments in the region sought to speed up those procedures including: San Diego, Vista, Chula Vista and Lemon Grove. She pointed out in Vista there was a "craft brewery-related revitalization" since the upcoming breweries triggered business attraction in the area. This is the case of Chula Vista where five new breweries opened along the main street "and now you're seeing restaurants coming […] there".

Hence, the presence of breweries in the urban fabric is very desirable because, "they represent an opportunity to develop gathering spaces and revitalize some neighbourhoods or places" according to the San Diego Council member interviewed. Notwithstanding, larger breweries are still classified as "light manufacturing". For these reasons, they are confined to industrial areas mainly because of sound, smell and pollution concerns.

As emerged in other studies [31], the public authority in San Diego supports economic development by harnessing the planning tools, mainly land use and zoning. In this respect, in 2015 the City Council updated the Land Development Code [32], introducing substantial improvements in favour of craft breweries. It introduced so called "Tasting rooms" within the Commercial Uses. According to the new measures, these are allowed "to offer tastings and sell beverages manufactured on the premises for on-site or off-site consumption". This offers a unique opportunity to locate micro-breweries or tasting rooms within the urban fabric where they can sell the beers produced within more dense and populated areas. At the same time, they can hold their large production sites in industrial parks where the land cost is much cheaper. By doing so, they can enlarge their customer bases, entrench relationships with the community and create more jobs.

However, the success of the craft beer industry has not only been prompted by the City Government. In fact, an array of actors played their roles in this game. For instance, the Higher Educational Institutions tailored specific programs for teaching how to brew beers. This is the case for both San Diego State University and University of California, San Diego which deliver courses fully recognized academic-wise and professional-wise. Additionally, there is a significant pursuit of innovation in that industry. This is the case of the frontrunner R&D centre, the White Labs, which brews yeast for all the breweries in San Diego. Finally, since 1997 the San Diego Brewer's Guild has been working as a common platform for the local breweries. On one side, the Guild has been working towards creating a shared calendar of public events related to the beers, namely fairs and exhibitions. On the other side, it helps to share knowledge about home-brewing. The Guild is listed among the External Stakeholder organizations the City government uses to refer [30].

Finally, the most remunerative industry related to craft brewing is the tourism sector. Several tour operators offer services for touring breweries. In fact, this sort of tourism service consolidates San Diego as the 8th city for number of tourists in the United States [30]. Additionally, many events regarding craft beers happen in San Diego, bringing more emphasis to the role of the local breweries as tourist attractions. In conclusion, the craft beer industry in the San Diego region shows a set of performance indicators which we can categorize as an "emerging" cluster from a life cycle perspective [33].

4 Conclusions

The concept of milieu provides an essential approach to understanding the territorial dimension of economic development. The collaborative environment and the material and immaterial relationships it establishes within the places in which the industry is located creates a place/industry identity which emerges as the most prominent feature characterizing the milieu itself. Furthermore, the milieu can lead to innovation through mutual learning processes and tacit knowledge among those parts of it.

In the interviews, desk study and fieldwork in San Diego, it clearly emerges that identity and collaboration are the main features characterizing the American craft breweries. Besides the high quality and innovative products, the shared behaviour towards collaboration and the promotion of the local dimension make San Diego craft breweries an idle case study in order to understand in depth the social innovation practices and spatial consequences of innovative milieus.

Acknowledgments. "The MAPS-LED project has received funding from the European Union's Horizon 2020 research and innovation programme under the Marie Skłodowska-Curie grant agreement No. 645651."

References

1. McCann, P.: The Regional and Urban Policy of the European Union. Elgar Editions, Cheltenham (2015)
2. Maillat, D.: From industrial district to the innovative milieu: contribution to an analysis of territorialised productive organizations. Louvain Econ. Rev. **64**(1), 111–129 (1998)
3. Aydalot, P.: Milieux Innovateurs en Europe. Group Européen sur les Milieux, Paris (1986)
4. Camagni, R.: Technological change, uncertainty and innovation networks: towards a dynamic theory of economic space. In: Regional Science, pp. 211–249 (1991)
5. Maillat, D., Quevit, M., Senn, L.: Réeseaux d'innovation et milieux innovateurs. In: Réseaux d'innovation et milieux innovateurs: un pari pour le développement régional, p. 6, GREMI-IRER-EDES, Neuchatel (1993)
6. Becattini, G.: Dal settore industriale al distretto industriale. Rivista di economia e politica industriale **7**(1), 7–21 (1979)
7. Stohr, W., Taylor, D.: Development from Above or Below. Wiley, Chichester (1981)
8. Brusco, S.: The Emilian model: productive decentralisation and social integration. Camb. J. Econ. **6**(2), 167–184 (1982)
9. Coffey, W., Polese, M.: The concept of local development: a stages model of endogenous regional growth. Papers Reg. Sci. Assoc. **55**, 1–12 (1984)
10. Garofoli, G.: Economic development, organization of production and territory. Revue d'economie industrielle **64**(1), 22–37 (1993)
11. Slee, B.: Endogenous development: a concept in search of theory. Options Mediterraneenes, no. 23 (1993)
12. Camagni, R.: The concept of innovative milieu and its relevance for public policies in European lagging regions. Papers Reg. Sci. **74**(4), 317–340 (1993)
13. Fromhold-Eisebith, M.: Innovative milieu and social capital - complementary or redundant concepts of collaboration-based regional development? Eur. Plan. Stud. **12**(6), 747–765 (2004)
14. Camagni, R.: Global network and local milieux: towards a theory of economic space. In: Conti, S., Malecki, E., Oinas, P. (eds.) Industrial Enterprise and its Environment: Spatial Perspective, pp. 195–216, Aldershot, Avebury (1995)
15. Brewers Association. https://www.brewersassociation.org/statistics/craft-brewer-defined/. Accessed 10 May 2017
16. Maillat, D., Lecoq, B.: New technologies and transformation of regional structures in Europe: the role of milieu. Entrepreneurship Reg. Dev. **4**(1), 1–20 (1992)

17. Governa, F.: La dimensione territoriale dello sviluppo socio-economico locale: dalle economie esterne distrettuali alle componenti del milieu. In: Magnaghi, A. (ed.) Rappresentare i Luoghi: Teorie e Metodi, pp. 309–324. Alinea Editrice, Firenze (2001)
18. Schnell, S., Reese, J.: Microbreweries, place and identity in the United States. In: Patterson, M., Hoalst-Pullen, N. (eds.) The Geography of Beer. Regions, Environment, and Societies, pp. 167–187. Springer (2014)
19. Flack, W.: American microbreweries and neolocalism: 'Ale-ing' for a sense of place. J. Cult. Geogr. **16**(2), 37–53 (1997)
20. Pike, A.: Introduction: brands and branding geographies. In: Brands and Branding Geographies, pp. 3–24. Edward Elgar, Cheltenham (2011)
21. Eberts, D.: Neolocalism and the branding and marketing of place by Canadian microbreweries. In: Patterson, M Hoalst-Pullen, N. (eds.) The Geography of Beer, pp. 189–199. Springer, Dordrecht (2014)
22. Karl-Strauss Brewing Company. https://www.karlstrauss.com/community/environment/. Accessed 15 May 2017
23. Craft Alliance. http://craftbrew.com. Accessed 2 July 2017
24. American Brewers Guild. http://www.abgbrew.com. Accessed 2 July 2017
25. Brewery Climate Declaration. http://allaboutbeer.com/article/breweries-sign-climate-declaration/. Accessed 8 July 2017
26. Arcadia Ales. http://arcadiaales.com. Accessed 20 June 2017
27. Freemont Brewing. https://www.fremontbrewing.com/community/. Accessed 6 2017
28. Hindy, S.: The Craft Beer Revolution. Palgrave Macmillan, New York (2015)
29. National University System Institute for Policy Research: San Diego Craft Brewing Industry: 2016 Update, San Diego, CA (2016)
30. The City of San Diego: The City of San Diego Economic Development Strategy 2014–2016, San Diego, CA (2014)
31. Bevilacqua, C., Spisto, A., Cappellano, F.: The role of public authorities in supporting regional innovation ecosystems: the cases of San Diego and Boston regions (USA). In: International Research Conference 2017: Shaping Tomorrow's Built Environment Conference Proceedings. Salford, UK (2017)
32. The City of San Diego. https://www.sandiego.gov/sites/default/files/legacy/development-services/pdf/industry/2014/mainordinance.pdf. Accessed 4 Dec 2017
33. Bevilacqua, C., Pronesti, G.: Clusters in designing S3-oriented policies. In: International Research Conference 2017: Shaping Tomorrow's Built Environment Conference Proceedings, Salford, UK (2017)

Innovation Districts as Turbines of Smart Strategy Policies in US and EU. Boston and Barcelona Experience

Bruno Monardo$^{(\boxtimes)}$ iD

Sapienza University of Rome, 00185 Rome, Italy
bruno.monardo@uniroma1.it

Abstract. Across US the most intriguing interpretation of 'Smart Strategies' and the emerging model that embodies the idea of recreating an innovative urban ecosystem is well represented by the concept of 'Innovation District', a 'geographic area where leading-edge anchor institutions and companies cluster and connect with start-ups, business incubators, and accelerators'. The city of Boston represents a paradigmatic case of flexible integration between urban economic redevelopment initiatives, changeable partnership architecture and exploitation of the potential of social innovation-related regeneration. The inspiration model of the 'Innovation District' of South Boston Seaport is referred to the 22@ Barcelona project, an initiative conceived in 2000 for regenerating an abandoned industrial site - 'El Poblenou'- with new thriving mixed use urban activities. This can be considered controversial and almost paradoxical given the fact that the Europe 2020 official Agenda is deeply committed in applying the Smart Specialisation Strategy for creating virtuous ecosystem in European urban regions following the US policy models and innovation clusters. The major challenge for an effective Smart Strategy style interpretation is related to the potential 'territorialisation' of urban redevelopment visions. The 'consciousness of places' with their local cultures can become a key-driver for embedded innovation. The 'place-based' approach allows to build virtuous regeneration projects including the potential of territorial 'dna' related to the local communities for identifying, recovering and increasing the values of local cultural specificities.

Keywords: Innovation Districts · Place based approach · Inclusion and equity

1 Introduction

Among the European Union present and future core policies, the issue of innovation represents the critical step between generating research ideas and turning them into viable actions that could help create growth, and foster sustainable regional and urban development, opening new seasons of prosperity.

The original engine of conceiving the 'Research and Innovation Strategy for Smart Specialization' EU policy (RIS3), within 'Europe 2020 Vision' is the necessity to cope with the long wave effects of the global economic crisis, tackling the rising research

© Springer International Publishing AG, part of Springer Nature 2019
F. Calabrò et al. (Eds.): ISHT 2018, SIST 100, pp. 322–335, 2019.
https://doi.org/10.1007/978-3-319-92099-3_38

and innovation gap in the European regions, as well as the increasing distance with the North American economy [1].

The European Cohesion challenge policy is rooted on the beneficial integration of three fundamental factors: the 'Smart Specialisation Strategy' concept enhanced by the 'cluster' policy, the high technology emphasis and the 'place-based' approach [2].

The main aim of these reflections is to highlight and support critically the idea that the 'Innovation District' phenomenon can be considered as the emerging interpretation model of an implicit 'smart strategy' at the local scale. In recent years across US, a new generation of redevelopment projects has been conceived in order to represent 'excellence poles' for rethinking the physical and social, as well as economic identity of blighted, strategically-located areas or deprived neighbourhoods, with the ambition of creating new thriving urban ecosystems. Since it is natural and informative for the EU to look at the American 'smart strategy' model, it is remarkable to underline the key example of the Innovation District phenomenon, in progress across US within the last decade, highlighting the significant differences that its diverse 'cultural styles' can suggest for Europe.

The idea of studying US policies to rethink effective strategies in a European context seems particularly appropriate, since the 'Smart Specialisation' concept is embedded in the 'transatlantic productivity gap' issue [3], due to the backward condition of many 'Old Europe' urban regions in using new technologies in order to support strategic economic sectors.

In the following section, a brief reconstruction of the pivotal EU requirement to tackle this gap is developed, going back to the roots of RIS3 policy, which was conceived by an expert group coordinated by Dominique Foray [4] and was incorporated into the Europe 2020 agenda as part of its tripartite goals of 'smart, sustainable and inclusive growth'.

Afterwards, it is briefly explored how RIS3 may be linked to US entrepreneurship and innovation policies, which in turn are strongly connected to the widespread application of the 'cluster theory', as it was re-conceived by Michael Porter in the early 1990s. From then on, significant best practices in the US have shown how the benefits derived from the implementation of the 'new' cluster theory have evolved, particularly in terms of economies of scale for urban agglomerations, stakeholder networks and an increase in the exchange of local knowledge.

The article then explores a significant fragment of policy and planning initiatives to implement the principles of research and innovation strategies in the Boston municipality, by analysing concisely the Seaport Innovation District planning initiative related to a complex cluster entanglement inspired by Barcelona @22 strategic project. The last part develops some preliminary reflections and remarks, discussing the positive and negative issues to be highlighted from the case experience of Boston and Barcelona, and reflecting on lessons that emerge from it with regard to the entailment on social and institutional innovation introduced by these new urban and regional regeneration models.

2   Smart Specialisation Strategies and Clusters

The European RIS3 platform defines 'Smart Specialisation' as "a place-based approach characterised by the identification of strategic areas for intervention based both on the analysis of the strengths and potential of the economy and on an Entrepreneurial Discovery Process (EDP) with wide stakeholder involvement". The 'Smart Speciali- sation' concept originally appears in the literature examining the 'productivity gap' between Europe and North America, a gap which had become evident since the midst of 1990s. After 1995 the information and communications technology sector (ICT) had boosted US productivity growth more than in the EU, where support for new tech- nologies for innovation had been scarce. The EU's poorer performance has been explained by multiple factors, such as a lower level of R&D investment [5], differences in industrial structure and the relatively slow pace of dissemination of new technologies across European economies [6].

In its 'Europe 2020' agenda, the EU incorporated a policy specifically aimed at tackling the productivity gap and launching a knowledge-intensive growth model: 'Research and Innovation Strategies for Smart Specialisation' (RIS3) [7]. This policy has been designed to promote local innovation processes in particular sectors and technological domains through a bottom-up identification of specific 'innovation pat- terns' [8].

RIS3 is based on four principles. First, that economic development is driven by knowledge and innovation, and in the long run is about 'true' economic regeneration that is not possible to plan *ex ante*; for this reason, it refuses to use the 'picking-the-winner' policy. Second, 'history matters', meaning that various EU regions have different potentials, institutional effectiveness, industrial specialisation and levels of knowledge; thus an analysis of each region's environment is vital. Third, the perspective of economic growth embraces a bottom-up approach. Fourth, the policy is demand-driven since it is derived from local potentials and local needs.

The RIS3 policy has thus embraced a 'place-based' approach [2] in order to identify the specifics that each region can utilize for successful innovation. This approach implies collaboration and sharing of information between local actors and all levels of government to enhance the 'grassroots' factors that create knowledge and transform it into sustainable innovation.

Local policy makers, universities and private entrepreneurs are recognised as key actors for promoting knowledge and innovation as the principal features for regional growth [9], whereas governments perform a strategic role in the production sphere, assigning greater importance to the involvement of local stakeholders and public/private coordination [10].

On one hand, these public policies are based on the concept that each region has its particular industrial and institutional history and that the local actors (entrepreneurs, policy makers, and civil society in general) should be involved in implementing regional development [11]. The policy process must therefore be inclusive and allow a large number of stakeholders to participate, in order to identify specific needs by using available resources.

On the other hand, the 'Entrepreneurial Discovery Process' (EDP) [3] is also being pursued; EDP consists of selecting and prioritizing fields and sectors where a cluster should be developed and where entrepreneurial activity unveils new domains for future specialisation. In the self-discovery process [12], both public and private sectors have to collaborate strategically, evaluating costs and opportunities and reducing the impact of necessarily imperfect information. Governments therefore have a prominent role - more important than merely the safeguarding of property rights - in avoiding corruption and guaranteeing economic stability.

The 'cluster' concept has been enjoying a surge in popularity over the past two decades, largely thanks to the work of Porter [13, 14], although early cluster theories date back to almost one hundred years ago [15]. In the Porter's conception, clusters are defined as "geographic concentrations of, and associated institutions (for example, universities, standard agencies, and trade associations) in particular fields that compete but also cooperate" [16].

At least three qualities in Porter's theory are strongly related to these reflections. First, the 'virtuous collision' between cooperation and competition, which concurrently creates pressure to innovate in the productive system; second: Porter's general definition of cluster as the encompassing of a broader range of regional agglomeration goes beyond the traditional Marshallian industrial district. Finally, and most notably, Porter "not only promoted the idea of 'clusters' as an analytical concept, but also as a key policy tool" [17] by explicitly including policy makers as key actors in fostering local economies. Indeed, policy makers took inspiration from the Porter's notion of the 'cluster' as a tool for promoting regional growth and competitiveness; this has led to "a proliferation of policies that seek to nurture and support cooperative relationships among firms and with other production-related agents" [18].

In general 'cluster policies' are conceived to build a territorial network, sort of 'platform' of local stakeholders (firms, institutions, public and private organizations, universities, technology transfer offices); territorial actors are encouraged to interact in formal and informal relationships in order to create, use and disseminate knowledge, enhance social trust and develop a harmonized vision for the future of the region. They operate below 'macro-level' polices, with the intention of improving the ecosystem for all firms through 'setting the table' activities [19], but above 'micro-level' policies, tailored to the need of individual firms [20].

While RIS3 policies have been conceived privileging the geographic and political reach of EU regional governments, cluster policies - or in more general terms, public initiatives aimed at fostering entrepreneurship in specific contexts - have shown their multi-scalarity in the passage from theory to practice, acquiring sense and significance not only at the regional level but also at the urban scale. Nowadays across the US, it is usual to discover 'place-specific' cluster policies [21], favouring a particular local area. Within the present general redevelopment policy of the city of Boston, the experience of the 'Boston Innovation District' (Sect. 3.1) can be deemed as paradigmatic. The rationale for such a geographic concentration can be ascribed to the broad objective of generating positive results in a designated area, which usually occur on a micro scale [22] and can be fostered by the coordination of a government entity that connects various companies. Cities provide a natural testing ground for these policies, thanks to a vast array of favourable conditions that can be found in urban areas: among them the presence of an educated and 'creative' workforce [23], a 'local supply' of entrepreneurs [24], a high concentration of private venture capital [25], a diversified business environment [26] and the provision of public infrastructure [21] are undoubtedly the most important. This is also why, for the purpose of these reflections, the analysis of US industrial and entrepreneurship policies is oriented to a specific urban context.

3 The Innovation District Phenomenon: Boston and Barcelona

The Greater Boston area is currently one of the most innovative locations in the US local-development landscape. This is thanks to its high agglomeration of educational institutions and industries, as well as its geography and infrastructure. As a result, the entire urban region is increasingly able to attract the interest of major investors and venture capitalists. The flourishing economic environment has positively impacted the economic growth of the metropolitan area, which is recording the highest rate of growth anywhere in the US [27]. Specifically, over the past thirty years, the cities of Boston and Cambridge - followed more recently by adjacent municipalities, such as Somerville and Charlestown - have implemented economic and urban policies that have turned the area into one of the most prosperous and vibrant zones in the nation. Public and private investments have therefore been made to boost sectors, such as education, financial services, life sciences and the high-tech industries, which today represent the main clusters that sustain the urban economy by creating jobs, as well as sustaining many other sectors.

The effects of these economic policies are felt in the territory via the spread of new development and renewal projects, which are changing the urban geography of the city by supporting the placement of innovation hubs within various neighbourhoods. The idea of creating an innovative urban ecosystem is embodied in the 'innovation-district' concept: a "geographic area where leading-edge anchor institutions and companies cluster and connect with start-ups, business incubators, and accelerators" [28].

Innovation districts are conceived as dense enclaves that merge the poietic and job potential of research-oriented anchor institutions and the high-tech start-ups in well-designed, amenity-rich residential, commercial and consuming environments. Creation, circulation and commercialization of new ideas are facilitated within these thriving atmospheres that leverage the intrinsic qualities of the virtuous urban context: physical proximity, relational density and dynamic identity.

The city of Boston represents a paradigmatic case of original and compelling integration between innovation policies and city redevelopment, thanks to the ongoing implementation of an explicit strategy whose core is the alignment between urban regeneration initiatives and exploitation of the potential of innovation-related, growing ecosystems. However, as it is concisely highlighted in the two following subsections, the strategy of spurring innovation within the city can risk to be interpreted more for promoting economic and real estate 'excellence poles' than for connecting disadvantaged populations to employment and educational opportunities.

3.1 Seaport Boston Innovation District

In 2004, the Boston Planning Development Agency (BPDA), then known as the Boston Redevelopment Authority (BRA), launched the 'LifeTech Boston' policy initiative, which can be seen as a significant incubatory step towards the eventual creation of a new redevelopment model, the first Innovation District in Boston. Its original mission was to foster the growth of Boston's life sciences and high-technology sectors by nurturing incumbent companies in the city and by attracting national and international business. It targeted three domains: biopharmaceutical, ICT and medical devices.

Its strategic goal was to attract new companies that were looking for favourable locations, by providing city services and identifying financial resources. In this activity, a network was created between trade and investment organizations, consulates, non-profits and public agencies. In particular, it worked with two stable partners: MassBio, a non-profit organization that represents and provides services and support for the life-science sector, and the Massachusetts Life Science Center (MLSC), a quasi-public agency tasked with implementing the Massachusetts State Life Sciences Act.

The most significant strand of the original strategy was later identified in the 'Boston Innovation District' (BID), a planning initiative launched in 2010 by the Menino administration and is still in progress; the project aims to create a complex neighborhood able to attract financers, resources and talent that mimics the success of 22@Barcelona, the most significant forerunning experience of all 'innovation district' models.

The BID project was conceived to redevelop the South Boston Waterfront, an underutilized area of 1,000 acres that previously hosted industrial activities and parking, and transform it into a thriving hub of innovation and entrepreneurship together with new residential, commercial and retail spaces (about 7.7 million sq. ft.) with a mixed-use configuration.

The BRA managed the project and provided partial funding for constructing new public spaces, building a network of private companies and using financial and planning tools within a PPP 'architecture' both to guarantee progressive implementation and to ease the cost burden of the project on the city's budget.

The public initiative has been actively involved in attracting both start-ups and more established companies, such as Vertex Pharmaceutical and more recently General Electric, both of which received significant tax breaks in return for setting up their new headquarters within the BID boundaries. Unique assets are concentrated in the dense redevelopment area, such as the world's largest start-up accelerator - 'MassChallenge' - and 'Factory 63', a significant experiment in innovative housing, providing private micro-apartments and public areas for working, gathering and organizing events (Fig. 1).

The BID vision has four main features, setting out general guidelines for how developments should be approached:

- *Industry-Agnostic*: the initiative is open to industries of every kind; this should allow for broad inclusivity of established companies and small enterprises, providing a framework for community engagement;

- *Clusters*: the BID's motto is "Work, Live, Play", which shows that the municipality hopes to attract amenities that would encourage entrepreneurs to spend more time in the district networking and socializing. This would bring entrepreneurs together into clusters to increase their proximity and density, making it easier for creative people in such a cluster environment to share technologies and knowledge. The city needs to retain talent through a working and living environment favourable to creativity and exchange; the creation of physical spaces that enable entrepreneurs to convene during and after work hours is an imperative for this municipal initiative, which leads to the recruitment of accelerators such as MassChallenge and the development of public meeting spaces such as District Hall;
- *Experimental*: the public administration is adopting an experimental framework characterized by expedited decision making and planning flexibility. The city's choice, reconfirmed by the present administration after Mayor Menino's original idea, aroused interest throughout the business community and created momentum for the public sector's efforts to attract developers, the creative industries, CEOs, entrepreneurs and non-profit organizations and to engage them in building a new community;
- *The City as Host*: differing from the 'university-as-host' scenario, as seen at MIT in Cambridge (Kendall Square), in the BID the city embodies the role of host institution. The identification of the innovation district as the flagship project in Boston means that the neighborhood will be free to develop organically, create momentum and allow innovation to spread throughout the city and its surroundings.

The centrepiece of BID is the District Hall, a large public space where innovators can meet, exchange ideas, explore potential synergies, finalize their creativity and come to concrete agreements. The building, opened in 2013, offers 12,000 sq. ft. of meeting space, and it is the result of a PPP between the BRA and private investors. The city also plans to add 1.2 million sq. ft. to the Massachusetts Convention Center, a major

Fig. 1. Boston seaport district master plan (Source: legacy.wbur.org).

focal point in the district, with a $1bn project to implement the construction of new private housing units.

3.2   22@Barcelona Innovation District

The project 22@ Barcelona emerges from the strategic will of the city government to transform an area occupied by abandoned factories, economic activities with reduced added value, and empty spaces that are scattered throughout the district. The project has been conceived within the context of the city of Barcelona's tradition in long-term strategic city planning and city model design. It is shaped around three axes: the physical, economic, and social regeneration of the urban context framed within the overall transformation of the east of the city together with the 'La Sagrera' station, the Olympic Village and the 'Forum'. It began in 2000 at a time of economic boom and as a result of the leadership exercised by the local government.

The implementation of the planning initiative has involved the creation of the '22@Barcelona' municipal company, which brought together efforts and hopes linked to the development of the project until 2011, the year of its dissolution. The 22@Network unites the interests of businesses installed in the area, headed first by the City Council of Barcelona, and more recently operating under the leadership of the private companies themselves.

The 22@ Innovation District is located in 'Poblenou', Sant Martí neighbourhood, characterized until the beginning of the transformation as a district that not only was called to renew the physical imprinting of the urban fabric, but also to emphasise the image of the 'compact city' trying to promote the urban regeneration of the area, both from the point of view of the existing and additional inhabitants and economic activities.

The Poblenou neighbourhood originally housed workers from the large companies located in the district. Historical activism, relational networks, and the proliferation of all types of civil society association activities contributed to a strong sense of neighbourhood in a large part of the population: the complexity of transformation in a socially and culturally vibrant area would have soon revealed the first disagreements and conflicts with citizens, unsatisfied and disappointed by the proposals and the lack of participation.

The @22 initiative covers about 200 hectares and 115 blocks of the XIX century original Ensanche Plan by Cerdà (Fig. 2); the core strategy has focused on the creation of clusters in competitive international companies and institutions, conceived as engines of the economic development of the entire Metropolitan area. These initiatives created clusters in five strategic sectors: Media, ICT, Biotech, Energy and Design. Besides the 4600 pre-existing residential private units, 4000 new subsidized housing units have been planned in the area (25% for rental). The Public Investment amount through a Special Infrastructure Plan (PEI) is about 180 million euros. According to the last available data (2015) the 'in progress' transformation shows that more than 1600 subsidized housing units and 40 ha for green areas (public and private) have been completed and approximately 14,000 m². have been constructed above ground level for the productive fabric facilities (the MediaTIC building and the business incubator Almogàvers Business Factory among others) and the neighbourhood fabric (CEIP Llacuna primary education centre or the Camí Antic de València Community Centre and Senior Citizens' Centre).

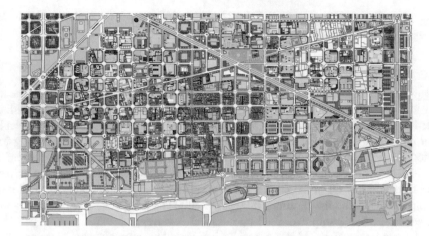

Fig. 2. Barcelona @22 district master plan (Source: 22barcelona.com).

The dynamic of the implementation process in 22@, planned since the end of '90s, was spoiled by the arrival of the economic and financial crisis. Despite that, a fundamental part linked to infrastructure was carried out before the slowdown of activity. Although 22@ is an area in which the preservation of jobs and the exporting dynamic of the businesses softened the effects of the crisis, the attraction of new companies was tackled by the paralysis of economic activity in the city as a whole.

More recently, and in a context of improvement in the behaviour of the economic activity, the 22@ district continues advancing towards the consolidation of innovation poles with a revisited model based on the collaboration between universities, enterprises and the public sector, by means of initiatives such as the Diagonal-Besòs Campus; at the same time the public private partnership conduction is trying to strengthen the attention towards the inclusionary sustainable dimensions, incorporating the social and solidarity economy into the transformation process of the productive model. However, despite the economic incentives mechanism for the private business attraction, nowadays more than 50% of 22@ land is still awaiting the completion of its transformation and the present municipal administration (headed by the Mayor Ada Colau) is rethinking the general strategy of the initiative.

4   What Innovation for Smart Strategies? Open Issues

As previously argued, Innovation Districts represent a new geography of economic development, amenity-rich, hyper-connected areas in city cores pursuing radical shifts from isolated suburban research parks developed in the 1970s and 1980s. Unlike the hyper-segregated business parks and residential districts that have been constructed for decades in many cities and suburbs worldwide, IDs include a range of distinctive features and contents which can be summarized into physical, economic and networking assets. The physical assets are represented by the public and privately owned spaces, designed and organized to stimulate new and higher levels of connectivity, collaboration and innovation; the economic ones are the firms, institutions and orga-nizations that drive, cultivate, or support an innovation-rich environment; networking assets focus on the relationships between actors (knowledge and administrative insti-tutions, firms, individuals) that have the potential to generate, sharpen, and/or accel-erate the advancement of ideas.

What is the interpretation model of 'smart strategies' emerging from the North American IDs phenomenon and from the urban redevelopment policy still in progress in the Boston area? Is it useful as a virtuous (or critical) lesson looking at the European Union context and its urban and regional policies?

Finding a single answer to these questions is not an easy task. Boston policy initiatives, the BID experience in particular and its inspiration model in Europe (Barcelona 22@) deliver an interesting set of lessons, key issues and suggestions related to at least three different dimensions of innovation: spatial, institutional and social.

Dealing with the physical spatial dimension, the BID urban planning blueprint doesn't look like following new sophisticated patterns, apart from the high tech research in the single architectural projects of the great companies in particular. In Barcelona the strong identity of the regeneration project is saved by the extraordinary relevance of the Ensanche Plan grid, conceived in the XIX century by Cerdà.

What is more significant can be related to the impact of the cluster choice and the proximity issue and their consequences on the new urbanity of the emerging districts and their ecosystem zooming backwards from the local to the urban scale. These experiences reveal the clear interconnection between clusters and the urban scale. Referring to Por-ter's theory, it would be inappropriate and substantially incorrect to investigate, follow and identify 'complete' clusters at a neighbourhood or municipal level since the assessment method adopted by Porter's team at Harvard Business School has been validated at state and county scales. Nevertheless, recent applied investigations on the possibility of recognizing significant cluster fragments at urban scale do deliver intriguing interpretations [29]. The Boston redevelopment initiatives show strong ties with a specific urban area, and the willingness to frame policy interventions within a wider spatial strategy of overall regeneration appears almost explicit. Physically dense concentrations of fragments and their ability to gain critical mass represent hotspots in the urban fabric, which 'topologically materialize' cluster fractals belonging to more com-plex and extended network systems. In general, in the 'Innovation District' phenomenon, the ideal objective of the regeneration strategy is the synergy between increased creative production and a high level of 'urbanity' associated with cross fertilization between public (and quasi public) spaces and actors who embody them.

As to the institutional dimension, the way in which 'innovation strategies' are interpreted in Boston, looking at what occurred in the evolution of the implementation process of Barcelona @22 after 2008, is linked to a 'flexible geometry' model that emphasizes the metamorphic synergies between the main actors of the classic 'triple and/or quadruple helix model'.

In particular, the City of Boston has proven to be sensitive to the 'institutional innovation' approach, playing a sophisticated role in tailoring adaptive partnerships among anchor institutions, investors, knowledge subjects and local communities. The Boston model represents the 'virtuous hybridization' of the various dimensions wisely mixed in the planning initiatives: from the overwhelming role of real-estate development to the increasing sensitiveness to local inclusion, from governance changeable profiles to socioeconomic quality. The dialectics generated by such contrasting approaches was a winning card: the BID experience demonstrates that locations, proximity, infrastructures, facilities and *ex ante* conditions still matter [30], and that the right choice of governance model plays a crucial role in the potential success or failure of such initiatives.

Given the considerable commitment and direct involvement of public institutions in the case of the BID, it is possible to interpret the evolution of the redevelopment strategy as the product of long-term planning and shared investment on the part of taxpayers, anchor institutions and private-sector partners. In a public-private/non-profit partnership, the cooperation between actors has been able to manage risk, mostly emphasizing the potential for private profit together with a recognized public benefit. Nevertheless, it is impossible to overlook the overwhelmingly business-driven philosophy of its policy and planning initiative.

The first lesson to be derived from the Boston model concerns flexibility in the stakeholder management, with an adaptive strategy based on entrepreneurial discovery/self-creation rather than on pre-conceived plans. One factor determining the success of these US initiatives, which at the same time coincides with some of the standard features of the RIS3 theory (entrepreneurial discovery, flexible strategy, elastic implementation) is versatility in the appropriate blending of 'stakeholders' from the urban zone, specifically public governmental institutions and local communities. In other words, it has a 'flexible geometry' approach in which roles can occasionally assume different identities and in which the boundaries between public and private initiatives are often blurred. By contrast, in the case of EU policy, these 'geometries' are likely to be shaped by a dominant, hierarchical, regional approach, in which the regions themselves catalyse and address the roles of the public and private actors that could potentially be involved in implementing and fostering innovation policies. And the first assessments about bottlenecks and weaknesses in the RIS3 implementation process [31] seem to confirm such a scenario.

Finally, innovation in its social equity dimension does not happen just because support is provided: the urban ecosystem as a whole has to be successfully reorganized and reinforced, including its physical and socioeconomic features. This is the biggest challenge faced by the current Boston administration after promoting a specific policy by locating 'Innovation Centers' in distressed neighbourhoods (the first one is Roxbury). It must discard traditional strategies in favour of doing something truly innovative: disrupting the patterns of inequality.

From a wider, EU-based point of view, the major challenge for an effective implementation of an RIS3 is to virtuously 'downsize' the role of industrial clusters, emphasizing the spatial and social sensitiveness to the redevelopment vision. A "consciousness of places" [32] is still crucial: a 'place-based' approach allows the creation of beneficial regeneration projects that are woven into the territorial 'DNA' of local communities in order to identify, recover and increase the value of local cultural specifics.

To avoid homogenization and perpetuation of income disparities, IDs can integrate targeted and high value methods of inclusion and policies for income redistribution — specifically, integrating quality educational programs including workforce development and access to appropriately priced housing and low cost properties for start up businesses — in order to support a sustainable increase in economic growth.

Unfortunately, existing IDs do not appear to be taking significant steps to ensure that the economic vitality and social benefits that come from an inclusive community are promoted. Already, calls of gentrification are surfacing. This may in part reflect a lack of true commitment to integrating policies that are truly aimed at preserving and promoting diversity.

IDs must necessarily be structured by a set of proactive policies that intentionally foster a socially and politically advantageous agenda for the many rather than the few. So, while the history of IDs is short, the evidence thus far seems to suggest that we are on track to perpetuate the trend to allow the districts to become, intentionally or not, an engine for reproducing the socioeconomic homogeny that the concept seeks to escape. This is the antithesis of the innovation IDs have the potential to achieve.

The use of the term ID, in and of itself, may be a catalyst for value increases [33]. In a number of cities, local stakeholders have applied the ID label to a project or area that lacks the minimum threshold of innovation-oriented firms, start-ups, institutions, or clusters needed to create an innovation ecosystem. Likely motivated by the quick return on investment, the profit motive for real estate developers to adopt the moniker seems clear.

Therefore it should be necessary to create and enforce intentional and conscious policies for inclusion and equity. Without these policies, existing community members will be displaced: in particular low income residents, as well as those cash strapped start up businesses that the districts seek to support and foster, will be driven from the community through market forces.

Promoting shared prosperity - making the economy work for all - is essentially about making capitalism inclusive. We cannot passively innovate our way out of exclusion and inequality. We have to summon the will to make the system work differently, work more fairly and inclusively. And we need to develop the rhetoric to convince those that only seek the quick economic return that a larger and more sustainable return is possible by integrating equitable policies on the front end.

The trend of IDs is, in many ways, a natural and organic response to shifting market dynamics and evolving technology. These districts offer multiple opportunities for neighbourhood revitalization, quality employment and poverty alleviation. Pursuing these will lessen the tensions between innovative and inclusive growth, which have emerged in many communities and, ultimately, offer a ladder of economic opportunity to low-income people.

Acknowledgments. This paper is related to the dissemination of the EU research project 'MAPS-LED' (*Multidisciplinary Approach to Plan Specialization Strategies for Local Economic Development*), Horizon 2020, Marie Sklodowska-Curie RISE, 2015-2019. The text is an author's personal evolution of his previous paper "What Interpretations for Smart Specialisation Strategies in European Urban Regions? Lessons from the Boston Area". In Bisello A., Vettorato D., Laconte P., Costa S. (Eds. 2018), *Smart and Sustainable Planning for Cities and Regions,* Springer.

References

1. Ark, B.V., O'Mahony, M., Timmer, M.: The productivity gap between Europe and the U.S.: trends and causes. J. Econ. Perspect. **22**(1), 25–44 (2008)
2. Barca, F.: An agenda for a reformed cohesion policy. A 'place-based' approach to meeting European Union challenges and expectations (2009). http://www.europarl.europa.eu/meetdocs/2009_2014/documents/regi/dv/barca_report_/barca_report_en.pdf. Accessed 4 Mar 2018
3. McCann, P., Ortega-Argilés, R.: Smart specialisation regional growth and applications to european union cohesion policy. Reg. Stud. **49**(8), 1291–1302 (2015)
4. Foray, D., David, P.A., Hall, B.H.: Smart specialisation. From academic idea to political instrument, the surprising career of a concept and the difficulties involved in its implementation. Ecole Polytechnique Féderale Lausanne: working paper 170252 (2011). http://pdfs.semanticscholar.org/29ad/6773ef30f362d7d3937c483003d974bc91c5.pdf. Accessed 4 Mar 2018
5. Falk, M.: What drives business R&D intensity across OECD countries? Appl. Econ. **38**(5), 533–547 (2006)
6. O'Mahony, M., Vecchi, M.: Quantifying the impact of ICT capital on output growth: a heterogeneous dynamic panel approach. Economica **72**(288), 615–633 (2005)
7. Camagni, R., Capello, R.: Regional innovation patterns and the EU regional policy reform: toward smart innovation policies. Growth Change **44**(2), 355–389 (2013)
8. Capello, R., Cappellin, R., Ferlaino, F., Rizzi, P.: Territorial Patterns of Innovation. Franco Angeli, Milano (2012)
9. Capello, R.: Smart specialisation strategy and the new EU cohesion policy reform: introductory remarks. Sci. Reg. **13**(1), 5–14 (2014)
10. Iacobucci, D.: Designing and implementing a smart specialisation strategy at a regional level: some open question. Sci. Reg. **13**(1), 107–126 (2014)
11. Coffano, M., Foray, D.: The centrality of entrepreneurial discovery in building and implementing a smart specialisation strategy. Sci. Reg. **13**(1), 33–50 (2014)
12. Hausmann, R., Rodrik, D.: Economic development as self-discovery. J. Dev. Econ. **72**(2), 603–633 (2003)
13. Porter, M.E.: The competitive advantage of notions. Harvard Bus. Rev. **68**(2), 73–93 (1990)
14. Porter, M.E.: Competitive advantage, agglomeration economies, and regional policy. Int. Reg. Sci. Rev. **19**(1–2), 85–90 (1996)
15. Marshall, A.: Principles of Economics. Macmillan, London (1920)
16. Porter, M.E.: On Competition. Harvard Business School Press, Boston (1998)
17. Martin, R., Sunley, P.: Deconstructing clusters: chaotic concept or policy panacea? J. Econ. Geogr. **3**(1), 5–35 (2003)
18. Aranguren, M.J., Wilson, J.R.: What can experience with clusters teach us about fostering regional smart specialisation. Ekonomiaz **83**(2), 126–145 (2013)
19. Lerner, J.: Boulevard of Broken Dreams. Princeton University Press, Princeton (2009)
20. Porter, M.E.: Clusters and Economic Policy: Aligning Public Policy with the New Economics of Competition. Harvard Business School Paper (2007)
21. Chatterji, A., Glaeser, E., Kerr, W.: Cluster of entrepreneurship and innovation. Innov. Policy Econ. **14**(1), 129–166 (2014)
22. Rauch, J.: Productivity gains from geographic concentration of human capital: evidence from the cities. J. Urban Econ. **34**(3), 380–400 (1993)
23. Florida, R.: The Rise of the Creative Class. Basic Books, New York (2002)

24. Glaeser, E., Kerr, W., Ponzetto, G.: Clusters of entrepreneurship. J. Urban Econ. **67**(1), 150–168 (2010)
25. Samila, S., Sorenson, O.: Venture capital, entrepreneurship and economic growth. Rev. Econ. Stat. **93**(1), 338–349 (2011)
26. Glaeser, E., Kerr, W.: Local industrial conditions and entrepreneurship: how much of the spatial distribution can we explain? J. Econ. Manage. Strategy **18**(3), 623–663 (2009)
27. Kahn, C.B., Martin, J.K., Mehta, A.: City of Ideas: Reinventing Boston's Innovation Economy: The Boston Indicators Report 2012. The Boston Foundation, Boston (2012)
28. Katz, B., Wagner, J.: The Rise of Innovation District: A New Geography of Innovation in America. Brookings Institution, Washington (2014)
29. Bevilacqua, L., Pizzimenti, P., Maione, C. (eds.): S3 Cluster Policy & Spatial Planning. Research Report, MAPS-LED project, EU Horizon 2020, Marie Sklodowska-Curie RISE (2017). http://www.cluds-7fp.unirc.it/. Accessed 4 Mar 2018
30. Lorenzen, M.: Social capital and localised learning: proximity and place in technological and institutional dynamics. Urban Stud. **44**(4), 799–817 (2007)
31. Capello, R., Kroll, H.: From theory to practice in smart specialisation strategy: emerging limits and possible future trajectories. Eur. Plan. Stud. **24**(8), 1393–1406 (2016)
32. Becattini, G.: La coscienza dei luoghi. Il territorio come soggetto corale. Donzelli Editore, Roma (2015)
33. Katz, B., Vey, J., Wagner, J.: One year after: observations on the rise of innovation districts. Brookings Institute (2015)

How Knowledge, Innovation and Place Work Together to Design Entrepreneurial Discovery Process: Insights from Maps-Led Project

Carmelina Bevilacqua⬛, Pasquale Pizzimenti⁽⊠⁾⬛, and Virginia Borrello

Mediterranea University of Reggio Calabria, 89100 Reggio Calabria, Italy
pasquale.pizzimenti@unirc.it

Abstract. The paper highlights the key findings of the research activities conducted under the MAPS-LED project, in examining how the place acquires a specific connotation in designing "tailored policy" for innovation and knowledge spill-overs. The contribution of the Project, in the current debate on the innovation policy, concerns the explanation of how territorial strategies can be part of regional innovation strategies for Smart Specialisation Strategies (S3). The MAPS-LED spatial oriented approach to US cluster highlighted the relevance of the urban dimension in concentrating knowledge resources and linking them to economic activities. The research focused on the occurrence of "innovation spaces" in the places characterized by the presence of Cluster, in order to identify specific urban areas (target areas) in which analyzing the interaction of cluster with the urban fabric, to make evident the emerging factor of a new demand of innovation-oriented physical transformation. Through the lens of the general framework of the Project, innovation spaces have been surveyed using on-line questionnaires distributed in different public and private innovation centers. The findings provide interesting insights on the role played by innovation spaces in the urban and the economic environment. Networking activities, services provided and the support of local place-based policies turn out to be key elements in spreading innovation in specific places and handler of the urban transformation demand. The quantitative approach to spatialize innovation joint with the qualitative approach through interviews led to connect Place, Knowledge and Innovation as main categories of output indicators to set the entrepreneurial discovery process as evidence-based and horizontal policy.

Keywords: Innovation spaces · Demand for urban transformation Entrepreneurial discovery process

1 Introduction

After almost 4 years since Smart Specialisation Strategies (S3) was introduced in the new agenda for Cohesion policy, the increasing importance of the cities in following and supporting the wave of innovation grounded on the Knowledge-based economy [1, 2] is contributing to underline the evolution path that S3 has taken, especially with respect cluster policy and cluster organizations. Beyond the concept of economic cluster related

© Springer International Publishing AG, part of Springer Nature 2019
F. Calabrò et al. (Eds.): ISHT 2018, SIST 100, pp. 336–345, 2019.
https://doi.org/10.1007/978-3-319-92099-3_39

to geographic concentration, the political context in which the activation of measures to enhance the creation and the development of cluster is partly connected to a complex governance, in which the City is acquiring the principal role. The MAPS-LED project focuses on these relationships to underline the comprehension of the context in design innovation policy according to regional innovation strategies for Smart Specialisation Strategies (S3). The process of investigation on linkages between space/place with innovation was conducted in seeking cognitive elements converged in the identification of a new concept of the urban dimension in the context of S3 in emphasizing the role of the Entrepreneurial Discovery Process (EDP). Academics and policy makers have been long discussed on the implementation of S3 at the regional level (RIS3), trying to identify the right path to implement a "tailored" innovation policy in driving the society towards a knowledge economy. The crucial element for making S3 implementation tailored to the context relates the activation of the Entrepreneurial Discovery Process (EDP) [3], which is triggered by entrepreneurial knowledge, the main ingredient of a process of smart specialisation [4]. However, EDP activation process, in the fact that embodies the peculiarity of the context (in terms of cultural, social and physical capital) and grasps the horizontal implication of S3, resulted problematic and different among EU regions, due to several risk factors such as the lack of preconditions for innovation, especially in lagging regions [5]. Some authors argue that localization economies characterized by the firms' concentration and their interaction - sharing of infrastructures, labor forces, suppliers and markets - combined with knowledge spill-overs, provide the pre-conditions for innovation [6]. This paper takes the perspective that the spaces in which innovation concentrates at urban level - namely innovation spaces - appear as catalysts for knowledge dynamics. They catch innovation and feed transformation processes towards a knowledge-based society through the convergence of entrepreneurial knowledge. The first section identifies the main characteristics of innovation spaces in the new wave of urban transformation characterized by the widespread of the innovation districts phenomenon in the transition towards the knowledge-based society. The second section focuses on the MAPS-LED project (Horizon 2020) research activities. Particularly, on the results deriving from on-line surveys provided to the users of the Cambridge Innovation Center aiming at analyze how the variety of knowledge can be embedded and driven by these innovation spaces. Findings, deriving from the Cambridge Innovation Center on-line survey, were grouped in three main drivers, — place, knowledge, and innovation - which synthesize the joint actions of public, private and community actors within innovation spaces. The capacity of innovation spaces to embed and drive knowledge in the current knowledge-based urban development phase can help in the comprehension of Entrepreneurial discovery process (EDP) activation elements. The case of the CIC shows how innovation spaces act as a catalyst for knowledge dynamics, stimulating a new wave of knowledge-based urban development. Such mechanisms can help in triggering entrepreneurial discovery process and expand innovation in deprived areas through public-private partnerships.

2 Innovation Spaces as Knowledge Convergence Hotspot in the Knowledge Based Urban Development Era

The shift from the industrial mass production to knowledge-intensive goods and services production generated novel economic development trajectories [7] in moving towards the knowledge-based society. Globalization is not a new but a changing phenomenon: the core drivers of globalization used to be trade in goods and capital flows, today, they are spurred on by rapid technological change, which is increasingly knowledge driven. In many EU and US cities, knowledge dynamics are sparking the interaction among actors involved in the production and use of innovation supported by local economic development strategies, urban policies, and planning tools. At the urban level, such dynamics are feeding a new wave of urban development contributing to the reshaping of cities through a re-interpretation of the relationship between innovation and space. This approach, defined as Knowledge-Based Urban development (KUBD) [7], is characterized by "*a new strategic development approach that involves management of value dynamics, capital systems, urban governance, development and planning*" [7] is commonly associated with the urban environments [8]. This phase is taking the shape of innovation districts: "geographic areas where leading-edge anchor institutions and companies cluster and connect with start-ups, business incubators and accelerators" [9] in physically compact, transit-accessible and mixed-used districts. Innovation Districts supply the growing demand of spaces for research, development and networking activities coming from the local entrepreneurial community, which plays a relevant role in innovation processes oriented both at the production or the use of innovation. In such spaces, the complex and dynamic community of innovators cluster, interact and connect with private and public actors. Since the late 2000s, innovation spaces grew constantly in US and EU cities [10]. The analysis of the literature pointed out how this is a rising topic in the academic and policy-makers debate. In the era of the continuous advancement of Information and Communication Technology, the face-to-face communication remains crucial [11] and innovation spaces appear to be the key places where knowledge can be shared and transferred. These spaces are "*physical manifestations of the current socio-economic and cultural forces*" [10] enabling and supporting innovation, and facilitating the creativity and critical thinking (of the participants) [12]. Wagner and Watch [10] identify eight main typologies of innovation spaces in the US: incubators, accelerators, co-working spaces, start-up spaces, innovation centers, maker spaces, research institutes, innovation civic hall. Here, innovation is boosted by bottom-up processes, stimulated and animated by the local entrepreneurial community which clusters and interacts proactively with the private and public sectors favoring the knowledge convergence. Others [13] emphasized the role played by local communities and the interaction among public and private actors, in boosting innovation in localized spaces of collaborative innovation (LSCI - hacker spaces, maker spaces, living labs, Fab labs and co-working spaces). These spaces involve the community empowering local innovation processes and creating a critical mass in specific places, where knowledge is continuously transferred and exchanged. In Europe, as stated by Foray [3], knowledge is fragmented and dispersed, making difficult the process of entrepreneurial knowledge convergence, which represents the basis of the entrepreneurial discovery process for S3 implementation. The MAPS-LED project preliminary findings highlighted how knowledge dynamics and innovation are included in economic development strategies, urban policies, and planning tools [14], in responding to the growing demand of transformation driven by such dynamics. The next section will briefly describe the rationale of the MAPS-LED project in analyzing the spatially oriented approach to US clusters, highlighting the relevance of the urban dimension in concentrating knowledge resources and linking them to economic activities. Particularly, it presents the results of an on-line survey [15] provided to the users of innovation spaces in the City of Cambridge (Ma, USA).

3 Entrepreneurial Knowledge Convergence in Innovation Spaces: The Cambridge Innovation Center

The MAPS-LED Research Project investigates how S3 can be implemented, with respect to the new agenda of Europe 2020, by incorporating a place-based dimension [16]. One of the preliminary findings of the project pertains to an innovative conception of the "place" as a key element for effective design and implementation of innovation-oriented policies [16]. From a spatial perspective, innovation can be explained by the occurrence of clusters, which provides a conceptual framework to describe and analyze important aspects of modern economies of different territorial dimensions [16]. The Cluster provides a conceptual framework to describe and analyze important aspects of modern economies [17] of different territorial dimensions. Cluster captures "the concept of dynamic location advantages" [18] in which "… local efficiency factors, like geographical and organizational proximity, external economies promoting a sort of industrial atmosphere, are overcome by more dynamic spatial elements like dynamic synergies and collective learning which explain innovation processes at the spatial level [18]. Furthermore, "when specialised and highly innovative small and medium-sized firms cluster in a particular area of the city, (…) an interesting question emerges on whether the innovative activities of these firms is more influenced by dynamic urbanization economies, i.e. by the more traditional advantages stemming from an urban atmosphere, (…) or by milieu economies, i.e. by collective learning of specialized knowledge, by specialization process of local specialised human capital [18].

The spatialization cluster methodology [16] led to consider the cluster, even with a physical configuration, as a proxy of innovation concentration because its occurrence is strictly connected to innovation, specialization, job creation [16]. The spatially oriented approach to US cluster highlighted the relevance of the urban dimension in concentrating knowledge resources and linking them to economic activities [16]. Clusters are featured by a high number of start-ups, which tend to locate in innovation spaces and attract R&D investments in generating innovative outputs [19].

This paper integrates the information deriving from the spatialization of clusters with an on-line survey provided to users of innovation spaces [16]. The main aim of the survey is to comprehend how innovation spaces are coming to the light as emerging factor of the new demand for innovation-oriented physical transformation. For the purposes of this paper, here are presented the results of the survey distributed to the users of the Cambridge Innovation Center (CIC). The CIC is a facility located in Cambridge that offers offices and co-working spaces for entrepreneurs and start-ups, as well as education programs, training and networking opportunity to increase workers' skills and connect innovators, venture capitalists, mentors, and big companies. The selection of the CIC as a case study came out from the spatialization of the Education and Knowledge Creation and the Business Services clusters in Cambridge [16], the presence of cluster-oriented policy initiatives and the urban regeneration initiatives in the area. The on-line survey was developed through the Survey Monkey web-tool [14] and distributed to the users via email. Since the survey was conducted online, and since it can be difficult to predict the number of people who have opened the emails containing the link to the survey, or who have read the posters that were hung on the bulletin boards, it hasn't been possible to calculate the actual response rate of the survey, but only the number of respondents to the survey itself. The respondents (i.e. the users of the innovation center) were divided in two main target categories: members and visitors. The members of the CIC are start-ups, entrepreneurs and companies that pay a membership fee to use the office spaces and the services that the center provides, while visitors are non-members who join the programs and events hosted in the Innovation Space, such as researchers, investors, entrepreneurs, etc. The total users reached by the survey are 53, the majority of them (71.7%) are CIC members.

The survey included 38 closed questions, which the authors have grouped into three main drivers: place, knowledge and innovation. The first is related to spatial factors (localization, proximity, attractiveness). The second is related to the activities and services provided by the CIC (network activities, co-working spaces, advice, financial support). The last provides information on the actors (company typology, business sectors, R&D activities, interaction with other companies) involved in the innovation process. The first set of questions relating to the "place" driver encloses those spatial factors - in terms of localization aspects, proximity and attractiveness - which constitute the reasons pushing companies to locate at the CIC. Here (see Fig. 1), are reported the answers of the members of the CIC about the attractiveness of the Kendal Square area (where the CIC is located) in terms of urban services. The participants rated the

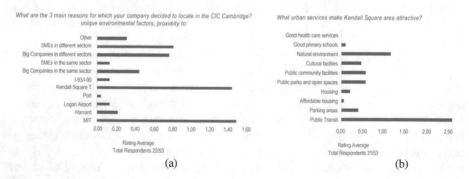

(a)   (b)

Fig. 1. Place driver. Proximity and attractiveness factors in Kendall Square area. Questions: What are the three main reasons for which your company decided to locate in the CIC Cambridge? (a) What urban services make Kendall Square area attractive? Rating options: 1st, 2nd, 3rd factor (b).

presence of public transit services as the most important feature for the attractiveness of Kendall Square, together with the proximity of the Charles River (natural environment) and public facilities (community facilities, parks and open spaces). The second important aspect relates to the proximity of the CIC to anchor institutions such the MIT as well as to public transportation facilities (subway station).

The second group of questions relates to the "knowledge" driver, investigating how innovation spaces act as a catalyst for knowledge transfer and exchange among users of the innovation center. This group of questions refers to activities taking place at the CIC and the typology of advanced and specialized services that the center provides.

Both the members and the visitors of the CIC indicated that the networking opportunity that the center offers, and the possibility to share ideas with other people and have feedbacks are some of the most important features of the CIC (see Fig. 2). In addition to these factors, the members of the CIC indicated that the wide range of options available for working, and the provision of business services and equipment are other important services that the center provides them, while the visitors highlighted that the training programs, workshops and conference that the CIC hosts are relevant services for the community.

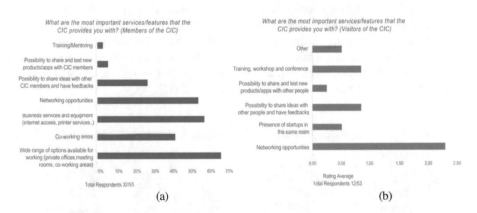

Fig. 2. Knowledge driver. Question: What are the most important services/features that the CIC provides you with? Members (a). Visitors (b).

From the results of the on-line survey (see Fig. 3), knowledge sharing appear to be key elements to feed innovation and competitiveness of companies located in innovation spaces. The members of the CIC highlighted the importance to interact with other companies (40,6%), research centers or innovation space (28,1%) and universities (18,8%) for the generation of new ideas through knowledge sharing and collaboration.

The need for interaction is also highlighted by the way companies intend to have access to the Key Enabling Technologies (KET). The 56,4% of the companies surveyed at the CIC intends to have access to KETs through the empowerment of contacts with universities (25%), public/private research centers and innovation centers (34,4%).

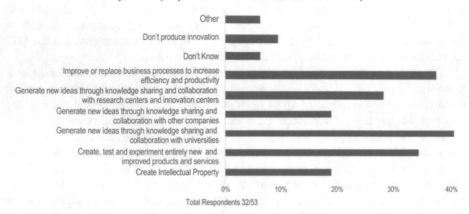

Fig. 3. Knowledge driver. Question: How does your company intend to be innovative and competitive?

The third group of questions relates to the "innovation" driver, investigating how innovation spaces contribute to the use/production of innovation. The innovation driver takes into account the actors operating in innovation spaces. In this case the typology of companies located at the CIC, the variety of business sectors, the research and development activity of companies, as well as the profile of the visitors that join the networking activities and the events held by the CIC (Fig. 4).

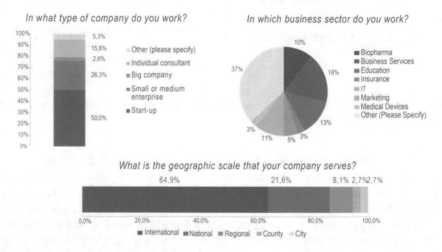

Fig. 4. Innovation driver. Questions: In what type of company do you work? In which business sector do you work? What is the geographic scale that your company serves? (Members of the CIC).

The 50% of the companies located in the CIC are start-ups. This confirms the general trend of the area, considered as the densest start-up area in the country. The start-ups which decided to locate at the CIC increased constantly in the last ten years, with a peak in 2015 of the 29, 7%. The non-members, or visitors, of the innovation space, are mainly people working for a star-up (33%), students/researchers/academics (33%) and individual professional (one-person company, 20%) (Fig. 5).

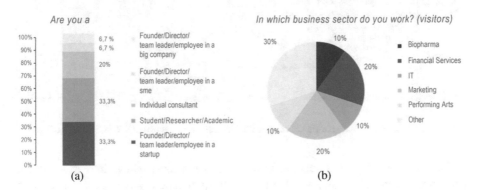

Fig. 5. Innovation driver. Questions: Are you a...(a) - In which business sector do you work? (Visitors of the CIC) (b).

The variety of business at the CIC was rated as one of the most important factors in pushing companies to locate at the CIC. However, only the 31,6% of the companies are focused on Research and Development. It can suggest a propensity to share information and knowledge rather than a specific focus on the production of innovation. The majority of members (93,5%) interact regularly with at least 1–5 companies (54,8%) and networking events are considered among the most important activities in favoring interaction at the CIC (56%) [15].

4   Conclusions

The results of the on-line survey show how innovation spaces are dynamic places where entrepreneurs interact with public, private sectors and communities in generating and spurring innovation. The City of Cambridge is supporting this new phase urban development with innovation-oriented economic development strategies and urban planning tools. Particularly, innovation spaces are a requirement for new development initiatives [14] in the Kendall Square area. In the case of the CIC, the complexity of knowledge dynamics as well the complexity of relationships among actors came out. However, the survey highlighted some rising negative effects. The urban development experienced in Kendall Square is generating gentrification phenomena characterized by housing and office spaces' unaffordability. Respondents have highlighted the following negative aspects: unaffordability of housing, unaffordability of space for businesses, increase of living costs. These elements suggest the need to focus also on "weak" and disadvantaged stakeholders through the development of specific programs.

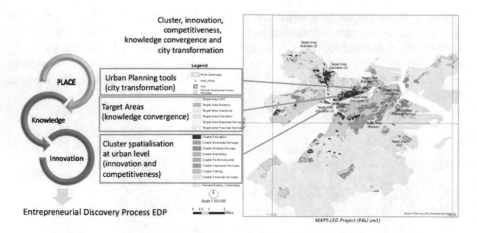

Fig. 6. Main categories of output indicators: Place, Innovation, Knowledge.

Data coming from the surveys have been synthesized taking into account three main key drivers: Place, Knowledge and Innovation. The drivers - place, knowledge, innovation - stem from the overlay mapping of cluster spatialization and zoning [16], which include innovation spaces and innovation initiatives [16]. The quantitative approach to spatialize innovation joint with the qualitative approach through interviews led to connect Place, Knowledge and Innovation as main categories of output indicators to set the EDP as evidence-based and horizontal policy. In those areas where clusters and innovation spaces occur, urban regeneration mechanisms empower the entrepreneurial convergence in specific places, featuring them as emerging factor in the current demand for innovation-oriented physical transformation. However, such processes call for the support of innovative financial instruments and products (from equity to loans) to stimulate the local credit access system and act as leverage for local economic development.

Acknowledgement. The MAPS-LED project has received funding from the European Union's Horizon 2020 research and innovation programme under the Marie Skłodowska-Curie Grant Agreement No. 645651.

References

1. Yigitcanlar, T., Bulu, M.: Urban knowledge and innovation spaces. J. Urban Technol. **23**(1), 1–9 (2016)
2. Szirmai, A., Naudé, W., Goedhuys, M.: Entrepreneurship, Innovation, and Economic Development. Oxford University Press, Oxford (2011)
3. Foray, D.: Opportunities and Challenges for Regional Innovation Policy, 1st edn. Routledge, New York (2015)
4. Foray, D., David, P.A., Hall, B.H.: Smart specialisation from academic idea to political instrument, the surprising career of a concept and the difficulties involved in its implementation. MTEI Working Paper-2011-001 (2011)

5. Capello, R., Kroll, H.: From theory to practice in smart specialisation strategy: emerging limits and possible future trajectories. Eur. Plan. Stud. **24**(8), 1393–1406 (2016)
6. Shearmur, R.: Are cities the font of innovation? A critical review of the literature on cities and innovation. Cities **29**, 9–18 (2012)
7. Yigitcanlar, T., Velibeyoglu, K.: Knowledge-based urban development: the local economic development path of Brisbane, Australia. Local Econ. **23**(3), 195–207 (2008)
8. Inkinen, T., Kaakinen, I.: Economic geography of knowledge-intensive technology clusters: lessons from the Helsinki metropolitan area. J. Urban Technol. **23**(1), 95–114 (2016)
9. Katz, B., Wagner, J.: The rise of innovation districts: a new geography of innovation in America. Metropolitan Policy Program at Brookings. Brookings Institute (2014)
10. Wagner, J., Watch, D.: Innovation Spaces: The New Design of Work. Anne, T., Bass, R.M. Initiative on Innovation and Placemaking at Brookings. Brookings Institute (2017)
11. Hall, P.: Cities of Tomorrow. An Intellectual History of Urban Planning and Design Since 1880, 4th edn. Wiley-Blackwell, Oxford (2014)
12. Bloom, L., Faulkner, R.: Innovation spaces: lessons from the United Nations. Urban Knowledge and Innovation spaces. Third World Q. **37**(8), 1371–1387 (2015)
13. Capdevila I.: Typologies of localized spaces of collaborative innovation. https://ssrn.com/abstract=2414402 or http://dx.doi.org/10.2139/ssrn.2414402. Accessed 3 Sept 2017
14. Bevilacqua, C., Pizzimenti, P.: Urban innovation-oriented policies and knowledge dynamics: insights from Boston and Cambridge, US. In: Talia, M. (ed.) Un Nuovo Ciclo Della Pianificazione Urbanistica Tra Tattica E Strategia-A New Cycle of Urban Planning Between Tactics and Strategies 2016, pp. 11–19. Roma-Milano, Planum (2016)
15. MAPS-LED Project on-line survey built with survey monkey website https://it.surveymonkey.com, developed and distributed to the members and visitors of the CIC in June 2017
16. Bevilacqua, C., Pizzimenti, P., Maione, C.: S3: cluster policy and spatial planning. Knowledge dynamics, spatial dimension and entrepreneurial discovery. Second Scientific report, MAPS-LED Project, Multidisciplinary Multidisciplinary Approach to Plan Smart specialisation strategies for Local Economic Development, Horizon 2020 -Marie Swlodowska Curie Actions-RISE-2014-grant agreement No 645651 (2017)
17. European Commission: The Role of Clusters in Smart Specialisation Strategies. Directorate General for Research and Innovation (2013)
18. Capello, R.: Milan dynamic urbanisation economies vs milieu economies. In: Simmie, J. (ed.) Innovative Cities. Spon Press, New York and London (2001)
19. Bevilacqua, C., Pronestì, G.: Clusters In Designing S3-oriented policies. In: Proceedings of the 13th International Postgraduate Research Conference 2017 (2017)

Land Value Capture by Urban Development Agreements: The Case of Lombardy Region (Italy)

Alessandra Oppio[1($\boxtimes$)], Francesca Torrieri[2], and Marco Bianconi[1,2]

[1] Politecnico di Milano, 20133 Milano, Italy
alessandra.oppio@polimi.it
[2] Federico II University of Napoli, 80138 Napoli, Italy

Abstract. This paper focuses on the interaction between land use planning and infrastructure provision, and offers an overview of the methods that are used in different development contexts to share the costs of local infrastructures and facilities between the public and the private sector. Over the last forty years or so, in the face of intense fiscal pressure and a drastic reduction in the transfer of resources from the central government, local authorities have increasingly been searching for alternative means to fund the provision of off-site infrastructures and facilities.

Starting from the analysis of the Urban Developments Agreements carried out in Lombardy Region over the last 15 years, the paper provides an overview of the surplus value capture mechanism, as it result from land use change and development with the aim of pointing out the issue of its allocation between public and private parties.

Keywords: Land value capture · Urban Development Agreements Lombardy region

1 Land Value Capture Mechanisms

1.1 The Notion of Land Value Capture

Land value capture refers to fiscal instruments through which public authorities can capture increases in properties values that are unrelated to actions of land owners [1]. When value increases depend on public policies, the reason for value capture is related to the costs of public investments, which advantage specific land owners rather than the entire community. Differently, in the case of rezoning decisions, land value increase for private owners is not only due to the special urban development they are going to foster, but it represents a kind of extraordinary gain endowed by public decisions.

In the past, large house-builders and other developers were in charge of providing on-site services (local estate roads, linkup to mains water and drainage, car parking, etc.), while the public sector was responsible for off-site provision - major roads, water supply, sewerage, and a range of other physical infrastructure and community facilities such as schools, hospitals, and green spaces. The expectation was that the State should meet most of those costs and then recoup them by means of general or local taxation.

© Springer International Publishing AG, part of Springer Nature 2019
F. Calabrò et al. (Eds.): ISHT 2018, SIST 100, pp. 346–353, 2019.
https://doi.org/10.1007/978-3-319-92099-3_40

However, over the last forty years, in the face of intense fiscal pressure and a drastic reduction in the transfer of resources from the central government, local authorities and their arm length agencies have increasingly been searching for alternative means to fund the provision of off-site infrastructure and facilities [2, 3]. On the one side, this led to a general preference for the adoption of user charges and hypothecated taxes over general taxation, but it also meant that as part of the development process, developers and/or landowners are increasingly asked to contribute to public goods that previously were provided by the State [4]. The privatisation of utilities and the contracting out of public services added a further dimension to this trend [5–7].

Countries with different administrative and fiscal traditions have different planning tools in place aimed at capturing land value. What almost all planning systems have in common is that they impose a levy on new development in order to fully or partially finance the provision of new local infrastructure, or the upgrading of the existing one [8, 9]. In some cases, standard charges are set, while in other instances contributions are negotiated within the framework of integrated programmes or complex partnership arrangements in addition to, or in lieu of, fixed tariffs. Forms of payment vary and can occur in cash, infrastructure or land. While some of those mechanisms have historically featured as a core component of regulatory regimes, and have therefore a long and checkered history, others have emerged relatively recently.

Starting from the description of the value capture mechanisms (Sect. 1.2), the paper focuses on negotiated exactions within Urban Developments Agreements carried out in Lombardy Region (Italy) over the last 15 years (Sect. 3). Finally, after an ex-post analysis of the incidence of public benefit gained by public administrations through the negotiation with private developers (Sect. 3.1), the issue of the allocation of the surplus land value between public and private parties is discussed (Sects. 3.2 and 4).

1.2 Types of Land Value Capture Mechanisms

Several forms of value capture mechanisms exist: (i) Impact fees; (ii) Joint developments; (iii) Property or land value taxes; (iv) Land banking; (v) Tax increment financing; (vi) Betterment levies; (vii) Development agreements) [1, 10–12]. The most common value capture mechanism is "Impact fees", that typically have to be paid by land owners as a contribution to infrastructures, which directly services their plots [13]. "Joint developments" are the second most common land value capture mechanism. Public authorities and private developers develop land jointly, thus sharing the resulting gains and the potential losses [14]. "Property or land value taxes" can be considered as a value capture mechanism when above all the property price on which they are based reflects market values rather than being updated according to a general index such as a house price index or the GDP deflator [15]. "Land banking" is the practice of making profits by reselling undeveloped or underdeveloped land [16]. It can be used effectively by public authorities, when land-use plans zone primarily those greenfield sites for development that are owned by publicly controlled land banks. "Tax increment financing" [17, 18] is the difference in tax revenues stemming from a potential investment project. This difference is generally used to finance the investment project. "Betterment levies" capture the increase in property values due to land rezoning or infrastructure provision for large areas allowed by public authorities.

Finally, the "Development agreements" or "Negotiated exactions" between local governments and developers are another common value capture mechanism. By this kind of agreements developers provide public services and/or financial contributions for obtaining planning/building permissions or rezoning decisions that allow more profitable development than the one defined by the urban plan. They are characterized by a contractual nature [19] thus allowing public authorities and developers to find specific solutions to specific problems by a flexible mechanism. This shift from 'taxation' to 'regulation' has been described as one of the key factors driving local government reform over the last two decades [4, 20]. The following sections will focus on the exactions negotiatied within Urban Development Agreement in Lombardy region (Italy) with the aim of sheding a light on the fairness of the exchange between public and private parties [21].

2 The Urban Development Agreements

2.1 The Terms of Negotiation

In the context of Urban Development Agreements, two main evaluation perspec-tives are generally considered: (i) the private developer's perspective focused on the viability of urban developments; (ii) the public perspective aimed at capturing a part of the surplus value generated by the new land uses. The first evaluation instance mostly deals with the gain envisaged by the developer's proposal rather than the return expected by implementing the traditional urban plan. The second is related to the amount of the surplus value that should balance the flexibility given to the private parties. On side, developers considers as legitimate the return given by their own proposal and entre-preneurial capabilities. Instead, on the other side, public authorities require a percentage of developers' return being it dependent on cross-synergies and cross-externalities created through investments and decisions both public and private [22].

Generally the two parties seem to have opposing interests at stake: the more one part gets, the less the other party gets and both they want as much as they can get.

3 The Case Study

3.1 The Urban Development Agreements in Lombardy Region

Since 2014 in Italy the minimum incidence of value capture for public authorities has been defined by a national law as a percentage of the capital gain obtained from the urban development (DL. 133 del 12/9/2014). Given this national line of action, each regional government shows a variety of interpretations and operational recommenda-tions defined at local level. In Lombardy Region, for example, the urban planning law prevails on the national one, thus the incidence of 50% is not entered in force and there are many different practices. In most of the cases decisions about the overall surplus value's allocation rule are taken case by case, according to variations of the functional programs defined over time. This high degree of flexibility allows to define specific agreements but at the same time it increases the risk that they are used ineffectively across different development projects.

The difficulty for the public authorities to estimate the adequacy of the value capture with respect to the overall surplus values generated by urban transformations and to sustainable growth objectives when they enter into negotiation with private parties is very common to the Urban Planning Agreements fostered in Lombardy Region.

This kind of negotiated planning has been firstly introduced in Lombardy region since 1986. The Urban Planning Agreements have enhanced urban transformations according to the following main features: (i) institutional cooperation between different government levels; (ii) subsidiarity; (iii) stakeholders' involvement for strategies and actions' definition; (iv) local private public partnership; (v) public investments' efficiency and effectiveness; (vi) functional mix; (vii) achievement of environmental and social objectives.

Among the large number of Urban Development Agreements carried out across the last seventies years, a sample of 15 case studies have been selected. They represent successful experiences since they have been definitively completed (Table 1).

Table 1. The Urban Development Agreements under evaluation: provinces and size.

	Urban Development Agreement	Province	Size (m2)	Volume (m3)
1	Abbiategrasso - *Area dismessa Ex Nestlès*	MI	17.527	40.000
2	Arcene - *Via Leopardi e Immobili Masciardi*	BG	95.100	53.000
3	Brivio - *Recupero area dismessa*	LC	9.955	22.000
4	Cernusco sul Naviglio - *Ex area Arcofalc*	MI	32.035	54.000
5	Cremona - *Ex Feltrinelli*	CR	283.440	124.431
6	Legnano - *Ex Opificio Cantoni*	MI	128.345	180.000
7	Nave - *Comparto Nave Centro P.A. 8/1*	BS	25.436	37.600
8	Rozzano - *Quinto de Stampi - Ex Cartiera*	MI	97.600	93.262
9	Rozzano - *Valleambrosia*	MI	86.204	78.762
10	Rozzano - *Rozzano vecchia*	MI	88.539	53.200
11	Segrate - *Cascina Ovi*	MI	44.290	57.222
12	Segrate - *Causa Pia*	MI	138.878	85.815
13	Temù - *Comparto 19, Lotto 71*	BG	11.493	11.176
14	Vimercate - *Via Mazzini n. 34*	MI	2.090	8.158
15	Vimodrone - *Comparto Nord/Ovest*	MI	225.403	329.080

By an ex-post analysis of the incidence of land value capture for public authorities across these 15 case studies [23], it has been possible to point out that the difference between the extra-contribution negotiated within the agreement and the obligations the private developer must pay to the public authorities according to the regional law and local regulations (L.R. 12/2005 art. 44) is generally positive, but extremely variable (See Graph 1). More in deep, the incidence of land value capture is higher when private functions are predominant, and this can be explained by the greater potential expected profit for the private developer. Nevertheless, a more accurate analysis of the agreements has revealed that very often the amount of land value capture is not proportional to the expected profits for the private, so that the negotiation seems to be done on a case by case approach.

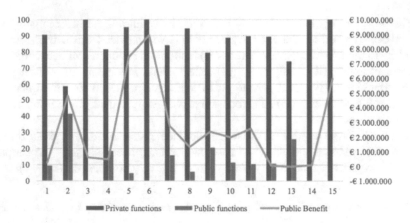

Graph. 1. Correlation between functional mix and Public Benefit.

3.2 Land Surplus Value: Measures, Power in Negotiations and Rules of Allocation

Given the flexible and contractual nature of Urban Development Agreements, the ex-post analysis has pointed out that the amount of public benefit is defined by a case by case approach with a very limited control on its fairness and appropriateness by public authorities. Since changes of use by planning permissions affect land value, part of this uplift will flow to landowners and to developers with respect to an appropriate level of return for compensating the investment risk.

Within this context, decision support systems are needed to evaluate the surplus value generated by the development, and hence to support local authorities to define how much they could ask to developers in the form of land, infrastructure components to be provided or commuted sums. The current surplus value models are based on the assumption that the value of a development project (or site) equates the monetary residual or surplus available once a site has been developed. More precisely, it is defined as a percentage of the difference between the land final value resulting from the urban development, transformation costs and expenses included, and its initial value [22, 24–28].

One of the most crucial as well as controversial issue is the land value that is used in the surplus evaluation models as threshold for estimating the overall value produced by urban developments. Many approaches suggest the use of market value of land, based on price signals from previous land transactions, rather than opting for different solutions such as existing value plus a given incentive. However, some critics claimed that such decision implicitly supports the transfer of increases in land value associated with planning permission to the landowner [29]. Such models embed generous assumptions related to the inducement needed to encourage landowners to release land for development. The critique moved to the market value approach is that such value is itself dependent on regulatory conditions. In other words, proposals to increase the benefit to be transferred to the local authority may appear unviable if the measure of surplus value is based upon transaction prices recorded in the past according to less stringent planning or environmental regulations. In the language of Ricardian rent theory, the market price of land does not determine the extent of planning charges, but rather it is the extent of planning charges that determines the market price of land.

4 Conclusions and Future Perspectives

This paper has critically reviewed the operation and the underlying assumptions related to the use of the land value capture mechanism applied in one regional context in Italy.

In the past much of the literature on development impacts and mitigation makes relatively simple assumptions about projects [30]. In reality, any technique used to estimate impact and decide upon mitigation is rooted in decisions belonging to the realm of public policy. In other words, the needs of the citizens triggering public expenditure, including the provision of new infrastructure, are needs generated not by only one, but rather by a complex mix of causes, and hence any decision on who will have to pay for it will necessary be grounded on technical evidence as much as in value judgements [31]. No process of refinement of the scientific evidence base has been used to assess the impact of development that can alter this basic fact. It follows that any consideration on the effectiveness and equity of a given exaction system is explicitly or implicitly grounded on a comparative assessment of the available alternatives. These encompass systems aimed at taxing capital gain (e.g. local property taxes or stamp duties on property sales), user pays systems, and transfers of central/regional government resources. In addition, at the opposite ends of the spectrum, there are always the options to either halt development altogether, or alternatively, to allow the development go ahead without any built-in mitigation measure.

More generally, changing circumstances force tax regimes to constantly adopt and adjust. If there would be a way to know the 'correct' value of a parcel of development land in a given context, the problem of deciding which instrument to use and which level of taxation to apply would vanish. While in the case of agricultural land, the 'right' price for a plot of land can be assessed in relation to its production value, in the case of urban land there are a number of intertwined aspects to be considered, the most important of which are definitely the building rights and obligations stemming from regulation. In Italy, the last thirty years were characterised by a growing flexibility allowed in managing plans; this occurred through the use of 'revisions' and 'modifications' procedures, and a more systematic use of negotiated development. It follows that, in order to be effective, land value capture practices have to cope with change.

References

1. OECD, The Governance of Land Use in OECD Countries Policy Analysis and Recommendations (2017)
2. Nelson, A.C.: Paying for Prosperity: Impact Fees and Job Growth, A Discussion Paper prepared for The Brookings Institution Center on Urban and Metropolitan Policy (2003)
3. Urban Land Institute: Infrastructure 2010: Investment Imperative. Ernst & Young, Washington (2010)
4. Altshuler, A., Gomez-Ibanez, J.: L'economia politica e le esalazioni sull'uso del suolo, in F. Curti Urbanistica e fiscalita locale, Maggioli Editore (1999)
5. Goss, S.: Making Local Governance Work: Networks, Relationships, and The Management of Change. Palgrave, London (2001)
6. Stoker, G.: Transforming Local Governance. Palgrave MacMillan, London (2003)
7. Karrer, F.: Il management dei servizi urbani tra piano e progetto. Officina, Roma (2008)
8. Hagman, D.G., Misczynski, D.J.: Windfalls for Wipe Outs: Land Value Capture and Compensation. American Planning Association, Washington (1977)
9. Reimer, M., Getimis, P., Blotevogel, H.: Spatial Planning Systems and Practices in Europe: A Comparative Perspective on Continuity and Changes. Routledge, London (2014)
10. Malme, J.H., Youngman, J.M.: An International Survey of Taxes on Land and Buildings. Lincoln Institute of Land Policy, Washington (1994)
11. Malme, J.H., Youngman, J.M. (eds.): The Development of Property Taxation in Economies in Transition: Case Studies from Central and Eastern Europe. World Bank Publications, Washington (2001)
12. Malme, J.H., Youngman, J.M.: The property tax in a new environment: lessons from international tax reform efforts, andrew young school international studies program public finance. In: The Challenges of Tax Reform in a Global Economy, Stone Mountain, Georgia (2004). http://www.issuelab.org/resource/property_tax_in_a_new_environment_lessons_fro m_international_tax_reform_efforts. Accessed 10 May 2016
13. Rosenberg, R.H.: Changing culture of American land use regulation: paying for growth with impact fees. SMUL Rev. **59**(177), 202–203 (2006)
14. OECD. OECD Principles for Public Governance of Public-Private Partnerships, Paris (2012). www.oecd.org/gov/budgeting/oecd-principles-for-public-governance-of-public-pri vate-partnerships.htm. Accessed 10 Mar 2017
15. Fensham, P., Gleeson, B.: Capturing value for urban management: a new agenda for betterment. Urban Policy Res. **21**(1), 93–112 (2003)
16. Van Dijk, T., Kopeva, D.: Land banking and central Europe: future relevance, current initiatives, western European past experience. Land Use Policy **23**(3), 286–301 (2006)
17. Levinson, D.M., Istrate, E.: Access for value: financing transportation through land value capture, Brookings Institute Metropolitan Infrastructure Initiative Series, Washington D.C. (2011)
18. Medda, F., Modelewska, M.: Land Value Capture as a funding source for urban investment: the Warsaw Metro System, Ernst & Young Better Government Programme Poland (2011)
19. Wegner, J.W.: Moving toward the bargaining table: contract zoning, development agreements, and the theoretical foundations of government land use deals. NCL Rev. **65**, 957 (1986)
20. Morphet, J.: Modern Local Government. SAGE Publications Ltd, London (2007)
21. Curti, F.: Lo scambio leale. Officina, Roma (2006)
22. Camagni, R.: Il finanziamento della città pubblica ». In: Baioni, M. (a cura di), La costruzione della città pubblica, pp. 39–56. Alinea, Firenze (2008)

23. Oppio, A., Torrieri, F.: Public and Private Benefits in Urban Development Agreements, Green Energy and Technology. Springer (In press)
24. Micelli, E.: Perequazione urbanistica. Pubblico e privato per la trasformazione della città. Marsilio, Venezia (2004)
25. Micelli, E.: La gestione dei piani urbanistici. Perequazione, accordi, incentivi. Marsilio, Venezia (2011)
26. Alterman, R.: Land use regulations and property values: the 'windfalls capture' idea revisited. In: Brooks, N., Donanghy, K., Knapp, G., Oxford Handbook on Urban Economics and Planning, Oxford: Oxford U.P., pp. 755–786 (2012)
27. Morano, P., Manganelli, B.: La stima delle aree fabbricabili a fini imu. Proposte per il superamento delle inadeguatezze nelle procedure messe a punto dai Comuni, Valori e Valutazioni **12**, 119–138 (2014)
28. Morano, P., Tajani, F.: The break-even analysis applied to urban renewal investments: a model to evaluate the share of social housing financially sustainable for private investors. Habitat Int. **59**, 10–20 (2017)
29. Crosby, N., Wyatt, P.: Financial viability appraisals for site-specific planning decisions in England. Environment and Planning C: Government and Policy (2016)
30. Healey, P., Ennis, F., Purdue, M.: Negotiating Development: Rationales and Practice for Development Obligations and Planning Gain. Routledge Spon Press, London (1995)
31. Bailey, S.J.: Local Government Economics: Principles and Practice. Palgrave Macmillan, London (1999)

The Transformative Power of Social Innovation for New Development Models

Martina Massari$^{(\boxtimes)}$ (iD)

University of Bologna, 40126 Bologna, Italy
martina.massari4@unibo.it

Abstract. The aim of the paper is to reflect on the link between social innovation and urban development. Starting from the recognition of social innovations as collective energies and intelligences, increasingly shaping urban
systems, the paper states that the growing dimension of these experiences
indicates the emergence of a phenomenon that deserves to be investigated, in
order to understand its innovation in organizational systems, design capabilities
and growth ambitions. Building on the assumption that social innovation
practices can become lever of transformation of the traditional city planning
approaches and practices, the paper highlights the crucial role of intermediate
place in fostering social innovation, replacing and integrating complex planning,
triggering processes of mutual institutional learning, and challenging public
authorities in rethinking their intervention in more adaptable forms. The paper
seeks to investigate, through the analysis of intermediate places of innovation,
how to promote the implementation of new urban models, which deepen the link
between social practices and urban and territorial development. The aim is to
explore the role of the specific intermediate places and their capacity to become
nexus for innovation in urban policies, planning tools and the territory, through
the analysis of two models of intermediate place, Urban Living Lab and
Community HUBs, which are useful bridges, able to answer to the transition
from social innovation to transformative innovation.

Keywords: Social innovation · Intermediate places · Service co-creation

1 Social Innovation and the Potential of Places

1.1 Introduction

Cities around the world are currently facing complex, varied and persistent challenges.
Well-known phenomena, such as urban saturation, climate change, structural crisis and
political turmoil, followed by a reduction of public resources, are juxtaposed with the
emerging pro-active role of different non-governmental actors [1] in the co-production
of public goods. This phenomenon is commonly defined as social innovation, a concept
with fuzzy boundaries, describing a combination of bottom-linked actions, by which
people find answers to social needs [2], that are not afforded by the market and not
anymore by the state, while building social cohesion and empowerment. Social
innovation is therefore an increasingly diverse form of private action with
public-interest purposes operating for the co-production of urban services and public

© Springer International Publishing AG, part of Springer Nature 2019
F. Calabrò et al. (Eds.): ISHT 2018, SIST 100, pp. 354–361, 2019.
https://doi.org/10.1007/978-3-319-92099-3_41

goods. According to Deleuze [3], social innovation is meant as a path-dependent process happening in places of opportunity in which local actors engage, facing various socio-economic problems within the urban space.

With these premises the paper seeks to draw a framework of assessment of some experiences of opportunity places which popularity stems from their ability to become nexus for social innovation in urban policies and to incubate new public policies. These 'intermediate places', such as urban agencies, Urban Living Labs, Innovation and Community HUBs, are proving to be successful because they act as interactive play-grounds in which the relationships between different actors generates practices that can be managed together with visions and strategies for a long-term cultural change perspective.

The contribution will assess whether the approaches operated by intermediate places can feed institutional approaches and orient it towards new development models more likely to meet socioeconomic, cultural and environmental needs of local communities.

1.2 Social Innovations as Emerging Socio-Spatial Elements

The multitude of episodes transforming the city is described with the 'buzzword' social innovation. Social innovation defines an ecosystem of practices of civic origin, exploring and experimenting "new ideas and solutions to meet social needs and at the same time creating new relationships of collaboration" [4]. They are depicted as new manifestations of participatory practices, able to achieve the power of a public policy [5], where tangible actions replace complex planning strategies. Social innovation is not a new concept, nevertheless, the growing number of initiatives is recently taking on such dimensions as to lead to the assumption that they are not only intervening in the resolution of their needs, but are laying the foundations for a new paradigm shift.

Social innovation has grown as an umbrella concept, including everything that emerges as an extra-planning activity in the city, bringing out the urgency to define a shared understanding, in order to subsequently clearly facilitate its promotion and up-scale. Among the different research approaches on social innovation, it is possible to distinguish some common features: the practical and engaging dimension, the civic focus, the collective action and the capacity to enrol in complex processes with a creative approach [6].

For the purposes of this article, social innovation is linked to urban and territorial regeneration [2, 7, 8] under the framework proposed by Frank Moulaert who reads it primarily as a research and development methodology and a path-dependent process, to tackle spatial and social challenges of the cities, proposing new contingent ideas. Social innovation becomes a social and territorial issue whose construction is strongly place-related and connected to the political and organisation forms [9]. The issues highlighted by Moulaert and other scholars provided the theoretical framework useful for reflecting on the effectiveness and power of such endogenous resources in fostering and activating innovation in urban planning routines. This approach is clear in the new 2014-2020 European programming and local policy agendas, in which the concept of competition is less evident than the capacity of the places to find internal resources for change. As Manzini states, the incorporation of social innovation in urban systems is becoming crucial in order to accept and move toward the direction of change, for mutual benefit [10].

Starting from this framework it is useful to reflect upon the skills, competencies, knowledge and learning needed, to nurture episodic innovation and bridging it towards an interaction with the urban policy-making process. The hypothesis is that specific innovation places are acting as intermediate levels providing the necessary skills and competencies, fostering episodic and micro innovation and working in a mutual exchange relationship with public institutions.

2 Places as Space-Society Interface

2.1 Intermediate Places as Innovation Bridges

The growing popularity of intermediate places as urban agencies, Urban Living Labs, Innovation and Community HUBs, stems from their ability to become nexus for social innovation in urban policies and in planning tools. From the Deleuzean-inspired model, emerges the interpretation of innovation as happening in places of opportunity [11] at the micro level, facilitating creative strategies and contaminations on a larger scale. The creative and mobilizing potential, which resides in the capital [12] of some specific places and contexts, hosting important innovation energies, could be a crucial factor in triggering long-term dynamics of social innovation for effective urban transformation. In many neighbourhoods several experiments are appearing, in the form of maintenance of public facilities, cooperation between businesses, management of the public green. This is much more than mere subsidiarity logic: local engagement, entrepreneurial skills, and the use of local knowledge, are new values that are being created thanks to the occasion for different actors to interact. As Crosta [13] stated in fact, the attention to urban user-oriented transformations is clearly characterised by the focus on their interaction: they are considered effective when generated as positive externalities of interaction between subjects [14]. With these premises it is possible to explore the characters of specific places, in fostering interactive connections between actors, resulting in co-created urban services that lead to the collective empowerment and ultimately an innovative urban governance model.

Urban Living Labs, Community Hubs, Policy Labs, City Service Hubs, Public Innovation Places, as open innovation spaces and context-driven environments, are some of the useful models able to integrate different levels of government and practices. These hybrid environments can become bridges between urban institutional mechanisms and micro-scale actions, aiming to be drivers of social and spatial innovation. They use circular, incremental and adaptive processes, shifting from projects to processes [15] where social innovation becomes one of the drivers of change.

The paper briefly delves into two models: Urban Living Lab [16] and Community HUBs [17] acting with different bindings and relationships with the public authorities and belonging to different institutional nature. The aim is to reflect upon the two diverse methodologies and their ability to produce new knowledge, skills, public goods and added value, according to the different local context.

2.2   Urban Living Labs and Community HUBs: How the Places Become Laboratories

Urban Living Labs. Living Labs, open innovation environments that put the end user at the centre of the process of co-production of goods and services [16], have long since extended their field of application from areas of experimentation of new technologies, to the co-creation of spaces, services, policies and urban processes.

After being adopted by the European Union, the methodology has recently evolved up to a third generation [18], oriented towards the transformation into platform of collaborative innovation, able to gather different stakeholders in a network of experimentations. This model, called Urban Living Lab (ULL), aims to offer an open and collaborative environment that considers the inhabitants as agents in the processes of transformation of the city and enables the exchange and co-creation of shared value in the city. A distinctive element of the ULL is the role of protagonist of the end user involved in systematic actions of co-creation, experimentation and evaluation. ULLs are both methodologies and physical places, where different energies of the territory meet, consolidated skills aggregate, and local knowledge is combined.

In Italy, ULLs often assume the characteristics of collaborative places and environments, organized around the contribution and mutual exchange of different urban actors. A recent example is Turin Living Lab-Campidoglio [19, 20] an initiative of the City of Turin by the former Executive Councillor for the Environment and the Chief Officer for Innovation and Economic Development, to promote, develop and test innovative solutions by local enterprises and associations, in a real-life context. An open call for projects and ideas declared the beginning of the Lab in 2016, with a following selection of 32 proposals to be implemented and tested in Campidoglio's neighbourhood. The project gave the opportunity to experiment with the community of the area and to test innovative forms of partnership with the ultimate aim of supporting research and innovation for public-interests purposes. Co-production of services and crowdsourcing approaches where used to define an inventory of needs, imagine new urban services and ultimately implement new processes and products in the neighbourhood. Through the process, the Lab was able to define a new policy context to support innovation by creating institutional arrangements able to allow the development and testing of innovative urban solutions by private actors and research bodies, in permanent exchange with citizens.

The public administration of Turin, through the Lab, tried to intercept innovation as an opportunity to answer to social challenges, presenting the possibility of negotiating objectives, strategies and policies, moving towards a logic of co-creation, yielding power while governing the process, discussing development goals and withdrawing old procedures to innovate policies.

Community HUBs. In the panorama of intermediate places, Community HUBs are emerging as hybrid places significantly influencing urban policies' implementation. Among the aims of the Community HUBs is the need to allow the most diverse actors to create the city, generating opportunities to collectively build urban policies. Community HUBs [17] are recently been defined as urban *ecotones*, places where different but close systems interact, where the place itself act as an intermediary between actors, public institutions, authorities and business. Their use of new channels of collaboration with administrations, as co-creation and hybrids methodologies, make them promising devices to intercept bottom-up and top-down levels [17]. The role they can play directly shaping spatial planning processes by providing implementation tools and practices and connecting communities, would help generate new governance models, shared knowledge, social and relational resources.

Among the first to be called Community HUB, the case of Kilowatt, hosted by the Serre of Giardini Margherita in Bologna, is an example of virtuous construction of urban identity, based on shared values and public interest objectives. Kilowatt defines itself as an environmental, social and cultural value-ideas accelerator, with a hybrid governance [22] that allows it to distribute the produced value, with communities and institutions, keeping the balance between economic sustainability and public services offer. The idea of Kilowatt is to answer to the evolution of job market and to the retraction of public welfare, offering a place where working spaces, urban local services and events, merge with collaboration and neighbourhood animation processes. In this intermediate place, several actors (professionals, business and associations) work together with the local community with the ambition to build a local proximity environment. The experience is structured as an enabler of social innovation with local community as a client, with the advantage to rely on a wider range of distributed resources among human and professional qualities of the community. Nevertheless, the added value of the operation derives from a 15 years contract with the local authorities (owners of the area) that enabled Kilowatt to promote long-term strategies and programming, allowing to externalise the value, while keeping the balance in economic sustainability.

3   Conclusions

The paper presents two model of intermediate places that could represent useful research and development resources for public administrations, and environments where new urban services and policies are tested, evaluated and improved before their implementation into a normative path. *The following* Table 1 *gives a framework of comparison between the elements highlighted in both cases.*

Both cases are developing an accountability path: the ULL increasing the accessibility to urban services and enabling the users by offering abilities, competencies to act in an autonomous way in the care of the urban goods; Kilowatt by co-producing public innovation from the bottom-up, starting from the local *milieu* [12], to connect initially separated resources with a strong attention to social inclusion. In the social innovation domain, they where able to give a territorial perspective, by creating new connections and bonds: Kilowatt, with a focus on the relationship people-local community; ULL, by enabling local vocations in a circular subsidiary perspective.

The two experiences differ mainly in the governance and the promoting actors. The case of Turin represents a top-down process promoted by the municipality in the attempt to modify an institutional path in an inclusive way [24]. The risk of this operation is to generate a reduction of responsibilities of the public in the matter of public services production, especially in some strategic sectors. On the other side, the

Table 1. Table of comparison Urban Living Lab and Community HUBs.

Project	Turin Living Lab	Kilowatt
Mission	Open Innovation to co-produce more inclusive services	Proximity to promote inclusion and social cohesion
Governance	Municipality promoter; profit and non-profit organisations	Multi-level partnership: direct management by cooperative
Target	Community members; companies	Community members, innovators, start-ups
Output	Improvement of quality and access to local services; public innovation	New urban participated welfare, innovative management models
Social Innovation	• Synergies with community; • Replication potential; • Modification of institutional paths	• Mutual exchange and recognition; • Value distribution; • New hybrid partnerships
Risks	Reduction public responsibility, lack of skills in civil servants	Elitism, exclusion of the actors that can't participate

management of Kilowatt, if handled only from the bottom-up, risks to exclude some specific actors, triggering elitism and generating cultural barrier for the access to the process. It must be avoided the risk that these places are experienced as exclusives, instead of capable of extending their effects to the whole community [23]. Therefore one aspect to be highlighted is the central role of the institutions, that should be able to look at social innovation practices from the outside, recognise the value, trigger some leverages and re-define themselves from the inside, getting involved in a mutual learning perspective. The public actor is also responsible for the transparency of the processes and the inclusion of all the possible beneficiaries of the city.

Furthermore, it is clear that in both cases, in order to establish integrated and adaptive processes, a flexible and dynamic involvement of different actors is needed, based more on open innovation, avoiding the exploitation of debate and public communication, to the detriment of the substance of the issues [21]. It is essential to avoid the 'sterilization' of the publics: a single target represents a problem for the success, the contamination and the reproduction of innovative urban productions. The impact of a social innovation on urban context is much more decisive and effective, if included in a community mobilization process useful to trigger a multiplication of energies.

The intermediate places have the ambition and the potential to represent a lever of development for public policies, but the aspect still to be negotiated concerns long-term sustainability that requires a preliminary planning activity and a continuous and effective monitoring and evaluation system. In the case of Turin, the survival of the process is linked to a political decision and will, and the risks are connected also to the lack of skills and capabilities of civil servants to manage complex and multi-level processes. What appears urgent is the need for developing new competencies for professionals and public servants, necessary to overcome the idea that social innovative actions themselves are sufficient to create value and to contribute to develop new urban innovation devices. In the case of Community-HUBs, the issue of economic sustainability is a central point. These places are often strongly reliant on short-term public funding that prevents high-value activities from being planned. It is therefore necessary to generate processes with a 'mixed supply chain' [17]: inside the intermediate places the institution should be forced to change and to recognise the value of these 'activators', while the activists must review their mechanisms in order to grow.

To conclude, in an urban scenario where the public actor is too big to engage in an effective manner with episodes of social innovation and too small to address substantial issues, intermediate places represent urban engines of production for the city. These places are able to intertwine multiple complex variables in a laboratory logic: handling different ingredients and intercepting multiple resources, building long networks and defining wide partnership. In order to bridge from social innovation to transformative innovation is necessary to start building a development vision, from the existing network of intermediate places from which to anchor new adaptive strategies and ultimately generating added value.

References

1. Rauws, W.: Civic initiatives in urban development: self-governance versus self-organisation in planning practice. Town Plann. Rev. **87**(3), 339–361 (2016)
2. Moulaert, F., Vicari Haddock, S.: Rigenerare la città. Pratiche di innovazione sociale nelle città europee. Collana Il Mulino/Ricerca, Bologna (2009)
3. Deleuze, G., Guattari, F.: A Thousand Plateaus: Capitalism and Schizophrenia (trans. B. Massumi). Athlone Press, London (1987)
4. Murray, R., Caulier-Grice, J., Mulgan, G.: The open book of social innovation. London, The Young Foundation and NESTA (2010). http://www.nesta.org.uk/publications/assets/features/the_open_book_of_social_innovation. Accessed 05 Sept 2017
5. Fareri, P.: Rallentare. Il disegno delle politiche urbane (edited by M. Giraudi). FrancoAngeli, Milano (2009)
6. TEPSIE. Social Innovation Theory and Research: A Guide for Researchers. Deliverable of: "The Theoretical, Empirical and Policy Foundations for Building Social Innovation in Europe" (TEPSIE), European Commission– 7th Framework Programme, 46 (2014)
7. Crosta, P.L.: Società e territorio, al plurale. Lo "spazio pubblico" – quale bene pubblico – come esito eventuale dell'interazione sociale. Foedus **1**, 40–53 (2000)
8. Evers, A., Ewert, B., Brandsen, T. (eds.): Social Innovations for social cohesion. Transnational Patterns and Approaches From 20 European Cities (2014). http://www.wilcoproject.eu/downloads/WILCO-project-eReader.pdf. Accessed 05 Sept 2017
9. Drewe, P., Klein, J.L., Hulsbergen, E.: The Challenge of Social Innovation in Urban Revitalisation. Techne Press, Amsterdam (2008)
10. Manzini, E.: Design, When Everybody Designs. An Introduction to Design for Social Innovation. MIT press, Boston (2015)
11. Hillier, J.: Straddling the post-structuralist abyss: between transcendence and immanence. Plann. Theory **4**, 271–299 (2005)
12. Camagni, R., Dotti, N.: Il sistema urbano. In: La crisi italiana nel mondo globale, Economia e società del nord (edited by Paolo Perulli e Angelo Pichierri), Einaudi, Torino (2010)
13. Crosta, P.L.: La politica del piano. Franco Angeli, Milano (1990)
14. Dvir, R.: Knowledge city, seen as a collage of human knowledge moments. In: Carillo, F.K. (eds.) Knowledge Cities, Approaches, Experience, Perspectives, pp. 245–272 (2006)

15. Ostanel, E.: Spazi fuori dal comune. Rigenerare, includere, innovare. FrancoAngeli, Milano (2017)
16. Concilio, G., De Bonis, L.: Smart Cities and planning in a Living Lab perspective. In: Campagna, M., De Montis, A., Isola, F., Lai, S., Pira, C., Zoppi, C. (eds.) Planning Support Tools: Policy Analysis, Implementation, and Evaluation Proceedings of the VII International Conference on Informatics and Urban and Regional Planning, INPUT FrancoAngeli, Milan (2012)
17. Calvaresi, C., Pederiva, I.: Community hub: rigenerazione urbana e innovazione sociale. In: Bidussa, D., Polizzi, E. (eds.) Agenda Milano. Ricerche e pratiche per una città inclusiva, Fondazione Feltrinelli, Milano (2016)
18. Leminen, S., Rajahonka, M., Westerlund, M.: Towards third-generation living lab networks in cities 7(11), 21–36 (2017)
19. Nesti, G.: Co-production for innovation: the urban living lab experience. Policy Soc. **4035**, 1–16 (2017)
20. Torinolivinglab. http://torinolivinglab.it/. Accessed 05 Sept 2017
21. De Leonardis, O.: Una questione d'inclusività. Urban@it Background Papers (2015). https://www.urbanit.it/wp-content/uploads/2015/09/BP_A_DeLeonardis.pdf. Accessed 05 Sept 2017
22. Venturi, P., Zandonai, F.: Imprese ibride. Modelli di innovazione sociale per rigenerare valore. Egea, Milano (2016)
23. Carta, M.: Reimagining Urbanism, Città creative, intelligenti ed ecologiche per i tempi che cambiano. ListLab, Trento-Barcellona (2013)
24. Martinelli, F.: Social innovation or social exclusion? Innovating social services. In: Franz, H.W., Hochgerner, J., Howaldt, J., Editorsthe, (eds.) Challenge Social Innovation. Potential for Business, Social Entrepreneurship, Welfare and Civil Society, Springer, Berlin (2012)

The Supportive City

Bianca Petrella(✉)

University of Campania Luigi Vanvitelli, 81100 Caserta, Italy
bianca.petrella@unicampania.it

Abstract. This paper does not describe the results of a specific research but presents a reflection on the possible urban scenarios based on sustainable development in all its possible aspects. In terms of city and territory as complex, open and dynamic systems, various aspects are tackled: from the need for participatory processes, to issues of urban security, to the coexistence of diversity, to illegal construction and so on.

Despite globalization and the tendency towards social and urban homologation, local contexts still present significant differences; they derive from the different historical stratifications, from the current social and economic conditions, from the local administrative capacity and efficiency. The skills, effectiveness and efficiency of government and governance are also different, not only in different countries but also within the same country. For the above reasons, the reflection, although of a general nature, makes particular reference to the conditions of urban systems in southern Italy, united by: urban degradation and widespread building, illegal construction, organized crime, recent immigration, deficient control of the territory, and so on.

Keywords: Participation · Security · Interculturalism · Public areas
Urban abusiveness

1 The Need to Participate in Choices

1.1 Government and Governance

The Agenda 21 [1] and the Habitat [2] Agendas affirm that active participation and governance are key areas to the future of the human settlements. The Italian Municipalities have activated several 21 Local Agendas, but nevertheless the democratic participatory processes are lagging behind [3]. The participation in the decisions should mark the transition from the form of government to that of governance. This means moving away from a "coherent form of authority, legitimate and exclusive place of power" [4] to a model of formulation and management of public policies; latter is characterized by coordinating and negotiating attitudes that sees the involvement of a plurality of cooperating subjects on possible alternatives.

The transition from government to governance has occurred in the last decades of the last century, when, on the one hand, the complexity and the segmentation of the social organization [5] has increased and, at the same time, the vertical and hierarchical decision-making model has entered into crisis.

© Springer International Publishing AG, part of Springer Nature 2019
F. Calabrò et al. (Eds.): ISHT 2018, SIST 100, pp. 362–371, 2019.
https://doi.org/10.1007/978-3-319-92099-3_42

The increase in electoral abstention is among the indicators of the breaking of the trust pact between public institutions and citizens [6]. In Italy, abstention takes a quite different form from most other European countries, in the sense that, even if the number of voters decreases slightly, the number of blank or spoilt ballot papers increases considerably. In the last parliamentary elections of 2013, abstention reached 28%; but if null votes are added, it is close to 32% [7]. Italian specificity derives from the strong moral sense of democratic participation in the vote, which, inculcated in society in the transition from monarchy to republican form, still remains actual.

The crisis of the traditional social pact also manifests itself with the increase in the search for greater autonomy of local powers (e.g., at different scales: the British Brexit, the Catalan question, the requests of the Northern and Central Italian regions) that would seem to configure the dissolution of the nation-state towards a scenario of many interacting polis: city-state. Although still far from this model of territory, the urban planner must be able to measure himself/herself against cities that can not longer be described as compact and articulated systems for predefined functions. The lack of ability to govern the (mostly qualitative) demand for transformation of complex urban systems has led to entities that are informal, disjointed, fragmented and have no certain boundaries [8]. In this scenario, the planner is no longer the one who elaborates a plan able to pursue the objectives decided by the political authority. The planner must use his technical skills to translate the alternatives into concrete projects. The different options derive from different public and private entities, which contribute to the construction of the shared design choice.

The scientific literature has produced numerous models of democratic participatory processes; they range from the Advocacy Planning of the Sixties [9] to the subsequent Metaplan [10], to the most recent Open Space Technology (OST) [11] and to the more typically Italian District Workshops and Public Debate [12]. In Italian participatory processes, the main difficulty to overcome is the significant diffidence towards the public institution and the bureaucracy that characterizes it; a further difficulty is the skepticism towards participatory methods, which initially, appear too simple to lead to actual results. It is precisely on these two aspects that we must initially work to build trust in the participatory process and this is why it has to make use of specialized figures [13].

The Public Debate [14], which has the weakness of having to intervene on a predefined project, is the most common practice in Italy but very often it is reduced to a mere formal fulfillment that it is difficult to produce future projects and visions, arising from comparison and the fusion of the ideas of the different actors. In Southern Italy, in particular, the professional figures of the facilitator and the mediator of conflicts are lacking; these skills must belong to all the technicians involved to ensure active listening and therefore the success and quality of the result.

2 Risks, Dangers and Security

2.1 Open Space Security

Designers are obviously required at all stages of the participation process in order to ensure the essential technical requirements of the result. They must make people understand why even some choices, apparently incomprehensible and initially not shared, are necessary to ensure environmental sustainability, comfort and safety of public, semi-public and private spaces [15–17].

The Charter of Megaride 94 states: "Town planning strategies will have to make the city safer, overcome physical inflexibility, and guarantee access and exit. To reach the city, move through it, leave it, and in this way share the city, are right that everyone must be able to enjoy" [18]. Safety conditions must be applied to the various levels of urban planning; depending on the scale of intervention, the appropriate actions must be implemented for the mitigation of danger and damage, caused by natural and technological risks [19] but also by street crime, that exploits poor planning, inefficient management and lack of maintenance of urban spaces. The latter, which are typical of many settlements, in the south of Italy, produce multiple negative effects as they limit the social life of the neighborhood, cause disaffection towards the common space, induce antisocial behavior that further increase the degradation, instead of opposing it and fight it [20]. Italy is very late on the problems of urban security [21].

Urban planning (at the various territorial scales in which it operates) can play a role of support to security at different levels of intervention. It should be noted that the rules of urban plans determine conditions of greater security (for example: distances between buildings, distribution system routes, protected road lanes, etc.) but can also increase the perception of security, creating places immediately understandable by those who he attends them [22].

The norm ENV 14383: 2003 "Prevention of crime - Urban planning and building design" indicates the necessity to implement the urban planning strategies, especially at the scale of the urban and building project [23].

2.2 Security and Diversity

ENV, however, neglects the aspect of safety that derives from the "sense of belonging" to a specific place and that immediately refers to the coexistence of cultural diversity in the same urban space [24]. A variety of cultures and ethnic groups characterize the current urban communities. Immigrants must share the same urban areas, formed in other times and for different needs from the present; this cohabitation, due to the original habits and mutual prejudices, risks creating conflicts between the indigenous group and the immigrant group and also among immigrant groups of different origins. As for all types of risk, also in this case, it is necessary to identify the source of danger; it must be sought, not in the immigrant community, but in the wrong urban organization that does not guarantee the quality of the coexistence of ethnic diversity. The organization of the common spaces of the cities is often also deficient for the other diversities that populate the cities. As Mumford [25] observed, as early as the 1940s, cities are designed for adult life and do not take into account the needs of children, as well as differences in gender, income, motor skills, etc. whose accessibility to all urban opportunities is not fully guaranteed.

The urban form is given by the relationships between building volumes, the related facades and streets, the squares on which they stand, but also the courtyards and hallways, visible from the outside, participate in the overall configuration, together with the most minute furnishing elements. All human beings are attracted or repelled by the physical, aesthetic and emotional experience of open urban spaces; if you feel comfortable in a public space, behaviors will spontaneously tend not only to protect it but also to work together to increase its beauty and usability. Whereas for autochthonous the sense of belonging is partly generated by the familiarity of spaces and architectures that express the social history of one's own culture [26], for the immigrant (especially if extra-European) this does not occur easily. An immigrant comes from other uses and customs, among which we must understand, in addition to the different form of urban spaces of origin, also the way in which they are enjoyed.

The relationship between immigrant and security must be investigated and resolved in at least two ways: from the point of view of the immigrant and from the point of view of the natives. For this last direction it can not be ignored that in Italy (induced mainly by media and by some political forces), the percentage of citizens who consider the immigrant a danger to security is increasing and therefore it is important that public and semi-public spaces are structured to favor the meeting and the possibility of exchange between different cultures. To this end, immigrants must be an integral part of the participatory processes activated for urban transformation decisions and, assisted by cultural mediators; they must be enabled to express their needs and particular sensibilities. The continuity of cultural exchanges generates interculturality that is an innovative social identity in which every diversity participating in the dialogue is enriched by the contribution of the other. Everyone, preserving its basic cultural heritage, builds, together with all the others, a new common culture on which to base the sharing of the same territory and cohabitation in the same urban place [27].

Urban syncretism is the fruit of both predisposed and conscious actions and spontaneous and informal actions. It is above all the result of time that allowed to sediment and make its own (recognizing itself in it) the innovative urban reality [28]. This was the case for the cities that we inherited from the past, so it will also be for those of tomorrow to which, even if desired, it can not be prevented from changing as a result of citizens, already foreigners, who live in them.

From sociological and psychological studies that have explored social conflict, especially from the Social identity theory [29], elements useful for the design of public urban spaces can be obtained (those spaces that inevitably have to be experienced together) to the design of private building volumes and to design the filter spaces between the public and the private part [30].

Social marginality, internal to economic relations, determines physical marginality. Both are realized with the expulsion from the center (social and urban), confining them, the hindering elements or, in any case, considered useless to the dominant interests; the realization of concretely intercultural and inter-class societies and cities is now an indispensable condition, unless one wants to abdicate and make conflict and insecurity endemic.

2.3   The Shared City

The public place's projects should encourage the meeting of all citizens: they must be areas for which both maintenance and management can be easily carried out and with minimal costs (including the active participation of the inhabitants). Public areas must not be rigidly determined but must be easily transformed and adaptable to the continuous changes in the demand produced by social change.

In support of the attention to be paid to the design of common areas and equipment for the community, it should be noted that ethnic and cultural diversity is an asset that can be evaluated economically [31]. They contributes significantly to the production of social capital: an urban territory with accessible associative conditions and the possibility of frequent meetings, it will certainly prove to be a fertile environment to raise and develop shared values and to contribute to the structuring of a new cohesive community. In fact, social capital does not only generate economic wealth but also social wealth and well-being, a condition which, in turn, is able to fuel economic capital and "civic welfare" [32].

Precisely because in the period of weak welfare the funds allocated to equipment and services for the community are increasingly scarce, it is necessary a careful estimation of the social and economic cost of a missed or inadequate intervention of realization and valorisation of equipment for the community and of urban quality as a whole.

Just as in the health sector, prevention is applied (also) to reduce the higher costs of subsequent treatments, in the same way it is necessary to intervene in advance, and to realize urban spaces and services (also) so as not to incur more burdensome expenses in the future due to social malaise and the consequences that it determines. An adequate supply of services to the marginal social categories helps to break down the conflict between groups that would no longer be forced to compete for access to scarce urban resources [33]. Since social capital is a productive factor, like physical and human capital, it is good that it comes into play in the cost-benefit analysis. In other words, we need to evaluate the benefits produced by the creation of new social networks and the strengthening of those already present that the creation of a new public space feeds. Conversely, we need to estimate the corrosion of social capital that the failure to create a suitable public space would determine, with damage to the local development of the economy and of society as a whole [34]. Durlauf and Fafchams [35] have summarized the three key points of social capital: it takes shape from informal organizational structures, based on social networks and associations; generates (positive) externalities for the (society); externalities are the product of a sharing of rules, values and mutual trust.

3   Urban Abusiveness

Interventions of urban regeneration can not ignore the presence of so-called informal cities. The forms and quotas of illegal construction manifest themselves in different ways and intensities in relation to local specificities, including among them the administrative, financial, managerial capacity and the effectiveness of territorial control by local authorities. The weakest points in Italy are the inefficient control of the urban spaces, the non-application of the laws and the specific laws are often confusing [36].

When the welfare state does not give an effective answer, alternative solutions are looked for independently. They range from caravans, to tents, to the occupation of proper or improper lodgings, to abusive constructions that can also become real agglomerations and are so widespread that they have specific denominations in each country (favelas, bidonville, barriadas, etc.) although the internationally most used word is slums. These are the sites where those who have not found a place in the official city have taken refuge. This happened because both the policy and the urban plans did not offer adequate housing for everyone.

Although in Italy the policies for social housing have been almost exhausted in the eighties of the last century [37], unlike other parts of the world, there are no large informal settlements, except for the "institutional, nomad and refugees camps" formed through the self-construction of real shacks.

The illegal building is present, in significant quantities. It is estimated that in Italy there are about 5 million abusive buildings and half of these are in the southern regions [38, 39]. The kinds of building violation range from buildings created without permits (but in compliance with planning regulations) to works not compliant with the building permit. It manifests itself both in a concentrated way (residential subdivisions) and in a scattered way, with single buildings, single-family or condominiums. Whatever the location and the size, the illegal building is characterized by a high urban degradation, marked by the lack of urban services, by the underdeveloped road and by the precariousness of network utilities (sewers, electricity, etc.). The urban degradation corresponds to the social degradation in that in these places a low-income population is concentrated, being also scarcely educated and in which the culture of illegality is rooted. Much of the urban abusiveness has been partly determined by the scarce amount of public housing and has been realized due to the lack of territorial control by local institutions. It is also for these reasons that, starting from 1985, three legislative provisions of building amnesty were issued, which were followed by regulatory measures aimed at redeveloping these types of settlements. The regenerative interventions of these places must act both on the specific part but above all, they must make it an active element of the urban system. To this end, and always with the active participation of the inhabitants, urban integration can contemplate (in addition to the necessary connections through infrastructures, transport services and articulation of green areas) the realization of symbolic elements aimed at providing that place with its own identity and recognisability, overcoming the anonymity and the original degradation.

The legitimate self-construction can replace the unauthorized self-construction, triggering the process of active participation of users in the management of the design, construction and management phases of the works. It can be practiced (in the associated and assisted form) both for the new building, and for the regeneration and restructuring of existing urban parts [40] and therefore also for the redevelopment of illegal settlements. Proceeding with this methodology, in addition to reducing costs, will lead to the construction of an urban space that meets the specific needs and, feeling the house and its surroundings as own, all users will help to keep the place in the best conditions and, therefore, to make it vital and secure.

What above expressed has to be extended to the other forms of housing precariousness in which many of the neighborhoods of social housing must be fully included. This is because the incompleteness, together with lack of control and public management, has led to a situation of degradation. Even if realized by the public institution and in a single solution, the conditions in which they are reduced lead to assimilate them to real slums for the conditions of decay, the high rates of crowding, the lack of basic services, etc. A further type of housing precariousness is that which occurs in particularly degraded areas of some historic centers. In these places, the "new" inhabitants are concentrated occupying unused residential buildings or settling down in fields facing the street, under the stairs, in attics and basements that is in rooms lacking the minimum requisites of habitability, where the immigrants have taken the place of the previous marginalized natives.

4 Conclusions

The urban regeneration is multidimensional and requires a multilevel governance acting simultaneously on the social, economic, cultural, urbanistic and building components. Beguinot [41] summarizes that we must act, at the same time, on the city of relations, on the city of stones and on the city of perception.

Social policies act primarily on the former while the shared urban planning must prepare the spatial component of the territory. The aim is that all citizens, in the kaleidoscope of all the cultural diversities, should be enabled to work and to have a social life in a smart and comfortable safe city.

With a view to an innovative governance, the cycle of urban policymaking [42] must obligatorily contemplate the full set of actions. It should be addressed as a whole approach, preparing the types of actions envisaged in the specific sectors: facilities, public spaces, social measures, etc. It must be very carefully with timetable and feasibility in an organic vision that prevents the realization by parts, or that prevents the execution of incomplete interventions. A minimalist solution is definitely preferable instead of a single portion of an ambitious project that risks not being completed, causing more damage than benefits. By further specifying the concept, it is not advisable to iterate many of the mistakes of the past. The Italian public housing were built with poor public utilities, and in some cases, public works were built and never been put into operation, because the activation costs have not been effectively assessed.

Accessibility is obviously a determining factor regarding the urban regeneration. If accessibility to resources is an integral part of human rights, accessibility to urban resources must be an integral part of citizens' rights. In order to satisfy that condition, and also help the integration and inclusion, the action must be performed both by the urban plan and by public policies. The former acts on spatial and temporal accessibility with a correct localization and proportioning of facilities and transport networks, while the latter must act with support for transport services and collective services and goods in general (economic and temporal accessibility).

Surely, the scale of the implementation plan, or even of the urban project, is more effective and incisive than the plan on a municipal and/or large area scale. It is in the implementation phase that the individual building works and the urban furniture are realized to create welcoming and safe spaces suitable for the functional and social diversity, by which the *terrain vague* are brought to new life [43]. Nevertheless, all the implementation plans must obviously be included in the general plan to ensure the juxtaposition and overall balance of the urban system.

It is necessary to conclude these reflections by recalling the Charter of Megaride 94 [18], which, promoted by Corrado Beguinot, was drafted with the participation of about six hundred urban operators from twenty-seven different countries. In ten simple principles, synthetically expressed, are set out the rules that every sustainable city must apply to with regard to the relationship: with nature, with intercultural, with social equity, with mobility and transport, with complexity, with technology, with the rehabilitation, reuse and revitalization of the already built, with risks and dangers of all kinds, with ancient and future beauty and, therefore, with time, understood as the ability to possess the memory of urban culture and translate it for future needs. In addition to the Megaride Charter, many others are national and international documents that indicate the necessary and correct ways of operating for the city and the territory; it would be enough just to remember the several Habitat Agenda [2] and Agenda 21 [1]. This means that neither the theoretical reflection nor the technical instrumentation of planning are lacking. However, it must not be believed that there are valid solutions for every context and social reality. It is the city itself that, based on its history, its past and, above all, the feeling of its citizens, old and new, must build its own path of progress devoted to dynamism and flexibility, showing itself capable of controlling all forms of discrimination or social marginalization.

Precisely in relation to such a "path", urban planning can also play an important as well, affecting strategic conditions for the life of the local population: this in accordance with the pattern of an open, supportive and secure city that Administration - with a view to good governance - must pursue.

References

1. UNEP, Agenda 21, United Nations Conference on Environment & Development, Rio de Janerio, Brazil, 3–14 June 1992
2. UN-Habitat, Habitat Agenda, United Nations Conference on Human Settlements, Istanbul, Turkey, 3–14 June 1996
3. Cabras, M.: Rigenerazione urbana: strumenti, politiche e possibilità per una nuova idea di città. Anci Lombardia, Milano, Italia (2017). http://www.risorsecomuni.it/. Accessed 01 Dec 2017
4. Bagnasco, A.: Tracce di comunità. Il Mulino, Bologna (1999)
5. Jessop, R.: Governance, fallimenti della governance e meta governante. In: Cavazzani, A., Gaudio, G., Sivini, S. (eds.) Politiche, governance e innovazione per le aree rurali, pp. 189–209. Edizioni Scientifiche Italiane, Napoli (2006)
6. Cuturi, V., Sampugnaro, R., Tomaselli, V.: L'elettore instabile-voto-non voto. Franco Angeli Edizioni, Milano (2000)
7. ISTAT, Elezioni e attività politica e sociale, 9. In: Annuario Statistico Italiano. Istat, Roma, Italia (2014)
8. Borja, J., Castells, M.: La città globale. DeAgostini, Milano (2002)

9. Davidoff, P.: Advocacy and pluralism in planning. J. Am. Inst. Planners **31**(4), 331–338 (1965)
10. Schnelle, W.E.: A Discursive Approach to Organizational and Strategy Consulting. BoD, Norderstedt, Deutschland (2008)
11. Owen, H.: Open space technology. Guida all'uso. Ferrari Editori, Rossano (2008)
12. Dioguardi, G.: Ripensare la città. Donzelli editore, Roma (2001)
13. Bobbio, L.: A più voci. Amministrazioni pubbliche, imprese, associazioni e cittadini nei processi decisionali inclusivi. Edizioni Scientifiche Italiane, Napoli (2004)
14. Pizzanelli, G.: La partecipazione dei privati alle decisioni pubbliche: politiche ambientali e realizzazione delle grandi opere infrastruttura. Giuffrè, Milano (2010)
15. Jeffery, C.R.: Crime Prevention through Environmental Design. Sage Publications, Thousand Oaks (1977)
16. Moeckli, D.: Exclusion From Public Spaces. A Comparative Constitutional Analysis. Cambridge University Press, Thousand Oaks (2016)
17. Palazzo, D.: NOOS - Not Only One Solution, An Urban Design Process. Mondadori Università, Milano (2008)
18. AA.VV.: Carta di Megaride 94. Città della pace, città della scienza. Giannini Editore, Napoli, Italia (1995)
19. Kendrick, T.: Identifying and Managing Project Risk. PMP, Amacom, San Carlos (2015)
20. Kelling, G.L., Wilson, J.Q.: Broken Windows: The police and neighborhood safety. Atlantic Monthly **3**, 29–38 (1982)
21. D'Onofrio, R., Trusiani, E.: Città, salute e benessere: Nuovi percorsi per l'urbanistica. Franco Angeli Edizioni, Milano (2017)
22. Ippolito, A.M. (ed.): La percezione degli spazi urbani aperti. Analisi e proposte. Franco Angeli Edizioni, Milano (2016)
23. AA. VV.: Pianificazione disegno urbano gestione degli spazi per la sicurezza. Manuale, European Commission Directorate General Justice, Freedom and Security (2008)
24. Kilbride, K.M. (ed.): Immigrant Integration: Research Implications for Future Policy. Canadian Scholars' Press, Toronto (2014)
25. Mumford, L.: La pianificazione per le diverse fasi della vita. Urbanistica **1**(4), 7–11 (1945)
26. Hauser, A.: Storia sociale dell'arte. Einaudi, Torino (1964)
27. Remotti, F.: L'ossessione identitaria. Laterza, Bari (2010)
28. Petrella, B., de Biase, C., De Salvo, V.: Syncretic design as solution for a new urban identity. In: Santos Cruz, S., Brandão Alves, F., Pinho, P. (eds.) Generative Places Smart Approchaes Happy People, pp. 601–625. Universidade do Porto, Porto (2015)
29. Tajfel, H., Turner, J.C.: An integrative theory of Intergroup Conflict. In: Austin, W.G., Worchel, S.: The Social Psychology of Intergroup Relations. Brooks/Cole Pub. Co., Monterey, California (1979)
30. Roccari, R.: Sicurezza urbana: analisi del legame tra ambiente costruito e criminalità. Exeo edizioni, Padova (2011)
31. Putnam, R.D.: La tradizione civica delle regioni italiane. Mondadori, Milano (1993)
32. Bruni L., Zamagni, S.: Economia civile. Efficienza, equità, felicità pubblica. Il Mulino, Bologna (2004)
33. Olzak, S.: The Dynamics of Ethnic Competition and Conflict. Stanford University Press, Stanford (1992)
34. Mutti, A.: Capitale sociale e sviluppo. La fiducia come risorsa. Il Mulino, Bologna (1998)
35. Stanford. https://web.stanford.edu/~fafchamp/soccaphandbook.pdf. Accessed 01 Dec 2017
36. Petrella, B., de Biase, C.: A typical Italian phenomenon: the unauthorized building. In: Proceedings of the Second International Conference on Advances in Civil and Structural Engineering - CSE the IRED, New York (2014)

37. Petrella, B.: L'edilizia residenziale negli ultimi quarant'anni. Due città emblematiche: Milano e Napoli. Fondazione Ivo Vanzi, Napoli (1989)
38. Legambiente: Dossier-L'Italia frana, il Parlamento condona, Edizioni Ambiente, Milano, Italia (2014)
39. ISTAT: La situazione del Paese. In: Rapporto annuale 2015. Istat, Roma, Italia (2015)
40. Bertoni, M., Cantini, A.: Autocostruzione associata e assistita. Progettazione e processo edilizio di un modello di housing sociale. Editrice Dedalo, Roma (2008)
41. Beguinot, C. (ed.): La città cablata un'enciclopedia. Giannini editore, Napoli (1989)
42. Bason, C.: Design for Policy. Routledge, London (2014)
43. de Solà Morales, I.: Terrain vague. Quaderns **212**, 38–39 (1996)

Quality Evaluation Method for Pediatric Hospital Buildings to Support the Territory

Francesca Primicerio[✉] and Giacomo Di Ruocco

University of Salerno, Fisciano 84084, Italy
fprimicerio@unisa.it

Abstract. Since ancient times the hospital has always been helping shape urban individuality, mainly affecting the territory development. In the best cases hospital has designed an intelligent system of synergies and integrations with other territorial and urban functions. The work focuses on the analysis in particular of pediatric hospital buildings. The objective of the research is to ensure higher quality and efficiency for these hospital buildings, in order to obstruct the ongoing obsolescence processes and guarantee an essential service to the territory. In the hospital design sector, the quality of the space is necessary not only to ensure more exceptional comfort for the users but has beneficial effects on the children healing process. Therefore the research proposes by using a multi-criteria methodology a method for assessing the quality of pediatric hospital structures at three different levels of investigation (Urban System, Building System, and Internal Space System). The originality of the work consists of recognizing the quality as a function of three specific criteria/objectives: *Humanization, Sustainability* and *Flexibility*. This new integrated multidisciplinary approach aims at representing a useful model to support systematically the appropriate design choices.

Keywords: Urban growth models · Pediatric hospital building
Quality evaluation · Methods and logical-operational tools · Humanization
Sustainability · Flexibility

1 Introduction

The analysis of some buildings in the Italian hospital context and above all in the pediatric sector brings out the poor quality of these structures. According to the Ministry of Health, in Italy there are 13 pediatric and/or maternal-child hospitals. The public health buildings are characterized by major deficiencies of functional nature and by the destruction of buildings. This category includes mainly the hospitals of Southern Italy, among the research objects, we can mention the "Santobono - Pausilipon" of Naples, the "Di Cristina" of Palermo and the "Giovanni XXIII" of Bari. The sites hosting these hospitals are inadequate if compared to modern construction technologies linked to the planimetric and structural flexibility of the healthcare architecture. They are buildings made of reinforced concrete that date back to the 70 s and in many cases have more or less pronounced deficits concerning seismic protection. This requirement, together with the total elimination of architectural barriers and the achievement of

© Springer International Publishing AG, part of Springer Nature 2019
F. Calabrò et al. (Eds.): ISHT 2018, SIST 100, pp. 372–383, 2019.
https://doi.org/10.1007/978-3-319-92099-3_43

microclimatic conditions adequate to the places of care, are the minimum characteristics essential for today's hospital. In most cases it is precisely these requirements listed not to be present in the structures of our country. Moreover, there is a strong deficiency also with regard to the system of plants, as in the case of the Santobono of Naples, where there is still the presence of traditional thermal power stations. These structures are typologically and technologically inadequate if compared to the current needs of users, regarding functionality, comfort, safety, performance. Today the buildings must relate to the people, to the environment and to the social context they belong to, which requires specific dimensional, functional and technological standards [1].

Several factors influence the dynamics of project development of hospital complex: discoveries in the medical industry, the new concept of "illness", the diffusion of new social and existential expectations, the transformations of the building process [2]. The development of the research takes into account the critical relationship between man-environment-building, in order to satisfy the needs of comfort and safety (thermo-hygrometric, acoustic, visual, correct choice of materials, the definition of internal and external spaces, integration of the hospital structure with the urban and territorial context). The design quality is one of the key elements for the success of hospital structure, and for the proper provision of the health service. That can lead to beneficial effects on the child's healing process [3].

The research focuses, in particular, on the attention of pediatric construction, as children represent particularly sensitive users. Some aspects of pediatric hospital can be assimilated to those of general hospital, with the exception of those dimensional, functional communicative characteristics that characterize the spaces for children.

Aiming at obstructing the obsolescence processes of these structures and giving back to the territory an essential service, we propose a method for assessing the quality of pediatric hospital structures by investigating three different scales. The work is still under completion. This research aims at supporting designers in design choices for improvement. The proposed methodology identifies three evaluation criteria (Humanization, Sustainability and Flexibility). For each of them the categories and indicators that compose the evaluation matrices are identified, as well as the final instrument.

2 Pediatric Hospital Buildings

2.1 The Relationship with the Context

Over the centuries, the concept of a place of care has changed in parallel with social, economic and cultural changes and with the evolution of medicine. Today, the hospital is an important urban element, with a collective character, which affects the entire territory [4]. The evolutionary cycle of the concept of hospital can be summarized in the following aspects:

Religious aspect in antiquity. In ancient times there were no real buildings for the care of the sick, these were housed in the temples and the care had a purely divine expression.

Charitable aspect in the Christian era. It was established that in every city the monasteries had to have a hospice for the poor, wayfarers and sick. The typological characteristics of medieval hospital buildings conform to the type of religious places and are easily identifiable in the *Cistercian Abbey of Fontain*, in the Yorkshire region (1132) and the *Abbey of San Gallo* in Switzerland (XIII century). The places for the cure were configured as places apart, outside the city, far from everyday life and from healthy people who could not come in contact with the sick [5].

Social aspect in the Renaissance period. It was understood that the hospital could no longer be considered a generic place but had to have its own functional, spatial and organizational characteristics that differentiated it from any other type of building. The fifteenth century sees the birth of grandiose nosocomial structures from the *Cà Grande* of Milan to the *Incurables* of Naples, to the *San Francesco* of Padua, to the *Ospedale Maggiore* of Cremona, etc. Numerous health care facilities for child care arose in Europe, alongside numerous initiatives to protect and promote childcare.

The *Ospedale degli Innocenti* (*Florence 1419–1427*) by Filippo Brunelleschi, one of the first structures for the health of children in Italy, is an emblematic model that has distinguished itself from the beginning for its particular social purpose. In its formal simplicity, it constitutes a symbol for the city and a representative work of the power of a generous government towards people [6]. Rigor and order are also found in the plant, within which it is possible to identify fundamental characteristics of Renaissance hospitals such as the courtyard and the porticos. The courtyard becomes a large square which with a simple language communicates with the people adapting itself to the evolutionary rhythms of the city [7].

Sanitary aspect in the modern era. The problems related to the excessive exploitation of hospital facilities, their position (often in densely populated historical centers), the lack of services, now could no longer be ignored. The *Lariboisiere Hospital in Paris* (designed in 1839 by M. P. Gauthier) constitutes the appearance of the modern type of pavilion hospital. The new philosophy, which inspired hospital buildings, was based on the utmost hygiene through ventilation, lighting, the separation of the buildings, the removal from the inhabited centers. As a rule, the hospitals in the pavilions were on one floor but soon the defects were evident (more land use and consequent higher costs were increased by the number of floors.) In the late nineteenth century the hospital structure in the halls began to enter in crisis, thus contrasting to the new solution of the "monoblock" hospital with vertical development. In the United States, among the first examples realized according to this new model we find the *Columbia Presbyterian Medical Center* in New York (1928) and the New York *Cornell Medical Center* (1933) located in the city center and contextualized in the skyline.

A hospital can affect and contribute to the redevelopment of a territory in many ways: by combining an inhomogeneous building fabric, constituting itself as a central pole of a surrounding space (in terms of size, morphological characteristics) and opening its interior to city life, etc.

A recent intervention which has been characterized as an element of strong urban redevelopment has been the *New Maternal-Infantile Hospital Gregorio Maranon* (*Madrid, 1996–2003, R. Moneo J. M. De La Mata*). This building is located in a very dense urban area. The monoblock type has made it possible to make the road system more functional and respect the uniformity of the surrounding buildings by redefining the margins of the insertion lot. The formal choices well suit the historical characteristics of the district and fulfill the functional recovery intention of the entire surrounding area.

The interventions of recovery and requalification of services such as hospitals not only make it possible to restore an essential service to the citizens but also a value to the history of the city, by participating in the development of the territory [8].

2.2   New Strategies: Between Innovation and Quality

Today the tendency towards hospital design has been activating different strategies to restore quality hospitals to people and the city.

Starting from a series of reflections on the contemporary relationship between hospital and city, the architect Renzo Piano together with the ex-minister Umberto Veronesi, for the Ministry of Health, has developed a strategy for a model of a high acute hospital with technological and assistance profile [9]. The study materialized with the drafting of guidelines (*D.M. 12 December 2000*) for the design and management of "highly complex technological and medium-sized" hospitals. The ten reforming principles of the ministerial document can be summarized as follows: **Humanization** (man must be placed at the center of health planning); **Urbanity** (the contemporary hospital must be urban and integrated with the surrounding territory and the daily life in which it is located); **Sociality** (the hospital must also be socially and culturally integrated with the environment that surrounds it); **Organization** (in particular by favoring the departmental model that allows an optimization of resources and facilitating collaboration between different healthcare professionals); **Interactivity** (the hospital must work in synergy with the other health facilities); **Appropriateness** (the hospital must be appropriately sized according to the services offered); **Reliability** (the professionalism of the operators and the environmental, hygienic and technological plant reliability are the basis of a quality service and a favorable perception of safety and protection by the patient); **Innovation** (the hospital is a constantly evolving structure and as such it must be flexible and adaptable to new needs); **Research** (scientific research is at the basis of human progress. Appropriate study spaces must be set up within hospitals); **Training** (the hospital must be considered a place of health training and education for all) [2]. Along with this decalogue, the functional layout of a hypothetical ideal hospital model was developed, divided into four blocks: hotel stay, hospital stay, operative block and emergency room.

An example of excellence in which we find many of the aforementioned design principles is the new *Meyer Pediatric Hospital* in Florence (2007). The solution proposed by the CSPE[1] design is based on a sensitive mimetic operation towards the

---

[1] CSPE, Centro Studi Progettazione Edilizia (Building Design Study Centre), was founded in Florence in 1975 by Antonio Andreucci, Paolo Felli and Romano Del Nord. In 1999 Giulio Felli and Corrado Lupatelli joined the practice as associates. Founded thirty years ago in Florence, CSPE is now one of Italy's leading architectural practice specialized in healthcare design and facilities for the public sector. Engaged in a wide spectrum of projects, it fosters innovation and supports academic research at national and international level.

surrounding landscape (through a design which follows the level curves) and towards the historic old health buildings (urbanity).

As indicated in the last Health Plan, Meyer has an important management task: it is a reference point for regional paediatricians through a network system integrated with other health services (interactivity). The realization of this new structure has been experienced by the hospital as to rethink the role and the services offered, as well as to experiment with new and important innovations in the health organization, in the construction and plant engineering technologies. The design took place in a global way, keeping in mind the specificity of the users' health and psychological needs at all stages. In this sense, the hospital-city relationship also changes by rethinking the health structure no longer closed on itself, like a small but open and continually changing city, configuring itself as a center of exchange open to the city from the social point of view (sociality).

The old building (Villa di Ognissanti) was destined for the university, the medical hotel, the administration and the medical clinics (research and training), while the specialized functions were included in the new construction. The high innovation of the Meyer Hospital consists of the relationship established between patient and architecture. The latter contributes to reducing environmental stress and ensure the functionality of the hospital complex, the hospital humanization, internal security, care quality, architectural and technological innovation, which improves the users well-being conditions. This artifact has been destined for hospital support services, while the specialized functions have been included in the new construction. Specifically in the villa were located the university, the health hotel, the administration and medical clinics.

The green invades all the building, being also present on the roofs of buildings. The relationship between the inside and the outside is continuous and uninterrupted, already from the entrance with the large window which allows you to observe the outdoor park (to therapeutic function) [10]. The plants have been organized in an exemplary manner to separate the different routes and traffic flows and facilitate orientation. The first two levels have been half buried. The three overall floors of the structure have been tapered and offset from each other, so as to create juts with large terraces. Despite of the building height, the inter-storey heights have been reduced to a minimum to improve the feeling of integration. The services and the church are located in the basement; on the ground floor the emergency room, the clinics, radiology, acceptance, pharmacy, commercial spaces and refreshment; and finally, on the first floor the operating rooms, the areas for intensive and specialist therapy.

The areas of hospitalization, all for children, have been planned partly on the first floor and on the whole second floor more panoramic (appropriateness). The child patient in this case is placed at the center of the entire design logic. The rooms[2] have been all arranged on one side, according to a segmented trend, and the services on the opposite side, while the typical areas for staff have been arranged along the route with a function of control. This solution avoids the presence of a narrow corridor, which instead turns into a livable space; the large rooms (24 m^2 for single and 30 for double rooms) are divided into two areas, one of which is more enclosed for the meetings with parents and

---

[2] Main reference law: D.P.R. 14/01/97.

doctors. Of particular interest for its implications on the humanization of the hospital is the Operation Block on the first floor, where the child can be accompanied by the parent up to the entrance of the operating room because a naturally covered lighted space was provided in front of each room pre-operative preparation takes place [11]. At the top the playroom is a place where colours dominate, with an iron glass structure for children's play and entertainment, in which the use of plastic shapes and particular lighting systems recovering Jean Piaget's cognitivist theories, which is necessary to promote children intellectual growth through sensorial development (appropriateness - humanizzation). Innovation is evident through the quality of energy saving technologies (innovation and reliability). There are photovoltaic systems and the bioclimatic greenhouse. A further bioclimatic system is made up of a high-insulation green cover (the solar sensors fixed to the roof which constitute a particular type of skylights formed by a polycarbonate shell able to direct the light avoiding an excessive number of rebounds), which contributes to climate comfort, as well as having a therapeutic function; also natural ventilation is used to guarantee an optimal microclimate, reducing the feeling of isolation. The set of all these solutions aimed at the maximum efficiency of the building organism was conceived by a team of specialists in environmental psychology, sociology, ergonomics, restoration, and landscape architecture, which symbolizes the high interdisciplinarity necessary for the design of the Third Millennium pediatric hospital.

3   A Quality Evaluation Tool for Pediatric Hospital Buildings

The study consists of: *Phase of knowledge* (State of the art analysis); *Operational phase* (Proposed evaluation method for the pediatric hospital building); *Application phase* (Case studies identification and application of the tool developed).

3.1   Operational Phase: Method

The proposed tool is based on multicriteria analysis (AHP method - Analytic Hierarchy Process) [12] which represents a large family of techniques able to take into account at the same time a variety of aspects of the problem that is being faced, both qualitative and quantitative. The main purpose of the tool is the rationalization of a decision-making process [13]. The methodological path that led to the definition of the instrument involved a series of steps (which will be explained in detail in the following paragraphs):

- Identification of the quality evaluation criteria;
- Identification of the criteria categories;
- Identification of the criteria indicators;
- Work out the three evaluation matrices (Humanization, Sustainability, Flexibility) and development of the weight allocation methodology;
- Assignment of weights to survey levels and to the categories of each criteria (step in development);
- Developed summary matrix and assignment of weights of the criteria (step in development);

The structuring of an evaluative problem takes place in a hierarchical form and implies that the elements - criteria, categories, and indicators are arranged in an ascending sense according to the level of abstraction: the elements placed higher up in the hierarchy are therefore abstract and general, while the more below concrete and particular.

The proposed evaluation tool is structured on three levels of scale, from the general to the particular, first investigating the external spaces around the building then the entire hospital and then the individual interior spaces. The complexity of the process is also closely linked to the problem of the observation scale. Depending on the scale there are different aspects to consider and with a different weight. At the first level (Urban System) the survey is carried out in relation to the hospital's setting up context (accessibility and public transport, possible interferences with the hospital area, etc.). The second level (Building System) investigates the quality of the open spaces of the hospital and its enclosure (paths, areas for children, signage, technological systems and energy saving). The third level (Interior Space System), in greater detail, investigates the individual internal departments (usability, functionality, materials and colors, etc.).

3.2   The Quality Evaluation Criteria

In this methodological hypothesis, during the first phase of knowledge, project experiences and state of the art were analyzed, which led to the identification of three criteria (*Humanization, Flexibility, and Sustainability*). These criteria are considered fundamental to obtain a quality product, both from a design point of view and as regards the sanitary functioning of the structure itself.

The trends analyzed, the study of examples of virtuous hospitals and inspections, and some reference regulations (*Guidelines of the D.M. 12 December 2000, Project Piano-Veronesi*) have formed the key to interpreting the needs of the pediatric hospital and together contributed to identifying these criteria. The three criteria identified summarize the 10 principles and address the multiple aspects established in the Plan Veronesi.

The identification of the "*Humanization*" criterion stemmed from the precise desire to focus attention on the needs and comfort of children and all hospital users. Therefore this concept blends the concepts of psychophysical well being and environmental psychology with the architectural quality and functionality of the hospital structure. Interior spaces designed for the needs of children, simple, clear and identifiable paths, furnishing elements that reinforce the value of the place of meeting and relational exchange [14]. The colour[3] and sound games, visual contact with the natural environment, represent salient aspects leading to an optimal management of the psycho-physical-social stress of the pediatric condition.

The identification of the "*Flexibility*" criterion arises from the knowledge that hos-pital buildings must be able to adapt to the multiple new needs. Furthermore, the building must survive over time thanks to the possibility of implementing several cycles of use of the building organism, reconfiguring, if necessary, the internal and external

---

[3] The colours act on the nervous system, triggering different effects. For example the warm colors have a calming effect.

structure and/or intervening in a simplified manner on the technological system that governs the space [15]. The aspect of flexibility is, therefore, a fundamental concept [16].

The term *"Sustainability"* embraces an expansive and complex concept. In this context, we intend to analyze only the aspects concerning environmental and social characteristics, to ensure that health structures must be able to be not only sustainable but above all sustainability witnesses.

3.3 Categories and Indicators

The categories and indicators represent the operational translation of the criteria, i.e., a way to express the objectives in such a way that they can be measured. In the summary image (see Fig. 1), shown below, the categories identified for each criterion are summarized.

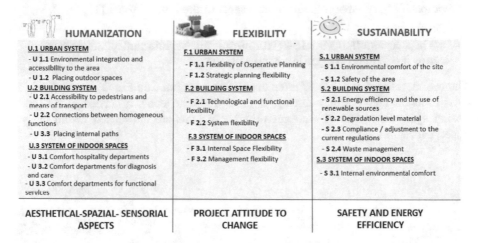

Fig. 1. Summary sheet of the categories.

To better understand what has been done, for reasons of complexity of the work, only the indicators of the Humanization criteria are shown below. This criterion represents, among the three, the most innovative.

- **U1 - URBAN SYSTEM:** U 1.1 - *Environmental insertion and accessibility to the area* (No alteration of the characters of the place, Easy to reach by public and private transport, Appropriate accesses to the area and differentiated by category of user); *U 1.2 Placing outdoor spaces* (Presence of suitable parking lots, Presence of suitable signs, Absence of architectural barriers, Presence of suitable equipped green areas, Presence of complementary external functions).
- **U2 - BUILDING SYSTEM:** U 2.1 - *Accessibility to pedestrians and means of transport* (Access to the building suitable and differentiated by category of user); *U 2.2 - Connections between the homogeneous functions* (Presence of functional-spatial relationships within the building, Recognition of functions within the

building); *U 2.3 - Placing internal paths* (Presence of internal paths suitable and differentiated by category, Absence of Architectural and Sensory Barriers, Presence of suitable signs).

– **U3 - INTERNAL SPACE SYSTEM:** *U 3.1 - Comfort hospitality departments* (Compatibility of inpatient environments with respect to the type of user, Comfort of the inpatient settings, Panoramic characteristics of hospitalization departments, Privacy and safety in hospital wards, Meeting the sanitary and plant health requirements in the wards); *U 3.2 - Comfort diagnosis and care departments* (Meeting the sanitary and plant engineering requirements in the diagnosis and treatment departments, Correct dimensioning of the spaces dedicated to the services of diagnosis and care and compatibility of the environments with respect to the type of user); *U 3.3 - Comfort departments for functional services* (Fulfillment of sanitary and plant engineering requirements in the departments dedicated to functional services, Correct dimensioning of spaces dedicated to functional services and compatibility of environments with respect to the type of user) [17].

3.4 Three Evaluation Matrices (Humanization, Sustainability, Flexibility) and the Weight Allocation Methodology

The tool developed consists of three evaluation matrices (see Fig. 2), respectively for each criterion mentioned, referring to three different levels of investigation, and a further synthesis matrix of the three criteria. Weights are introduced within the proposed evaluation tool.

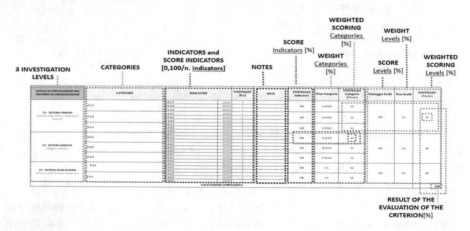

Fig. 2. Example of the structure of one of the three criteria evaluation matrices.

In practice, the weights measure, through numerical values, the priorities assigned to the various aspects of the problem and for this reason they have never absolute value but only relative [18]. To guarantee a certain degree of objectivity in identifying these and then to the whole instrument, the following were used: the participatory survey technique; the "PCT" Paired comparison technique (Saaty method) [19]. We proceeded to define the weight allocation methodology. Therefore, three (computerized) questionnaires were prepared for each criterion to be administered in the case of humanization to patients/carers and medical staff (of the Department of Medicine of the University of Salerno) and in the other two cases to technical designers. Therefore, the potential users and potential technicians interested in strengthening these structures were asked to express a preference based on the comparison of the indicators in pairs. The phase of restitution of results (and therefore of weights) is still in development. The results obtained from the questionnaires will be transferred to the specially elaborated double entry tables, where a numerical value is assigned equal to: **1** (when the indicator is considered more important than the other); **0.5** (when the two indicators are considered of equal importance); **0** (when the indicator is considered less important than the other). In this way we will determine the relative weights.

3.5 Summary Matrix and Weights of the Criteria

Quality is understood (in this study) as a function of humanization, flexibility and sustainability. The synthesis matrix puts the three criteria into a system. The synthesis matrix of the three criteria makes possible to extract the total value (from 0 to 300) of the quality, obtained from the sum of the results of the matrices of each criterion. This value will be inserted in a measurement scale, represented by symbols, which allows an intuitive classification of the quality (see Fig. 3).

CRITERIA	SCORE	WEIGHT	WEIGHTED SCORING
HUMANIZATION	0-100	x	x
FLESSIBILITY	0-100	x	x
SUSTAINABILITY	0-100	x	x
TOTAL QUALITY ASSESSMENT			0-300

SCORE
0 - 59 = ★   60 - 119 = ★★   120 - 179 = ★★★
180 - 239 = ★★★★   240-300 = ★★★★★

Fig. 3. Example of synthesis matrix.

The assessment of sustainability concerns the aspects concerning safety and respect for the environment, for this reason, certainly more weight will be attributed. On the other hand, humanization, which investigates functional-aesthetic aspects, which are very important for guaranteeing a good quality of life for users, will have a lower weight compared to the criterion of sustainability, but higher than that of flexibility. From the study, the latter, in fact, turns out to be less incident on overall comfort. The proposed tool is preliminary because, this last step (assigning weights to the criteria) is still under way and experimentation.

4 Conclusion

The last phase (Application phase) is in development. Structured the evaluation model can be applied to a concrete case study that will see the opportunity to verify the validity of the product and indicate the appropriate meta-design improvement proposals. For the application of the tool we are oriented in the choice of a critical case study of Campania. The application phase includes 3 steps: *Data sensing*; *Data evaluation* (application of the tool); *Improvement proposals* (design choices). This work, in this regard, aims at providing greater comfort to users and a better livability of the hospital with particular attention to the needs of children. The redevelopment project of the health facilities, if thought of the different scales, can represent an opportunity to give back a better life quality to the whole territory, not only to users, but to the entire territorial context where it is located.

References

1. Greco, A., Morandotti, M.: Edilizia ospedaliera. Esperienze e approfondimenti per una progettazione consapevole. Alinea Editrice, Milano (2011)
2. Campolongo, S.: Edilizia ospedaliera. Approcci metodologici e progettuali. Hoepli, Milano (2006)
3. Dall'Olio, L.: L'architettura degli edifici per la sanità. Officina Edizioni, Roma (2000)
4. Petrilli, A.: Il testamento di Le Corbusier. Il progetto per l'ospedale di Venezia. Marsilio Editore, Venezia (1999)
5. Hassenpflug, V.: Ospedali Moderni. Editrice Internazionale, Arte e scienza, Roma (1964)
6. Masciadri, I.: Ospedali in Italia. Tecniche Nuove, Milano (2011)
7. Meoli, F.: Innovazione organizzativa e tipologica per l'ospedale. In: Carrara, G. (a cura di) collana: Architettura e Tecnologia. Gangemi Editore, Roma (2015)
8. Carbonara, P.: Architettura Pratica. Utet, Firenze (1971)
9. Morandotti, M.: Modelli progettuali per l'edilizia ospedaliera. TCP, Pavia (2001)
10. Nenci, A.: Profili di ricerca e intervento psicologico-sociale nella gestione ambientale. FrancoAngeli, Roma (2008)
11. Meoli, F.: L'architettura dell'ospedale a 15 anni dall'Art. 20 L.67/88. Criteri per la valutazione della qualità progettuale complessiva. Palombi Editori, Roma (2006)
12. Saaty, T.L.: How to make a decision: the analytic hierarchy process. Eur. J. Oper. Res. **48**, 9–26 (1990)
13. Voogd, H.: Multicriteria evaluation for urban and regional planning. Pion Limited, London (1983)
14. Delle Fave, A., Marsicano, S.: L'umanizzazione dell'ospedale. Riflessione ed esperienze. FrancoAngeli (2004)
15. Sicignano, E., Petti, L., Di Ruocco, G., Scarpitta, N.: A model flexible design for pediatric hospital. In: 5th INTBAU International Annual Event - Heritage, Place, Design: Putting Tradition into Practice on Putting Tradition into Practice: Heritage, Place and Design, pp. 1464–1472. Springer International Publishing, Milano (2017)
16. Cellucci, C., Di Sivo, M.: Habitat contemporaneo. Flessibilità tecnologica e spaziale. FrancoAngeli, Milano (2016)

17. Prodi, F., Stocchetti, A., Boccadoro, S. (a cura di): L'Architettura dell'ospedale: aspetti tecnico-sanitari. Alinea, Firenze (1992)
18. Kwangsun, Y., Ching-Lai, H.: Multiple Attribute Decision Making. Springer, Heidelberg (1981)
19. Saaty, T.L.: Decision Making for Leaders: The Analytic Hierarchy Process for Decision in a Complex World. New Edition RWS Publications, Pittsburgh (2001)

Unauthorized Settlements: A Recovery Proposal of Villaggio Coppola

Claudia de Biase$^{(\boxtimes)}$, Luigi Macchia$^{(\boxtimes)}$, and Sharon Anna Somma$^{(\boxtimes)}$

University of Campania Luigi Vanvitelli, 81031 Aversa, Italy
sharonannasomma@libero.it

Abstract. In Italy, Campania is known to be one of the regions with the highest number of illegal settlements. Specifically, such illegal settlements occur mainly in the Napoli and Caserta provinces. Castel Volturno is one of the municipalities has been recognized to play a negative role in this current issue. Castel Volturno, is the area of intervention for the application of a recovery plan for illegal settlements. An analysis of the planning tools in force and preliminary tools exiting, analysis of urban analysis through detailed study of the Kevin Lynch techniques, identification of performance elements on Villaggio Coppola, before moving on to the proposal for the new unauthorized settlement recovery plan. The paper focuses on understanding how to solve the problem of illegal settlements through different possibilities provided by the law; and how to create desirable solutions for administrations and citizens to improve the territory that is often defaced by the careless hand of man, which risks compromising a cultural landscape of inestimable value.

Keywords: Unauthorized settlements · Renovation · Urban planning

1 Illegal Settlements

1.1 The Phenomenon of Illegal Settlements

By Italian Law A building is considered illegal when a complete buildings or part of them are built or extended in the absence or partial derogation from the necessary regulatory approvals. The building TU (P.D. no. 380/01), identified three types of illegal building [1]:

- The realization of a building without permission to built;
- The realization of a completely different building from that provided in the proposed project;
- The realization of a work with essential variations.

This paper is the result of collective work of the authors as well as individual personal as specified: *The phenomenon of illegal settlements* (Sharon Anna Somma), *Castel Volturno planning tools* (Luigi Macchia), *An analysis methodology and Villaggio Coppola* (Claudia de Biase), *A proposal PRUA for Villaggio Coppol*a (Sharon Anna Somma), *Abstract and Conclusion* can be attributed to both.

© Springer International Publishing AG, part of Springer Nature 2019
F. Calabrò et al. (Eds.): ISHT 2018, SIST 100, pp. 384–391, 2019.
https://doi.org/10.1007/978-3-319-92099-3_44

In the recent years, have been proposed different classifications of the kinds illegal urbanism, but they are those of two authors have been taken into consideration how most important classification: Marcelloni and Fontana in 1988; and CRESME in 2000. The phenomenon of illegal settlements is characterized by shapes, nature and purposes. First of all, the proposal by Marcelloni and Fontana identified, a shape, and nature of illegal building, such as, radical modification of use destinations in residential building, building production areas for agricultural production, a high number of secondary houses, heavy renovation. While CRESME has identified purpose of the illegal building, about, necessary illegal building, refers to the need to provide a house as a place indispensable for life highlighting serious individual economic difficulties; and speculative illegal building, a different vision respect to necessity building. This paper will be exclusively about unauthorized settlements or better about the illegal building of neighborhoods or lotting.

1.2 Castel Volturno Planning Tools

In Italy, Campania is known to be one of the regions with the highest number of illegal settlements; in fact the illegal building are about 63.3%, namely about 70.000 of the building compared to the authorized ones (ISTAT data, 2015) [2]. Specifically, such illegal settlements occur mainly in Napoli with about 86.000 requests presented, and Caserta provinces [3]. Castel Volturno, located on the coast, has been recognized to play a negative role in this current issue. Currently, the plans in force related to Castel Volturno are: the Regional Plan (PTR), the Territorial Plan (PTCP), the hydrogeo-logical plan (PAI), and plans of management SIC-ZPS areas. The PTR of Campania region, approved with the regional law L.R. n°13/2008, defines specific parameter, addresses and strategic contents. Castel Volturno belongs to the *Ambiente Insediativo* (AI) n°1 (Piana campana), in the *Sistema Territoriale di Sviluppo* (STS) F1 zone (Litorale domitio) and in the *Campo Territoriale Complesso* (CTC) n°8 (Litorale domitio). The PTCP of Caserta, approved in 2012, recognises 6 framework areas. Castel Volturno belongs to the framework named Litorale Domitio. Castel Volturno also belongs to PAI, introduced with regional law L.R. n°1/2012, managed by the Authority of Bacino of the Liri-Garigliano-Volturno. The ZPS-SIC areas of common interest which are part of the municipality of Castel Volturno are SIC area *Pineta di Castel Volturno* (IT8010020), SIC area *Pineta di Patria* (IT8010021), SIC area *Foce Volturno-Variconi* (IT8010028) and ZPS area *Variconi* (IT7010018).

1.2.1 An Analysis Methodology
The transformation that affects the territory determines a visible change on a geo-graphical scale mostly since the 60s. This current issue is due to absence of regularity, resulting in unruly building sprawl; following this, Castel Volturno territory has been particularly affected. Today, the regularity tool in force, is a Perimeter of the urban settlements introduced by the law L.N. n°765/67 that defines as a town all that falls within the perimeter. According to this law, every building that is outside the perimeter is illegal. Furthermore a first distinction is made in different areas with different intended uses compared to the plan (Fig. 1).

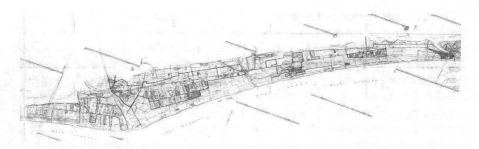

Fig. 1. Perimeter of urban settlements since 1967.

Overlapping of the Perimeter within the meaning of L.N. n°765/67 and the perimeter of the areas covered by the PUC. It has been hard to quantify the right number of the illegal buildings, being about 40 years after perimeter; furthermore, the spread of catastrophic events such as earthquake in the 70s and 80s had led to a strong expansion, with the result of an improvement of the sprawl with the progress of the building towards the maritime state-owned area. For instance, only in 2003, it has been stipulated a Program Agreement proposing interventions for the productive valorization of local resources, likewise safeguarding and conserving natural resources and rebalancing the territory through the restructuring of urban subsets. However, there are no references to interventions for the reduction of an illegal use of land. Since 2011, preliminary proposal of PUC has been presented in compliance with art. 23 paragraph 3 of L.R. n°16/2004 [4] which has identified unauthorized settlements proposing for these the methods of urban and building recovery through the formation of implementation plans. A preliminary proposal of PUC, allowed the verification of requests for amnesty that in case of full compliance with the provisions of the PUC, the possibility of being able to obtain the amnesty. In municipality of Castel Volturno, the presence of different types of illegal building has been ascertained. To date, more than 11000 requests for amnesty have been submitted in conjunction with L.N. n°47/1985, L.N. n°724/1994 and L.N. n°326/2003 [5] to obtain the so called *Condono*. Meanwhile, among these requests just a low number has been examined; furthermore, several ones aren't remediable due to the connection with buildings built on public lands, or burdened by civic, landscape and hydro-geological restrictions. To these practices of about 70/80 illegal building more for each year should be added (Fig. 2).

From overlapping inside the (see Fig. 4), the increase of the building density from the 60s up became apparent. This results to be outside the perimeter not subjected to verification of conformity with respect to the municipal territory. To date, the illegal actions has caused a slow decline of the whole territory. The development of this phenomenon has probably been caused by the fact that the territory has always been deprived of a planning tool approved in a definitive form, if not a perimetration of the urbanized areas on the basis which permits were issued for some constructions, as for the production plants. Moreover, in the various instruments proposed by administrations, a repressive intervention has never been mentioned for buildings which do not present any possibility of being condoned or restored for legal reasons.

Fig. 2. Overlapping of the Perimeter within the meaning of L.N. n°765/67 and the perimeter of the areas covered by the PUC preliminary recovery plan 2011

1.3   Villaggio Coppola

Villaggio Coppola was born as an unauthorized settlement, emblematic example of urban sprawl, a private seaside village [6]. A village that stands on a privately-owned land belonging, according to a Regio Decreto of 1911, to Cecere family [7].

Until the end of the Second World War, the territory remained uncultivated and characterized on a legal level by 1933 Land Law (Legge Fondiaria)[1], then, the end of 50s great economic it was transformed: from agricultural land to a country with a tourist vocation. In 1962 Coppola village (which was named after Coppola brothers) it was organized a first large-scale parcelling on the family-owned land that fell within the municipal area of the highest tourist interest due to the natural values of the coast (Fig. 3).

The start of works to build Village Coppola – Pinetamare, dates back to 7th December 1965.

This project which was inspired by the urban "model" of tourist housing settlement of Baia Domitia, proved to be different especially in the territorial distribution, distorting the natural landscape of Caserta; in fact, the pinewood next to the village, declared as whole and subject to landscape restrictions, was soon swallowed up by speculative constructions. The area by Villaggio Coppola covers about 950.000 m^2 including the stretch of the SS7 quarter Domitiana. With regards of the identified area, the coastal strip with building sprawl and illegal land use affects about 538.000 m^2 of which 114.000 m^2 were found to be privately owned. The remaining 424.000 m^2 consist of land belonging to the maritime state property and the state forestry. On May 16, 2001, the demolition of the first of the eight towers started; between the months of February and August 2003 the other seven towers were demolished; these abatements assumed a fundamental symbolic meaning of the strong presence of the State in a territory battered with regards of building abuses and a difficult social situation [8].

---

[1] Nuove norme per la bonifica integrale art. 22 R.D. 13 febbraio 1933 n. 215, through the structure of the reclamation areas of the district in which they are known or more plots, the concession holder of the sector Consortium proceeds, for the purposes of reclamation, according to a reception plan, to the meeting of plots of men, to sell each owner, in exchange of his land, to pay one or more land for the purpose of reclamation.

Fig. 3. First Plan of the extension of Villaggio Coppola 1962.

1.4   A Proposal PRUA for Villaggio Coppola

The ability to generate an environment that creates meaningful connections is related to the characteristics of *legibility* and *imageability* [9] and is essential for the survival of every urban ecosystem to determine the creation of a strong identity of the place. Generally, an urban system must be legible, capable to lead to the identification of a structure, through sensorial signals defined and perceived by the observer, arbitrarily selected by the community and finally manipulated by urban planners, by constantly relating the image of the city with form, quality and objectives.

The identity of Castel Volturno differs from the territorial context for the urban form that has had a remarkable development, from the 60s to today. From a first reading of the territory emerges what follows [10]:

- *Strong points*: Pinewood which extends the stretch from Baia Verde to Marina di Licola; the nature reserve of the Variconi oasis; the delta of the Volturno River and the Regi Lagni, the seafront position; the presence of a tourist harbour; the fast infrastructure that connects the cities of Rome, Naples and Caserta; the building and urban fabric that defines the typical building typologies;
- *Weak points*: the great extent of the unused agricultural area; the strong degradation; the same harbour currently incomplete; the lack of iron infrastructure and a good mobility system for connections with the cities of Campania and Lazio; the lack of equipment, the integration of the different ethnic groups existing on the territory.

Safety, for example, is one of the aspects to consider for the improvement of living conditions, which has not been taken into account since the construction of Villaggio Coppola. Crime events, of low and high degree, are the aspects to be evicted in order to rebuild a good settlement. From the ensemble of all the elements is clear that the characteristic of this part of the territory is the absolute lack of services and equipment for socialization and aggregation; this lack is due to the randomness of urban residential stratifications and is illegal use. The lack of a public space, a key element for individual and social well-being, able to support multi-cultural integration and a better quality of life for citizens of every age and ethnicity in the Castilian context. This kind of analysis also provides the identification of five performances dimensions, such as nodes, edge, district, path and landmark [9] (see Fig. 4). These features are able to show to the citizens perception with respect to the places of the territory. The Keywords of

Proposal PRUA for Villaggio Coppola are the following: recovery urban settlements, demolition, and reuse and regeneration of forsaken buildings with "zero-cost land use". The choice of the area of intervention derives from the conditions of the territory, such as: imposing weight of the phenomenon of illegal building, uncontrolled growth of settlements without primary and secondary urbanizations and the absence of an existing regulatory instrument that regulates the territory. The objective is to operate through the proposal of a Recovery Plan for the unauthorized settlements for Villaggio Coppola, referring to the art. 29 of the L.N. n°47/1985. The main objective is to reduce and count the abuse, while where possible, to recover illegal buildings; it considers a greater comfort for the settlement, through the primary and secondary re-urbanization and redevelopment through works and services to favor a integration to the citizens. In order to hypothesize an intervention on the territory, an analysis of the building fabric and of their conditions was carried out (degraded, incomplete or decommissioned), as well as two types of interventions to be applied: restoration of the building, and the demolition of unsinkable buildings. For this purpose, every feasibility study was necessary for each intervention (see Fig. 5). Demolition interventions (45 terraced houses, 22 residential buildings, and 3 of different destinations) are planned on the coastal strip, as they fall within the owned-state area, and within the coastal strip (300 from the shoreline line), with the constraint of unbuilding this intervention provides for the identification of new public spaces. Regarding the recovery, there are two ways of intervention: recovery of urban voids or brownfields with change of intended use; restoration of the existing building, characterized by changes of intended use. In two areas new buildings will be shared, within which the families residing in the demolished buildings will be relocated. The typical building is structured on two floors which are expected to be given to each family; made with sustainable materials that conform to the climatic and territorial context.

One of the fundamental principles of a recovery plan is to encourage the direct participation of citizens in the "self-recovery" of the neighborhoods; hence, they contribute to those who have to realize new volumes, through mechanisms of rebalancing between the sale of public areas and the payment of extraordinary contributions. In addition, the project includes a new waterfront design and aspires to a new identity for Villaggio Coppola, from an illegal village to a tourist center, downsizing the existing urban fabric, through the concept taken from the road stratification. All the material of the demolished buildings will be recycled and used as an element of urban furniture, arranged along the new promenade in order to create a real artistic with sculptures or on the facades of buildings to be recovered falling in the residential area; among the most common materials used, flooring or coatings. Out of the total number of buildings on the beachfront, 10 are excluded from the demolition, which is instead expected to be recovered, as they are new buildings of which 4 are decommissioned, and some of the remaining are in degraded conditions. One of the reasons why demolition isn't planned, is the presence of green spaces that can be used as a new access point from the inside.

This also allows the perception of visual cuts from which the sea can be seen from the pinewood. The recovery of abandoned, degraded, and unfinished buildings consists of consolidation in structural and aesthetic terms, through the use of sustainable materials, compliant with the territory, starting from an intervention on the façade for

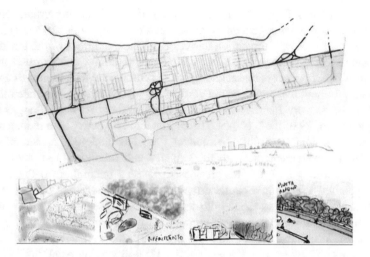

Fig. 4. Perceptive Analysis of Villaggio Coppola

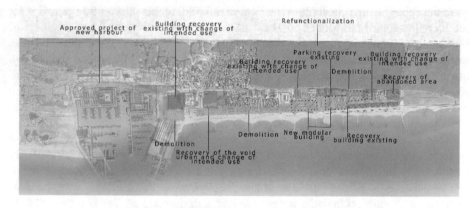

Fig. 5. Juridical table of the feasibility study.

which the homogenization of all the buildings on which color bands will be applied using recycled materials. As for unfinished buildings, we are currently testing the hypothesis that a complete recovery may occur according to the criteria of sustainability. This may assume a different role from that of the original project, providing to the citizen a common place to socialize and integrate. In conclusion, the new waterfront proposal is based on a restyling of the driveway, reorganizing a promenade for colors, recalling those that make up the territory, which identifies the different functions of the city: residential area, touristic area and new green area.

2 Conclusion

The scientific discussion, at national and European level, on the unauthorized building and settlements, is agreed on an assumption: the soil is an exhaustible resource for the environment and the landscape, therefore, central in urban and territorial planning and management. The paper, therefore, focuses on the recovery of this large amount of heritage, in a sustainable key, which leads to understand how to solve the phenomenon of illegal settlements through the recovery proposal plan the chosen area, that Villaggio Coppola. In addition, the problems of the territory that emerged during the work, have led to the hypothesis of a wide and justified actions:

- Demolition: (1) with rebuild, achieving an improvement of the primary and secondary urbanization; (2) without rebuild, provides for the identification of new public spaces.
- Recover of abandoned area, and restyling of green areas.
- Acquisition of assets of some building with change of use destination.

The desirable recovery project for Villaggio Coppola in addition to urban redevelopment and construction, using sustainable materials and the use of alternative energy, should also promote the removal of physical and social barriers resulting from the illegitimate development of the building. Environmental sustainability is visible in the provision of new green areas for the community and, Social sustainability is achieved through the promenade, and then, economic sustainability is substantiated by the provision of a reduced number of demolitions - economically burdensome - and in the prediction of a building and urban recycling for private owners.

References

1. de Biase, C., Losco, S.: Up-grading illegal building settlements: an urban-planning methodology. In: International Conference Green Urbanism, pp. 455–456 (2016)
2. ISTAT, Rapporto BES, pp. 221–222. Streetlib, Roma (2015)
3. de Biase, C., Forte, F.: Abusivismo nella provincia di Caserta: prime riflessioni su aspetti urbanistici ed estimativi. In: METE, vol. 3, pp. 1–10. Officine Grafiche (2013)
4. de Biase, C.: Un toolkit per le piccole e grandi trasformazioni urbane, Aracne (2007)
5. Losco, S., de Biase, C.: Insediamenti abusivi e pianificazione urbanistica in Campania. Urbanistica Informazioni **269–270**, 82–84 (2016)
6. Ricci, M.: Rischio paesaggio, International journal of Urban Culture, Edizioni Scientifiche Italiane, p. 47, Napoli (2008)
7. De Jaco, A.: Inchiesta su un comune meridionale: Castelvolturno, Editori Riuniti (1972)
8. Il Villaggio Coppola Pineta Mare. http://www.casertace.it/. Accessed 21 Nov 2017
9. Lynch, K.: The Image of the City. Edizioni MIT, Press (1960)
10. Lynch, K.: Progettare la città: la qualità della forma urbana. Edizioni Eta Libri (1990)

The New Urban Governance. The Case of Madrid: Between Utopia, Desire and Reality

María José Piñeira Mantiñán[1]([⊠]), Rubén Camilo Lois González[1], and Jesús Manuel González-Pérez[2]

[1] University of Santiago de Compostela, 15782 Santiago de Compostela, Spain
mariajose.pineira@usc.es
[2] University of the Balearic Islands, 07122 Palma (Mallorca), Spain

Abstract. The article analyzes the degree of success achieved by the municipal governments that came to power in Spain after the 2015 elections. In the face of corruption, vulnerability and the neo-liberal growth model that has prevailed since the mid-1980s, they promote a new model of urban governance based on the principles of efficiency, transparency, equity and citizen participation. Principles that laid the foundations of their electoral programs. After two years, it is time to assess if this desire has been fulfilled. Through the Madrid case study this paper will provide an overview of how local authorities are simultaneously addressing the dual pressures of reducing debt as well as restoring the welfare state and the trust of the population. We will explore to what extent they have managed to fulfill their promises, the main problems they have had, how the opposition perceives their urban policies and if they continue to have the support of associations and institutions that originally endorsed them.

Keywords: Multilevel governance · Participatory democracy · Madrid

1 The New Models of Urban Governance

Since the beginning of the twenty first century the governance of cities has encouraged a wide range of academic works focusing on the novel models of urban governance, the relationships between governments at different scales and the municipalities response to the social and financial effects of the crisis [1–5]. These issues are especially relevant in large Spanish cities, and particularly in Madrid [6, 7]. In fact, the capital has suffered from the excesses in the application of a neo-liberal program at a local level which sought to encourage growth at all costs and create a globalized business city. Although some of these objectives were achieved by the consolidation of large multinational companies (with presence in Europe and Latin America), and the generation of wealth and employment since 1990, this development was abruptly discontinued by the 2008 financial crisis. The real estate market that sustained speculative and continuous economic growth plummeted, thousands of people lost their jobs, could not deal with their debts, and the feeling of discontent and bleakness spread to all levels of society. The conservative municipal government continued with its plans of urban aggrandizement (candidacy for the 2024 Olympic Games, investment in large public works, etc.), and detached itself from the will of a population that demanded quick and new solutions to

© Springer International Publishing AG, part of Springer Nature 2019
F. Calabrò et al. (Eds.): ISHT 2018, SIST 100, pp. 392–401, 2019.
https://doi.org/10.1007/978-3-319-92099-3_45

their problems. As a result of this process, a new alternative left political party, grouped in the Ahora Madrid candidacy, won the 2015 local elections taking over the city mayor's office. Since then Ahora Madrid have tried to realise a new urban project with utopian proposals (favouring a participatory-assembly management model), important changes in the objectives of urban development and a socio-territorial redistribution of government priorities [7, 8]. They have managed to put forward new programs, although in some cases they have faced major challenges in a scenario that will be analyzed under the contemporary governance perspective.

The first attribute of the municipal governments of change emerged in 2015, and Madrid as an outstanding example, is the assumption of the principles of new urban governance [9, 10]. It is about networking, taking into account the opinions of organized social actors, both the classics (neighbourhood associations, merchants, etc.) and those born in the heat of anti-crisis protests (Anti-evictions Platform, those affected by bank fraud, neighbourhood youth movements, etc.) [8, 11]. In this way, governance becomes less hierarchical and segmented according to the areas of competence, and is horizontally integrated (thanks to citizen consultation), connecting problems that corresponded to specific sectors (for example, linking the fight against poverty and social policies with housing) [12, 13].

Without a doubt, and as we have expressed in previous works [7], the governing approach has substantially changed. However, it is challenging to realise all the objectives of the electoral program in local governance when the communication with the (conservative) regional government, and especially the national (also conservative) government, is very difficult, moreso when an austere fiscal policy freezes the city council funds in the banks as guarantee. In parallel, with a media environment where the opinions of the opposition parties are broadly publicised, as a way of defying some substantial changes.

In any case, the principles of new governance have somehow been applied and a series of terms are based on the daily practices of the municipality [14]. Sustainability is among them, with actions promoting sustainable mobility, such as limiting access to the city center by private car, and urban ecology policies improving the quality of life. The principle of subsidiarity had been used by previous municipal governments and now the proximity of the administration to the problems of the people is needed. Administrative decentralization processes have been confirmed, particularly at neighborhood level, complemented by the principles of participatory local democracy. Along with equity and efficiency, but above all transparency in management, an imperative for those urban political movements that grew and won fenomenal support in their fight against political and financial corruption. Transparency applied to the management of public budgets, in the hiring processes and in the salaries and income of political representatives, which should be of public knowledge. All this, closely linked to the principles that fuelled the 15-M protests, the indignant's movement, the birth of the political party of the alternative left *Podemos* and behind the *Ahora Madrid* candidacy, connects with the classic demand of the right to the city, citizen empowerment and the reinforcement of citizen participation, continually defended in the protests that took place in the worst of crisis and in the form of governing when the crisis began to recede (in general, since 2010) [15].

These updated and consensual government schemes have somehow been applied in Madrid, although there have been many challenges to their realisation. Despite this, Madrid continues to be a very representative example of the recent debate on the new urban governance in Spain. In the first place, because it is the main Spanish city and, therefore, the country's main big city that supported an alternative left government in an impressive movement of change spurred on by the 2015 local elections. Together with the capital, Barcelona, Valencia, Zaragoza, Cádiz, Badalona, Sabadell and A Coruña are all examples of this political trend, all of them with incontestable urban importance. Secondly, and this is a novelty, the victory in Madrid has meant a complete break with the neo-liberal and developmental government models that the Popular Party (Partido Popular) mayors applied for more than twenty years. This framework is similar to Valencia or Cádiz, but can not be extrapolated to the other cases already mentioned. In addition, breaking into a capital city and Autonomous Region traditionally dominated by the right, with traditional government systems, has created an environment of greater pressure by the media and supported by the political and business elites against a new urban government often labelled as radical. In fact, urban governance has been easier in peripheral cities (such as in Catalonia or Galicia), where new political movements have been associated with established identity or nationalist forces. In Madrid, the intense participatory experience fostered by the 15-M Movement, together with a prominent public figure at the head of the mayor's office, has prompted groundbreaking ways of governing the daily problems of people at a local level. In this context, this work has a double objective. On the one hand, to study the main obstacles that the current Madrid City Council has had in developing new urban policies. On the other, analyzing the degree of acceptance of this new management and governance model by the association networks and main citizen platforms.

2   Methodology

In previous works we approached the study of new models of urban governance in Madrid and Barcelona from the analysis of statistical variables, the study of government programs and the allocation of municipal budgets [7]. After three years of municipal government, we wanted to evaluate the achievements, the weaknesses in the execution of the political programs and the applicability of the new governance models based on the use of qualitative methodologies. This being the semi-structured interview with key characters the main method used. Between ten and fifteen questions were posed to different types of key people: political leaders (Deputy Director of the Department of Decentralization and Territorial Action and Technical General Secretary of the Department of Culture of the City of Madrid), citizen platforms (responsible for La Ingobernable-Social Centre of Urban Commons and the Anti-evictions Platform (PAH)), and Neighbour Associations through the Regional Federation of Neighborhood Associations of Madrid (FRAVM).

The questions revolved around six main variables: (a) policies to fight vulnerability and social exclusion; (b) citizens' resilience measures; (c) citizen participation actions; (d) new socially just city model vs. old models dominated by strategic and competitiveness objectives; (e) the use of space, with its contrasts and inequalities, in public policies; (f) multilevel governance. Note that the same questions have not been asked to all the interviewees, but have been adapted to their area of expertise and where they demonstrate greater knowledge. This allows us to obtain more extensive and valuable information on the programmed topics. The interviews lasted an average of one hour and took place over three days in June 2017. The results of these interviews are compared with some of the most important theoretical contributions of multilevel governance, as a means to evaluate the success of the new urban policies implemented by the Madrid City Council.

3 The Evolution of Urban Policies in Madrid. From Strategic Planning to Intervention in Disadvantaged Neighborhoods

When the first democratic local councils were formed in March 1979, urbanism became a fundamental part of local politics. Neighborhood associations contributed to this recognition, at least in large cities, where many of their actions against the struggle during the Franco dictatorship related to an urban problem [16]. With the arrival of the first local governments of democracy, these associations weakened resulting in a loss of influence reminiscent of what is happening with citizen movements today. The first democratic city councils were governed by leftist parties so the neighborhood associations had to rethink their role, closer to collaboration than to opposition.

3.1 Urban Policies in Madrid During the Democratic Period (1979–2015)

In the last five decades Madrid has had seven mayors. During the first ten years the socialist party (PSOE) governed, retaining many socialist leftist principles. The right (PP) led Madrid during the longest period, 26 years, a period of real estate bubble and, in recent years, of crisis and social cuts, surrendering to the principles of an entrepreneurial and neo-liberal city. According to Fuente and Velasco [17], urban policies in Madrid are divided into three stages.

– First democratic governments (1979–1988): the first socialist mayor defended the idea that local policies allowed a better transmission of citizen needs to democratic institutions, by their proximity. As a consequence of this new vision, coalitions between the municipal government and the social fabric, the one that had led the neighborhood movements since the 1960s, emerged [17]. The city model was designed based on four main axes: (a) compact city (General Plan of Urban Planning -PGOU-, Special Plan of Protection and Conservation of Historic-Artistic Buildings, Neighborhood Program in Remodeling and Immediate Action Plans); (b) green areas (the surface is doubled); (c) reinforcement of the public sector (more public employment, the Metro ceases to be a private company); (d) implementation of neighborhood participation and decentralization processes (Rules for Citizen Participation, Sectoral Councils for Participation, decentralization of Madrid in 18 districts). The reference was the new PGOU (1985) that was projected as an unconventional "bottom up" process and from the parts to the whole [16]. At the end of this period, the social model under construction began to evolve towards the

search for competitiveness. In 1988 PROMADRID was created, developing a diagnosis of Madrid with the idea of becoming the basis of the strategic plan.
- The consolidation of a management city (1989–2003): the strategic planning became the structure of the new model of the 1990s city. In 1992, Madrid was designated "European Capital of Culture". Published under the title Madrid Futuro: Strategic Plan of Madrid [18], a model linked to external projection, urban competitiveness, the attraction of high-level added investments was proposed, with the ultimate goal of consolidating Madrid as an European regional capital. Citizen participation was replaced by major strategic decisions; the compact city (PGOU 1985) gave way to an increasingly diffuse city based on population and functional decentralization (PGOU 1997). This model was based on large public investments for the improvement of transport and connectivity infrastructures: M-45 ring road, Barajas Plan (airport), Puerta de Atocha Station, Madrid-Sevilla High Speed Train (AVE), South Bus Station, etc. [17]. The modernization of cultural heritage in the urban center (Reina Sofía Museum National Art Center, Thyssen Bornemisza Museum, National Music Auditorium, Casa de América, etc.) and the construction of mega-facilities and singular urban pieces in the new tertiary peripheries (IFEMA exhibition grounds at the Campo de las Naciones, Municipal Conference Center, International Trade Center etc.). Fundamental actions for a city obsessed with competitiveness.
- Towards the global city (2003–2015): there was no change of model but the idea of a global competitive city was deepened. Faced with a type of political leadership (first stage) and managerial (second), the business approach triumphed since 2003 [17]. This stage included the years of the most intense housing bubble (1996–2007) and the beginning of the crisis. In 2002, the Madrid Municipal Promotion Company (esMadrid.com) was created for the provision of services related to the cultural, tourist, economic and business promotion and dissemination. A promotion related to the candidacy of Madrid to host the Olympic Games of 2012, 2016 and 2020. In this same line, in 2007, the Madrid Global Office was created. The main urban actions were not carried out based on a new PGOU. The project is trusted as smaller and more flexible at urban level. Investments are concentrated in the central areas of the city, in those most visible, most easily promoted and privileged neighborhoods. These are the cases of the Remodeling Project on Serrano Street, the Recoletos-Prado Axis Remodeling Plan, the so-called Operation Rio, the unfinished Operation Chamartín and the construction of the Cuatro Torres complex. With the constant commitment of Madrid as an Olympic city, investments in large infrastructures were not abandoned: the tunnelling of the M-30 and the construction of a new airport terminal. Madrid became global, but increasingly unequal.

3.2 The Change. In Search of Social Justice in the City (2015–2019)

The rise of global Madrid, so meticulously planned since the beginning of the 1990s, results in a sudden fall with serious social and urban consequences. The crisis was parallel to a double movement of scale: the international financial crisis and the crisis in the Spanish real estate cycle [19]. In 2010, 24.4% of the municipal population lived in poly-vulnerable areas or areas of integral vulnerability [20].

Since 2015 the government of Madrid is led by Ahora Madrid, a confluence movement comprised of left parties: Podemos, Ganemos Madrid, Izquierda Unida and Equo. The mayor is an emeritus judge and former member of the Communist Party of Spain, who defined her political project as a "popular unity citizen candidacy". The program of Ahora Madrid included many of the demands of the 15-M Movement and meant the beginning of a new urban agenda that focuses on the search for social justice (2015–2019). Its arrival to power was understood in the context of increasing urban impoverishment, the inequalities intensified by the crisis, the policies of economic adjustment and social cuts. The strategies of the neo-liberal city were rejected. The competitiveness and urban promotion became secondary. The construction of large infrastructures and mega-equipment were no longer priorities. And the distribution of centralities, in terms of public investment by neighborhoods, prioritized the most disadvantaged periphery districts with greater parts of the population at risk of social exclusion. The 2016 and 2017 budget allocation prioritized, on the one hand, the less privileged and peripheral neighborhoods, located mostly south of the municipality of Madrid. Usera, Puente de Vallecas and Villaverde increased their budget between 32% and 38% in the 2015 and 2017 period while the most central and northern districts, with higher rents, had lower growth or, as in the case of Salamanca, negative (2015–17: −8.1%). On the other hand, the social issue has been crucial. In 2017, social spending increased 22.2% compared to 2016. All these changes take place at an urban scale and are not related to regional (Autonomous Community) or metropolitan level policies. On the one hand, the Autonomous Community is governed by the Popular Party, which has ruled the municipality for the last 26 years and its legacy is being rejected by the current municipal government. On the other hand, although there is an extensive literature on the metropolitan region of Madrid, its limits are not legally defined and there is not a metropolitan government. Furthermore, interest in metropolitan planning has been lost due to the lack of funding to address macro-projects. As a consequence, planning and intervening in the traditional consolidated city and its vulnerable neighborhoods are prioritized.

Based on a political ideology focused on the reinforcement of the public and the guarantee of basic services, we highlight two types of actions. First, the right to housing, the strengthening of neighborhood culture and the fight against urban poverty, inequality and evictions: the creation of an anti-eviction office (Office of Mortgage Intermediation) and the stoppage of evictions by the Municipal Housing Office. Secondly, the commitment to a new model of urban governance, decentralized, transparent and participatory: transparency in budgets and government actions (Transparency Portal); decentralization towards the districts (their budgets increased 11.42% in 2017); citizen empowerment and participatory budgets (the 2017 budget integrates projects decided by citizens to a value of 60 million euros).

In this stage a new reading of the city takes place and it is committed to a closer urbanism, with small-scale actions. Instead of focusing on global competitiveness based on urban projects, as was the case in the previous stage, now the neighbourhood is concentrating attention. The purpose of this paper is not to analyze the spatial functions of the new government. However, a number of projects in the 2015–2019 plan are worth mentioning. To highlight: (a) promotion of decent housing (extension of the public park for rental housing, and rehabilitation of homes and buildings);

(b) planning based on general interest (comprehensive urban regeneration of vulnerable areas); (c) implementation of an efficient and sustainable mobility plan (promotion of public transport, pedestrian mobility and use of bicycles); (d) territorial balance (investments in the improvement and construction of cultural and sports facilities; comprehensive neighborhood improvement plans); and (e) regeneration of the urban space.

4   The Conflict Between Desire and Reality

Even with the interesting precedent of the first socialist government of Tierno Galván, the arrival to power of Ahora Madrid has probably been the most important change in the model of urban governance of the capital during all of democracy. However, the theory (or program) that sustains it faces multiple obstacles.

The first are internal, between the political parties that support it. *Podemos*, with independent actors like Mayor Carmena, and *Ganemos Madrid* driven by 15-M activists. The second one is the distance with their social bases. Thus, groups and associations such as the anti-eviction or the right to decent housing movement (PAH), the urban commons movement and the Regional Federation of Neighborhood Associations of Madrid (FRAVM) are critical of the new government because of the lukewarmn of some reformist measures, which, for example, have not solved the serious residential problems generated by the crisis, having been left out of government actions the mortgage intermediation office or the negotiation with the banks on the issue of housing vulture funds. In this sense, in the interviews held, it is noted that the housing issue is key for these groups. On the one hand, they affirm that in Madrid there is no housing alternative that allows a bilateral relationship with banks and owners. The volume of housing offered by the SAREB -entity created to help the reorganization of financial institutions troubled by their excessive exposure to the real estate sector- and social housing funds -with moderate prices for families in a more vulnerable situation- has been very low, some 15,000 in Spain (2013–2016). On the other hand, the supply of social rent housing promoted as a result of the implementation of a Territorial Re-balancing Fund has been insufficient, while the illegal occupation of housing has been denied. A problem that in working-class neighborhoods in the south such as Villaverde, Usera, Vallecas or Carabanchel, affects entire streets, where 50% of buildings have been illegally occupied for more than seven years. At times, this generates ghettos and violent confrontations with "legal" residents.

The takeover of municipal power has revealed the local trap [2]. In some cases, there has been a real concern from the City Council to reduce the vulnerability of neighborhoods, but the solution exceeds the local level (infrastructure, mobility) and must be managed at metropolitan, regional or national level. In others, the action of the new government is limited by excessive control when accepting new initiatives and the fear of altering some rules of the game converted into legal norms by the regional and Spanish administrations.

Decentralization and the fight against social inequality are two of the main programmatic proposals of the new governance model. Regarding the first, most interviewed experts and groups' representatives agree that it is at a very early stage and is not achieving the expected success. The creation of the District Boards (a new autonomous body from which to make city policies and achieve greater proximity between government and citizens) and the Local Forums as instruments of citizen participation, have developed differently depending on the neighborhoods. In a similar order, they consider that the fight against inequality has not reached the desired results either. Despite the fact that a study on the social vulnerability of neighborhoods has been carried out and a Territorial Rebalancing Fund has been created to help disadvantaged people, there is a lack of qualified staff to implement the projects.

The implementation of multilevel governance, in which urban policies and projects were designed from the ground up, although guaranteed through the Local Forums and its working groups, soon began to show signs of weakness. The fact that any citizen could participate in the working groups and propose improvement projects for their neighborhoods was a novelty. And having 60 million euros to do it through the so-called participatory funds generated great expectation. However, from the FRAVM certain inconsistencies were observed in the process. The fact that citizens' initiatives had to be submitted in writing, digitally signed and voted on-line was a bias in the neighborhoods that ended up proposing more improvement projects and initiatives receiving more votes. While those newly created neighborhoods such as the Ensanche in Vallecas, where there is a young adult population, with IT knowledge and more active in the associative movements, have been awarded a large number of grants; other neighborhoods, such as Villaverde de Abajo, severely affected by evictions and unemployment, have had barely any initiatives because a significant part of the population has no studies or access to new technologies and does not even dare to speak out about improvements that seem very basic and do not have the consistency of an urban project (improving street cleaning, sidewalks, paving of streets, caring for green areas, collecting solid waste, etc.).

5 Conclusion

Ahora Madrid has introduced new initiatives and programs in the way of governing the city: transparency, decentralization, closeness of the administration to citizens, etc. However, we consider that the municipal government does not have yet a complete model of the city and that its interest is more focused on resolving some critical issues such as housing, social vulnerability and pollution. Furthermore, implementation of some of the proposals included in the electoral program is not totally successfull: sometimes legislation limits the initiatives; sometimes they are not able to measure how many stuff or resources are needed to carry out some of these proposals. Likewise, the groups interviewed (PAH, urban commons and the FRAVM) are not pleased with the procedure or with the results. They consider that at the present stage the relationship between them and the local council can be defined as tension-utopia: (a) contending that although the citizen movements played a key role in the design of the electoral program of Ahora Madrid, they are not currently having the level of participation they deserve on the execution of urban policies and the desired political project; (b) noting that no drastic measures have been taken to curb the impact of evictions and vulture funds; (c) denouncing that the city council does not want to face the corruption interwoven by the previous governments that consented to the cession of public buildings, for example, to private foundations with no public tender process. Finally, although they acknowledge that the City Council has made an effort to help the neighborhood associations, the associative networks are very weak. We may recall what happened in the 1980s, the relationship between the first democratic local councils governed by the left and the neighborhood associations. Even so, the Madrid platforms and associations interviewed continue to support the government of Manuela Carmena, since although the problems are many and not all the objectives have been achieved, neither in time nor in form, the change is substantial compared to the previous municipal governments and a return to the old days is not desirable. In any case, some questions arise in the short-term: how future local governments will deal with the legacy received, especially if they belong to another political party? Are new policies well accepted by population? And if the economy recovers, would this fact lead to the recovery of the previous neoliberal model, i.e. megaprojects?

Acknowledgments. This article fits into the projects "New models of government of the cities" (CSO2016-75236-C2-1-R) & "Crisis and vulnerability in Spanish island cities". (CSO2015-68738-P), both financed by Ministry of Economy and Competitiveness.

References

1. Agnew, J.: The territorial trap: the geographical assumptions of international relations theory. Rev. Int. Polit. Econ. **1**(1), 53–80 (1994)
2. Purcell, M.: Urban democracy and the local trap. Urban Stud. **43**(11), 1921–1941 (2006)
3. Fanstein, S.: The Just City. Cornell University Press, Ithaca (2010)
4. Iglesias, M., et al.: Políticas urbanas en España. Grandes ciudades, actores y gobiernos locales. Icària, Barcelona (2011)
5. Lake, R.W.: The subordination of urban policy in the time of financialización. In: DeFillipis, J. (ed.) Urban Policy in the Time of Obama, pp. 45–64. University of Minnesota Press, London (2016)
6. Blanco, I., Griggs, S., Sullivan, H.: Situating the local in the neoliberalisation and transformation of urban governance. Urban Stud. **51**(15), 3129–3146 (2014)
7. González, J.M., Lois, R.C., Piñeira, M.J.: The economic crisis and vulnerability in the spanish cities: urban governance challenges. Proc. Soc. Behav. Sci. **223**, 160–166 (2016)
8. Cabreirizo, C., Klett, A., García, P.: De alianzas anómalas a nuevos paisajes políticos. Madrid, Lavapiés y otras geografías de lo común. URBS, Revista de Estudios Urbanos y Ciencias Sociales **5**(2), 163–178 (2014)
9. Kooiman, J.: Societal governance: levels, models and ordrs of social-political interaction. In: Pierre, J. (ed.) Debating Governance, pp. 138–166. Oxford University Press, Oxford (2000)
10. Lange, P., et al.: Governing towards Sustainability - conceptualizing modes of governance. J. Environ. Plann. Policy Manage. **15**(3), 403–425 (2013)
11. Brenner, N., Theodore, N.: Preface: from the new localism to the spaces of neoliberalism. Antipode **34**(3), 41–347 (2012)
12. Davies, J.S.: Local governance and the dialectics of hierarchy, market and network. Policy Stud. **26**(3–4), 311–335 (2005)
13. Ramesh, M., Fritzen, S.: Transforming Asian Governance: Rethinking Assumptions, Challenging Practices. Routledge, London (2009)

14. UNHabitatGlobal Campaign on Urban Governance. http://mirror.unhabitat.org/content.asp?typeid=19&catid=25&cid=2097. Accessed 26 Nov 2017

15. Lois, R.C., Piñeira, M.J.: The revival of urban social and neighbourhood movements in Spain: a geographical characterization. Erde **146**(2–3), 127–138 (2015)

16. Terán, F.: Historia del urbanismo en España III. Siglos XIX y XX. Cátedra, Madrid (1999)

17. Fuente, R., Velasco, M.: La política urbana en Madrid: un relato provisional. Geopolítica(s) **3**(1), 35–59 (2012)

18. PROMADRID: Madrid Futuro: Plan Estratégico de Madrid. Ayuntamiento de Madrid, Madrid (1993)

19. Rodríguez, M., García, B., Muñoz, Ó.: Del Madrid global a la crisis urbana. Hacia la implosión social. In: Observatorio Metropolitano de Madrid (eds.) Paisajes devastados. Después del ciclo inmobiliario: impactos regionales y urbanos de la crisis, pp. 123–177. Traficantes de Sueños, Madrid (2013)

20. Dirección General de Revisión del Plan General: Revisión del Plan General. Diagnóstico de ciudad, vol. II. Ayuntamiento de Madrid, Madrid (2012)

Isovalore Maps for the Spatial Analysis of Real Estate Market: A Case Study for a Central Urban Area of Reggio Calabria, Italy

Pierfrancesco De Paola[1]([⊠]), Vincenzo Del Giudice[1],
Domenico Enrico Massimo[2], Fabiana Forte[3], Mariangela Musolino[2],
and Alessandro Malerba[2]

[1] Federico II University of Naples, 80125 Naples, Italy
pierfrancesco.depaola@unina.it
[2] Mediterranea University of Reggio Calabria, 89100 Reggio Calabria, Italy
[3] Luigi Vanvitelli University of Campania, 81031 Aversa, Italy

Abstract. Generally, with reference to the geographical variability of real estate values, the observed variables may have non-linear relationships with the response variable. For this reason it's possible to combine kriging techniques with additive models to obtain the geoadditive models. In this paper a geoadditive model based on penalized spline functions has been applied, in order to obtain improvements respect to usual Kriging techniques and to provide a spatial distribution of real estate unitary values for a central area of the city of Reggio Calabria (Italy). This is the first preliminary phase for the verification of the robustness of the real estate sample, or for the subsequent individuation of progressive real estate sub-samples, for to detect and to identify possible potential market premium in real estate exchange and rent markets for green buildings.

Keywords: Post carbon city · Green buildings · Geoadditive models
Semiparametric regression · Real estate market analysis

1 Introduction

In this study a geoadditive model characterized by penalized spline functions has been implemented, in order to analyze the spatial distribution of real estate unitary market prices in the urban area central of Reggio Calabria (Italy). This step is the first preliminary phase for the verification of robustness of the real estate sample, for the subsequent individuation of progressive real estate sub-samples, for to detect and to identify a possible potential market premium in real estate exchange and rent markets for green buildings [23–32].

Geoadditive models have many advantages, even in small local real estate markets, being able to analyze the variation of real estate values in an area of interest [1–6]. In addition, these models allow to predict, quantify and locate in real time where and how

---

The authors contributed equally to the study.

© Springer International Publishing AG, part of Springer Nature 2019
F. Calabrò et al. (Eds.): ISHT 2018, SIST 100, pp. 402–410, 2019.
https://doi.org/10.1007/978-3-319-92099-3_46

real estate values vary in urban context, with possibility to correlate these variations with any phenomenon or economic effect (for example, delimitation of micro-zones, modeling of locational variables, delimitation of areas with homogeneous values, etc.) [7–29].

The combined use of penalized splines with techniques of spatial statistics (Kriging) allows to obtain spatial maps with high reliability on which to base any decisions related to urban investments.

Kriging is a regressive technique used for geostatistic spatial analysis that allows to interpolate a variable in the space minimizing the mean square error. Generally, knowing the variable value in some points of space, it is possible to determine the variable value in other points for which there are no measures, this through a weighted average of known values. The weights depend on the spatial relationship of measured values in the range of unknown point (for which no information is available). For the calculation of weights a semi-variogram is often used, it is a graph capable of putting in relation the distance between two points and the semi-variance value among the measurements made respect to these two points. Substantially, a semi-variogram exposes in a qualitative and quantitative mode the spatial autocorrelation between two points [1–5].

In Kriging the spatial interpolation is based on the autocorrelation of variable, considering that a variable vary continuously in space. Currently Kriging is often used in the implementation of Geographic Information Systems (G.I.S.), which represent computer systems able to produce, manage and analyze the spatial data associating one or more alphanumeric descriptions to each geographical element [7, 8]. In particular, the determination of interpolation areas can currently be done through the use of exponential, gaussian, linear, rational or spherical functions. As alternative to these functions the use of geoadditive models with penalized spline functions has been used in this paper in order to achieve significant improvements in forecasting for surfaces interpolation [4, 5].

2 Model Specification

The complexity of relationship between real estate unitary prices and explanatory variables has conducted to implementation of a geoadditive model.

Generally, geoadditive models are composed by a semi-parametric additive component, which it serves to express the relationship between model's non-linear response and explanatory variables, and a model with linear mixed effects that expresses the spatial correlation of observed values [1].

The first component involves a low rank mixed model representation of additive models. For simplicity we shall present the case of two additive components. Suppose that (s_i, t_i, y_i), $1 \leq i \leq n$, represent measurements on two predictors s and t for the response variable y, in this case the additive model is:

$$y_i = \beta_0 + f(s_i) + g(t_i) + \varepsilon_i \tag{1}$$

where f and g are unspecified smooth functions of s and t respectively. Therefore, if we define u_+ to equal u for $u > 0$ and 0 otherwise, a penalized spline version of model (1) involves the following functional form [4, 5, 10]:

$$y_i = \beta_0 + \beta_s \cdot s_i + \sum_{k=1}^{Ks} u_k^s (s_i - \kappa_k^s)_+ + \beta_t \cdot t_i + \sum_{k=1}^{Kt} u_k^t (t_i - \kappa_k^t)_+ + \varepsilon_i \qquad (2)$$

In Eq. (2) there is the penalization of the knot coefficients u_k^s and u_k^t, where $\kappa_1^s, \ldots, \kappa_{ks}^s$ and $\kappa_1^t, \ldots, \kappa_{kt}^t$ are knots in the s and t directions respectively. The penalization of the u_k^s and u_k^t is equivalent to treating them as random effects in a mixed model [10].

Setting $\beta = (\beta_0, \beta_s, \beta_t)^T$, $u = (u_1^s, \ldots, u_{ks}^s, u_1^t, \ldots, u_{kt}^t)^T$, $X = (1 \, s_i t_i)$ with $1 \le i \le n$, $Z = (Z_s | Z_t)$, with:

$$Z_s = \left[(s_i - \kappa_k^s)_+ \right]_{1 \le i \le n, 1 \le k \le Ks}, \quad Z_t = \left[(t_i - \kappa_k^t)_+ \right]_{1 \le i \le n, 1 \le k \le Kt} \qquad (3)$$

penalized least squares is equivalent to best linear unbiased prediction in the mixed model:

$$y = X\beta + Zu + \varepsilon; \ E\begin{pmatrix} u \\ \varepsilon \end{pmatrix} = 0; \ \mathrm{cov}\begin{pmatrix} u \\ \varepsilon \end{pmatrix} = \begin{bmatrix} \sigma_s^2 \cdot I & 0 & 0 \\ 0 & \sigma_x^2 \cdot I & 0 \\ 0 & 0 & \sigma_\varepsilon^2 \cdot I \end{bmatrix} \qquad (4)$$

Model (4) is a variance components model since the covariance matrix of $(u^T \varepsilon^T)^T$ is diagonal. The variance ratio $\sigma_\varepsilon^2 / \sigma_s^2$ acts as a smoothing parameter in s direction. Penalized spline additive models are based on low rank smoothers, as defined by Hastie [2], considering that linear terms are easily incorporated into the model through the $X\beta$ component.

At this point we can incorporate a geographical component by expressing kriging as a linear mixed model and merging it with an additive model such as model (4) to obtain a single mixed model (defined as geoadditive model).

Universal kriging model for (x_i, y_i), $1 \le i \le n$ (y_i are scalar and x_i represent geographical location included in R^2 domain) is [1]:

$$y_i = \beta_0 + \beta_1^T x_i + S(x_i) + \varepsilon_i \qquad (5)$$

where $S(x)$ is a stationary zero-mean stochastic process and ε_i are assumed to be independent zero-mean random variables with common variance σ_ε^2 and distributed independently of S. Prediction at an arbitrary location x_0 is done through the following expression:

$$\hat{y}(x_0) = \hat{\beta}_0 + \hat{\beta}_1^T x_0 + \hat{S}(x_0)$$

Then for a know covariance structure of S the resulting equation is:

$$\widehat{y}(x_0) = \widehat{\beta}_0 + \widehat{\beta}_1^T x_0 + c_0^T \left(C + \sigma_\varepsilon^2 I\right)^{-1}\left(y - \widehat{\beta}_0 - \widehat{\beta}_1^T x\right) \tag{6}$$

where:

$$C = \left(\text{cov}\{S(x_i), S(x_j)\}\right)_{1 \leq i,j \leq n}$$
$$c_0^T = \left(\text{cov}\{S(x_0), S(x_i)\}\right)_{1 \leq i \leq n}$$

For the implementation of Eq. (6) we can use:

$$\text{cov} = \{s(x), S(x')\} = C_\theta(\|x - x'\|) \tag{7}$$

where $\|v\| = \sqrt{(v^T v)}$ and C_θ is a term of a Matérn covariance function. The complete formulation of C_θ term corresponds to:

$$C\theta(r) = \sigma_x^2(1 + |r|/\rho)\exp(-(r)/\rho) \tag{8}$$

Equation (8) is the simplest member of the Matérn family and ρ term can to be choose with following rule to ensure scale invariance and numerical stability [1]:

$$\widehat{\rho} = \max_{1 \leq i,j \leq n}\|x_i - x_j\| \tag{9}$$

For all aspects and matters above reported, a geoadditive model can be described, substantially, as a single linear mixed model as follow:

$$y_i = \beta_0 + f(s_i) + g(t_i) + \beta_1^T \cdot x_i + S(x_i) + \varepsilon_i \tag{10}$$

It we put $X = \left(1\, s_i t_i x_i^T\right)$ with $1 \leq i \leq n$ and $Z = (Z_s|Z_t|Z_x)$, where Z_s and Z_t are defined by Eqs. (3) and $Z_x = Z\Omega^{-1/2}$ with:

$$X = \left(1\, x_i^T\right)_{1 \leq i \leq n}$$

$$Z = \left[C_0\left(\|x_i - \kappa_k\|/\rho\right)_{1 \leq k \leq K}\right]_{1 \leq i \leq n}$$

$$\Omega = \left[C_0\left(\|\kappa_k - \kappa_{k'}\|/\rho\right)_{1 \leq k,k' \leq K}\right]$$

$$C_0(r) = (1 + |r|)\exp(-|r|)$$

The model has this representation:

$$y = X\beta + Zu + \varepsilon \tag{11}$$

where:

$$E\begin{pmatrix} u^s \\ u^t \\ \tilde{u} \end{pmatrix} = 0; \mathrm{cov}\begin{pmatrix} u \\ \varepsilon \end{pmatrix} = \begin{bmatrix} \sigma_s^2 I & 0 & 0 & 0 \\ 0 & \sigma_t^2 I & 0 & 0 \\ 0 & 0 & \sigma_x^2 I & 0 \\ 0 & 0 & 0 & \sigma_\varepsilon^2 I \end{bmatrix} \qquad (12)$$

Model (10) can be extended to incorporate linear covariates through the $X\beta$ term. The extension to more than two additive components is straightforward.

3 Empirical Analysis

In this section are presented the results of theoretical model described in previous section for a sample of real estate data.

For a central neighborhoods of the city of Reggio Calabria (Italy), n. 187 sales of residential units located in a limited geographical area have been observed in a period of twenty years about. Obviously here are presented the results for a test designed to verify the validity of the model proposed. The residential units have the same building types and are included into a homogeneous urban area in terms of services and infrastructural qualification. For each property the real estate market price and the amounts of some real estate characteristics are known, as shown in Table 1.

Table 1. Variable description.

Variable	Description
Real estate unitary price (UPRICE)	expressed in Euro/1000
Geographic coordinates (XCOORD, YCOORD)	expressed with longitude and latitude
Property's age (EPOCA)	expressed retrospectively in no. of years
Sale date (DAT)	expressed retrospectively in no. of years
Number of services (BATH)	no. of services in residential unit
Positional Variable (ZONE)	expressed with a score scale (from 1 to 5, passing from more central areas to peripheral areas)
Maintenance (MAIN)	expressed with a score scale (1 for bad conditions, 2 for mediocre conditions, 3 for good conditions, 4 for optimal maintenance state)
Floor level (LIV)	no. of floor level of residential unit

On the basis of data the following semi-parametric model has been implemented:

$$UPRICE = LIV + MAIN + BATH + ZONE + DAT + f(EPOCA) + f(XCOORD, YCOORD)$$

Results and main indices of model verification are presented in tables and graphic that follow. The determination of knots for the spatial component and its geographical coordinates are identified by the space filling algorithm, implemented in default.knots.2D function library of R Software [3]. The model (13) was therefore estimated by the Re.M.L. method using the spm library of R software.

The estimates of effects in the non-linear model have been significant by values of freedom degrees (*df*) and smoothing parameters (*spar*). The values of obtained predictions are consistent with observed data, also analysis of residuals has not shown any abnormality in its structure. In examined area the spatial distribution of real estate unitary prices clearly shows how the geographical component affect the prices of sampled properties (Table 2).

Table 2. Statistical description of variables.

Variable	Std. Dev.	Median	Mean	Min	Max
UPRICE	0.30	0.95	0.99	0.41	2.14
XCOORD	0.01	15.65	15.65	15.63	15.68
YCOORD	0.03	38.09	38.09	38.02	38.19
EPOCA	14.46	19.00	22.78	2.00	73.00
DAT	5.35	10.00	9.56	0.00	23.00
LIV	1.38	3.00	3.42	1.00	7.00
BATH	0.53	2.00	1.66	1.00	4.00
ZONE	0.69	5.00	4.58	2.00	5.00
MAIN	0.90	3.00	3.07	1.00	4.00

The main result of the interpolation is a thematic map depicting the real estate unitary values in the urban context considered, in which blue and red colors represent unitary values, respectively, lowest and highest values (see Figs. 1 and 2).

Fig. 1. Reggio Calabria map with indication of the urban context considered.

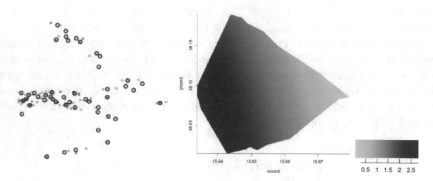

Fig. 2. Spatial distribution of real estate unitary prices, knots placement and location of housing units.

4 Conclusion

The geoadditive models are an effective vehicle for the analysis of spatial real estate data and other applications where geographic point data are accompanied by covariate measurements.

In this paper the spatial distribution of unitary real estate prices has been analyzed for a central urban area of Reggio Calabria, such experimentation has allowed to verify the reliability of the proposed model. The results obtained by the application of the proposed model suggest that geoadditive models can be successfully used for the prediction and spatial distribution of unitary real estate market values. This result may aid for the verification of progressive real estate sub-samples in able to detect and to identify possible potential market premium in real estate exchange and rent markets for green buildings [23–32].

The objectives pursued with the theoretical model proposed are many and varied, such as the study of different segments of local real estate markets, or even the prediction and interpretation of the phenomena related to the genesis of rewards of position, with particular reference to problems of transformation and investments for urban areas affected from projects or action plans, and in order to optimize the choices of use of goods and resources.

References

1. Kammann, E.E., Wand, M.P.: Geoadditive models. Appl. Stat. **52** (2003)
2. Hastie, T.J.: Pseudosplines. J. Roy. Stat. Soc. Ser. B **58** (1996)
3. Wand, M.P., French, J.L., Ganguli, B., Kammann, E.E., Stuadenmayer, J., Zanobetti, A.: SemiPar 1.0 R package (2005). http://cran.r-project.org. Accessed 14 Oct 2017
4. Wand, M.P.: Smoothing and Mixed Models. Comput. Stat. **18** (2003)
5. Ruppert, D., Wand, M.P., Carroll, R.J.: Semiparametris regressions. Cambridge University Press (2003)

6. Morano, P., Locurcio, M., Tajani, F., Guarini, M.R.: Fuzzy logic and coherence control in multi-criteria evaluation of urban redevelopment projects. Int. J. Bus. Intell. Data Min. **10**(1), 73–93 (2015)
7. Guarini, M.R., D'Addabbo, N., Morano, P., Tajani F.: Multi-criteria analysis in compound decision processes: the AHP and the architectural competition for the chamber of deputies in Rome (Italy). Buildings **7**(2), 38 (2017)
8. Antoniucci, V., Marella, G.: Immigrants and the city: the relevance of immigration on housing price gradient. Buildings **7**, 91 (2017)
9. Antoniucci, V., Marella, G.: Small town resilience: housing market crisis and urban density in Italy. Land Use Policy **59**, 580–588 (2016)
10. Del Giudice, V., De Paola, P.: Spatial analysis of residential real estate rental market. In: d'Amato, M., Kauko, T. (eds.) Advances in Automated Valuation Modeling. Studies in System, Decision and Control, vol. 86, pp. 9455–9459. Springer (2017)
11. Del Giudice, V., De Paola, P., Forte, F.: The appraisal of office towers in bilateral monopoly's market: evidence from application of Newton's physical laws to the Directional Centre of Naples. Int. J. Appl. Eng. Res. **11**(18), 9455–9459 (2016)
12. Saaty, T.L., De Paola, P.: Rethinking Design and Urban Planning for the Cities of the Future. Buildings **7**, 76 (2017)
13. Del Giudice, V., Manganelli, B., De Paola, P.: Depreciation methods for firm's assets. In: ICCSA 2016, Part III, LNCS, vol. 9788, pp. 214–227. Springer (2016)
14. De Ruggiero, M., Forestiero, G., Manganelli, B., Salvo, F.: Buildings energy performance in a market comparison approach. Buildings **7**, 16 (2017)
15. Del Giudice, V., De Paola, P., Manganelli, B.: Spline smoothing for estimating hedonic housing price models. In: ICCSA 2015, Part III. LNCS, vol. 9157, pp. 210–219. Springer (2015)
16. Ciuna, M., Milazzo, L., Salvo, F.: A mass appraisal model based on market segment parameters. Buildings **7**, 34 (2017)
17. Del Giudice, V., De Paola, P., Manganelli, B., Forte, F.: The monetary valuation of environmental externalities through the analysis of real estate prices. Sustainability **9**(2), 229 (2017)
18. Del Giudice, V., De Paola, P., Cantisani, G.B.: Rough set theory for real estate appraisals: an application to directional district of Naples. Buildings **7**(1), 12 (2017)
19. Del Giudice, V., De Paola, P., Cantisani, G.B.: Valuation of real estate investments through fuzzy logic. Buildings **7**(1), 26 (2017)
20. Del Giudice, V., De Paola, P., Forte, F.: Using genetic algorithms for real estate appraisal. Buildings **7**(2), 31 (2017)
21. Del Giudice, V., Manganelli, B., De Paola, P.: Hedonic analysis of housing sales prices with semiparametric methods. Int. J. Agric. Environ. Inf. Syst. **8**(2), 65–77 (2017)
22. Del Giudice, V., De Paola, P., Forte, F., Manganelli, B.: Real estate appraisals with bayesian approach and markov chain hybrid monte carlo method: an application to a central urban area of Naples. Sustainability **9**, 2138 (2017)
23. Massimo, D.E.: Green Building: characteristics, energy implications and environmental impacts. case study in Reggio Calabria, Italy. In: Coleman-Sanders, M. (ed.) Green Building and Phase Change Materials: Characteristics, Energy Implications and Environmental Impacts, vol. 01, pp. 71–101. Nova Science Publishers (2015)
24. Massimo, D.E.: Valuation of urban sustainability and building energy efficiency: a case study. Int. J. Sustain. Dev. **12**(2–4), 223–247 (2009)
25. Massimo, D.E.: Emerging issues in real estate appraisal: market premium for building sustainability. In: Aestimum, pp. 653–673 (2013)

26. Massimo, D.E.: Valutazione del rapporto tra insediamento ed energia. Il mercato immobiliare come driving force della sostenibilità urbana. In: Atti della 32° Conferenza Italiana di Scienze Regionali, AISRe. Il ruolo delle città nell'economia della conoscenza, Milano (2011)
27. Massimo, D.E., Musolino, M., Barbalace, A.: Stima degli effetti di localizzazioni universitarie sui prezzi immobiliari. In: Marone, E. (ed.) La valutazione degli investimenti pubblici per le politiche strutturali. Firenze University Press, Firenze (2011)
28. Massimo, D.E.: Stima del green premium per la sostenibilità architettonica mediante Market Comparison Approach, Valori e Valutazioni (2010)
29. Massimo, D.E., Del Giudice, V., De Paola, P., Forte, F., Musolino, M., Malerba, A.: Geographically weighted regression for the post carbon city and real estate market analysis: a case study. In: Calabrò, F., Della Spina, L., Bevilacqua, C. (eds.) Local Knowledge and Innovation Dynamics Towards Territory Attractiveness Through the Implementation of Horizon/E2020/Agenda2030. Springer, Berlin (2018). ISBN 978-3-319-92098-6
30. Del Giudice, V., Massimo, D.E., De Paola, P., Forte, F., Musolino, M., Malerba, A.: Post carbon city and real estate market: testing the dataset of reggio calabria market using spline smoothing semiparametric method. In: Calabrò, F., Della Spina, L., Bevilacqua, C. (eds.) Local Knowledge and Innovation Dynamics Towards Territory Attractiveness Through the Implementation of Horizon/E2020/Agenda2030. Springer, Berlin (2018). ISBN 978-3-319-92098-6
31. Spampinato, G., Massimo D.E., Musarella, C.M., De Paola, P., Malerba, A., Musolino, M.: Carbon sequestration by cork oak forests and raw material to built up post carbon city. In: Calabrò, F., Della Spina, L., Bevilacqua, C. (eds.) Local Knowledge and Innovation Dynamics Towards Territory Attractiveness Through the Implementation of Horizon/E2020/Agenda2030. Springer, Berlin (2018). ISBN 978-3-319-92098-6
32. Malerba, A., Massimo D.E., Musolino, M., De Paola, P., Nicoletti, F.: Post carbon city: building valuation and energy performance simulation programs. In: Calabrò, F., Della Spina, L., Bevilacqua, C. (eds.) Local Knowledge and Innovation Dynamics Towards Territory Attractiveness Through the Implementation of Horizon/E2020/Agenda2030. Springer, Berlin (2018). ISBN 978-3-319-92098-6

The Rise of the Co-creative Class: Sustainable Innovation-Led Urban Regeneration

Claudia Trillo$^{(\boxtimes)}$ (iD)

University of Salford, Salford M5 4WT, UK
`C.Trillo2@salford.ac.uk`

Abstract. While it is clear that cities are the ideal setting for the rise of the creative class, it is not yet clear how to detect emerging innovation in cities and how to implement it in a sustainable way. This paper builds on and moves forward a research project undertaken on Smart Specialisation Strategy and starts to operationalize a previously theorized novel approach to assess the emerging innovation in cities, based on unveiling factors, tools and triggers allowing cities to enable sustainable innovation-led urban regeneration. The case study of Media City UK has been used to test whether some of the findings emerged from the previous studies would be suitable to allow understanding the potential of an emerging innovation hub and its potential sustainability. Findings from Media City UK show that: conventional indicators can be successfully complemented with dynamic proxies capable to capture the changing nature of innovation; geographical boundaries should be set up as dynamic edges capable to capture the rationale of social innovation in the open networks. It is also recommended to explicitly include equity among the goals to be pursued through innovation-led urban regeneration strategies, to ensure that *social innovation* -and not just *innovation*- allows achieving urban inclusive growth.

Keywords: Sustainable innovation-led urban regeneration · Inclusive growth
Social innovation · Co-creative class · Smart specialisations

1 Introduction

The new role played by creativity in shaping the everyday life, society and spaces has been clearly supported by evidence and led to identify a real and proper new social force steering contemporary economics, the so called *creative class* [1]. Further studies recently demonstrated the correlation between unconventional innovation and density of the cities [2, 3], showing that although patents are more likely to be produced in suburban areas, unconventional innovation is happening in the densest urban areas. Because of that, more *place-based* policies should be consequently pursued. It could sound like a good news for social scientists that more face-to-face interaction is needed in order to produce innovation at least at its earlier stage, since this confirms that human ties -rather than formalized structures- work better in oiling the mechanism of collaboration and co-production. This could suggest to build on and move forward the concept of *creative class* towards the idea of the rise a new *co-creative* class. It has also been suggested, following the line of thoughts of emphasizing the role of

© Springer International Publishing AG, part of Springer Nature 2019
F. Calabrò et al. (Eds.): ISHT 2018, SIST 100, pp. 411–421, 2019.
https://doi.org/10.1007/978-3-319-92099-3_47

supercooperators in changing evolutionary patterns [4] and drawing from qualitative data gathered in some innovation districts, that dense urban patterns could be interpreted as a cognitive infrastructure for innovators [5], where the intensity and quality of positive feedback loop in the production of new ideas generated by *co-creators* are maximized by the spatial proximity and related physical factors such as mixed use, public transit, availability of incubators, accelerators, civic centers, social incubators, HE anchors, all framed within spatially managed urban regeneration initiatives [6]. Creative entrepreneurs, willing to cooperate and produce shared knowledge, prioritize places featured by: density, accessibility, availability of shared spaces that make good and cooperative actions *frequent and observable* [7]. This is possible not only if urban environments are dense, but also if they are shaped in a way -familiar to urban designers- that encompasses concepts such as walkable urbanism, smart growth, new urbanism, placemaking.

However, it has been also highlighted how innovation *per se* is not a driver for sustainable growth [8–10], in contrast, it may even act as catalyst for inequality. As advocated by the United Nations (Habitat III) with the Sustainable Goal 11: Sustainable Cities and Communities: Make cities inclusive, safe, resilient and sustainable, inclusive growth should be a pillar of any urban development strategy and plan. This principle is interrelated with the concept of social innovation, which focuses on *social* progress rather than on progress in itself and escapes the easy route of technocracy in innovation in favor of a value-laden vision of the change. The *process* matters, not just the goal. The innovation *ethos* should lead to tackle societal challenges and spill over the benefits to the wider community, not just to produce new patents. The rationale of such a vision could be regarded as intertwined with the paradigm of the Economics for the Common Good [11, 12], whose origin dates back to the Enlightening and coincides with the rise of Civic Economic, taught since 1754 by Antonio Genovesi in Naples. Applied to the concept of innovation districts and implemented within innovation-led urban regeneration strategies, this humanistic vision of innovation implies that sustainable innovation-led urban regeneration strategies should be planned and designed by taking in consideration (1) the social negative and positive potential spillovers, (2) the way in which the geography of inequality will be influenced by a given change, (3) the unbalance stemming from a *destructive creation* or from an even positive shock in the system.

Apart from putting forward traditional approaches to tackle social inequalities through spatial planning (additional services, more education, social housing...), current planning theory and practice is still missing a clear and comprehensive approach to this issue. All the aforementioned instruments are usually considered as remedial actions reflecting a sort of philanthropic attitude rather than real and proper economic tools for generating more economic success. Despite of the long-lasting debate about the Third Way, urban planning approaches, instruments and tools still reflect an understanding of the two concepts of "social equity" and "urban economic development" as if they were clashing forces dividing society and to be considered respectively as triggers for costs (the social) or as drivers for financial incomes (the economic). This is mainly due to a gap in the discipline that is far to be filled. Unfortunately, it is still difficult to exactly assess the economic damages produced by unequal development, however attempts have been made. Moreover, while it is clear that cities are the ideal setting for the rise of the (co-)creative class, it is not yet clear how to detect emerging innovation in cities and how to implement it in a sustainable way. Part of the problem is due to the misalignment between the rapid pace of innovation and the obsolete set of conventional data which planners and policy makers usually rely on, in order to make their decision. It would exceed the scope of this paper to provide solutions to all those issues in a comprehensive manner. This paper represents a small contribution towards this direction. It aims at showing how a novel approach to data visualization in spatial planning would help mapping innovation and supporting decision makers in the identification of spatial gaps in the geography of the (social) innovation, which would otherwise remain undetected and unresolved. This has been achieved by adopting a single case study research strategy, focusing on an innovation district -Media City UK, MCUK- located in Salford, Greater Manchester Area. Following this section focused on the problem statement (1), the following sections will: (2) clarify in which way the current European place- based strategies for innovation relate to the European urban agenda, to set up the link between economic development strategies and sustainable urban development; (3) discuss the empirical findings from the case study and (4) offer insights to policy makers and planners willing to achieve sustainable innovation-led urban regeneration.

2   Place-Based Strategies and Urban Agenda: The EU Perspective

The debate between place-based and place-blinded strategies has been informing the international policy making agendas for many years [13, 14]. Following the rationale of supporting place- based development strategies, the European Commission has recently introduced the so called Smart Specialization Strategy (S3 or RIS 3). This has been achieved by including the S3 within the regional development policies for the current programming period 2014–2020 through a quite strong regulatory approach, by means of the so-called ex ante conditionalities, meaning that the adoption of a dedicated strategy for S3 is a precondition for regional authorities in being able to access to some funds. In a nutshell, S3 is an innovative policy concept which emphasizes the principle of prioritisation in a vertical logic, non-neutral. It is based on 5 principles: (1) Granularity (the level had to be not too high); (2) Entrepreneurial discovery (entrepreneurs in the broadest sense discover and produce information about new activities); (3) Priorities will not be supported forever, since the strategies need to be adaptive and flexible; (4) S3 shall be an inclusive strategy and (5) has an experimental nature, meaning that - within a certain degree- a risk-taking approach is acceptable in public policies [15]. The development of the policy strategies and methods suggested by the European Commission in terms of spatial approach to the Smart Specialisation have been so far influenced quite explicitly by the Cluster Theory [16]. This theory has been turned in the US into a real proper national project, clustermapping. The EU has introduced a similar platform shadowing the US one. However, the complexity of the S3 concept is broader that a merely economic driven approach. Policy documents, political debate and academic literature seem to converge on the idea that a necessary driver for place-based development is social capital, which can be enabled by Social Innovation [17].

A EC granted research project, Multidisciplinary Approach to Plan Smart Specialization Strategies for Local Economic Development - MAPS LED, allowed to conclude, with respect to the social dimension, that S3 boost sustainable innovation- led urban regeneration under the following conditions: (1) the overall spatial urban ecosystem is innovation-supportive [5], (2) development is people-driven rather than place-based [18], (3) smartness allows achieving systemic innovation rather than mere technological innovation [19, 20]. Furthermore, S3 should explicitly pursue inclusive growth as part of the overall strategic agenda.

The pivotal role played by innovation in S3 paves the way to building a clear link between S3 and places where innovation tends to happen more frequently, i.e., cities [1] and in particular it suggests focusing on the urban hotspots of the innovation process, the innovation districts as defined by Katz and Wagner [21]. Previously, Katz and Bradley [22] had clarified that, though the territorial patterns of innovation could emerge in form of clusters whose spatial essence could be understood at regional scale [16], nevertheless "our open, innovative economy increasingly craves proximity and extols integration, which allow knowledge to be transferred easily between, within, and across clusters, firms, workers, and supporting institutions. (...) The avanguard of these megatrends is largely found not at the city of metropolitan scale (...) but in smaller enclaves, what are increasingly being called innovation districts". Swinney recently highlighted the nexus between clusters and cities [23], through an interpretation of cities as a specific type of cluster, where multiple industries operate benefitting from co-location. If concentrated urban hotspots are more likely to host unconventional innovation, wouldn't be wiser to prioritize few selected neighborhoods and concentrate efforts to leverage the emergence of new innovation districts that could trigger a wider economic development? Indeed, this would require a perfect alignement between urban and economic development strategies. To explore whether this is happening at European policy level, the current S3 policies have been checked against the priorities of the EC urban agenda and viceversa.

The current framework steering the urban agenda in the European Union -known as "Pact of Amsterdam"- has been signed in May 2016 and explicitly underpins the Habitat III goals of sustainable urban development and inclusive growth. In Sect. 2 (Priority Themes and cross-cutting issues of the Urban Agenda for the EU) it is clearly asserted that "the complexity of urban challenges requires integrating different policy aspects to avoid contradictory consequences and make interventions in Urban Areas more effective, (hence), ... the partnerships shall consider the relevance of the following cross-cutting issues for the selected priority themes: ... (12.3) Sound and strategic urban planning (link with regional planning, including 'research and innovation smart specialisation strategies' (RIS3), and balanced territorial development), with a place-based and people based approach." The Pact of Amsterdam explicitly mentions S3 as a key element that should be incorporated in the urban strategies, including a section entirely focused on "Civil Society, Knowledge Institutions and Business", which recognizes "the potential of civil society to co-create innovative solutions to urban challenges". Although social innovation is not explicitly recalled in this section, supporting (social) co-creation is strongly recommended. The path forward on how to incorporate innovation in sustainable urban transformation is therefore clear. Place-based and people-based strategies, nurtured by the civic society through co-creation, should be prioritized on place-blinded strategies based on a technocratic approach.

While the Pact of Amsterdam explicitly recognizes the role of S3 within the New Urban Agenda, S3 weakly recognize the strong tie they should have with cities. The Guide for the S3 [15] includes an annex specifically focused on S3 delivery instruments and horizontal approaches, but does not include cities nor the urban agenda. Though a network of nine European cities is currently sharing knowledge on how bridging the gap between S3 and cities through an Urbact project (InFocus), still little evidence were found on the awareness that due to the nature of S3, in particular due to the emphasis on the entrepreneurial discovery, they should be prioritized in denser urban areas featured by clear factors as listed in the previous section. This gap is partly due to the persisting difficulty to turn the multi-level governance mantra into an easily implementable and scalable planning practice, but also to the difficulties in elaborating evidence-based ground-breaking initiatives. For example, still institutional evaluations and political decisions on innovation are deemed accountable on the basis of data which cannot keep up with the pace of innovation. The following section discusses a project developed in partnership with the City of Salford that allowed understanding the potential of unconventional data to support decision making on an innovation district.

3 Innovation and Inclusive Growth Through Data Visualisation

Manchester is possibly one of the best example of European post-industrial cities that were able to reinvent their future in a very successful manner thanks to both knowledge economy and advanced innovation. A flagship regeneration project that is paradigmatic with this respect is MediaCity UK - MCUK, located in an area which corresponds to the former docklands of the city of Manchester, directly connected through the sea by the Manchester Canal. Following the closure of the industrial activities in 1982, the entire area of Salford Quays went through a systematic regeneration, still ongoing. One of the key-triggers of the local economic of the Salford Quays area is represented by the relocation to this area of a real and proper hub for the production of digital and media innovation, including companies of the caliber of BBC and ITV. The University of Salford opened a new campus in the area, whose educational activities are intertwined with the industry neighbors. This regeneration process has turned an area that 30 years ago was still a no man's land into a thriving engine for not just local but also regional growth. According to the Salford City Council [24], in 2015 Salford Quays hosted 3,500 residents and 900 businesses supporting over 26,000 jobs – around 23,000 more jobs than were lost when the docks closed. The recent improvements to the local transport network, the Metrolink, allowed a direct tram connection between the area and the Manchester city center, contributing to make it a walkable place. Hotspots for innovation - incubators and accelerators- are growing in the area, such as The Landing, The Greenhouse, as well as the University of Salford itself. A previous research paper [6] presented a grid for the assessment of spatial factors related to successful innovation-led urban regeneration, which had been developed within the MAPS LED

Table 1. Assessment grid for innovation – driven development applied to MCUK.

Spatial factor	Trigger	Spatial tool	Ecosystem
Dense and walkable urban environment. *The area is extremely dense and shows a high degree of walkability*	**Public local authorities:** *The SCC is leading the process*	**Mixed use:** *The plan is mixed-use* **Public transit:** *The Metrolink expansion improved connectivity*	**Proactive local public authorities** *The SCC is leading the process very effectively.*
Spatially identifiable hotspots. *Highly significant public spaces are numerous in the area and accelerators and incubators are growing (Greenhouse, Landing)*	**Private companies, HE, Public authorities** *Many private stakeholders boosted the development of the area, complemented by public institutions*	**Incubators Accelerators (both public and private)** *Present*	**Anchor companies Anchor institutions Champion(s)** *Very strong presence of private companies of international calibre (BBC, ITV)*
Local2local and local2global networks. *Actors such as the University of Salford could provide support to the activation of networks*	**Private companies, HE, Public authorities** *Strong presence*	**Spatial proximity of the local2local networks** *Not detected (December 2017)*	**Local and global incubators/accelerators** *International linkages of the Network of incubators/accelerators not detected*
Spatially identifiable civic innovation centres and socially driven incubators *Not detected (December 2017)*	**Public local authorities NGOs** *Not detected (December 2017)*	**Regeneration initiatives.** Salford Quays regeneration plan; **Civic centres** *Not detected;* **Social incubators** *Not detected*	**Active local authorities and communities** *Scarce presence of community groups*
Spatially identifiable anchor HE institutions: *University of Salford opened the new campus MCUK in the heart of the area*	**HE Institutions** *University of Salford*	**HE anchors HE accelerators and incubators** *The University of Salford has its own incubator*	**Starts-ups generated by the HE institutions** *The University of Salford incubator has generated starts up*

project. Table 1 shows the application of the assessment grid to MCUK, suggesting that although some elements are missing, the urban ecosystem of MCUK features the majority of the factors which made other urban spaces supportive of innovation-driven economic development.

The successful identification of the majority of the factors suggests that the MCUK spatial pattern is supportive of an innovation-driven economic development strategy, while the grid highlights some gaps on social innovation and civic engagement. Indeed, the contribution of MCUK to pursuing inclusive growth at the wider scale remains controversial. Pockets of deprivation still exist along the border of the regeneration area, hosting a population still plagued by low levels of education, poor health and economic conditions. The aspiration of mapping the potential of MCUK in triggering inclusive growth and overcoming the social gaps between a fast growing core and a lagging behind fringe suggested to look into both conventional and unconventional data. The potential of these latter was suggested by some studies developed by the MIT Senseable City Lab and by some attempts to understand urban inequality through data generated by social media [25]. In the MCUK exploratory investigation, a web based GIS platform developed by the University of Salford Think Lab in collaboration with Mirrorworld, namely City Data Explorer, has been used to visualize conventional data, complemented by most up to date data elaborated by two main stakeholders of the MCUK area, the University of Salford and the Salford City Council. The SPIN Unit further integrated these data with a set of unconventional data elaborated through social media datasets. The selection of unconventional data has been driven by their availability, rather than by theoretical consideration. The main goal was to depict the most up to date socio- economic appreciation of the area and its context. Figure 1 shows the concentration of companies located in MCUK, within the context of the surrounding neighborhoods (Fig. 1). The MCUK area is surrounded by a fringe where the level of deprivation is high (red in the map). Figure 2 overlays the postcode of the University of Salford alumni to the current location of the business (Fig. 2). The map shows a strong presence of alumni around the MCUK companies, leading to the hypotheses that

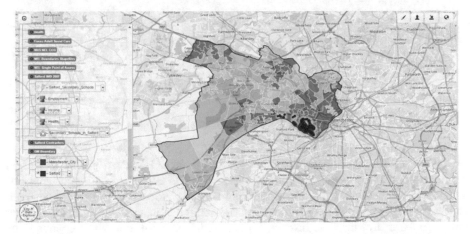

Fig. 1. Location of the companies in MCUK and social deprivation index of the surrounding.

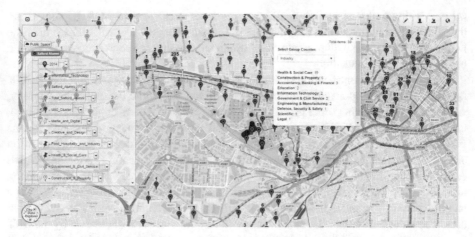

Fig. 2. Location of the University of Salford alumni in relation to the companies in MCUK.

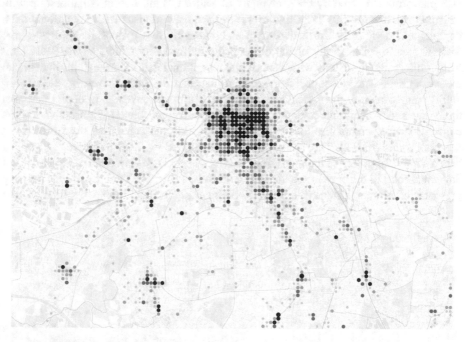

Fig. 3. The complexity of economic activities according to real time data.

MCUK is retaining educated workforce within the area. However, in the pockets of deprivation surrounding MCUK, due to the gaps in the educational level, it would be difficult to catch up with the emerging economic development. The more the innovation process is boosted in the innovation district, the more this latter should be better linked to the rest of the city. Instead, further studies showed how MCUK tends to be more isolated than connected. To be able to understand better the current spatial economic dynamics of MCUK within the Greater Manchester Area, unconventional data have been chosen on the basis of their free availability. By drawing from Four Corners, Instagram, Twitter, a rich set of maps and graphics have been elaborated by the SPIN unit. As an example, Fig. 3 shows the density and frequency of use of public places and services, determined through a previously tested methodology based on indicators drawn from social platforms open access data [26].

4   Findings and Conclusions

The rapid pace of innovation in MCUK can be considered a case of successful innovation- led urban regeneration. However, a pattern on inequality is also evident, with lagging-behind deprived pockets surrounding the shining innovation districts. The previous section showed how social and economic dynamics appear aligned with the current process of economic development. However, further studies based on unconventional data gathered with no restrictions in terms of administrative boundaries, show how two main "active corridors" are emerging, one leading from the Manchester City Center (main circular area in the middle) to South following the axis of the University of Manchester (the Oxford corridor), another leading from the Manchester City Center to East towards the Crescent campus of the University of Salford (corresponding to the Chapel Street regeneration corridor). Instead, MCUK (West side of the city center, position easily identifiable thanks to a bend in the Manchester Canal) appears quite insulated.

This observation would recommend to intensify mixed used development along the corridor that connects MCUK to the Manchester City Center, rather than focusing on a mere residential development for areas. This would also allow creating more economic opportunities for a still deprived area surrounding the innovation district, creating job places suitable to attract different educational levels.

The testing of the assessment grid and mixed datasets in terms of honing effectiveness on decision making has been performed by an actual stakeholders' engagement process, delivered through a series of meetings and workshops over a 6 months period, during which experts in charge of making strategic decisions were able to discuss the data. The approach was considered valuable and further stakeholders, following the project dissemination event, showed interest in applying the methodology.

The exploratory investigation developed in MCUK allowed demonstrating that (1) conventional indicators can be successfully complemented with dynamic proxies capable to capture the changing nature of innovation; (2) geographical boundaries should be set up as dynamic edges capable to capture the rationale of social innovation in the open networks, in order to ensure that external spillovers are properly considered. Finally, it is recommended to explicitly include the goal of inclusionary growth among the goals to be pursued through the urban transformation, to ensure that social innovation -and not just innovation- allows achieving a balanced change in the urban socio-economic and material structure of cities.

Acknowledgements. The paper builds on and puts forward some findings developed within the MAPS-LED project, funded by the European Commission under the Horizon 2020 Program, Project ID: 645651. The author wishes to thank the University of Salford for the support gained from the Higher Education Innovation Funding (HEIF), which allowed testing the framework against a real-world case study.

References

1. Florida, R.: The rise of the Creative Class. Basic Books, New York (2002)
2. Berkes, E., Gaetani, R.: The geography of unconventional innovation. https://economicdynamics.org/meetpapers/2015/paper_896.pdf. Accessed 06 Dec 2017
3. Florida, R.: How Innovation Leads to Economic Segregation. CityLab, 24th October 2017. https://www.citylab.com/life/2017/08/the-geography-of-innovation/530349/. Accessed 02 Dec 2017
4. Nowak, M.: Supercooperators: Altruism, Evolution, and Why We Need Each Other to Succeed. Free Press, New York (2011)
5. Trillo, C.: Smart specialisation strategies as drivers for (Smart) sustainable urban development. In: Ergen, M. (ed.) Sustainable Urbanization. InTech. https://www.intechopen.com/. Accessed 06 May 2016
6. Trillo, C.: Towards an assessment methodology for smart specialisation strategies: sustainable local development. In: IRWAS, vol. 149. University of Salford (2017)
7. Rand, D.G., Yoeli, E., Hoffman, M.: Harnessing reciprocity to promote cooperation and the provisioning of public goods. Policy Insights Behav. Brain Sci. **1**(1), 263–269 (2014)
8. Berkes, E., Gaetani, R.: Income segregation and rise of the knowledge economy. https://sites.northwestern.edu/. Accessed 13 Dec 2017
9. Florida, R.: The Geography of Innovation. CityLab, 3rd August 2017. https://www.citylab.com/life/2017/08/the-geography-of-innovation/530349/. Accessed 01 Dec 2017
10. Devaney, C., Shafique, A., Grinsted, S.: Citizens and inclusive growth, RSA in collaboration with JRF. https://www.thersa.org/globalassets/pdfs/reports/rsa_citizens-and-inclusive-growth-report.pdf/. Accessed 06 Dec 2017
11. Zamagni, S.: L'Economia del Bene Comune. CittaNuova, Roma (2007)
12. Tirole, J.: Economics of Common Good. Princeton University Press, Princeton (2017)
13. Barca, F., McCann, P., Rodríguez-Pose, A.: The case for regional development intervention: place-based versus place-neutral approaches. J. Reg. Sci. **52**(1), 134–152 (2012)
14. Monardo, B., Trillo, C.: Innovation strategies and cities. Insights Boston Area Urban. **157**, 154 (2016)
15. EC: RIS3 GUIDE, Joint Research Center, Smart Specialisation Platform (2011)
16. Porter, M.E.: Location, competition and economic development: local clusters in a global economy. Econ. Dev. Q. **14**(1), 15–20 (2000)
17. EU: Guide to Social Innovation (2013)
18. Devaney, C., Trillo, C.: Spinning the wheel and switching on the lightbox. Towards a novel evaluation for Smart Specialisation. Urbanistica **157**, 125–128 (2016)
19. Agbali, M., Arayici, Y., Trillo, C.: Taking the advancement of sustainable smart cities seriously: the implication of emerging technologies. Tamap J. Eng. **2017** (2017)
20. Agbali, M., Trillo, C., Arayici, Y., Fernando, T.: Creating smart and healthy cities by exploring the potentials of emerging technologies and social innovation for urban efficiency lessons from the innovative city of Boston. Int. J. Urban Civ. Eng. **11**(5), 617–627 (2017)

21. Katz, B., Wagner, J.: The Rise of Innovation Districts: A New Geography of Innovation in America. Brookings Institution Press, Washington D.C. (2014)
22. Katz, B., Bradley, J.: The Metropolitan Revolution: How Cities and Metros Are Fixing Our Broken Politics and Fragile Economy. Brookings Institution Press, Washington D.C. (2013)
23. Swinney, P.: How do we encourage innovation through clusters? In: Center for Cities. http://www.centreforcities.org/. Accessed 06 Dec 2017
24. SCC: Salford City Council. https://www.salford.gov.uk/mediacityuk/. Accessed 06 Dec 2017
25. Shelton, M., Poorthuis, A., Zook, M.: Social media and the city: rethinking urban socio-spatial inequality using user-generated geographic information. Landsc. Urban Plann. **142**, 198–211 (2015)
26. Baeza, J., Cerrone, D., Mannigo, K.: Comparing two methods for urban complexity calculation using the Shannon-Wiener Index. WIT Trans. Ecol. Environ. **226**, 369–378 (2017)

A Novel Approach for Establishing Design Criteria for Refugees' Shelters

Rania Abumaradan[1]([⊠]) [iD] and Claudia Trillo[2] [iD]

[1] Petra University, Amman 961343, Jordan
raburamadan@uop.edu.jo
[2] University of Salford, Salford M5 4WT, UK
C.Trillo2@salford.ac.uk

Abstract. There is currently growing interest amongst the international community of policy makers about finding solutions for accommodating refugees, who currently concentrate mainly in countries located in the MENA region such as Syria, Yemen, Iraq, Palestine and Libya. Man-made disasters were the main cause for displacing people, whether inside their countries or forced to cross the border. This paper discusses the impact on refugees produced by not adequate accommodations and gathers refugees' inputs on customizing a suitable place for living. The paper focuses on a case study, the Al Zaatari camp in Jordan, which is the most recent camp of MENA region. The case study is developed through an extensive primary data collection, including interviews with refugees in the Al Zaatari camp. Refugees provided the researchers with their inputs by sharing their needs and desires in a new situation. Further insights were provided by involving NGOs experts and their insights and experiences, through an iteration cycle within the design science method. The main finding from the research is a new perspective opposite to a standardized approach to refugees' needs, built through listening to what refugees have to say on how to improve their lives in camps.

Keywords: Refugees' camps · Al Zaatari · Settlement criteria

1 Introduction

Accommodating refugees is becoming a crucial concern for the global community. The United Nations High Commissioner for Refugees [UNHCR] [1] reported that non-government organizations indicate there are 6,000,000 refugees distributed between 60 camps in Middle East and North Africa and 4.8 million registered refugees in camps in Asia and Africa. Refugees are accommodated in secured and protected shelters in the host country. In order to mitigate any risks and hazards, refugees are sheltered in the quickest time possible. According to the Sphere handbook [2, 3] in facts a shelter is a place to provide security, safety and protection from the climate changes.

However, it is also central ensuring human dignity and sense of community in order to support refugees to recover from the impact of disasters. Local governments and humanitarian organizations such as UNHCR reported that there are many areas around

© Springer International Publishing AG, part of Springer Nature 2019
F. Calabrò et al. (Eds.): ISHT 2018, SIST 100, pp. 422–432, 2019.
https://doi.org/10.1007/978-3-319-92099-3_48

the world suffering from a lack of attention on the refugees' conditions, leading to major challenges in many sectors such as healthcare, education, economic, cultural and social aspects.

This paper discusses the impact on refugees produced by not adequate accommodations and gathers refugees' inputs on customizing a suitable place for living. The paper focuses on a case study, the Al Zaatari camp in Jordan, which is the most recent camp of MENA region. The case study is developed through an extensive primary data collection, including in depth interviews with refugees in the Al Zaatari camp. Refugees provided the researchers with their inputs by sharing their needs and desires in a new situation. Further insights were provided by involving NGOs experts and their insights and experiences. The main finding from the research is a new perspective opposite to a standardized approach to refugees' needs, built through listening to what refugees have to say on how to improve their lives in camps.

Some authors [4–6] discussed the challenges of refugee settlement in a short time, showing conflicts between temporary and permanent solutions. Temporary shelters are usually the most practical solution for accommodating refugees. However, temporary shelter units have drawbacks in terms of meeting the physical and non-physical needs of refugees. At the beginning, refugees are directed to settle in shelters which are seen as suitable to meet basic needs in an emergency situation. Emergency shelters are often provided in the form of tents which can last for up to two years, after which weather conditions affect the durability [7]. After this time, refugees should be moved to semi-permanent or permanent prefab shelters to obtain stability and reach their satisfaction. In this regard, studies into temporary shelters are divided in two groups: program management of urban planning strategies and program design management. Previous examples in affected countries such as Japan, Turkey and Iran have shown that the two programs aim to build temporary shelters in less time and cost. A variety of strategies have been employed in pre-disaster and post-disaster programmes, such as informal and formal strategies. While the formal strategy leads to standardization of factors and to one single approach on technology, the informal strategy takes advance of a variety of technologies. However, the drawbacks of the informal sector can lead to conflicts with the local community, because it is often unregulated and does not take into account planning strategy, health protection considerations, sanitation, sewage and other factors [8].

This paper examines the establishing of adequate shelters for refugees by considering the refugees' perspective within a formal strategy applicable globally. This approach is pursued by developing a list of criteria that guide the process of accommodating refugees as shown in Fig. 1. Criteria control the linkage between shelters implementation and design proposal of adequate shelters.

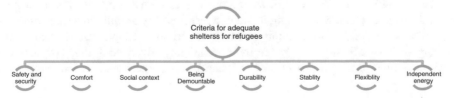

Fig. 1. List of criteria

2   Previous Studies of Accommodations

The literature review shows examples of accommodating affected people by natural and man-made disasters; several organizations presented solutions to settle affected people in a new location. This paper discusses the possibility of providing adequate shelters based on criteria derived by refugees in their living context. Despite abundant examples of temporary shelter on global level in different affected locations such as Iran, Turkey, Indonesia and India, the literature review did not show examples that can provide an immediate shelter also aimed to provide adequate place for living in a broader sense.

In Turkey, a bottom-up strategy was applied where the government has offered strategy and program planning. Turkey's efforts did not meet the needs of refugees with regular prefabricated concrete slab units. In spite of the planning program for such units, the shortage was to cover social safety and stability by supporting gathering spaces for refugees in a camp.

Another example is located in Cape Town, where a project offered shelters made from recycled materials and created by domestic skills as shows in Fig. 2. The project aimed to combine standardization and local skills of variety and multiplicity by creating cluster organization of units [8]. However, the drawbacks of the informal sector without formal supervision led to a nasty community that caused living difficulties, health problems and poor hygiene.

Fig. 2. Informal shelter in Cape Town (Source: www.google.jo)

Another solution utilized in disaster situations is compacting concrete as a honeycomb concrete panel which is light and as a result is easy to build in a short time [9]. In terms of durability, honeycomb is based on origami design that is used in the military shelters. Military shelters are more technically advanced in terms of materials used and structure. A number of soft wall accordion shelters such as tent accordions have been developed, also closed structured units [9] can come from a number of building industries that have provided emergency shelters and their structural components as shown in Fig. 3.

Fig. 3. Military folded shelter (Source: www.inhabitat.com)

In 2009–2010, following the Indian Ocean tsunami, India, Thailand, Indonesia and the Habitat for Humanity (HFH) repaired homes by providing a core house that used recycled materials and offered the potential for future extension [10].

However, no single example exists from international institutions and other organizations showing explicit attention for social and cultural needs of refugees and solutions are usually based on providing tents as an emergency solution and developing prefabricated shelters regardless of local context conditions.

The number of refugees around the world has risen dramatically over the past 20 years and the agencies involved in providing vast quantities of shelters for different climate regions have consequently suffered in terms of their efficiency. Refugees need to be accommodated immediately in an emergency or temporary shelter as a quick response to their desperate situation of being homeless [11]. Whilst tents may provide an immediate solution, many practices, unfortunately, have inappropriately used them for more than a few months. Generally, a tent does not meet refugees' needs in terms of offering privacy, space for families or social activity. UN-HABITAT& IFRC [12] present a planning strategy point of view, showing how tents are an inadequate answer compared with a local shelter model built by refugees; consequently such refugees often build new tents beside the old ones to more adequately meet their needs. A tent is not an ideal solution for accommodating refugees for more than a few months; it is clear that tents cannot withstand harsh weather conditions in the long term and they provide insufficient space for social and daily activities as shown in Fig. 4. The type of material a tent is manufactured from is another negative element, often making them inflexible and inadequate in terms of thermal comfort.

Prefabricated shelters are an inadequate solution for refugees because of time and cost limitations that negatively affect the efficiency of shelters [13]. Whilst prefabricated shelters have the potential to be a flexible and easy solution in the early staging of program strategy, the difficulties of adaptability to the climate conditions (especially in

Fig. 4. Challenges of units' performance and services (Source: www.google.jo)

harsh climate conditions such as cold, or hot climates) mean that it would not be a perfect solution as it would interfere with the technology inside the shelter, such as insulation and mechanical ventilation.

3 Refugees Needs-Criteria for Refugees' Requirements

Shelters allow meeting basic needs, however, it has been noticed that a gap exists between what refugees need for better living conditions and what organizations provide [14]. Refugees, as any other people, experience a critical situation particularly when they arrive to a host community, however, while it may be enough at the first stage providing them with an emergency shelter, at a later stage stress factor and psychological suffering require more complexity of needs satisfactions. Accordingly, the paper presents the relation between shelter as one of basic needs and its impact on refugees' satisfactions, where shelter will be presented through criteria, incorporating the refugees' perception on what the criteria should meet. Although a shelter is an object that protects people from external dangers, it is also a place to experience dignity, identity and sense of belonging, thus it must be appropriate for people to obtain satisfactions of stability, security and participation in the host community. Refugees' shelters are an important quality of life factor for establishing safe and secure feeling and determining characteristics of comfort, control, stability, environment resistance, and social and cultural privacy, refugees' experiences of daily living by creating adequate spaces for different activities such as women gathering, cooking and washing.

The paper proposes criteria for managing design performance justifying them through interviews with refugees and fieldwork. Criteria are developed by a list of requirements created from primary data collected in the Al Zaatari camp.

A brief discussion of the criteria follows.

Safety and security criteria need to be meet both in case of natural or man-made disasters. Refugees do not move in one route, they follow unplanned movements to achieve their needs which include safety and security [15].

Comfort is about a person's well-being and is influenced by external factors such as temperature, lighting, noise, as well as it is affected by individual conditions (e.g. age and gender) [16].

Social context holds multiple meanings and definitions and refers to what people bring to a place to establish their social and cultural characteristics. Researchers indicate to social context as place attachment or spirit of place because following the act of sharing daily activities, cognitive experiences, emotions and beliefs in a certain piece of land [17]. Although refugees in camps try to bring their meaning of home by modifying the built environment, they are limited by insufficient thermal resistance and structure effectiveness of the shelters [18].

Being *demountable* refers to the ability to take apart a structure and build it again within a period of time. The Institution of Structural Engineers [19] describes a demountable structure as a temporary structure based on a type of material, structure, cost and time. The requirements of demountable structures are distributed between a numbers of factors which are time determined, flexibility of modernity, use of light-weight materials and components of the structure. A demountable structure is a solution for quick and easy protection compared with conventional temporary structures. In this regard, the ability to use such structures many times requires more attention from a design point of view.

Stability leads to the route that refugees follow when they decide to go to a certain place rather than others. Besides the demand of leaving a dangerous area, refugees would choose a place that is similar to their social and cultural background, which means they would look to neighboring countries with similar characteristics to the original one [20].

Durability is defined as the ability and capability of design components to provide safe construction. Durability cannot be separated from flexibility; it defines levels of materials and structure strength, external and internal factors such as weather conditions and building operation and maintenance. Durability addresses the long term use of shelters [21]. Many factors are related to durability and must be taken into consideration such as; temperature, solar radiation, humidity, wind, and dust.

The need for *flexibility* is to reduce wasted time and materials by changing used spaces and increasing building longevity. For this reason, in emergency situations, flexibility and modularity is required due to the temporary situation. Flexibility is required by UNHCR in a rapidly changing and moving world [22]. Flexibility, rapidity, simplicity and security are needed in camps. Flexibility refers to circumstances that appear throughout the operation, such as adding elements to a shelter because of expanded families or repairing shelters.

Independent constant energy is the capacity of a physical system to do work or produce a change. Young [23] defines energy as coming from natural resources of the environment such as sunlight, wind movement, rain and wave motion [16].

4   The Research Methodology

The research methodology is based on a single case study research strategy, developed through fieldwork, informal interviews, semi-structured interviews with refugees and experts in the field. The paper focus on Al Zaatari camp in the north of Jordan as shown in Fig. 5, to understand refuges needs and their settlement requirements. Al Zaatari is the most recent camp in Middle East region and the second biggest camp around the world. The researchers conducted interviews with refugees in two stages; stage one included informal interviews with five refugees and the stage two included twenty four semi structured interviews with refugees, in both stages the interviews were conducted in their context. Qualitative data were analyzed through Nvivo, number of nodes and codes have been categorized to understand the priorities from the refugees' stories, data outcomes allowed developing a list of refugees' requirements and were revisited against the literature and experts' opinions.

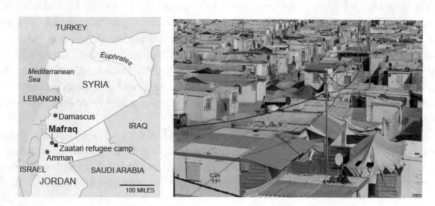

Fig. 5. The Al Za'atari camp (Source: www.google.jo)

The interview questions focused on the organizations consideration of climate difficulties based on users' views, refugees provided different perspectives to achieve comfort more related to their traditional understanding. The researchers integrated different points of views from the literature and the fieldwork by applying the rationale of finding intersections between the criteria and the interviewees' perspective. Interview questions also consider if social aspects are incorporated in existing shelters, investigating whether only technical solutions or also social solutions can be considered important. In short, the paper explores the ability of existing shelters in Al Zaatari camp of providing adequate space for living, by drawing from the refugees experiences matched against the criteria.

5   Discussion

The literature review shows how a number of criteria are followed by NGOs and local governments in emergency situations for the establishment of camps globally. The Sphere Project's [2] shows the core standards and minimum principles as shown in Fig. 6 including: protection, core standards, water supply and sanitations, food security, shelter and settlement and health action.

UNHCR [22] focuses on two categories which are; shelter and settlement and non-food items (bedding, clothing and household items) as shown in Fig. 7.

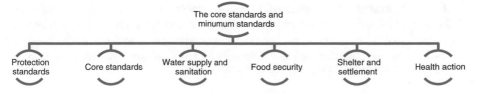

Fig. 6. The core standards of Sphere project for refugees

Fig. 7. The standards of UNHCR for refugees

Although international organizations are involved in managing programs for the settlement of refugees in emergency situation, it is clear that such organizations concentrate on standards and minimum principles valid for a short period of time that are not applicable longer than a few months. Moreover, refugees as users are not involved in the definition and implementation of programs and failure in understanding their needs leads to poor standardization and prioritization.

This paper contributed to solve this problem by filling the gap in the knowledge and creating a set of eight criteria by combining literature and refugees' requirements collected through interviews in Al Zaatari camp. The list could help organizations and local governments when accommodating refugees by providing adequate shelter and appropriate living condition in a camp. The following diagram Fig. 8 shows the point of view of refugees as resulted from the interviews in what they consider adequate or inadequate, with respect to the social considerations.

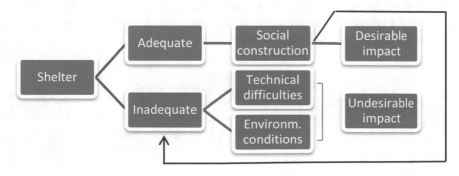

Fig. 8. The impact of inadequate and inadequate shelter

Figure 9 presents the impact of technical aspects and relation with inadequate shelters and how criteria could steer the process of achieving satisfactory shelters. In short, the list of criteria, which has been developed giving a central role to the refugees' perspective and complementing their view with experts' opinions, allows making better decisions and supports NGOs and local government.

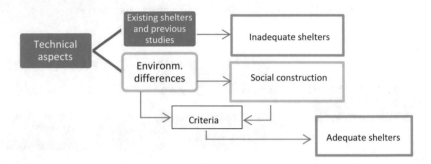

Fig. 9. The impact of technical aspects

6   Conclusion. Inadequate Standardizations and Criteria

The paper highlights challenges in offering adequate shelters to refugees. It achieves this goal by assessing the gaps between the findings from the literature review and primary data collected by interviewing the users (refugees). The paper shows how inadequate practices of institutions and humanitarian organizations towards refugees' settlement led mainly to standardized solutions for settlement refugees, which lack considerations of the differences regarding social aspects and environment challenges in camps. The paper finally offers a list of eight criteria developed through the involvement of the final users in the discussion. The connection of literature review and previous practices with the requirements expressed by refugees in the fieldwork gives a foundation to improve a proposed list of criteria that might be a solution for the settlement of refugees in adequate shelters globally. Institutions and local governments concentrate mainly on protecting refugees' life from danger immediately, however, after a while safety and protecting life are not enough and other factors should be taken in account. Social aspects and environmental specificities are still not well considered, while criteria offer a comprehensive way to connect social aspects and environmental challenges through technical requirements as shown in Fig. 10.

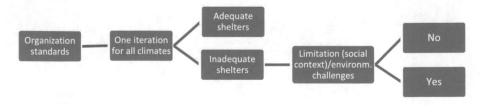

Fig. 10. The impact of social context and environment challenges on criteria

To conclude, the lack of attention on social and cultural aspects, as well as on environmental challenges, led to inappropriate solutions of refugees' shelter. This paper offers a set of criteria that allows appropriate stakeholders prioritization appropriate, prevent standardizations of shelter solutions and customize appropriate shelters based on differences and variation of local context, applicable globally.

References

1. United Nations High Commissioner for Refugees. Middle East and North Africa. UNHCR Global Appeal, Geneva (2014–2015)
2. The Sphere Project: Humanitarian Charter and Minimum Standards in Humanitarian Response. The Sphere Project, Geneva (2011)
3. Bulley, D.: Inside the tent: community and government in refugee camps. Secur. Dialogue 45(1), 63–80 (2014)
4. Hadafi, F., Fallahi, A.: Temporary housing respond to disasters in developing countries - case study: Iran-Ardabil and Lorestan Province Earthquakes. World Acad. Sci. Eng. Technol. 4(6), 1258–1264 (2010). International Science Index 42
5. Arslan, H., Cosgun, N.: Reuse and recycle potentials of the temporary houses after occupancy: example of Duzce, Turkey. Build. Environ. 43(5), 702–709 (2008)
6. Quarantelli, E.L.: Patterns of sheltering and housing in US disasters. Disaster Prev. Manag. 4(3), 43–53 (1995)
7. Jabr, H.: Housing conditions in the refugee camps of the west bank. J. Refugee Stud. 2(1), 75–87 (1989)
8. Lizarralde, G., Root, D.: Ready-made shacks learning from the informal sector to meet housing needs in South Africa. CIB World Build. Congr. 87, 2068–2082 (2007)
9. Bradford, N.M., Sen, R.: Multiparameter assessment guide for emergency shelters: disaster relief applications. J. Perform. Constr. Facil. 19(2), 108–116 (2005)
10. Schilderman, T.: Adapting traditional shelter for disaster mitigation and reconstruction: experiences with community-based approaches. Build. Res. Inf. 32(5), 414–426 (2004)
11. Davis, I.: Shelter After Disaster. Oxford Polytechnic Press, London (1978)
12. United Nations Human Settlements Programme: International Federation of Red Cross and Red Crescent Societies. Shelter Project (2009). UN-HABITAT, IFRC (2010)
13. United Nations High Commissioner for Refugees: Displacement. The new 21st Century Challenge. UNHCR global trends, Geneva (2013)
14. Heywood, F.: Understanding needs: a starting point for quality. Hous. Stud. 19(5), 709–726 (2004)
15. Clarin, M.: Climate refugees, refugees or under own protection? A comparative study between climate refugees and refugees embraced by the United Nations Refugee Convention (Unpublished dissertation). Karlstad University, Karlstad (2011)
16. Fuchs, M., Hegger, M., Stark, T., Zeumer, M.: Energy Manual: Sustainable Architecture. Birkhäuser, München (2012)
17. Cross, N.: Designerly Ways of Knowing. Springer, London (2006)
18. Sinan, M., Sener, M.C.A.: Design of a post disaster temporary shelter unit. A/Z ITU J. Fac. Architect. 6(2), 58–72 (2009)
19. Institution of Structural Engineers: Temporary demountable structures: Guidance on procurement, design and use. The Institution of Structural Engineers (IStructE), London (2007)
20. Maslow, A.H.: A theory of human motivation. Psychol. Rev. 50(4), 370–396 (1943)

21. Bradford, N.: Design optimization of composite panel building systems emergency shelter applications (Unpublished Ph.D. thesis), University of South Florida, South Florida (2004)
22. United Nations High Commissioner for Refugees: Handbook for Emergencies. UNHCR Headquarters, Geneva, Switzerland (2007)
23. Young, W.R.: Photovoltaic applications for disaster relief. Florida Solar Energy Center, Florida (2001)

A Comprehensive Proposition of Urbanism

With Potential Applications on Users' Urban Cognitive-Mapping Users' Generated Urban Designs

Mohammed Ezzat[(⊠)]

German University in Cairo, New Cairo City 11432, Egypt
mohammed.ezzat@guc.edu.eg

Abstract. "Urbanism could be comprehended using three perspectives labelled Visual, Emotional, and Rational" is the comprehensive proposition the paper introduces, asserts its validity, and analyses its potential utilizations. The proposition assures not only that there are three relativistic perspectives that could sufficiently help in having a proper understanding of Urbanism, but it also assures that these three relativistic perspectives are contrasting. By contrasting we mean that having an understanding of one perspective would entail a closer understanding of the others. Urbanism is comprehensible by perceiving its immense notions using the three perspectives, the repository of these perceived notions is accumulated by a companion comprehensive model. There is a desperate need for such comprehensive, highly abstract, proposition rather than relying solely on fragmented concrete low-level propositions for the analysis of complex subject like urban identity. Holistic abstraction is indispensable for proper understanding and prediction of users' urban cognition depending on their urban identity, and hence for developing socially and culturally sustainable urban designs. For instance, the successfulness of any empirical computational simulation of the users' urban Cognition-mapping is highly dependent on such comprehensive proposition. Additionally, using the comprehensive model as a standard monolithic tool of analysis would enrich the social comparative studies. A main task of the paper is to examine the validity of the comprehensive proposition by analysing the exhaustive understanding of important urban notions using the three perspectives. During that examination, characteristics of the three perspectives would be disclosed and accumulated using the companion comprehensive model. Our findings conform to theoretical and philosophical ontological foundations.

Keywords: A unifying model of Urbanism · Visually driven Urbanism
Rationally driven Urbanism · Emotionally driven Urbanism

1 Introduction

In our earlier research efforts for defining people's urban identity [1], we were faced with the reality that there is a desperate need for comprehensive propositions in the field of Urbanism. These comprehensive propositions could be thought of as part of the meta-definition of the field, and consequently, their existence would facilitate the

© Springer International Publishing AG, part of Springer Nature 2019
F. Calabrò et al. (Eds.): ISHT 2018, SIST 100, pp. 433–443, 2019.
https://doi.org/10.1007/978-3-319-92099-3_49

communication and representation not only internally between the urban notions themselves but also between Urbanism and the other complex fields, in our case, it was the theories and practices of the field of identity. The difficulties that we faced prior to defining the proposed comprehensive proposition, and its associated comprehensive model, were numerous and the analytical results were less reliable. For example, the urban identity empirical analysis of any sampled group would always be done using granular propositions form different complex fields. That low-level of analysis would always yield less reliable results that are specific to the case-study. Any serious scientific efforts of generalization, explanation, or mainly prediction are not viable beyond the locality of the sampled group. To tackle these limitations, we found that the low-level granular propositions should be coupled with more abstract comprehensive propositions that would facilitate the nominal structuring and grouping of these granular propositions into a simple and cohesive comprehensive whole. A supportive explanation of the desperate need for such comprehensive propositions maybe materialized by the field of the philosophy of science, which mandates the coupling between granular constituent statements and the abstract holistic propositions for any discipline to be scientifically substantiated [2].

We started our search for such comprehensive proposition by navigating possible different dimensional definitions of Urbanism [3, 4]. Afterward, the discovered critical urban dimensions were decomposed into a multitude of keywords. An intuitive process of clustering, relating, and organizing these keywords yielded the persistent existence of three extremes between which the other variations of any urban notion started to distribute based on their degrees of relevance to the related three extremes. These three extremes were labelled Visual, Emotional, and Rational. Additionally, the other variations that were relatively distributed between the three extremes were recognized as mere syntheses of the three relevant extremes [5]. The conducted intuitive process for discovering the comprehensive proposition is known as holism by reduction. Although, based on that reductionism, we intuitively believed in the sufficiency and the contrast between the three perspectives, a fair part of the paper would be exhausted trying to trace the validity of that intuitive conclusion.

Therefore, we propose the abstract reduction of urbanism into three perspectives, by which Urbanism would then be categorized and related. The universe of Urbanism is a continuum of notions, ideas, and values. These notions would be coherently explored by the three perspectives. The proposed categorical and relational representation of Urbanism by the three perspectives would entail a better coherent understanding of its immense notions, and consequently a better discoverability of each notion's variations.

This means that it is the categorical exploration of urban variations that the introduced model is devised for. As a matter of fact, ontological propositions are mere categorical tools. Discovering the variations' of Urbanism is synonymous with analysing the multitude versatility of people's urban identities. The structured comparative studies between these versatile identities should enhance the definition, prediction, and analysis of the complexity of people's urban identity. A main task of the companion comprehensive model is to mnemonically and coherently maintain the repository of the discovered characteristics of the three perspectives and their synthesized variations.

Improving the Understanding of Urbanism by enhancing variations' discoverability, categorically and relationally by the proposed perspectival model, has major foundations in philosophy. For instance, Nietzsche concluded that the perspectivist ontology, alternatively known as perspectivist epistemology, is the main way of having an ontological abstraction of any science [6]. Meaning that any science should be reduced to several atomic perspectives that are then responsible for defining all the individuals of that science, and that is also the same views of Plato and Kant that may be summarized as "The Universal is a Precondition for Knowing the Individual" [2]. A direct consequence of that perspectival/individual dependency is that knowledge could be represented as syntheses of variations [2, 5, 7]. However, the slogan of "Information is a difference that makes the difference" is probated by Wu Kun's theory of the ontology of human cognition [8]. These foundational theories conform to our cognitive-mapping approach by representing the two main urban cognition actors, the users and the urban constructs, as empirical instances of Urbanism. Urbanism's variations' representation would enable us to reliably parametrize and characterize these two urban instances. The colour theory also discovers millions of colour variations based on three basic colours.

Via the forthcoming exploration of Urban movements, theories, and constructs, we would sense the contention between the three perspectives. We would also sense that urban data could be satisfactorily classified by these three perspectives. we would first examine the comprehensive proposition against the two fundamental urban notions of place and design and then intuitively against other urban notions (Fig. 1(1)). Then we would test our hypotheses of the *sufficiency* and the *contrast* of the three perspectives. during that examination, each perspective's discovered variations of the selected urban notions would disclose characteristics of the three perspectives. The accumulation of the discovered variations/characteristics in the companion model is the concluded practical descriptor of Urbanism. The examination should support the *contrasting* hypothesis between the perspectives. but why the contrasting property is that important? Although the practicality of the proposition is highly dependent on the contrasting property, there is a unanimous agreement in the field of the philosophy of science that the constituent parts of any comprehensive proposition should be dependent [7]. Our argued dependency relationship is that the three perspectives are *contrasting*. In brevity, without the contrasting property, the argued comprehensiveness of the proposed proposition could be compromised.

2 Examining the Comprehensive Proposition and Defining the Corresponding Comprehensive Model

The examination's main role is to test the assumption that these three perspectives are *contrasting* and that the chosen urban notions, "Place", "Design", and the others, could be *sufficiently* relatively comprehended using merely these three perspectives. Any variation beyond the three perceived extremes is a mere synthesis between any/all of these three extremes [5]. We will conduct this examination by firstly intuitively discovering the three extreme variations of urban notions, and then investigating theoretically the same notion. A comparison between the intuitive and the theoretical discoveries would be analysed. The intuitive discovery is done based on the assumption that the three perspectives are *contrasting*. The comparison between the contrasting intuitive discovery and the rigor theoretical interpretation is meant to test the assumed contrasting quality of the three perspectives. The two main founding definitive notions of Urbanism of "Place" and "Design" were selected to conduct such examination.

The Urban Notion of "Place"

The context of discourse, from which the comprehensive proposition would be extracted, is the atmospheric reflections of urban constructs. Therefore, it is not a surprise to perceive and analyse our first, foundational, urban notion of "Place".

The Contrasting Relativistic Perception of the Urban Notion of "Place"

The Visual Perspective. The visual perspective is about a higher philosophy that is mostly not related to daily activities. This philosophy shapes each corner of the visual perspective's figurative products. The users are experiencing and living that philosophy, their activities and their social behaviours are always considered as subordinate to what that deriving philosophy dictates. The visual designs are produced by artists, poets, or musicians who find or compose the higher philosophy and get devoted to it until the conclusion of their detailed products. Place is a direct spatial consequent of that philosophy. The sophisticated, pure, dreamy, timeless, beyond-reality places are exemplary visual products. Not only function and form that follow design, but also all the aspects of the created design are born from a determining spirit.

The Emotional Perspective. Activities hosted inside the space, their interaction between each other and between them and the hosting space, and the corresponding messages between the hosting space from one side and the hosted activities and the social norms from another is what this perspective's place is all about. Place exists, celebrates, motivates, gets its identity, and is enriched by its hosted activities. Liveness, warmth, cultural enrichment, and joy are amongst the main features of its evolved places.

The Rational Perspective. Integrity and truthfulness is the slogan that derives that perspective. That integrity is always in dialog with the direct consensus logical existential interpretations and explanations. Functionality is an important exemplary mandated quality by the logical existential paradigm. Truthfulness is the main source of beauty.

The tendency of explaining and interpreting applies even for any related conception of amusement and beautification, which is an inherently needed criterion of architecture.

The Theoretical Investigation of the Urban Notion of "Place"

The Visual Perspective. although the Le Corbusier's saying "Architecture is the play of forms- wise, correct, magnificent-the play of forms in the light" may be a good description of this perspective. The Bauhaus's earlier manifesto was what this perspective really stands for. The Bauhaus's earlier manifesto rested on the spirit, which is the shaping higher philosophy, that should artfully unify all the elements of the final products.

These elements may include architecture, landscaping, interior design, product design etc. the international style, which is mostly rational, was crafted by the leaders of the Bauhaus. The international style was remembered generally for its monolithic Marxism vision of the world and also for its disregard of the immediate or broader contextual and social realities. The two prominent postmodernism styles that took over after the fall of that style were known for their plurality and multiplicity that brought complexity to their designs, against the prior monolithic approach. They implemented multiplicity either by plural historical references or by deconstructing any nominal or physical structure in a way that enriches the users' experience and the contextual communication. Both of the movements appreciated societal and uniqueness values. Contemporary wise, the industrial and digital revolutions facilitated the unprecedented complexity of the design process and the uniqueness of any part of the final design, and that uniqueness is a main characteristic of visual designs.

The Emotional Perspective. Camillo Sitte, at the end of the nineteenth century, was known for his initiation of refollowing the model of the naturally developed historical European towns and cities, at the time Europe was highly visually oriented. Siena is a good example of that naturally developed urban cities over hundreds of years. The city is regarded as one of the most visited touristic attractions in Florence, Italy. It is known as one of the most pedestrian-friendly cities in the world. It has a rich, organic, and individualistic qualities. Its central Plaza Del Campo, a UNESCO's declared World Heritage Site, is a source of inspiration for humanitarian and urban studies. There are plenty of contemporary thinkers and urban movements that praise these naturally developed urban cities. For example, Bryan Lawson praised the concept of places as a glorification of their users' social behaviours [9]. Places can conduct such glorification by stimulating, securing, and reflecting the identities of their users. Christopher Day admired the sensitive blending between the traditional socially developed urban constructs and their immediate cultural and environmental surrounding [10]. he devised a new designing methodology called "consensus design" that is done cooperatively between the users to produce natural organic designs. This "consensus design" process could be established under minor supervision of architects. Francis Tibbalds was known for his "making people-friendly towns" theme [11]. This theme was the base for the ten commandments of architecture Tibbalds envisioned in correspondence to Prince Charles challenge. It appreciated values like adaptability, sophistication, joy, mixed uses and activities, respecting human scale, and pedestrians' freedom to walk about. The New Urbanism movement emphasised on the two pillars of "traditional neighbourhood design" and "Transit-oriented development", as procedural implementations for mimicking the naturally developed Urbanism.

The Rational Perspective. Despite that some important rational features were historically evident in the religiously driven Gothic style, like the truthfulness of the structural exposition and the traces of irregularity of their plans to reflect the innate functionality, contemporary rationality was initiated by the end of the seventieth century. It was inspired by the functional interpretations of the Biological, Mechanical, Gastronomic, and Linguistic analogies. According to philosophy, a rational person is the one who is disposed to intend to do what he believes he ought to do. This philosophical definition could be translated to the values of integrity and truthfulness. After the fall of the international style, which is more or less a rational movement, other rational movements took place. These movements may include the theoretical Archigram movement; the High-tech architecture, e.g. The Pompidou Centre; and recently The Neo-futurism movement, e.g. Norman Foster. Rationalists' truthfulness invalidated the use of ornaments and the concealment of structural elements or building services. it also advised the truthful expression of materials' reality, their corresponding construction techniques, and slogans like "Form adheres to utilization".

The Urban Notions of "Design"

Now that we have a basic understanding of the nature of the three perspectives' perception of the founding urban notion of place, the repository of the companion comprehensive model would be enriched by accumulating the other perceived urban notion of design (Fig. 1(1)). The more notions perceived by the three perspectives, the better understanding of the three perspectives we have, and consequently the easier it is to perceive more notions by them. This escalating understanding behaviour should have a decisive contribution to the model's knowledge syntheses. Firstly, we would examine three notions of "Design". These notions are "Product/Process", "Top-Down/Bottom-Up", and "People/Professional". We would then intuitively perceive other urban notions. The accumulated notions are to support the *contrasting/ sufficiency* arguments.

The Contrasting Relativistic Perception of the Urban Notions of "Design"

The Visual Perspective. The visual designs are spatial products shaped by a professional authority. That authority exercises a fair amount of time composing and defining the deriving philosophy; and It accordingly chooses proper sources of inspiration that would be then interpreted to lay the specifications of its prospect design. Therefore, a visual design is a specifications' prescribed, macro-scale, top-down product. The granular social activity is considered as a consequent of the higher deriving philosophy. The uniqueness of the visual designs is due to the uniqueness of its fashioning artist, and that uniqueness is an implied highly valued criterion.

The Emotional Perspective. The emotional designs are evolving processes between the two parties of the hosting place and the hosted activities. Emotional places emerge and evolve based on the communicative messages, either generated or received, between these two parties. Hence, the emotional design is a micro-scale process. The harmonious relationship between the activities and the surrounding, the existing social and environmental tissue, is the highest value that derives all the facets of the design. Therefore, it is a bottom-up complex evolving process that keeps interacting over time. People's complex designs integrate well with this ongoing naturally guided process.

The Rational Perspective designs are mesoscale and mechanistic. Usefulness, Practicality, and Integrity are amongst the dominant traits that derive all the facets of the design. People's complex designs could potentially be abstractly integrated.

The Theoretical Investigation of the Urban Notions of "Design"

We would try to interpret more urban constructs in the eyes of the three perspectives to have even clearer insights during the analysis of the additional urban notions.

The Visual Perspective. We may mention three more urban movements/constructs for the sake of clarity. They are mostly visual rather than anything else due to Their remarkable picturesque and symbolic qualities. They are all driven by precise specifications determined by higher philosophies. For example, Urbanism during and Dictatorship period in Europe, in the first half of the twentieth century, was meant to emphasize on the dictatorial regimes' socio-political programs and to reflect their image of strength, power, efficiency, and superiority. The city became a public product of art, Urbanism emphasized on their imageability by unifying all the facets of art (including architecture, painting, sculpture, photography, furniture, and landscape design) under a single umbrella. Colossal buildings, the inhumane scale for reflecting their philosophy, and prominent axial paths are amongst the common visual specifications. Traces of this paradigm are found in plenty of examples including the built well-visited "Via Imperiale" in Rome, currently known as "Via Cristoforo Colombo", that connects the Colosseum with Piazza Venezia; and also the unbuilt plans of the "World Capital Germania" in Berlin. The same fingerprints exit in Romanian cities, like the city of Pompeii, and in Renaissance cities, like the cities of Palma Nova and Neuf-Brisach. The Liberal Monument, a group led by famous historians and architects, suggested the utilization of a few dispersed formalized centres for resolving and contradicting the anarchy of the sprawl. These visual formalized centres should be not bottom-up but top-down, not every day but elite, and not fabric but figure; and that matches the intuitive findings.

The Emotional Perspective. We may mention two more urban constructs. The first is two different people's urban designs of the sprawl, in the developed countries, and the spontaneous settlements, in the developing countries. Both of them represent the massive, people's naturally designed, contemporary urban extensions. Although many considers both of them as chaotic constructs that may have plenty of problems/side-effects that need immediate rectification, others tend to praise their implicit complexity, naturality, and sophistication [12]. For example, some studies showed that more than half of the USA's population are inhibiting sprawling areas. The Favela is a national image that has plenty of touristic visitors and that has pervasive existence in the movies' industry. if we try to deduce the commonalities between these people's designs, the historical Sumerian and Islamic cities, and the aforementioned natural emotional urban developments, we would conclude the same intuitive emotional findings. These constructs are sophisticated procedural interactions between their constituting agents, physical or nominal. These designs are micro-level bottom-up evolving processes.

Discussion

The aforementioned urban theories, movements, or constructs, which could be used to fairly represent Urbanism, could be satisfactorily classified by the three perspectives, and that should support the *sufficiency* argument of the three perspectives. Additionally, the intuitive relativistic perception of the urban notions was done based on the three perspectives' *contrasting* assumption. If we to compare the intuitive understanding with the theoretical interpretations of the aforementioned urban data, we would strongly sense the validity of our main *contrasting* hypothesis between the three perspectives.

We may also sense the raw power of accumulating the urban notions and the coherent collaborations between them in enhancing the intuitive discoverability of urban variations. For instance, we may briefly intuitively perceive new urban notions of Rationality, Societal behaviourism, and individuality; and then incorporate them in composing several Plausible statements about the three perspectives as the following:

- The visual designs are about a higher philosophy that gets shaped, and its conceptual context gets defined, by a professional artistic authority. That philosophy depends on narratives to lay expressive specifications of a figurative product. This implies the uniqueness, the top-down, and macro-level qualities of these designs.
- The emotional designs are evolving processes. the interactions between the diverse agents, nominal or physical, derive these processes. These micro-scale ingredients procedurally interact for the creation of the whole. traces of plurality, diversity, harmony with the surrounding, complexity, and sophistication are evident in these designs.
- The rational designs are logical and meso-scaled. coherence or abstraction between the other perspectives is acquired. Rationalism generates notions of its own as well.

Some may disagree with these statements, but we should keep in mind that though they are variations by their own, they are meant to be extreme variations that synthesize the other possibilities. Now, we would examine the third argued pillar of the proposed model, which is Knowledge syntheses (Fig. 1(3)). For instance, Although The Deconstructivism was mostly practiced based on visual preferences, its definition departed visuality in favour of certain emotional features like respecting the human scale and being context sensitive. Other syntheses examples may include: micro agents may evolve for laying specifications of a product rather than shaping a process, or else, these agents should be rationally expressed and each has apparent marks of his own; an authority may lay the specifications for the agents to evolve toward a final product, the involved agents and the authority's rules should be rationally physically marked and their abstraction should shape the final product; a mechanic could be approved and then utilized partly as a specification and partly as a process, for both of the agents and the authority; a design may start from a meso-scaled abstract definition of the needed functional unit, then the involved agents and the overall specs would be laid simultaneously; etc. We may also observe that the proposed comprehensive proposition was derived from, and tested against, the context of the atmospheric reflections of urban constructs; and that context is thought to be of primal relevance to the paper's main usage of users' urban cognition. The need for other contexts of discourses, and consequently other comprehensive propositions, is subject to practical needs. As prior attempts of urban unifying models: Henry Lefebvre described unification as the desperately needed linkage for his planned, perceived, and used knowledge spaces [13]; Patrik Schumacher analysed urbanism as an Autopoietic system, a self-evolving sub-social system [14]; The three factored David Kolb's general theory of Urbanism defined the aspects of contemporary complex places [12]; while the Complexity theory of cities offered the definition of Urbanism as a chaotic system and simulated cities accordingly [15].

3 Potential Applications

Predicting the intended users' urban cognition and incorporating it into the design process would enhance our opportunities in Tailoring users' generated, socially sustainable, and culturally sustainable designs. To do so, we need to have a basic understanding of the nature of the users' complex urban cognition. The proposed comprehensive model, the three perspectives' accumulated perceived urban notions, is expected to play a vital role in discovering their different conceptions and impressions of urban constructs and as well in describing their urban preferences. If we compare their urban impressions against their preferences, we could then continually predict their recommendations in the vast variations and possibilities of the urban design's decisions. Users perceive urban constructs based on their cognitive urban identity. The cognitive urban identity term is coined for specializing the cognitive analysis solely in relation to urban constructs, and that is why the proposed comprehensive model is derived from and tested against influential urban constructs. Depending on their urban cognition, they understand urban constructs by decomposing them hierarchically into smaller physical features, similar to the way Kevin Lynch city's image analysis, and then they recognize their corresponding impressions against the most significant features. Their conceptions and their impressions are all done relatively based on their different identities. These concluded impressions could be then compared with their preferences to reach the efficient equivalent decision (Fig. 1(2)). The comprehensive model is meant to set the context of discourse of that whole cognitive process.

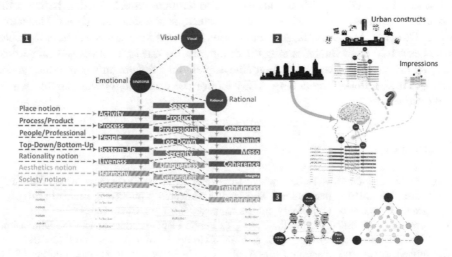

Fig. 1. (1) the introduced accumulated urban notions perceived by the three perspectives. (2) the main two actors of the cognition process, the users and the urban construct, empirically instantiated by the comprehensive model. (3) knowledge syntheses representation.

The aforementioned users' cognitive-mapping model could be simulated computationally using Agent-Based modelling ABM. ABM is generally devised as a social simulative apparatus that tackles the complex nonlinear evolutions and interactions of the society's constituents on both of the smaller scale and on the larger synergetic scale. We envision that an empirical ABM could simulate the users' cognitive-mapping as a tool based on the comprehensive model. The envisioned simulative tool would have several agents that represent the intended users and also the urban constructs. The users' agents are empirically generated depending on actual users' questionnaires, interviews, and observations. The urban constructs' agents would represent the magnitude of the physical features of any proposed urban construct. These decomposed urban constructs' features would be compared with the simulated users' agents' cognition to structure a recognized hierarchical significance of these decomposed features. Characterizing and parametrizing the users' agents as structured instances of urbanism, by the comprehensive model, is similar to the known techniques of empirical agent-based modelling [16].

4 Conclusion

In the paper, we investigated the possibility of a Triadic perspectival ontological representation of Urbanism paralleled with the syntheses of knowledge for facilitating the coherent discoverability of the variations of urban constructs' characteristics. By doing so, we present a basic definition of Urbanism to the other analytical universes, like the universe of Identity. It may seem, at the first glance, that exhaustive understanding of Urbanism for instantiating both of the intended users, as a hierarchically structured instances of Urbanism, and their recognized urban constructs as a long shot. However, this could be a very robust approach for mapping the complex users' cognition in sole relation to Urban constructs. The harmonious relationship between the two urban instances, the users and the constructs, is the key for "Users' Generated Urban Design". And accordingly, maintaining the harmony between the two urban instances would be facilitated during the design process due to the proposed categorical tool. In future work, we would perceive a plethora of more urban notions, and a consequent agent-based modelling ABM tool of the users' cognitive-mapping would be developed.

References

1. Ezzat, M., El-khorazaty, T., Salama, H.: Urban identity using the six dimensions of urbanity. In: Archtheo 2016 X International Theory of Architecture Conference, Istanbul (2016)
2. Agazzi, E.: Scientific Objectivity and Its Contexts. Springer, Cham (2014)
3. Carmona, M.: Tiesdell, S., Heath, T., Oc, T.: Public Places - Urban Spaces the Dimensions of Urban Design. Routledge Taylor & Francis Group, London & New York (2010)
4. Cuthbert, A.R.: Understanding Cities Methods in Urban Design. Routledge, London (2011)
5. Nakamori, Y.: Knowledge Synthesis: Western and Eastern Cultural Perspectives. Springer, Tokyo (2016)

6. Welshon, R.C.: Perspectivist Ontology and de re Knowledge. In: Nietzsche, Epistemology, and Philosophy of Science: Nietzsche and the Sciences II, pp. 39–46. Springer (1999)
7. Esfeld, M.: Holism in Philosophy of Mind and Philosophy of Physics. Springer (2001)
8. Kun, W., Brenner, J.E.: An informational ontology and epistemology of cognition. Found. Sci. **20**, 249–279 (2015)
9. Lawson, B.: The Language of Space, 1st edn. Architectural Press, Oxford (2001)
10. Day, C.: Spirit & Place, 2nd edn. Architectural Press, Elsevier, London (2002)
11. Tibbalds, F.: Joining it all together. Urban Des. Group **3**, 3–5 (1980)
12. Kolb, D.: Sprawling Places, 1st edn. The University of Georgia Press, Athens (2008)
13. Lefebvre, H.: The Production of Space. Basil Blackwell Ltd., Oxford (1974)
14. Schumacher, P.: The Autopoiesis of Architecture: A New Framework for Architecture, vol. 1. Wiley, London (2011)
15. Portugali, J.: Complex Artificial Environments: Simulation, Cognition and VR in the Study and Planning of Cities. Springer, Heidelberg (2006)
16. Smajgl, A., Barreteau, O.: Empiricism and agent-based modelling. In: Empirical Agent-Based Modelling-Challenges and Solutions, pp. 1–26. Springer, New York (2014)

A Typology of Places in the Knowledge Economy: Towards the Fourth Place

Arnault Morisson[(✉)] [iD]

Mediterranea University of Reggio Calabria, 89100 Reggio Calabria, Italy
arnault.morisson@unirc.it

Abstract. In the knowledge economy, the rise of new social environments is blurring the conventional separation between the first place (home), the second place (work), and the third place. The paper aims to construct a typology of places in the knowledge economy. The research methodology is based on an exploratory case study approach investigating the social environments in Paris (France) that don't fit in the traditional typology of places. The paper finds that the new social environments in the knowledge city can combine elements of the first and second place (coliving); of the second and third place (coworking); and of the first and third place (comingling). Furthermore, the combination of elements of the first, second, and third place in new social environments implies the emergence of a new place, the fourth place. The paper contributes to understand how the knowledge economy is changing the nature of places in the global post-industrial city.

Keywords: Third place · Knowledge economy · Clusters · Innovation districts Creative economy · Reinventing Paris

1 Introduction

In November 2014, the City of Paris launched the Reinventing Paris call for innovative urban development projects to transform 23 sites into showcase architectural projects [1]. In February 2016, a jury comprised of international and national experts selected 22 projects to reinvent Paris [1]. The projects included coliving spaces, coworking spaces, FabLabs, shared living spaces, makerspaces, living labs, live-work spaces, flexible spaces, rooftop bars and restaurants, incubators, hotels, and hostels [1]. The process of reinventing Paris didn't start with the call for projects. Indeed, since the late 2000s, as in any other global post-industrial cities, Paris has experienced the rise of a multitude of new social environments [2, 3].

Oldenburg and Brissett [4] add to the traditional dichotomy of first place (home) and of second place (work) the concept of third place, which they define as places "where people gather *primarily* to enjoy each other's company" (p. 269). The dominant third places are the coffee shops, the bar, beauty parlors, general stores, and the community centers [5]. The role of the third place has always played an essential role in the diffusion of ideas and knowledge. The insurance company, Lloyds of London, first operated out of a coffee house [6]. In Silicon Valley in the 1970s and 1980s, the most famous informal spaces-such as Walker's Wagon Wheel Bar and Grill in Mountain

© Springer International Publishing AG, part of Springer Nature 2019
F. Calabrò et al. (Eds.): ISHT 2018, SIST 100, pp. 444–451, 2019.
https://doi.org/10.1007/978-3-319-92099-3_50

View-have probably disseminated more ideas than conventional seminars [7]. In the knowledge economy, new social environments are deliberately being created combining places in order to facilitate networking and the exchange of knowledge.

The paper explores the rise of the new social environments in Paris (France). The new social environments studied are bounded in the city of Paris. The research conducted for this paper is based on three sources of data: semi-structured interviews, secondary data, and direct observations. The paper finds that the new social environments in the knowledge city can combine elements of the first and second place (coliving); of the second and third place (coworking); and of the first and third place (comingling). Furthermore, the combination of elements of the first, second, and third place in new social environments implies the emergence of a new category of place, the fourth place. The paper contributes to the understanding how the knowledge economy is changing the nature of places in the global post-industrial city.

2 The Knowledge Economy and Places

In the 1990s, capitalist countries started to undergo an economic transition towards post-Fordism, or knowledge-based economies [8, 9]. The knowledge-based economy recognizes the importance of knowledge as the driver of productivity and economic growth, emphasizing the role of information, technology and learning in economic performance [10, 11]. In the knowledge economy, technological innovation is essential for economic prosperity. The academic literature provides, both across nations and over time, a solid theoretical background linking technological innovation to the progress of countries, regions, cities, and firms [12–16]. The transition from the mass production to knowledge-based economy, which was instigated by disruptive technological innovations, namely Information and Communication Technologies (ICT), has had profound consequences on the social, organizational, and institutional structures [17–22]. The knowledge economy has reshaped, for instance, the whole organization of production [23, 24]; the nature of learning and the innovation process [25–27]; social relationships [28, 29]; and the nature of work [30–35].

Since the late 2000s, new social environments-such as hacker spaces, maker spaces, Living Labs, FabLabs, shared living spaces, coliving, and coworking spaces-have been emerging in the post-industrial cities [2, 3]. The emergence of new social environments is the result of concurrent trends fostered by the knowledge economy, namely, the integration of work and personal life [32, 36]; the importance of informal networks [37, 38]; the importance of tacit knowledge [39–42]; the millennials' preference to live in urban centers [43]; and overall new organizations of work [30, 31, 34, 35].

3 Methodology

The research methodology is based on a single significant and concept sampling case-study approach, using primary and secondary data. The author uses case studies "out of the desire to understand complex social phenomena" [44]. The purpose of this case study is to explore the new social environments emerging in the knowledge economy. The case selected is the city of Paris (France). The paper investigates a contemporary phenomenon that has not been fully examined, in which the researcher has no control on the actual phenomenon [45]. A qualitative approach is, as a result, the most appropriate method [45, 46].

The research conducted for this paper is based on three sources of data: semi-structured interviews, secondary data, and direct observation. In total, 9 semi-structured interviews were conducted in different social environments that combine two or more elements of Oldenburg's and Brissett's [4] typology. The people interviewed were founders or employees working for the social environments. The stakeholders were selected according to their strong knowledge and diverse perspectives on the phenomena studied [47]. The interviews aimed to investigate how founders and employees perceived their spaces to be in the traditional typology of places. The secondary data used for the research included articles in newspapers, such as *Les Echos*, *Le Figaro*, and *Le Monde*; articles in news websites, such as Rude Baguette and TechCrunch; and the websites of the studied social environments. The researcher also conducted formal and informal observations in May, June, and September 2017. First, formal observations through planned visits with representatives of the studied places, and second, non-participatory informal observations in order to uncover network dynamics and local buzz. In total, the researcher conducted about 18 h of formal and informal observations in order to examine how the new spaces investigated were utilized by their users.

The data was examined in an analytic deductive manner. That is, the qualitative analysis was first deduced using Oldenburg's and Brissett's [4] typology of first, second, and third place. Alongside this deductive phase of analysis, the researcher looked, in an inductive manner at patterns in which two or more categories of Oldenburg's and Brissett's [4] typology were combined in the newly created social environments. Typologies are "classification systems made up of categories that divide some aspect of the world into parts along a continuum" [48]. Validation is achieved through prolonged engagement, persistent observation, and triangulation in order to "assure that the right information and interpretations have been obtained" [49]. The newly proposed typology allows readers to make decisions regarding transferability.

4 Case Study – Paris, France

Paris (France) is a wealthy global city-region, which has a population of 12,5 million inhabitants generating a GDP of USD 818 billion in 2015, that represents a GDP of around USD 65,500 per capita [50]. Although, Paris is lagging behind its peers on patenting and innovation activities, the city concentrates France's research and innovation activities and is one of the leading innovation hubs in the world [50]. In 2014, Anne Hidalgo, Mayor of Paris, launched the Reinventing Paris call for innovative urban development projects in order to accelerate Paris' transition into the knowledge economy [1]. Reinventing Paris engages 150,000-m^2 construction of new space, which contains within it, 1,300 apartments, 60,000-m^2 of offices and coworking spaces, 4 hotels, 3 hostels, and 26,000-m^2 of green spaces [1]. The paper does not have the ambition to list all the places that combine two or more of Oldenburg's and Brissett's [4] typology, but rather to provide examples of emerging categories that fit the proposed typology of places in the knowledge economy.

4.1 The Combination of the First and Second Place – The Coliving Space

The combination of the first place (home) and the second place (work) is the coliving space. In 2016, the startup *HackerHouse* was created in Paris. The startup provides shared accommodations and working areas to entrepreneurs. As of 2017, the *HackerHouse* has four apartments that each have different focuses, such as virtual reality, blockchain, software development, and design. The *HackerHouse* also organizes events and meetups for the public and for the "Hackers" living in the *HackerHouses*. In one *HackerHouse* for instance, the amenities include shared bedrooms with bunk beds for up to 10 people, two bathrooms, a kitchen, and common areas with large working tables, TV screens, white boards, and high-speed internet. In Reinventing Paris, the *Bains Douches & Co's* in the 15th district includes two buildings, one building dedicated to shared apartments and one building dedicated to coworking space [1]. The function of coliving space and to create an atmosphere conducive to living and working under the same roof.

4.2 The Combination of the Second and Third Place – The Coworking Space

The combination of the second place (work) and the third place is the coworking space. *Hubsy, Anticafé, Craft Coffee Shop*, and *10H10* are coworking cafés in which people pay an hourly or daily flat rate in order to work in the coffee place while having access to free drinks and food. In contrast with traditional café place (third place), where the primarily function is "to enjoy each other's company" [1], the co-working cafés' function is to work and network. Coworking places such as *WeWork La Fayette* in the 9th district have a coffee shop and a pop-up restaurant at the center of its coworking space. *The Bureau* in the 8th district is a luxury coworking place that has a high-end restaurant and bar open only to its residents and the residents' guests. *Numa* and *Cool and Workers*, in the 2nd district and the 11th district, respectively, each open their first floors, which include coffee shops, to residents and guests. The startup *BlaBlaCar*, a collaborative car-sharing company, provides its workers with workstations but also common areas with a café. The function of coworking space is to create an atmosphere conducive to work and network in order to favor the exchange of knowledge and to foster collaboration opportunities.

4.3 The Combination of the First and Third Place – The Comingling Space

The combination of the first place (home) and the third place is the comingling space. In Reinventing Paris, *Le 29 Hôtel (très) Particulier* in the 17th district, mixes shared apartments and shared common spaces offering to the residents and guests cultural events and concierge services [1]. *Mama Shelter* and *Mob Hotel* are hotels that not only offer the traditional bedrooms but also shared common areas in order to favor networking and mingling between the guests. The *Allure* residential project in the ZAC Clichy-Batignolles will provide its residents with shared spaces, a shared rooftop with a shared kitchen, and concierge services. The function of the comingling spaces is to favor social interactions and networking opportunities between its residents.

4.4   The Combination of the First, Second, and Third Place – The Fourth Place

The combination of the first place (home), second place (work), and the third place is the fourth place. Opened in 2017, *Station F* is a 34,000-m^2 innovation center that combines restaurants, bars, a post office, fablab, and 3,046 working desks for 1,100 startups [51]. In 2018, *Station F* will open *Home*, a 100 shared apartments residence for the entrepreneurs and knowledge workers working at *Station F*. In Reinventing Paris, the *Stream Building* offers *Zoku* mini lofts with shared spaces, co-working space, a shared rooftop, bars, and restaurants. The *Stream Building* is divided into four categories: Stream Work, Stream Play, Stream Eat, and Stream Play. The fourth place blurs the frontier, within the same space, of the first (home), second (work), and third place making the space, a place in itself. The function of the fourth place is to foster networking, to promote mingling, and to favor collaboration, face-to-face interactions, and the exchange of tacit knowledge.

5   Discussion

The knowledge economy is changing the traditional typology of place. The Reinventing Paris call for innovative projects has been a catalyst project for the city of Paris to conceive and imagine the "Places of Tomorrow" in the knowledge city. Global cities such as New York, London, San Francisco, and Paris are experiencing the emergence of new spaces that are making the traditional typology of spaces and places obsolete. The typology proposed in this paper suggests that places in the knowledge economy are increasingly overlapping, which ultimately produces a fourth place (c.f. Fig. 1). The

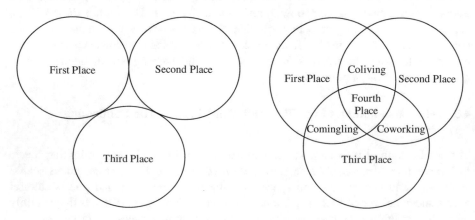

Fig. 1. Places before (left) and in (right) the knowledge economy.

frontier between the traditional first, second, and third place is increasingly unclear, leading to the creation of new spaces and a new place. The emergence of the fourth place, coworking, comingling, and coliving spaces highlight the importance of tacit knowledge, social interactions, networks, and the spatial dimension of innovations in the knowledge economy.

6 Conclusions

A new typology of place is emerging in the knowledge economy. The new typology stresses the blurring frontier between the traditional first, second, and third places. In post-industrial global cities, like Paris, new social environments are emerging. These spaces are often merging elements of two or more places, leading to new spatial categories such as coliving, coworking, and comingling spaces. This paper proposes a fourth place in addition to the first, second, and third place. The fourth place combines elements of the first, second, and third place, making it a place in itself. In the fourth place, the frontier between social and private dynamics, work and leisure, networking and social interactions, and collaboration and competition are blurry, making it the place for the knowledge economy. Policymakers should favor the creation of the fourth place and new social environments such as coliving, coworking, and comingling spaces by promoting mixed-use zoning that incorporates different dimensions of places and by providing incentives in cities where the market is too weak to foster the creation of these new spaces. Further research should investigate the concept of the fourth place.

Acknowledgments. This article is part of the MAPS-LED research project, which has received funding from the European Union's Horizon 2020 research and innovation programme under the Marie Skłodowska-Curie grant agreement No. 645651.

References

1. Pavillon de l'Arsenal: Réinventer Paris. Editions du Pavillon de l'Arsenal, Paris (2016)
2. Botsman, R., Rogers, R.: What's Mine is Yours: How Collaborative Consumption Is Changing the Way We Live. Collins, London (2011)
3. Gandini, A.: The rise of coworking spaces: a literature review. Ephemera 15(1), 193 (2015)
4. Oldenburg, R., Brissett, D.: The third place. Qual. Sociol. 5(4), 265–284 (1982)
5. Oldenburg, R.: The great Good Place: Café, Coffee Shops, Community Centers, Beauty Parlors, General Stores, Bars, Hangouts, and How They Get You Through the Day. Paragon House Publishers, Vadnais Heights (1989)
6. Knight, R.V.: Knowledge-based development: policy and planning implications for cities. Urban Stud. 32(2), 225–260 (1995)
7. Castells, M.: The Rise of the Network Society. Blackwell Publishers, Oxford (1996)
8. Amin, A.: Post-Fordism: A Reader. Blackwell Publishers, Oxford (1994)
9. Drucker, P.F.: The discipline of innovation. Harvard Bus. Rev. 76(6), 149–157 (1998)
10. OECD: The Knowledge-Based Economy. OECD Publishing, Paris (1996)
11. Powell, W.W., Snellman, K.: The knowledge economy. Annu. Rev. Sociol. 30, 199–220 (2004)

12. Fagerberg, J.: International competitiveness. Econ. J. **98**(391), 355–374 (1988)
13. Freeman, C.: Technical innovation, diffusion, and long cycles of economic development. In: Vasko, T. (ed.) The Long-Wave Debate, pp. 295–309, Springer, Berlin (1987)
14. Rosenberg, N.: Innovation and Economic Growth. OECD Publishing, Paris (2004)
15. Schumpeter, J.A.: The Theory of Economic Development: An Inquiry into Profits, Capital, Credit, Interest, and the Business Cycle. Transaction Publishers, New Jersey (1934)
16. Solow, R.M.: Technical change and the aggregate production function. Rev. Econ. Stat. **39**, 312–320 (1957)
17. Castells, M.: The Informational City: Information Technology, Economic Restructuring, and the Urban-Regional Process. Blackwell Publishers, Oxford (1989)
18. Freeman, C.: The Nature of Innovation and the Evolution of the Productive System in Technology and Productivity. OECD, Paris (1991)
19. Geels, F.W.: Technological transitions as evolutionary reconfiguration processes: a multi-level perspective and a case-study. Res. Pol. **31**(8), 1257–1274 (2002)
20. Kemp, R.: Technology and the transition to environmental sustainability: the problem of technological regime shifts. Futures **26**(10), 1023–1046 (1994)
21. Perez, C.: Structural change and assimilation of new technologies in the economic and social systems. Futures **15**(5), 357–375 (1983)
22. Perez, C.: Technological revolutions, paradigm shifts and socio-institutional change. In: Reinert, E. (ed.) Globalization, Economic Development and Inequality: An Alternative Perspective, pp. 217–242. Edward Elgar, Cheltenham (2004)
23. Florida, R., Kenney, M.: The new age of capitalism: innovation-mediated production. Futures **25**(6), 637–651 (1993)
24. Piore, M.J., Sabel, C.F.: The Second Industrial Divide: Possibilities for Prosperity. Basic Books, New York (1984)
25. Chesbrough, H.: Open Innovation: The New Imperative for Creating and Profiting from Technology. Harvard Business School Press, Boston (2003)
26. Gibbons, M., Limoges, C., Nowotny, H., Schwartzman, S., Scott, P., Trow, M.: The New Production of Knowledge: The Dynamics of Science and Research in Contemporary Societies. Sage, Thousand Oaks (1994)
27. Rothwell, R.: Towards the fifth-generation innovation process. Int. Mark. Rev. **11**(1), 7–31 (1994)
28. Klinenberg, E.: Going Solo: The Extraordinary Rise and Surprising Appeal of Living Alone. Penguin, City of Westminster (2012)
29. Turkle, S.: Alone Together: Why We Expect More from technology and Less from Each Other. Basic Books, New York (2012)
30. Kalleberg, A.L.: Nonstandard employment relations: part-time, temporary and contract work. Annu. Rev. Sociol. **26**(1), 341–365 (2000)
31. Kalleberg, A.L.: Flexible firms and labor market segmentation: effects of workplace restructuring on jobs and workers. Work Occup. **30**(2), 154–175 (2003)
32. Lewis, S.: The integration of paid work and the rest of life. is post-industrial work the new leisure? Leisure Stud. **22**(4), 343–345 (2003)
33. Terranova, T.: Free labor: producing culture for the digital economy. Soc. Text **18**(2), 33–58 (2000)
34. Trilling, B., Fadel, C.: 21st Century Skills: Learning for Life in Our Times. Wiley, San Francisco (2009)
35. Smith, V.: New forms of work organization. Annu. Rev. Sociol. **23**(1), 315–339 (1997)
36. Himanen, P.: The Hacker Ethic. Random House, New York (2010)
37. Nardi, B.A., Whittaker, S., Schwarz, H.: It's not what you know it's who you know. First Monday **5**(5) (2000)

38. Saxenian, A.: Regional Advantage: Culture and Competition in Silicon Valley and Route 128. Harvard University Press, Cambridge (1994)
39. Bathelt, H., Malmberg, A., Maskell, P.: Clusters and knowledge: local buzz global pipelines and the process of knowledge creation. Prog. Hum. Geogr. **28**(1), 31–56 (2004)
40. Gertler, M.S.: "Being there": proximity organization, and culture in the development and adoption of advanced manufacturing technologies. Econ. Geogr. **71**(1), 1–26 (1995)
41. Rodríguez-Pose, A., Crescenzi, R.: Mountains in a flat world: why proximity still matters for the location of economic activity. Cambridge J. Reg. Econ. Soc. **1**(3), 371–388 (2008)
42. Von Hippel, E.: "sticky information" and the locus of problem solving: implications for innovation. Manag. Sci. **40**(4), 429–439 (1994)
43. Florida, R.: The Rise of the Creative Class: and How Its Transforming Work Leisure Community and Everyday Life. Basic Books, New York (2002)
44. Yin, R.K.: Case Study Research: Design and Methods. Sage Publications, Newbury Park (1994)
45. Eisenhardt, K.M.: Building theories from case study research. Acad. Manag. Rev. **14**(4), 532–550 (1989)
46. Creswell, J.W.: Qualitative Inquiry and Research Design: Choosing Among Five Approaches. Sage Publications, Thousand Oaks (2013)
47. Eisenhardt, K.M., Graebner, M.E.: Theory building from cases: opportunities and challenges. Acad. Manag. J. **50**(1), 25–32 (2007)
48. Patton, M.Q.: Qualitative Evaluation and Research Methods. Sage Publications, Thousand Oaks (2015)
49. Stake, R.E.: Multiple Case Study Analysis. Guilford Press, New York (2013)
50. Parilla, J., Marchio, N., Trujillo, J.L.: Global Paris: Profiling the Region's International Competitiveness and Connections. Brookings Institute, Washington (2016)
51. Dillet, R.: A walk around Station F with Emmanuel Macron. TechCrunch. https://techcrunch.com/2017/07/01/a-walk-around-station-f-with-emmanuel-macron/. Accessed 21 Nov 2017

Cluster Identification: The Case of United Kingdom

Massimiliano Ferrara and Claudio Massimo Colombo[(⊠)]

Mediterranean University of Reggio Calabria, 89100 Reggio Calabria, Italy
claudio.colombo@unirc.it

Abstract. The explosion of cluster phenomenon has called for the identification of industrial agglomerations in the UK. The authors have been pointed out only those industries that have higher rates of innovation and better economic results as a function of their presence - defined by the European Union as emerging industries – given that these are fields in which new cross-sectoral linkages are most likely to emerge. This paper, following the methodological approach developed by Porter and adapted to the European Union by the European Cluster Observatory, set two benchmark parameters – Location Quotient and Economic Specialisation - in order to detect the emergence of business agglomerations. The mapping result suggests that United Kingdom has a wide spectrum of cluster of emerging industries and that Berkshire, Buckinghamshire and Oxfordshire is the most innovative region in the whole country.

Keywords: Business agglomeration · Cluster · Competitiveness in UK

1 Introduction

Business agglomeration is one of the most popular form of cooperation that arose during the last decades in Europe and elsewhere [1, 2]. According to the definition given by Porter [3], "clusters are geographic concentrations of interconnected companies, specialized suppliers, service providers, and associated institutions in a particular field that are present in a nation or region". That is to say, it is a tool for promoting and sustaining competitiveness, innovation, growth on local, regional and national levels. In the majority of cases, a cluster initiative influences the strategy of enterprises, and it increases their competitiveness because of the delivered added value. As emerged by the Maps-Led research project, "the cluster, even with a physical configuration, has been considered as a proxy of innovation concentration because its occurrence is strictly connected (by definition from the Porter's model) to innovation, specialization and job creation" and "spatializing cluster acquires the meaning to spatialize innovation" [4].

Given the central role of the cluster in innovation and knowledge dynamics, the objective of this paper is to identify business clusters in United Kingdom. This is because, based on the innovation performances ranked by the European Innovation Scoreboard [5], the United Kingdom is an "Innovation Leader" at European level. Moreover, as stated by Swords [6], regional development in the United Kingdom derives directly from the work of Porter. In fact, the same Porter led a study financed by

© Springer International Publishing AG, part of Springer Nature 2019
F. Calabrò et al. (Eds.): ISHT 2018, SIST 100, pp. 452–461, 2019.
https://doi.org/10.1007/978-3-319-92099-3_51

the British government, in order to provide recommendations for the implementation of a cluster development program in that region [7].

This paper is divided in 5 parts. After the introduction, Sect. 2 focus on the porterian model for cluster mapping and in the translation of this tool at European level. Section 3 explains the methodology used by the authors for identifying the emerging industries clusters in United Kingdom. Section 4 describes the data gathering process and the application of Location Quotient and Economic Specialisation ratios. The last section illustrates the results and the richness of the different cluster categories operating in the country.

2 Porterian Cluster Mapping

The most recognized method for identifying cluster, based on a quantitative strategy, has been made by Porter and the Institute for Strategic Competitiveness at the Harvard Business School [8]. In particular, they created a model that allowed using proxies that translated the cluster definition by Porter [3] in quantitative terms. The mapping focused on spatial concentration of economic activities in a given territory and on the relationships among different sectors through an analysis of correlations in terms of employment and of input-output links. The industrial agglomerations thus detected reflect the presence of spillovers, common interests or connections such as the sharing of labor pools, skills or even technological cooperation. This methodology has enabled 51 so-called traded clusters to be identified, which also account for one third of the total U. S. labor force [9].

The methodology on which the American model is based, developed by Porter, has been translated into Europe thanks to the Center for Strategy and Competitiveness at Stockholm School of Economics, which in 2006 has established the European Cluster Observatory (ECO) [10]. Scholars have taken into account territorial differences and European peculiarities to identify traded clusters. As a result, also for Europe 51 clusters categories have been identified and, thanks to an innovative methodology developed by ECO that takes into account four different statistical dimensions, they managed to capture the strongest existing cross-industry linkages and spillovers. Emphasis was placed on employment specialization, size, employment productivity and employment growth. Employment specialization is defined by European Commission [11] as "the relative measure" that "indicates how much stronger a region is in a cluster category than would be expected given its overall size, compared to the average employment size in the specific cluster category across all regions"; size refers to the number of employees and establishments in the geographical area of interest; productivity is measured by the wages paid; employment growth represent the dynamism because it is a measure that capture if a cluster benefit from cluster effects and it is measured by employment growth and the presence of fast-growing firms [11]. However, the core objective has been to identify clusters of emerging industries, i.e. those industries that have higher rates of innovation and better economic performances as a function of their presence [8]. In fact, these are fields in which new cross-sectoral linkages are most likely to emerge. In order to achieve this objective, the analysis followed three specific stages. "First, the strongest current cross-sectoral linkages are identified, drawing on a traditional cluster mapping analysis. Second, the cluster categories of traded industrial sectors are broadened and in some cases merged, to capture an additional layer of weaker linkages beyond the stronger linkages within a given cluster category. Third, a final list of the ten strongest emerging industries are selected from the wider group of candidate emerging industries that have shown the highest economic dynamism in the recent past" [8]. The final result is made up of 10 clusters, each one consisting of specific industrial sectors [12].

For the purposes of this paper, we will consider the ten Emerging Industries as a cluster category of reference for the spatial identification of the most promising clusters of UK. This is because, emerging industries are the clusters in which new cross-sectorial linkages are most likely to stand out due to "a path-dependent process of related diversification" [13].

3   Methodology for Mapping the Emerging Industries in UK

The methodology developed in this section has its origin in the aforementioned work of ECO [8, 11, 12]. As a starting point, the definition of cluster for the ten emerging industries has been analyzed. Each emerging industry – Advanced Packaging, Bio-pharmaceuticals, Blue Growth Industries, Creative Industries, Digital Industries, Environmental Industries, Experience Industries, Logistical Services, Medical Devices, Mobility Technologies – is composed by several economic activities identified by the NACE name. NACE can be defined as the statistical classification of economic activities in the European Community and it provides a four-digit code that identifies the nature of economic activities. The next step has been to spot the presence of the ten Emerging Industries in the UK. To achieve this goal has been used the Location Quotient (LQ) for the variable "Employment", the "indices of specialisation most commonly used to identify the presence of clusters within a specific geographic locale" [14]. The location quotient indicate the degree of concentration of a cluster in a particular region relative to its concentration in the national economy [15]. It is used to understand if the employment of an industry in a region is above or below the average.

Therefore, the LQ_E can be shown as:

$$LQE = \frac{\frac{E_{ir}}{E_r}}{\frac{E_{in}}{E_n}} \tag{1}$$

where:
   E_{ir} = is the industry employment in region r
   E_r = is the overall employment in region r
   E_{in} = is the industry employment in nation n
   E_n = is the overall employment in nation n

When the LQ_E has a value above one, it indicates that the emerging industry has a concentration of employment above the national average. However, as commonly agreed in the literature, LQ that has employment as a variable cannot in any way help to discern between a region with an above-average LQ, but made up of a single very large company, and a region with an above-average LQ and many companies employing the same number of workers [2, 15]. Thus, given the fact that the degree of concentration of firms above the average is synonymous of inter-firm interactions and spillover effects [16], the Location Quotient for the variable "Firms" could help capture the existence of a cluster in the emerging industry. Using the denomination by Paton, Del Castillo and Barroeta [17], in this case the LQ$_F$ is defined as Economic Specialisation (CE).

So, the CE$_F$ is:

$$CEF = \frac{\frac{F_{ir}}{F_r}}{\frac{F_{in}}{F_n}} \qquad (2)$$

where:
  F_{ir} = is the industry firms in region r
  F_r = is the overall firms in region r
  F_{in} = is the industry firms in nation n
  F_n = is the overall firms in nation n

Also in this case, when the CE$_F$ has a value above one, it indicates that the emerging industry has a concentration of firms above the national average.

Drawing inspiration from the approach employed by ECO, the assessment of each UK region has been carried out through a two-stage process. In the first step, it has been assessed which regions have fulfilled the following condition:

LQ$_E$: this index of specialisation has to be equal to 2 or more so to explore outperforming regions. As a matter of fact, a location quotient equal to 2 means that the given industry is represented by a 100% bigger share of employment in the given region than the industry's share of employment on the level of all regions [11].

At this point, having stated that the CE is another indicative element of clustering, in the second step it has also been ascertained which regions fulfil the following requirement:

CE$_F$: in this instance, as stated in the literature [18], the coefficient has to be equal to 1.10 or more.

This means that each region of the United Kingdom that has been tested must comply with the dual criteria – LQ and CE – in order to be recognized the presence of a cluster category.

4 Data and Results

Regarding the geographical units, the focus has been set on NUTS 2 level. The UK territory is divided into 40 areas [19]. Nevertheless, due to the lack of data regarding some variables (e.g. employment) in Northern Ireland, this area has been omitted from the study.

Equation (1)

With respect to the statistics about employment, have been used the dataset available in Business Register and Employment Survey of the UK Office for National Statistics for the year 2015 [20]. Data have been aggregated for SIC Codes in a way to reflect the composition of each Emerging Industry. SIC Codes are the "Standard industrial classification of economic activities" that provide a four-digit code that identifies the nature of economic activities. Given the fact that UK SIC Codes are identical to the NACE codes used by the ECO for identifying clusters, it has not been necessary to make variations of the datasets obtained.

The application of Eq. (1) has resulted in the identification of the presence of a total of 12 emerging industries in 8 regions that have a LQ equal to 2 or more. From the Tables 1, 2 and 3 is possible to notice that the best performing regions are Berkshire, Buckinghamshire and Oxfordshire, that met the requirement in the Biopharmaceuticals, Digital Industries and Medical Devices industries, and West Midlands and North Eastern Scotland that met the requirement in Advanced Packaging, Mobility Technologies, Blue Growth Industries and Environmental Industries respectively.

Table 1. UKJ1 location quotient

	Biopharmaceuticals	Digital industries	Medical devices
Berkshire, Buckinghamshire and Oxfordshire	2.92	2.29	2.06

Source: Own elaboration

Table 2. UKG3 location quotient

	Advanced packaging	Mobility technologies
West Midlands	2.18	2.14

Source: Own elaboration

Table 3. UKM5 location quotient

	Blue Growth industries	Environmental industries
North Eastern Scotland	3.61	4.82

Source: Own elaboration

Eventually, Leicestershire, Rutland and Northamptonshire, Herefordshire, Worcestershire and Warwickshire, East Anglia, Inner London – West, and Outer London - West and North West fulfilled the criterion in 1 emerging industry each one.

Regarding the Emerging Industries, Advanced Packaging, Biopharmaceuticals, and Mobility Technologies are active in 2 regions; Blue Growth Industries, Creative Industries, Digital Industries, Environmental Industries, Logistical Services, and

Medical Devices are located in 1 region; and Experience Industries is not concentrated in any region.

Equation (2)
Concerning Economic Specialisation, have been used the dataset available in Uk Business Counts of the UK Office for National Statistics for the year 2016 [21]. A different year from Eq. (1) has been chosen due to the absence of geographical data for the same year. It is assumed that for the study to be executed this does not result in a loss of significance. Also in this case, data have been aggregated for SIC Codes in a way to reflect the composition of each Emerging Industry following the methodology used for Eq. (1).

The application of Eq. (2) has resulted in the identification of the presence of the CE coefficient in a total of 120 occurrences (Figs. 1, 2, 3, 4, 5 and 6).

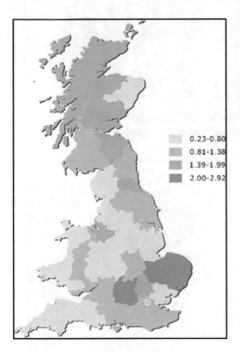

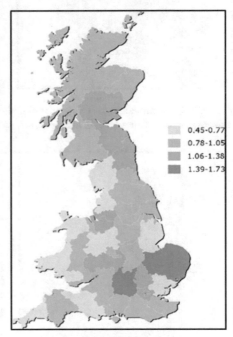

Fig. 1. LQ benchmark in the UK for the biopharmaceutical cluster. Source: Own elaboration

Fig. 2. CE benchmark in the UK for the biopharmaceutical cluster. Source: Own elaboration

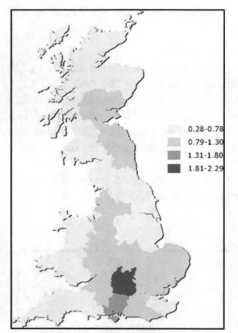

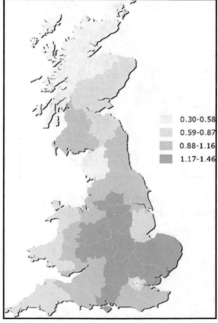

Fig. 3. LQ benchmark in the UK for the digital industries cluster. Source: Own elaboration

Fig. 4. CE benchmark in the UK for the digital industries cluster. Source: Own elaboration

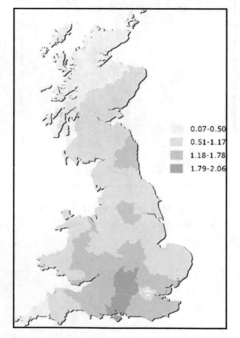

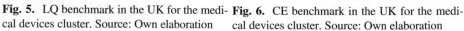

Fig. 5. LQ benchmark in the UK for the medical devices cluster. Source: Own elaboration

Fig. 6. CE benchmark in the UK for the medical devices cluster. Source: Own elaboration

Table 4. NUTS 2 level regions LQ and CE benchmark parameters

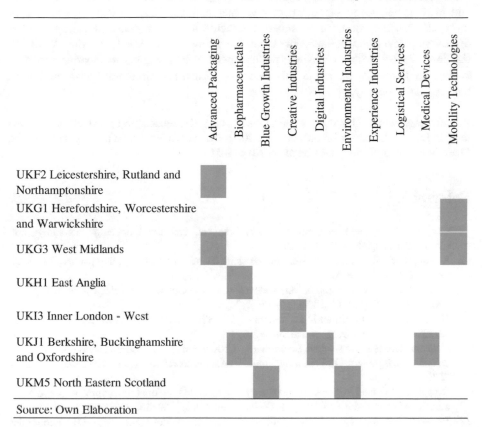

	Advanced Packaging	Biopharmaceuticals	Blue Growth Industries	Creative Industries	Digital Industries	Environmental Industries	Experience Industries	Logistical Services	Medical Devices	Mobility Technologies
UKF2 Leicestershire, Rutland and Northamptonshire	■									
UKG1 Herefordshire, Worcestershire and Warwickshire										■
UKG3 West Midlands	■									■
UKH1 East Anglia		■								
UKI3 Inner London - West				■						
UKJ1 Berkshire, Buckinghamshire and Oxfordshire		■			■				■	
UKM5 North Eastern Scotland			■			■				

Source: Own Elaboration

5 Conclusions

Given the established benchmark parameters, from the Table 4 is possible to notice that in 7 regions there are a total of 11 clusters of emerging industries and that Berkshire, Buckinghamshire and Oxfordshire region is the place with the higher number of business agglomerations. In fact, it has Biopharmaceuticals, Digital Industries and Medical Devices industries agglomerations. Therefore, given that clusters are proxy of innovation, it can be argued that this region is the most innovative in the whole country.

Moreover, Advanced Packaging, Biopharmaceuticals and Mobility Technologies are the only business clusters of emerging industries that are located in two regions respectively. However, despite the wide spectrum of clusters in the different regions, it should be stressed that two emerging industries – Experience Industries and Logistical Services - are not active in the list of clusters found throughout the UK.

In conclusion, cluster mapping in the UK regions must be an incentive for national and local authorities to better target their interventions. Thus, in regions with one or more clusters, the commitment must be directed towards a phase of entrepreneurial discovery process that enhances the competitive advantages already existing. While, in the case of areas without clusters, policies need to be designed to identify new specialization domains. This is the only means of achieving a structural transformation of the regional economy.

Acknowledgements. This work is part of the MAPS-LED research project, which has received funding from the European Union's Horizon 2020 research and innovation programme under the Marie Skłodowska-Curie grant agreement No 645651.

References

1. Cooke, P.: Regional innovation systems clusters, and the knowledge economy. Ind. Corp. Change **10**(4), 945–974 (2001)
2. Martin, R., Sunley, P.: Deconstructing clusters: chaotic concept or policy panacea? J. Econ. Geogr. **3**(1), 5–35 (2003)
3. Porter, M.: Location competition, and economic development: local clusters in a global economy. Econ. Dev. Q. **14**(1), 15–34 (2000)
4. Bevilacqua, C., Pizzimenti, P., Maione, C.: S3: Cluster Policy & Spatial Planning. MAPS-LED Project, Reggio Calabria (2017)
5. Growth: European Innovation Scoreboard - Growth - European Commission (2017). http://ec.europa.eu/growth/industry/innovation/facts-figures/scoreboards_en. Accessed 27 Sept 2017
6. Swords, J.: Michael Porter's cluster theory as a local and regional development tool: the rise and fall of cluster policy in the UK. Local Econ. **28**(4), 369–383 (2013)
7. Porter, M., Ketels, C.: UK Competitiveness: moving to the Next Stage. DTI Economics Paper (2003)
8. European Commission 2014: European Cluster Panorama 2014. http://eco2.inno-projects.net/2014-10-15-cluster-panorama-d1.4a.pdf. Accessed 5 Sept 2017
9. Delgado, M., Porter, M., Stern, S.: Defining clusters of related industries. J. Econ. Geogr. **16**(1), 1–38 (2015)
10. Crawley, A., Pickernell, D.: An appraisal of the European cluster observatory. Eur. Urban Reg. Stud. **19**(2), 207–211 (2011)
11. European Commission 2016: European Cluster Panorama 2016. http://ec.europa.eu/DocsRoom/documents/20381. Accessed 5 Sept 2017
12. European Commission: Methodology and findings report for a cluster mapping of related sectors (2014). http://ec.europa.eu/DocsRoom/documents/16527/attachments/1/translations/en/renditions/pdf. Accessed 5 Sept 2017
13. Neffke, F., Henning, M., Boschma, R.: How do regions diversify over time? Industry relatedness and the development of new growth paths in regions. Econ. Geogr. **87**(3), 237–265 (2011)
14. Spencer, G., Vinodrai, T., Gertler, M., Wolfe, D.: Do clusters make a difference? Defining and assessing their economic performance. Reg. Stud. **44**(6), 697–715 (2009)
15. Resbeut, M., Gugler, P.: Impact of clusters on regional economic performance. Competitiveness Rev. **26**(2), 188–209 (2016)

16. Glaeser, E., Kerr, W.: Local industrial conditions and entrepreneurship: how much of the spatial distribution can we explain? J. Econ. Manag. Strategy **18**(3), 623–663 (2009)
17. Paton, J., Del Castillo, J., Barroeta, B.: Regional economic transformation: the role of clusters in specialised diversification (2014)
18. Del Castillo, J., Paton, J.: Cluster identification and analysis: four regional cases in Spain. ERSA Conference Papers (2011)
19. Ec.europa.eu.: National structures (EU) – Eurostat (2017). http://ec.europa.eu/eurostat/web/nuts/national-structures-eu. Accessed 7 Sept 2017
20. Nomisweb.co.uk.: Nomis - Official Labour Market Statistics (2017). https://www.nomisweb.co.uk/query/construct/summary.asp?mode=construct&version=0&dataset=172. Accessed 7 Sept 2017
21. Nomisweb.co.uk.: Nomis - Official Labour Market Statistics (2017). https://www.nomisweb.co.uk/query/construct/summary.asp?mode=construct&version=0&dataset=142. Accessed 11 Sept 2017

The Neutrality Between "Us" and "Others", a Framework for Sustainable Social/Cultural Urban Development

A Tool of Analysis and a Goal for Urban Intervention

Mohammed Ezzat[✉]

German University in Cairo, New Cairo City, Egypt
mohammed.ezzat@guc.edu.eg

Abstract. A conceptual theoretical manifestation of "Us/Others" relationship could be traced in fields like identity, where a demarcated boundary is somehow apparent, and societal inclusion of minorities. However, the paper perceives the "Us" and "Others" relationship as a persistent epistemological and psychological relationship that needs explanation. "Us" can't exist apart from the "Others". "Others" is not necessarily a person; it is whatsoever beyond "Us". Consequently, the paper recognizes the "Us/Others" relationship as a point of departure and as an essential step for the achievement of a universal unifying whole that disregards boundaries and sets a common platform of communication that sustains everyone's conception of Urbanism. For such a universal whole to exist, the boundaries between "Us" and "Others" need to diminish into neutrality. Such neutrality would be conducted based on the abstract understanding of both of "Us" and "Others". Otherness is a kind of enrichment of the universal whole rather than a weakening factor, and it could be the missing soul of Globalization. As a matter of fact, the universal whole is a monolithic meaning of infinite Otherness's shapes. During the paper, Otherness would be handled as a tool of analysis for instrumenting the cultural/social sustainable metropolitan/urban transformations. Otherness would be analytically represented as hierarchical instances of the structured continuum of Urbanism, which is disclosed by a specially devised comprehensive tool of analysis. Such monolithic representation of Urbanism and their analytical instances constitute the proposed framework and is expected to comparatively and coherently support the top-down and bottom-up neutral urban analyses and interventions. The paper would firstly exhaust the metaphysical definition of "Us" and "Others", then secondly would present the possible intervening policies for promoting neutrality as the universal form of Urbanism.

Keywords: Neutral designs · Otherness · Universal urban designs
Unifying model of urbanism

© Springer International Publishing AG, part of Springer Nature 2019
F. Calabrò et al. (Eds.): ISHT 2018, SIST 100, pp. 462–470, 2019.
https://doi.org/10.1007/978-3-319-92099-3_52

1 Introduction

Understanding the identity of any sampled group implies defining the significant reflections of the values, beliefs, qualities, and the alike on that group. However, defining such reflections entails the implicit definition of the "Others", which is the complementary of this set. That complementary set of the discovered reflections is the suggested identity of the "Others". For instance, when an individual belongs to a group, then that group is "Us" and anything beyond that group is "Others". This dilemma between "Us" and "Others" always persists as a needed analytical phenomenon for designing any sustainable social/cultural urban development, especially in historical sites' developments or the sites in their vicinity, where "Us" is clearly established and any change by "Others" is highly susceptible. The relation between "Us" and "Others" can either be cooperative, both of them cooperate in producing the ultimate reflections on the sampled group, or it could be segregated, both of them are clearly desperately confined. In the paper, we would focus on the neutral relationship when a clear distinction between "Us" and "Others" diminishes. Neutrality is, therefore, dependent on the definition of "Us" and "Others". In other words, neutrality is relative to the sampled group and not an absolute phenomenon. To define that grey neutral area between "Us" and "Others", we need to have a basic understanding of both of "Us" and "Others". During the paper, we would rely on a comprehensive model of Urbanism [1] to coherently and theoretically define "Us" and "Others" (Fig. 1). The comprehensive model states that the continuum of Urbanism could be coherently understood using three perspectives labelled Rational, Visual, and Emotional. We would first investigate the metaphysical definition of "Us" and "Others", then we would analyse the different possible approaches for promoting neutrality.

But, is the Otherness merely the paper's creation? Magnifying the phenomenon of the otherness would inspire the artful transcendence that is beyond the realistic singular perception of the person. Additionally, unleashing the ability to philosophize the universe is a major impact of analysing the otherness. Such philosophization would otherwise be trapped in the singular eye of the perceiver [2]. Pursuing what is beyond the lived and the experienced is the way of visualizing and theorizing the otherness, and great poets had sensed the power of such salvation [3]. Many contemporary authors of literature engaged in the debate about the concept of the Utopia of the otherness. Such Utopian Otherness is thought to be the needed soul of the socioeconomic Globalization [4], and that is the fundamental position the paper stands for.

2 Neutrality as a Tool of Analysis

In parallel research effort [1], we define a specially devised comprehensive tool of analysis of Urbanism. That tool represents Urbanism as a categorically related set of traits and characteristics. By using such tool, that comprehensively and coherently enumerates the characteristics of urban constructs, we would be able to describe both of "Us" or "Others" as variations of these abstract characteristics (Fig. 1). Nevertheless, discovering the different characteristics of the otherness should be coupled with answering a metaphysical question about the significance of such analysis. The answer may come from the field of the philosophy of science [5, 6]. An ontology of the information theory, originally introduced by Wu Kun, reduces the interconnected concepts of mind, knowledge, epistemology, perception, cognition, and information into the sole concept of differentiation [5]. Such a doctrine's slogan is "a difference that makes a difference". Our proposed concept of the Otherness corresponds with that minimal representation of the difference. The concept of Otherness is also an influential theory of social representation [7].

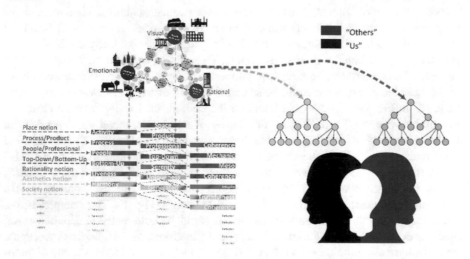

Fig. 1. The comprehensive tool of Urban analysis for representing the "Us/Others" instances.

However, the significance of analysing the otherness mandates knowing what does "Us" or "Others" stand for? And who does each represent? There must be categorizing criteria, by which, "Us" starts to stratify from the "Others". These categorical criteria could for instance be:

- Spatial criteria: like here and there spatial qualities.
- Group belonging: like belonging to this group, that group, or to both.
- Having specific characteristics: like religion, ethnicity, education, etc.

Departing from this simple stratifying categorical representation of the otherness, we may constructively define the complexity of the otherness as:

1. There is no such a single criterion of the otherness's stratification, but rather there are immense amounts of categorical criteria.
2. These immense criteria prioritize and relate, hierarchically, differently based on the context of discourse.
3. Each of these stratifying criteria could extend over different levels of a plethora of different meanings. This multi-levelled representation implies the decomposability of "Us" or "Others" into smaller constituting groups.

4. This decomposability principle entails that "Us" or "Others" could be abstractly represented in rather the more detailed multi sub-groups representation (Fig. 1).

The utilization of that abstract representation of "Us", or the dually "Others", against their constituting sub-groups may suggest the following:

1. The abstract representation of "Us" would presume the direct understanding of its dual, the "Others".
2. The accumulation of the multi sub-groups constituting "Us" or "Others" into a holistic abstract presentation of "Us" or "Others" is an identity-related decision.

One of the expected positive contributions of the proposed neutrality concept, the universal Urbanism, is that it may overcome the deficiencies and the drawbacks inherent in the aforementioned criteria's categorical stratification. The proposed otherness that mainly depends on the hierarchical representation of Urbanism, using the comprehensive tool of Urbanism, would act as a translation from any other negative criterion of stratification into the more commonly acceptable criteria of Urbanism. It would create a common platform, which implies a monolithic meaning of mutual acceptance, for any other criterion to be democratically communicated. In the following section, we would try to investigate the possibilities of spreading neutrality in any sampled group.

3   Neutrality, as an Intervening Goal, to Achieve

Promoting neutrality is the only intervening sustainable approach. Imposing neutrality, as a mere political top-down decision, would be an act of disregard to the inherent complexity of any community. Additionally, assuming a reliable priori understanding of such complexity could be a presumptuous assessment that can be easily challenged. Any presumed intervention or understanding should be done by the community itself or, at least, by close collaboration with it. Then, the collaborative approach is the only viable method of intervention. Therefore, the top-down intervening actions should be supported by the emergent, bottom-up, neutral cooperation between "Us & Others". That cooperative bottom-up evolution of neutrality, and consequently for shaping the infinite shapes for the universality, could take any of the following several forms:

1. Nominal cooperation: in the nominal cooperation, the reflected meaning of any proposed new urban construct should be neutral. In this approach, the world of semantics is the point of interest. It is a tailored phenomenological neutrality. It is similar to the perceived knowledge space of Henri Lefebvre's planned, perceived, and used trilogy of place [8]. This means that, a single physical urban construct should reflect neutral meanings to "Us" and to "Others" (Fig. 2(2)). It is a single construct that simultaneously represent "Us/Others". This cooperative nominal approach may be attained by the following:
   a. Highlighting on the commonalities between "Us" and "Others", and utilizing them as main formative criteria for the proposed monolithic neutral urban construct that positively reflects everyone's preferences.

   b. Perceiving "Us" from the "Others" eyes, or mutually perceiving "Others" from the "Us" eyes for the creation of the monolithic neutral that suits everyone.

2. Spatial physical cooperation: in the physical cooperation approach, multiple materialized urban constructs that differently representation both of "Us" and "Others" would be composed together for the creation of the whole (Fig. 2(1)). The composition can take one of the following shapes:

   a. Equally dispersed composition: it is more apparent to recognize "Us" and "Others" as a monolithic holistic mixer rather than separate constructs.

   b. Asymmetric dispersed composition: it is more attainable to recognize each of "Us" or "Others" as separate constructs rather than recognizing them as a mixture.

3. Hybridized approach: a hybridization between the previous approaches is done by:

   a. Mixing nominal and physical urban constructs.

   b. Planning multiple interwoven physical constructs, that either reflect "Us" or "Others"; then add a layer of abstraction to produce the monolithic nominal cooperative construct that suits everyone. Meaning, abstractly producing nominal single construct out of multiple physical ones.

The "Us/others" users' experience of place is not static but dynamic, and their reflections are an exposure-based, time-dependent experience. The selectivity of which policy to promote for neutrality is part of the sampled group's identity, and its successfulness depends on the locality of the targeted group.

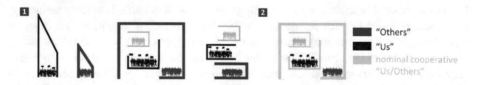

Fig. 2. (1) Different formal compositions between "Us/Others" constructs, (2) a nominal cooperative representation.

3.1 Implementing the Neutrality's Promoting Policy

Urban top-down intervention, in already shaped historical and cultural urban constructs, should be highly sensitive to their social, cultural, and physical conditions; and additionally, it should have the knowledgeable intelligence and insight for their prospect future changes. Any intervention, for promoting neutrality, shouldn't come from a day and night decision, but rather it should develop and evolve sensibly and intelligibly. The refusal of sudden change and the appraisal of evolutional one can be implemented by planning dispersed pilot urban nuclei as inception points of any intervention, mainly for feedback enquiry (Fig. 3). These nuclei should have the following characteristics:

4. They have the resilience, responsiveness, and adaptability for the current and futuristic conditions of their hosting communities.

5. They have to be on continual change, which should be energized by the bottom-up feedbacks and positively accept top-down political tailored interventions as well.

Therefore, these nuclei could be thought of as living organisms that have the mental and memorial capacity to be autonomous entities, and they have a specific goal to achieve, which is promoting neutrality. However, as much as they develop naturally, they should embrace direct, professionally planned, interventions as well. For These direct professionally planned interventions not to undermine the main goal of the proposed evolutional nature of the nuclei, they should be aware of the following:

1. The spatial extents and limits for such direct intervention.
2. The qualitative limitations, and their foreseeable social/cultural consequences.

And accordingly, any direct professional intervention would not drift the proposed nuclei form their goal-oriented natural development. These proposed dispersed nuclei's evolution would normally span the different phases of:

1. The inception phase: A pilot miniature projects should be dispersed prior to imposing any top-down intervening measurements. Their role is to collect the data needed for the coordination between the bottom-up emergence and the possible top-down interventions.
2. The maturity phase: in this phase, the characteristics of the nuclei are clearer, and the top-down intervening decisions are easier to analyse and to predict their consequences.

All of the aforementioned variations of Intervening methods and abstract Shapes of Intervention (Figs. 1 and 3) could be utilized as predictive/analytical tools during the emergence of the proposed nuclei.

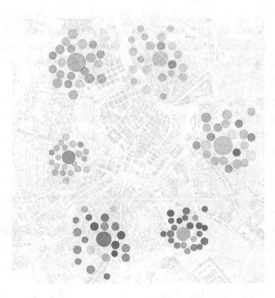

Fig. 3. The desperate naturally self-organizing pilot nuclei for the top-down intervention.

Discussion: The proposed intervening dispersed nuclei may misleadingly seem presumptuous, experimental, or even lack proper theoretical foundations. A parallel concept of the nuclei is introduced to the field of Urbanism not only as an explanation of the non-residential core [9], the nucleus of a neighbourhood, but also as a tool for modelling the spatial land-use Change in even the complex emergence of the sprawl or the spontaneous settlements [10]. Such modelling is viable based on the multiple-nuclei theory that recognizes the megacities as the agglomerative growth of dispersed points of attraction rather than as the emergence of a singular central Business District [11]. Such nuclei agglomerative growth could easily be observed in any studied city [12]. Therefore, our proposed self-organizing nuclei pilots are robust tools for analysis. They are observed vastly in the ongoing urban transformation processes and they are already utilized as simulative tools of Urbanism.

4   The Framework by the Comprehensive Model

So far, we introduced the techniques needed for defining and maintaining neutrality, but we don't know how to understand the fundamental notion of "Us", or mutually "Others". We are lacking the proper tools for representing "Us" and "Others", which are the building blocks of neutrality. In a parallel research effort [1], we specifically devised a comprehensive tool for understanding Urbanism intended for representing "Us/Others". This comprehensive tool was developed based on exhaustive analysis of urban constructs, historical and contemporary ones. That analysis yielded the possibility of abstractly and coherently categorizing the urban constructs' characteristics underneath three urban perspectives. The approach of holistically categorizing urban constructs is closely related to exhaustive exploration of the known atmospheric reflections of any urban construct.

The categorical three perspectives were labelled Visual, Rational, and Emotional. Urban constructs characteristics' variations were perceived coherently by the three perspectives. The conducted procedure for assuring that perspectival representation of the urban constructs is known, in the philosophy of science, as holism by reduction [6, 13, 14]. We may briefly describe the three perspectives as:

- *The visual perspective* is about a higher philosophy that is mostly not related to daily activities. This philosophy shapes each corner of the visual perspective's figurative products. The users are experiencing and living that philosophy, their activities and their social behaviours are always considered as subordinate to what that philosophy dictates. Visual designs are produced by artists, poets, or musicians who produce that higher philosophy. The manifesto of the Bauhaus, the city as an art museum of the dictatorship period at Europe at the first half of the twentieth century, and the Romanian/renaissance cities are all examples of that perspective.
- *The emotional perspective* is about the Activities hosted inside the space, their interaction between each other, and the interaction between them and the hosting space. Place exists, celebrates, motivates, gets its identity, and is enriched by its hosted activities. Liveness, warmth, cultural enrichment, and joy are amongst the main features of its evolving places. The sprawl, the spontaneous settlements,

the naturally developed cities of Europe, and the Sumerian/Islamic cities are examples of that perspective.

- *The rational perspective.* Integrity and truthfulness is the slogan that derives that perspective. Truthfulness is the main source of beauty. The tendency of explaining and interpreting applies even for any introduced conception of amusement and beauty, which is an inherently needed criterion of architecture. The gothic architecture, though religiously driven, the international style, the high-tech movement, and the Neo-futurism movement, i.e. Norman Foster, are rational products.

5 Discussion

Is promoting neutrality an act of ethicality? When we introduced the concept of the universal, we meant the monolithic meaning vs. infinite shapes. Neutrality is the social/cultural sensible variation of the crude, though economically successful, globalization. Neutrality recognized the monolithic form of the different. So, what is wrong in having crude boundaries between "Us" and "Others"? Why should we break the boundaries of localities? Or simply, what is the gain of doing so? These questions can be easily answered as a comparison between the harsh delimiters between "Us" and "Others" against the proposed locality-sensitive neutrality's reflections of the following:

- The universality is the monolithic reflected meaning of tolerance, acceptance, cooperation, mutual respect, richness of the diversity in life, openness, identity as a boundary of inclusion rather than a boundary of exclusion, and the artful transcendence/philosophization that is beyond the singular person's view of the world.

Single meaning vs. multiple infinite shapes is what the proposed neutrality stands for, and that is what we mean by relativistic non-absolute users-dependent neutrality.

6 Conclusion

For promoting neutrality, we need to justify any top-down transformative action by the bottom-up feedbacks. These feedbacks are acquired by planned dispersed pilot nuclei and they should enhance our ever-changing understanding of both of "Us" and "Others". We presented the possible intervening policies that should be tightly related to these feedbacks. Neutrality should enhance our quality of life, as a community, by adding certain needed social values to the existing repository of values and by broadening our artful/philosophized conceptions of the world. The specially devised comprehensive continuum of Urbanism is the paper's proposed framework of neutrality. In future work, we would select similar urban constructs from various parts of the world and we would then utilize the proposed neutral framework for analysis.

References

1. Ezzat, M.: A Comprehensive Proposition of Urbanism: With Potential Applications on users' urban cognitive-mapping and users' generated urban designs. In: International Symposium New Metropolitan Perspectives (ISTH2020), Reggio Calabria (2018, in Press)
2. Desmond, W.: Art, Origins, Otherness: Between Philosophy and Art. State University of New York Press, Albany (2003)
3. Fernández-Medina, N.: The Poetics of Otherness in Antonio Machado's Proverbiosy Cantares. University of Wales Press, Cardiff (2011)
4. Arenas, F.: Utopias of Otherness. University of Minnesota Press, London (2003)
5. Kun, W., Brenner, J.E.: An informational ontology and epistemology of cognition. Found. Sci. **20**, 249–279 (2015)
6. Esfeld, M.: Holism in Philosophy of Mind and Philosophy of Physics. Springer, Dordrecht (2001)
7. Augoustinos, M., Riggs, D.: Representing Us and Them: constructing white identities in everyday talk. In: Social Representations and Identity: Content, Process, and Power, pp. 109–130. Palgrave Macmillan, New York (2007)
8. Lefebvre, H.: The Production of Space. Basil Blackwell Ltd, Oxford (1984)
9. Mehaffy, M., Porta, S., Rofe, Y., Salingaros, N.: Urban nuclei and the geometry of streets: the emergent neighborhoods' model. Urban Des. Int. **15**, 22–46 (2010)
10. Abdullahi, S., Pradhan, B., Al-sharif, A.A.: Spatial land use change modeling techniques. In: Spatial Modeling and Assessment of Urban Form. Analysis of Urban Growth: From Sprawl to Compact Using Geospatial Data, pp. 171–185. Springer, Berlin (2017)
11. Torrens, P.M.: How Cellular Models of Urban Systems Work (1. Theory) (2000)
12. Berry, B.J.L.: The Human Consequences of Urbanisation. Macmillan Press, London (1973)
13. Agazzi, E.: Scientific Objectivity and Its Contexts. Springer, Cham (2014)
14. Welshon, R.C.: Perspectivist Ontology and De Re Knowledge, in Nietzsche, Epistemology, and Philosophy of Science: Nietzsche and the Sciences II, pp. 39–46. Springer (1999)

Industrial Districts as Cities. Supra-Local Governance in the Sassuolo Ceramics District

Cristiana Mattioli[(✉)] [iD]

Politecnico di Milano, 20133 Milan, Italy
cristiana.mattioli@polimi.it

Abstract. The paper discusses the plural notion of metropolitan territories in Italy by focusing on industrial districts (IDs), which are not included in the national urban agenda. IDs represent today not only important productive systems inserted in global value chains, but also peculiar urban conurbations formed by processes of territorial coalescence, thus requiring adequate representation and attention. In the Sassuolo ceramics district, some changes and challenges have recently stimulated new reflections among local institutions and have finally led to the proposal of the 'city-district' idea, which has actually encountered many difficulties and resistances. The paper critically reflects on the feasibility, necessity and possible configuration of this new governance system, firstly by showing that some relevant issues for the competitiveness and sustainability of the ID have already reached a supra-local scale, thus demanding common strategies and policies of territorial improvement and regeneration; secondly, by giving suggestions for its effective implementation.

Keywords: Industrial districts · Supra-Local spatial planning · Sassuolo

1 Introduction: The Role of Italian Industrial Districts

Territorial policies in Italy have recently shown a kind of contradictory and complementary configuration. On the one hand, a limited number of strategic 'metropolitan cities' have been identified, following a hierarchized and top-down vision of national territorial government (Law 56/2014). However, the homogeneous definition of 'metropolis' hardly fits with the Italian plural urban and socio-economic configuration, and with contemporary, post-metropolitan territorial processes, reaching regional extension [1–3]. On the other hand, important contemporary issues (such as climate change, hydrogeological risk, depopulation and abandonment, etc.) and new economic trends (tourism, soft economy, etc.) have enlightened again inner areas [4].

Both positions leave aside 'intermediate regions' characterised by diffused and polycentric urbanisation and industrialisation, which attract the attention of some EU development policies [5]. Among them, industrial districts (IDs) (or industrial production zones) represent a peculiar situation for the strong connection between economy, society, and territory [6]. In IDs the industrialisation appeared in tardive and original forms, delineating an alternative model of development for the Fordist one.

IDs still occupy a prominent position within national system. Their population is large; their extension covers more than 20% of national surface and, above all, their

© Springer International Publishing AG, part of Springer Nature 2019
F. Calabrò et al. (Eds.): ISHT 2018, SIST 100, pp. 471–478, 2019.
https://doi.org/10.1007/978-3-319-92099-3_53

economy is crucial for national GDP [7]. In fact, they account for more than 50% on national export and host many leading, innovative companies [8].

Facing globalised market, knowledge economy and internationalisation, IDs have profoundly changed, sometimes even in divergent ways [9]. They are more differentiated than before, for what concerns industry, population and territorial configurations. Multi-level networks have broken the original self-restraint configuration, opening IDs to global value chains and competitive regional systems [10, 11].

From a territorial standpoint, IDs originated from commercial and industrial towns surrounded by a countryside organised in a sharecropping system; thus, these territories show both urban and rural elements. Due to their polycentric organisation and extensive urbanisation that is constantly expanding since the 90s', they are today peculiar urban conurbations, formed by processes of territorial coalescence [12]. IDs are thus cities, even though without an adequate political representation.

2 Sassuolo Ceramics District: Essays of Supra-Local Governance and Planning

The Sassuolo ceramics ID is located in the Emilia Romagna region and is the biggest conurbation of the foothill area, gathering eight Municipalities in the two Provinces of Modena and Reggio Emilia (about 180,000 inhabitants). In productive terms, its firms represent 80% of Italian domestic production of ceramic tiles and employ more than 18,000 people.

At the end of the XIX century, Sassuolo became an important centre specialised in ceramics tiles production, but it was only after World War II that industrial development arose. Plants rapidly grew in number and spread all over the territory, shaping a continuous conurbation [13] and largely exploiting the land. Soon enough, environmental and health problems appeared. As a response, local administrations, trade unions and banks started to cooperate to find urgent and coherent solutions, in particular fostering industry modernisation and compatibility.

This configuration consolidated in the 80s' with the introduction of the 'Comprensorio', a new government level between the Municipality and the Province. Inside this framework, a 'PTCC - Piano Territoriale di Coordinamento Comprensoriale' (District Territorial General Plan) was elaborated (1983) with the aim to promote a more balanced, sustainable and differentiated socio-economic development. The PTCC proposed an 'integrated, polycentric and metropolitan system' characterised by widespread and diverse services, an efficient network of mobility, land preservation, industry improvement. The plan never reached an operative level. However, it paved the way for several experiences of cooperation, which also involved private actors: two European studies on transport in the 90s'; the agreement on local goods logistics (1997); the PRUSST program for territorial regeneration (early 2000); the EMAS European certification (2003); the recent formation of the Unions of Municipalities that act on common public services efficiency and rationalisation.

The supra-local dimension is indeed not new in the ceramics ID. However, despite its functional, economic and social internal integration, the system continues to be managed on local basis, therefore in a fragmented and barely efficient way.

To overcome this situation, during the last municipal elections (2014) the left-wing party (Partito Democratico) started a reflection on a common territorial strategy proposing the 'city-district' vision. The 'city-district' is intended as a supra-local structure oriented towards a new sustainable model of development. Led by a dedicated department established in Sassuolo, the project involved at first seven Municipalities in the foothill area, where the majority of industrial plants are located. Facing sectorial crisis, the supra-local strategy is mostly aimed at the creation of a more diverse and integrated local economic system. For this reason, the first collective initiative was the creation of a new territorial brand called 'Terra Maestra', which enhances local attractiveness by offering itineraries related to industrial and cultural tourism, agri-food and automotive excellences (such as, Parmigiano Reggiano and Ferrari).

3 Objectives and Methodology

The Sassuolo ceramics district represents an interesting case-study for analysing the evolution of Italian IDs, in particular in relation to supra-local governance system and spatial planning. The 'city-district' idea highlights the necessity to identify a supra-local system (eventually a more structured administrative entity) able to tackle critical issues that single Municipalities cannot face alone, for budget and expertise lacks or simply because of their limited territorial authority.

The paper aims to critically reflect on the feasibility, necessity and configuration of this new governance level, by concentrating on its territorial and urban effects. It will discuss in particular three inter-municipal emergent issues (firms reorganisation, mobility and logistics system, and habitability), selected accordingly to their relevance for local administrations and stakeholders (received through interviews) and inside local debate (newspapers, citizens' committees, municipal decisions, etc.), as well as through an accurate and continuous fieldwork conducted over the last 5 years.

4 Inter-municipal Emergent Issues

4.1 The Consolidation of Leading Companies

Despite the still relevant role of horizontal integration [14], the Sassuolo ceramics ID has evolved into a productive system mainly controlled by big business groups, consolidated through merges and acquisitions [15]. Due to the historic distribution of industrial plants, these processes led to the formation of multi-plant factories spread on the territory regardless of municipal boundaries. This organisation has revealed all its limits facing technical innovation, which on the contrary tends to concentrate in selected, specialised spaces in reasons of its high costs. Also, during the recent crisis, the production decrease caused a generalised underuse of industrial plants and work-force. As a consequence, leading companies have entered a period of great spatial transformation [16], integrating two more general trends: the evolution toward "Industry 4.0" (ICT, innovation, etc.) and the internalisation of tertiary functions (R&D, marketing, product customisation, etc.) [8].

On the one hand, they are reorganising their production system by centralising it in few places, mainly for logistical reasons. Consequently, some factories are abandoned, while others (the most central, accessible and visible ones) are consolidated. This polarisation has clearly exacerbated local rivalry related to attractiveness policies, job offer and tax payment, reinforcing at the same time the bargaining power of few entrepreneurs, in particular in case of expansion on greenfield.

On the other hand, workplaces are today more clean and multifunctional [17]; indeed, their qualified spaces accommodate also welfare facilities and activities open to the territory.

Although being locally rooted, leading companies have entered global networks, precociously undertaking internationalisation [18]. Today they have showrooms in Milan, annually organise the world leading trade show in Bologna – where the Ceramics Research Institute is also located – and find added-value resources in a space that is no more restrained to ID, nor to medium-sized cities located nearby, but rather enlarged to metropolitan scale [19].

4.2   Uncoordinated Top-Down Actions for Logistics and Mobility

Whereas common infrastructural strategies promoted in the 90s' have certainly improved local situation (by creating an east-west connection and a link road to Modena), some of their previsions remain unrealised. Meanwhile, industrial system has deeply changed, especially during the crisis. Despite the production decrease, the territory continues to suffer from heavy congestion, as distribution system is today more and more fragmented. The ceramics ID is also reshaping into an international logistics hub serving European markets with a sophisticated supply chain [20].

For these reasons, the demand for new infrastructures is still strong. However, local administrations seem to have few relevance in the decision-making process. Strategies are developed at supra-local government levels.

Recently, the Italian Minister for Infrastructure went to Sassuolo to relaunch the construction of a new highway, designed for the first time in the 80s'. The 15 km-long infrastructure would connect Sassuolo with the national highway system (A1 Milano-Napoli) and northern-Europe (via A22 Brennero highway) by running along the Secchia river, the primary ecological corridor of the territory, where preservation and leisure activities coexist with high environmental impact functions (such as gravel pits, transport and logistics small hubs). Besides environmental issues, the highway is today locally questioned for its costs and effective need, as even entrepreneurs point out its extreme delay and its probable financial unsustainability. More in general, the highway promotes an obsolete logistics model, strictly connected to road transport.

On the contrary, important resources have been allocated in the last years for railway improvement, making its use greater than the average in other productive sectors. A new railway yard is ready to open, except the link road has not been built yet and the Province of Reggio Emilia has suddenly erased from its planning instruments the prevision of a railway connection to the ceramics ID. Similarly, the new Piano

Urbano della Mobilità Sostenibile (Sustainable Urban Mobility Plan) – in definition – includes only the four Municipalities in the Province of Modena.

In a period of resource scarcity and urgent environmental problems (hydrogeological risk, pollution, etc.), it is difficult to see how (uncoordinated) planning decisions can link local development to new major works, rather than improving the efficiency and capacity of existing, sustainable infrastructures. It is therefore no surprising that many local committees organise to contrast this resource-consuming tendency.

4.3   A Fragmented, Degraded Territory

The territory represents today an important, competitive and distinctive asset for companies, both as innovation cluster [8] and liveable space.

In Sassuolo (as in other IDs), the poor quality of places becomes a critical issue in terms of territorial attractiveness and habitability. Actually, present urban conditions are the result of sectorial (sometimes even contradictory) spatial strategies, inward-looking urban projects and a planning process that has simply juxtaposed objects without paying attention to the space between them or supra-local dimension [21]. This situation has worsened during the crisis period.

Local rivalry has become particularly intense on commercial distribution, as it often represents the unique possibility for Municipalities to collect land taxes and recover former industrial, polluted sites. The diffusion of so many medium-sized shopping malls is however unsustainable and unable to produce urban and social quality. Indeed, it promotes a consumerist way of life - in which social encounters and public spaces are connected only to shopping experience - based on private car mobility, which is translated into functional spaces surrounded by wide parking lots.

The resource scarcity has also severe impacts on the rich equipment of public facilities. Whereas industrial companies' initiatives can increase local offer, especially in industrial areas, existing welfare spaces require today extraordinary maintenance and adaptation to standards. Due also to negative demographic trends and new lifestyles, some public services are or will be soon reduced and centralised in capital cities, where the majority of cultural and leisure activities are located, therefore creating new territorial hierarchies and inequalities.

These dynamics evidently question traditional territorial polycentrism - namely the dispersion of services and functions -, fostering the recognition of wider 'metropolitan territories', where complementarities must be strengthened, duplications avoided, and internal connections better organised [22].

5   The 'City-District' as Territorial Project

Many issues are thus demanding for a more comprehensive territorial strategy. Hence, the definition of an inter-municipal governance system cannot be postponed later on.

If the Sassuolo ceramics district (as other Italian territories) wants to compete on a global level, particular attention has to be addressed to its territorial quality, with the aim to produce a more diverse, efficient, attractive and liveable territory, for its inhabitants, companies, and the new high-skilled workforce. Consequently, IDs improvement is not only an industrial issue, but also a territorial one.

Adding new objects is no longer enough. It is necessary to demonstrate commitment to the protection, improvement and reconnection of social fixed capital by transforming, innovating and reforming existing urban materials [23]. To this end, the definition of a long-term, shared vision of spatial development is of the utmost importance.

First of all, in order to control land take, promote urban regeneration and put an end to the inefficient and damaging municipal rivalry, it is fundamental to review and limit local expansions, reframing them inside an inter-municipal 'ecology of additions and subtractions' [24]. In this way, extensions and compensations could be managed at a supra-local level, by favouring volumes transfer from fragile areas and drosscapes reclamation.

A common strategy could also guide the identification of some territorial 'structural' elements and figures, a starting point for a systemic reform of the city-district [25]. Some suggestions may be envisioned moving from the above-mentioned transformations and trends.

Some companies have already engaged directly in projects of territorial qualification by working inside and outside the factory. However, being random and isolated, these interventions are detached from what stands beside them, becoming pure landmarks. Private initiatives need to be oriented to construct a landscape that is more coherent to production quality. Moving from companies' visibility and enlargement demands, it is possible to coordinate actions for the improvement of industrial roads and areas, as well as city surroundings, also by applying to the existing urban fabric some of the EIPs indications [26, 27].

For what concerns transport issue, both logistics and collective mobility reorganisation require common strategies, able to reinforce local negotiating power. Railway, in particular, can fulfil a dual task. It may lay the foundations of an extended and more equal urban system, in which town centres and services are efficiently connected, congestion is reduced, and accessible urban tissues are valorised and densified. Also, it may insert the 'city-district' in a wide regional system, by connecting it to nearby cities and the new high-speed train station of Reggio Emilia Mediopadana.

Finally, local habitability may be improved with a system of public and open spaces, connecting and including welfare facilities [28]. Water streams may become environmental corridors and elements of continuity inside urban and rural areas, integrating issues of hydrogeological safety and common use, while interstitial portions of countryside may be transformed into 'agricultural parks' with the active cooperation of local farmers.

6 Conclusion

The 'city-district' has encountered many obstacles during its implementation, due especially to the lack of a strong political will. However, without a radical cultural transition, IDs risk to become mere concentrations of multinational companies, detached by the increasingly degraded context. Or, they risk following development paths imposed by global logics, regardless the quality of local territory.

To avoid this scenario, the paper suggests not to define a new level of government, which has shown to be inefficient in the past, but rather to set a multi-level governance structure for enhancing territorial regeneration through cooperation among administrations and private actors. Indeed, the territory represents the link between local firms (competitors) and may be used to maintain and establish relations of cooperation, thus reinventing the original integration between economy, society and territory.

Evidently, each actor and each intervention may refer to a long-term shared vision. In this sense, it is crucial to collectively choose one among the many possible futures and invest to realise it [22].

However, to be effective, the paper argues that strategic vision must assume the character of a territorial project, which can be implemented and adjusted over time through flexible, multi-level and public-private alliances.

In a period of resource scarcity, the financial aspect is not insignificant, either. On the one side, few concrete, integrated territorial projects can become opportunities to rationalise and aggregate resources, by jointly participate in national and European announcements. On the other side, companies' involvement in territorial regeneration can be enhanced by defining new instruments and procedures (such as fiscal measures) for private benefits. In this way, the still relevant economic capacities of local industry may effectively be valorised for collective aims [29].

Finally, the Sassuolo case-study shows that IDs can represent a testing room for the definition of supra-local and multi-level governance systems, by which to start a sustainable reform of the existing city-territory and a more general reflection on Italian territorial, administrative reorganisation.

References

1. Indovina, F.: Metropoli territoriale e sviluppo economico-sociale. Economia e Società regionale 109(1), 43–61 (2010)
2. Balducci, A., Fedeli, V., Curci, F.: Post-Metropolitan Territories. Looking for a New Urbanity. Routledge, London and New York (2017)
3. Tewdwr-Jones, M., McNeill, D.: The politics of city-region planning and governance. Reconciling the national, regional and urban in the competing voices of institutional restructuring. Eur. Urban Reg. Stud. 7(2), 119–134 (2000)
4. Barca, F., Casavola, P., Lucatelli, S.: A Strategy for Inner Areas in Italy: Definition, Objectives, Tools and Governance. Materiali UVAL 31. Roma (2014)
5. McCann, P., Ortega-Argilés, R.: Smart specialization, regional growth and applications to European Union cohesion policy. Reg. Stud. 49(8), 1291–1302 (2015)
6. Becattini, G.: Industrial Districts: A New Approach to Industrial Change. Edward Elgar Publishing Ltd, Cheltenham (2004)
7. Pertoldi, M.: Landscapes of production: an investigation into italian industrial clusters. Rev. Hist. Geogr. Toponomastics II, 3–4, 57–68 (2007)

8. Corò, G., Micelli, S.: I distretti industriali come sistemi locali dell'innovazione: imprese leader e nuovi vantaggi competitivi dell'industria italiana. Economia Italiana **1**(1), 1–24 (2007)
9. De Marchi, V., Grandinetti, R.: Industrial districts and the collapse of the Marshallian model: looking at the italian experience. Competition Change **18**(1), 70–87 (2014)
10. Scott, A.J.: Global City-Regions. Trends, Theory, Policy. Oxford University Press, New York (2001)
11. Soja, E.W.: Accentuate the regional. Int. J. Urban Reg. Res. **39**(2), 372–381 (2015)
12. Calafati, A.G.: Economie in cerca di città. La questione urbana in Italia. Donzelli, Roma (2009)
13. Piccinato, G.: Urban landscape and spatial planning in industrial districts: the case of Veneto. Eur. Plan. Stud. **1**(2), 181–198 (1993)
14. Porter, M.E.: The Competitive Advantage of Nations. Free Press, New York (1990)
15. Brioschi, F., Brioschi, M.S., Cainelli, G.: From the industrial district to the district group: an insight into the evolution of capitalism in Italy. Reg. Stud. **36**(9), 1037–1052 (2002)
16. Mattioli, C.: Emergenti, ma non isolate. Medie imprese e territorio nella metamorfosi del distretto di Sassuolo. Territorio **81**, 105–109 (2017)
17. Rappaport, N.: Vertical Urban Factory. Actar Publishers, New York (2015)
18. Russo, M.: Processi di innovazione nei distretti e globalizzazione: il caso di Sassuolo. In: Tattara, G., Corò, G., Volpe, M. (eds.) Andarsene per continuare a crescere. La delocalizzazione internazionale come strategia competitiva, pp. 281–308. Carocci, Roma (2006)
19. Micelli, S., Rullani, E.: Idee motrici, intelligenza personale, spazio metropolitano: tre proposte per il nuovo Made in Italy nell'economia globale di oggi. Sinergie **84**, 127–156 (2011)
20. Russo, M.: Distretto industriale e servizi di trasporto. Il caso della ceramica. Franco Angeli, Milano (1990)
21. Lanzani, A.: Shrinking cities, cities in crisis. Post-planning reflections. In: Lapenna, A., Younès, C., Rollot, M., D'Arienzo, R. (eds.) Ressources Urbaines Latentes, pp. 269–290. Metis Presses, Genève (2016)
22. Rullani, E.: Lo sviluppo del territorio: l'evoluzione dei distretti industriali e il nuovo ruolo delle reti di città. Economia Italiana **2**, 427–472 (2009)
23. Viganò, P.: Elements for a theory of the city as renewable resource. In: Fabian, L., Giannotti, E., Viganò, P. (eds.) Recycling City, pp. 13–24. Giavedoni, Pordenone (2014)
24. Easterling, K.: Subtraction. Stenberg Press, Berlin (2014)
25. Lanzani, A., Merlini, C., Zanfi, F. (eds.): Recycling Industrial Districts. Settlements, Infrastructure and Landscape in Sassuolo. Aracne, Roma (2016)
26. Lowe, E.A., Moran, S., Holmes, D.: Fieldbook for the Development of Eco-industrial Parks. Indigo Development for US-EPA, Oakland (1996)
27. Conticelli, E., Tondelli, S.: Eco-industrial parks and sustainable planning: a possible contradiction? Adm. Sci. **4**, 331–349 (2014)
28. Munarin, S., Tosi, M.C.: Welfare Space. On the Role of Welfare State Policies in the Construction of the Contemporary City. List, Trento (2014)
29. Secchi, B.: Per un'agenda urbana e territoriale. In: Calafati, A.G. (ed.) Città tra sviluppo e declino. Un'agenda urbana per l'Italia, pp. 5–12. Donzelli, Roma (2014)

Measuring Urban Configuration:
A GWR Approach

Konstantinos Lykostratis[1]([⊠]), Maria Giannopoulou[1],
and Anastasia Roukouni[2]

[1] Democritus University of Thrace, 67100 Xanthi, Greece
klykostr@civil.duth.gr
[2] Delft University of Technology, 2628 CD Delft, Netherlands

Abstract. The relationship between accessibility, as a measure of the fixed location of the property, and land value is well recognized. Space Syntax theory was developed as a set of tools to analyze relationships between structures and functions of cities introducing accessibility measures of the urban grid. Even though admitted that location parameters comprise the most influential factors of urban property value, accessibility measures of centrality, based on network (integration and choice) quantifying urban morphology, have gained little attention in land value literature, despite the fact that urban grid morphology has a crucial role in property market structure. Recently, there has been growing interest for spatial statistics which count for spatial aspects of phenomena such as the land market. The frequently used OLS regression adopted for statistical inference on variables influencing a phenomenon is mostly inefficient for comprehension of spatial phenomena mainly due to instabilities caused by spatial autocorrelation and fixed parameter assumption. GWR extends the classic regression model by allowing spatially varying coefficient estimations while also accounting for spatial autocorrelation. The goal of the research presented herein is to estimate which spatial accessibility radius better explains objective land value and to explore local spatial relationships between geometric accessibility and land values using GWR, Space Syntax theory and GIS techniques, in Xanthi city, a medium sized city in Northern Greece. This research has led to inferences concerning the importance of Space Syntax geometric accessibility in the interpretation of land values, with local patterns of accessibility influence emerging.

Keywords: Accessibility · Space Syntax · GWR · Objective land value

1 Introduction and Background

The relationship between accessibility and land values is well recognized. From location theory and von Thunen's model, to monocentric and polycentric city models [1–3], and amenity-based theory [4], accessibility as a measure of the fixed location of a property [5] is an important parameter defining urban economy and life quality [6, 7] often used in human geography [8]. Recently, there has been growing interest for statistics which count for the spatial aspects of phenomena such as the land market [9]. Counting for these spatial aspects has a crucial role in the comprehension of such kind

© Springer International Publishing AG, part of Springer Nature 2019
F. Calabrò et al. (Eds.): ISHT 2018, SIST 100, pp. 479–488, 2019.
https://doi.org/10.1007/978-3-319-92099-3_54

of phenomena and their differentiations [10]. Accessibility is hard to be defined and measured [11, 12] summarize measures of centrality based on network and graph theory. One of the approaches is based on Space Syntax theory. Accessibility as seen through Space Syntax theory is also known as spatial accessibility [13] or a special case of geometric accessibility different than classic measures (geographic accessibility) stated above [14]. Space Syntax theory was developed by [15] as a set of tools to analyze relationships between structures and functions of cities.

According to natural movement theory [16] open spaces' layout is responsible for people moving or understanding space [17]. The open spaces layout forming the urban grid, is represented through an axial map and then turned into an axial graph, the latter interpreted through graph theory [18]. The centrality measures generated by the graph quantifying accessibility for every segment of the network or the relationship between a network segment to all other segments of the network, are integration and choice. Integration measures mean depth (depth here should be sensed as the number of lines travelled to move from the origin line to the destination line) of a segment to every other segment of the network [19]. Hence, an axial line is highly integrated when it can be easily reached form other lines of the network, making integration an accessibility measure of the line [20]. Choice measures the participation degree of a line in all shortest paths between all network lines [20]. Hillier and Lida [21] summarize three notions of travel: topological (based on fewest turns), geometric (based on least angle change) and metric (based on least length). Network relationships are based on topology and geometry [21] and direction [20]. Spatial analysis based on Space Syntax theory can be conducted for the grid as a whole examining the relationship between every segment of the network to all other segments but can also be conducted for grid parts defined by depth (in Space Syntax terms).

Even though generally admitted that location parameters comprise the most influential factors of urban property value, geometrical accessibility which quantifies urban morphology has gained little attention in land value literature [22] despite the fact that urban grid morphology has a crucial role in property market structure [23]. Research on this field relates space syntax measures with office rent values [23, 24] and residential property values, revealing through regression models positive statistically significant relationships between housing price and global integration but also negative and statistically significant relationships between housing price and choice [25–27]. Corresponding inferences are figured for global integration in regression models accounting housing tax bands [28, 29]. As far as local integration is concerned positive relations with housing values have been reported [22, 25, 30], although Chiaradia et al. [28] recorded contrary results, probably due to different samples and different grid structure [27]. It is worth noting that coefficient differentiation has been recorded in correlations and regression models applied to different sub regions of the same sample [22, 26, 31], or even both positive and negative values of the choice measure co-existed in the same city [43].

The frequently used OLS regression, usually adopted for statistical inference on variables influencing a phenomenon, is mostly inefficient for comprehension of spatial phenomena mainly due to instabilities caused by spatial autocorrelation [32–34]. Moreover, classic regression assumes that coefficients derived from model calibration are fixed throughout the sample and geographic area [10]. Maddala [35] notes parameter differentiation both spatially and over time in geographic phenomena, such as property value as underlined above. In case there is evidence of spatial hetero-geneity, local statistic methods should be applied [36] in order to overcome troubles related to coefficient bias, reduced proportions of the variation explained by models calibrated with global methods, and spatial autocorrelation [37]. GWR [38, 39] extends the classic regression model by allowing spatially varying coefficient estimations, also accounting for spatial autocorrelation [10]. GWR is applied for every observation based on neighboring observations defined by kernel selection (fixed or adaptive) according to observation distribution [10, 37, 40].

Fig. 1a. Research area.

Fig. 1b. Objective land value.

2 Methodological Framework and Case Analysis

Data used in this research refer to 7415 land parcels covering the entire city of Xanthi, a city of 55000 inhabitants in Northern Greece. The urban tissue (Fig. 1a) consists of distinct sections with a particularly interesting variety in form and density, which is followed in general terms in its consecutive extensions: traditional parts with a coherent organic tissue, newer extensions with rectangular grid or normal geometries with great variety in the size of blocks [31]. Parcels which constitute the spatial level of the analysis along with blocks forming Xanthi's urban grid were modeled in ArcGIS in order to be joined with necessary descriptive data. Objective value as set by Greek Ministry of Finance was chosen as the land value. In Greece property taxes and taxes on transfers, inheritance and donations are based on objective value which is determined by parcel condition, capitalization potential (defined by floor space ratio) and parcel location (in terms of having front facing commercial road). Moreover, transaction prices and housing prices are based on objective value. Thus, the objective value was calculated for every parcel. Space syntax measures were calculated using DepthmapX [41, 42]. Segment angular analysis [20] was chosen in order to calculate integration and choice for different metric radii (global, 200, 400, 600, 800, 1000, 1200, 1500, 2000, 2500, 3000, 3500, 4000 m). Analysis results (global and local integration and choice values) were assigned for each land parcel to its nearest segment using spatial join ArcGIS techniques.

In order to estimate which accessibility radius (in Space Syntax terms) better explains objective value, 13 regression models were constructed (1 for each variable set, integration and choice) each one responding to different radius (1 global model and 12 local models):

$$y = ax_1 + bx_2 \tag{1}$$

a being integration and b choice. Regarding linearity for the OLS calibration, some of the variables were log-transformed. The correlation matrix led to models 2 and 3 be discarded due to low correlation ($r < 0,2$) between choice and land value. The rest of the models were calibrated with OLS, results presented in Table 1. Multicollinearity tests based on VIF values [43] and spatial autocorrelation test based on Moran's I [44, 45] were run, results presented in Table 1.

The best model (based on R-squared and AIC Criterion) explaining the objective value of parcels was then calibrated with GWR for possible local relations to show up. Because of the density and uniform distribution of observations the chosen kernel was Gaussian fixed, size based on AICc minimization [38]. AIC is also used to compare the GWR and OLS model. Moreover, following [46] (F3 criterion) parameters were checked for statistical significance of non-stationarity. Using local VIF values the model was also checked for local multicollinearity [47, 48], which if present can lead to unreasonable signs and big standard errors. For the analysis described above the gwmodel [49, 50] was used. Results presented in Tables 2, 3 and 4. Lastly, the Moran's I index is calculated for the GWR model and is compared to the OLS.

3   Results and Discussion

The results from the OLS regression indicate that both integration and choice measures are statistically significant variables with integration being more influential to land value and no multicollinearity issues reported. Integration has a positive relation in contrast to choice's negative relation. The analysis shows that metric radius 1500 m is the level of accessibility that best explains objective value (54% proportion of variation), but spatial autocorrelation of residuals is a problem not to be disregarded. An equal interval classification of the objective values is presented in Fig. 1b, with the lowest values appearing at the traditional parts and the highest ones at the city centre. The spatial analysis results based on Space Syntax theory seem to coincide with city's aspects. Figure 2b illustrates the segments of the network that are mostly traversed in any possible travel choice. These segments are actually the parts of the network that link the traditional parts of the city with the newer extensions also having high business and retail activity. Equivalent comments can be made for Fig. 2a where the city centre is obvious along with the segregated areas.

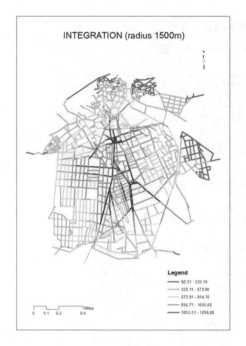

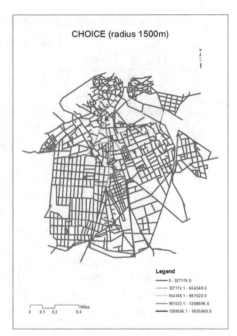

Fig. 2a. Integration (radius 1500 m) measures.   **Fig. 2b.** Choice (radius 1500 m) measures.

The 1500 m radius model is then calibrated with GWR with impressive results. Comparing the AIC value of the two models, the GWR model is definitely superior while according to related tests no local multicollinearity is detected. Furthermore, both measures show statistically significant spatial non-stationarity (F3 criterion), indicating that GWR is the appropriate approach for the issues discussed herein. One of the important advantages of GWR is the ability of mapping the results [10, 46]. Integration coefficient mapping is presented in Fig. 3a. As it is noticed the relation is positive with the highest values detected in areas with high integration values, consistent with previous theory [22]. One of the interesting points is the region on east where integration is much more capitalised probably because of the notable degree of segmentation. One of the most interesting findings is reported in Fig. 3b. As it is noticed the city's central district negatively values choice in contrast with the southwest region where the opposite condition occurs. Interpreting those results from a residential property angle, the high choice values, often related to vehicular flows [51], and noise and traffic congestion [25], as it is the case in the city centre, negatively influence values. On the contrary, the southwest region is characterized by high Floor Space Area index indicating strong housing and population density, along with some observed retail uses. Positive influence of choice on (housing) values is also reported in [22], in regions with relative features showing that residents are tolerant in noise and traffic. Moran's I Z-score for GWR model residuals is impressively reduced to 191.47 but still a problem.

Table 1. OLS Analysis results, Multicollinearity and Spatial autocorrelation test.

Model	Coef		Robust t	R2	Adj. R2	AIC	VIF	Moran's I
	Unstd	Std						
1	0,002*	0,708*	72,379	0,449	0,449	5687,859	1,262	291,58*
	0,000*	−0,094*	−7,396				1,262	
4	0,003*	0,673*	64,675	0,394	0,394	6397,236	1,293	347,72*
	−0,021*	−0,111*	−9,273				1,293	
5	0,002*	0,724*	77,622	0,467	0,467	5445,241	1,270	339,87*
	−0,017*	−0,101*	−9,417				1,270	
6	0,002*	0,758*	85,064	0,511	0,511	4803,238	1,271	330,72*
	−0,017*	−0,106*	−10,752				1,271	
7	0,002*	0,772*	86,800	0,531	0,531	4496,055	1,269	326,04*
	−0,017*	−0,109*	−11,545				1,269	
8	0,002*	0,779*	87,411	0,540	0,540	4354,766	1,261	322,20*
	−0,017*	−0,113*	−12,663				1,261	
9	0,002*	0,748*	76,993	0,489	0,489	5132,044	1,280	318,04*
	−0,017*	−0,122*	−12,984				1,280	
10	0,002*	0,708*	71,204	0,449	0,449	5686,884	1,260	301,00*
	0,000*	−0,095*	−7,277				1,260	
11	0,002*	0,702*	71,139	0,444	0,443	5761,158	1,258	294,97*
	0,000*	−0,091*	−7,098				1,258	
12	0,002*	0,708*	72,178	0,450	0,449	5680,417	1,258	292,08*
	0,000*	−0,093*	−7,303				1,258	
13	0,002*	0,71*	72,561	0,451	0,451	5656,722	1,262	292,26*
	0,000*	−0,096*	−7,547				1,262	

* = statistical significance at p = 0.001 level

Table 2. Non-stationarity significance test.

Variables	F3 statistic	Numerator DF	Denominator DF	Pr(>)
Intercept	344,917	2120,113	7387,8 <	2,2e−16*
INT1500	463,258	3089,790	7387,8 <	2,2e−16*
LNCH1500	93,145	705,909	7387,8 <	2,2e−16*

* = statistical significance at p = 0.001 level

Table 3. GWR results, OLS-GWR comparison.

Variables	Coefficients			
	OLS	GWR		
		Min	Median	Max
INT1500	0,002*	−2,18E–05	1,08E−03	0,0028
LNCH1500	−0,017*	−5,33E–02	−1,29E−02	0,0378
AIC	4354,766	−3043,825		
R²	0,540	0,831		
Adj. R²	0,540	0,829		

* = statistical significance at p = 0.001 level

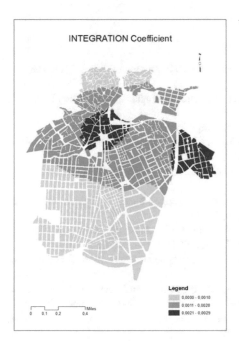

Fig. 3a. GWR integration coefficient.   **Fig. 3b.** GWR choice cefficient.

Table 4. Local multicollinearity diagnostics.

	INT1500 VIF	LNCH1500 VIF	Local CN
Min.	1,051	1,051	6,224
1st Qu.	1,238	1,238	10,553
Median	1,385	1,385	13,146
Mean	1,389	1,389	13,928
3rd Qu.	1,534	1,534	17,198
Max.	1,875	1,875	28,651

4 Conclusions and Perspectives

This research has led to inferences concerning the importance of Space Syntax geometric accessibility in the interpretation of land values. This observation is much stronger when integration and choice are used in locally estimated regressions with varying coefficients such as GWR. One of the most interesting outcomes is that in radius 1500 m spatial accessibility measures, form indices which explain residential property land.

Moreover, it was shown that using the GWR method spatial autocorrelation problems are reduced, while local patterns of accessibility influence emerge. An interesting idea for future research could be to focus on more parcel data to be added to the analysis along with other accessibility measures in order to achieve a higher degree of reliability.

References

1. Alonso, W.: Location and Land Use: Toward a General Theory of Land Rent. Harvard University Press, Cambridge (1964)
2. Mills, E.S.: Studies in the Structure of the Urban Economy. Johns Hopkins Press, Baltimore (1972)
3. Muth, R.: Cities and Housing. University of Chicago Press, Chicago (1969)
4. Brueckner, J.K., Thisse, J.F., Zenou, Y.: Why is central Paris rich and downtown detroit poor? An amenity-based theory. Eur. Econ. Rev. **43**(1), 91–107 (1999)
5. Orford, S.: Valuing locational externalities: a GIS and multilevel modelling approach. Env. Plan. **29**(1), 105–127 (2002)
6. Nilsson, P.: Natural amenities in urban space - a geographically weighted regression approach. Landscape Urban Plan. **121**, 45–54 (2014)
7. Spinney, J., Maoh, H., Kanaroglou, P.: Location and land values: comparing the accuracy and fairness of mass appraisal models. Can. J. Reg. **37**(1/3), 19–26 (2014)
8. Chin, H.C., Foong, K.W.: Influence of school accessibility on housing values. J. Urban Plan. Dev. **132**(3), 120–129 (2006)
9. Bencardino, M., Nesticò, A.: Urban sprawl, labor incomes and real estate values. In: Borruso, G., et al. (eds.) ICCSA 2017. LNCS, vol. 10405, pp. 17–30. Springer (2017)
10. Fotheringham, A.S.: Geographically weighted regression. In: Fotheringham, A.S., Rogerson, P.A. (eds.) Spatial Analysis, pp. 243–254 (2009)
11. Handy, S.: Planning for accessibility: theory and practice. In: Levinson, D.M., Krizek, K.J. (eds.) Access to Destinations, pp. 131–147. Elsevier Publishers (2005)
12. Porta, S., Crucitti, P., Latora, V.: The network analysis of urban streets: a primal approach. Env. Plan. B Urban Analytics City Sci. **33**(5), 705–725 (2006)
13. Charalambous, N., Mavridou, M.: Space syntax: spatial integration accessibility and Angular Segment Analysis by Metric Distance (ASAMeD). In: Hull, A., Silva, C., Bertolini, L. (eds.) Accessibility Instruments for Planning Practice. COST Office, pp. 57–62 (2012)
14. Jiang, B., Claramunt, C., Batty, M.: Geometric accessibility and geographic information: extending desktop GIS to space syntax. Comput. Env. Urban Syst. **23**(2), 127–146 (1999)
15. Hillier, B., Hanson, J.: The Social Logic of Space. Cambridge University Press, Cambridge (1984)

16. Hillier, B., Penn, A., Hanson, J., Grajewski, T., Xu, J.: Natural movement: or, configuration and attraction in urban pedestrian movement. Env. Plan. **20**(1), 29–66 (1993)
17. Karimi, K.: A configurational approach to analytical urban design: space syntax methodology. Urban Des. Int. **17**(4), 297–318 (2012)
18. Penn, A., Turner, A.: Space syntax based agent simulation. In: Schreckenberg, M., Sharma, S.D. (eds.) Pedestrian and Evacuation Dynamics, pp. 99–114. Springer, Berlin (2002)
19. Hillier, B.: Space is the Machine: A Configurational Theory of Architecture. Space Syntax, London (2007). E-edition
20. Hillier, B., Vaughan, L.: The city as one thing. Prog. Plan. **67**(3), 205–230 (2007)
21. Hillier, B., Lida, S.: Network and psychological effects in urban movement. In: International Conference on Spatial Information Theory, COSIT 2005. LNCS, Ellicottville, NY, United States, vol. 3693, pp. 475–490. Springer, Heidelberg (2005)
22. Xiao, Y., Webster, C., Orford, S.: Identifying house price effects of changes in urban street configuration: an empirical study in Nanjing. China Urban Stud. **53**(1), 112–131 (2016)
23. Desyllas, J.: Berlin in Transition: Using Space Syntax to analyse the relationship between land use, land value and urban morphology. In: Major, M.D., Amorim, L., Dufaux, D. (eds.) Proceedings of the First International Space Syntax Symposium, pp. 04.1–04.15. University College London, London (1997)
24. Enström, R., Netzell, O.: Can space syntax help us in understanding the intraurban office rent pattern? Accessibility and rents in Downtown Stockholm. J. Real Estate Fin. Econ. **43**(4), 548 (2011)
25. Law, S., Karimi, K., Penn, A., Chiaradia, A.: Measuring the influence of spatial configuration on the housing market in metropolitan London. In: Kim, Y.O., Park, H.T., Seo, K.W. (eds.) Proceedings of the Ninth International Space Syntax Symposium, pp. 121.1–121.20. Sejong University Press, Seoul (2013)
26. Matthews, J.W., Turnbull, G.K.: Neighborhood street layout and property value: the interaction of accessibility and land use mix. J. Real Estate Fin. Econ. **35**(2), 111–141 (2007)
27. Shen, Y., Karimi, K.: The economic value of streets: mix-scale spatio-functional interaction and housing price patterns. Appl. Geogr. **79**, 187–202 (2017)
28. Chiaradia, A., Hillier, B., Barnes, Y., Schwander, C.: Residential property value patterns in London: space syntax spatial analysis. In: Koch, D., Marcus, L., Steen, J. (eds.) Proceedings of the 7th International Space Syntax Symposium, pp. 015.1–015.12. Royal Institute of Technology (KTH), Stockholm (2009)
29. Narvaez, L., Penn, A., Griffiths, S.: Space Syntax Economics: decoding accessibility using property value and housing price in Cardiff, Wales. In: Greene, M., Reyes, J., Castro, A. (eds.) Proceedings of the Eighth International Space Syntax Symposium, Santiago de Chile, pp. 1–19 (2012)
30. Topcu, M., Kubat, A.S.: Computers, the analysis of urban features that affect land values in residential areas. In: Koch, D., Marcus, L., Steen, J. (eds.) Proceedings of the 7th International Space Syntax Symposium, pp. 026.1–026.9. Royal Institute of Technology (KTH), Stockholm (2009)
31. Giannopoulou, M., Vavatsikos, A.P., Lykostratis, K.: A process for defining relations between urban integration and residential market prices. In: Calabrò, F., Della Spina, L. (eds.) Procedia - Social and Behavioral Sciences, vol. 223, pp. 153–159. Elsevier (2016)
32. Dubin, R.A.: Spatial autocorrelation: a primer. J. Hous. Econ. **7**(4), 304–327 (1998)
33. Pace, R.K., Barry, R., Sirmans, C.F.: Spatial statistics and real estate. J. Real Estate Fin. Econ. **17**(1), 5–13 (1998)
34. Pace, R.K., Gilley, O.W.: Using the spatial configuration of the data to improve estimation. J. Real Estate Fin. Econ. **14**(3), 333–340 (1997)
35. Maddala, G.S.: Econometrics, 2nd edn. McGraw-Hill, New York (1977)

36. Haining, R.: Spatial Data Analysis Theory and Practice. Cambridge University Press, Cambridge (2003)
37. Fotheringham, A.S., Brunsdon, C., Charlton, M.: Geographically Weighted Regression. Wiley, Chichester (2002)
38. Fotheringham, A.S., Charlton, M.E., Brunsdon, C.: Geographically weighted regression: a natural evolution of the expansion method for spatial data analysis. Env. Plan. A **30**(11), 1905–1927 (1998)
39. Duarte, C.M., Tamez, C.G.: Does noise have a stationary impact on residential values? J. Eur. Real Estate Res. **2**(3), 259–279 (2009)
40. McMillen, D., Redfearn, C.L.: Estimation and hypothesis testing for nonparametric hedonic house price functions. J. Reg. Sci. **50**(3), 712–733 (2010)
41. Turner, A.: Depthmap 4, A Researcher's Handbook. http://www.vr.ucl.ac.uk/depthmap/handbook/depthmap4.pdf. Accessed 21 Nov 2004
42. Varoudis, T.: DepthmapX – Multi-platform Spatial Network Analyses Software. https://github.com/varoudis/depthmapX. Accessed 09 May 2012
43. Pedhazur, E.J.: Multiple Regression in Behavioral Research, 3rd edn. Harcourt Brace, Orlando (1997)
44. Cliff, A.D., Ord, J.K.: Spatial Autocorrelation. Pion, London (1973)
45. Cliff, A.D., Ord, J.K.: Spatial Processes: Models and Applications. Pion, London (1981)
46. Leung, Y., Mei, C.L., Zhang, W.X.: Statistical tests for spatial nonstationarity based on the geographically weighted regression model. Env. Plan. A **32**(1), 9–32 (2000)
47. Wheeler, D.C.: Diagnostic tools and a remedial method for collinearity in geographically weighted regression. Env. Plan. A **39**(10), 2464–2481 (2007)
48. Wheeler, D., Tiefelsdorf, M.: Multicollinearity and correlation among local regression coefficients in geographically weighted regression. J. Geogr. Syst. **7**(2), 161–187 (2005)
49. Gollini, I., Lu, B., Charlton, M., Brundson, C., Harris, P.: GWmodel: an R package for exploring spatial heterogeneity using geographically weighted models. J. Stat. Softw. **63**(17), 1–50 (2015)
50. Lu, B., Harris, P., Charlton, M., Brundson, C.: The GWmodel R package: further topics for exploring spatial heterogeneity using geographically weighted models. Geo-spat. Inf. Sci. **17**(2), 85–101 (2014)
51. Turner, A.: From axial to road-centre lines: a new representation for space syntax and a new model of route choice for transport network analysis. Env. Plan. B Urban Anal. City Sci. **34**(3), 539–555 (2007)

Planning for Antifragility and Antifragility for Planning

Ivan Blečić[1]([⊠]) and Arnaldo Cecchini[2]

[1] University of Cagliari, 09124 Cagliari, Italy
ivanblecic@unica.it
[2] University of Sassari, 07041 Alghero, Italy

Abstract. We argue that antifragility is a valuable and contentful goal for planning, distinct from resilience. We present a possible conceptualisation and delineate the essential properties of an antifragile planning, its affinities with the capability approach, and discuss the possible sources of its legitimacy within the conception of a liberal-democratic state. Hence the suggestion to incorporate antifragility into both the methodology and the content of planning.

Keywords: Antifragile planning · Complexity · Antifragility · Resilience

1 Introduction

The special complexity of territorial systems, of which cities are historically but the most plentiful navels, adds trickiness to the already burdensome responsibility of planning. Other than making "hard" predictions about the evolution of such systems unfeasible in principle, it also brings about greater uncertainty of action, raising both operational and deontological concerns for planning [1–6].

To address these concerns, we here advance the proposal that the concept of antifragility offers a contentful and legitimate partial goal for planning. To appraise the merits of the proposal, we confront it with the emerging approach of urban resilience, and argue that (territorial, urban) antifragility: (i) is distinct from (territorial, urban) resilience; (ii) may be elected as a more general, superordinate goal for governing territorial systems; and (iii) that for some of its tenets it can become a (partial) condition of legitimacy for planning, beyond mere policy preferences. To make this last point we must look for the possible sources of legitimacy for antifragile planning within the framework of a liberal-democratic theory of state. Here we circumscribe the scope of our effort: without necessarily ruling out other options, we ought to construct its compatibility and legitimacy within liberal-democratic theoretical framework, for antifragile planning to be applicable and operative here and now.

© Springer International Publishing AG, part of Springer Nature 2019
F. Calabrò et al. (Eds.): ISHT 2018, SIST 100, pp. 489–498, 2019.
https://doi.org/10.1007/978-3-319-92099-3_55

2 Complexity Redoubled

The complexity of territorial systems as social systems is double: they are complex in the "simple" sense of being large "many-body systems" [7], composed of many parts (elements, components) interacting non-linearly, and which by way of these interactions and feedbacks exhibit forms of spontaneous order and emergent properties at higher levels of hierarchy; but the complexity of territorial systems is redoubled for they include a special kind of interacting elements, "agents", endowed with free agency and autonomy, which we cannot but presuppose as an operational and moral (if not ontological) assumption.

This *redoubled complexity* is what makes the (long-term) evolution of territorial systems unpredictable in principle; the effects and ultimate outcomes of planning policies, actions and regulations uncertain; and the political responsibility of planning practice messier.

2.1 Fragile, Robust, Resilient, Antifragile

Among attempts of conceptual overhaul of planning theory to deal with this state of affairs, one of the most prominent, and thriving, approach is that of urban resilience: broadly speaking, building the capability of urban systems to absorb shocks, perturbations, volatility, and to bounce back to its prior equilibrium.

While there are still notable operational issues to the application of the concept of resilience in public policy making [8], we want here to appraise the essential general properties and merits of resilience by situating it within a more general examination of the modes in which things (objects, organisms, institutions, systems) "respond" to uncertainty, to events, perturbations, stressors, volatility, disorder – in short, to time.

Here, the conceptual triad "fragile-robust/resilient-antifragile" proposed by Taleb in his 2012 book *Antifragile: Things that Gain from Disorder* [9] helps clarify things. Following Taleb, the essential property of fragile things is that time can only harm them, because perturbations can only damage, break or destroy them, and never benefit them. If things are robust or resilient, they are essentially indifferent to time, because perturbations (of up to a certain intensity) do not affect them in the case of robustness, or they absorb, bounce back and recover from perturbations (of up to a certain intensity) in the case of resilience.

Antifragility is different. Taleb proposes to calls a thing (object, organism, institution, system) antifragile if it can from time - from events, perturbations, stressors, volatility, disorder - also gain, get stronger, improve, evolve, better adapt.

The usefulness of the conceptual triad fragile-robust/resilient-antifragile goes beyond its analytical relevance. The problem of efficacy of the classical planning arises, among other things, out of its tight, overly demiurgic, reliance on *hard* predictions (to predict with accuracy and precision what exactly will and has to happen?, when?, where?, with that intensity?, as a consequence of what planning action?). If, in the light of the redoubled complexity of territorial systems, hard predictions are indeed hard, must we not then settle for something *less*? The Talebian triad points at such a different possible attitude of the planning practice: rather than pretending to obtain and rely on hard predictions, we could, and cannot but, settle for something we may call *weak* predictions: the point is not to accurately and precisely predict what will happen, when and where, but to explore what are the system's responses to perturbations, volatility, even low-probability events, in other words to detect its fragility, robustness, resilience or antifragility. To planners, such a *weak prediction* is, as Taleb rather convincingly lays out, more accessible a practice than making hard prediction.

3 Antifragile Planning

3.1 Is Antifragility a Desirable Goal for Planning?

Although the notion of antifragility in Taleb is richer, its essential general property of "gaining from disorder", and the distinction we've drawn between fragility, robustness, resilience and antifragility, are consequential enough to let us lay out the argument for antifragility as a contentful goal for planning, distinct from resilience.

We share the view that urban resilience *is* a justifiable policy goal. Kolers [10] has fairly persuasively shown that, within a liberal framework – both "classical" (Lockean) and "equalitarian" (Rawlsian) – resilience can and should be incorporated as a necessary condition of legitimacy and justice. By extension, planning as a field of public policy ought to incorporate resilience into both methodology and content.

But here we want to make two observations. First, in point of terminological clarity and etymological pedantry, it seems to us often not everything that gets called resilient is indeed resilient. True, within the planning debate the notion of resilience has been extended beyond the mere idea of bouncing back to the previous equilibrium, so, not rarely, one notices the need to supplement the term "resilience" with adjectives "adaptive", "evolutionary", "transformative", or similar, which strictly speaking pertain to antifragility. Therefore it seem to us appropriate to advance a modest suggestion to sharpen up our common conceptual apparatuses, if only to enrich our vocabulary and to make a more perspicuous use of it, and to start saying "antifragile" when essentially we talk about antifragility, and "resilient" when we talk about resilience.

This hope for terminological clarity leads us to our second and more substantial observation. Although there is more to antifragility, and its conceptualisation by Taleb is richer, to draw the distinction with resilience we want here to focus on one essential property of antifragility: optionality, the property of having options (possibilities, rights, entitlements, capabilities to do, to have, to become, to change the course of action, to reverse prior decisions, ...), but not obligations. Indeed, optionality is a fundamental difference between antifragility and resilience: while at the core of antifragility, optionality *in the strong sense* is absent in resilience. We say in the strong sense because we must be careful with the distinction we are making. Indeed, the goal of resilience may be pursued for things (institutions, services, infrastructures, environmental and ecological systems) that are valuable not only for, say, human well-being, but also for providing certain optionality to people. But in this weak sense, resilience *at its face value* does not contemplate the possibility and the desirability that these "things which provide optionality" (institutions, services, infrastructures, systems) evolve, improve, gain with time, perhaps even in terms of their purpose of providing optionality. For that, the goal of antifragility needs to be put to work.

3.2   What Would an Antifragile Planning Be Like?

We can now attempt to delineate the essential properties of an antifragile planning (for a more extensive treatment, see [6]). One could assign two different meanings to the idea of pursuing the goal of antifragility in planning. One is to intend it as the application of the goal of antifragility to the *object of planning* (what to do, and what better avoid doing, in order to make territory more antifragile, and less fragile). This way of understanding antifragile planning can be called "planning for antifragility".

The second way to understand antifragile planning is to think of applying the goal of antifragility to the *planning itself*, to its practices and outputs (how to make more antifragile and less fragile the territorial planning decisions and policies, their decision-making processes, procedures, policies, regulations, management, and indeed plans themselves). We can call this second way of understanding antifragile planning "antifragility for planning".

The two meanings are not operationally unrelated. An antifragile planning policy (e.g. designed in such a way to be sufficiently open-ended and flexible, and to embed sufficient optionality in the system, in order to be able to embrace and benefit from the emergence of opportunities, spontaneous social energies, entrepreneurship, and from yet unpredictable, or low-probability, circumstances) could both advance the goal of antifragility of the territorial system it acts upon, while being *itself* antifragile.

Therefore, while the distinction is analytically useful, and there could be operational separation between first-order criteria for "planning for antifragility" on one hand, and for "antifragility for planning" on the other, what we propose to call antifragile planning should be antifragile in both senses. Conversely and by extension, a "fragile planning" is fragile in both these senses: it fragilises the territorial systems it acts upon, and its processes, outputs and actions are in themselves fragile.

Let us then attempt to lay out a possible set of conditions and criteria for an antifragile planning.

Fragilisers. Since other than what do to, antifragile planning is much about what better avoid doing, we must first identify factors, practices, attitudes, modes of seeing and intervening on social systems in general, and on territorial systems in particular, which may fragilise them. Here is a (non exhaustive) repertoire of such *fragilisers*:

1. *Plans and policies based on fragile predictions.* This is a general and catch-all category of fragilisers, implicit in many of the more specific ones that follow. It is rooted in the idea that the purpose and task of planning is to strongly determine specific spatial and social arrangements and dispositional outcomes, and that therefore planning requires *hard* predictions of what the city will be like and how it will evolve on the mid- to long-run. Such predictions cannot but be obtained through models highly sensitive to parameters, and thus structurally fragile. If territorial systems are complex as we said, and therefore in many aspects intrinsically unpredictable, and if the efficacy of our decisions and actions are grounded on *hard* and hence fragile predictions, then the decisions and actions are themselves fragile, in both senses we have given it above.

2. *Excess of centralisation*-cum-*micromanagement.* A policy preference may make centralisation justifiable, say, for the efficacy and expediency of pursuing certain

policy goals, for example equity or uniformity of conditions, opportunities or outcomes. It however threats to fragilise and to jeopardise the antifragility of a system when the aim is not only to indicate general frame of reference for individual action, to grant rights, to provide universal public goods, and to solve collective action problems, but also to micromanage the workings of the system, in all its parts.

3. *Efficiency* and *optimisation*. To be clear, there are cases where it may be desirable to make more efficient and to optimise certain sub-systems, services, processes, but this is an uncontroversial effort *only if* their purpose is uncontroversial and their performance is measurable on a single criterion. Is such a teleology a feasible goal to pursue in the case of territorial systems and of cities in general? In other words, is it meaningful to ask what the general purpose of a territorial system, or a city, might be? Shouldn't we rather think of territorial systems as a common platform for the "heterogony of ends", as a shared spatial, social, cultural and economical context within which different subjects chose and pursue their different objectives, ends and life plans? There is a further fragilising power behind the thrust for efficiency and optimisation at all costs. When focused only on the immediate (short-run, first-order) effects, such efforts may trim down optionalities, remove safeguards and protective redundancies, reduce the opportunities for exaptations, and thus shrink the possibility of evolutionary adaptations, of change of uses, of embracing the multiplicity and heterogeneity of ends and needs - present and future.

4. *Specialisation*. Excess of specialisation is a special case of trimming down the optionality, for it makes the system fragile to external perturbations and reduces its capacity to learn and adapt to changes in the environment, to become something else, to have a new life.

5. *Simplification* and *standardisation*. Not taking into account the (redoubled) complexity and the possibility of counterintuitive behaviour of the system, related to its capacity for autopoiesis and to the feedback effects.

6. *Lack of consensus building*, and the *crumbling of the "cement of society"*, in its various forms, such as excessive economic inequality when it spills over into the inequality of real opportunities and capabilities, undermining the social cohesion from within. This is a noteworthy point in our discussion of antifragile planning, since social cohesion must not be understood as a stationary state of "harmony", but a dynamic, ultimately precarious, outcome of conflicts, reciprocal accommodations, and partisan mutual adjustments.

The Three Planes of Antifragile Planning. The perspective of antifragility allows us to tentatively distinguish three operational planes for the planning practice: (i) the *via negativa*, (ii) the shared vision and (iii) the space of the project.

Via negativa. Building on a Taleb's idea, with *via negativa* we intend a system of mainly negative general rules providing "external" restrictions which delimit the space of possible actions and prohibitions. Largely nomocratic in nature [11, 12], they follow the logic of the *via negativq* insofar as they do not directly predetermine the outcomes, do not impose performative behaviours, do not indicate what to do and what has to be done, but mainly what is forbidden, what should not be done. But the concept of *via negativa* also refers to the removal of what could be harmful and of the superfluous which wastes social and human energies [13], from counterproductive constraints to procedural and normative superfetation.

A *prima facie* content of prohibitions of the *via negativa* in planning is to be looked for among the fragilisers we described before. For that reason, the *via negativa* constitutes a core of an antifragile planning, and its political legitimacy may be (a least partially) independent from mere policy preference. However, under the tenet of *via negativa* there is also space for policy options subject to democratic deliberation and collective decision making. The idea of *via negativa* does not necessarily imply a withdrawal to a planning miniarchism or the maintenance of the status quo in the legal and institutional conditions for individual behaviour. There is rather a Lockean twist to it: as long and the rules and regulations preserve principles and forms of generality and superordination, and retract from short-term contingencies and conveniences, the *via negative* does not exclude the possibility of structural transitions and (even radical) "changes of regime".

Shared Vision. A planning firmly tied to hard predictions is fragile, but a planning which does not "tend towards future" and does not "create future" is a contradiction in terms. It is not unreasonable to think that a political community would want to think about its future, at least on an time horizon accessible to one's imagination of three to four generations, and that it would want to avoid undesirable futures.

This is the task of planning, for which it needs a deliberation on a *reasonably shared vision* of the desirable scenarios and those to avoid, through an effective strategic decision. In essence, antifragility-wise, such a shared vision is a concrete declination, in a precise historic moment, based on available resources, of the set of different freedoms which compose the right to the city. For that purpose, we hold it can be argued that the Senian capability approach [14] can provide a theoretically and practically feasible ethical-normative focus, compatible with the goal of antifragility and with the a-teleological conception of territory and city, offering a necessary flexibility for autonomous action and project design (of individuals, economical and social subjects, institutions), and thus making the territory more antifragile, without with that fragilising individuals, indeed expanding their optionality. We do not have space here to go into more details (but see [6, 15]), however, besides providing a normative framework to cast the content of antifragile planning into, we will see in the next section how the capability approach can also play the pivotal role in providing the basis for the legitimacy of the goal of antifragility in planning.

The Space of Project. Finally, there is the plane of flexible and variable actions which permit the free expression of individuals in social forms they choose; their actions and decisions only being subordinated to the respect on the restrictions of the via negativa and consistent with the shared vision.

4   On Legitimacy of Antifragile Planning

While making the case for resilience as a necessary condition of legitimacy and justice, Kolers [10] denies this status to antifragility. We object.

Kolers' argument that antifragility is illiberal is grounded on the fact that anti-fragility is not just contextual but also level-relative: "[t]he citizens' affairs cannot all be anti-fragile, because in many cases the anti-fragility of some involves capitalizing on the fragility of others. And the state or community cannot itself be anti-fragile because part of its function is to absorb some of its citizens' fragility." Now, it is true that complex systems are often organised on levels of hierarchy (from micro to macro), and Kolers is of course right to point out that, as Taleb makes it explicit, antifragility on one level of hierarchy often emerges at the expense of fragility of the components at lower levels.

But perhaps Kolers' assessment is too hasty. Taleb does not really attempt, nor was that his aim in *Antifragile*, to work out a feasible political theory (compatible with liberalism). But, even if he were to lay down one with excessive shades of social Darwinism (up to jeopardising essential liberal principles), we are not bound to be fundamentalist Talebians, nor should that be the most fruitful, let alone the only possible political theory.

In essence, we object that the level-relative nature of antifragility poses an insur-mountable obstacle for the construction of a workable (liberal-democratic) legitimi-sation. In passing, we must point out that liberal political theory is not that unfamiliar with dealing with level-relative concerns. For instance, is the two-tier structure of the Rawlsian principles of justice not operating precisely with such concerns in mind, when the lexical priority of the liberty over the difference principle wards off certain treatments on the "lower" level (violation of equal right to basic liberties) even at the expense of maximising the performance on the "higher" level (the share of primary goods for the least advantaged)?

Let us then sketch out the argument for the legitimacy of antifragility (in planning).

4.1   Antifragility of What to What?

So, yes, there can be specific dispositional relationships between the (anti)fragility of a system and that of its sub-systems and components. A system is antifragile when it gains from the fragility of its components (such antifragility can for instance be obtained when the system incorporates mechanisms of learning, advances via trials-and-errors, improves through tinkering, local experiments and failures).

Here indeed arises the core problem with antifragility as a political ideal, since some of the components of the systems we are talking about are individuals, social groups and human communities, whose fragility can be a legitimate concern of the state, and whose treatment (liberties, well-being, security) is a valuable goal, let alone being at an unconstrained disposal to be fragilised, even right off sacrificed, for the sake of the "greatest antifragility for the greatest number" at the higher level.

So the question is: at what level should antifragility per pursued? Or, to paraphrase Carpenter *et al.* [16], antifragility of what to what? We have until now defined anti-fragility in terms of "gaining from disorder", but on the level of planning and public policy, it is not uncontroversial what should be the informational focus to measure the "gain", since in the case of social systems we are bound to deal with individual subjectivity and plurality. This question can have a positive variant (at what level of the system's hierarchy does the gain show up?, who gains?, at the expenses of whom or of what?), and a normative variant (who should gain?).

So, in the case of cities, one could raise the question if their antifragility as "organisms" may be at the expense of the "survival" of (some) individuals and social groups. We could give two answers. First is that sometimes happens: in the evolution of cities social groups and functions disappear, get transformed, change their nature, and there could be limits to what a public policy can and is admissible to do about it. The second answer, though, is that a city without its people does not exist, and that people are called citizens, a term with a series of implications, both on the individual level, and in point of social cohesion.

To compare the desirability of antifragility vs. resilience, we must address the question posed by Carpenter *et al.* "of what to what" [16]. The desirability of resilience depends if the system to be made resilient is functional and valuable as-is (if the status quo is bad, say, inequitable, making it resilient may not be a valuable goal), and what its permanent disruption would bring about (for instance, does the disruption, say, bring about greater inequities). That is why, to justify resilience as a legitimate political goal, Kolers' answers the question of "resilience of what to what" by clarifying that it should regard "the resilience *of* reasonably just social systems, *to* disruptions that could significantly reduce quality of life and undermine democracy, [and therefore it] would seem to be of important moral concern" [10].

But, taking into consideration our previous point on optionality in antifragility, similar argument may also be deployed in favour of antifragility, as soon as we look a little bit further from the mere issues of ecological and environmental resilience. Our key point is this: one can pursue the goal of antifragility for valuable systems, by endowing them with optionality and asymmetry of gains versus harms (even, say, in terms of equity), to make them evolve favourably in the face of uncertainty. In this sense antifragility can be a meaningful and legitimate normative goal.

This is where the capability approach can fruitfully come to our assistance. Besides being tightly related to the idea of optionality, the capability approach points us to the possible road to address the question of "antifragility of what to what". We could almost give it a Rawlsian twist, and say that - first principle - to the classical individual rights and protections (which better be resilient) we should add individual capabilities (whose extension better be antifragile); and then - second principle - to pursue the goal of systemic antifragility compatible with the first principle. Because through capabilities, the system is made antifragile by endowing individuals with optionality, this in part means to make individuals themselves more antifragile. So it seems to us possible to acknowledge that there is a favourable terrain for synergies, constructive interaction and positive feedbacks between the two principles. (For a more extended discussion of urban capabilities and their link to antifragile planning see [6]).

We must admit there is a limit to the rawlsian analogy. Given that capabilities are differentiated, contentful and purpose-oriented, they are in principle not as universal as the basic rights, and are not as purpose-neutral as the rawlsian conception of the primary goods, so there may be trade-offs to be had when determining the set of capabilities, and if necessary, which individual capabilities should be prioritised over which, and over what systemic optionality. Indeed. But this observation pertains to the objections raised to the capability approach in general, so be we allowed not to wage this debate here, other than suggesting that these trade-off may be the proper object of the *plane of the shared vision*, subject to public debate and democratic deliberation.

5 Conclusions

We have sketched out the content of an antifragile planning and discussed how its legitimacy may be worked out within the framework of a liberal-democratic theory of state. Our argument was that some of tenets of antifragile planning can be embedded in the more general necessary conditions of legitimacy of planning, while other may be object of policy preferences and goals, subject to public debate and deliberation.

Before ending, we want to ward off any antifragility fundamentalism. From all we say does not follow that cities should be only made of antifragile things and subsystems. Cities and territories as systems are composed of many things, elements, components and subsystems, some fragile, some robust, some resilient and some antifragile. And there are many good, sufficient and efficient reasons for this state of things. That is why antifragility isn't always the goal to pursue *for everything, in every case and at any cost*. There are in fact fragile elements and systems, which at least in part own their *raison d'être*, value, importance or beauty, to their fragility; sometimes fragility is pursued for the charm and allure of the delicate, ephemeral, frangible, weak. Then there are objects and systems which are robust, and they better be designed and build that way, for reasons of functionality, economy, usability and simplicity. Then there are resilient systems, altogether functioning satisfactorily: in the strict sense, urban resilience is a desirable feature of valuable systems in response to adverse events, to environmental disasters, infrastructure failures and social crises [17, 18]. All in all, there are many objects and subsystems within territories and cities which are fragile, robust, resilient and antifragile, and as such are necessary and useful for its survival, good functioning, and human well-being.

That being said, we hold that antifragility can be elected as a more general and superordinate horizon for a territorial system, and that then, within it, it can and should include many subsystems and things which are, from case to case, resilient, robust, fragile, antifragile.

By introducing the notion of antifragility and by contrasting is with resilience, our primary purpose here was to point towards a way to broaden the general coordinates of the debate on planning under complexity and uncertainty. We acknowledge that much theoretical, operational and empirical groundwork has yet to be covered, especially on the contextualisation and application of the conceptual framework to different planning problems and trade-offs, for ours to ultimately prove to be a valuable contribution to the debate.

References

1. Portugali, J.: Self-Organization and the City. Springer, Berlin (1999)
2. Batty, M.: Cities and Complexity. MIT Press, Cambridge (2005)
3. Innes, J.E., Booher, D.E.: Planning with Complexity. Routledge, London (2010)

4. De Roo, G., Hillier, J., Van Wezemael, J. (eds.): Complexity and Planning. Ashgate, Farnham (2012)
5. Portugali, J., Meyer, H., Stolk, E. (eds.): Complexity Theories of Cities Have Come of Age. Springer, Berlin (2012)
6. Blečić, I., Cecchini, A.: Verso una pianificazione antifragile. Come pensare al futuro senza prevederlo. FrancoAngeli, Milano (2016)
7. Andersen, P.W.: More is different. Science 177(4047), 393–396 (1972)
8. Duit, A.: Resilience thinking: lessons for public administration. Public Adm. 94(2), 364–380 (2015)
9. Taleb, N.N.: Antifragile: Things That Gain from Disorder. Allen Lane, London (2012)
10. Kolers, A.: Resilience as a political ideal. Ethics Policy Environ. 19(1), 91–107 (2016)
11. Moroni, S.: Rethinking the theory and practice of land-use regulation: towards nomocracy. Plan. Theory 9(2), 137–155 (2010)
12. Moroni, S.: Why nomocracy: structural ignorance, radical pluralism and the role of relational rules. Prog. Plan. 77(2), 46–59 (2012)
13. Moroni, S.: Libertà e innovazione nella città sostenibile. Ridurre lo spreco di energie umane. Carocci, Roma (2015)
14. Sen, A.K.: The Idea of justice. Harvard University Press, Cambridge (2009)
15. Blečić, I., Cecchini, A., Talu, V.: Capability approach in urban quality of life and urban policies: towards a conceptual framework. In: Maciocco, G., Johansson, M., Serreli, S. (eds.) City Project and Public Space. Springer, Berlin (2013)
16. Carpenter, S., Walker, B., Anderies, J.M., Abel, N.: From metaphor to measurement: resilience of what to what? Ecosystems 4, 765–781 (2001)
17. Pickett, S.T.A., Cadenasso, M.L., Grove, J.M.: Resilient cities: meaning, models, and metaphor for integrating the ecological, socio-economic, and planning realms. Landscape Urban Plan. 69, 369–384 (2004)
18. Musco, F., Zanchini, E. (eds.): Il clima cambia le città. Strategie di adattamento e mitigazione nella pianificazione urbanistica. FrancoAngeli, Milano (2014)

Investment Property in Rental: Profitability and Risk Analysis

Franco Prizzon[(⊠)] [iD] and Andrea Cullino

Politecnico di Torino, 10125 Turin, Italy
prizzon@polito.it

Abstract. Until 25/30 years ago it was not essential to study the investments in detail to understand where to deposit own savings: the growing markets and the economic boom favored the success of a large portion of investments on the market and in particular for the housing market, where the investment risk was particularly low. In recent years, however, the economic landscape has reversed and there is an increasing need for specific analyses that seek to try to estimate the future development of the value of own tangible property and the assessment of potential income. This work focuses on the risk and profitability of rental real estate investments, looking at average return values according to the different uses of buildings and identifying an average risk value, and then comparing them with securities investments such as ten-year BTPs.

Keywords: Real estate investments · Building Risk-Profitability
Turin Buy-Rent

1 Investment and Investor

Before entering the heart of the topic, it is necessary to specify some methodological elements adopted in order to make the work as close as possible to a real investment case [1, 2]. First of all, the work is oriented to the average investor, which has relatively modest capital, and that operates according to the logic of "good saver". The category of "average investor" refers to the vast majority of Italian investors, who wish to exploit their savings through small-scale investments. In this way the scope of the investment is reduced: small to medium sized buildings, individual family units or small buildings destined to commercial activity; this excludes industrial sheds, rural buildings and large buildings.

Finally, the average investor carefully assesses where it is better to invest [3], concentrating on the search for an easily leasable property, for instance a residential apartment, shop or car park, not located in stately buildings nor inside buildings with valuable finishes or particular locations: the investor is interested about common property, a competitive price, so the investor can rent it for a sum equal to the average rental value. Therefore, the research will focus the attention on buying-selling values and renting out residential houses, shops and car boxes[1].

---

[1] The basic values used as the starting data for the analysis were collected from the OMI database, Observatory of the Real Estate Market, of the Agenzia delle Entrate, that is based to real value of Buy-Rental investment. The OMI is an official source that allows you to consult data on the prices of real estate values and leases throughout Italy.

© Springer International Publishing AG, part of Springer Nature 2019
F. Calabrò et al. (Eds.): ISHT 2018, SIST 100, pp. 499–506, 2019.
https://doi.org/10.1007/978-3-319-92099-3_56

The center of our investigation is the territory of Turin and the neighboring municipalities, exactly the 34 municipalities belonging to the first and second territorial areas near the city with the addition of some territories considered particularly significant. The municipalities the same municipalities will then be subdivided into three internal zones: center, semi-center and periphery (with the exception of those in which real estate values do not change significantly depending on the area, for which these specific subdivisions would be superfluous).

2   The Analysis

In Turin, instead, the information were collected according to the typical subdivision into 26 districts [4, 5] (Fig. 1).

	DISTRICT / ZONE	BUILDING TYPE	MARKET VALUE (€/mq)		RENTAL VALUE (€/mq x mese)		AVARAGE MARKET VALUE (€/mq)	AVERAGE RENTAL VALUE (€/mq x mese)	GROSS RETURN
			min	max	min	max			
TORINO	Centro	Civil apartments	2100	2850	8,0	12,0	2475	10,0	0,048
		shops	2600	5200	17,9	28,5	3900	23,2	0,071
		Garage	2400	2750	19,2	25,6	2575	22,4	0,104
	Valentino	Civil apartments	1850	2750	6,0	9,1	2300	7,6	0,039
		shops	1200	1650	7,4	10,0	1425	8,7	0,073
		Garage	2500	2700	19,9	24,0	2600	22,0	0,101
	San Salvario	Civil apartments	1400	1950	5,9	9,0	1675	7,5	0,053
		shops	1000	1800	7,7	15,0	1400	11,4	0,097
		Garage	1600	2400	14,2	21,2	2000	17,7	0,106

Go on …

	DISTRICT / ZONE	BUILDING TYPE	MARKET VALUE (€/mq)		RENTAL VALUE (€/mq x mese)		AVARAGE MARKET VALUE (€/mq)	AVERAGE RENTAL VALUE (€/mq x mese)	GROSS RETURN
			min	max	min	max			
BRANDIZZO	Cenetrale	Civil apartments	820	1250	3,5	4,8	1035	4,2	0,048
		shops	760	1050	4,1	5,8	905	5,0	0,066
		Garage	800	1150	4,5	6,2	975	5,4	0,066
BRUINO	Cenetrale	Civil apartments	1150	1500	4,8	6,6	1325	5,7	0,052
		Garage	810	1150	4,8	6,7	980	5,8	0,070
CAMBIANO	Cenetrale	Civil apartments	810	1150	3,8	5,0	980	4,4	0,054
		shops	730	1100	5,6	6,6	915	6,1	0,080
		Garage	930	1100	5,4	6,6	1015	6,0	0,071
	Semicentrale	Civil apartments	830	1650	5,1	7,1	1240	6,1	0,059
		shops	730	1100	5,5	6,6	915	6,1	0,079
		Garage	930	1100	5,4	6,6	1015	6,0	0,071
	Perferia	Civil apartments	830	1250	4,1	5,9	1040	5,0	0,058
		shops	790	1150	5,5	6,6	970	6,1	0,075
		Garage	940	1150	5,4	6,6	1045	6,0	0,069

Go on …

Fig. 1. Gross Return by District and Zone, Gross Return = (Value* of rentals × 12 months)/ Market value*

A first analysis of the data collected already reveals significant differences in return: it is not uncommon to find areas where the profit is 4–5 times higher than others.

In any case, what the entrepreneur assesses when he makes an investment is always the net profit, which mean the difference between the gross return value and the expenses incurred for the same investment.

Therefore, the analyses include the costs necessary for the management and the maintenance of the investment [6] (amount for the contract, energy report of the building, ordinary overhaul and replacement of the boiler, repairs, replacement, painting, etc.), general accounting (administrative and business consultant costs), insurance and property tax (In this situation we decided to apply 30% net because our commercial investigation concerns a wide range of investors.). Expenses also include the vacancy quote: when the contract terminates, the lessor loses the rent of that property for a period, due to the search time lapse necessary to find a new tenant. The expectation of this transition period is approximately 2 months. In the current context the tenant change house not later 4 years, so we estimate that the owner of the accommodation loses two of monthly rent every 4 years. We prefer to estimate the defaulting tenant as a risk component in our analyses.

The analysis of the data shows that there is a deep gap between the profits: we find not-ordinary data but we have to explain them for a correct interpretation.

First of all, it seems necessary to point out that the data obtained are the result of surveys on large large-scale (statistical basis), approximations, even if limited, and estimates, so they should make the calculation not perfectly in line with reality [7].

Moreover, it is right to remember the we are talking about statistics, so not all the cases of buying-renting a property are close to the data that we have found. This happens because, in practice, there is not only the law of the market, of the statistical average, of "good business", but often the law of feeling, chance and necessity. It is therefore easy to find investments with significantly lower returns in reality. We could consider the contexts in which the lessor owns a rented property, with high maintenance and management costs, but he keeps it in spite of everything due to affective reasons, for example. We have to consider also the opposite cases: it should be equally normal that there is higher profits of investment. We could consider, for example, the circumstances in which the tenant, in love with the aesthetics of the property or that has not the material time for the research of similar house, is willing to pay an out-of-market figure, too high compared to the average of the area and period. Similarly, when the lessor is able to pay property lower than the market value, the return is higher than the average returns [8, 9]. As you know, the estimation of an investment, whatever it could be, requires comparisons, theoretical reasoning and careful statistical forecasts. However, these factors can't lock the reasoning, because they are implicit variables in final evaluation. *Small Parenthesis: The zone by zone tables with the calculation results are not entered for not protract the text. It is considered more important to think about average values as they have a lower level of error.*

A first important distinction immediately emerges from the net return values: it would not be correct to put at the same level the investment in the purchase and renting of properties with different uses. In fact, it can be observed that box profits, if compared to the returns of the apartments, have on a much higher average value: an average of 3.5%, far from the average value of residential houses, which is 1.8%.

We couldn't say the same things about the stores, which are located in an intermediate area between the two previous aforementioned: the average of all the values is around 3.4%. These differences are justified by market reasons. Boxes, for example, have average higher per square meter thanks to their limited size (We refer to 14–17 m^2): in fact their rental quotes are accessible to a large part of the population. Although the price per square meter is theoretically disadvantageous if we compare it to other properties, Renting box's cost is lower, so there is a broad range of customers.

Instead the stores, maintain lower cost/sq.m compared to the cost/sq.m of boxes, as we have already said before about measurement, but their economic value is included in a higher range than the accommodation's class (on average more than one percentage point). There are many reasons with which we could explain what we have just said, below the most important ones:

- They lodge business activity which produces income;
- The manager of the commercial activity should deduct part of the rent from the fees;
- They are often located in transit ways, city centers, quarters and locality and, therefore, generally in areas where the property has a higher market value per square meter;
- It is more risky investment: in fact, in recent years, the average replacement period of the lessee has decreased considerably and so, in order to amortize the costs of the months in which the property remains vacant, the monthly fee is raised, even if the increase is a bit considerable.

Therefore it is necessary to deal with these two main topics, civil dwellings and shops, in different ways.

Concerning the first ones mentioned, there is average return value of 1.8% for all the municipalities of Turin' district and an average return of 1.8% for the city. As you could see, these averages are not different from each other.. However, this fact was predictable: renting the house is one of the highest (if not the most wasteful) expenses that a family has to support. Accordingly, it is advisable to choose carefully, considering what the market offers and what are the most competitive prices: however, prices do not always follow criteria linked to the value of the property, but they tend to be levelled towards the lowest average value, because the risk is that it is not possible to rent at prices considered fair by the owner.

3   The Comparison: Risk and Profitability

In recent years, due to the economic crisis, the trend of the BTPs has been extremely volatile; this depends on fears of a breakdown in the euro zone and doubts about the sustainability of Italian public debt. To date, values seem to have stabilized in view of increasing investor confidence and the slight growth of national markets. The trend in yield over the last decade of the 10-year BTP with maturity has fluctuated between a minimum of 1.53% and a maximum of 7.261% (nov. 2011). The table below shows the current values (updated to 11/11/2017) of ten-year government bonds. The average net return on these securities is 1.91% points. For the purpose of assessing the rental investments, it was decided to consider the investment in ten-year BTPs as' zero-risk'. It is clear that there is no investment with risk equal to zero. However, for practical purposes, it is necessary that we consider a way to find compromise between risks and returns of the investment. At the moment, the safest investments on the market are those that target government bonds.

Risk is defined as the variability (or volatility) of investments, obviously including potential losses as unexpected gains. The risk is, of course, always present in the market, but what changes is its perception by investors or, in other words, their confidence in the investment. Savers perceive the risk especially when it shows itself in the form of loss. Closely linked to the perception and evaluation of financial risk is the expected return on the investment: the higher the risk, the more the investor requires remuneration, premium for this risk. Risk is a key investment variable [10, 11]. In fact, Each person is characterized by a specific propensity to financial risk, which represents the level of individual tolerance to the possibility that the value of their investment should fluctuate more or less significantly over time. When you decide to invest in the markets, whatever they are (financial, real estate, etc.), it is necessary to consider this factor and choose the instrument, which suits the preferences and needs of each individual in the best way. Inasmuch as this work is not aimed at one single individual, the assessment will not take into account the many personal aspects involved in the choice of investment. We could define the concept of personal aspects as propensities towards certain types of investment, even if they are not apparently the most advantageous if we consider purely mathematical and statistical calculations, dictated by preferences, habits and individual needs, that influence their decision. For example, there are people used to exclude real estate purchases as the possibility of future investments due to habit or fear, leaning towards investments such as real estate or commodities, even if perhaps less profitable or more dangerous. These decisions are made because a strong emotional component such as serenity and satisfaction influences the choice of the type of investment.

If we consider the return given by the BTPs, the choice of any other investment, if made in rational way, could start from this return, to whom we have to add risk premium that could supply the higher danger of investment if we compared to, although very low (near 0), which the risk which depends on State's credit instruments (Table 1).

Table 1. Comparison Net Return and Risk - Ten-year BTPs

Net return	Risk
1.95%	Null value (0%)

It is still to be emphasized that economic inflation is the prolonged increase in the average price level of goods and services in determined period of time, which generates decrease in the purchasing power of the currency. Therefore, if inflationary period could be expected in the coming years, this consideration will be reflected in the net profits on State's bonds and part of the percentage will be used to cover the difference in purchasing power that my capital loses each year. Foreseeing inflationary developments in recent years, we could consider one percentage point as the reference value in calculation purposes. We have also to consider two parameters of return-risk in real estate investment. Regarding the first one, remembering that the investigation concerns only Turin and its districts, the average value is 2.9% points per annum of the initial investment to purchase the property. Considering default risk, real estate investment does not foresee the possibility of losing completely the initial capital invested [12]. The only way this could happen is that the property ceases to exist physically, which is unlikely situation in nature. Therefore, the invested capital does not risk default, but on the contrary is a victim of the market, since the value of the property increases or decreases the purchase/sale price according to market trends.

However, real estate investment presents another risk, which affects profitability, making it equal to zero: the arrears. In order to quantify it, it was conducted survey on sample of 132 rented properties in Turin and its suburbs. The owners' interviews showed that 18.2% of tenants have been evicted in the last year due to the insolvency of the landlord's rent. Instead, 12.9% complains of delays and irregular payment times, but then recovered in short time. Newspapers and magazines of sector, considering national data, speak about higher percentages, with around 42.0% of property owners complaining about the failure to pay rent or, in any case, delays in payment after a few months. In northern Italy the situation seems to be better, indeed in this area the values are around 23.0%. In order to continue our reasoning, 20% is taken as the reference percentage, the result of the comparison between the value emerging from the specialized publications and the valor of the interviews. However, it is to consider that the percentage inherent in delays is not taken into consideration, because they are not considered as loss, but only profit's postponement of a few months, therefore actually negligible. The risk for an owner not to receive rent for a longer or shorter period is assumed to be 20%. However, this risk cannot be compared directly with the risk of other investments, because, as I have already mentioned, the risk is not calculated on default, but on profitability. In fact, the risk is the return of about 3% on investment and not for its entire duration, but only for a period of time that, considering the bureaucratic timing, varies from a minimum of 3 months to a maximum of 19 (in case of the tenants are minors or over 65 years old, the times could be further extended). At the end of this period, the lodging could be reoccupied, but another time reappears the risk of 20% not receiving income for a more or less short period of time. Therefore, the output data of this analysis give these percentages as a result, as regards real estate investment is concerned (Table 2):

Table 2. Comparison Net Return and Risk - Rent & Buy Investment

Net return	Risk
2.9%	20% (on return)

It is good, also for this investment, to try to think about how inflation affects and how it modifies its values. We could separate this topic into two parts, the first one about the capital invested and the other one about the annual profitability of this investment.

The capital invested is not directly influenced by the cost of living. In real estate, at the time of the initial investment, it is like if you exchange money for the real estate. So from that moment I no longer own my initial capital, but property whose value increases or decreases according to the market. The discourse concerning profitability is different. Although this is a percentage of the invested capital, it could be increased every year with the increase in ISTAT by a value very close to that of the life price. However, this applies only to shops and boxes. For housing, in fact, since the formula of "cedolare secca" has been chosen for the calculation, it is not possible to apply increases until the end of the contract which, in our case, is 8 years. The following example tries to explain these arguments in terms of amount in order to understand how much the ISTAT increase will affect housing. An initial capital of €100,000 is invested for the purchase of an apartment. This, according to the chosen formula, gives an average annual net profitability of 2.9%, which is equivalent to €2,900.

After eight years that same apartment will have profit of 23,200€. If the owner had been able to apply the ISTAT increase and thus adjust the rent according to high cost of living, the return after 8 years would have been €24,028. The low percentage influence and the possibility, at the end of the contract, to adjust the rent on the basis of expensive life allows, in order to simplify the calculation, to consider zero the influence of inflation also on the profit of the housing and therefore has no influence on the entire investment of purchase and rent. Comparing the following tables, the choice of investment, if guided only by the logical return-risk, would not seem difficult and would prefer to choose ten-year BTPs, given the good relationship between risk and return (Table 3).

Table 3. Comparison between Rent & Buy Investment and Ten-year BTPs

	Net return	Risk
Ten-year BTPs	1.9%	Null value (0%)
Rent & Buy Investment	2.9%	15% (on return)

In this way, the investment of rent would exclude from the choice, because it is not profitable compared to the risk incurred. If, however, this table is modified with the above inflationary considerations, we could be see that all debt securities and credit instruments are decreasing in profitability because, as already explained, part of it is used to cover expensive life. The table would then be represented with these new values to compare (Table 4):

Table 4. Comparison between Rent & Buy Investment and Ten-year BTPs – with inflation

	Net return (-inflation)	Risk
Ten-year BTPs	0.9%	Null value (0%)
Rent & Buy Investment	2.9%	15% (on return)

As you could see, the landscape has changed. The profitability of debt securities has fallen and now has values close to one percentage point.

Then there is still one aspect that we could consider. In this historical-territorial context, the property market has experienced and is still experiencing a period of severe crisis, with as consequent the drastic drop in house prices compared to the 2007 maximum values. Therefore, in addition to the risk of losing the return, there is another danger on the property that coincides with its loss of value, which would coincide with the loss of part of the initial capital invested.

From the analysis of the latest sales' data it appears that the curve is changing its inclination and prices are lightly reversing the trend. This affects the overlying tables. The return on investment in real estate would be positively affected by this recovery forecast as the return on investment would no longer come only from renting out the property, but also from the value's increase of the property compared with the time of purchase. It is easier to understand this argumentation if I provide an example. The usual accommodation, already considered in the previous example, costs today 100,000 € and annually promises net yield of 2,900€ (23,200€ at the end of the 8 year contract). The property at the expiry of the 8-year contract with the tenant will have increased its value by percentage value equal to the growth of the real estate market in these years. This increase will therefore have an impact on the return of investment, which will have on average net return of 2.9% on rent plus the percentage growth from the purchase value of the property, all without any risk of default.

4   Concluding Considerations

Summarizing, if we read the data in this key and take as true all the considerations previously made, the investment in rent seems to prevail over the debt securities considered. I think it's useful now to specify that there are many involved variables on which there is no certainty, such as inflation (we can't know the trend nor the future value), the real estate market (the value of the property could also decrease throughout the period by declining and completely modifying all the previous valuations), taxes (increases or future decrees could substantially change the results), the risk of arrears (in the future it could increase or decrease, upsetting the results).

References

1. Brueggeman, W., Fisher, J.: Real Estate Finance & Investments (Real Estate Finance and Investments). McGraw-Hill/Irwin, Boston (2006)
2. Montagnana, M., Prizzon, F., Zorzi, F.: Qualitative models for analysing housing market dynamics. Pap. Reg. Sci. **69**(1), 153–165 (1990)
3. Schlarbaum, G.G., Lewellen, W.G., Lease, R.C.: Realized returns on common stock investments: the experience of individual investors. J. Bus. **51**(2), 299–325 (1978)
4. OICT. http://www.oict.polito.it. Accessed 12 Apr 2017
5. OMI. http://www.agenziaentrate.gov.it/wps/content/nsilib/nsi/aree+tematiche/osservatorio +del+mercato+immobiliare+omi. Accessed 12 Apr 2017
6. Fregonara, E., Giordano, R., Ferrando, D.G., Pattono, S.: Economic environmental indicators to support investment decisions: a focus on the buildings' end of life stage. Buildings **7**, 65 (2017)
7. Massari, M., Zanetti, L.: Valutazione. Fondamenti teorici e best practice nel settore industriale e finanziario. McGraw-Hill, Milano (2008)
8. Giliberto, M.: Equity real estate investment trusts and real estate returns. J. Real Estate Res. **5** (2), 261–263 (1990)
9. Ibbotson, R.G., Siegel, L.B.: Real estate returns: a comparison with other investments. Real Estate Econ. **12**(3), 234–242 (1984)
10. Naranjo, A., Ling, D.C.: Economic risk factors and commercial real estate returns. J. Real Estate Financ. Econ. **14**(3), 287 (1997)
11. Ross, S.A., Zisler, R.C.: Risk and return in real estate. J. Real Estate Financ. Econ. **4**(2), 175–190 (1991)
12. Piantanida, P., Rebaudengo, M.: The construction sector crisis in Italy: any strategy for small and medium-sized builders? In: SGEM 2017 Conference Proceedings, Book 5, vol. 2, pp. 759–766, 24–30 August 2017

A Mass Appraisal Model Based on Multi-criteria Evaluation: An Application to the Property Portfolio of the Bank of Italy

Leopoldo Sdino[1], Paolo Rosasco[2(✉)], Francesca Torrieri[3], and Alessandra Oppio[1]

[1] Polytechnic of Milan, 20133 Milan, Italy
[2] University of Genoa, 16123 Genoa, Italy
rosasco@arch.unige.it
[3] University of Naples Federico II, 80121 Naples, Italy

Abstract. This paper presents an application of multicriteria evaluation to select the property characteristics in order to estimate the most probable market value of a large public property portfolio. The methodology proposed, based on the involvement of key actors of the decision process, aim to support the decision process of value judgment in a more flexible way overcoming the difficulty presented by econometric models due to the scarcity of a large sample data; it is referred to a multi-parameter estimated model and tested on a large property portfolio owned by the Bank of Italy. The application has shown that this type of procedures can be a reliable tool to analyse the real estate values and solve estimation problems concerning consistent real estate assets on which analytical methods and regression models are hardly applicable due to the scarcity of data.

Keywords: Mass appraisal methods · Real estate value
Multi-criteria evaluation

1 Introduction

In the context of Mass Appraisal many methodology have been developed for the evaluation of a large number of real estate assets, with specific reference to experimental studies on econometric statistical methods such as:

- *hedonic price method*: mainly used to quantify the effects caused by social and urban factors on the value of properties as well as by location factors environmental factors [1–5];
- *artificial neural networks*: mainly used to forecast property values and for market segmentation [6–10];
- *spatial analysis method*: used to verify, through GIS applications, the weight of the variable "accessibility" [11–15];
- *ARIMA models (autoregressive integrated moving average models)*: used to explain the prices of residential properties in relation to the macro-economic variables [16–19].

© Springer International Publishing AG, part of Springer Nature 2019
F. Calabrò et al. (Eds.): ISHT 2018, SIST 100, pp. 507–516, 2019.
https://doi.org/10.1007/978-3-319-92099-3_57

As noteworthy, the application of these methodologies depend on the availability of market data on market prices or rents and moreover on quantitative and qualitative information of property characteristics. Is well known that the value of an asset depends on its characteristics and on the weights that these characteristics assume in the formation of market values [20–22]. The lack of transparency of the real estate market, especially in the Italian context, and the insufficient of sample data available leads to experiment new approaches to Mass Appraisal evaluation more oriented to the decision process.

The search for a new methods that reflects the real decision-making process of the market players has led to the proposal, also in the field of real estate evaluation, of approaches borrowed from the decision theory. The aim is to approximate the utility functions expressed by the actors operating in the real estate market.

Curto and Simonotti [23] identifies in the context of decision theory useful theoretical and methodological supports of qualitative and quantitative methods to support estimative judgment. In particular, the author proposes the application of the Analytic Hierarchy process (AHP) method, as a tool capable of define a scale of values and quantifying qualitative real estate variables; recently, other authors apply the multi-criteria technique [24, 25] and the fuzzy theory [26, 27] in real estate case studies. The common assumption is that the estimation judgment can be interpreted as a deductive decision process, supported by quali-quantitative methods developed in the aim of decision theory.

In the present paper a multi-criteria approach will be tested for the evaluation of a large public property portafolio owned of the Bank of Italy. In particular, the proposed approach aims to estimate the weight of different characteristic based on the consultation of key actors in the decision making process.

2 Materials and Methods

The case study refers to the appraisal of the market value of a part of the income earning property owned by the Bank of Italy being transferred to the "Società Italiana di Iniziative Edilizie e Fondiarie" (SIDIEF S.p.A.).

The property portfolio is located in different area of the Italian territory and consists in investment assets of guarantee fund reserves of the retirement pension of the Bank of Italy's staff (TQP); it includes about 6,300 units of which about 80% is residential properties and the remaining 20% having different destinations. The properties are mainly located in the city of Rome (83%), Naples (7%), Aquila (3%), Campobasso, Catania, Como and Salerno in Italy.

The total floor area of the entire properties (excluding the properties located in the city of L'Aquila affected by the earthquake in 2009) is 340,000 commercial sqm so distributed: 292,500 sqm for residential destination, of which 24,000 sqm for office and about 23,500 sqm for commercial use. At 31 December 2012, 88% of these assets (excluding properties located in L'Aquila) were rented (90% for residential use, 70% for other uses); the reference value of the entire property assets is 435,000,000 €[1].

The approach proposed for the selection of real estate characteristics in order to the evaluate the entire portfolio; due to the large scale, is coherent with the first point within the basic methodological requirements of the mass appraisal procedure that are:

1. identification of the most significant property characteristics;
2. collection of market data and property characteristics of assets;
3. choice of the evaluation model in order to reflect and simulate the market.

Regard to the first point "identification of the most significant property characteristics" different extrinsic and intrinsic characteristics where considered due to the different locations, floor areas, destination, typology and status of maintenance.

The evaluation method used to estimate the property portfolio refers to the Adjustment grid method and multicriteria analysis for the assessment of the weights assigned to the different property characteristic.

Starting from the average market value (V_{med}), published by the "Observatory on Property Values" of the Italian Tax Authority, different adjustment was introduced considering two weighted coefficients called "Status coefficient" (KS) and "Market coefficient" (KM): the KS coefficient was introduced to keep into account the properties characteristics; the KM coefficient was introduced to keep into account the market conditions (use of property, number of sales and purchases within real estate market, prices trend, etc.).

The value (Vi) of each property is so estimated[2]:

$$V_i = V_{med} \cdot KS_i \cdot KM_i \tag{1}$$

In the next paragraph each step of the methodology will be describe.

2.1 The Identification of Property Characteristics and Their Weight

The KS coefficient takes into account the extrinsic, intrinsic and technological characteristics of properties. Each characteristic was defined using a bottom up approach or with the help of panel of expert in the field of real estate assets management of the Bank of Italy, and classified in two groups (or panels): "Administration" and "Technicians".

The first part of the evaluation process concerned the identification of the quality level of each property characteristic (21); this selection was based on a check list fill in during a site inspection. This first step led to the assignment of a score, on a scale of values from 0, corresponding to "poor" and 5 corresponding to "excellent" quality level of the characteristic. Based on the outcome of the site inspection and their professional experience, the members of the two panels define a list of most important real estate

---

[1] The value was reported in the documents attached to the Financial Statement of the Bank of Italy at 31 December 2012.

[2] The value of the coefficient KM can vary from 0,9 (worst market conditions) to 1,1 (better market conditions).

characteristics that influence the market value. The selected characteristics are divided into 3 groups that identify an equal number of coefficients to make up the KS value, in particular:

1. *extrinsic coefficient* (Ke): keeps into account the following characteristics: accessibility, social context and level of services in the neighbourhood where the property is located;
2. *intrinsic coefficient* (Ki): keeps into account, in relation to the type of real estate, the state of conservation of the constructive elements, such as structure, roof, external curtain walls, windows and doors, etc.;
3. *technological coefficient* (Kt): it keeps into account, in relation to the type of property, the state of conservation of the heating plant and other amenities, such as water distribution network, electric system, lift, etc.

The characteristics identify by the expert is reported in Table 1.

Table 1. Property characteristics selected by the panel to appraise the property value

Extrinsic	Intrinsic	Technological
Social context	Internal design	Lift
Services	Vertical structure	Thermal plant
Pollution	External curtain walls	Energy qualification
Green areas	Door and windows	Electric system
Proximity to the city centre	Roof	Concierge service
Public means of transport	Horizontal structure	Other plants
Parking areas		
Exposure/Proximity to natural beauties/monuments		
Stores		

The following step regarded the determination of the impact (or weight) of each individual characteristic on the real estate value. Through the technique of pair ways comparison derived from the Analytic Hierachy Process (AHP)[3] method defined by Saaty [28], each member of the panels has compared each characteristic with all other ones and according to the level of preference. A score has been assigned on the base of the following scale of measurement[4] derived from the Saaty scale:

- 1: in case of two different characteristics have the same level of importance;
- 2: in case of slight predominance (minimum difference);
- 3: in case of average predominance (average difference);
- 4: in case of strong predominance (maximum difference).

---

[3] Within the AHP - a multicriteria technique - the pair ways comparison is used to define the weights of the criteria and the impact of each solutions on the criteria.

[4] The original scale of measurement established by Saaty vary from 1 to 9.

The eigenvalues extracted from the matrix of pair ways comparison and the following standardization of values allow to calculate the weights of each characteristics and assign the KS coefficient to each property, shown in an aggregate way and for the two groups of the panel, are indicated in Table 2.

Table 2. Real estate characteristics and related percentage weight

Characteristic	%	Characteristic	%
Internal design	7.3	Thermal system	4.7
Social context	7.0	Energy qualification	4.3
Lift	6.6	Roof	4.3
Services	5.8	Horizontal structure	4.2
Pollution	5.5	Parking area	4.1
Green areas	5.3	Electric system	4.1
Vertical structure	5.2	Proximity to natural beauties/monuments	3.8
Proximity to the city centre	5.1	Stores	3.7
External curtain walls	5.0	Concierge services	2.5
Public means of transport	5.0	Other systems	1.9
Door and windows	4.7		

As Table 2 shown, that all the characteristics identified have some importance in the definition of the properties values. Looking at the aggregate results, the difference between the maximum and the minimum percentage weights is 5.4%; the extrinsic characteristics account for more than 45% of the property value, while the intrinsic characteristics account for 30.7% and the technological characteristics account for 24% (Fig. 1).

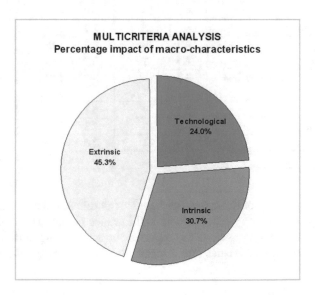

Fig. 1. Percentage weight of groups of property characteristics.

Looking at the judgment expressed by the two group of experts (Table 3) we can notice (Fig. 2):

- the weight assigned to technological characteristics is unchanged and for the two groups it accounts for 24% of the property value;
- the extrinsic characteristics that account for more than 45% of the property value, are reduced to 36.8% for the "Technicians" group and increase to over the half of the property value for the "Administration" group (51.4%).

Table 3. % impact of characteristics on the property value. Comparison between the "Technicians" and "Administration" groups.

Characteristics	Assigned value percentage					
	Tot. (%)	Rank	Tech. (%)	Rank	Adm. (%)	Rank
Internal design	7.3	1	5.4	6	8.6	1
Social context	7.0	2	4.8	9	8.6	2
Lift	6.6	3	7.2	4	6.1	5
Services	5.8	4	4.7	10	6.5	4
Pollution	5.5	5	3.2	18	7.1	3
Green areas	5.3	6	4.3	13	6.1	6
Vertical structure	5.2	7	8.4	1	2.9	18
Proximity to the city centre	5.1	8	4.5	11	5.6	7
External wall curtains	5.0	9	7.4	2	3.3	16
Public means of transport	5.0	10	5.0	7	4.9	9
Doors and windows	4.7	11	5.0	8	4.5	10
Thermal system	4.7	12	3.8	14	5.3	8
Energy qualification	4.3	13	4.4	12	4.2	13
Roof	4.3	14	5.9	5	3.1	17
Horizontal structures	4.2	15	7.2	3	2.1	20
Parking area	4.1	16	3.8	15	4.4	12
Electric system	4.1	17	3.7	16	4.4	11
Proximity to natural beauties and monuments	3.8	18	3.3	17	4.1	14
Stores	3.7	19	3.1	19	4.1	15
Concierge service	2.5	20	2.3	21	2.6	19
Other systems	1.9	21	2.6	20	1.4	21
TOTAL	*100*		*100*		*100*	

In Fig. 3 a clustering of the properties based on all different coefficients is reported.

To sum up, the property has extrinsic and intrinsic characteristics that are above average. The values below average are reported for assets being currently renovated. The technological characteristics are higher than the average levels of ordinary properties.

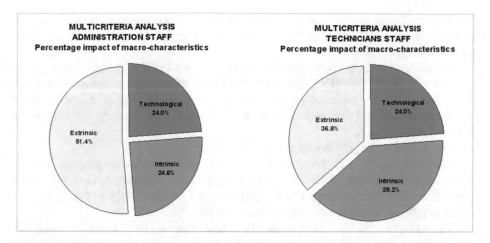

Fig. 2. Percentage (%) impact of the categories of characteristics on the value of properties Comparison between the "Administration" and "Technicians" groups

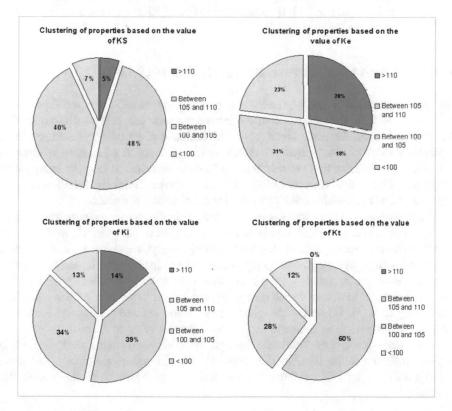

Fig. 3. Clustering of properties based on the values of the coefficient KS and the sub-coefficients Ke, Ki and Kt

The maintenance condition of these properties is above the average, this means a good management of the properties by the Bank of Italy, which has focused on their enhancement over time.

For each property the KS coefficient is calculate multiplying the score assigned to each characteristic with its weight; the KS value is shown with reference to the 100 index assigned to an "average" property on the real estate sub-market with value equal to the medium value (Vm) registered by the "Observatory on Property Values" of the Italian Tax Authority.

The variation range is between 111.4^5 and 85.6.

Most properties have better characteristics than an ordinary property on the sub-market of reference; within the 67 real estate properties estimated, 62 (93%) have a coefficient KS > 100. More than 52% (35) have a KS coefficient higher than 105 and a remaining value of 5% is even higher than 110 (Fig. 3).

The individual coefficients in which KS is broken down (Ke, Ki and Kt) are shown with reference to the 100 index assigned to the average property on the sub-market of reference.

The variation range of Ke is between 117.3 (assigned to a real estate property located in Naples) and 94.9 (assigned to a real state property located in Rome).

The variation range of Ki is between 114.0 and 88.6 (assigned to a real state properties located in Rome).

3   Conclusions and Future Research Perspectives

The paper proposed a quali-quantitative methodology based on a multi-parameter estimated model for the evaluation of the large real estate property portfolio owned by the Bank of Italy. In accordance with the indications given by the international appraisal standards, mass appraisal develops the forecast in relation to a group of property characteristics that are selected due to their significance in order to assess the value; they vary within segments of the real estate market and their identification is an important aspect to obtain reliable results from estimation activities.

When statistic models cannot be used – such as multiple regression models – since it is impossible to create considerable samples of properties or when the characteristics of properties are heterogeneous, the multi-criteria analysis can be a suitable tool to select and weight meaningful property characteristic. In the case under analysis the involvement of professionals on the real estate market helps to select property characteristics and determine their percentage impact on the value of a property portfolio comprising about 6,300 units. The characteristics and percentage impacts have been used in a multi-parameter estimate model to assign the value to each of them.

The values have been validated through the comparison with market prices of similar properties in the same market segment and with values (minimum and maximum) published by "Observatory on the Real Estate Market" of the Italian Tax Authority.

---

[5] The values are assigned to a real estate properties located in Rome.

The application has shown that this type of procedures can be a reliable tool to analyse the value creation functions within the real estate market segments and solve estimation problems concerning large real estate assets on which analytical estimates and regression models are hardly applicable.

References

1. Manganelli, B., De Paola, P., Del Giudice, V.: Linear programming in a multi-criteria model for real estate appraisal, Lecture Notes in Computer Science (including subseries Lecture Notes in Artificial Intelligence and Lecture Notes in Bioinformatics), vol. 9786, pp. 182–192 (2016)
2. Blomquist, G., Worley, L.: Hedonic prices, demand for urban housing amenities and benefit estimates. J. Urban Econ. **9**(2), 212–221 (1981)
3. Graves, P., Murdoch, J.C., Thayer, M.A., Waldman, D.: The robustness of hedonic price estimation: urban air quality. Land Econ. **64**(3), 220–233 (1988)
4. Janssen, C., Soederberg, B., Zhou, J.: Robust estimation of hedonic models of price and income for investment property. J, Property Invest. Finan. **19**(4), 342–360 (2001)
5. Morancho, A.B.: An hedonic valuation of urban green areas. Landscape Urban Plan. **66**(1), 35–41 (2003)
6. Borst, R.: Artificial neural networks: the next modelling/calibration technology for the assessment community? Property Tax J. **10**(1), 69–94 (1991)
7. Collins, A., Evans, A.: Artificial Neural networks: an application to residential valuation in the U.K. J, Property Valuat. Invest. **11**(2), 195–204 (1994)
8. Worzala, E., Lenk, M., Silva, A.: An exploration of neural networks and its application to real estate valuation. J. Real Estate Res. **10**(2), 185–201 (1995)
9. Cechin, A., Souto, A., Aurelio, M.: Real estate value at Porto Alegre City using artificial neural networks. In: Sixth Brazilian Symposium on Neural Networks Proceedings, pp. 237–242, 22–25 November 2000
10. Ge, X.J., Runeson, G., Lam, K.C.: Forecasting Hong Kong housing prices: an artificial neural network approach. In: Proceedings of International Conference on Methodologies in Housing Research (2003)
11. Anselin, L., Getis, A.: Spatial statistical analysis and geographic information systems. Ann. Reg. Sci. **26**, 19–33 (1992)
12. Griffith, D.A.: Advanced spatial statistics for analysing and visualizing geo-references data. Int. J. Geograp. Inf. Syst. **7**(2), 107–124 (1993)
13. Zhang, Z., Griffith, D.: Developing user-friendly spatial statistical analysis modules for GIS: an example using ArcView. Comput. Environ. Urban Syst. **21**(1), 5–29 (1993)
14. Theriault, M., Des Rosiers, F.: Combining hedonic modelling, GIS and spatial statistics to analyze residential markets in the Quebec Urban Community. In: Proceedings of the Joint European Conference on Geographical Information, EGIS Foundation, The Hague, The Netherlands, vol. 2, pp. 131–136 (1995)
15. Levine, N.: Spatial statistics and GIS: software tools to quantify spatial patterns. J. Am. Plan. Assoc. **62**(3), 381–390 (1996)
16. Kim Hin, D.H., Calero Cuervo, J.: A cointegration approach to the price dynamics of private housing. J. Property Invest. Finan. **17**(1), 35–60 (1999)
17. Sivitanides, P., Southard, J., Torto, R.G., Wheaton, W.C.: The determinants of appraisal-based capitalization rates. Real Estate Finan. **18**(2), 27–38 (2001)

18. Chang, Y., Ko, T.: An interactive dynamic multi-objective programming model to support better land use planning. Land Use Policy **36**, 13–22 (2013)
19. Iacoviello, M.: Consumption, house prices, and collateral constraints: a structural econometric analysis. J. Hous. Econ. **13**(4), 304–320 (2004)
20. Forte, C., De, Rossi B.: Principi di Economia ed Estimo. Etas Libri, Milano (1974)
21. Sdino, L. (a cura di): Contributi e riflessioni economiche, estimative, finanziarie per le professioni immobiliari, in Atti del 1° Corso per Agenti Immobiliari. Tecnocopy, Genova (1998)
22. Sirmans, G.S., Benjamin, J.D.: Determining apartment rent: the value of amenities, services and external factors. J. Real Estate Res. **4**(2), 33–43 (1989)
23. Curto, R., Simonotti, M.: Una stima dei prezzi impliciti in un segmento del mercato immobiliare di Torino, Genio Rurale 3 (1994)
24. Breil, M., Giove, S., Rosato, P.: A Multicriteria Approach for the Evaluation of the Sustainability of Re-use of Historic Buildings in Venice. In: IDEAS Working Paper Series from RePEc (2008)
25. Giove, S., Rosato, P., Breil, M.: An application of multicriteria decision making to built heritage. The redevelopment of Venice Arsenale. J. Multi Criteria Decis. Anal. **17**(3–4), 5–99 (2010)
26. Bagnoli, C., Smith, H.C.: The theory of fuzzy logic and its application to real estate valuation. J. Real Estate Res. **16**(2), 169–200 (1998)
27. Bonissone, P.P., Cheetham, W.: Financial applications of fuzzy case-based reasoning to residential property valuation. Fuzz-IEEE **1**, 37–44 (1997)
28. Saaty, T.L.: The Analytic Hierarchy Process. McGraw Hill, New York (1980)

The Other Side of Illegal Housing.
The Case of Southern Italy

Giuseppe Guida[✉]

Department of Architecture and Industrial Design, Luigi Vanvitelli University
of Campania, 81031 Aversa, Italy
giuseppe.guida2@unicampania.it

Abstract. In the last decade, Italian territorial policies have become increasingly characterized by a broad laissez-faire approach, which negates the value and the importance of urban planning and of common goods.

The point of departure for the hypothesis formulate in this paper is the ascertainment of an assault on landscape and of an increase in soil consumption. The latter is closely related to the fluctuations of the property markets during the crisis, which are particularly acute in prime areas or in "legally difficult" areas. This tendency is particularly visible in Southern Italy, where economic development has been historically weak. There, the crisis has made some territories available for particularly remunerative real estate investments.

As a consequence, urban planning is seen as a hurdle to be overcome with derogations, agreement-based policies, public-private partnerships, housing plans, as well as with past and ongoing emergencies.

Data on coastal areas, cities, and towns shows an increase of soil consumption due to dwelling, which is in turn connected to the structural crisis of local public finances and to the connected depletion risk for high-quality landscapes and their identity.

Against this background, this paper looks at the other side of unauthorized development, i.e. on the housing stock which, for different reasons, has been legalized. In doing so, it suggests possible solutions for the complex issue of building amnesties.

Keywords: Illegal housing · Periurban · Urban policies

1 Introduction

Urban and real estate dynamics have long been connected to the economic cycle and, more broadly, to local and global financial systems.

In Italy, the period of structural crisis (2008–2015) and the subsequent slow recovery have strongly impacted on the housing market and on soil consumption rates. In 2016, not all economic sectors have grown: growth was negative in the agricultural sector (−3% after growing by 4,2% in 2015, which was an extraordinary crop year) and in the public services sector (−2%). A positive but moderate growth was registered in the service sector (2,5%). Higher growth rates have been registered in the industrial sector (5,2%, after years of decline). Industrial growth was supported by good production results, especially in the building sector (8,7%, after the 9,6% of 2015), which

© Springer International Publishing AG, part of Springer Nature 2019
F. Calabrò et al. (Eds.): ISHT 2018, SIST 100, pp. 517–524, 2019.
https://doi.org/10.1007/978-3-319-92099-3_58

is quickly recovering after the decline of the last decade [1]. This growth is related to the renewed attention towards medium-scale cities, as testified by the Programma Periferie (a requalification plan for suburbs), and towards larger cities, as testified by the PON Metro programme. Other factors behind this positive trend include the implementation of the Masterplans for Italy's newly instituted metropolitan areas and for the city of Taranto, as well as a better integration between the national government and municipal authorities. This is suggested by the fact that these Masterplans are based on a mutual agreement and the national government's improved ability to replace regional authorities in cases of inertia [1].

According to a previous analysis of [2], the economic and financial conditions of Italian municipalities are often critical: out of 445 municipal bankruptcies registered in the country between 1989 and 2011, 353 (i.e., 79%), were declared in the South of the peninsula. About 31% of municipalities of Calabria has declared bankruptcy. Campania follows suit with a rate of 20,5%. Here, however, the number of citizens suffering from these financial cracks is higher: almost two million people, or almost the 33,5% of the total regional population. The same took place in 35 municipalities in Puglia (13,5%), 24 in Sicily (6%), 19 in Basilicata, 18 in Abruzzo, 14 in Molise and 3 in Sardinia. The remaining 92 municipalities that declared bankrupt are mainly located in Lazio (43) and Lombardy (15). The rest are located in Emilia Romagna (8), Marche (6), Piedmont (5), Umbria (4), Tuscany (4), Liguria (4), and Veneto (3). In the examined period, we can calculate a total of almost 3 million and 750 thousands Italian citizens affected by municipal bankruptcy.

Seen against the backdrop of the general crisis faced by the country and in consideration of the length of the recession, these data only partially reflect the fragile situation of local authorities in Southern Italy. The research hypothesis of the paper is that public administrations "adapt" to such economic conjunctures [3] in different manners: they stretch the current regulations, undertake borderline initiatives or implicitly encourage illegal practices. The effects of these transformations are especially visible in the outskirts, where the urban and peri-urban sprawl require solutions to restore adequate living and housing conditions.

A growing distrust towards the Western economic system, and towards the development strategies for the South inspired by it, is encouraging the elaboration of more rapid alternatives. These, however, are not necessarily connected to development programmes, but are rather aimed at guaranteeing a rapid inflow of funds and thus to make daily administration possible for cash-strapped local authorities. Possibly, these alternatives allow the realisation of important projects as well as the nourishing of political contacts and networks.

Cersosimo and Donzelli [4] analyse the ebb and flow of several "waves" of economic growth policies: forced industrialization; social protection through mass hiring in public sectors, and hyper-growth of (irrational and inefficient) infrastructure projects. Similarly, the "new" development opportunities produced in the 1990s and connected to local economic districts have been met with increased scepticism. Already before the recession, industrial projects in the south were in crisis. Equally, the economic/urban "programmes" (the "Integrated Territorial Projects" and the "Territorial Pacts"), whose

EU funds (i.e. the Regional Operative Programmes) had begun to rekindle the sec-ondary and tertiary sectors in the south. This particularly applied to the most "fluid and dynamic" areas [5] of the country, where more solid development dynamics seemed to have taken root. There are several ways to intervene, and many of these are not codified and only implemented locally. However, other policy tools are wide-ranging and, for this reason, particularly problematic. Chief among these is the so-called Piano Casa, a national framework on housing that regional authorities must further detail and implement.

The different regional laws implementing the Piano Casa have resulted in the production of unclear and contradictory laws, with the effect of slowing down proce-dures. Therefore, the Piano Casa appeared unable to provide the "economic recovery" or the "requalification of the existing heritage" that are mentioned in the title of the agreement between the national government and the regions and in the title of regional laws. On the contrary, the Piano Casa has turned out to be a means to undermine significant part of the environmental protection law.

Most of the regional laws implementing the Piano Casa in Southern Italy and beyond have extended (if compared to the more reasonable previous agreement between the national government and the regions) the possibilities of constructions even to areas placed under environmental protection. In Lazio, for example, the pro-visions of Piano Casa can be applied in the following protected areas: 47 ha in the Insugherata Park, 28 hectares in the Tenuta dei Massimi, and 31 hectares in Monte Mario, for a total of 106 hectares. Still in Lazio, the Piano Casa can be implemented in regional parks that are not affected by hydrogeological management plans. With the strategic use of "B zones" identified by the relevant laws, it is possible to build on 29 hectares of the Decima Malafede natural reservation and on 40 hectares in the Mar-cigliana park (data provided by Legambiente). The consequences of this are also visible in the Veio park: 16 hectares in the municipality of Campagnano, 39 hectares in the municipality of Sacrofano and 252,5 hectares in the municipality of Rome, for a total of 307,5 hectares. There are problems also in the park of Bracciano, where the area of development in the B areas identified by the relevant law, between Anguillara Sabazia, Bracciano and Trevignano Romano is equal to a total of 1.076 hectares. Practically, only lake areas remain protected. Regions like Campania and Calabria, even if their local laws officially exclude certain areas from the application of the Piano Casa, in practice allow this to happen with the use of derogations and, as stated above, by issuing unclear laws that rely on the interpretation of municipalities. The Regional Law 25/2009, the Basilicata region introduced a new legal figure: the "architectural land-scape protection law". This might allow interventions that would otherwise be banned under the pre-existing legal framework on landscape protection.

The case of the Sardinia region is particularly telling. On January 20th, 2012, the regional government approved an amendment to its Piano Casa provisions, which had been published in the regional law n. 4 of 23/2009 (called *Extraordinary dispositions for economic support through construction and for the promotion of strategic inter-ventions and development plans*). The national government challenged this amendment and remarked that art. 7. subsection 1, f., permits interventions "in derogation to the provisions of building codes and urban planning measures of the city" and in derogation to "the current regional regulations". The national government commented that "in matters pertaining to the implementation of the Piano Casa, [the amendment] unconstitutionally degrades landscape protection to the same legal rank of urban planning and construction codes" [6]. Also in this case, the regional government of Sardinia attempted to stretch the possibilities afforded by Piano Casa, also in a patently illegitimate manner.

Lastly, while the Puglia region seems geared towards a virtuous path and to be determined to avoid such unrefined tampering of landscape protection laws, Sicily and Campania only refer to landscape protection law for areas characterised by a complete construction ban, which are generally small in size and not relevant.

A similar argument can be made about the "Conferenza di Servizi" (a consultation between local authorities and local stakeholders). This tool has now become the standard procedure to agree upon and define broad-ranging derogations to landscape protection laws. Making matters worse, the scope of such consultations was broadened in the national taxation measured passed in 2010 (Law 122/2010) and was partly extended to documents and procedures concerning cultural and landscape heritage [7].

The approach introduced by the European Landscape Convention (implemented in Italy with the Code for cultural and landscape heritage) is equally problematic. By privileging the social perception of landscape and the participation of the affected communities, the Convention surely paves the way for the shared and participative creation of territorial charters and thus for the fruition of landscapes and of common goods. By doing so, however, it also shifts the responsibility for landscape protection on the fragile local dimension [8]. This approach, indeed, requires an active civic society that is able to combine local expertise with expert knowledge through participative democracy initiatives that enable communities to reflect on how to best transform their living and social environment. In short, new forms of learning [3] and of participation would be necessary to increase the sense of place of the affected communities. However, the reality in southern Italy is far from this ideal case and, broadly speaking, "land" and "property" are still seen as private goods, as opposed to relational goods. In the same vein, regulations and protection laws are perceived as a violation of freedom, and are thus obstructed or circumvented. This further nudges local authorities, who often depend on consensus for their re-election, to damage the landscape.

The motive of so many derogations to landscape protection is clearly that of enabling profit making in valuable areas, for both private and public finances. The ultimate outcomes of this process are the depletion of the territory, the irrational dispersion of human settlements [9], the impoverishment of valuable landscapes, a long-term worsening of development perspectives of the affected territories and the positive externalities that would otherwise derive from the correct management of the landscape (Fig. 1).

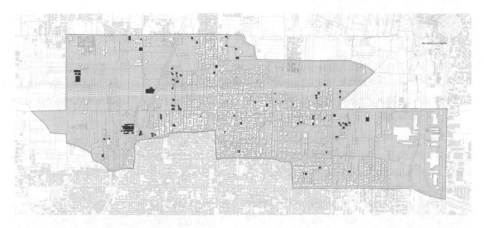

Fig. 1. Municipality of Capodrise (CE): localisation of dwelling settlements (source: Giuseppe Guida, Giovanni Bello, 2017).

2 Norm and Form

According to the latest report produced by the National Council for Economics and Labour and by the Italian statistical institute, the highest intensity of unauthorised development in Italy is found in Campania, where nearly one third of all the housing built in the last ten years is unauthorised. In the whole country, the same figure amounts to 20%, whereas in the north of the country only 5% of the housing was built illegally.

If one includes housing developed legally, but which was built in derogation of ordinary legislation, the results are even more dramatic. In the last ten years, nearly 30% of the legal housing has been built following ad-hoc regulations that created exceptions to the ordinary urban and landscape planning tools. Chief among these derogations is the Piano casa, which the Campania regional authority (which is tasked with the management of the territory) has adopted and extended for two more years. In a similar vein, the Campania region has invented (and reinvented) laws such as Law 19/2001, which expedited paperwork for building applications and thus removed a significant hurdle to construction in areas under landscape protection laws. The millions cubic meters of underground garage in the Sorrento peninsula, either completed or in construction and under legal review, will remain its outrageous symbol.

The first consequence of the current crisis is the introduction of EU policies aimed at containing the public debt. These policies negatively affected the investment levels of local authorities such as regions and municipalities. The spending ability of the Italian municipalities was thus severely limited. In order to keep providing public services, local authorities began to look for different ways to finance their investments and to use new instruments. These include:

- Project financing, where privates carry out an investment that is deemed to be in the public interest through an ad-hoc company, expecting profits through the management of the financed facility;

- Public property leasing, which consist in the construction of public works through a developer and a financer. The latter will eventually own the facility and rent it out to the relevant authority;
- The sale of publicly owned estate that authorities no longer intend to keep;
- Debt management tools that allow authorities to lower their debt or to spread its repayment to a longer period of time. This sometimes entails significant risks linked to possible interest rates hikes (e.g. derivatives).

Along these factors, which have entered the standard toolbox public institutions, a new one is emerging: the usage of valuable areas (protected by landscape conservation laws) for residential, service, and leisure purposes. This is often achieved with the tools mentioned above and, most importantly, by drawing ad-hoc legislation that waives urban and landscape regulations. This tool is perhaps even more dangerous than those listed above, both in terms of social costs and of irreversible effects on the territory. Furthermore, their financial benefits are uncertain and only hypothetical, much unlike the economic returns of the involved privates.

It is thus not a coincidence that, according to the most recent report of the Italian environmental protection agency on soil consumption [10], Campania ranks highest for soil consumption in the first 10 km from the coastline. In Campania, about 18% of this portion of land is thus developed, also on steep slopes (5,4%) and in areas characterised by hydraulic and seismic danger (8,5% and 10,4%, respectively). All these values are expected to grow. This kind of soil consumption has direct and indirect effects on the landscape, on the ecosystems, on biodiversity and on essential factors of climatic and hydro-geological regulation.

This problematic process connects different actors and different dynamics: real estate investors, politics, unauthorised development, organised crime, lack of controls, as well as a variegated and spatially dispersed set of satellite activities. Urban planning, in consequence, constitutes a hurdle to be bypassed with ad-hoc legislation, consultation with local stakeholders, public-private partnerships on "fast-track" public projects, self-construction, and regional Piano Casa.

In this scenario, which is only the last stab at the legitimacy of urban and territorial planning, southern Italy is most intensely exposed to and threatened by predatory building speculation and investment capitals. Data show an increase of soil consumption on coastal areas, in cities and towns, which exposes landscapes and places to the risk of impoverished landscapes and identity loss.

3 Possible Solutions

This paper presents the preliminary results of an ongoing research. Data portray a worrisome tendency towards the deregulation of soil use, which in turn reduces the scope and the efficacy of planning - a state of affairs that might be difficult to reverse. The problems highlighted before, and the ongoing legal production on soil use is at odds with Jonas' call for responsibility, which would highlight the urgency of contrasting the indisputable state of emergency that looms over local authorities [11].

Planning may thus become a useless exercise, particularly in its attempts at achieving long-term strategies, a balanced development, the requalification and the protection of the landscape. The less stringent and more permissive regulations remain in place, such as building codes (called *Ruec*) and technical standards (*Nta*), while the overall legal framework and the vision for the territory, as well as controls and protection of its urban-environmental equilibrium become weaker.

In southern Italy, the absence of structural support mechanism for the economy determines a more intense usage of ad-hoc regulations and derogations on behalf of both the national and regional legislative bodies. This sets off a vicious circle that impoverishes the landscape, alters the housing market, and limits territorial development. This process is less visible in large urban centres such as Naples, Palermo, or Bari, than it is in medium-small cities, on hillside and on coastal areas, especially when these are sparsely populated but have very high housing prices.

The enormous amount of illegal housing or housing built after derogations to the standard planning and landscape provisions has grown in the last decades: about 200 thousand regularisation requests sent after the building amnesties of 1985, 1994, and 2003 (although this should not be applicable in the Campania region). Of these requests, about 85 thousand were produced in Naples alone. Despite a special procedure to process these requests, about half of them remain to rot in the cramped rooms of the Regularisation Office, aptly located in a semi-abandoned building in the public housing area of Ponticelli.

The main challenge consists in finding a way to avoid that the responsibility for requalification, regularisation, and requalification measures only weigh on municipalities. A possibility might be the creation of a public, independent agency [12]. Such an agency might help negotiate shared solutions or, where necessary, expedite the demolition of illegal housing and couple such interventions with territorial development and requalification strategies.

References

1. Svimez: Rapporto 2017 sull'economia del Mezzogiorno. Il Mulino, Bologna (2017)
2. Svimez: Rapporto annuale Svimez (2012) sulla finanza dei comuni. http://www.svimez.it. Accessed 15 Sept 2017
3. Donolo, C.: Disordine. L'economia criminale e le strategie della sfiducia. Donzelli, Roma (2001)
4. Cersosimo, D., Donzelli, C.: Mezzo giorno e mezzo no. Realtà, rappresentazioni e tendenze del cambiamento meridionale. Meridiana **26**(27) (1996)
5. Trigilia, C.: Il Sud in mezzo al guado. Meridiana **31** (1998)
6. Edilizia urbanistica. http://www.ediliziaurbanistica.it/pf/testo-news/21687/Piano-Casa-Sardegna-impugnato-dal-Governo. Accessed 15 Sept 2017
7. Settis, S.: Paesaggio, Costituzione, Cemento. La battaglia per l'ambiente contro il degrado civile. Einaudi, Torino (2010)
8. INU: Rapporto dal Territorio 2010. INU Edizioni, Roma (2011)
9. Guida, G.: Immaginare città. Metafore e immagini per la dispersione insediativa. Franco Angeli, Milano (2011)

10. Ispra: Consumo di suolo, dinamiche territoriali e servizi ecosistemici. Roma (2017)
11. Guida, G.: La Regione trasforma il paesaggio in un cantiere, La Repubblica/Napoli 22(02) (2012)
12. Curci, F., Formato, E., Zanfi, F. (eds.): Territori dell'abusivismo. Come uscire dall'Italia dei condoni. Donzelli, Roma (2017)

The Participation Process in Urban Revitalization Projects

Yakup Egercioglu$^{(\boxtimes)}$

Izmir Katip Celebi University, 35620 Izmir, Turkey
yakupegercioglu@gmail.com

Abstract. Izmir is the third most populous city in Turkey and one of the oldest settlements on the Mediterranean basin. It is vital to protect Izmir's cultural heritage, integrate historically significant areas into modern life, and promote sustainable urban development. To achieve this, the Izmir Metropolitan Municipality has implemented the multidisciplinary "**Izmir Tarih (History) Project**", and the participants include local administrations, national public institutions, local public enterprises, non-governmental organizations, the private sector, funding institutions and the local community. This paper discusses the participation process implemented in Izmir and certain projects whose pros and cons will be determined.

Keywords: Izmir History Project · Urban revitalization projects
Participation

1 Introduction

Located in western Turkey, Izmir is the country's third most populous city and is one of the oldest settlements on the Mediterranean basin. Archaeological excavations have shown that Izmir was first established at the Yeşilova Mound in the Bornova district and was inhabited continuously from approximately 8500 to 4000 BC [1]. In 333 BC, under the rule of Alexander the Great, Izmir was re-established on the slopes of Mount Pagos, which today is known as Kadifekale [2]. Until the 16th century, Izmir was a small port town located on the outskirts of Kadifekale. From the 17th century onward, it developed as an important harbour city and expanded towards the coastline [3]. Its development continued through the 19th century and Izmir eventually became a cosmopolitan city with a population consisting of Muslim Turks, Jews, Armenians, Greeks and Levantines [4]. Over the course of the 20th century, the historical centre of Izmir underwent many changes for a variety of reasons. First, the Izmir fire of 1922 caused widespread damage to urban areas. After the 1922 fire, housing demand increased and reconstruction of damaged areas became a priority [5]. After the 1950s, rural-urban migration in western Turkey led to important socio-political and socio-economic changes; Izmir was affected by this migration more than many other cities in western Turkey. During this period, historic city centres faced urban sprawl and illegal housing because most residents began to move out to new, developing areas of the city, while the slopes of Kadifekale, the traditional residential area of the city, became home to migrant communities [6] (see Fig. 1). The Izmir Tarih Project, which is also a branding

© Springer International Publishing AG, part of Springer Nature 2019
F. Calabrò et al. (Eds.): ISHT 2018, SIST 100, pp. 525–533, 2019.
https://doi.org/10.1007/978-3-319-92099-3_59

project of the city, is a long-term endeavor and includes a large geographic area. The Project covers approximately 275 ha, with approximately 1500 listed buildings and 8500 offices included within its borders [8]. The main objective of the project is to improve the historical consciousness of Izmir's residents, thus making them more aware of their city's cultural heritage. The project also aims to revitalise the Kadifekale-Agora-Kemeraltı historical axis by preventing further decline in value and by maximizing its tourism potential as part of the city's branding process. Given these aims, the project's strategic priority is to include all stakeholders in the process. These features distinguish the project from other revitalization approaches that do not invite the public to contribute to the process. In contrast, the Izmir Tarih Project includes many participants: local administrations and central public institutions have leadership roles; the private sector and funding institutions have investment roles; civil public institutions provide support; architects, archaeologists, planners and sociologists provide expertise; and the local community also contributes. The project could not succeed without the participation of the local public.

Fig. 1. Locations and a panorama of neighborhoods in 19th century in Izmir [7].

2   The Izmir Tarih (History) Project

In the literature, public participation, community participation and citizen participation words are being used interchangeably to address participation of inhabitants or users. Citizen participation has been debated by various scholars, however, as Desai [9] argues there is no single definition or single method about participation. All definitions and conceptualizations of participation aim to improve power of citizens in decision-making. In 1990s, participation was defined as a process of turning citizens' ideas into initiatives and actions [10]. In urban case Creighton [11] argued that citizens who involved in participation were supposed to affect or interact with the end product. Thus, citizens became an attention point in participation cases. On the other hand Mitchell, [12] made a link between participation and social sustainability; and argued that participation of local people could provide long-term sustainability of urban revitalization. In similar a vein, UN defined the term as a mutual relation between citizens and participation; participation is creation of opportunities which provide contribution to members of community, give opportunity to influence the development process and share the results of the process equally [13]. In another words, it is argued that participation of the community could increase the quality of the product which will be used by the community. Arnstein said that lack of power distribution in participation causes inefficiency [14]. In other words, for citizens, being participant without having power in decision making, makes them passive participants. Arnstein illustrated a participation ladder based on eight rungs.

The Izmir Metropolitan Municipality carried out two workshops in the 2014 and 2015 when launching and developing the Izmir Tarih Project. First workshop included second sub-region Havra District, fifth sub-region Hotels' District and eleventh sub-region Anafartalar Street second stage. Second workshop included first sub-region Agora, thirteenth sub-region housing fabric, fourteenth sub-region second housing fabric and sixteenth sub-region Kadifekale-Ancient Teathre. There were three sessions for both workshops: Identification of condition mapping and route map. Participants' attendance also have great importance in workshops. Based on the characteristics of these regions and on previous academic research, the area of the Izmir Tarih Project has been divided into nineteen sub-regions to improve planning/design strategies for each one.

These division decisions took into account the density of each region's functions, including residential, commercial, and religious buildings, as well as archaeological sites. Thus, each sub-region is hosting different renovation, revitalization and restoration projects based on its own particular characteristics; this entails both local problems and opportunities. **Tarkem** is a pioneer and a unifying institution that intends to improve cooperation and dialogue among the public, the private sector and civil public institutions. This is generally neglected in revitalization projects in Turkey (see Fig. 2).

Fig. 2. The nineteen sub-regions of project [15].

As a result, operational guidelines for the sub-regions, with the strategic goal of encouraging tourism, were established by the participants. Operational guidelines, based on the results of the workshops, were grouped into four categories by the moderators. The first category is that of **general criteria** for the studied sub-regions. This describes the aims of the operations. General criteria for the sub-regions also can be understood as the goals to be accomplished throughout the region. The second category is that of **operations related to more than one unit (plot)** in each sub-region. These operations are based on the evaluation multiple plots in the sub-region that share similar designs. Third, **micro-operations** are mainly related to the restoration of certain historic buildings in the sub-regions. All of these historical buildings were also selected by the participants. Finally, there are **operations that should be implemented in entire sub-regions** and are mainly related to problems like transportation and lighting, which are the responsibility of municipalities [15]. Results of the workshops recommended that many operations be carried out in the sub-regions. To facilitate prioritization, the operational guidelines were adapted into questionnaires and sent to all workshop participants. Priorities in the studied sub-regions were determined according to the responses of the participants. After this determination of operational guidelines and priorities, operational plans were published as a book and sent to all participants and to other potential stakeholders [16]. Finally, a meeting called a **Dialogue Conference** was arranged to announce the priorities of the operations and to facilitate possible public-private partnerships. Some of these projects, which will be explained in the following sections, have already begun, and new operational plans for the other sub-regions are currently being developed in workshops (Fig. 3).

Fig. 3. A view of workshops in Izmir History Project workshop building.

3   Process of Operation Plans of the Izmir Tarih Project

Strategic approaches in the operation plans involve all areas in the related sub-region. It is expected spontaneous improvement for areas that out of the related sub-region with the influence of other areas. Detailed operation plans have six main phases.

1. **Targets for region**: First phase includes determination of main features, values, strengths and weaknesses of the region. In this context, targets can be clealy identified.
2. **Operations to be carried out for more than one unit**: Projects that have comprise only a single building could not be sufficient to reach determined targets. Instead, region needs more complex project that include more than one unit together.
3. **Microoperations**: There are many single building project or public space project proposals. This phase aims reusing buildings with a new or authentic function.
4. **Operations to be carried out for whole region**: Suggestions that related with whole region is determined by participants. These suggestions focus on preservation oriented plan, transportation, lightning, infrastructure and traffic.
5. **Inspection**: It is evaluated that targets in first phase whether achieved or not. This is a control phase and aims constant dialog with actors.
6. **Monitoring:** Feedbacks are one of the most important elements of the project. Monitoring the process and determining problems that hinder the project are the last but most efficient phase [15] (Fig. 4).

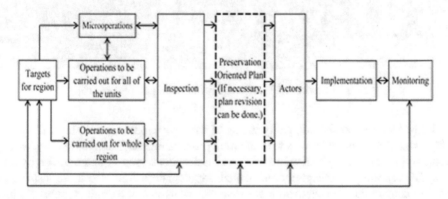

Fig. 4. Operation plan working scheme [15].

The Participant Role: The most important features of this project are multi-disciplinarily and participation. These features differ the project from other revitalization approaches that don't have an opportunity to public for contribution in process. The project couldn't accomplish without any participation from the public of the region. The main actors of the İzmır History Project can be listed as:

- Local administrations (Izmir Metropolitan Municipality and Konak Municipality)
- Central public institutions (Ministry of Culture and Tourism, Ministry of Development, Ministry of Economy)
- Local public institutions (Universities, İzmir Regional Directorate of Foundations, İzmir Culture and Tourism Directorate)
- Civil public institutions (Izmir Chamber of Commerce, Kemeraltı Tradesmen Association, Union of Chamber of Merchants and Craftsmen, Associations and Foundations etc.)

- Private sector (TARKEM-Historic Kemeraltı Construction Investment Trading Joint Stock Company etc.)
- Supporter & Fund provider institutions (National and International Funding, Embassies, Izmir Development Agency.)
- Participants (citizens) (Fig. 5)

Fig. 5. Location of 1. Sub-region and important project [16].

Izmir Metropolitan Municipality is one of the most productive administrations and it has been carried out a lot of a revitalization projects recently. In addition to planning preparations, institutions that related about historical conservation was founded by both Izmir Metropolitan Municipality and Konak Municipality. And also these institutions provide cooperation between different disciplinarians and economic support. Expropriations are the most vital stage in renovation-revitalization works. Izmir Metropolitan Municipality carried out numerous expropriations in Izmir historical city center to start and develop Izmır History Project.

Universities also have an important role in the process such as prepare the planning activities, restoration projects and doing researches etc. Studies are done with the students about conservation areas and it is a big chance for brainstorming that is useful producing projects. Izmir Regional Directorate of Foundations and Izmir Culture and Tourism Directorate as a local public institutions take a responsibilities about historical fabrics and monuments. Public and civil public institutions participating in the project also have an important impact. For instance Kemeraltı Tradesmen Association which was founded voluntarily by tradesmen, Izmir Chamber of Commerce and Union of Chamber of Merchants and Craftsmen etc. Kemeraltı Tradesmen Association was founded several times before. It always aims to be a symbol of a tradesmen's problems and requests, and also it provides a relationship between local administration and tradesmen. The association contributes to advertise of Kemeraltı.

TARKEM as a private sector support is a multi-partner joint stock company that was found by 116 partners to preserve and develop urban assets of Izmir. It started its studies on 19 November 2012 to regulate, preserve and hand down to the future generations. The main aim of the company is production of the necessary rehabilitation-renovation policy, contribution to increase living standards and improving space quality depending on the area of the space character with evaluating the investment and resources effectively. The other purpose is introducing Izmir to "Old Town" concept that European cities such as Prague, Budapest, Lisbon have [16] (Fig. 6).

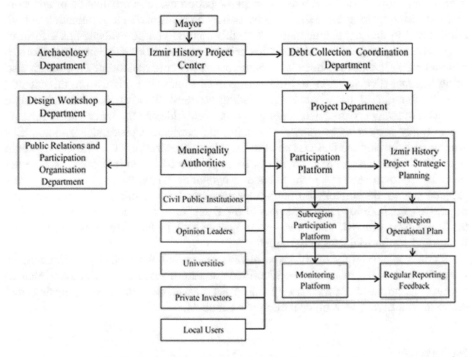

Fig. 6. Izmir History Project organization scheme [16].

As mentioned above, integration of the project with public is one of the prior decisions. It becomes clear that local authority should provide an opportunity to citizens to involve them in decision making process. This involvement should be made by using appropriate participation tools which can provide expression of their thoughts freely. These thoughts should be included in decision making.

4 Conclusion

Conservation, restoration and revitalization projects in historically significant environments are generally carried out by both municipalities and the private sector. However, while some gentrification may be accepted, residents of these areas should benefit the most from these projects. In the last few years, the Izmir Metropolitan Municipality has been including local residents in decision making and design processes not only in historical areas, but also in other parts of the city. With this goal in mind, the Izmir Tarih Project has also been encouraging the participation of a diversity of actors in these processes. This paper has presented the methodology and operational steps of the Izmir Tarih Project as a new participation model. Participation of all actors in the decision making and design processes may be evaluated as a combination of both administration and communication. Thus, this method can be evaluated as a form of governance and may enrich relationships between communities and their municipalities as local authorities. Workshops with participants, the core method of the project, also bring together diverse actors to discuss important topics. For this reason, this method may be thought of as a consensus-building strategy. Tourism has been the main objective of many urban revitalization projects worldwide and the Izmir Tarih Project is not the only project to be carried out using the participant process. For example, a similar participation method was used in the urban regeneration of the Raval District in Barcelona. The same method could not be implemented in Izmir because each settlement has its own set of characteristics and dynamics. Thus, the methodology of this project had to be suitable for Izmir specifically. As a result of this process, the objectives, methodology and strategies of the Izmir Tarih Project were evaluated using the participant process. However, the project is still in its early stages and can be thought of as the trial of a new model.

Thus, the possibility of mistakes always exists and the methodology should be revised when these mistakes are identified. In future research, the authors plan to monitor the processes of the Izmir Tarih Project, especially after implementation, and compare this model with other participation models.

References

1. Akurgal, E.: Bayraklı Kazısı Ön Rapor. Türk Tarih Kurumu Basımevi, Ankara (1950)
2. Baykara, T.: Izmir Şehri ve Tarihi. Ege Universitesi Matbaası, Izmir (1974)
3. Ulker, N.: XVII. Ve XVIII. Yüzyıllarda Izmir Şehri ve Tarihi. Akademi Kitabevi, Izmir (1994)
4. Maeso, J., Lesvinge, M.: Smyrna in 18th and 19th Centuries: Western Perspectiv. Mas Printery, Izmir (2013)
5. Serçe, E.: Tazminattan Cumhuriyete Izmir'de Belediye. Dokuz Eylül Yayınları, Izmir (1998)
6. Tekeli, I.: Izmir Tarih Projesi Tasarım Stratejisi Raporu. Izmir Tarih Proje Yayınları, Izmir (2015)
7. Beyru, R.: The City of Izmir in 19th Century. Literatur Publishing, Istanbul (2011)
8. Tekeli, I.: Kentleşme ve Kentsel Dönüşüm. Tarih Vakfı Yurt Yayınları, Istanbul (2014)

 9. Desai, V.: Community Participation and Slum Housing: A Study of Bombay. Sage Publications, New Delhi (1995)
10. Hillier, J., Healy, P.: Critical Essays in Planning Theory. Ashgate, Farnham (2008)
11. Creighton, J.: The Public Participation Handbook: Making Better Decisions Through Citizen Involvement. Wiley, San Francisco (2005)
12. Mitchell, D.: The Right to the City: Social Justice and the Fight for Public Space. The Guilford Press, New York (2003)
13. United Nations: Earth Summit: Agenda 21. UN, New York (1993)
14. Arnstein, S.: A ladder of citizen participation. J. Am. Inst. Plan. **35**(4), 216–224 (1969)
15. Kutlu, G.: Agora, Kadifekale Operasyon Planları. Izmir Tarih Proje Yayınları, Izmir (2015)
16. Kutlu, G.: Havralar Bölgesi, Anafartalar Caddesi Operasyon Planları. Izmir Tarih Proje Yayınları, Izmir (2015)

An Evaluation Framework
for Resilience-Oriented Planning

Gabriella Esposito De Vita[1]([⊠]) [iD], Roberta Iavarone[2] [iD],
Antonia Gravagnuolo[1] [iD], and Ines Alberico[3] [iD]

[1] CNR - National Council of Research, IRISS, Naples, Italy
g.esposito@iriss.cnr.it
[2] CNR - National Council of Research, IBAF, Porano, Italy
[3] CNR - National Council of Research, IAMC, Naples, Italy

Abstract. The ability of a system to absorb, recover from and successfully adapt to stressing circumstances can be defined as "resilience". To make cities more resilient toward natural disasters, several international initiatives recommend to consider the risk management not only in emergency conditions.

The identification of characteristics that make resilient the cities toward natural disasters and the connection between the resilience goals and the risk management phases (mitigation, preparedness, response, recovery) are the main purposes of the framework implemented in the present work. At this aim, we considered the city as complex, dynamic, self-organizing system, continuously changing under the pressure of perturbing internal or external factors. The framework, structured in a Geographic Information System, is useful at different territorial management scales and can host many types of data. Starting from a critical review of international frameworks, focused on the resilience, four *drivers* (economic, social, environmental and institutional), several *driver descriptors* (number: 15) and *sub-drivers* (number: 36) were identified to improve the resilience and to manage the territory during the risk management phases. This frame allows to overcome the sectorial approaches of territorial management promoting the integration of resilience goals (prevent, prepare for, cope with, respond to, and recover from) and of risk management phases (mitigation, preparedness, response, recovery) into ordinary planning tools.

Keywords: Urban resilience · Urban regeneration · Risk management

1 The City Resilience and the Disaster Risk Management

The resilience concept rooted in ecology describes the capacity of a natural system to adsorb shocks from disturbances and to turn in equilibrium state [1]. Later, it was adopted in urban planning because of some similarities with the antrophic system [2, 3]. At present, resilience is a largely debated concept because of its overlapping with sustainability notion [4–6]. Several authors consider the resilience enhancement as a means to improve the system sustainability [7–10], others invert this relation and describe the sustainability as a factor contributing to resilience [11–17]. Furthermore,

© Springer International Publishing AG, part of Springer Nature 2019
F. Calabrò et al. (Eds.): ISHT 2018, SIST 100, pp. 534–546, 2019.
https://doi.org/10.1007/978-3-319-92099-3_60

resilience and sustainability are considered to have different aims that can be complementary or competitive [18–22].

The resilience concept expresses the capacity of a system to withstand, respond to, adapt and prepare more readily to shocks and stresses to emerge stronger after tough times, and live better in good times [23]. It should not be referred to a static equilibrium, because urban systems can change and become different from their original conditions, in response to strains [24, 25].

The Agenda 2030 for Sustainable Development [26] advocates the development of safe, inclusive, resilient and sustainable cities (Goal 11), by protecting and safeguarding the world's natural and cultural heritage (Target 11.4), reducing the number of deaths and damages caused by disasters (Target 11.5), and increasing the number of cities adopting and implementing integrated policies and plans towards resilience to disaster and holistic disaster risk management (Target 11.b).

The needs pointed out in the Agenda 2030 [26] can be considered the result of a long-lasting process characterized by the enunciation of many important concepts. The International Decade for Natural Disaster Reduction [27] recommended to integrate disaster-mitigation programs, land use and insurance policies for disaster prevention and to establish education and training programs to enhance community preparedness. The Yokohama Strategy and Plan of Action for a Safer World [28] and the International Strategy for Disaster Reduction [29] highlighted that managing disaster in the emergency condition alone is not sufficient, because it yields temporary results at a very high cost, thus underlining that all the phases of risk management should be considered integral parts of policies and urban planning at regional, national and international scales.

All these recommendations were received by Hyogo Framework for Action (2005-2015) which stressed the importance of innovative and pro-active approaches to involve people in all stages of disaster risk reduction [30]. It also pointed out the need of building a culture of safety and resilience at all levels, improving international collaboration on resilience issues and allocating proper resources to reduce the impacts of natural hazards. Moreover, only recently the Sendai Framework for Disaster Risk Reduction (2015–2030) has emphasized the different concepts of disaster management and disaster risk management [31]. Disaster management is applied when events have already occurred while disaster risk management assumes the implementation of actions that allow the city to cope with dangerous events and reduce their effects. The latter was seen as part of a multidimensional and systemic framework including also the preservation of natural and cultural heritage in the face of natural hazards since its historic, aesthetic, social, scientific and spiritual values for past, present, and future generations.

Furthermore, urban resilience studies show different priorities across global, regional, city, community and facilities levels [32]. On the global scale, collective measures for the management and protection of ecological systems are detected and the importance of their ability to cope with crises is emphasized [33–35]. On the regional scale, scientific works pay attention on the stability and diversification of urban economic structures to cope with unknown risks and pressures [36–38]. On the city scale, studies focus on policy management and propose strengthening the institutional arrangement to guarantee the adoption of elastic measures [39, 40]. On the community scale, studies highlight the importance of providing basic material conditions for residents, such as sufficient water, healthcare and dwellings, stressing the importance of the insurance benefits of employment [41–43]. On the facilities scale, studies suggest to guarantee traffic and communication infrastructure to ensure their immediate availability in emergencies [44, 45].

To improve the cities resilience, we propose a framework named *"Resilience and Disaster Risk Management"* (hereafter, *RDRM*) that overcomes the sectorial approaches of territorial management and promotes the integration of resilience and disaster risk management into territorial plans (e.g. Regional Territorial Planning, Provincial Territorial Planning, Municipal territorial planning).

The *RDRM* includes 4 *drivers* (economic, social, environmental and institutional), 15 *driver descriptors* and 36 *sub-drivers*. The latter was grouped according to the resilience goals (prepare for, respond to, withstand, adapt to) and for each one the role for one or more risk management phases was identified (mitigation, preparedness, response, recovery). In the following paragraphs we described: (a) the concepts of drivers and sub-drivers for an urban system and a summary of resilience sub-drivers proposed by scientific work and international framework and (b) the drivers and sub-drivers useful to improve the resilience and reduce natural risks. Moreover, the relations between resilience goals (prepare, respond, withstand, adapt) and risk management phases (mitigation, preparedness, response, recovery) were established.

2 City Resilience and Natural Risks

Disasters caused by natural events and environmental stresses such as climate change have a critical impact on the cities, affecting both natural and built environments, and thus economy and society. To reduce the consequences of these events, the resilience assessment become essential. Several works analyzed the resilience of city system without considering external disturbances; some of them considered a single resilience category (social, economic, environmental, institutional, infrastructural) and tested their methodology through case study [46–49], others used a systemic approach highlighting the difficulty in controlling and comparing a large numbers of variables [50–52]. Few studies considered concurrently environmental, social, economic, and institutional categories and the occurrence of dangerous events [53–55] (Table 1). Moreover, a large part of studies focused on the reduction or mitigation of natural risks [56, 57], others concerned the post-disaster emergency planning caused by sudden, large and catastrophic events [3] while few studies defined policies and actions dedicated to the prevention [58].

In the present work, the city was considered an urban ecosystem, that include both nature and humans in a largely built environment [59], controlled by several drivers and sub-drivers. The drivers can be defined as a superior complex phenomena governing the direction of the ecosystem change, which could have both of human and nature origin. The anthropogenic drivers are based on economic, social and political fundamental needs like food, health, clean water, employment, energy. Natural drivers on the other hand are majorly independent from anthropogenic causes and could be referred to as "force majeure", like for example earthquakes, volcanic eruptions or tectonic drift [60]. The sub-drivers are tools to "measure" the resilience level of cities (e.g. for environmental driver: Adequate Green Infrastructures are realized to cope with natural hazard, Buildings are robust and safe to cope with natural hazards).

Table 1. Summary of driver suitable for the urban resilience assessment. Scale of work, and case study are reported. The symbol "x" indicates the lack of information. (from Iavarone et al. 2017; modified)

Scale	Resilience driver	Case study	References
County	Social	United States Counties	[46]
Magisterial District	Economic	Magisterial districts of South Africa	[47]
Territorial	Systemic	x	[50]
County, metropolitan statistical area	Systemic	Southeast region counties (United States); Gulfport-Biloxi, Charleston and Memphis metropolitan statistical areas	[53]
Regional, County	Systemic	x	[51]
County	Community	82 counties of the state of Mississippi	[48]
City	Social economic	Caldas da Rainha-eniche (Portugal); Évora – Elvas (Portugal)	[49]
City	Systemic	x	[52]
District	Systemic	Quetta, Belucistan (Pakistan)	[54]
City	Systemic	Delhi (India)	[55]

According to [61] and several frameworks widely tested worldwide [62], 4 drivers and 26 sub-drivers of resilience for city exposed to dangerous events were identified as potential tools for the resilience assessment and the disaster risk management. All sub-drivers are part of a "RDRM" framework extensively described in Iavarone et al. (this volume). The core is a tree structure that relates the sub-drivers of resilience with disaster risk management phases. In detail, the sub-drivers were firstly classified according to the necessities of resilience goals (prepare, respond, withstand, adapt) and then their capacity to reduce the effects of hazardous events, in both long or short term, was used to establish the link with mitigation, preparedness, response, recovery phases of disaster risk management.

3 Resilience and Disaster Risk Management

An urban environment in a good resilience status is characterized by: (a) innovative and diversified economic activities (economic driver); (b) inclusive society, active citizen networks and accessibility to opportunities, infrastructures and services (social driver); (c) sustainability of urban development, natural resources and the availability of adequate infrastructures (environmental driver) and (d) clear leadership and long-term vision, proper resources at local scale, governments cooperation and openness to participation (institutional driver) (Table 2).

Table 2. Economic, social, environmental and institutional drivers and driver descriptors are listed.

Driver	Note	Driver Descriptor
Economic	All productive activities, trade and services in a specific territory are considered. This driver considers economic sectors (primary, secondary and tertiary) and their diversification, the level of innovation and creativity, vitality of entrepreneurial environment, skills and education of workforce. All of them influence the overall capacity of response of a community to adverse events	E1 - Employment and workforce E2 - Entrepreneurship E3 - Local productivity
Social	This driver includes socio-economic characteristics of population such as age, gender, employment, education, income, health and wellbeing. It also considers the access to communication means, transport means and health services; also, social cohesion and cultural aspects such as the relationship of community with its environment. All these characteristics influence the inclination of a community to recover from adverse events	S1 - Socio-economic S2 - Services (communication, transport, health) S3 - Socio-cultural
Environmental	This driver considers the characteristics of the natural and built environment, land uses and infrastructures. It also includes natural and cultural heritage as element of particular value, but also as factor of increased vulnerability if adequate mitigation measures are not implemented	En1 - Natural environment En2 - Built environment En3 - Infrastructures (streets, energy, ICT) En4 - Land use
Institutional	This driver considers the capacity of institutions to manage the territory, also in case of natural dangerous events, through urban/territorial planning tools, risk management tools and emergency management tools. It includes financial resources available at local level for risk management, and the capacity of institutions to be open and inclusive, promoting active participation of the community to emergency planning and risk management decision processes	I1 - Leadership and long-term vision I2 - Territorial management I3 - Institutional collaboration I4 - Financial resources I5 - Citizens' engagement

Specifically, we recognize 7 economic sub-drivers, 8 social sub-drivers, 12 environmental sub-drivers and 9 institutional sub-drivers (Table 3). All these sub-drivers were also considered vital tools to limit the consequences of hazardous events and thus they were allocated, according to the needs and the programming time (long or short term), in one or more disaster risk management phases (mitigation, preparedness, response and recovery).

Table 3. Structure of drivers and sub-drivers of urban resilience related to risk management. For driver descriptor see the Table 1.

Driver	Driver descriptor	Sub-driver	References	Goals of resilience	Risk management phase
Economic	E1	The population in working age is actively employed	[47–49, 51, 53]	Adapt	Recovery
		Workforce has diverse skills	[62]	Adapt	Recovery
		Workforce is employed in sectors useful to cope with natural hazards	[46, 49, 51]	Respond, Withstand, Adapt	Response, Recovery
		Industries are diverse	[62]	Adapt	Recovery
	E2	Innovation takes place	[62]	Prepare for, Adapt	Mitigation, Preparedness, Recovery
		The entrepreneurial ecosystem is vital	[64]	Adapt	Recovery
	E3	Adequate stock of primary resources is ensured	[47]	Respond, Withstand, Adapt	Response Recovery
Social	S1	Society is inclusive	[62]	Adapt	Recovery
		Resources are equally distributed	[46–53]	Respond, Withstand, Adapt	Response, Recovery
	S2	People have access to communication devices	[51–53]	Prepare for, Adapt	Preparedness, Response
		People have access to transport means	[62]	Respond, Withstand	Response
		People have access to health services	[50, 51, 53]	Respond, Withstand	Response
	S3	People recognize and feel proud of their city's identity	[65]	Prepare for, Adapt	Mitigation, Recovery
		The Heritage Community takes care and valorises cultural heritage/landscape	[65]	Prepare for, Adapt	Mitigation, Recovery
		Citizen's organizations are active in the community	[62]	Prepare for, Respond, Withstand, Adapt	Mitigation, Preparedness, Response, Recovery

(*continued*)

Table 3. (*continued*)

Driver	Driver descriptor	Sub-driver	References	Goals of resilience	Risk management phase
Environmental	En1	Adequate Green Infrastructures are realized to cope with natural hazard	[66]	Respond, Withstand	Response
		Natural heritage is preserved	[67]	Adapt	Recovery
	En2	Buildings are robust and safe to cope with natural hazards	[50, 53]	Prepare for, Respond, Withstand	Mitigation, Response
		Inhabited areas are not overpopulated	[50, 51, 53]	Prepare for, Respond, Withstand, Adapt	Mitigation, Response, Recovery
		Buildings are covered by insurance	[50, 51, 53]	Adapt	Recovery
		Adequate equipped and safe spaces are available for emergency	[53]	Prepare for	Preparedness
		Cultural heritage/landscape is well-conserved	[67]	Prepare for, Adapt	Mitigation, Recovery
		Cultural heritage/landscape is safeguarded from natural hazards	[52]	Prepare for, Adapt	Mitigation, Recovery
	En3	The infrastructures are well distributed over the territory	[46, 51, 53]	Prepare for, Respond, Withstand, Adapt	Mitigation, Response, Recovery
		The infrastructures are well maintained	[62]	Prepare for, Respond, Withstand, Adapt	Mitigation, Response, Recovery
	En4	Degraded areas are absent or in phase of regeneration	[47, 50]	Prepare for, Adapt	Mitigation, Recovery
		The city/territory carries out urbanization rates	*	Prepare for, Adapt	Mitigation, Recovery

(*continued*)

Table 3. (*continued*)

Driver	Driver descriptor	Sub-driver	References	Goals of resilience	Risk management phase
Institutional	I1	Leadership and long-term vision are clear/include learning from past natural events	[47]	Prepare for, Adapt	Mitigation, Preparedness, Recovery
	I2	Urban planning is regulated	[50, 51]	Prepare for, Adapt	Mitigation, Recovery
		Risk management plans are available and periodically updated	[51, 52]	Prepare for, Adapt	Mitigation, Preparedness, Recovery
	I3	Local governments cooperate with regional and national governments	[62]	Prepare for, Respond, Withstand, Adapt	Mitigation, Response Recovery
		Local governments, institutions and civil society organizations cooperate	[68]	Prepare for, Respond, Withstand, Adapt	Mitigation, Preparedness, Response, Recovery
	I4	Adequate financial resources are available at municipality level	[68]	Prepare for, Adapt	Mitigation, Recovery
	I5	People are informed about the natural hazards that may affect the city	[68]	Prepare for	Preparedness
		People are able to apply emergency plans directives	[68]	Prepare for, Respond Withstand	Preparedness, Response
		Government is open and citizens' participation takes place	[62]	Prepare for, Adapt	Mitigation, Preparedness, Recovery

The star symbol (*) shows the sub-driver proposed in the present work.

Mitigation focuses on the effects of hazardous events, it aims to reduce the interaction of human, property and environment with dangerous events applying structural and non-structural actions. The latter are programmed for a long-term and in ordinary condition [63]. An example of a sub-driver useful in this phase is: "*buildings are robust and safe to cope with natural hazard*", it pertains to "*prepare*", "*respond*" and "*withstand*" resilience goals (Table 3). The knowledge of how many buildings are far from this condition is fundamental to improve the resilience through the reduction of weak buildings exposed to adverse events.

Preparedness addresses the process for developing and maintaining capabilities for the whole community in both pre and post event through the education, outreach and training of the population. An example of a sub-driver useful in this phase is labelled *"People have access to communication device"*, it is related to *"prepare"* and *"adapt"* resilience goals (Table 3).

Response addresses the actions taken in the immediate aftermath of a dangerous event to save and sustain lives, meet basic human needs (food, shelter, clothing, public health and safety), reduce the loss of property and limit the effect on critical infrastructure and on environment. After the solution of the immediate emergency issues, the focus shifts to planning actions aiming at restoring property, utilities, stabilization of public services and conclusion of clean-up process. An example of a sub-driver useful in this phase is *"the infrastructures are well distributed over the territory"*, it pertains to *"prepare"*, *"adapt"*, *"respond"* and *"withstand"* resilience goals (Table 2).

Recovery encompasses both short-term and long-term efforts for the rebuilding and revitalization of affected communities, aiming at the return to a degree of physical, environmental, economic and social stability. Example of sub-drivers are: *"Adequate stock of primary resources is ensured"* in a short-term, it pertains the *"respond"*, *"withstand"* and *"adapt"* resilience goals while *"leadership and long term vision are clear/include learning from past natural events"* in a long-term, it concerns the *"prepare"* and *"adapt"* resilience goals (Table 2).

4   Conclusive Remarks

In the present work, the city environment is considered an ecosystem and the concept of resilience ruled out for natural environment was applied to urban environment.

We proposed the use of drivers and sub-drivers for assessing the resilience of cities exposed to natural hazard and ménage the territory during the four phases of risk management. In detail, economic, social, environmental an institutional drivers and 36 sub-drivers were described. The single sub-driver indicates the optimal conditions that cities should have to cope with natural disaster risk and its measure allows to identify the gap between this status and the real condition. Moreover, this framework overcomes the sectorial approaches of territorial management and considers the resilience and the risk management as parts of ordinary territorial planning tools. Future research will focus on the identification of quantitative and qualitative indices useful to rank the territory in different resilience classes. The framework is flexible and thus provides the possibility to draw several maps, one for each sub-driver, and to visualize the spatial distribution of zones with different resilience values. Moreover, it allows the summary of maps offering the possibility to identify in a simple way the areas needing interventions to improve the resilience.

References

1. Holling, C.S.: Resilience and stability of ecological systems. Ann. Rev. Ecol. Syst. **4**, 1–23 (1973)
2. Folke, C.: Resilience: the emergence of a perspective for social–ecological systems analyses. Global Environ. Change **16**(3), 253–267 (2006), https://doi.org/10.1016/j.gloenvcha.2006.04.002. Accessed 21 May 2017
3. Davoudi, S., Shaw, K., Haider, L.J., Quinlan, A.E., Peterson, G.D., Wilkinson, C., Fünfgeld, H., McEvoy, D., Porter, L., Davoudi, S.: Resilience: a bridging concept or a dead end? "Reframing" resilience: Challenges for planning theory and practice, interacting traps: resilience assessment of a pasture management system in northern Afghanistan, urban resilience: What does it mean in planning practice? Resilience as a useful concept for climate change adaptation? The politics of resilience for planning: A cautionary note. Plan. Theo. Pract. **13**(2), 299–333 (2012)
4. Redman, C.L.: Should sustainability and resilience be combined or remain distinct pursuits? Ecol. Soc. **19**(2), 37 (2014)
5. Pizzo, B.: Problematizing resilience: Implications for planning theory and practice. Cities **43**, 133–140 (2015)
6. Marchese, D., Reynolds, E., Bates, M.E., Morgan, H., Clark, S.S., Linkov, I.: Resilience and sustainability: similarities and differences in environmental management applications. Sci. Total Environ. **613–614**, 1275–1283 (2018)
7. Milman, A., Short, A.: Incorporating resilience into sustainability indicators: an example for the urban water sector. Glob. Environ. Change **18**, 758–767 (2008)
8. Walker, B., Pearson, L., Harris, M., Maler, K.-G., Li, C.-Z., Biggs, R., Baynes, T.: Incorporating resilience in the assessment of inclusive wealth: an example from South East Australia. Environ. Resour. Econ. **45**, 183–202 (2010)
9. Saunders, W.S.A., Becker, J.S.: A discussion of resilience and sustainability: land use planning recovery from the Canterbury earthquake sequence. New Zealand. Int. J. Disaster Risk Reduct. 14, 73–81 (2015)
10. Jarzebski, M.P., Tumilba, V., Yamamoto, H.: Application of a tri-capital community resilience framework for assessing the social–ecological system sustainability of community based forest management in the Philippines. Sustain. Sci. **11**(2), 307–320 (2016)
11. McEvoy, D., Lindley, S., Handley, J.: Adaptation and mitigation in urban areas: synergies and conflicts. Proc. Inst. Civ. Eng. Munic. Eng. **159**, 185–191 (2006)
12. Chapin, F.S., Kofinas, G.P., Folke, C.: Principles of Ecosystem Stewardship: Resilience-Based Natural Resource Management in a Changing World, pp. 1–409 (2009)
13. Avery, G.C., Bergsteiner, H.: Sustainable leadership practices for enhancing business resilience and performance. Strategy Leadersh. **39**(3), 5–15 (2011)
14. Closs, D.J., Speier, C., Meacham, N.: Sustainability to support end-to-end value chains: the role of supply chain management. J. Acad. Mark. Sci. **39**, 101–116 (2011)
15. Ahi, P., Searcy, C.: A comparative literature analysis of definitions for green and sustainable supply chain management. J. Clean. Prod. **52**, 329–341 (2013)
16. Bansal, P., DesJardine, M.R.: Business sustainability: it is about time. Strategic Organ. **12**(1), 70–78 (2014)
17. Saxena, A., Guneralp, B., Bailis, R., Yohe, G., Oliver, C.: Evaluating the resilience of forest dependent communities in Central India by combining the sustainable livelihoods framework and the cross scale resilience analysis. Curr. Sci. **110**, 1195 (2016)
18. Derissen, S., Quaas, M.F., Baumgaertner, S.: The relationship between resilience and sustainability of ecological-economic systems. Ecol. Econ. **70**(6), 1121–1128 (2011)

19. Fiksel, J., Goodman, I., Hecht, A.: Resilience: navigating toward a sustainable future. solutions, pp. 1–13 (2014)
20. Lizarralde, G., Chmutina, K., Bosher, L., Dainty, A.: Sustainability and resilience in the built environment: the challenges of establishing a turquoise agenda in the UK. Sustain. Cities Soc. **15**, 96–104 (2015)
21. Lew, A.A., Ng, P.T., Ni, C. (Nickel)., Wu, T. (Emily).: Community sustainability and resilience: similarities, differences and indicators. Tour. Geogr. **18**, 18–27 (2016)
22. Meacham, B.J.: Sustainability and resiliency objectives in performance building regulations. Build. Res. Inf. **44**, 474–489 (2016)
23. The Rockefeller Foundation, ARUP: City Resilience Framework, New York (2015)
24. Ahern, J.: From fail-safe to safe-to-fail: sustainability and resilience in the new urban world. Landscape Urban Plan. **100**(4), 341–343 (2011)
25. Carpenter, S., Walker, B., Anderies, J.M., Abel, N.: From metaphor to measurement: resilience of what to what? Ecosystems **4**(8), 765–781 (2001)
26. United Nations: Transforming our World: the 2030 Agenda for Sustainable Development. (2015)
27. United Nations: The International Decade for Natural Disaster Reduction – IDNDR (1989). http://www.un.org/documents/ga/res/44/a44r236.htm. Accessed 15 Mar 2017
28. United Nations: Yokohama Strategy and Plan of Action for a Safer World. World Conference on Disaster Reduction, Yokohama, Japan (1994)
29. United Nations: International Strategy for Disaster Reduction (1999). http://www.unisdr.org/files/resolutions/N0027175.pdf. Accessed 15 Mar 2017
30. UNISDR: Hyogo Framework for Action 2005–2015: Building the Resilience of Nations and Communities to Disasters In Extract from the final report of the World Conference on Disaster Reduction. World Conference on Disaster Reduction. Kobe, Hyogo, Japan (2005). https://doi.org/10.1017/cbo9781107415324.004. Accessed 15 Mar 2017
31. UNISDR: Sendai Framework for Disaster Risk Reduction 2015–2030. Sendai, Miyagi, Japan (2015)
32. Zhanga, X., Li, H.: Urban resilience and urban sustainability: what we know and what do not know? Cities **72**, 141–148 (2018)
33. Kareiva, P., Watts, S., McDonald, R., Boucher, T.: Domesticated nature: Shaping landscapes and ecosystems for human welfare. Science **316**(5833), 1866–1869 (2007)
34. Leichenko, R.: Climate change and urban resilience. Curr. Opin. Environ. Sustain. **3**(3), 164–168 (2011)
35. Woolhouse, M.E.J., Rambaut, A., Kellam, P.: Lessons from Ebola: improving infectious disease surveillance to inform outbreak management. Sci. Transl. Med. **7**(307), 307rv5 (2015)
36. Stone, R.: Lessons of disasters past could guide Sichuan's revival. Science **321**(5888), 476 (2008)
37. Barata-Salgueiro, T., Erkip, F.: Retail planning and urban resilience - an introduction to the special issue. Cities **36**, 107–111 (2014)
38. Toubin, M., Laganier, R., Diab, Y., Serre, D.: Improving the conditions for urban resilience through collaborative learning of Parisian urban services. J. Urban Plan. Devel. **141**(4) (2015)
39. Barthel, S., Parker, J., Ernstson, H.: Food and green space in cities: A resilience lens on gardens and urban environmental movements. Urban Stud. **52**(7), 1321–1338 (2015)
40. Beilin, R., Wilkinson, C.: Introduction: Governing for urban resilience. Urban Stud. **52**(7), 1205–1217 (2015)

41. Braun-Lewensohn, O., Sagy, S.: Community resilience and sense of coherence as protective factors in explaining stress reactions: comparing cities and rural communities during missiles attacks. Community Ment. Health J. **50**(2), 229–234 (2014)
42. Vallance, S.: An evaluation of the Waimakariri district council's integrated and community-based recovery framework following the Canterbury earthquakes: implications for urban resilience. Urban Pol. Res. **33**(4), 433–451 (2015)
43. Mehmood, A.: Of resilient places: planning for urban resilience. Eur. Plan. Stud. **24**(2), 407–419 (2016)
44. Chang, S.E., McDaniels, T., Fox, J., Dhariwal, R., Longstaff, H.: Toward disasterresilient cities: Characterizing resilience of infrastructure systems with expert judgments. Risk Anal. **34**(3), 416–434 (2014)
45. Testa, A.C., Furtado, M.N., Alipour, A.: Resilience of coastal transportation networks faced with extreme climatic events. Transp. Res. Rec. **2532**, 29–36 (2015)
46. Cutter, S., Boruff, B., Shirley, W.: Social vulnerability to environmental hazards. Soc. Sci. Q. **84**(2), 242–261 (2003)
47. Naudé, W.A., Mcgillivra, Y.M., Rossouw, S.: Measuring the Vulnerability of Sub-National Regions. Oxf. Dev. Stud. **37**(3), 249–276 (2009)
48. Sherrieb, K., Norris, F., Galea, S.: Measuring capacities for community resilience. Soc. Indic. Res. **99**, 227–247 (2010)
49. Gonçalves, C., Marques da Costa, E.: Framework and Indicators to Measure Urban Resilience. AESOP-ACSP Joint Congress, Dublin (2013)
50. Normandin, J.M., Therrien, M. C., Tanguay, G.A.: City strength in times of turbulence: strategic resilience indicators. In: Proceedings of the Joint Conference on City Futures, Madrid, 4–6 June (2009)
51. Peacock, W.G., Brody, S.D., Seitz, W.A., Merrell, W.J., Vedlitz, A., Zahran, S., Harriss R. C., Stickney R.R.: Advancing the resilience of coastal localities: developing, implementing and sustaining the use of coastal resilience indicators: a final report. Hazard Reduction and Recovery Center, Texas A&M University, College Station (2010)
52. Cutter, S.L.: The landscape of disaster resilience indicators in the USA. Nat. Hazards **80**, 741–758 (2016)
53. Cutter, S.L., Burton, C.G., Emrich, C.T.: Disaster Resilience Indicators for Benchmarking Baseline Conditions. J. Homel. Secur. Emerg. Manage. **7**(1) (2010)
54. Ainuddin, S., Routray, J.K.: Earthquake hazards and community resilience in Baluchistan. Nat. Hazards **63**, 909–937 (2012)
55. Prashar, S., Shaw, R., Takeuchi, Y.: Assessing the resilience of Delhi to climate-related disasters: a comprehensive approach. Nat. Hazards **64**, 1609–1624 (2012)
56. UNISDR: Making cities resilient report (2012)
57. Vale, L.J., Campanella, T.J.: The Resilient City: How Modern Cities Recover from Disaster. Oxford University Press, Oxford (2005)
58. Pizzo, B., Fabietti, V.: Environmental risk prevention, post-seismic interventions and the reconstruction of the public space as a planning challenge. An introduction. IJPP-Ital. J. Plan. Pract. **3**(1) (2013)
59. McPhearson, P.T., Feller, M., Felson, A., Karty, R., Lu, J.W.T., Palmer, M.I., Wenskud, T.: Assessing the effects of the urban forest restoration effort of MillionTreesNYC on the structure and functioning of New York citi ecosystems. Cities Environmane **3**(1), 1–21 (2010)
60. Oesterwind, D., Rau, A., Zaiko, A.: Drivers and pressure- untangling the terms commonly used in marine science and policy. J. Environ. Manage. **181**, 8–15 (2016)

61. Iavarone, R., Gravagnuolo, A., Esposito De Vita, g., Alberico, I.: Resilient city, an approach to copo with natural hazard. In: XV Forum Internazionale, Le vie dei Mercanti, World Heritage and Disasters, Napoli and Capri, Italy (2017)
62. OECD: Resilience cities (2016)
63. FEMA: Developing and Maintaining Emergency Operations Plans. Comprehensive Preparedness Guide 101, Version 2.0. (2010)
64. Ewing Marion Kauffman Foundation. Measuring an Entrepreneurial Ecosystem (2015)
65. European Union. Cultural Heritage Counts for Europe project - CHCfE. Full report (2015)
66. EEA. https://www.eea.europa.eu/themes/sustainability-transitions/urban-environment/urban-green-infrastructure/indicators_for_urban-green-infrastructure. Accessed 18 Feb 2018
67. ISTAT. http://www.istat.it/en/well-being-and-sustainability/well-being-measures. Accessed 25 Nov 2017
68. United Nations: Disaster Preparedness for Effective Response. Guidance and Indicator Package for Implementing Priority Five of the Hyogo Framework (2008)

Re-signification Processes of the Productive Heritage for a Renewed Urban Quality

Chiara Corazziere(✉) ⓘ

Mediterranea University of Reggio Calabria, 89100 Reggio Calabria, Italy
chiaracorazziere@gmail.com

Abstract. The following essay is a summary of what has been produced in the activities of the research grant *Productive heritage: research of records left on the territory* in the broader research project *The importance of the company in the development of society: how to read and enhance the cultural heritage inherited from productive activities*. The aim of the research is to develop a strategy that has as its objectives the valorization, promotion and communication of the cultural heritage inherited from the disused productive activities which, today, require be *re*-meant to take on a renewed role in the contemporary fabric.

The analysis carried out on a sample of nine case studies allows to investigate regenerative processes whose effectiveness can already be assessed in terms of the ability to assimilate the contradictions characterizing the physical condition of the productive heritage, justified on past logics, compared with the current urban, social, economic context. This capacity, deduced from the comparison between what emerged from the interpretative reading of the sites and the synthesis of the design logic already adopted, is articulated according to some design criteria useful for the subsequent definition of guidelines; these, declined according to the five light actions, *reading, mapping, enhancing, re-generating, innovating*, build a methodology useful to trigger those *re*-signification processes capable of assigning a renewed urban quality to large areas of the urban fabric, of generating common spaces and therefore sure of everyday life, to guarantee the presence of places dedicated to new working, culture, and welfare communities.

Keywords: Re-signification processes · Productive heritage · Urban quality

1 Introduction

The following essay is a re-elaboration of what has been produced by the author so far in the activities of the research grant entitled *Productive heritage: research of records left on the territory* in the broader research project *The importance of the company in the development of society: how to read and enhance the cultural heritage inherited from productive activities*. The research carried out is aimed at developing a strategy that, starting from the cultural identity and interpretative reading of the reference territorial context, has as its specific objectives the valorization, promotion and communication of physical realities that, thanks to the identity memory that characterizes

© Springer International Publishing AG, part of Springer Nature 2019
F. Calabrò et al. (Eds.): ISHT 2018, SIST 100, pp. 547–554, 2019.
https://doi.org/10.1007/978-3-319-92099-3_61

them and the opportunity to take on a new role in the contemporary context, can certainly be considered cultural heritage.

The bibliographic study on the broader theme of the industrialization of the national territory in the eighteenth and nineteenth centuries, had as its first outcome the identification of a repertoire of productive activities among those most conditioning, at the time of their creation, the urban, environmental, social, economic context of implantation and that today demand to be re-meant to assume a renewed role in the contemporary fabric.

We can be said, therefore, that where processes of re-signification of the productive heritage have been initiated, conceived within broader dynamics of innovation and promotion of the next context, experiments can be observed whose effectiveness can already be assessed in terms of urban regeneration, environmental and of the landscape, of occupational repercussions, of innovation and social inclusion, of an educational and safe city [1].

On the contrary, where the road to mere recovery, the freezing of values has been followed, museological projects have been created, which today appear to be monuments of themselves decontextualized because they do not conceive the historical narration on the logic of contemporary communication and on the request of a more and more diversified public or, on the contrary, they propose destinations of use completely divorced from the context and from the original matrix of the spaces and, therefore, are not characterized by any originality.

All the investigated realities, therefore, are united by having lost their original function, but only some see in attributing to the achievement of a future goal of sustainable development on a creative basis the possibility of becoming a cultural heritage to be handed down to future generations.

Among the latter and based on re-signification processes already being applied, according to more or less advanced stages, a survey sample was defined consisting of the following case studies: Progetto Manifattura (Rovereto_TN), OGR (Torino), Piazza dei Mestieri (Torino), Area PoP Up (Ravenna), CAOS (Terni), NEXT SNIA Viscosa (Rieti), Centrale Montemartini (Roma), ExFadda (San Vito dei Normanni_BR), Reali Ferriere ed Officine (Mongiana_VV).

The sample is significant because it is possible to draw design ideas that are not only repeatable in similar contexts; re-elaborated according to a structured methodology, in fact, they can trigger those re-signification processes able to assign a renewed urban quality to large areas of the urban fabric, to generate common and therefore safe spaces for everyday life, to guarantee the presence of places dedicated to new working, culture and welfare communities.

2   Assimilate the Contradictions to *Re*-signify Permanence

The research already carried out shows that the possibility of the heritage investigated to become an inheritance depends on the ability to assimilate the contradictions characterizing the physical condition, justified on past logics, compared with the urban, social, economic and contemporary context [2]. This capacity is declined according to some *changes of state* defined by *design criteria* and deduced from the comparison between what emerged from the interpretative reading of the sites and the synthesis of the design choices already underway.

2.1 Open Economic System *vs* Closed Economic System

The "closed circuit" economic systems that characterized the productive activities of the past - for ex. culture/mineral extraction-processing raw material-product packaging - are replaced today by open systems in which the actors are both public and private and in which they generate new production methods based on the gathering of the community around common interest centers, on the sharing of knowledge, on cooperative learning, on open processes of collaboration and co-generation of the project already in its conceptual and/or realization phases [3].

Autonomous with respect to strategic spatial locations and the need to procure raw materials and natural sources, these systems are part of wider processes of urban regeneration and environmental recovery and are based on potential networks - from a cultural, tourism, production and social point of view - of the territory.

2.2 System of Values *vs* System of Measures

The investigated production plants, whatever the type, worked as devices designed to package a product, according to a certain quantity - measure - established by the capacity of the instrumentation and the manpower used.

Nowadays, regenerated disused plants become devices in which to generate processes and ideas, before products, in which the human component is no longer a measurable element [4], but an indispensable condition for launching creative and innovative production formulas such as start-ups, thematic incubators, laboratories, smart environments, narrating museums (see Fig. 1).

2.3 Indeterminacy of the Work Space/Portable Workspace *vs* Coincidence Architecture/Function

The coincidence of architecture with a precise function - characteristic of all the investigated production activities - has created "unchangeable" spaces if we think of them solely to replicate any traditional production process. The continuous changes that have invested the economic models and the consequent modifications of the social map, the decommissioning of some production spaces and their abandonment, has meant that this coincidence is less and that the work space can be understood today as flexible, changing, shared, a meeting place for ideas, or even virtual and, therefore, "portable" [5].

2.4 Re-use of the Existing One *vs* Creation of New Buildings

The recovery of the disused production areas today assumes a new centrality because it allows to face contemporary problems common to all the cities, such as, among many, the reuse of abandoned or underutilized assets and the contrast to land consumption [6].

Fig. 1. Procession *of Reparationists* by William Kentridge, work created specifically for the central court of the OGR, Turin.   To realize his work, Kentridge was inspired by the former industrial and working vocation of the place. The sculptural collection, in black steel, is composed of a procession of 15 figures that alludes to the work of repairing trains and bodies (ph V. Gioffrè).

The temporary use and the realization of easily reversible interventions is the preferred formula to start, in a relatively short time, the use of an area already regenerated as we proceed with the realization of the overall project which usually involves medium to longer times [7].

Among the analyzed case studies, the legal formulas adopted are different, starting from lease rentals, the free loan for use, the self-determined rate, the barter of skills.

2.5   Instances of the Actual Demand *vs* Investors

The new users of the disused production assets are no longer developers and private investors, but subjects involved in various activities. Cultural, creative initiatives, profit and non-profit companies, social gathering places, museum installations and educational paths, social inclusion activities and corporate social responsibility operations are the key to developing alternative processes to enhance the productive heritage destined, otherwise, underuse or abandonment.

The usual strategies promoted exclusively by the logic of the market, therefore, are reversed and restarted by the demands of the actual demand; the production of the *object* is renounced in favor of the construction of a regenerative process.

2.6   Co-working and Co-living Modalities *vs* Rigid Social Hierarchy

The productive process characterizing traditional enterprises has always been founded on a rigid hierarchy that saw the involvement of a few thinkers in the face of a large working class, according to a model in which there was a rigid coincidence between the social system and the economic system.

Most of the regenerative processes investigated see, today, the involvement of a community of actors engaged in different activities, equally collaborating in co-generating ideas and projects, up to the implementation phases, selected on the basis of the contribution offered and not on the logic of social or economic importance [8].

The sharing of the workspace is an expression of this logic in which all the skills are put in place for the pursuit of a common goal and are goods for exchange to take advantage of the possibilities to reside, even if for limited periods, pertaining to temporary activities and/or linked to laboratory and artistic formulas, within the regenerated spaces.

3   Reading, Mapping, Enhancing, *Re*-generating, Innovating

A further scan of some experiments already underway in the investigated sites has made it possible to derive five actions - *reading, mapping, enhancing,* re-*generating, innovating* - on which to base a hypothesis of design principles/guidelines also in order to verify the possibility of synthesizing a flexible approach, which can be declined from the *general* to the *local*, conceived according to general principles, but capable, at the same time, of being adaptable and coherent with the cultural identity of the specific case.

3.1   Reading_Research the Testimony on the Territory

It is the objective of this direction to facilitate the search for some elements that characterize the heritage inherited from production activities; this in order to clearly outline the characters of a physical reality that does not specifically refer to the category of industrial archeology, but that thanks to the importance of identity memory that characterizes it and the potential to take on a new role in the contemporary context, can be considered certainly cultural heritage.

Lost the original function, the achievement of a future goal of sustainable development on a creative basis necessarily passes through the search for some characters, both material and immaterial that are, together, surviving testimony on the territory and memory of a more or less recent past [9], but also elements on which to map future interpretative maps and all subsequent steps of the regenerative process.

Therefore, in this first action, what is proposed is an objective reading, aimed at composing almost a taxonomy of the productive activities, through their determination, that is the recognition and identification, and a classification, that is the description and the collocation in a complex system.

3.2   Mapping_Interpret the Permanence to Identify New Vocations

It is the objective of this address to invite a systemic view of the context in which the production reality has operated; for this purpose the mapping action does not refer to a simple listing of elements but rather to an empirical and perceptive observation of complex systems structuring the context starting from what remains, from the permanence, material and immaterial, referable to the activity production ceased.

At the same time, mapping is an interpretation of the constituent elements of the contemporary context - not only, therefore, of infrastructures and connections - as a basin of potential users and stakeholders, of already existing attractors, of already active networks, of social, cultural and economic dynamics already in progress.

This in order to draw a complex territorial framework from which to draw inspiration to imagine a new life for the productive heritage compatible and consistent with the vocations of the context and in which to isolate the nodes on which to implement a system design choice with the capabilities and territorial potentialities, in the logic of complementarity, of co-development rather than competition [10].

3.3   Enhancing_Integrating Needs and Vocations

It is the objective of this address to interrelate the mapping of real, tangible and intangible heritage with that relating to human potential. For this purpose, the valorization action refers to the ability to establish the most correct dynamics on which to establish a new dialogue with the territory, making on the one hand the problems closely linked to the productive heritage, and on the other recognizing in the context those potentials capable of supporting useful strategies to the wider urban context.

If understood as a moment of integration between needs and vocations of the territory, the valorization is, therefore, a necessary condition to move from an object to a relational approach, from a logic aimed at transforming the collective assets in financial value to that of investing in social capital. and transformation of the problem areas of the city into opportunities for new working, culture, welfare communities [11].

3.4   *Re*-generating_Meeting the New Questions and Lifestyles

It is the objective of this direction to give shape to the vision, to identify for the productive heritage a new vocation deduced from the opportunity to recover the abandoned or underutilized permanence and the ability to trigger a process of regeneration, re-activation and involvement of the entire territory, starting from a new synthesis of cultural, social, economic-productive resources and offering the community a set of tools for the use of the territory.

Working in the perspective of regeneration, in this case, necessarily means confronting a more complex context than the one in which the productive activity was originally established, in which it compose articulated and multidimensional intervention systems, capable of responding to multiple objectives, to new questions and lifestyles expressed by the city and the communities; all these concerns can no longer be resolved exclusively in terms of urban form and design [12].

In this sense, the recovery of disused production areas assumes a new centrality because compared to the past, when the intervention could be limited to the timely redevelopment of the site, today the area or the artefact must be understood as objects projected into the future, inserted into wider dynamics innovation and promotion able to address issues common to all cities such as, for example, the containment of land consumption and giving priority to the reuse of unused heritage and the now standardized need to involve the community.

There are two main reasons: the first, of a dimensional nature given by the potential of a reuse project of a production area to radically transform a large urban and territorial geography; the second of a vocational nature given the ability of a disused productive heritage to catalyze a large community of actors.

This in order to make a trend reversal in which the community plays the main role: no longer rely on a traditional approach in which an expert chooses the solution between different alternatives but recognizes that you can use the cognitive, economic, social resources distributed between multiple and complementary actors. What can be drawn from a widespread knowledge that manifests itself in different forms and expressions with respect to which the role of the expert becomes relative, able to initiate a design vision strongly rooted in the context but capable, at the same time to create connections with the external.

3.5   Innovating_Search the Balance Between Constraints and Opportunities

It is the objective of this direction to stimulate the innovation of the policies and services on which it depends, in virtue of new social demands and services, the fallout, in terms of urban quality, of a project of re-signification of the productive heritage and its competitive ability.

Quality, understood as a point of balance between constraints and opportunities, as well as a synonym of urban safety against a risk that in addition to environmental and seismic is in our case more markedly cultural, has become a prerequisite as well as a success factor to start regenerative processes above all to return a functional qualification to old containers within pre-existing fabrics and for which any intervention must aim to intercept real questions, propose new lifestyles, anticipate trends, create opportunities for synergy between different economies [13].

In this logic the regeneration of the productive heritage can also be a vehicle of social innovation insofar as it contributes to building a model of city that aims to reduce distances, to mix populations and to exploit resources to give new answers to problems such as the insertion in the world of work, social integration, active aging, the weakening of identity memory.

4   Conclusions

The proposed methodology does not want to suggest a *procedure* but rather should be understood as a set of *shared recommendations*, of orientation for those - public administrators, private actors, scientific community - interested in the common goal of a correct reading and an effective re-signification, according to a replicable and gradual approach to the heritage inherited from production activities.

The guidelines formulated are suggestions of practices developed on the knowledge acquired during the research activity carried out up to now but constantly updated and able to verify, *in itinere*, the effectiveness of the path undertaken and to be modified, if necessary, in progress.

The *modus operandi* formulated, therefore, has as its objective the triggering of regenerative processes starting from local potentials, therefore designed *ad hoc*, but which is able to address, at the same time, even those common and transversal themes to the different territorial contexts in which there exists a heritage erected by a productive activity for which to imagine, starting from the reading and interpretation of the permanence, innovative formulas of valorization.

References

1. Braae, E.: Beauty Redeemed. Recycling Post-industrial Landscapes. Birkhäuser, Basel (2015)
2. Latz, P.: Rust Red. Landscape Park Duisburg-Nord. Hirmer, Munich (2016)
3. Brunetta, G., Moroni, S. (eds.): La città intraprendente. Comunità contrattuali e sussidiarietà orizzontale. Carocci Editore, Roma (2011)
4. Ricci, M.: Nuovi Paradigmi. List, Trento (2012)
5. Marini, S., Bertagna, A., Gastaldi, F. (eds.): Architettura, città, società. Il progetto degli spazi del lavoro. Università Iuav di Venezia, Venezia (2012)
6. Marini, S., Corbellini, G. (eds.): Recycled Theory: Illustrated Dictionary. Quodlibet, Macerata (2016)
7. Inti, I., Cantaluppi, G., Persichino, M.: Temporiuso. Manuale per il riuso temporaneo di spazi in abbandono. Altreconomia, Milano (2014)
8. Campagnoli, G.: Riusiamo l'Italia. Da spazi vuoti a start-up culturali e sociali. Il Sole 24 Ore Editore, Milano (2014)
9. Augé, M.: Macerie e Rovine. Il senso del tempo. Bollati Boringhieri, Torino (2004)
10. Latouche, S.: Breve trattato sulla decrescita felice. Bollati Boringhieri, Torino (2008)
11. Associati, T.A.M. (ed.): Taking care. Progettare per il bene comune. BeccoGiallo, Padova (2016)
12. D'Arienzo, R., Younès, C.: Recycler l'urbain. Pour une écologie des milieux habités. MētisPresses, Genève (2014)
13. Fontanari, E., Piperata, G. (eds.): Agenda RE-CYCLE. Proposte per reinventare la città. Il Mulino, Bologna (2017)

Temporal Dynamics of Land Values and Determinants

Antonio Nesticò⓪ and Massimiliano Bencardino$^{(\boxtimes)}$ ⓪

University of Salerno, 84084 Fisciano, Italy
mbencardino@unisa.it

Abstract. The temporal evolution of land values is conditioned by the reference macroeconomic framework and, more strictly, by the socio-demographic and productivity characteristics of the territorial study area. If these dependencies are widely recognized in the literature, on the other hand, the correlation levels between the variables at stake, especially in relation to the parameters of the survey area, are not investigated in quantitative terms.

With the present paper, we intend to establish the measure of the diachronic correlation between market values of the land (vineyards, olive groves and irrigated crops) and variables able to affect the mechanisms of price formation.

Geographic Information Systems (GIS) are used for the processing of cartographic representations useful for visualizing the spatial distribution of data, which pertain to the vast area of the Province of Salerno (Italy). Correlation curves return the measures searched for the time interval from 2000 to today, allowing to highlight the effects of the structural crisis in the years after 2008.

The entire analytical path starts from the collection of information and construction of the datasets, passes through the selection of parameters and for the processing of the maps and can establish quantitatively the levels of correlation between market values and territorial variables. Thus, it provides useful elements for future research, aimed at defining an explanatory model of the temporal trend of mercantile appreciations of agricultural soils.

Keywords: Land values · Temporal correlation · Economic model
Geographic Information Systems

1 Land Values and Explanatory Variables: Diachronic Analysis and Cartographic Representation

The market values of agricultural land are spatially correlated both with the income produced by workers in the territory, and with the population density, able to express the urbanization level of the area. This is demonstrated in a study on market appreciation for the year 2015 of vineyards, olive groves and irrigated arable crops in the Province of Salerno, which includes 158 municipalities for a total area of 4,952 km^2 [1–4]. It's about

---

The contribution to this paper is the result of the joint work of both authors to which the paper has to be attributed in equal parts.

© Springer International Publishing AG, part of Springer Nature 2019
F. Calabrò et al. (Eds.): ISHT 2018, SIST 100, pp. 555–563, 2019.
https://doi.org/10.1007/978-3-319-92099-3_62

cultural destination that are widespread in the survey area and are crucial for the total value of agricultural outputs.

With this paper the research is developed in order to investigate the temporal correlation levels in the period 2000–2015 for the three variables: land value, per capita income and resident population [5–9].

Market information is given by the Italian Council for the Agricultural Research and the Analysis of Agrarian Economy (CREA). The data on income and population are respectively extracted from the data warehouse of the Ministry of Economy and of the Italian Statistical Institutes (Istat) and have the Municipality as territorial unit.

The CREA provides the minimum V_{min} and maximum V_{max} market values of the different crops for each of the 17 agricultural regions in which the Province of Salerno is subdivided. According to the Italian Statistical Institutes (Istat), the agrarian region is constituted «by groups of municipalities according to homogeneous rules of territorial continuity in relation to certain natural and agricultural characteristics and, subsequently, aggregated by altitude». Figures 1, 2 and 3 show the temporal trend of average market values (V_{min} + V_{max})/2 for vineyards, olive groves and irrigated arable crops.

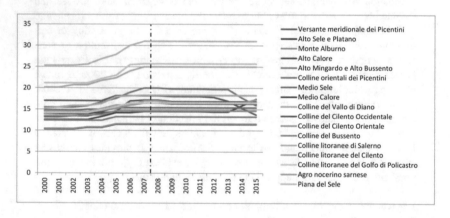

Fig. 1. Temporal trend of the land value (vineyards) in the agricultural regions.

The graph in Fig. 4 also represents the per capita income values, aggregated to the scale of the agricultural regions starting from the data collection on a municipal level, to allow an immediate comparison with the market values. An extract from the reference dataset is in Table 1.

From the observation of Figs. 1, 2, 3 and 4 two salient elements emerge:

1. for all three crops, the land values show temporal trends that tend to rise and are completely comparable to each other. In particular, a rather pronounced growth in the period 2000–2008, is followed by a substantial stagnation of prices in the subsequent time frame 2008–2015, as a result of the structural crisis that has invested the economy not only national;

2. the trend of per capita incomes tends to follow the one of the values, so to express accentuated levels of correlation between the variables at stake.

Table 1. Per capita income for agrarian region in the years 2000–2015.

Agricultural region	2000	2001	...	2007	2008	...	2014	2015
Versante merid. dei Picentini	5,021	5,610	...	7,494	7,358	...	7,719	7,979
Alto Sele e Platano	4,523	4,805	...	6,677	6,611	...	6,933	7,238
Monte Alburno	5,604	6,013	...	7,747	7,866	...	8,255	8,393
Alto Calore	5,504	5,882	...	7,643	7,789	...	8,047	8,598
Alto Mingardo e Alto Bussento	4,455	4,816	...	6,718	6,492	...	6,956	7,308
Colline orientali dei Picentini	5,615	6,106	...	8,271	7,961	...	8,789	9,016
Medio Sele	5,198	5,560	...	7,333	7,219	...	7,659	7,945
Medio Calore	5,398	5,773	...	7,439	7,407	...	7,977	8,198
Colline del Vallo di Diano	5,408	5,864	...	7,701	7,407	...	7,938	8,196
Colline del Cilento Occidentale	5,231	5,814	...	7,819	7,539	...	8,213	8,414
...	...	...		...	...		...	...
Colline litoranee di Salerno	8,347	9,005		11,399	9,931	...	11,826	11,921
Colline litoranee del Cilento	5,654	6,212		8,453	8,046		9,058	9,147
Agro nocerino sarnese	5,593	5,894		7,736	7,284		8,064	8,292
Piana del Sele	5,865	6,350		8,530	7,966		8,754	8,913

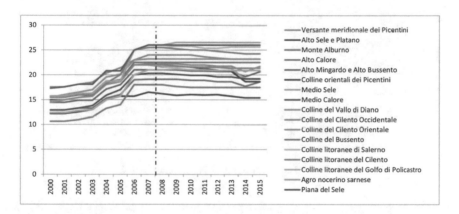

Fig. 2. Temporal trend of the land value (olive groves).

With regard to the first point, the year 2008 marks a different time trend of the parameters. The vertex of the income curves for the year 2008 (see Fig. 4) clearly separates the values *ante* 2008 from those *post* 2008, differentiating between the previous growth phase and the subsequent stagnation. The vertex is more accentuated for high-income areas (Coastal hills of Salerno, Hills of Eastern Cilento, Sele's Plain), presumably marked by speculative phenomena. The *post* 2008 crisis is also highlighted by the Italian Institute of Agrarian Economy (INEA), moreover commenting on the effects of agricultural policy actions: «The demand for land is particularly weak because of the economic crisis, while the supply is struggling to adapt to the new quotations and it is waiting for an improvement of the conjuncture. In essence, it is reiterated the lack trading activity which, in confirmation of what happened in the

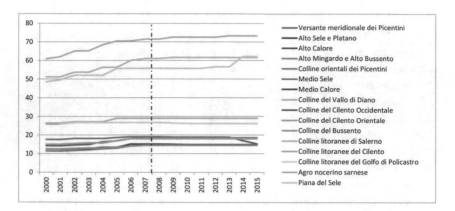

Fig. 3. Temporal trend of the land value (irrigated arable crops).

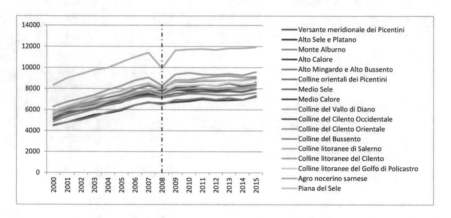

Fig. 4. Temporal trend of per capita income.

urban real estate market, has recorded a rapid contraction in recent years. The new reform of the common agricultural policy does not seem to have given any significant effect on land values, since the changes in the aid mechanisms mainly affect the relationship between landowners and tenants» [10, 11].

The different variations both of values and incomes in the two periods *ante* 2008 and *post* 2008 also emerge from the eight cartographic elaborations presented in Fig. 5 implementing Geographic Information Systems. The differential terms exclude the year 2008, marked by a vertex in the time curves.

The three maps a2, b2 and c2 show the substantial market stagnation in the period following the 2008, where the prevalence of light yellow expresses modest or nil variations in values, to which correspond equally modest increases in income (Fig. 5 d2) at most up to +7.67%.

On the contrary, the market value increases shown in Figs. 5a1, b1 and c1 are accentuated, although with differences. In fact, for vineyards and olive groves three

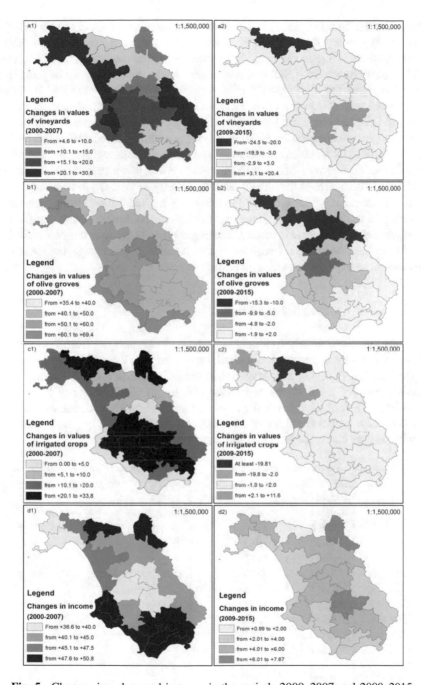

Fig. 5. Changes in values and incomes in the periods 2000–2007 and 2009–2015.

factors affect the increments: (1) the distance from the urban area; (2) the proximity to the coast; (3) journey times up to the main transit arteries. Instead, for the irrigated arable crops there is a reversed behavior respect to the three indicated factors, above all respect to the distance from Salerno, reasonably due to the lower level of profitability, which encourages the improvement of agricultural areas less appreciated in terms of geographical location. On the other hand, in the same von Thünen's model the areas destined to irrigated crops are more distant than the residential zone [12, 13].

The comparison between 2000–2008 values (maps Fig. 5a1, b1 and c1) and incomes (Fig. 5d1) expresses a marked correlation in coastal agrarian regions, in particular for vineyards and olive groves. Moving from south to north, this is true for the coastal Hills of the Gulf of Policastro, of the Cilento and for the Sele's Plain. On the contrary, there is not a high correlation in the coastal Hills of Salerno, located further North. This finds an explanation in the complex dynamics that tourism of the Amalfi Coast and the urban economy of Salerno generate in this agrarian region [14–18]. Lower correlation is finally found in the inner agricultural regions, where significant percentage increases of the values are not always associated with equal increases in incomes.

2 Temporal Correlation Levels

The correlations qualitatively highlighted in the previous paragraph are now evaluated in quantitative terms. Since there is a similar trend for the 17 agrarian regions of study, then the correlation levels measurement is made by the provincial average values. Figure 6 shows the trend of the population residing in the Province of Salerno from 2000 to 2015. The construction of the Figs. 7, 8 and 9 allows to estimate the diachronic correlations of the market values of vineyards, olive groves and irrigated arable crops both with the per capita income and the population.

Year	Population	ΔP
2000	1,075,127	
2001	1,072,611	-2,516
2002	1,072,856	+245
2003	1,076,957	+4,101
2004	1,082,657	+5,700
2005	1,083,365	+708
2006	1,081,855	-1,510
2007	1,086,276	+4,421
2008	1,089,417	+3,141
2009	1,090,030	+613
2010	1,091,984	+1,954
2011	1,092,876	+892
2012	1,092,574	-302
2013	1,093,453	+879
2014	1,105,485	**+12,032**
2015	1,108,509	+3,024

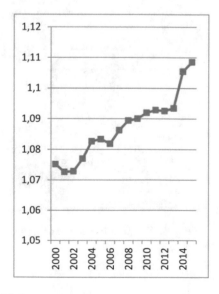

Fig. 6. Population trend in the Province of Salerno (elaborations from Istat data).

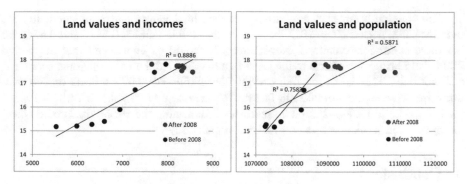

Fig. 7. Temporal correlation for the vineyards.

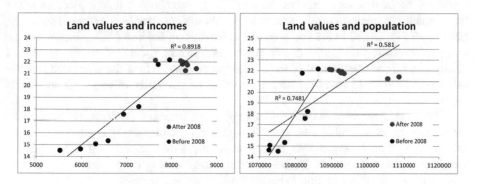

Fig. 8. Temporal correlation for the olive groves.

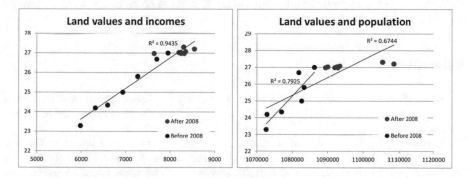

Fig. 9. Temporal correlation for the irrigated crops.

For all three cultural destinations, there is a strong correlation between land values and per capita income, as shown by the R^2 measures of 0.8886, 0.8918 and 0.9435 respectively for vineyards, olive groves and irrigated arable crops.

The correlation appears less strong between land values and population. In fact, over the entire period 2000–2015 the R^2 measures fall to 0.5871, 0.5810 and 0.6744 for the three cultural destinations. Nonetheless, it should be considered that the period 2008–2015 brings the effects of the crisis that has affected the entire national economy, and not only. This is a time frame whose instability inevitably reflects on the relationships between the variables. In fact the R^2 increase up to 0.7583, 0.7481, 0.7925 if read in the period 2000–2008 *ante* recession, confirming the strong diachronic correlation between market values and resident population.

3   Conclusions

The characterization of the functional relation able to explain the formation mechanisms of the land values imposes the preliminary selection of independent variables that exhibit significant correlations with the dependent market variable. The study, developed with regard to the vast area of the Province of Salerno (Italy), shows that the temporal evolution of both the per capita income, representative of the territory's ability to produce wealth, and of the resident population follow the same trend that in the period 2000–2015 connotes the market values of vineyards, olive groves and arable crops.

The diachronic analysis uses Geographic Information Systems (GIS) for the elaboration of cartographic representations able to visualize the data evolution. From this interpretation emerges also the effect of the structural crisis that has heavily weighed on the economy in the period after 2008, causing an exogenous shock that makes the temporal correlation between land values and population unstable.

The measurement of correlation levels confirms the results in quantitative terms. From these we can start further research with the purpose to generalize the functional relationships between agricultural land values and socio-economic variables through the writing of a value function.

References

1. Bencardino, M., Granata, M.F., Nesticò, A., Salvati, L.: Urban growth and real estate income: a comparison of analytical models. In Gervasi, O., et al. (eds.) ICCSA 2016, Part III. LNCS, vol. 9788, pp. 151–166. Springer, Switzerland (2016)
2. Bencardino, M., Nesticò, A.: Urban sprawl, labor incomes and real estate values. In: Gervasi, O., et al. (eds.) ICCSA 2017. LNCS, vol. 10405, pp. 17–30. Springer, Cham (2017)
3. Bencardino, M., Nesticò, A.: Demographic changes and real estate values. A quantitative model for analyzing the urban-rural linkages. Sustainability 9(4), 536 (2017)
4. Bencardino, M., Nesticò, A.: Spatial correlation analysis among land values, income levels and population density. In: ISTH 2018 (2018)
5. Gregor, H.F.: Geografia de la Agricoltura. Vieens Vives, Barcelona (1970)
6. Orefice, M.: Estimo, 1st edn. UTET, Torino (1984)
7. Masini, E., Barbati, A., Bencardino, M., Carlucci, M., Corona, P., Salvati, L.: Paths to change: bio-economic factors, geographical gradients and the land-use structure of Italy. Environ. Manag. 61, 116–131 (2017)

8. Bencardino, M.: An estimate of land take in municipal planning of the Campania region. In: Gervasi, O., et al. (eds.) ICCSA 2017. LNCS, vol. 10408, pp. 73–88. Springer, Cham (2017)
9. Greco, I., Bencardino, M.: The paradigm of the modern city: SMART and SENSEable cities for smart, inclusive and sustainable growth. In: Murgante, B., et al. (eds.) ICCSA 2014, Part II. LNCS, vol. 8580, pp. 579–597. Springer, Heidelberg (2014)
10. Povellato, A., Bortolozzo, D. (a cura di): Indagine sul mercato fondiario in Italia. Rapporto regionale 2013. INEA Istituto Nazionale di Economia Agraria, Roma (2014)
11. Povellato, A., Bortolozzo, D., Longhitano, D.: Il mercato fondiario. In: Annuario dell'agricoltura italiana, Vol. LXVII, Cap. VIII. INEA Istituto Nazionale di Economia Agraria, Roma (2014)
12. Sinclair, R.: Von Thünen and urban sprawl. Ann. Assoc. Am. Geogr. **57**, 72–87 (1967)
13. Jones, A.P., McGuire, W.J., Witte, A.D.: A reexamination of some aspects of Von Thünen's model of spatial location. J. Reg. Sci. **18**(1), 1–15 (1978)
14. Blake, A.: The dynamics of tourism's economic impact. Tour. Econ. **15**(3), 615–628 (2009)
15. De Mare, G., Manganelli, B., Nesticò, A.: Dynamic analysis of the property market in the city of Avellino (Italy): the Wheaton-Di Pasquale model applied to the residential segment. In: Murgante, B., Misra, S., Carlini, M., Torre, C., Nguyen, H.Q., Taniar, D., Apduhan, B. O., Gervasi, O. (eds.) ICCSA 2013. LNCS, Part III, vol. 7973, pp. 509–523, Springer, Heidelberg (2013)
16. Calabrò, F.: Local communities and management of cultural heritage of the inner areas: an application of break-even analysis. In: Gervasi, O., et al. (eds.) Lecture Notes in Computer Science, vol. 10406. Springer (2017)
17. Della Spina, L., Lorè, I., Scrivo, R., Viglianisi, A.: An integrated assessment approach as a decision support system for urban planning and urban regeneration policies. Buildings **7**, 85 (2017)
18. Napoli, G., Giuffrida, S., Trovato, M.R., Valenti, A.: Cap rate as the interpretative variable of the urban real estate capital asset: a comparison of different sub-market definitions in Palermo, Italy. Buildings **7**(3), 80 (2017)

A Multicriteria Economic Analysis Model for Urban Forestry Projects

Maria Rosaria Guarini[1](✉) (iD), Antonio Nesticò[2] (iD),
Pierluigi Morano[3] (iD), and Francesco Sica[1] (iD)

[1] Sapienza University of Rome, 00185 Rome, Italy
mariarosaria.guarini@uniroma1.it
[2] University of Salerno, 84084 Fisciano, Italy
[3] Polytechnic University of Bari, 70126 Bari, Italy

Abstract. The urban green areas represent a strategic resource for the contemporary city sustainable development. In addition to aesthetic and recreational functions, their presence contributes to increase the environmental quality level by improving the microclimate, preserving the biodiversity and promoting the territory economic growth. However, the interventions execution designed to provide the built areas of the so-called urban forests is rarely indicated as a priority action in the urban spaces planning because often a different allocation of available resources is preferred. In this work, starting from the definition of a indicators set useful for expressing the not only financial, but also social, cultural and environmental components of value of the projects for urban forestry, the aim is to build a multi-criteria economic analysis protocol purposeful at predicting the correct funds distribution between initiatives for realization of urban forests at a district scale. The characterization of the model is carried out through the logical-mathematical tools of the Operational Research developed according to Continuous Linear Programming principles.

Keywords: Urban forestry · Economic evaluation · Sustainable development
Multi-criteria decision analysis · Continuous Linear Programming

1 Urban Forestry Projects for the Sustainable Redevelopment of the Territory

During the period after the Second World War, there have been in the world poorly controlled city transformation processes that have returned over time a highly uneven territory characterized by areas with «unprecedented population growth, lack of maintenance due to institutional breakdowns, unauthorised/unregulated construction works and nonchalant behaviours on the part of the citizens» [1]. In Italy since the second half of the years '80 it has been tried to remedy these problems by proposing redevelopment actions that have not always been possible to realize also for the

---

The contribution to this paper is the result of the joint work of the four authors to which the paper has to be attributed in equal parts.

© Springer International Publishing AG, part of Springer Nature 2019
F. Calabrò et al. (Eds.): ISHT 2018, SIST 100, pp. 564–571, 2019.
https://doi.org/10.1007/978-3-319-92099-3_63

economic-financial resources scarcity available to recover run-down areas with a view to sustainable development of territory [2–7] also thorough public-private partnership [8–10].

In Europe starting with the Leipzig Charter on Sustainable European Cities (2007), the European Community directives define specific objectives (*Sustainable Development Goals*) to be pursued also with the recovery of "unhealthy" urban areas [11], placing the focus not only on the built, but also on society, economy, culture and environment, seen as essential components for the integrated development of territory [12]. In essence, it is suggested to be simultaneously pursued objectives linked to the natural (biodiversity, resilience, bio-productivity), economic (fulfilment of the primary needs, strengthening of the equity, increase of goods and services) and social (cultural diversity, institutional sustainability, social justice, participation) system through projects aimed at regenerating city portions according to urban sustainability logics [13–15].

In this direction, the realization and preservation of green areas in the urban consolidated contexts – from this the designation of *urban forest* – helps to «support positive economic, social and environmental links between urban, peri-urban and rural areas by strengthening national and regional development planning», with the aim of «make cities inclusive, safe, resilient and sustainable» [16]. In fact, forestry operations include actions turned to satisfy needs with not only environmental, but also social, economic and cultural character, returning a unitary territory vision led according to multidimensional logic able to consider both financial aspects than of extra-monetary character [17]. In relation to the need of requalifying degraded urban areas, the urban forestry interventions offer eco-systemic services indicative of the urban green land multifunctionality based on the interaction between «Environment-Society-Economy» [18–20].

The relationship between «Environment-Society-Economics» paves the way for the definition of a analysis model able to evaluate through multicriteria logic the overall effects which are generated on the built by such interventions [21].

The aim of this paper is to outline an evaluation methodology that takes into account the multidimensional character of urban forestry initiatives through:

1. definition of the criteria able to measure the project capacity to achieve objectives of the environmental, economic and social system;
2. attribution of performance indicators for each criterion;
3. description of the optimisation model for allocating available resources between proposed interventions.

When the problem consists in allocation of resources between initiatives aimed at the recovery of degraded urban areas through urban forestry actions, it is useful the use of the Operational Research optimization algorithms able to take into account more criteria through the writing of an algebraic expression called *objective function* [22]. In the following it is illustrated:

– in the Sect. 2 the indicators set useful to express the multidimensional character of the urban forestry initiatives based on the indications present and shared in International and European documents;

- in the Sect. 3 the financial and economic analysis model, which allows to solve decision-making problems of resources allocation between investment alternatives through a logical-operational approach using mathematical expressions written according to Continuous Linear Programming (CLP) principles [23, 24].

Finally, the results of the study and possible developments of the proposed methodology are given in the Sect. 4.

2 The Multicriteria Logic for the Economic Evaluation of Urban Forestry Actions

As already said, a indicators set of both environmental and social, economic and cultural nature should be identified to define urban forestry interventions with the aim of promoting the sustainable planning of the territory.

In the literature, the indicators employed for the evaluation of the effects produced by urban forestry interventions mainly concern aspects of ecological-environmental nature and they interest rarely economic, social and cultural issues [25–27].

The Food and Agriculture Organization (FAO) identifies 10 Key Issues to be considered in the design phase as evaluation criteria: 1. Human Health and Well-Being, 2. Climate Change, 3. Biodiversity-Landscapes, 4. Economic Benefits-Green Economy, 5. Risk Management, 6. Mitigation Land-Soil Degradation, 7. Water-Watersheds, 8. Food-Nutrition Security, 9. Wood Security, 10. Socio-Cultural Value [28].

In the present text a breakdown of the FAO key issues is proposed in three different sectors respectively of environmental, economic and socio-cultural nature on the basis of the eco-systemic services typological classification [29].

The Table 1 summarizes the existing relationships between the eco-systemic services types (criteria), the key issues proposed by FAO (sub-criteria) and performance indicators chosen on the basis of the cases addressed in the literature [30–33]. The evaluation model proposed in the paper is defined by considering these elements. The choice of indicators is carried out taking into account the evaluative aspects of both the area *status quo* to be requalified, as well as the social, cultural and financial consequences arising from the proposed urban forestry action. It is provided a summary indicator - the *Total Expected Value* (TEV) - which allows to express the effect produced by the single project on the intervention area. It is obtained summing the value the indicators have taken depending on the data collected.

In order to orient the choice between investment alternatives (param COSTS) subject to budget constraint (param BUDGET), the model is built for the optimal allocation of resources made according to urban quality levels to be achieved in perspective of recovering particularly degraded areas (set AREAS). The urban forestry initiatives designed for this objective are evaluated on the basis of performance indicators presented in Table 1 (set INDICATORS). Here, the income indicator can be one of those traditionally used in the Cost-Benefit Analysis, i.e. the Net Present Value, the Internal Rate of Return, the Payback Period. So, a synthetic index (param TEV) adapt to express the goodness of the proposed project is defined.

Table 1. Indicators set for urban forestry projects.

Ecosystem services[a]	Key issues[b]	Indicators set
Environmental services	Mitigation land	Green areas
	Soil degradation	
	Climate change	Territorial
		Fragmentation index
	Biodiversity-landscapes	Air pollution level
	Water-watersheds	
Economic services	Economic benefits	Income indicator
	Green economy	
	Risk management	
Socio-cultural services	Human health and well-being	Destined areas to recreational activities
	Food-nutrition security	
	Wood security	
	Sociocultural value	No of new workers

Legenda: [a]McVittie A. et al., 2013 [b]FAO, 2016

The problem of public resources allocation among n areas to be requalified subject to urban forestry actions is examined. An evaluation model is suggested for the allocation of financial resources (set RATE) to be used for the urban quality level improvement.

A Mathematical Programming Language (AMPL) software is used to structure the model. The model is defined in the CLP logic (see Sect. 3). CLP can be an effective tool to define the amount of resources to be allocated to urban forestry actions for the city integrated development. In the aforementioned logic, functional relationships of the problem to be solved find synthesis in linear mathematical expressions implemented according to the CLP optimization algorithms. The simplex algorithm is selected for this case, which seems better able to solve resources scheduling problems according to multicriteria logic.

3   Allocation of Resources According to Economic Optimization Models

The use of multicriteria evaluation systems resolves the problem of resources distribution between areas subject to requalification projects [34]. Specifically, each area represents an investment choice D_i, whose characteristics are expressed with criteria provided through appropriately selected performance indicators. These define in a unique way the project of the i-th area of which the Total Expected Value is obtained from the linear combination of the values assumed by each indicator.

The multiobjective evaluation problems – such as that of our interest – can be solved by considering a mathematical model in which: the decision D_i to allocate a part of the resources x_i to the i-th area (A_i) depends on the TEV value assumed by the i-th project; the combination $(P_{AREA_i}, GOAL_j)$ expresses the project capacity P_{AREA_i} to pursue the j-th objective (GOAL); $Ps(GOAL_j)$ represents the weight of the j-th GOAL and $TEV_{i\_in}(P_{AREA_i})$ is the Project's Total Expected Value linked to the AREA$_i$.

The TEV$_{fin}$ deriving from the proposed project is measured through a coefficient ε_i act to express the increase in uniform value ΔTEV_i of the i-th areas:

$$TEV_{i\_fin} = TEV_{i\_in} + \Delta TEV_i = TEV_{i\_in} + (TEV_{i\_in} \times \varepsilon_i) \tag{1}$$

Using perequative logics for the estimation of ε_i, the following mathematical expression is considered:

$$\Delta TEV_{(i-1)} = \Delta TEV_i = \Delta TEV_{(i+1)} \tag{2}$$

where the increase in the value of the i-th area is equal to that of other intervention areas.

The financial resources rate x_i required for the increase of i-th area TEV value is defined according to TEV_{i_fin} and the investment cost estimated for each area.

The problem therefore consists in the estimation of the ε_i through an optimization model of the type:

$$\begin{cases} \max \sum_{i=1}^{n} \varepsilon_i \times TEV_{i\_in}(P_{AREA_i}) = \sum_{j=1}^{m}\sum_{i=1}^{n} \varepsilon_i \times Ps(GOAL_j) \times TEV_{i\_in}(P_{AREA_i}, GOAL_j) \\ \varepsilon_i > 0 \\ \sum_{i=1}^{n} \varepsilon_i = 1 \\ TEV_{(i-1)\_in} \times \varepsilon_{(i-1)} = TEV_{i\_in} \times \varepsilon_i = TEV_{(i+1)\_in} \times \varepsilon_{(i+1)} \end{cases} \tag{3}$$

The model is implemented with AMPL syntax by using the linear programming continues algorithms (Table 2).

In summary, the n areas to be recovered (set AREAS) are evaluated according to m indicators (set INDICATORS). A rate of resources (set RATE) is established for each area according to the available budget (param BUDGET).

The PARAMETERS section declares the numerical values that define the Total Expected Value of the project on the i-th area (param TEV).

Defined the unknowns of the problem (var $\varepsilon\{i$ in RATE$\} >= 0$), the objective function is written as it follows:

$$\text{MAXIMIZE OBJECTIVE}: \text{ SUM } \{I \text{ IN AREAS, IN RATE}\} \text{ TEV}[I] \times \varepsilon[I] \tag{4}$$

Finally, it is important the CONSTRAINT about the increase in the value of the i-th areas:

$$\text{s.t.}(subject\ to)\ \text{constraint\_0}: \text{ TEV}_{(i-1)} \times \varepsilon_{(i-1)} = \text{TEV}_i \times \varepsilon_i = \text{TEV}_{(i+1)} \times \varepsilon_{(i+1)} \tag{5}$$

Table 2. The optimization model

SETS
set AREAS;
set RATE;
set INDICATORS;
PARAMETERS
param TEV{i in AREAS};
VARIABLES
var ε\{i in RATE\} $> = 0$;
OBJECTIVE FUNCTION
MAXIMISE objective: sum\{i in AREAS, in RATE\} TEV[i] $\times$ ε[i];
CONSTRAINTS
s.t. constraint_0: $TEV_{(i-1)} \times \varepsilon_{(i-1)} = TEV_i \times \varepsilon_i = TEV_{(i+1)} \times \varepsilon_{(i+1)}$

Depending on the ε_i obtained values, the amount of financial resources x_i to be allocated to the n economic units for the increase in value, is deduced considering the investment cost of the project provided for the i-th area compared to the available budget.

4   Conclusions

The intervention policies adopted in the city don't always achieve useful objectives for urbanised territory sustainable development. The lack of a systemic vision between natural and built environment induces to keep in little consideration the execution of urban forestry projects particularly suitable for the redevelopment of the built. In view of this, it is necessary to use analysis methodologies proper to evaluate interventions of such species in the light of the multiple effects produced on the entire urbanized fabric.

A useful evaluation protocol is suggested based on the multi-criteria decision analysis approach in order to promote forestry actions in the cities through an equitable allocation of available funds.

The proposed model allows to determine the amount of resources to allocate to areas to be recovered through urban forestry projects evaluated with multicriterial logics. Particularly, the adopted analysis scheme is suitable for the investment plans construction from public subjects to attain fixed quality levels, not only of financial nature, but also of social, environmental and cultural character. For this reason, multidimensional indicators capable of measuring forestry initiatives under multiple aspects in view of urban sustainability are taken into account. The indicators are selected in the light of the main literature references.

The AMPL programming environment is used for the model construction because it allows to write in simple form the evaluation protocol through linear algebraic functions expressed according to Continuous Linear Programming principles. Applications to case studies are in progress, with the aim of verifying the analysis instrument effectiveness, its adaptability to the multiple and different real situations, as well as to outline future research prospects.

References

1. Jack, M.W., Coles, A.M., Piterou, A.: Sustainable project management in urban development projects: a case study of the Greater Port Harcourt City Development Project in River State, Nigeria. Sustain. Dev. Plan. VIII **210**, 209–219 (2017)
2. El Din, H.S., Shalaby, A., Farouh, H.E., Elariane, S.A.: Principles of urban quality of life for a neighbourhood. HBRC J. **9**, 86–92 (2013)
3. Graziano, T.: Riconversione funzionale, verde urbano e gentrification: dalla promenade palntée di Parigi alla High Line di New York. Riv. Geogr. Ital. **121**(1), 45–60 (2014)
4. Hodson, M., Marvin, S.: Intensifying or transforming sustainable cities? Fragmented logics of urban environmentalism. Local Environ. **22**, 8–22 (2017)
5. Guarini, M.R.: Self-renovation in Rome: Ex Ante, in Itinere and Ex Post Evaluation. In: Gervasi, O., et al. (eds.) 16th International Conference on Computational Science and its Applications - ICCSA 2016. LNCS, vol. 9789, pp. 204–218 (2016)
6. Morano, P., Locurcio, M., Tajani, F., Guarini, M.R.: Fuzzy logic and coherence control in multi-criteria evaluation of urban redevelopment projects. Int. J. Bus. Intell. Data Min. **10**(1), 73–93 (2015)
7. Bencardino, M., Nesticò, A.: Demographic changes and real estate values. A quantitative model for analyzing the urban-rural linkages. Sustainability **9**(4), 536 (2017)
8. Directive 2014/23/UE of the European Parliament and Council in 26 February 2014 on the award of the concession contracts. http://eur-lex.europa.eu/legal-content/EN/TXT/PDF/?uri=CELEX:32014L0023&from=EN. Accessed 04 Mar 2018
9. Guarini, M.R., Battisti, F.: Benchmarking multi-criteria evaluation: a proposed method for the definition of benchmarks in negotiation public-private partnerships. In: Murgante, B., et al. (eds.) 14th International Conference on Computational Science and Its Applications - ICCSA 2014. LNCS, vol. 8581, pp. 208–223 (2014)
10. Morano, P., Tajani, F.: Decision support methods for public-private partnerships: an application to the territorial context of the Apulia Region (Italy). In: Green Energy and Technology, pp. 317–326 (2017)
11. Vitali, W., Fini, G., Bovini, G.: L'Agenda per lo Sviluppo Urbano Sostenibile (2017)
12. Leipzig Charter on sustainable European cities, Extracted from CdR 163/2007 EN-COM/SAB/lc, Lipsia (2007). http://ec.europa.eu/regional_policy/archive/themes/urban/leipzig_charter.pdf. Accessed 04 Mar 2018
13. Gonzalez, M.J.G.: Planning, urban sprawl and spatial thinking. Eur. J. Geogr. **8**(1), 32–43 (2017)
14. Yan, J., Shen, Y., Xia, F.: Differentiated optimization of sustainable land use in metropolitan areas: a demarcation of functional units for land consolidation. Sustainability **9**, 1356 (2017)
15. Nesticò, A., Galante, M.: An estimate model for the equalisation of real estate tax: a case study. Int. J. Bus. Intell. Data Min. **10**(1), 19–32 (2015)
16. ONU: Trasformare il nostro mondo: l'Agenda 2030 per lo Sviluppo Sostenibile (2015)
17. Konijnendijk, C., Kjell, N., Randdrup, T., Schipperini, J.: Urban Forest and Trees. Springer, Amsterdam (2005)
18. Farber, S.C., Costanza, R., Wilson, M.A.: Economic and ecological concepts for valuing ecosystem services. Ecol. Econ. **41**, 375–392 (2002)
19. Bateman, I.J., Mace, G.M., Fezzi, C., Atkinson, G., Turner, K.: Economic analysis of ecosystem service assessments. Environ. Resour. Econ. **48**(2), 177–218 (2010). https://doi.org/10.1007/s10640-010-9418-x

20. Votsis, A.: Planning for green infrastructure: the spatial effects of parks, forests, and fields on Helsinki's apartment price. Ecol. Econ. **132**, 279–289 (2017). https://doi.org/10.1016/j. ecolecon.2016.09.029

21. Munda, G.: Social multi-criteria evaluation for urban sustainability policies. Land Use Policy **23**, 86–94 (2006). https://doi.org/10.1016/j.landusepol.2004.08.012

22. Kaiser, M.G., El Arbi, F., Ahlemann, F.: Successful project portfolio management beyond project selection techniques: understanding the role of structural alignment. Int. J. Project Manag. **33**, 129–139 (2014). https://doi.org/10.1016/j.ijproman.2014.03.002

23. Chiveco, E.: Integration of linear programming and GIS for land-use modelling. Int. J. Geogr. Inf. Sci. **7**(1), 71–83 (1993). https://doi.org/10.1080/02693799308901940

24. Jeroen, A.C.J.H., Eisinger, E., Heuvelink, Gerard B.M., Stewart, J.T.: Using linear integer programming for multi-site land-use allocation. Geogr. Anal. **35**(2), 148–169 (2003). https:// doi.org/10.1111/j.1538-4632.2003.tb01106.x

25. Clark, J.R., Matheny, N.P., Cross, G., Wake, V.: A model of urban forest sustainability. J. Arboric. **23**, 17–30 (1997)

26. Dobbs, C., Escobedo, F.J., Zipperer, W.C.: A framework for developing urban forest ecosystem services and goods indicators. Landscape Urban Plan. **99**, 196–206 (2011). https://doi.org/10.1016/j.landurbplan.2010.11.004

27. Kenney, W.A., Van Wassenaer, P.J., Satel, A.L.: Criteria and indicators for strategic urban forest planning and management. Arboric. Urban Forest. **37**(3), 108–117 (2011)

28. FAO: Guidelines on Urban and Peri-urban Forestry. Forestry Paper, 178 p. (2016)

29. McVittie, A., Hussain, S.: The Economics of Ecosystems and Biodiversity – Valuation Database Manual (2013)

30. Sheppard, S.R., Meitner, M.: Using multi-criteria analysis and visualisation for sustainable forest management planning with stakeholder groups. Forest Ecol. Manag. **207**, 171–187 (2005)

31. Gough, A.D., Innes, J.L., Allen, S.D.: Development of common indicators of sustainable forest management. Ecol. Ind. **8**, 425–430 (2008)

32. Woodall, C.W., Amacher, M.C., Bechtold, W.A., Coulston, J.W., Jovan, S., Perry, C.H., Randolph, K.C., Schulz, B.K., Smith, G.C., Tkacz, B., Will-Wolf, S.: Status and future of the forest health indicators program of the USA. Environ. Monit. Assess. **177**, 419–436 (2011). https://doi.org/10.1007/s10661-010-1644-8

33. Östberg, J., Delshammar, T., Wiström, B., Nielsen, A.B.: Grading of parameters for urban tree inventories by city officials, arborists, and academics using the Delphi method. Environ. Manag. **51**(3), 694–708 (2013). https://doi.org/10.1007/s00267-012-9973-8

34. Nesticò, A., Sica, F.: The sustainability of urban renewal projects: a model for economic multi-criteria analysis. J. Property Investment Finance **35**(4), 397–409 (2017). https://doi. org/10.1108/JPIF-01-2017-0003

Spatial Correlation Analysis Among Land Values, Income Levels and Population Density

Massimiliano Bencardino$^{(\boxtimes)}$ and Antonio Nesticò

University of Salerno, 84084 Fisciano, Italy
mbencardino@unisa.it

Abstract. The correlation between market values of agricultural land and infrastructural, socio-demographic and productivity characteristics is widely recognized. In the present paper, the Authors intend to establish how the endowment of infrastructures, the levels of income and the demographic density are able to affect the land values of the corresponding territory.

The analysis considers the market values of the vineyards, olive groves and irrigated crops, distributed throughout the Province of Salerno (Italy) and relevant for the overall value of agricultural production of the study area. The real estate appraisals, related to the last available survey of 2015, are provided by the Council for Agricultural Research and Analysis of the Agrarian Economy (CREA), for each of the 17 agricultural regions in which the Province is divided. They are at first cartographically represented through Geographic Information Systems (GIS) and then analyzed according to the geo-location both of the main urban centers and of the infrastructural network that characterizes the area. So, the distribution of values is explained in the light of established theories of Economic Geography.

Then, the market values of agricultural soils are synchronously correlated with the surveys on the taxable income and the population density, respectively extracted from the data warehouse of the Ministry of Economy and Finance and of Italian Statistical Institute (Istat). Through the data processing and the thematic maps building, the correlations between the involved variables are quantified, in order to establishing functional relations that can be extended to other similar areas.

Keywords: Land values · Economic model · Demography
Territorial planning · Geographic Information Systems

1 Introduction and Aim of the Work

The market values of agricultural soils depend on the characteristics of location, attributable to the specific geographical area, and the peculiarities of the individual property, as explained by the estimative literature [1–3]. The first are affected by planning policies and by economic and socio-demographic factors [4–8], so that the

---

The contribution to this paper is the result of the joint work of both Authors, to which the paper has to be attributed in equal parts.

© Springer International Publishing AG, part of Springer Nature 2019
F. Calabrò et al. (Eds.): ISHT 2018, SIST 100, pp. 572–581, 2019.
https://doi.org/10.1007/978-3-319-92099-3_64

mercantile appraisals of agricultural land can be considered able to summarize the effects of complex territorial dynamics [9, 10]. But what are the main economic and socio-demographic indicators that influence the average purchase price and what are the correlation levels?

With the aim of answering this question, three parameters are taken into account: (1) the market value of agricultural soils with specific crop destination; (2) the average taxable income, representative of the economic capacity of taxpayers; (3) the population density, able to express the urbanization level. The complexity of the examining problem is thus analyzed in a simplified scheme, against the number of potential determining variables.

The following paragraph describes the distribution of market values of vineyards, olive groves and irrigated crops in the Province of Salerno, which includes 158 municipalities for a total area of 4,952 km^2. These are the most frequent crop destinations in the survey area and relevant for the overall value of agricultural production. In the 3th paragraph, the spatial distribution of income and population density is examined, and in the 4th, the correlation levels are quantified, deriving the results of the quantitative analysis and outlining future research.

2 Land Values and Characteristics of the Territory

The study is developed on the data set of the Council for Research in Agriculture and the Analysis of Agricultural Economics (CREA), which divides the entire national territory into agricultural regions (see excerpt in Table 1).

Table 1. Excerpt from the dataset of land values.

Cod.	Agricultural regions		Altimetric zone	Crop type	k€/ha.
6501	Versante merid. dei Picentini	1	Inland mountain	Irrigated crops	15.2
6501	Versante merid. dei Picentini	1	Inland mountain	Olive groves	23.2
6501	Versante merid. dei Picentini	1	Inland mountain	Vineyards	13.7
6502	Montagna Alto Sele e Platano	1	Inland mountain	Irrigated crops	14.5
6502	Montagna Alto Sele e Platano	1	Inland mountain	Olive groves	15.4
6502	Montagna Alto Sele e Platano	1	Inland mountain	Vineyards	11.5
6503	Monte Alburno	1	Inland mountain	Irrigated crops	n.v.
6503	Monte Alburno	1	Inland mountain	Olive groves	17.5
6503	Monte Alburno	1	Inland mountain	Vineyards	11.6
...	...	...	...	...	...
6512	Colline del Bussento	3	Internal hill	Irrigated crops	14.5
6512	Colline del Bussento	3	Internal hill	Olive groves	22.5
6512	Colline del Bussento	3	Internal hill	Vineyards	14.7
6513	Colline litoranee di Salerno	4	Coastal hill	Irrigated crops	61.6
6513	Colline litoranee di Salerno	4	Coastal hill	Olive groves	25.6
6513	Colline litoranee di Salerno	4	Coastal hill	Vineyards	31.0
...	...	...	...	...	...

Returning to the definition of the National Institute of Statistics (Istat), the agrarian region is constituted «by groups of municipalities according to rules of homogeneous territorial continuity in relation to certain natural and agricultural characteristics and, subsequently, aggregates by altimetric zone» .

The CREA provides the minimum and maximum market value of the different qualities of cultivation, for each of the 17 agricultural regions in which the Province of Salerno is subdivided. Through GIS tools, a cartographic representation of the mean market value is produced, for vineyards, olive groves and irrigated crops. Figures 1, 2 and 3 show the distribution of values on the territory. On the same maps the main road arteries, namely the A2 Salerno-Reggio Calabria, the A3 Naples-Salerno, the SP430 Cilentana and the SS517 Bussentina, are reported.

The shown distribution of values seems to respond to the von Thünen's model, according to which the real estate appraisals decrease from the center to the periphery, by increasing the distance.

It must be emphasized that, today, the von Thünen's logic is no longer fully capable of describing the spatial organization criteria of agricultural production activities. In fact, the model – elaborated in the first half of the nineteenth century – refers to a profoundly different economic phase, in which the primary sector represented the most relevant in the formation of wealth and the impact on the profit of transport costs was considerably greater [11]. Instead, in the present phase of globalization, the territories can no longer be considered isolated and independent entities from the global networks [12].

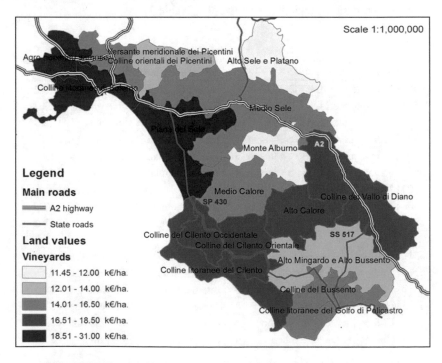

Fig. 1. Land value of vineyards in the province of Salerno (year 2015).

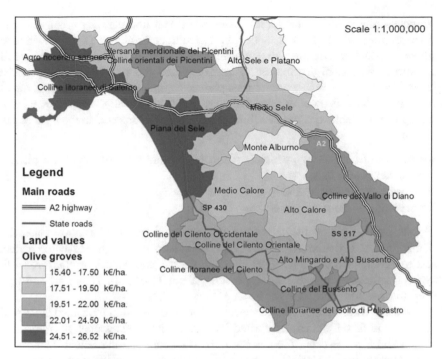

Fig. 2. Land value of olive groves in the province of Salerno (year 2015).

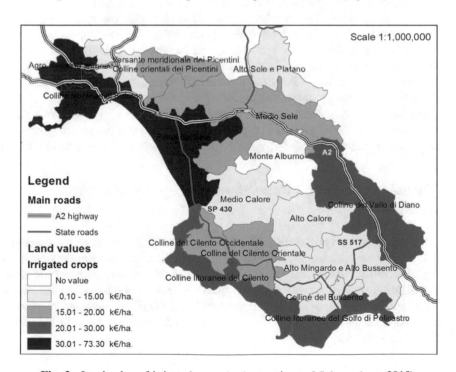

Fig. 3. Land value of irrigated crops in the province of Salerno (year 2015).

Nevertheless, the model of the German economist can find partially empirical correspondence when it's applied to a simple agricultural economy [13], or where the small-scale agriculture continues to produce the same crops in an inertial way. According to the above, the spatial distribution of market values at a regional scale can still be explained both by the incidence of transport costs and by other intrinsic territorial characteristics. This is exactly what happens in the case study, where for all three cartographic representations the distribution of values is interpreted in the light of three conditions:

1. the distance from Salerno, which is the main city and the most important place for the exchange of goods;
2. the coexistence of coastal strip and internal areas;
3. the network of the main roads.

So, the average mercantile appraisals of agricultural soils tend to grow going from the southernmost agricultural regions (High Mingardo and High Bussento) to those closest to the capital Salerno (Coastal hills of Salerno), with values of vineyards, olive groves and irrigated crops that go respectively from 12.0–14.0, 17.5–19.5 and 0.10–15.0 up to over 30 thousand € per hectare.

In the same way, an increase in values is shown moving from the inland areas to the coastal ones. In fact, if the High Sele and Platano has land values ranging from 11.5 to 17.5 thousand € per hectare, the Coastal hills of Cilento and the Sele's Plain respectively assume values of: 16.5–18.5 and 18.5–31.0 thousand € per hectare for vineyards, 22.0–24.5 and 24.5–26.5 for olive groves, 20.0–30.0 and 30.0–73.3 for irrigated crops.

Finally, the previous two conditions are associated with the spatial anisotropy, generated by the main road arteries and already recognized by von Thünen as a variant to the radially decreasing income model [14, 15]. In fact, a double directrix of higher market prices, given by the A2 Salerno-Reggio Calabria highway and the Cilentana coastal road, is recorded. Likewise, a peak of values is detected in the Agro nocerino-sarnese, strongly connected to the A2 and A3 highways.

Therefore, the accessibility is still a key variable if translated in terms of an advanced economy, no longer just interpreted as transport costs but probably as speed of connection to the markets and dynamism of the socio-productive contexts [16–19].

3 Taxable Income and Population Density. Synchronic Analysis

For a more detailed analysis of the land value function, the informations represented in Figs. 1, 2 and 3 are correlated with the level of taxable income and with the population density. The relative numerical terms, useful for the GIS elaborations and cut at the municipal scale, are respectively extracted from the data warehouse of the Ministry of Economy and of the Italian Statistical Institutes (Istat). Then, they are aggregated to the scale of the agrarian regions to allow more immediate comparisons with the market values (Figs. 4 and 5).

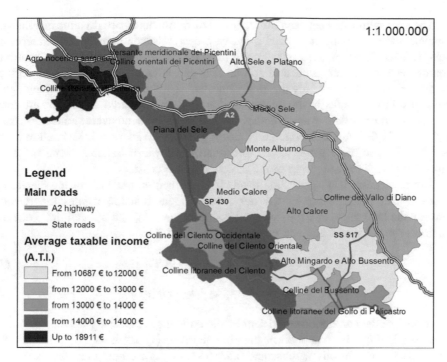

Fig. 4. Average taxable income for each agricultural region (year 2015).

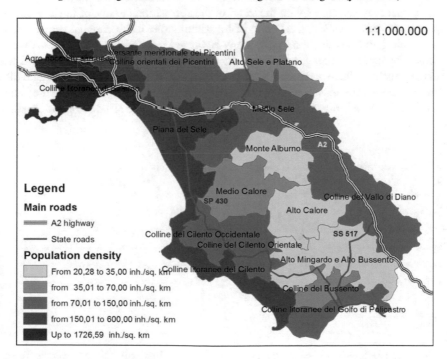

Fig. 5. Population density for each agricultural region of the province (2015).

The correspondence between the territorial distributions of both taxable income and population density with the spatial trend of property values immediately emerges in Figs. 1, 2 and 3. So, it turns out that the three elements (the distance from the urban area of Salerno, the coexistence of coastal strip and internal areas and the network of the main roads) affect not only the land values but also the economic and demographic indicators. For example, in the internal region of High Mingardo and High Bussento, among the most distant from Salerno, lowest levels both of average income (from 10,687 to 12,000 €) and population density (from 20.28 to 35.00 inh./km^2) are recorded. Instead, they progressively grow moving towards Salerno, along the most important road lines and going in the direction of the coast.

The measurement of correlation levels is performed on the basis of the regression model, which better interprets the data. So, a linear function defines the relation between the land value and the average taxable income (1), and a logarithmic one the relation between land value and population density δ (2):

$$Land\ valueV = a_1 + b_1 \cdot [A.T.I.] \tag{1}$$

$$Land\ valueV = a_2 + b_2 \cdot \log(\delta) \tag{2}$$

The values of the correlation coefficient R^2 are quite high, especially for the vineyards, as shown in Figs. 6, 7 and 8. In the first function (1) R^2 is equal to 0.7827 for vineyards, 0.5089 for olive groves and 0.5356 for irrigated crops (Figs. 6a, 7a, 8a); in the (2) R^2 is respectively 0.7054, 0.8365 and 0.6416 (Figs. 6b, 7b, 8b).

But what reasons can explain the trend of the correlation curves, of linear type with respect to the variables V and ATI and of logarithmic one if the variables are V and δ?

According to Sinclair (1967): «In many advanced industrialized parts of the world, the basic forces determining agricultural land use near urban areas are associated with urban expansion. Where these forces are in operation, the agricultural pattern quite often is one of increasing intensity with distance from the city, quite the reverse of the pattern generalized by Von Thünen's theory» [11]. In fact, the land use responds to the pursuit of

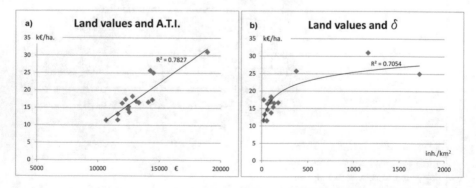

Fig. 6. Correlation curves between the land values of the **vineyards** and the Average Taxable Income ATI (a) and between the land values and the population density (b), year 2015.

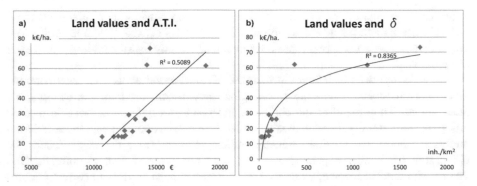

Fig. 7. Correlation curves between the land values of the **olive groves** and the ATI (a) and between the land values and the population density (b), year 2015.

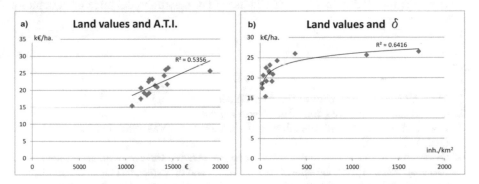

Fig. 8. Correlation curves between the land values of the **irrigated crops** and the ATI (a) and between the land values and the population density (b), year 2015.

differential rent. Thus, if the income from agricultural lands is underestimated, the allocation of the soil resource necessarily shifts towards urban transformation [20–25].

Therefore, it is possible to affirm that, in the fringe areas, the increasing population density, due to a more compact urbanization, is not associated with an equal increase in the mercantile value of agricultural soils. This phenomenon is confirmed by a gradually weaker growth in agricultural values with increasing population density. So, the relation between V and δ is approximated through a logarithmic function better than with a linear function.

4 Conclusions

The complex territorial dynamics depend on a large number of closely related variables. Assuming that the market value of soils summarizes the effects of multiple processes, in the present paper the logical connections and the correlations between land values, taxable income and population density are investigated. The vast and heterogeneous area of the Province of Salerno (Italy) is taken as a reference. Here, coastal surfaces and inland areas with different altitudes coexist, a dominant city core and an articulated road network are present and differently distributed levels of wealth are expressed.

Considering the disciplinary principles of the Appraisal as well as Economic Geography models, Geographic Information Systems are used for the cartographic elaborations. They manifest the ability of independent variables, i.e. taxable income and population density, to explain land values. The results obtained with the correlation analysis support the qualitative interpretation gathered from the maps. Then, the elaborations rationally express that linear functions are sufficient to correlate land values with taxable income, while logarithmic curves must be used to correctly return the correlation between land values and population density.

The check of the results in other territorial contexts, as well as the use of additional parameters, provide interesting research perspectives, to define a general function of the value for the interpretation of the mechanisms that determine the price of agricultural soils.

References

1. Medici, G.: Principi di Estimo, 5th edn. Calderini, Bologna (1977)
2. Orefice, M.: Estimo, 1st edn. UTET, Torino (1984)
3. Codice delle Valutazioni Immobiliari. Italian Property Valuation Standard. Tecnoborsa (2011)
4. Bencardino, M., Greco, I., Ladeira, P.R.: The comparative analysis of urban development in two geographic regions: the state of Rio de Janeiro and the campania region. In: Murgante B. et al. (eds.), ICCSA 2012, Part II, LNCS, vol. 7334, pp. 548–564. Springer, Heidelberg (2012
5. Morano, P., Locurcio, M., Tajani, F., Guarini, M.R.: Fuzzy logic and coherence control in multi-criteria evaluation of urban redevelopment projects. Int. J. Bus. Intell. Data Min. 10(1), 73–93 (2015)
6. Della Spina, L., Lorè, I., Scrivo, R., Viglianisi, A.: An integrated assessment approach as a decision support system for urban planning and urban regeneration policies, Buildings 7(85) (2017)
7. Nesticò, A., Sica, F.: The sustainability of urban renewal projects: a model for economic multi-criteria analysis. J. Property Invest. Financ. 35(4), 397–409 (2017)
8. Zambon, I., Serra, P., Bencardino, M.C., Carlucci, M., Salvati, L.: Prefiguring a future city: urban growth, spatial planning and the economic local context in Catalonia. Eur. Plan. Stud. 25, 10(3), 1797–1817 (2017)
9. Bencardino, M., Nesticò, A.: Urban sprawl, labor incomes and real estate values. In: Borruso, G., et al. (eds.) ICCSA 2017. LNCS, Vol. 10405, pp. 17–30, Springer, Cham (2017)
10. Bencardino, M., Nesticò, A.: Demographic changes and real estate values. a quantitative model for analyzing the urban-rural linkages. Sustainability 9(4), 536 (2017)
11. Sinclair, R.: Von Thünen and Urban Sprawl. In: Annals of the Association of American Geographers, vol. 57, pp. 72–87. Taylor & Francis (1967)
12. De Matteis, G., Lanza, C., Nano, F., Vanolo, A.: Geografia dell'economia mondiale, 4th edn. UTET, Novara (2010)

13. Conti, S., De Matteis, G., Lanza, C., Nano, F.: Geografia dell'economia mondiale, 2nd edn. UTET, Torino (1993)
14. Gregor, H.F.: Geografia de la Agricoltura. Vieens Vives, Barcelona (1970)
15. Jones, A.P., McGuire, W.J., Witte A.D.: A reexamination of some aspects of Von Thünen's model of spatial location. J. Reg. Sci. **18**(1), 1–15 (1978)
16. Greco, I., Cresta, A.: From smart cities to smart city-regions: reflections and proposals. In: Borruso, G., et al. (eds.) ICCSA 2017. LNCS, vol. 10405, pp. 282–295, Springer, Cham (2017)
17. Calabrò, F.: Local communities and management of cultural heritage of the inner areas. an application of break-even analysis. In: Gervasi, O., et al. (eds.) LNCS, vol. 10406. Springer (2017)
18. Dolores, L., Macchiaroli, M., De Mare, G.: Sponsorship for the sustainability of historical-architectural heritage: application of a model's original test finalized to maximize the profitability of private investors. Sustainability **9**(10), 1750 (2017)
19. Napoli, G., Giuffrida, S., Trovato, M.R., Valenti, A.: Cap rate as the interpretative variable of the urban real estate capital asset: a comparison of different sub-market definitions in Palermo, Italy. Buildings **7**(3), 80, 1–25 (2017)
20. Sali, G.: Intorno alla rendita fondiaria dei suoli agricoli, Ce.S.E.T.. In: Atti del XXXIX Incontro di Studio, pp. 133–144, Firenze University Press (2010)
21. Ferrara, A., Salvati, L., Sabbi, A., Colantoni, A.: Soil resources, land cover changes and rural areas: towards a spatial mismatch? Sci. Total Environ. **478**, 116–122 (2014)
22. Nesticò, A., De Mare, G.: Government tools for urban regeneration: the cities plan in Italy. A critical analysis of the results and the proposed alternative. In: Murgante, B., Misra, S., Rocha, A.M.A.C., Torre, C., Rocha, J.G., Falcão, M.I., Taniar, D., Apduhan, B.O., Gervasi, O. (eds.) ICCSA 2014, Part II, LNCS, vol. 8580, pp. 547–562, Springer, Heidelberg (2014)
23. Bencardino, M.: Dinamiche demografiche e consumo di suolo negli ambienti insediativi della Regione Campania, 1st edn. Libreriauniversitaria.it Edizioni, Limena PD (2017)
24. De Mare, G., Nesticò, A., Macchiaroli, M.: Significant appraisal issues in value estimate of quarries for the public expropriation, Valori e Valutazioni 18, 17–23. DEI Tipografia del Genio Civile, Roma (2017)
25. Rizzo, S.L., Smeghetto, F., Lucia, M.G., Rizzo, R.G.: Sprawl dynamics in rural–urban territories highly suited for wine production. mapping urban growth and changing territorial shapes in North-East Italy. Sustainability **9**(116), 1–20 (2017)

Land Consumption and Urban Regeneration. Evaluation Principles and Choice Criteria

Francesca Salvo, Massimo Zupi$^{(\boxtimes)}$, and Manuela De Ruggiero

University of Calabria, 87036 Rende, Italy
massimozupi@vodafone.it

Abstract. One of the main problems in the debate around land consumption concerns the need to address actions of urban regeneration towards the restoration of consumed land rather than mere containment of land consumption. The poor appreciation of the value related to those transformations aimed at recovering free areas, imposes an estimative remark aimed at the definition of tools able of comparing the benefits related to ecosystem services and those related to ordinary transformations. A key role needs to be held by local communities that, conveniently involved in the assessment, can help addressing urgencies and emergencies of the studied environment, and at the same time actively participate to the process of territorial, social and cultural transformation.

Keywords: Land consumption · Environmental sustainability
Highest and Best Use

1 Land Consumption Law in Italy

The enactment of the Land Consumption Government bill by the Chamber of deputies (12 May 2016) has had great emphasis on national press, stimulating a large debate aimed at identifying the limits and the shortcomings of this law.

In particular, ISPRA (Istituto Superiore per la Protezione e la Ricerca Ambientale) taked notice of the definition of land consumption, different than the one made by European Union. This definition, that misses out a large number of areas (public utility services both on a local and general level, infrastructures and priority settlements, areas for the improvement of existing productive activities, enclosed premises, all those actions connected to agricultural activities), could represent an obstacle to the reduction of land consumption, as well as a limitation to monitoring processes [1].

When we talk about soil consumption, one of the main problems are the measurement methods. In fact, we find ourselves in a field in which the survey instruments also play a decisive role: detection systems, data return methods, etc. [2].

In reference to the complexity of the measurement, there are several projects at the European and national level aimed at defining solid methodologies and comparable data for the measurement of land use; among the others, the European Corine Land Cover (CLC) project, the monitoring conducted by ISPRA in collaboration with the National System for the protection of the environment and the LEAC (Land and Ecosystem Accounts) methodology, set up by the European Agency for the Environment, are worth mentioning. However, completely satisfactory results do not match a

© Springer International Publishing AG, part of Springer Nature 2019
F. Calabrò et al. (Eds.): ISHT 2018, SIST 100, pp. 582–589, 2019.
https://doi.org/10.1007/978-3-319-92099-3_65

very wide range of monitoring methods. In fact, they are often dissimilar and produce inhomogeneous, hard-to-compare measurements of the consumed land [3].

From a strictly disciplinary point of view, the weakness of support and promotion devices concerning the objectives of urban reuse and regeneration has been underlined. The few strategies summoned up by the law around this important topic are still too vague and unenterprising [4].

Currently, the government bill is being discussed in the Environment and Agriculture Commissions of the Senate (almost 600 days have passed since enactment by Chamber of deputies) and the attention on this topic is heavily faded. However, environmental associations (FAI, LIPU, WWF, Legambiente) and the National Institute of Urbanism are asking strongly for its approval before the upcoming end of this Legislative period.

Therefore, it is useful to make some remarks about some of the aspects that concern a strictly operative interpretation of the principles mentioned by the Law.

2 Reduction of Land Consumptions Restoration of Consumed Land

Beyond the specific contents of the law and the subsequent limits to its application, in the operational established procedure, there always is the possibility of a more conservative interpretation of the rules and a more innovative way, according to general principles and the specific objectives expressed in it. In particular, the Law properly addresses land consumption "reduction", referring to the need of limitation to further land consumption. However, the provision does not go against the possibility to foster and improve actions aiming at reducing land already consumed, by re-naturalizing impaired areas.

In this perspective, it is useful to start from the definition of urban regeneration addressed by the Government Bill that refers to "a coordinated combination of urban, building and socio-economic actions in consumed areas, including those actions aiming at fostering the settlement of urban agriculture activities, such as urban gardens, teaching gardens, social gardens and shared gardens. Activities that pursue the objective of replacement, reuse and requalification of the built environment, in a sustainable and land consumption reducing way, by addressing transformation operations in built areas, increasing of the ecologic-environmental potential, reducing water and energy consumption and providing proper primary and secondary services" [5].

The need of improvement of the ecological-environmental potential could represent an important bargaining lever to end the land consumption reduction logic (or at the latest compensation logics) and to achieve the definition of effective actions of restoration of consumed land. From this point of view, the initiatives of urban regeneration could free themselves from an approach deeply linked to the concept of reuse (not by chance the two terms are used together in the Law) unable to go beyond the objectives, the aspirations and the results of "urban renewal" that is essentially a process of physical transformation [6], of "urban development", the aims of which remain unclear, and "urban revitalization", that, although expressing the need of an action, does not identify a method, an approach and above all the subjects to which it is directed [7].

Two groups of actions can orientate this new form of urban regeneration, centered on the improvement of the ecological-environmental potential. On the one hand, actions aiming at climate change adaptation of urban settlements could be promoted: green and blue infrastructures; public spaces shaped to act as water storages when storms occur, avoiding the flooding of the surrounding areas; the enlargement of floodplains by watercourses, helping lamination in case of freshet [8]. On the other hand, the extent and health of ecosystems could be improved in order to safeguard their biodiversity and the services they provide, coherently with the 2020 European Strategy for the conservation of biodiversity [9].

3   Environmental Sustainability vs Economic Sustainability

The main obstacle to the possibility to orientate urban regeneration initiatives towards the restoration of consumed land can be detected in the lack of comprehension of the intrinsic economic value that these types of actions embody. Nowadays, we are used to thinking exclusively in terms of market and real estate income, therefore urban transformations are guided only by economic assessments that fulfil these aspects. By contrast, it is essential to prove that, by introducing in decisional processes the environmental benefits assured by free land, we can guarantee a consistent reduction of land consumption and general savings to the community and to public finances, due to the safeguarding natural resources.

Although it is renowned that soil sealing has a certain environmental cost given by the decay of eco-systemic functions and by the altering of ecologic balance, it is clear that restoration of consumed land represents a value, also in economic terms.

However, the assessment of environmental benefits cannot ignore the involvement of the communities that are directly interested by these events of transformation. Only by a direct investigation, made of specific exploration methods, of the particular needs and demands of citizens, we can highlight the direct and indirect utility expressed by the great part of eco-systemic services that a land ensures.

However, this type of approach is involved in the debate around urbanism that centres on topics such as citizenship, common goods and public space. Nowadays, contemporary civil society has the urge to participate, choose, organize and produce its own well-being. This urge is expressed by the replevin of the right to use common goods that need to be taken from private or corporative uses, and handed back to self-arranged communities. The so called collaborative urbanism represents an attempt to channel the new energies of which we dispose, new local based practices, new sense of belonging.

A context such as this one, is the necessary assumption to regeneration actions centred on the improvement of environmental and ecological potential, as long as they are supported and guided by a rigid assessment of costs and benefits, able to take into consideration the benefits given to the soil.

4 Evaluation Approaches and Choice Criteria of Territorial Transformations

The topic of the evaluation of ecological and environmental resources is heart-felt and discussed in the field of assessment.

The consciousness that intangible resources have an intrinsic value, not necessarily extra-monetary, has defined several study perspectives aiming at measuring and evaluating the quality of these resources.

Part of this context need to be considered the assessment perspectives for what concerns urban regeneration and, in particular, those of the restoration of consumed land [10].

In fact, the policies of territorial transformation need to be built on scientific, economic and assessment tools, able to render the correct economic value that urban transformations bear, not only in an ordinary or speculative way, but also those aiming at safeguarding eco-systemic services.

In theory, all real estates are subject to transformation and enhancement, that, in terms of urban and territorial transformations, concern mostly conservation, recovery, requalification of existing immovable complexes, when these goods have architectonic, historic, environmental implications or, on the contrary, demolition and reconstruction for those reale estates that are obsolete and lacking of value.

Recent environmental emergencies force us to think about the opportunity to imagine territorial transformations free from speculative pressure, taking into account also the possibility of restoring consumed land and reducing construction.

It is obvious that whatever building action involves the entire construction sector, real estate and, more in general, the wellbeing of a community, with backlashes on the economic profitability of real estates but also on the added social and cultural value of these transformations.

The necessity of mediating environmental urges and the connected economic sustainability, calls upon an assessment remark that has to give to planners the possibility of choosing between possible transformation actions and demonstrating the opportunity offered by the choices made [11, 12].

Since the transformation process causes a change in the value of immovable resources, the choice among possible transformations needs to be guided by transformation value, in the means of the value of the resources to be transformed defined according to their economic susceptiveness, considering technical, legal and budget constraints.

According to assessment theory, transformation value is an economic syncretic aspect derived from the combination of market price and cost, being conceptually equivalent to the variance between the expected market value of the transformed immovable and the transformation cost:

$$\begin{aligned} \textit{Transformation value}\,(\textit{input}) \\ = \textit{value of the transformed good}\,(\textit{output}) \\ - \textit{cost of transformation} \end{aligned} \tag{1}$$

The transformation cost is defined as the recovery, requalification, construction, demolition and reconstruction cost, according to the type of actions required by the transformation.

In accordance with its distinctive features, transformation value can be used as a criterion or tool for the detection of the most suitable territorial transformations (not only in monetary terms but also in extra-monetary ones).

In fact, transformation value is the most adequate economic marker to identify the *Highest and best use* (HBU), that is the most convenient between the transformation values related to the possible uses that the immovable can host, including also the current use.

HBU performs a counselling role, identifying the most profitable transformation and destination, from the investor's point of view (both public or private, individual or collective).

In this perspective, the assessment methodology, through the transformation value, is an indispensable instrument to orientate territorial and urban planning choices, and mostly in the evaluation of costs and benefits related to the different possible land uses and for regeneration and safeguard policies.

In its most traditional meaning, transformation value can be easily evaluated in ordinary contexts of recovery of existing buildings or in the demolition of immovable complexes that do not have any intrinsic or potential value, and for the subsequent reconstruction, that are respectively calculated as:

$$
\begin{aligned}
&\textit{Transformation value}_1 = \textit{Value of the recovered building} - \textit{Recovery costs} \\
&\textit{Tansformation value}_2 = \textit{Value of the new building} - \textit{Demolition costs} - \textit{Construction} \quad (2) \\
&\textit{costs.}
\end{aligned}
$$

The terms of these equations can be gauged according to the instruments provided by international assessment standards (*market oriented, cost approach, income approach*), thanks to the immovable data of which we dispose for comparison.

More complex is the definition of the transformation value for cases, not yet largely diffused, of clearance of land in order to safeguard eco-systemic systems, biodiversity and environmental benefits assured by free areas and by the recovery of land once consumed.

Although the assessment of transformation value follows the classical definition

$$
\begin{aligned}
\textit{Transformation value}_1 = \textit{Value of cleared area} - \textit{Demolition costs} \\
-\textit{New accommodation costs.}
\end{aligned} \quad (3)
$$

The most complex question is that related to the assessment of the value of the cleared areas.

The assessment of cleared areas requires an approach that is able to mediate between quality and quantity, beauty and efficiency, to reconcile when possible the different, and often contrasting, elements of which the good is made.

From this point of view, the assessment can and has to provide instruments able to support the planner in his project choices through the definition of a complex value that wants to determine the "value of a quality".

The evaluation of eco-systemic services refers to ecological and cultural values [13, 14] and, more generally, is based on the assessment of the total economic value [15], defined as the sum of the value of all the fluxes of the services that the natural capital generates, suitably updated.

There are many values that need to be considered in the economic assessment of the eco-systemic services:

– Use value, related to the utility provided by consumers;
– Option value, related to the will to assure a service in future;
– Existence value, related to the possibility of preserving a service from its destruction;
– Legacy value, that refers to the possibility of using a certain service for future generations.

The total economic assessment related to eco-systemic services has to be defined by appealing to non-monetary assessments (environmental impact assessment), or to monetary evaluations referring to real market (conventional and assessment methods) or to the notion of surplus of the consumer (noted preferences method - travel cost method, oedometric method - and declared preferences method - contingent assessment).

The objective is the definition of values not necessarily related to an effective use of resources or services, and, when it is not possible to state a connection with consumables or the value of private goods, it is necessary to appeal to contingent assessment by faking a non-existing market.

The strong point of this approach is ascribable to the possibility of tautly evaluating the total economic value in all its components (use and non-use value) on one side, and to the methodology based on a participative approach of the involved community, on the other.

It is obvious the contingent assessment approach needs to be suitably planned, case by case, for each community and each specific assessment requirement, mediating environmental and ecological requests and the economic convenience assessment.

The direct involvement of the community, that has the possibility to express itself in the contingent assessment, offers the possibility to highlight the emergencies and necessities of the community and, at the same time, allows citizens to gain awareness on the value related to eco-systemic benefits granted by land.

In fact, the economic tool offered by the assessment approach, gives the planner the possibility to concretely demonstrate that, in certain circumstances, the economic value related to the restoration of consumed land can exceed that derived from a common building transformation, fostering also the developing of a political and urban culture based on active participation.

5   Conclusions

The definition of the value of cleared land, made possible thanks to the assessment of the total economic value, is important for a precise study of the general context of the possible transformations of dismissed areas, in order to identify the Highest and Best Use.

The possibility of using an economic parameter able to quantify in monetary terms the benefit related to the clearance of consumed land, that can be compared to that derived from the most common transformation actions, can help urban policies to exceed a speculative vision normally oriented to building transformation, safeguarding the eco-systemic functions and ensuring the ecologic balance.

By overtaking a vision oriented by the concept of reuse and embracing one based on urban regeneration aiming at safeguarding and promoting the ecological and environmental services, represents an obligatory challenge for current urban policies.

From this point of view, a central role needs to be played by communities that, suitably involved with awareness of its environmental responsibility, can evidence the capability of participation, active citizenship, sympathy, responsibility, all essential to the promotion of a correct territorial, cultural and social transformation.

References

1. Munafò, M., Marinosci, I., Riitano, N.: Disegno di legge sul contenimento del consumo del suolo e riuso del suolo edificato. In: Rapporto ISPRA 2016, Consumo di suolo, dinamiche territoriali e servizi ecosistemici (2016)
2. Bonora, P. (ed.): Atlante del consumo di suolo. Per un progetto di città metropolitana. Il caso Bologna, Baskerville, Bologna (2013)
3. Bencardino, M.: Demographic changes and urban sprawl in two middle-sized cities of Campania region (Italy). In: Murgante, B., et al. (eds.) ICCSA 2015, Part IV. LNCS, vol. 9158, pp. 3–18. Springer, Cham (2015)
4. Arcidiacono, A., Di Simine, D., Oliva, F., Ronchi, S., Salata, S.: La dimensione europea del consumo di suolo e le politiche nazionali. INU Edizioni (2017)
5. Disegno di Legge 2039, Contenimento del consumo del suolo e riuso del suolo edificato, testo approvato dalla Camera dei Deputati (2016), http://www.regioni.it/cms/file/Image/upload/Aula_120516.pdf. Accessed 06 Dec 2017
6. Couch, C.: Urban Renewal: Theory and Practice. Macmillan, London (1990)
7. Musco, F.: Rigenerazione urbana e sostenibilità. FrancoAngeli, Milano (2009)
8. Lenzi, S., Filpa, A.: Nuove prospettive per il riuso delle aree dismesse; ospitare gli interventi per l'adattamento climatico degli insediamenti urbani. In: Rapporto ISPRA 2016, Consumo di suolo, dinamiche territoriali e servizi ecosistemici (2016)
9. COM, 244, EU Biodiversity Strategy to 2020 (2011). http://ec.europa.eu/environment/pubs/pdf/factsheets/biodiversity_2020/2020%20Biodiversity%20Factsheet_EN.pdf. Accessed 06 Dec 2017
10. Morano, P., Tajani, F., Locurcio, M.: GIS application and econometric analysis for the verification of the financial feasibility of roof-top wind turbines in the city of Bari (Italy). Renew. Sustain. Energy Rev. **70**, 999–1010 (2017)

11. Morano, P., Tajani, F.: Decision support methods for public-private partnerships: an application to the territorial context of the Apulia Region (Italy). In: Stanghellini, S., et al.: Appraisal: from Theory to Practice, Green Energy and Technology, pp. 317–326 (2017)
12. Morano, P., Tajani, F.: Evaluation of the financial feasibility for private operators in urban redevelopment and social housing investments. In: Gervasi, O., et al. (eds.) Computational Science and Its Applications, Part III. LNCS, vol. 9788, pp. 473–482 (2016)
13. Gómez-Baggethun, E., Martín-López, B., García-Llorente, M., Montes, C.: Hidden values in ecosystem services. A comparative analysis of preferences outcomes obtained with monetary and non-monetary valuation methods. In: Paper Presented at DIVERSITAS OSC2 Biodiversity and Society: Understanding Connections, Adapting to Change, 12–16 October, Capetown (2009)
14. Christie, M., Fazey, I., Cooper, R., Hyde, T., Kenter, J.O.: An evaluation of monetary and non-monetary techniques for assessing the importance of biodiversity and ecosystem services to people in countries with developing economies. Ecol. Econ. **83**, 67–78 (2012)
15. Heal, G.M., Barbier. E., Boyle. K., Covich, A., Gloss, S., Hershner, C., Hoehn, J., Pringle, C., Polasky, S., Segerson, K., Shrader-Frechette, K.: Valuing Ecosystems Services: Toward Better Environmental Decision-making. National Research Council, Washington D.C. (2005)

Land Value Hot-Spots Defined by Urban Configuration

Konstantinos Lykostratis$^{(\boxtimes)}$ and Maria Giannopoulou

Democritus University of Thrace, 67100 Xanthi, Greece
klykostr@civil.duth.gr

Abstract. Location characteristics and accessibility are widely used in real estate research as they are considered to predominantly shape property values. Space syntax is a well-known methodology for urban analysis, elaborating accessibility measures of the urban grid. Even though urban morphology has a crucial role in property market structure, Space Syntax centrality measures integration and choice, which are used to quantifying urban space accessibility, have gained little attention in land value literature. Recently, there has been a growing interest for spatial statistics which embrace all tools needed for studying spatial phenomena such as the land market, so that spatial relationships and spatial patterns can be interpreted. The proposed methodological framework of the present paper attempts to explore the relationship between land value and geometric accessibility measures (global integration and choice), applying a spatial approach which combines LISA indicators and Space Syntax theory in order to visualize local correlation patterns. In order to be tested and validated, the framework is applied to Xanthi, a medium-sized city in Northern Greece. Results indicate significant relationships between geometric accessibility and objective land values, along with differences in the emerging spatial relationship patterns between different parts of the city.

Keywords: Accessibility · Space Syntax · LISA · Objective land value

1 Introduction and Background

Paraphrasing Kiel and Zabel's [1] view about houses, for land property at large, it can be stated that urban land property is by definition an immovable commodity, hence its value is shaped predominantly by location characteristics. Orford [2] notes that location characteristics, which are property-unique, are synonym to accessibility. Accessibility measures are widely used in real estate research as they define urban economy and quality of life [3, 4]. Space syntax is a well-known methodology of urban analysis, which uses computer techniques to analyze urban configuration.

Accessibility is hard to be defined and measured [5]. In land value research it is mostly modeled as the distance from business centers, proximity to public transport and services such as schools and retail, or potential for jobs (these kinds of measures are used e.g. in [4, 6–8]). Studying centrality as a spatial procedure [9], Space Syntax theory elaborates accessibility measures of the urban grid in order to analyze urban space morphology and to describe relationships between built space and its social

© Springer International Publishing AG, part of Springer Nature 2019
F. Calabrò et al. (Eds.): ISHT 2018, SIST 100, pp. 590–598, 2019.
https://doi.org/10.1007/978-3-319-92099-3_66

aspects [10, 11]. Space Syntax analysis takes into account topological dependence of parts which compose urban open spaces, spatial arrangement of which, defines urban grid's structure. According to Space Syntax developers open space network is a continuous system that can be divided in segments [12], thus allowing urban grid to be represented as an axial map [13]. Axial maps are studied through axial graphs which are analyzed using graph theory [14]. Space syntax measures quantifying urban space accessibility are based on axial graph analysis.

The centrality measures calculated and used in this research are integration and choice, describing the relationship between an axial line to all other axial lines of the system [9] in terms of topology, geometry [15] and direction which define the notion of distance [16]. Integration measures mean depth (depth here should be considered as the number of lines travelled to move from the origin line to the destination line) of a segment to every other segment of the network [17], reflecting how far a segment is to all other segments, forming a measure of proximity [18], a measure reflecting how central a space is to all other spaces of the network [19]. Choice measures the participation degree of a line in all shortest paths between all network lines [16]. Different analysis approaches have been proposed based on the notion of distance in shortest paths [15]. Spatial analysis based on Space Syntax theory can be executed for the grid as a whole examining the relationship between every segment of the network with all other segments (global parameters) but can also be executed for grid parts defined by depth (in Space Syntax terms) (local parameters).

Even though urban morphology has a crucial role in property market structure [20], accessibility defined by Space Syntax has gained little attention in land value literature [21]. Studies relating Space Syntax with housing values [22] indicate statistically significant positive relationships with global integration besides statistically significant negative correlations with choice in models using housing prices [23–25]. Corresponding inferences are figured for global integration in regression models accounting housing tax bands [26, 27]. Bivariate correlations reveal that global integration is positively related to land values [28]. Moreover, Entsröm and Netzell [29] conclude that global integration helps substantially in office rent value interpretation, revealing a pattern [20]. Furthermore, conflicting results on variable influence is noticed, probably due to different research areas and spatial layout [27].

Recently, there has been growing interest for spatial statistics in the description of spatial phenomena such as the land market [30, 31], originating from the always growing use of spatial data. In contrary with a-spatial data, spatial data involve the spatial aspect as in the case of distances or measurements of phenomena with spatial reference [32, 33]. Spatial statistics embrace all tools needed for studying spatial phenomena, so that spatial patterns and spatial relationships can be interpreted. In this context, Local Indicators of Spatial Association (LISA) have been proposed [34], quantifying the degree to which spatial correlation between two variables exists at local scale, while allowing for statistical significance testing [35]. LISA can help identify hot-spots [36] sub-areas of the research area as they produce discernible clusters [37]. Local indicators are generally used in order to identify clusters arising from global correlation [38], possible locations of specifically significant clustering [36].

The proposed methodological framework of the present paper attempts to explore the relationship between land value and geometric accessibility measures, applying a spatial approach which combines LISA indicators and Space Syntax theory in order to visualize local correlation patterns. In order to be tested and validated, the framework is applied to Xanthi, a medium-sized city in Northern Greece [22].

2 Methodological Framework and Case Analysis

The data used in this research refer to 7415 land parcels covering the entire city of Xanthi, a city of 55000 inhabitants in Northern Greece. The urban tissue of Xanthi (Fig. 1a) consists of distinct sections with a particularly interesting variety in form and density: traditional parts with a coherent organic tissue, newer extensions with rectangular grid or normal geometries with great variety in the size of blocks [16]. Parcels which constitute the spatial level of the analysis along with blocks forming Xanthi's urban grid were modeled in ArcGIS in order to be joined with necessary descriptive data. The objective value as set by the Greek Ministry of Finance was selected to be used as the land value. This was decided due to the fact that, in Greece, property taxes and taxes on transfers, inheritance and donations are based on objective value which is determined by parcel condition, capitalization potential (defined by floor space ratio) and parcel location (in terms of having front facing commercial road). Moreover, transaction prices and housing prices are based on objective value. Thus, the objective value was calculated for every parcel. Space syntax measures were calculated using DepthmapX [39, 40]. Segment angular analysis [16] was chosen in order to calculate space syntax accessibility measures. Analysis results (global integration and choice values) were assigned for each land parcel to its nearest segment using spatial join ArcGIS techniques.

Bivariate global Moran's I (1) as an index of spatial correlation between the variables [35] was adopted in order to assess the existence of spatial clusters. Then, bivariate local Moran's I (2) was used, exploring spatial patterns between geometric accessibility measures and objective land value. These local indicators are defined as:

$$I_{ab} = \frac{z_a W z_b}{n} \tag{1}$$

$$I_{ab}^i = z_a^i \sum_J w_{ij} z_b^j \tag{2}$$

where w is the row-standardized spatial weights matrix, z_a is the standardized value of objective land value (z_a^i being the standardized value at location i) and z_b is the standardized value of accessibility (z_b^j being the standardized value of neighboring location j). The weight matrix was based on the nearest neighbors approach with 20 neighbors chosen as nearest, a common choice in real estate analysis. Statistical significance of the bivariate relationship is based on a permutations approach with 999 random permutations constructed. GeoDa (version 1.12.0) software [41] was used for analysis and ArcGIS for map visualization.

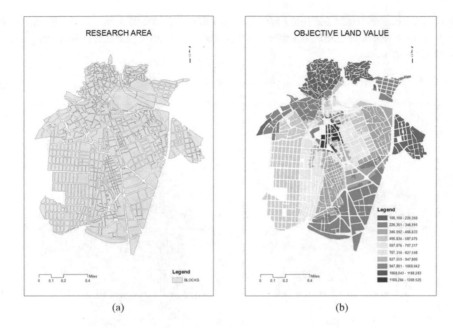

Fig. 1. (a) Research area. (b) Objective land value.

3 Results and Discussion

An equal interval classification of the objective values is presented in Fig. 1b, with the lowest values appearing in the traditional parts and the highest ones in the city centre. For Fig. 2a and b, equal interval classification was chosen with blue shades indicating lower values of the accessibility variables rising up gradually to beige and finally red shades indicating higher values. Integration map (Fig. 2a) illustrates that the city centre along with principal road axes and axes with retail activity are highly integrated, in contrast with the obviously segregated areas surrounding city's main urban core. Spatial analysis results are consistent with previous work [29] relating high integration values with gridiron urban tissue and low levels with cul-de-sacs. In this case, traditional city parts with organic tissue are the ones with the lowest integration values. Figure 2b illustrates the segments of the network that are mostly traversed in any possible travel choice. These segments are actually the parts of the network that link the traditional parts of the city with the newer extensions, also having (as proposed by Law et al. [27]), high business and retail activity.

Spatial correlation between global integration and land value is reported in Table 1. Both cases have a bivariate global Moran's I > 0 with z-values > 100 at p = 0.001 level, indicating the existence of positive spatial correlation and spatial clusters. As already stated Local indicators of association (LISA) identify the location and type of association between variables. LISA results are presented in maps 5 and 6, showing that a spatial pattern exists for both accessibility parameters at local scale. In general, significant observations for integration and choice at p < 0.05 level are 63% and 52% respectively. Results indicate differences in the emerging spatial relationship pattern between city parts.

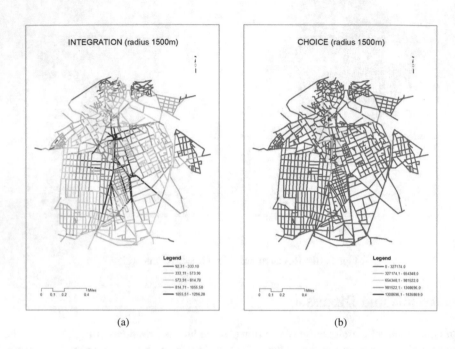

Fig. 2. (a) Integration (radius 1500 m) measures. (b) Choice (radius 1500 m) measures.

Table 1

Variables	Moran's I	sd	Z-value	p-value
Integration	0,5917	0,0029	201,7121	0,001
Choice	0,2718	0,0026	105,7353	0,001

Figure 3a illustrates the existence of both positive and negative relationships between integration and objective values. Low integration and objective land values are located in northern and eastern Xanthi, such as the traditional city parts, which are mostly residential areas with low Floor Space Index (FSI) (FSI is used as measure of density). Significant high integration and high objective value regions are concentrated in inner locations mostly around city centre and roads with high commercial retail activity and high FSI. Of particular interest are the LH clusters (low objective value and high integration) mostly located in the southern city parts between highly integrated road axes.

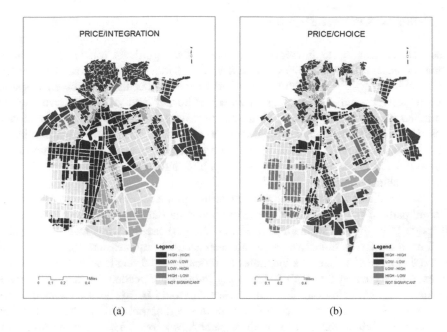

(a)   (b)

Fig. 3. (a) LISA analysis for Integration. (b) LISA analysis for Choice.

These clusters could be divided in two parts: one located by the city centre, related to high integration values, high density (expressed through FSI) but small lots, and the other located in the southernmost part of the city where although high integration values are concentrated, FSI is medium.

LISA analysis results for choice and objective land values are presented in Fig. 3b. Deep red colored (HH) areas indicate that the higher the choice, the higher the land values. This region coincides with the city centre, which is marked by strong commercial and business activity and road axes which comprise retail activity cores. In segregated areas (LL) in northern, southern and eastern Xanthi, low objective values are related with low choice values. These are residential areas defined by low building density (low FSI, low building height, detached houses, small lots). Of special concern is the LH region located in the southern part of the traditional core where objective values relate negatively with choice. Grid complexity justifies higher choice values, while this part of the traditional core although sharing the same density conditions stated above, is an emerging core of eating/leisure activity. Finally one more cluster (HL) is arising in the southwestern part of the city where higher objective values are noticed along with lower choice values. This region (mostly a residential area), unlike the traditional core, is characterized by high building and population density (high FSI and flats) with an urban layout following a gridiron pattern.

4   Conclusions and Perspectives

This research has led to inferences concerning the significant relationship between geometric accessibility and objective land values. Moreover, it is more than evident that combining spatial association tools and Space Syntax theory allows for spatial relation clusters to emerge, helping the interpretation of land values. Recognizing the parameters that form land value is an important process regarding real estate activities, loans, infrastructure financing, investments or even determination and distribution of taxes. Understanding the relationship between those parameters and land value can be very useful for urban planners as well as public and private investors in all urban decision-making processes.

Results indicate differences in the emerging spatial relationship patterns between different parts of the city. One of the most interesting outcomes is that although some city parts (city centre, northern and eastern Xanthi) are showing respective spatial relationships between objective value and each accessibility measure, for all other city parts only one of the measures illustrates significant spatial relationship with objective values, thus reflecting that both accessibility indexes are needed for land value interpretation. An interesting idea for future research would be to focus on spatial correlation analysis of integration and choice measures in different radii, probably shedding this way more light to the land value phenomenon. Moreover, analysis of multivariable models with Space Syntax measures, parcel data and neighborhood structure variables could be applied to show the magnitude of each variable.

References

1. Kiel, K.A., Zabel, J.E.: Location, location, location: the 3L approach to house price determination. J. Hous. Econ. **17**(2), 175–190 (2008)
2. Orford, S.: Valuing locational externalities: a GIS and multilevel modelling approach. Environ. Plann. **29**(1), 105–127 (2002)
3. Nilsson, P.: Natural amenities in urban space - a geographically weighted regression approach. Landscape Urban Plann. **121**, 45–54 (2014)
4. Spinney, J., Maoh, H., Kanaroglou, P.: Location and land values: comparing the accuracy and fairness of mass appraisal models. Can. J. Reg. **37**(1/3), 19–26 (2014)
5. Handy, S.: Planning for accessibility: theory and practice. In: Levinson, D.M., Krizek, K. J. (eds.) Access to Destinations, pp. 131–147. Elsevier Publishers (2005)
6. Asabere, P.K., Huffman, F.E.: Zoning and industrial land values: The case of Philadelphia. Real Estate Econ. **19**(2), 154–160 (1991)
7. Cervero, R., Duncan, M.: Neighbourhood composition and residential land prices: Does exclusion raise or lower values? Urban Stud. **41**(2), 299–315 (2004)
8. Kok, N., Monkkonen, P., Quigley, J.M.: Land use regulations and the value of land and housing: an intra-metropolitan analysis. J. Urban Econ. **81**, 136–148 (2014)
9. Cutini, V.: Centrality and land use: three case studies on the configurational hypothesis. Cybergeo: Eur. J. Geogr. (2001). Systems, Modelling, Geostatistics, document 188
10. Bafna, S.: Space syntax: a brief introduction to its logic and analytical techniques. Environ. Behav. **35**(1), 17–29 (2003)

11. Teklenburg, J.A.F., Timmermans, H.J.P., Van Wagenberg, A.F.: Space syntax: standardised integration measures and some simulations. Environ. Plann. **20**(3), 347–357 (1993)
12. Hillier, B., Hanson, J.: The Social Logic of Space. Cambridge University Press, Cambridge (1984)
13. Hillier, B., Penn, A., Hanson, J., Grajewski, T., Xu, J.: Natural movement: or, configuration and attraction in urban pedestrian movement. Environ. Plann. **20**(1), 29–66 (1993)
14. Penn, A., Turner, A.: Space syntax based agent simulation. In: Schreckenberg, M., Sharma, S.D. (eds.) Pedestrian and Evacuation Dynamics, pp. 99–114. Springer, Germany (2002)
15. Hillier, B., Lida, S.: Network and psychological effects in urban movement. In: International Conference on Spatial Information Theory, COSIT 2005, Ellicottville, NY, United States. LNCS, vol. 3693, pp. 475–490. Springer, Heidelberg (2005)
16. Hillier, B., Vaughan, L.: The city as one thing. Prog. Plann. **67**(3), 205–230 (2007)
17. Hillier, B.: Space is the Machine: A Configurational Theory of Architecture, E-edition. Space Syntax, London (2007)
18. Karimi, K.: A configurational approach to analytical urban design: Space syntax methodology. Urban Des. Int. **17**(4), 297–318 (2012)
19. Law, S., Stonor, T., Lingawi, S.: Urban value: measuring the impact of spatial layout design using space syntax. In: Kim, Y.O., Park, H.T., Seo, K.W. (eds.) Proceedings of the Ninth International Space Syntax Symposium, pp. 061.1–061.14. Sejong University Press, Seoul (2013)
20. Desyllas, J.: Berlin in transition: using space syntax to analyse the relationship between land use, land value and urban morphology. In: Major, M.D., Amorim, L., Dufaux, D. (eds.) Proceedings of the First International Space Syntax Symposium, pp. 04.1–04.15, University College London, London (1997)
21. Xiao, Y., Webster, C., Orford, S.: Identifying house price effects of changes in urban street configuration: An empirical study in Nanjing, China. Urban Stud. **53**(1), 112–131 (2016)
22. Giannopoulou, M., Vavatsikos, A.P., Lykostratis, K.: A process for defining relations between urban integration and residential market prices. In: Calabrò, F., Della Spina, L. (eds.) Procedia - Social and Behavioral Sciences, vol. 223, pp. 153–159. Elsevier (2016)
23. Law, S., Karimi, K., Penn, A. Chiaradia, A.: Measuring the influence of spatial configuration on the housing market in metropolitan London. In: Kim, Y.O., Park, H.T., Seo, K.W. (eds.) Proceedings of the Ninth International Space Syntax Symposium, pp. 121.1–121.20. Sejong University Press, Seoul (2013)
24. Matthews, J.W., Turnbull, G.K.: Neighborhood street layout and property value: The interaction of accessibility and land use mix. J. Real Estate Financ. Econ. **35**(2), 111–141 (2007)
25. Shen, Y., Karimi, K.: The economic value of streets: mix-scale spatio-functional interaction and housing price patterns. Appl. Geogr. **79**, 187–202 (2017)
26. Chiaradia, A., Hillier, B., Barnes, Y., Schwander, C.: Residential property value patterns in London: space syntax spatial analysis. In: Koch, D., Marcus, L., Steen, J. (eds.) Proceedings of the 7th International Space Syntax Symposium, pp. 015.1–015.12. Royal Institute of Technology (KTH), Stockholm (2009)
27. Narvaez, L., Penn, A., Griffiths, S.: Space syntax economics: decoding accessibility using property value and housing price in Cardiff, Wales. In: Greene, M., Reyes, J., Castro, A. (eds.) Proceedings of the Eighth International Space Syntax Symposium, pp. 1–19. Santiago de Chile (2012)
28. Min, K., Moon, J., Kim, Y.: The effect of spatial configuration on land use and land value in Seoul. In: Kubat, A.S. Ertekin, O., Guney, Y.I., Eyubolo, E. (eds.) Proceedings of the Sixth Space Syntax Symposium, pp. 080.1–080.16, Istanbul (2007)

29. Enström, R., Netzell, O.: Can space syntax help us in understanding the intraurban office rent pattern? Accessibility and rents in downtown Stockholm. J. Real Estate Finance Econ. **43**(4), 548 (2011)

30. Bencardino, M., Granata, M.F., Nesticò, A., Salvati, L.: Urban growth and real estate income. a comparison of analytical models. In: Gervasi, O., et al. (eds.) ICCSA 2016. LNCS, vol. 9788, pp. 151–166. Springer, Cham (2016)

31. Bencardino, M., Nesticò, A.: Urban sprawl, labor incomes and real estate values. In: Borruso, G., et al. (eds.) ICCSA 2017. LNCS, vol. 10405, pp. 17–30. Springer International Publishing, AG (2017)

32. Casetti, E.: The expansion method, mathematical modeling, and spatial econometrics. Int. Reg. Sci. Rev. **20**(1–2), 9–33 (1997)

33. Fotheringham, A.S., Brunsdon, C., Charlton, M.: Geographically Weighted Regression. Wiley, Chichester (2002)

34. Anselin, L.: Local indicators of spatial association-LISA. Geograph. Anal. **27**, 93–115 (1995)

35. Anselin, L., Syabri, I., Smirnov, O.: Visualizing multivariate spatial autocorrelation with dynamically linked windows. In: Anselin, L., Sergio Rey, S. (eds.) New Tools for Spatial Analysis. CSISS, Santa Barbara (2002)

36. Getis, A.: A history of the concept of spatial autocorrelation: a geographer's perspective. Geograph. Anal. **40**(3), 297–309 (2008)

37. Griffith, D.A.: Spatial autocorrelation. In: Kitchin, R., Thrift, N. (eds.) International Encyclopedia of Human Geography, pp. 308–316. Elsevier (2009)

38. McIlhatton, D., McGreal, W., Taltavul de la Paz, P., Adair, A.: Impact of crime on spatial analysis of house prices: evidence from a UK city. Int. J. Hous. Markets Anal. **9**(4), 627–647 (2016)

39. Turner, A.: Depthmap 4, A Researcher's Handbook. http://www.vr.ucl.ac.uk/depthmap/handbook/depthmap4.pdf. Accessed 22 Apr 2004

40. Varoudis, T.: DepthmapX – Multi-platform Spatial Network Analyses Software. https://github.com/varoudis/depthmapX. Accessed 14 Nov 2012

41. Anselin, L., Syabri, I., Kho, Y.: GeoDa: an introduction to spatial data analysis. Geograph. Anal. **38**(1), 5–22 (2006)

Commons and Cities. Which Analytical Tools to Assess the Commons' Contribution to the Economic Life of the Cities?

Maria Patrizia Vittoria[✉]

Institute for Research on Innovation and Services for Development,
National Research Council (CNR), Naples, Italy
p.vittoria@iriss.cnr.it

Abstract. Cities often manifest strong differences in their de-facto informal political cultures. Do these differences have anything to do with their own performance levels? Moreover, if so, what are the precise mechanisms by which they affect performance? Looking at the case of the urban Commons, particularly the Italian Social Centres started at the end of '70s, as a proxy of urban informal political culture, and revisiting the mainstream business economics and its conceptual consequents, the present paper surveys the main questions and inherent analytical tool sets. The aim of the paper is to find a good framework for further analysis on Commons and metropolitan perspectives. For the institutionalists, the Commons have to be studied as alternative governance modes laying beyond the State and Market. While, following the dynamic capabilities view they could be treated as evolutionary inter-organizational solutions with a potential in terms of creativity and entrepreneurship. The conclusion summarizes the cases for analysing urban Commons with the dynamic capabilities approach.

Keywords: Urban Commons · Firm's theories · Local development

1 Introduction

The Commons' issue spreads, as a matter of sociology and political science since the emergence of political opposition movements to the *fin du siècle* neo-liberal ideas. These movements was in general attempted to criticize the world's mercification to the benefit of extending State intervention [1]. In Italy, since the late '70s, their political action (mostly carried out by the Social Centers) was based on the extensive involvement of the citizens and on the material occupation of abandoned urban sites (such as ex-factories, warehouses, schools, and historical buildings) [2–4]. Other studies, concerning law and legislation, have recognised that following different patterns of social interaction, dependent on specific urban identities, some initiatives were able to push the local public authorities to recognise the 'civic use' of resources, introducing a new model of 'participatory administration' [5].

In the field of the economic analysis, the Commons issue can be treated in different ways depending by the selected methodological approach. In brief, we can distinguish two basic approaches. One follows the traditional focus of the political economics.

© Springer International Publishing AG, part of Springer Nature 2019
F. Calabrò et al. (Eds.): ISHT 2018, SIST 100, pp. 599–605, 2019.
https://doi.org/10.1007/978-3-319-92099-3_67

Where the production of goods, distribution of the social product, and allocation of resources are the economic system's main functions and the feudal-mercantilist, capitalistic and socialist one are the main response modes. Here, the Commons issue has been seen and discussed as a model of production [6] and new infrastructural paradigm, able of managing trust until the creation of a new currency [7]. By another side, a second approach overcomes the idea, supported by the traditional one, of the firm as a mere decision-maker of prices and quantities for a fixed quantity of products. Here, starting from the crucial Penrose's scientific contribution, the firm has been considered as an organization. A pool of (internal and external) resources implying complex competitive strategies, no more bounded with a single market, able to adopt different development logics, and finally able of influencing the surrounding economic environment.

Since the emergence of these ideas, there was a growing flow of research contributions, where a complete review [8] suggests distinguishing them in two basic addresses: managerial and behaviorist. It is in this field of analysis and on the contributions that have been produced on the specific theme of the Commons that we would like to focus our attention. Here, in the late 1990s, starting from the managerial theories of the firm and its consequents, Elinor Ostrom provided the most comprehensive theoretical contribution to the analysis. She focused on the rules that could make effective the investments at the margin between private and public interests. Her major contribute was in reaching a kind of Commons' heuristics demonstrating the limits of a framework (mostly grounded in the transaction cost economics) based on production costs and human actors' (bounded) rationality.

In the meanwhile, less has been said on the how some Commons' organisational ability have improved the social value creation processes at both a firm and urban levels. In our view, following the scientific Ostrom's suggestions and the theoretical advances made in general in the field of firm's theories there is a good chance of upgrading the analytical tool set.

In trying to recollect the main argumentations at the base of this conceptual transition, the present paper contributes to calibrate the emerging answers about the future of the contemporary urban commons within a selected organisational framework. In our view, the proposed analytical tool kit will better fit the upcoming Commons policy needs with the local economic transformation.

2   Treating the Commons Issue with Firm's Theories. From the Institutional 'Structure of Efficient Governance' Towards 'Purposefully-Participative Creative Action'

How the research approach in studying the Commons could change according to the new theories of the firm? To what extent is it useful to insist on the search for the possible rules that govern them when the analysis should give information about their dynamic creative and innovative organizational capabilities?

Far from the idea of covering the whole survey of the firm theories produced in the last decades, this section proceeds around the fil-rouge developed by the Ostrom's basic assumptions and her research strategy.

In particular, the early Ostrom's concerns were about finding the institutional arrangement to overcome the crucial dilemma between selfish interests and the use of resource in common. Imposing the idea of economic actors rationality, she adopted the game theory as analytical tool [9].

This research strategy has been justified and dates back to the Coase [10] conception of the nature of the firm. As it is widely known, in the Coase view, firms and markets were seen as alternative modes of governance, the choice between them made to minimize transaction costs. The boundaries of the firm were set by bridging transactions into the firms so that at the margin the internal costs of organizing equilibrated with the costs associated with transacting in the market.

Initiated by Coase seminal paper, a substantial literature has emerged on the relative efficiencies of firms and markets. This literature, greatly expanded by Williamson [11, 12] and others, has come to be known as Transaction Costs Economics (TCE). It analyzes the relative efficiencies of markets and internal organisation, as well as intermediate forms of organisation such as strategic alliances.

Contractual difficulties associated with asset specificity are at the heart of the relative efficiency calculations in TCE. When specific assets are needed to support efficient production, then the preferred organisational mode is internal organisation. Vertical and other forms of integration are preferred over contractual arrangements when efficient production requires investors to make irreversible investments in specific assets. The structures used to support transactions are referred to as governance modes. Internal organization (doing things inside the firm) is one such governance mode.

Thus, the TCE accepted that to think of the firm and markets as alternative modes of governance useful. Relatedly, the selection of what organise internally depends on the nontradability of assets in term of transaction costs.

To have a view of the conceptual evolutions to the Coase and Williamson's TCE approach, we need to draw attention on the transaction costs basic conceptualisation. This expression refers to costs that must be incurred to make an exchange, a contract or an economic transaction in general; they represent the costs of using the market. In fact, relationships that economic agents establish on the market would be at zero cost only if the information were perfect, complete, distributed symmetrically, there was no uncertainty and the contractors were perfectly rational. In general, however, agents must incur costs for establishing relationships (research costs of the contractor and brokerage services), for the negotiations, for the definition and drafting of contracts (the costs of consultants and lawyers) and for the control of compliance with the agreements (costs for monitoring the activities of the parties and enforcement). The magnitude of these costs depends on the type of transactions and in particular: (a) the degree of specificity of the physical and human capital involved; (b) uncertainty; and (c) the frequency of transactions.

High transaction costs do not necessarily imply that market exchanges are impossible. Rather, it implies that the idea of perfect markets (i.e. the market as efficient coordination of economic assets) and spot contracts (as a tool for fixing the exchanging price) has been replaced by organisational strategies based on behavioural control.

Thus, the analysis of how institutions are formed, how they operate and change, and how they influence behaviour in society has become the Ostrom's (2005) major subject of inquiry.

To these issues the early critics involved converting the notion of assets' non-tradability entirely into the concept of transaction costs. Initially, Alchian and Demsetz [13], tried to operationalize the concept of transaction costs with 'technological non-separabilities[1], while Williamson (1985) focused on 'specific assets[2].

With the Nelson and Winter (1982) evolutionary approach behavioural routines and path-dependencies are involved in the dominant conception of the nature of the firm. The memory of the firm is, therefore, the basis of its behaviour. When necessary (i.e. when the firms outcomes become non-satisfactory), the firm seeks new routines both within and outside of itself. The search could lead to the introduction of new process-products or even at the exit of the firm from the market. The main concern of the firm is no more the optimal production structure but to organise itself and fit the changing competitive environment.

Thus, the optimizing interacting agents to obtain a market equilibrium have disappeared. In addition, a comparison between companies in the same sector through their cost structures becomes meaningless.

When, recovering the seminal contribution of Penrose (1959), a dynamic resource-based view of the firm was attempted, further reasons (beyond the asset specificity) why assets are considered 'not tradable' were introduced. David Teece – the main sponsor of the theory - explains the assets' uniqueness through some examples: "...the land on the corner of Park Avenue and 59th Street in New York City rarely comes onto the market. The ability to write highly creative and efficient software for computer operating systems is not widely distributed. Brands that signal particular values (e.g. Lexus) are likewise thinly traded" [14]. Moreover, he firstly introduced the concept and discussed the inherent organisational implications of assets co-specialization. "The assets that are co-specialised to each other need to be employed in conjunction, often inside the firm. This isn't the emphasis of Coase, Alchian and Demsetz, or of Williamson. Assembling co-specialised assets inside the firm in the Dynamic Capabilities (DC) framework is not done primarily to guard against opportunism and recontracting hazards, although in some cases that may be important. Instead, because effective coordination and alignment of these assets/resources is difficult to achieve through the price system, special value can accrue to achieving good alignment within the firm. This is different from what Barnard (1938) has suggested with his emphasis on the functions of the executive as rooted in cooperative adaptation of a conscious and deliberate kind. Here the focus is on the 'orchestration' of co-specialised assets by strategic managers. It is *pro-active* [purposefully and/or participative] process designed to: (1) keep co-specialised assets in value-creating co-alignment, (2) select new co-specialised assets to be developed through the

---

[1] This view introduced to the idea of the firm as 'team production function'.

[2] The central assumption in this view was that the purpose of organizations is to reduce the transaction costs. Where 'market' is the most suitable situation in conditions of certainty, non-specificity of capital, and occasional transactions, 'hierarchy' is much more effective in opposite situations.

investment process, and (3) divest or run down co-specialised assets that no longer help yield value. Rather than stressing opportunism, the emphasis in DC is on change processes, *inventing and reinventing* the architecture of the business, asset selection, and asset orchestration" [14].

The DC framework is centrally concerned on the strategic management function, which transcends the question of optimal firm boundaries. Value can be created by astutely organising assets both inside and outside the firm. This view, concerning the strategic management function, which transcends the question of optimal firm boundaries, disclosed the way for the 'networked organisations' described as characterised by lateral or horizontal patterns of exchange, interdependent flows of resources, and reciprocal lines of communication [15]. The opportunity to create competitive advantage and rents from alliance relationships and the key factors driving them have been discussed in Dyer and Singh's [16] article on the relational view of the firm. Using the Teece definition of DCs, relational capability can be viewed as a type of dynamic capability with the capacity to purposefully create, extend, or modify the firm's resource base, augmented to include the resource of its alliance partner. The relational capabilities are a precondition for firms to access the benefits from their network ties.

Moreover, Alchian, Demsetz, and Williamson have all emphasised opportunistic free riding. Indeed, their human actors are assumed to be boundedly rational, self-interested seeking, opportunistic. While, the DC framework adds other (arguably less ubiquitous) traits of human nature: (1) intrapreneurship and entrepreneurship, and (2) foresight and acumen.

In the DCs relational view of the firm, formal contracts should effectively protect the interests of each side, and there is a greater likelihood of equity-based relationships when there are high levels of asset specificity on each side. Here, the new element of effective governance is the importance of informal safeguards (such as trust and reputation) in protecting the interests of each side from opportunism. The effective governance entails the choice of the appropriate mix of formal and informal safeguards to govern the partnering relationships.

Comparing the TCE and DC views of the firm emerge two different analytical perspectives and methodologies. Like resource-based theory, the DC view is content-oriented. A central concern has been its connection to wealth creation and capture [17]. It is based not on mainstream economics but on evolutionary economics [18]. It claims also for research on strategy process and a much wider set of disciplines, including sociology, psychology, decision science and political science.

3   Conclusion

The overview of the firm's theories since the Coase seminal paper and until the most recent dynamic resource-based view produced by Teece and other colleagues (widely recognized as the DC framework) together with our experience in the field allow us evidencing the cases for analysing urban Commons with the DC approach.

Our experience on the field is grounded in a case-study research conducted on 30 social centers located in the Neapolitan urban area. The empirical analysis has been conducted in the April-July and September-December 2016 fieldwork periods [19].

Main evidences have testified that beyond the burdens deriving from the initial political conflict (between the political activists and local authorities), which is established for the occupation of the site, many stable organisations are able to return with subsistence expenses through the organisation of cultural events. In these cases, we have noted that the self-financing (often aided with specific on-line crowd funding campaigns) is the inherent organisational ability. A good level of sensitivity with the context, in general and a specific strategy of resources acquisition are in the focus of some of them. In practice, the relational skills come into play when a relationship of trust is established with the neighbourhoods' residents. Inclusive environments to the residents and free social and legal services are the main forms of local social interaction.

Thus, if we share with the evolutionary theorists the idea that every firm (organisation) is different from another, it follows that we can assume that every Commons is different from another. Looking at a single case searching the inherent conditions of internal efficiency, an explorative progressive-stages analysis can be designed.

As suggested by the DC view, a great influence in the firm's efficiency construction is in its external relational abilities. Moreover, the organisational ability to engage and maintain an external informal network in knowledge-based domains is a possible source of value creation. This statement applies to not for profit as well as for profit organisations.

In the Commons case, individuals motivated by political engagement are more likely to be inclined to make altruistic decisions and to embark on bonds based on trust. Long distance on-line knowledge networks are easily established with other activists and high personalities in different fields. Due to these links, they have potential advantages (for example, the speeding up the feedbacks) in different knowledge management situations.

In going to step up a level in geographical scale examining the space of city, the free group membership and grassroots participation (as organizational issues) at the base of the Commons process of creation can be considered as a 'giving voice' process at collective urban preferences. The Commons' ability in giving voice at collective preferences as well as their recognition as appropriate mix of formal and informal safeguards governing effectively the urban challenges (cfr. par. 2 in the text) issues are, in parallel, interesting the last local development theory [20] and policy [21, 22]. As known, the smart specialization policy maker is interested in the Entrepreneurial Discovery Process (EDP). The EDP importance is related to the recognition that the government does not have innate wisdom or the ex-ante knowledge about future priorities.

Often entrepreneurial individuals that are well-placed to explore and identify new activities often will not have sufficient connections to marketing and financing sources, reducing their incentives to enter in the process; likewise, "discovery" opportunities may be constrained due to the large-scale investments required by some projects, and in particular by the spillovers that are specific to knowledge driven investments. Thus, making explorative analyses of the local urban Commons following the dynamic resource-based view could lead to discover co-ordination externalities and new potential market opportunities.

References

1. Dardot, P., Laval, C.: Del Comune, o della Rivoluzione nel XXI secolo. Derive-Approdi, (2015)
2. Dines, N.: What are 'Social centers'? a study of self-managed occupations during the 1990s. Transgressions A J. Urban Explor. **5**, 23–39 (2000)
3. Dines, N.: Tuff city. Berghahn Books, New York (2012)
4. Mudu, P.: Resisting and challenging neoliberalism: the development of Italian Social Centers. Antipode **36**(5), 917–941 (2004)
5. Micciarelli, G.: Le teorie dei Beni Comuni al banco di prova del diritto. La soglia di un nuovo immaginario istituzionale, Politica&Società **1**, 123–142 (2014)
6. Fumagalli, A.: Sfruttamento e sussunzione nel capitalismo bio-cognitivo. Derive e Approdi (2017)
7. Fumagalli, A., Braga, E.: La sfida dell'istituzione finanziaria del Comune. Derive e Approdi (2015)
8. Zamagni, S.: Economia Politica. Teoria dei Prezzi, dei Mercati e della Distribuzione. NIS, La Nuova Italia Scientifica (1987)
9. Ostrom, E.: Governing the Commons. The Evolutions of Institutions for Collective Actions. Cambridge University Press, Cambridge (1990)
10. Coase, R.: The nature of the firm. Economica **4**, 386–405 (1937)
11. Williamson, O.: Markets and Hyerarchies: Analysis and Antitrust Implications. Free Press, New York (1975)
12. Williamson, O.: Economic Institutions of Capitalism. Free Press, New York (1985)
13. Alchian, A.A., Demsetz, H.C.: Production, information costs, and economic organization. Am. Econ. Rev. **62**, 777–795 (1972)
14. Helfat, C.E., Finkelstein, S., Mitchell, W., Peteraf, M.A., Singh, H., Teecem, D.J., Winter, S.: Dynamic capabilities. In: Understanding the Strategic Change in Organizations. Blackwell Publishing, Malden (2007)
15. Staw B., Cummings, L.L.: Research in Organizational Behavior, Research in Organizational Behavior, pp. 295–336 (1990)
16. Dyer, J.H., Singh, H.: The relational view: cooperative strategy and sources of interorganizational competitive advantage. Acad. Manag. Rev. **23**(4), 660–679 (1998)
17. Teece, D.J., Pisano, G., Schuen, A.: Dynamic capabilities and strategic management. Strateg. Manag. J. **18**(7), 509–533 (1997)
18. Nelson, R.R., Winter, S.: An Evolutionary Theory of Economic Change. Harvard Business Press, Cambridge (1982)
19. Vittoria, M.P., Napolitano, P.: Comunità informali come 'luoghi creativi' e drivers di produttività urbana. Il caso dei Centri Sociali a Napoli, Riv. Econ. del Mezzog. **1** (2017)
20. Hausmann, R., Rodrik, D.: Economic development as a self-discovery. J. Dev. Econ. **72**, 603–633 (2003)
21. Foray, D., David, P.A., Hall, B.H.: Smart Specialization. From academic idea to political instrument, the surprising career of a concept and the difficulties involved in its implementation, Lausanne (2011)
22. Foray, D.: Smart Specializations: Challenges and Opportunities for Regional Innovation Policies. Routledge (2015)

The Creative City: Reconsidering Past and Current Approaches from the Nomocratic Perspective

Stefano Cozzolino[(✉)] [iD]

ILS - Research Institute for Regional and Urban Development,
Karmeliterstraße 6, 52064 Aachen, Germany
stefano.cozzolino@ils-forschung.de

Abstract. This paper reconsiders the idea of the creative city in light of the nomocratic perspective and provides critical reflections on past and current practices. The main argument is that planning should not be based on direct intervention to create a creative city (for instance, by supporting certain economic activities or boosting the attractiveness of specific neighborhoods or places), but rather it should enable its spontaneous emergence by creating the conditions in which creativity can be expressed and experimented in space by the largest part of civil society.

Keywords: Creative city · Nomocratic planning · Spontaneous order

1 Introduction

In the literature, there are different conceptions of what makes a city creative and different ideas regarding the nature of city processes that give rise to what we generally refer to as a creative city [1]. As a consequence, in planning theory and practice, we see different approaches for its development and promotion. This paper considers two theoretically opposite planning approaches. In the first case, the city is conceived of as a simple, plannable entity [2]. In this approach, planning is an instrument to achieve a pre-determined idea of the (creative) city. Its function is mainly positive in the sense that it actively drives urban changes toward the politically preferred state of affairs (for instance, through land-use plans, incentives, events, and so on). In the second approach, the city is conceived of as complex system based on self-organizing orders and the unplannable nature of many urban aspects is taken into account [2–4]. In this approach, planning is not used instrumentally to reach a specific state of affairs but as a general framework that welcomes unpredictable actions and emergent socio-spatial configurations. Its function is mainly negative (i.e., it mainly avoids the emergence of negative externalities) and does not directly determine urban changes [5]. Moroni [6] defines the first approach as teleocratic planning and the second as nomocratic planning. In the teleocratic approach, "planning intervention is fundamental […] to establish a desired overall state of affairs through deliberate coordination of the contents of the (private) independent urban activities". In this case, the role of public authorities is that of "generating order through specific, purpose-dependent directives and instructions".

© Springer International Publishing AG, part of Springer Nature 2019
F. Calabrò et al. (Eds.): ISHT 2018, SIST 100, pp. 606–614, 2019.
https://doi.org/10.1007/978-3-319-92099-3_68

Conversely, "with the nomocratic approach, public authorities regulate the actions of the private actors (via abstract and general rules that apply equally to everyone) and plan solely their own actions (in particular, to guarantee certain kind of infrastructures). [...] This works around the notion of spontaneous order and on a rediscovery of the ideal of the rule of law". In particular, the author maintains that "spontaneously generated order is desirable for generating new powers that would otherwise not exist."

In brief, assuming the teleocratic approach, becoming a creative city is possible through top-down interventions and political coordination aimed at pre-determined socio-spatial configurations. Instead with nomocratic planning, the creative city becomes a space in which spontaneous and unpredictable actions can be pursued and developed without having any specific socio-spatial configurations in mind [7]. Starting from an overview of past and current theories and practices (Sect. 2), the paper argues that the way in which planning policies deal with the creative city is today widely teleocratic and rarely nomocratic. It also discusses the problems and limits of current practices (Sect. 3) and re-evaluates the conception of the creative city in light of the nomocratic perspective (Sect. 4). Then, it provides certain devices to revise contemporary planning policies (Sect. 5).

2 The Creative City in Practice

For over thirty years, the idea of the creative city has been discussed and explored by planners. Introduced in the 80s, it became widely debated in the 90s with Charles Landry's book The Creative City: A Toolkit for Urban Innovators [8]. Initially, the idea of the creative city was proposed in response to the predominating old urban renewal practices that were frequently focused on the development of the so-called urban hardware (i.e., the built environment) rather than the urban software (i.e., culture, social capital, etc.). The main goal was to shift orthodox approaches to policies and practices to be focused more on people and their networks by providing the means to foster local creativity and culture. Charles Landry, like many others, believed this approach to be more effective in regenerating cities, communities, and urban economies. Essentially, what the author wrote - and what he still stands for today - is similar to what Jane Jacobs has been sustaining and successfully defending for all her life: cities are made first of all by people, not objects [2]. However, in contrast with Jacobs's argument, who recognizes the complexity and spontaneity of urban processes (especially regarding entrepreneurial discoveries) [9], urban creativity started to be coordinated and planned as though it was a simple object by means of statutory, top-down public interventions.

Around the idea of the creative city, a movement of thinkers began to recognize the importance of creative infrastructure and social capital as driving forces for urban regeneration and economic revitalization [10]. Governors began to support this idea and believed that regeneration processes could be induced and boosted through planning interventions targeted at the development of creative cultural environments and clusters and at the promotion of specific segments of creative or cultural economies [11–13].

Two main phases of planning practices are recognizable in the development of creative cities. The first phase started roughly in the late 80s. Public policies and financial investments mainly dealt with the built environment in order to generate global cultural milieus [14]. In the literature, this phase is largely labeled as Post-Fordism, which refers to the period when big industrial plants and mass producers of goods left the cities, providing an opportunity for re-urbanization processes in the name of cultural economies [15]. Classic examples are the large-scale transformations generated by urban development corporations (for instance, during Margaret Thatcher's government in the UK) [16], the transformations inspired by the European Capital of Culture program [17], or the various punctual constructions of post-modernist buildings and landmarks aiming at reproducing the so-called Bilbao effect [18]. As Marcuse sustains [19] "Since the mid-1990s, [...] planners have increasingly turned to arts and culture as development tools. Performing and visual arts centers, festivals, public art, artist live/work buildings, artists' centers and community cultural centers have been planned, invested in and built as ways to revitalize emptying downtowns and attract tourists."

The second phase began with the financial contraction of the last 10–15 years, which provoked a general shift from direct public real estate investments to the pro-motion of global networking and the facilitation of local creative industries. In this phase, planning interventions did not focus as much on the built environment and physical transformations but prevalently on local economies and cities' brand strategies [20, 21]. Simply to give an idea, a recent example is the program UNESCO Creative Cities Network, which identified 70 cities from all over the world according to their particular potential in various cultural and creative economic sectors (such as music, literature, folk art, design, media arts, foods, and cinema) considered to be strategic in generating positive impacts on the cities' images and attractiveness. (Consider that many other examples can be provided by looking at municipal and local city policies).

During both the first and the second phase, planning interventions had some common features: first, they aimed to regenerate and strengthen cities' attractiveness by reinforcing their images and global competitiveness; second, they tried to create or support specific, vibrant urban environments (often, policies were connected to specific neighborhoods); third, they were often driven by public, top-down interventions and investments; fourth, they worked at the local level in order to gain results at the global level. In consideration of these common characteristics, the main goal of planning has been the development of creative milieus [22] which, in theory, should enable expansion of certain economic sectors, improve places deemed to be strategic, as well as encourage the growth of cities' economies through side-effects and chain reactions.

All this is undertaken with two main ambitions in mind: the revitalization and regeneration of specific places so that they can become more competitive, and the attraction of what Richard Florida defines as the creative class [23, 24]. In brief, planning interventions for creative cities are expected to stimulate virtuous economic cycles, repositioning cities in the global ranking and regenerating specific areas and cultural activities.

Today, planning policies rely heavily on strategies of this kind. Striking cases of such attempts are the so-called citadels. Citadels of culture, arts, music, and science - to name a few - are typical examples of planning interventions in our times. These interventions are often not limited to slogans, but they are also formalized in land-use plans under different names, for example that of cultural district [25].

3 Problems and Limits of Current Practices from the Nomocratic Perspective

A first general observation is that the wide debate on creative cities highlights certain - sometimes hidden - crucial aspects of cities' functionality (for instance cultural and creative aspects). In fact, when creativity and culture are manifested in cities, they represent a symptom of diffused economic and social well-being. In this regard, the use of rankings, which are currently very common, is an interesting practice to compare different cities and their creative propensity. However, it is decisive to distinguish the evaluation of urban creativity from urban policies and interventions that aim at actually developing urban creativity. In fact, differently from the usefulness of rankings (to which we can agree or disagree upon, but they still remain documents that do not directly affect cities), planning interventions may pose several ethical and practical questions. In particular, from the nomocratic perspective, the promotion of specific planning interventions for the development of a creative city raises at least three main issues.

3.1 Crises of the Rule of Law

In both the current and past approaches for developing creative cities - yet different - it is hard, if not inherently impossible, to respect the rule of law and offer equal treatments to all urban actors and urban spaces through planning rules that are impartial (i.e., laws cannot be applied equally to all) and stable (i.e., rules cannot be written in ways that provide reliable expectations). This because planners are asked to select strategic urban areas and economic sectors at the expense of other areas and economic sectors that are in this way excluded. In other words, politicians and majorities must arbitrarily decide "winners and losers." In addition, following a political "opportunistic" approach, planning becomes more inclined to change strategies any time new social-economics trends and political opportunities for urban transformations emerge.

Thus, considering such temporality, planning is apt to become temporary instead of stable, preferring short-term objectives instead of long-term projections (note that if planning rules change too quickly, this provokes serious problems of uncertainty) [26]. Moreover, in this approach, planners need to select a favorite group of "players" and treat them differently (in this regard, we must consider that even the choice to invest in one economic sector instead of another induce to unequal treatments and, at the end, still to the selection of advantaged groups of interests).

3.2 Interventionism and Political Decisions

Making cumulative interventions in the market in order to favor specific areas and economic sectors on the long-run may create price distortions and deteriorate existing self-organizing orders. Unequal treatments or unexpected incentives induce urban actors to readapt their behaviors to public governors' decisions by dismantling local contextual knowledge, expectations, and plans. This creates an inefficient use of decentralized forces and knowledge [27]. From this perspective, it is worth underscoring that planning interventions are not implemented in a vacuum but always upon real social processes that are the result of self-coordination and self-organization. Moreover, interventionism (i) "impinges directly on the underlying norms, such as respect for liberty, that support self-organization itself" [28]; (ii) increases public decision-makers propensity to support political action at the expense of free enterprise; (iii) makes dis-intervention difficult (i.e., it affects local knowledge in such a way that it makes it easier to generate further planning interventions) [29]; and (iv) turns what should be restricted to private economic decisions into political decisions. Overall, this produces outcomes favorable to interest groups participating to political debates, which may not be in the public interest [4]. Therefore, this approach induces urban actors to spend more time than they need to build political consensus around their initiatives instead of market consensus with the probability to build gray coalitions and conflicts of interest [30].

3.3   Planned Gentrification

Another important point is that urban regeneration policies in the name of the creative city have often initiated gentrification processes [31]. For instance, D'Ovidio [15] maintains that "arts, and culture in general, have been systematically used in order to revamp the urban environment and to promote neighborhood development. [...] This is achieved with high social costs. [... The] displacement of artists and [disadvantaged] people [comes with the] increased value of real estate." This to say that there is a trade-off between promoting the regeneration of particular neighborhoods and the protection of existing social capital. Undoubtedly, improving places affects their attractiveness and, therefore, their affordability. The real ethical problem comes when such processes are not the result of spontaneous orders but, instead, the intentional target of public policies. To quote Tallon [16], "gentrification can be encouraged through leisure and cultural regeneration, the staging of 'middle-class' events, and the development of flagship property-led buildings. [However] their relevance to local populations is often questionable." Beside this, the implementation of centrally driven regeneration policies often induces the serial replication and homogenization of the global urban landscapes and the formation of a "culture of nowhere" [32].

4   Reconsidering the Creative City in the Light of the Nomocratic Perspective

Current and past approaches have some questionable limits and pose certain ethical and practical problems. By assuming a nomocratic perspective, this section underscores three points that should help us to re-evaluate and expand the existing idea of the creative city without abandoning it. (This is obviously a small selection of relevant aspects as other components can be taken into consideration).

4.1   Creativity Is Not an Elitist Issue

Creativity cannot be considered as an elitist issue nor as something enclosed in few economic sectors or types of activities (for instance, as indicated by UNESCO, music, literature, folk art, design, media arts, foods, and cinema). Rather, creativity must be considered to be a transversal and potential quality of any urban action that requires entrepreneurship and inventiveness [33]. In this sense, the first device is that of moving from the idea of a creative city as a space offering certain creative and cultural services or functions to the idea of a creative city as a space where it is possible to undertake and develop a plurality of creative actions. Governments cannot know in advance how and when unpredictable creative innovations will benefit society as a whole [27]. Hence, the room for creativity can be expanded by means of an institutional framework that allows a large space for experimentation as well as people ready to utilize it freely and productively: "expanding the creative class means creating an environment in which [...] impediments to engaging in creative activity are low" [34]. In doing so, society can benefit from continuous processes of trial and errors, reaching a level of creativity and innovation otherwise unachievable. This openness, however, also requires the tolerance to accommodate failures and the emergence of social-spatial configurations that are not always appreciated by the political majority. In other words, it may be that certain places and economic sectors will succeed more than others despite the lack of political consensus.

4.2   Many Relevant Factors Are Unplannable

As suggested by Richard Florida [23], one peculiarity of creative environments is that they attract other creative people. Nevertheless, the emergence (or not) of such creative environments depends on factors that are largely uncontrollable and therefore hardly plannable [35]. In fact, according to Holcombe [4] the conditions favoring the development of creative environments can be of different kinds: "some are purely a matter of geography (for instance, climate, proximity to beaches, etc.); other are the products of planning interventions (such us parks, transportation networks, public space); others are market-driven rather than policy-driven (for example, nightlife, bars, restaurants, diversity in population and lifestyles, and so on)". The last group of conditions in particular cannot be provided directly by means of public action. These amenities become available as diversified economies grow and accumulate with the passage of time [36]. Their existence is the product of self-organization and free-enterprise. Such an evolutive process takes times and is incremental. Moreover, it requires the development of many small actions of separate private individuals. Thus, the emergence of creative and cultural environments cannot happen overnight nor can it be directly planned and then created from scratch all at once: as in many other kinds of urban phenomena, time and uncertainty are two fundamental aspects that cannot be excluded from our discourse.

4.3   Limits and Potential of Planning Rules

From a planning perspective, when thinking of creativity, two main types of action are of interest [37]: one relates to how agents use a certain space, and the other concerns how agents transform a space. In the former case, changes mainly relate to the activities carried out in a particular building or urban space. In the latter case, changes mainly concern the built environment itself. Creativity can be expressed in both cases if land-use plans and building codes welcome it. In Steward Brand's words [38] "in buildings, evolution is always and necessarily surprising. You cannot predict or control adaptivity. All you can do is make room for it".

In this sense, one of the greatest efforts planners must undertake is providing planning conditions and rules that allow for a wide range of experimentation regarding the possibility to carry out different uses (or the re-use of building and open spaces) and consenting to adjust the physical environment to unknown up-coming needs. The kinds of processes and changes that I have in mind are mainly small and highly connected to practical circumstances. Therefore, promoting the creative city should mean first of all expanding and diffusing the opportunity for creative activities to take place exactly - but not exclusively - where creativity is not expected to emerge.

5   Final Remarks

Today, the idea of the creative city is still important but needs to be reconsidered. Creativity cannot be foreseen nor planned. In order for it to be manifested, flexibility and experimentation are required, not rigid planning. "The [real] question is to generate the conditions [... such] that everyone can become creative" [39]. Conversely, today's practices provide space for "creative" planners and politicians but not so much for the creative initiatives of ordinary citizens. From this perspective, even though the expression "creative city" is usually used to indicate certain urban environments generally perceived as creative or vibrant, it is crucial to underscore that creativity remains first of all a quality of actions and does not strictly refer to (physical) spaces (which obviously may condition the development of new creative activities or attract creative people to different degrees). Moreover, what we consider to be creative clusters are, most of the time, the result of a combination of complex self-organizing orders and not so much of planning interventions [18]. This does not mean that creative cities do not need planning, but a different approach that at least takes into consideration the risks of excessive interventionism (Sect. 3), as well as certain emergent and hardly plannable aspects and peculiarities of urban creativity (Sect. 4). In conclusion, contrary to what we saw in the last decades, planners should be less focused on directly developing a creative city and more careful in providing a planning framework that welcomes a plurality and diffused creative actions. This means accepting that creative cities as we commonly envision them (for instance, attractive and full of cultural activities, bars, art galleries, vibrant environments, and so on) cannot be the direct result of strong, top-down public interventions, but are rather a consequence of certain conditions (only in minimal part directly controllable, such as public infrastructures and planning rules) that enable the spontaneous emergence of entrepreneurship and diffused creativity.

References

1. Landry, C., Bianchini, F.: The Creative City. Demos, London (1995)
2. Jacobs, J.: Death and Life of Great American Cities. Random House, New York (2007)
3. Alfasi, N., Portugali, J.: Planning rules for a self-planned city. Plan. Theor. **6**(2), 164–182 (1961)
4. Holcombe, R.: Cultivating creativity: market creation of agglomeration economies. In: Andersson, D.E., Andersson, E., Mellander, C. (eds.) Handbook of Creative Cities. Edward Elgar Publishing, Northampton (2011)
5. Cozzolino, S., Buitelaar, E., Moroni, S., Sorel, N.: Experimenting in urban self-organization: framework-rules and emerging orders in Oosterwold, Cosmos + Taxis **4**(2), 49–59 (2017)
6. Moroni, S.: Rethinking the theory and practice of land-use regulation: towards nomocracy. Plan. Theor. **9**(2), 137–155 (2010)
7. Andersson, D.E.: Creative cities need less government. In: Andersson, D.E., Andersson, E., Mellander, C. (eds.) Handbook of Creative Cities. Edward Elgar Publishing, Northampton (2011)
8. Landry, C.: The Creative City: A Toolkit for Urban Innovators. Earthscan, London (2012)
9. Jacobs, J.: The Economy of Cities. Random House, New York (1969)
10. Scott, A.: The Cultural Economy of Cities. Sage, London (2000)
11. Griffiths, R.: Creative enterprise and the urban milieu. In: Paper presented at the AESOP Annual Congress, Vienna 13–17 July (2005)
12. McCarthy, J.: Quick fix or sustainable solution? Cultural clustering for urban regeneration in the UK. Leisure Studies Association, Eastbourne (2007).
13. Mommaas, H.: Cultural clusters and the post-industrial city: towards the remapping of urban cultural policy. Urban Stud. **41**(3), 507–532 (2004)
14. Montgomery, J.: Cultural quarters as mechanisms for urban regeneration. Part 1: conceptualising cultural quarters. Plan. Pract. Res. **18**(4), 293–306 (2003)
15. D'Ovidio, M.: The Creative City Does Not Exist: Critical Essays on the Creative and Cultural Economy of Cities, vol. 2. Ledizioni, Milan (2016)
16. Tallon, A.: Urban Regeneration in the UK. Routledge, London (2010)
17. Kunzmann, K.: Culture, creativity and spatial planning. Town Plann. Rev. **75**(4), 383–404 (2004)
18. Plaza, B.: Evaluating the influence of a large cultural artifact in the attraction of tourism: the Guggenheim Museum Bilbao case. Urban Aff.Rev. **36**(2), 264–274 (2000)
19. Marcuse, A.: Cultural planning and the creative city. In: Paper presented at the Annual American Collegiate Schools of Planning Meetings, Ft. Worth, Texas (2006)
20. Ashworth, G., Kavaratzis, M.: Cities of culture and culture in cities: the emerging uses of culture in city branding. In: Hass, T., Olsson, K. (eds.) Emergent Urbanism: Urban Planning & Design in Times of Structural and Systemic Change, pp. 73–79. Ashgate, Oxon (2016)
21. Kavaratzis, M., Ashworth, G.J.: City branding: an effective assertion of identity or a transitory marketing trick? Tijdschrift voor economische en sociale geografie **96**(5), 506–514 (2005)
22. Hall, P.: Creative cities and economic development. Urban Stud. **37**(4), 639–649 (2000)
23. Florida, R.: Cities and the Creative Class. Routledge, London (2005)
24. Ray, P.H., Anderson, S.R.: The Cultural Creatives: How 50 Million People are Changing the World. Harmony Books, New York (2000)
25. Frost-Kumpf, H.: Cultural Districts: The Arts as a Strategy for Revitalizing Our Cities. Americans for the Arts, Washington (1998)

26. Brennan, G., Buchanan, J.M.: The Reason of Rules. University Cambridge Press, Cambridge (2008)
27. Hayek, F.A.: The use of knowledge in society. Am. Econ. Rev. **35**(4), 519–530 (1945)
28. Ikeda, S.: Urban interventionism and local knowledge. Rev. Austrian Econ. **17**(2–3), 247–264 (2004)
29. Ikeda, S.: Interventionism and the progressive discoordination of the mixed economy. Adv. Austrian Econ. **5**, 37–50 (1998)
30. Chiodelli, F., Moroni, S.: Corruption in land-use issues: a crucial challenge for planning theory and practice. Town Plann. Rev. **86**(4), 437–455 (2015)
31. Smith, N.: New globalism, new urbanism: gentrification as global urban strategy. Antipode **34**(3), 427–450 (2002)
32. Kunstler, J.H.: The Geography of Nowhere. Touchstone, New York (1993)
33. Kirzner, I.M.: Entrepreneurial discovery and the competitive market process: An Austrian approach. J. Econ. Lit. **35**(1), 60–85 (1997)
34. Holcombe, R.: Planning and the invisible hand: allies or adversaries? Plan. Theo. **12**(2), 199–210 (2013)
35. Easterly, W., Freschi, L., Pennings, S.: A Long History of a Short Block: Four Centuries of Development Surprises on a Single Stretch of a New York City Street. Conference, NYU Development Research Institute (2015)
36. Desrochers, P., Leppälä, S.: Creative cities and regions: the case for local economic diversity. Creativity Innov. Manage. **20**(1), 59–69 (2011)
37. Cozzolino, S.: The city as action. The dialectic between rules and spontaneity. Doctoral dissertation, Milan Polytechnic (2017)
38. Brand, S.: How Buildings Learn: What Happens After They're Built? Penguin, New York (1994)
39. Moroni, S.: Land-use Regulation for the Creative City. In: Andersson, D.E., Andersson, E., Mellander, C. (eds.) Handbook of Creative Cities. Edward Elgar Publishing, Northampton (2011)

Landscape as Driver to Build Regeneration Strategies in Inner Areas. A Critical Literature Review

Stefania Oppido[✉] [iD], Stefania Ragozino [iD], Donatella Icolari [iD],
and Serena Micheletti [iD]

Institute for Research on Innovation and Services for Development,
National Research Council of Italy (IRISS-CNR), 80134 Naples, Italy
s.oppido@iriss.cnr.it

Abstract. In the scientific debate and political strategies, solving the gap between core and non-core areas is becoming pivotal for achieving a more balanced territorial development. The emerging issues in current scientific literature regard how to reverse marginalisation trend in inner areas by carrying out approaches able to enhance the territorial capital in an endogenous development perspective. For this purpose, strategies for these areas have to look carefully at their strengths, such as a low pollution rate, a more direct access to natural resources and a rooted local identity. In many of these contexts, landscape is a specific asset that should be enhanced as driver to trigger new development dynamics, to achieve local attractiveness and competitiveness starting from multidimensional values recognised to the landscape.

In this perspective, collaborative processes can be developed, aimed at involving local communities and stakeholders in recognising landscape resources and values as starting point to enhance place identity in marginalised contexts. Therefore, the interpretation phase should develop a dynamic and collective process, supported by interdisciplinary methods and tools to improve territorial capital and, at the same time, strengthening social cohesion, in order to lay the groundwork for co-planning and co-designing processes.

Keywords: Inner areas · Landscape · Place-based Regeneration

1 Introduction

The ongoing research activities deal with development strategies in inner areas for reversing marginalisation trend, within the research project "Place-based Regeneration Strategies and Policies for Local Development", at the National Research Council of Italy. The aim is contributing to cope with the main challenges of inner areas through

---

Within the unitary work of the research group, S. Oppido developed §1, S. Micheletti §2, S. Ragozino §3 and D. Icolari §4.

© Springer International Publishing AG, part of Springer Nature 2019
F. Calabrò et al. (Eds.): ISHT 2018, SIST 100, pp. 615–624, 2019.
https://doi.org/10.1007/978-3-319-92099-3_69

place-based strategies based on the identification/recognition of the landscape as driver to trigger new territorial development dynamics.

Starting from the main issues of the scientific international debate and the political agenda about inner areas, the research group is developing an Action Research process in the selected case of Alta Irpinia, pilot area for the Campania Region (Italy) within the Italian National Strategy for Inner Areas. In the preliminary analysis of the case, civic activism and bottom-up initiatives emerged. They are focused on the touristic reuse of the historical Avellino-Rocchetta Sant'Antonio railway, re-named the "Irpinia landscape railway" from the local associations, considering it as "vehicle" for narrating the local heritage. The current research phase deals with the definition of the "arena" for a collaborative process aimed at supporting community and local actors in the construction of a shared narrative of their territory "through the train". The general goal is to consider the landscape as a driver for a widespread regeneration process of the inner area [1, 2].

The issue is relevant because the peripheralisation process is causing damages in many territories, both in terms of loss of identity and increasing environmental risks. Depopulation processes, decay, land maintenance reduction, and abandonment of the cultural heritage and landscape make these areas less and less attractive for people and traditional business [3]. In a longer-term perspective, these conditions will lead to spiralling-down effects [4, 5] and determine a "territorial degeneration", with environmental and social costs [6, 7]. This process of decline is improved by the progressive reduction of public services, the loss of employment opportunities, which in turn lead to further depopulation.

In order to reverse these negative trends, the current debate highlights the need to look beyond the conceptualisation of traditional dichotomies between "leading" and "lagging" regions, to explore new research paths and policies able to enhance specific values of inner areas - such as environmental quality, people's well-being, cultural and symbolic values of landscape [8] - focusing on endogenous development and bottom-up approaches [5].

For these reasons, territorial capital enhancement and social cohesion strengthening are keywords in the scientific and cultural debate as well as in the political agenda for inner areas, because they can become the driving forces to contrast abandonment and to reduce social exclusion. However, in practice these objectives are still poorly implemented. In a widespread perspective, the goal is to achieve a more balanced territorial development and social equity between core and non-core areas [9–11].

Out of the global development routes, inner territories have often preserved their diversity, strong sense of belonging and local identity [8]. Their territorial capital includes excellent or everyday landscapes whose protection, management and planning can contribute to the consolidation of local identity, economic growth and job creation [12]. Therefore, landscape as specific asset should be enhanced to promote territorial development and to achieve local attractiveness and competitiveness in inner areas, based on territorial diversity and respect for the "uniqueness of place".

In this perspective, as promoted by the Italian National Strategy for Inner Areas, a place-based approach could be addressed [13, 14], focusing on local communities as key actors for a regeneration process based on awareness of the place, identification of local tangible and intangible resources and their values [15–17]. Within this approach, a community who recognise its values can determine a process of "*patrimonializ-zazione*", which means interpreting the territorial heritage as element to strengthen the local identity and to trigger socio-economic development [18]. This process can contribute to rebuild ties between community and territory, and among different local actors, by activating self-recognition and self-organisation of development processes.

For this purpose, it is necessary to develop a collaborative process, starting from multidimensional values recognised to the landscape. The analysis will include not only physical elements but also collective memories and meanings. It will be implemented through interpretative tools, involving local communities, aimed at narrating the territory and its heritage, as support for the planning process [19]. Furthermore, the involvement of local communities and socio-economic stakeholders, on the one hand, allows to bring out local needs as a central element in development strategies, and, on the other hand, it can facilitate the implementation of decision processes by sharing responsibilities through co-planning and co-designing methods [2].

The paper includes three sections following this introduction; (Sect. 2) a short description of marginalisation processes in inner areas and the main strategies implemented in Italy; (Sect. 3) a critical review of territorial development approaches; (Sect. 4) finally, a practical framework to achieve collaborative processes to involve local key actors in recognising landscape resources and values as starting point to support territorial strategies.

2   Inner Areas as European and Italian Research Field

The inner areas include places geographically and relationally distant from supply centers for essential services. Since the II post-war period, the economic attractiveness of urban contexts has changed the historic balance between cities and peripheries in favour of the first, especially in terms of services, investments and employment. Physical distance and exclusion from socio-economic and political dynamics caused depopulation and desertification, triggering interregional migrations to urban areas [7, 20, 21]. Depopulation has progressively determined the deterioration of natural and agricultural resources with environmental risks, abandonment of historical heritage, and expenditure of territorial capital [2, 7, 22].

Starting from the 80s, scholars initiated a debate and European and national governments took care of inner areas in order to reduce unbalance, social inequality and costs caused by abandonment of these territories and to enhance their features. Only in 2008 strategies for territorial cohesion clarified that, modifying the perspective on "periphery", territorial differences can be a strength for sustainable development [23]. In fact, with systemic actions, local interventions aim at development and growth processes by strengthening territorial balance. EU directives focus on multilevel governance models and strategies – from European to local scale – based on principles of vertical and horizontal subsidiarity, therefore urging interaction among different local government institutions, and between public institutions and communities [24, 25].

In Italy, peripheral areas crisis has begun since '30s with depopulation of mountain areas, but the phenomenon of inner areas started to be endorsed in '50s by economic studies related to southern Italy issues [26]. During '90s, local development strategies

(Territorial Pacts, Integrated Design, and Leader Approach) have been activated, and in 2014 Social Cohesion Agency promulgated the Italian Strategy for Inner Areas (*SNAI*) to contrast the process of decline. Current national territory is represented by 60% of inner areas (52,7% of municipalities) in which only one quarter of population lives (13,540 million) [7, 22]. The *SNAI* purpose is to reverse the demographic trend by acting on negative conditions (low accessibility, reduction of services, depopulation), by promoting natural and cultural heritage enhancement, by improving local community well-being conditions, and by enhancing employment [7, 20]. For these purposes, reuse and enhancement of unused or underused buildings and heritage can represent the starting point of regeneration processes.

Within the *SNAI*, 65 area-projects across the Italian regions to realise the National Strategy at local level and one pilot area per region to implement the strategy have been defined. The areas have been identified through a methodological study based on the Italian municipalities which represent centres of basic services, and by classifying other municipalities in four classes, determined by the distance (in minutes) from the centre offering services: belt areas (less than 20 min), intermediate areas (between 20–40 min), remote areas (40–75 min), ultra-remote areas (more than 75 min) [7].

Inner areas policies are supported by European funds such as ERDF, ESF, EAFRD, EMFF 2014–2020 and Italian Stability Law funds, invested by municipalities with multilevel actions: the former to protect and enhance environmental, cultural and productive heritage (comprising sustainable tourism), the latter to achieve purposes about quantity and quality of essential services (education, mobility and health) [7].

As suggested by Green Paper [23], the *SNAI* promotes place-based strategies but suggests centre-local implications [25]. Based on co-design principles, *SNAI* stimulates multilevel governance integrated with participatory processes by local networking because "Services developed with local input are more likely to reflect the most important needs […]. National priorities continue to play a part in the process, but there is an implicit negotiation that allows all the parties to have a say in the final outcome" [23: 89]. Furthermore, according to ordinariness principle, the *SNAI* considers inner areas as a laboratory for local development to enhance environmental, cultural and socio-economic resources by making implementable and replicable processes in other territories, even central ones [25].

3   Critical Review of Territorial Development Approaches

The gap between core and non-core areas is increasing in all Europe and is becoming pivotal in the scientific debate and political strategies [27]. The emergent theme is the "inner periphery", which mainly includes rural areas or small towns defined through their lower rate of industrial production than core areas, and a high rate of population loss. This latter, beyond being serious in itself, has a relevant impact on communities and economies of these places [28] by affecting with a spiralling-down effect the shaping and evolution of regional and local identity [4, 29]. These areas suffer for several disadvantages among which their incapacity to attract investments, mainly because they are not competitive like a core area is - unfortunately they are not desirable for traditional investors because they are unable to make use of eventual investments, to create jobs and to develop innovation [8]. Most of the time, they are studied and approached through models based on development trajectories of core areas, while there would be the need to approach them taking into account specific issues and peculiarities of their local resources. Inevitably, market-led approaches and traditional growth models are unable to catch these specific issues, for which it is necessary to rethink the concepts of *development* and *growth*. The hoarding of resources and capitals, or generically economic indexes, cannot be intended as the only way to understand and approach these places. Strategies for these areas have to look carefully at their positive elements, such as a low pollution rate, a more direct access to natural resources and a rooted local identity - characteristics that core areas do not have [8].

What should be done if industrialisation or smart specialisation are not the keystone for these development processes? What could be different options for non-core areas? With specific regard to the phenomenon of the peripheralisation, scientific debate is going beyond the difference between productive and less productive areas. It gave the floor to a wider concept based on power relationships between core and non-core areas centred on cultural perceptions and evolution of local economies [8, 27, 29, 30].

This reflection transforms the issue of "competitiveness" into "territorial/local competitiveness" [14, 31, 32], which is "interpreted as residing in creativity rather than in the pure presence of skilled labour; in local trust and a sense of belonging rather than in pure availability of capital; in connectivity and relationality more than in pure accessibility; in local identity, beyond local efficiency and quality of life" [15: 1368]. In order to better understand this concept, it is necessary to introduce the consolidated concept of *territorial capital* – the "competitive potential of a given territory" [32: 1387]. This latter consists of traditional, material and functional elements to be combined with non-material, cognitive and relational elements, and helps us to claim that these particular areas need complex and multidimensional approaches for their development. Scholars approached this concept since 2000 [34, 35] and deepened it from different points of view by stressing more and more on its narrative potentialities [36] and its nature as resource intended also as an element from which it is possible to promote territorial development [37]. Mainly, these discourses including themes as *endogenous growth* and *territorial diversity* have become pivotal since there has been talk of the role of intangible resources within sustainable development.

At European level, there is a consolidated idea about the usefulness of policies to develop territorial capital, and for each region a specific territorial capital has been recognised, which "is distinct from that of other areas and generates a higher return for specific kinds of investments than for others, since these are better suited to the area and use its assets and potential more effectively" [37: 1].

This critical literature review revealed the recognition of the rich potential of these inner areas in opposition to the codification of them as non-competitive areas – the concept of territorial capital highlights the richness and not the lack of these territories. Studies, experimentations and practices on these areas identify the territorial capital as an occasion to understand, develop and test new approaches to improve territorial development. In spite of this, although political interest at European level is pushing in this direction [11, 23, 39], current territorial development practices have not yet been well-established on this cognitive base. So much so that among main instances promoted by scientific debate there is the lack of a strong focus on landscape identity and the recognition that this last can improve effectiveness of policies [40].

The Italian Strategy for Inner Ares identified more of these topics as pillars to be used to build a systemic framework approaching the peripheralisation and marginalisation of these areas. Specifically, place-based approach [7, 13, 16, 33, 41] is the selected strategy to manage development, growth and regeneration trajectories by taking into account local assets, landscape identity, and inclusive governance.

4   Collaborative Processes Identifying Landscape as Development Resource

The last part of this contribution proposes a practical framework based on a collaborative approach in which the landscape plays a role of catalyst for territorial strategies. The fieldwork, developed following an Action Research process, is the Alta Irpinia inner area in the Campania Region. Here, evidence from the bottom-up initiatives carried out in the last years highlights the role of social actors in promoting landscape as specific asset of local identity, starting from the reuse of the historical Avellino-Rocchetta Sant'Antonio railway for the narrative of the landscape. Bottom-up activities through the train have increased a new interaction among local actors and have created a new narrative of places, facilitating a physical and intellectual access to territorial resources. This element represents the starting point to develop a collaborative and co-design process aimed at preserving local memory and promoting cultural identity. The final goal is a regeneration strategy based on connectivity between different territorial actors, "aimed at the *re-appreciation* of meanings and values of place, the *re-grounding* in cultural and ecological place based assets, the *re-positioning* in new and hidden economies" [38: 34].

For this purpose, an interpretation phase should be developed through a dynamic and collective process supported by interdisciplinary methods and tools. This approach considers the territorial resources as result of dynamic and collective cognitive and material processes, determined by the interaction between communities and environment. Local identity is recognised in cultural landscape as depositary of an historical memory built on mutable, fragile and hidden relationships. In this perspective, the interpretation and communication process of cultural landscape promote conservation and new knowledge production [42, 43].

This recovery and reconstruction process addresses and retells a large range of narrative knowledge - histories, memories, personal and collective tales, experiences, traditions, local know how, iconographic, cartographic and photographic sources and historical data - in which landscape is identified as narrative medium, meaning and marketing of place [25]. Specifically, in inner areas this participatory interpretation of cultural heritage and resources, and open narratives of landscape can make innovative forms of sustainable community-oriented management based on the enhancement of local attractiveness by improving access to sites and touristic services [44]. This process aims at identifying values and culture of the territories to define community identity, and to reuse local cultural knowledge and practices as tools for promoting new strategies for sustainable development based on local distinctiveness (handcrafting traditional gastronomy, agri-food traditions) and also related to creative industries and creative economy [45].

In this perspective, the landscape narrative is an effective tool of communication and sharing of local identity, it reinforces social cohesion, keeps alive cultural practises and creates place meaning through participatory processes as cultural mapping and co-design, identified as a process to trigger development strategies based on specific values [46, 47]. Indeed, cultural mapping intended as a social practice is based on involvement of different territorial stakeholders and communities to create, especially in rural and inner areas, proximity between local communities and cultural heritage and to plan community services, communication systems and community-based bottom-up activities. The cultural mapping, as economic, social and community development tool, can be driven by top-down experts, as well as bottom-up and community-centred approaches. It is developed through quantitative and qualitative methodologies, in order to collect different information on territorial assets and hidden resources [48]. Top-down approach uses quantitative indicators and fits to a national dimension; conversely, bottom-up approaches and qualitative methodologies are an effective tool for local contexts to investigate and to preserve intangible and immaterial assets and dynamic character of the environment. Alternative approaches combine quantitative indicators and qualitative methods to make cultural mapping an integrated development tool [49, 50].

The planning process of a cultural mapping must answer to the question whether to elicit an internal or external response as the main strategic design objective [51].

The knowledge produced by the cultural mapping process can be used for new spatial representations and visual interfaces focused on community memories and experiences [52] and it can be managed by innovative multimedia and multimodal frameworks to highlight place meanings and to enhance place identity communication [53].

The potentialities of collaborative processes emerge also in initiatives of territorial enhancement and management based on bottom-up networks [54] such as Local Action Groups (LAGs), *Comunità di Fiume e di Paesaggio* (River and Landscape Communities) and network of cultural and research institutions for the preservation of the historical memory and local identity through educational, cultural and economic activities. Forms of self-representation of social resources and cultural and territorial capital [55, 56] are ecomuseums conducted through participatory tools such as participatory inventories, community and landscape maps to recognise the territorial heritage [57].

Starting from this framework, IRISS researchers are carrying out an interpretation phase of the Irpinia landscape in order to highlight the proactive role of local community in the representation and enhancement of local identity by a learning and interaction approach consistent with Action Research process.

Acknowledgment. The research activity is developed within the research project "Place-based Regeneration Strategies and Policies for Local Development", coordinated by G. Esposito De Vita at the IRISS-CNR.

References

1. Oppido, S., Ragozino, S., Micheletti, S.: Reusing heritage: activist planning for place-based regeneration processes. In: José Antunes, F., Simões, J.M., Morgado, S., Marques da Costa, E., Cabral, J., Ramo, I.L., Silva, J.B., and Baptista-Bastos, M. (eds.) AESOP Annual Congress - SPACES OF DIALOG FOR PLACES OF DIGNITY: Fostering the European Dimension of Planning, pp. 2236–2241. Universidade de Lisboa, Lisbon (2017)

2. Oppido, S., Ragozino, S., Micheletti, S.: In land-scapes: sharing responsibilities to regenerate publicness and cultural values of marginalised landscapes The historical railway Avellino-Rocchetta Sant'Antonio of Alta Irpinia. Urbani Izziv (2017)

3. Lang, T.: Socio-economic and political responses to regional polarisation and socio-spatial peripheralisation in Central and Eastern Europe: a research agenda. Hung. Geogr. Bull. **64**, 171–185 (2015)

4. Emery, M., Flora, C.: Spiraling-up: Mapping community transformation with community capitals framework. Community Dev. **37**, 19–35 (2006)

5. Küpper, P., Kundolf, S., Mettenberger, T., Tuitjer, G.: Rural regeneration strategies for declining regions: trade-off between novelty and practicability. Eur. Plan. Stud. **26**, 1–27 (2017)

6. Sassen, S.: Expulsions: Brutality and Complexity in the Global Economy. Harvard University Press (2014)

7. Barca, F., Casavola, P., Lucatelli, S.: A Strategy for Inner Areas in Italy: Definition, Objectives, Tools and Governance (2014)

8. Kinossian, N.: Planning strategies and practices in non-core regions: a critical response. Eur. Plan. Stud. **26**, 1–11 (2017)

9. Commission of the European Communities: New Partnership for Cohesion: Convergence, Competitiveness, Cooperation: Third Report on Economic and Social Cohesion, Brussels (2004)

10. OECD: Productive regions for inclusive societies (2016)

11. ESPON: Polycentric Territorial Structures and Territorial Cooperation (2016)

12. Europe, C.O.: European landscape convention. In: Report and Convention (2000)

13. Atkinson, R.: Territorial capital, attractiveness and the place-based approach: The potential implications for territorial development. Territ. Cohes. Eur. 70th Anniv. Transdanubian Res. Inst., 297–308 (2013)

14. Camagni, R., Capello, R.: Regional competitiveness and territorial capital: a conceptual approach and empirical evidence from the European Union. Reg. Stud. **47**, 1383–1402 (2013)

15. Stephenson, J.: The cultural values model: an integrated approach to values in landscapes. Landsc. Urban Plann. **84**, 127–139 (2008)

16. Barca, F.: An agenda for a reformed cohesion policy. A place-based approach to meeting European Union challenges and expectations (2009)

17. Becattini, G.: La coscienza dei luoghi. Il territorio come soggetto corale. Donzelli, Roma (2015)

18. Poli, D.: Il patrimonio territoriale fra capitale e risorsa nei processi di patrimonializzazione proattiva. In: Meloni, B. (ed.) Aree interne e progetti d'area, pp. 123–140. Rosenberg e Sellier, Torino (2015)

19. Magnaghi, A.: Montespertoli. Le mappe di comunità per lo statuto del territorio. Alinea Editrice, Firenze (2010)

20. Lucatelli, S.: Strategia Nazionale per le Aree Interne: un punto a due anni dal lancio della Strategia. Agriregionieuropa **12**, 4–10 (2016)

21. Meloni, B.: Aree interne, multifunzionalità e rapporto con la città. Agriregionieuropa **12**, 61–65 (2016)
22. Oppido, S., Ragozino, S., Micheletti, S.: Riuso del patrimonio ferroviario (non) dimenticato e processi di rigenerazione. Avellino - Rocchetta Sant'Antonio: il treno irpino del paesaggio. In: Urbanistica è/e azione pubblica. La responsabilità della proposta - Società Italiana degli Urbanisti, Roma 12–14 giugno 2017, Rome (2017)
23. Commission of the European Communities: Green Paper on Territorial Cohesion Turning territorial diversity into strength - SEC(2008) 2550 (2008)
24. OECD: Strategies to Improve Rural Service Delivery. OECD, Paris (2010)
25. Storti, D.: Aree interne e sviluppo rurale: prime riflessioni sulle implicazioni di policy. Agriregionieuropa **12**, 65–69 (2016)
26. Barbera, F.: Il terzo stato dei territori: riflessioni a margine di un progetto di policy. Meloni B., a cura di, Aree interne e Progett. d'area. Torino Rosenb, Sellier (2015)
27. Lang, T., Henn, S., Ehrlich, K., Sgibnev, W.: Understanding Geographies of Polarization and Peripheralization: Perspectives from Central and Eastern Europe and Beyond. Springer (2015)
28. Brown, D.L., Argent, N.: The impacts of population change on rural society and economy, Routledge Int. Handb. Rural Stud. **85** (2016)
29. Dax, T., Fischer, M.: An alternative policy approach to rural development in regions facing population decline. Eur. Plann. Stud. **26**, 1–19 (2017)
30. Wirth, P., Elis, V., Müller, B., Yamamoto, K.: Peripheralisation of small towns in Germany and Japan – Dealing with economic decline and population loss. J. Rural Stud. **47**, 62–75 (2016)
31. Camagni, R.: Regional competitiveness: towards a concept of territorial capital. In: Seminal Studies in Regional and Urban Economics, pp. 115–131. Springer (2017)
32. European Commission: Territorial state and perspectives of the European union, scoping document and summary of political messages (2005)
33. Hambleton, R.: Place-based leadership: a new perspective on urban regeneration. J. Urban Regen. Renew. **9**, 10–24 (2015)
34. Camagni, R.: On the concept of territorial competitiveness: sound or misleading? Urban Stud. **39**, 2395–2411 (2002)
35. Organisation for Economic Co-Operation and Development - OECD: Territorial Outlook., Paris (2001)
36. Camagni, R., Caragliu, A., Perucca, G.: Territorial capital, relational and human capital, PRIN - Capitale Territ. Scenar. quali-quantitativi di superamento della Cris. Econ. e Finanz. per le Prov. Ital. **36** (2011)
37. Tóth, B.I.: Territorial capital: theory, empirics and critical remarks. Eur. Plann. Stud. **23**, 1327–1344 (2015)
38. Horlings, L.G.: Connecting people to place: sustainable place-shaping practices as transformative power. Curr. Opin. Environ. Sustain. **20**, 32–40 (2016)
39. European Commission Committe on Spatial Development: ESDP European Spatial Development Perspective, Towards Balanced and Sustainable Development of the Territory of the European Union (1999)
40. Loupa-Ramos, I., Bernardo, F., Carvalho Ribeiro, S., Van Eetvelde, V.: Landscape identity: implications for policy making. Land Use Pol. **53**, 36–43 (2016)
41. Pugalis, L., Bentley, G.: Place-based development strategies: possibilities, dilemmas and ongoing debates. Local Econ. **29**, 561–572 (2014)
42. ICOMOS: The Florence Declaration on Heritage and Landscape as Human Values (2014)
43. UNESCO: Third UNESCO World Forum on Culture and Cultural Industries (2014)

624   S. Oppido et al.

44. European Commision: Towards an integrated approach to cultural heritage for Europe. Brussels, Brussels (2014)
45. Buratti, N., Ferrari, C.: La valorizzazione del patrimonio di prossimità tra fragilità e sviluppo locale. Un approccio multidisciplinare. FrancoAngeli, Milano (2011)
46. Duxbury, N., Garrett-Petts, W.F., MacLennan, D.: Cultural mapping as cultural inquiry. Routledge (2015)
47. Soini, K., Birkeland, I.: Exploring the scientific discourse on cultural sustainability. Geoforum **51**, 213–223 (2014)
48. Longley, A., Duxbury, N.: Introduction: Mapping Cultural Intangibles (2016)
49. Freitas, R.: Cultural mapping as a development tool, City. Cult. Soc. **7**, 9–16 (2016)
50. Reed, M.S., Fraser, E.D.G., Dougill, A.J.: An adaptive learning process for developing and applying sustainability indicators with local communities. Ecol. Econ. **59**, 406–418 (2006)
51. Chiesi, L., Costa, P.: Making territory through cultural mapping and co-design. In: How community practices promote territorialisation. In: Dessein, J., Battagl., E., Horlings, L. (eds.) Cult. Sustain. Reg. Dev. Theor. Pract. Territ. Routledge (2015)
52. ICOMOS: The ICOMOS Charter for the interpretation and presentation of cultural heritage sites. Int. J. Cult. Prop. **15**, 377–383 (2008)
53. Duxbury, N., Saper, C.: Introduction: Mapping Culture Multimodally (2015)
54. Cerquetti, M.: Building Bottom-Up Networks for the Integrated Enhancement of Cultural Heritage in Inner Areas. Towards New Paths (2017)
55. Ribaldi, C., De Varine, H.: Le radici del futuro. Il patrimonio culturale al servizio dello sviluppo locale. Econ. della Cult. **16**, 274 (2006)
56. De Varine, H.: Patrimoni e territori. Territorio (2014)
57. Garzena, P.: SIGNIUM. Mappe di comunità. La rivista dell'ecomuseo del biellese (2004)

Navigating Neo-liberal Urbanism in the UK. Could a Social Entrepreneur Be Considered an Activist Planner?

Stefania Ragozino(✉) iD

Italian Research Council CNR, Institute for Research on Innovation
and Services for Development IRISS, 80134 Naples, Italy
s.ragozino@iriss.cnr.it

Abstract. Based on the classification developed by Tore Sager about recognised modes of activist planning, this article offers a reflection on the figure of the social entrepreneur intended not only as a *bridge actor* but also as a *proactive actor* within the planning process in a political and economic scenario strongly affected by the consequences of neoliberism. The work proposes an empirical approach aiming to deepen an Anglo-Saxon experience of a social enterprise acting in one of the 39 deprived areas recognised by the New Deal for Communities and involved in its regeneration process. Evidence from the fieldwork permits reflection on the potential role of the social entrepreneur in a time of austerity urbanism in the UK.

Keywords: Austerity urbanism · Activist planning · Social enterprise

1 Introduction

Neoliberal governance has been strengthened with the 2008 crisis. Austerity measures have long been part of neoliberal repertoire that constructs governmental downsizing, reinvigorating free markets and individual liberty. These trends are structuring a new system of urban politics, an *urbanisation of neoliberal austerity* that has a direct and long-term impact on the city, of which Localism is one of the most pernicious consequences [1]. New forms of planning react to neoliberal politics and austerity urbanism, starting from a worldwide civic movement that addresses wider issues of planning [2, 3]. Among this boundless sphere in which are included recognised forms of resistance, grassroots initiatives and more performed groups of professionals, Tore Sager classified *activist planning modes* that are «unconcealed and recognised by the government, and including efforts of both lay and professional planners» [4].

In the Sager's work as in this one, *neoliberalisation* is intended as «dynamic processes through which neoliberalism is continuously being transformed» especially affecting planning process through evolving market-led trajectories [5]. Taking into account civic activism as «a reaction to lay-offs, foreclosures, hard-handed welfare reforms and democratic deficit in handling recession» [4] and the significant influence of successive waves of neoliberalisation on urban movements, activism planning could be defined as going beyond protest, and as being aimed at developing an alternative

© Springer International Publishing AG, part of Springer Nature 2019
F. Calabrò et al. (Eds.): ISHT 2018, SIST 100, pp. 625–634, 2019.
https://doi.org/10.1007/978-3-319-92099-3_70

planning proposal [2]. Sager intends the activist planner not necessarily as a planning professional, she is not linked to one particular ideology, although she resists neo-liberal policies; she is interested in going beyond possible institutional participatory process to start grassroots-initiatives and, when the institutional decision-making process is not sufficiently inclusive and transparent, is engaged in the preparation of alternative or supportive plans. Although she may be positioned differently to the formal planning process or supporting one of the conflicting parties, she is not necessarily anti-government. Importantly, she may have an intermediate position between parties as a mediator or by being affiliated with an NGO, association or group interested in facilitating planning within the community. The crucial point for a successful activist planner is having a vibrant network of actors outside of the formal planning. Sager recognised six modes of activist planning, which are presented in Table 1.

Table 1. Recognised modes of activist planning (Source: Sager, 2016).

	Government planner	Civil society planner
Loyal to group or community	Official partisan planning	Advocacy planning, community-based activist planning
Committed to strategic cause	Equity planning, inside issue advocacy	Radical planning, critical-alternative initiatives
Committed to relational cause	Public activist mediation	Intermediate activist planning

Sager built this classification scheme taking into account the position of the activist planner and the rationale of the action. Referring to the full article for details [4], here we are interested in the last cell, *Intermediate activist planning*, in which the author deepened the unbalanced power relations among territorial and urban actors and their influence in the planning process. In this model, NGOs, third sector local organisations, and planning-related NGOs are identified as *affiliated* to the activist planner, with a role of being the *bridge actor* in the context capable of altering power relationships within the planning process [6]. In a wider literature review, it has been observed that these actors - identified under the umbrella concept of social economy - are actively engaged in urban regeneration processes and construction of more effective urban governance models [7, 8]. Among these, the social entrepreneur, putting together social with entrepreneurial aspects, is the one strongly deputed to intervene on urban context. A rich scientific literature regarding social enterprise considers it as an *organisation with social purposes* [9, 10], in addition a more context-related review includes a specific territorial mission of «transforming adverse environmental conditions into enabling ones to improve the lives of their target beneficiaries» [11], and a «more direct influence [...] on urban planning» [12].

After these considerations, is it possible to consider the social entrepreneur not only as *bridge actor* but also as a *proactive actor* engaged in the planning process? Could the social entrepreneur be intended as an active planner? Could he/she alter power relations of the planning process in order to make it less complicate and faster?

This paper explores these questions in a case of urban renewal in the UK because it seems a significant setting to test whether social entrepreneur are able to pursue their social goals and to influence the flailing planning process, despite the strong marketization of public policy in the country. There the effects of the economic crisis on planning are considered to have been more severe than elsewhere [5], just think to rooted tensions between central and local government, as well as the Localism Act (2011), which aimed at obscuring but sustaining neo-liberal policies [13].

The paper presents one experience of a Community Interest Company involved in a heritage-led regeneration process in Plymouth (Devon, UK) in order to reflect on the potential role of the social entrepreneur as activist planner responding to neoliberal drift. This experience has been studied through the qualitative approach of mixed-methods case study [14], structured as follows: (1) Analysis of the current planning scenario in Plymouth; (2) Thematic site visits; (3) N. 8 semi-structured interview with engaged actors in urban development project (snowball approach); (4) Interaction with the social cultural enterprise RIO (site visit, data sharing and elaboration).

Following this introduction, this paper presents (Sect. 2) the selected experience of UK in the field of planning and social enterprise. Then the paper presents (Sect. 3) the case of RIO in Plymouth, and finally (Sect. 4) a reflection about the research questions.

2   Neo-liberal Urbanism and Social Entrepreneurship in UK

Recent political re-orientations represent a dramatic step backwards for UK's well-established planning tradition [4, 15]. These setbacks revolve around three policies: (1999) political and institutional devolution; (2000) regional spatial planning; (2010) neighbourhood planning. These consecutive changes may be better understood within the framework of *neoliberalisation*, in which material values are emphasised [5]. Additionally, the Localism Act (2011, Cameron government) was aimed at the devolution of decision-making powers from central government control to individuals and communities. Its reported objective was to enable communities to address local needs through the new tier of planning, the Neighbourhood Planning (2012). At the same time, it weakened the role of planners and local government representatives. These contradictions have the paradoxical effect of creating a new form of centralism, whereby *empowered* communities are under threat of seeing their autonomy revoked by the central government [13, 16]. In this scenario, the line between public and private responsibilities is blurred [4].

One of the basis of Localism is the ideology of Big Society, «a form in which a new round of roll-back neoliberalism is enabled by public sector cuts, dismantling state institutions and the privatization of health services» [17]. This favours the diffusion of social enterprises in order to facilitate welfare reform and discipline third sector. As early as 2002, the Blair government accelerated the debate on social enterprise by launching the Social Enterprise Coalition and by implementing the Social Enterprise Unit. With the Companies Act (2004), a new legal form of social enterprise was created - the Community Interest Company (CIC). «CICs are limited companies which operate to provide a benefit to the community they serve. They are not strictly "not for profit", and CICs can, and do, deliver returns to investors. However, the purpose of CIC is primarily one of community benefit rather than private profit. Whilst returns to investors are permitted, these must be balanced and reasonable, to encourage investment in the social enterprise sector whilst ensuring true community benefit is always at the heart of any CIC» [18]. Just over ten years after the establishment of the CIC, 10,000 units were registered in UK, which listed 68,000 social enterprises. Collectively, these contributed £24 billion to the national economy in turnover and realised profits for £240,000 [19]. In these last years, this Anglo-Saxon approach to social enterprise – the most consolidated in Europe – has emphasised entrepreneurship and market logics. The European model, conversely, is more oriented to community and more combative of market-led trajectories [20].

Although the rationale of social enterprise is the creation of social value, such enterprises must balance "value creation" with "value capture" [21]. For this reason, social enterprises are understood here as "hybrid" organisations, because they cannot be classified as either private, public or non-profit organisations. Rather, a social enterprise spans their respective boundaries, bridges different institutional levels and faces conflicting institutional logics [22–24]. Social enterprise is included in social economies - hybrid by definition - because it engages market, public bodies, and funds economies and people, which normally work independently or through negotiation [17]. It is defined also as "particular dynamic" within third sector, «a tool for building bridges between distinct components of the third sector» [24], and «a crossroads element between cooperative and non-profit organisation» [25]. Compared to cooperatives, social enterprise pursues interests that are more widely shared. Compared to non-profit organisations, it has a higher propensity to risk-taking. «The pursuit of financial sustainability and social objectives requires the generation of sufficient revenue to invest in business activities at the same time as maintaining investment in social projects to create social value and drive forward social change» [22]. The challenge of social enterprises is to find the way in which the social value creation is strictly linked to, or better integrated with, the achievement of economic stability or economic objectives [9]. For these reasons, social impact finance is organised so that economic return creates social value [26]; it is precisely the level of hybridisation that can affect the economic return.

In the following paragraph, the author describes the role of the social enterprise RIO within the regeneration process in the neighbourhood of Devonport with a focus on the selected topics in order to be able to reflect on the research questions.

3   Planning Initiatives and RIO's Role in Devonport

In the wider UK scenario, author selected the city of Plymouth because it is a hot spot for social enterprise activity with the most active social enterprise networks in the country [27]. It is a city located 310 km from London; with 267,700 inhabitants, it is the second largest city of the southwest region after Bristol. Its reputation is generally positive because it is the greenest city in the UK and tourism industry is growing, it holds the ninth largest university and one of the main commercial port of UK and, after London, has the highest numbers of post-war listed buildings [28]. The economy of the city depends primarily on the defence and public sectors. The presence of the Royal Navy in the neighbourhood of Devonport, meant that the city avoided both the first wave of deindustrialisation and the correspondent investments for redevelopment. This has postponed its development in comparison to cities of the UK.

The current neighbourhood of Devonport was a city built in 1690 as Royal Navy's worker settlement. In 1800, its population exceeded that of two neighbouring towns, Plymouth and Stonehouse, as well as its economic and political power spread as demonstrated from its built heritage still existing. In 1914, Devonport became a peripheral neighbourhood of the new city of Plymouth and this caused a loss of its political and commercial function. During WWII, Plymouth has been severely destroyed, the Royal Navy in Devonport being a primary target of the German air force. Although in the 1960s the Royal Navy helped maintain high employment rates, most of Devonport contained post-war low-quality houses, as well as historical built heritage that has not been properly preserved. So that Devonport became one of the most deprived areas of UK, developing the reputation of *no-go area*.

In 2000, Devonport was included in thirty-nine-long list of deprived areas that would benefit from a fund of £50 million for a ten-year regeneration project as part of the New Deal for Communities (NDC). In parallel, the Conservation Areas programme considered Devonport as a potential resource for the entire city of Plymouth for its forty-two buildings included in the statutory list of Buildings of Special Architectural or Historic Interest. During 2004, the Devonport Regeneration Community Partnership had drawn, after extensive consultation with the community, the Devonport Development Framework, a plan for the physical development of the area (in addition to the NDC framework). This is considered to be an extremely complex and ambitious series of interlinked projects [29]. In 2006, the City Council of Plymouth produced the Devonport Area Action Plan, based on NDC's themes and Conservation Areas' objectives, focusing on the demolition of 565 properties with the simultaneous construction of more than 1100 new homes and conservation projects for historical buildings and parks (Devonport Guildhall and Column, Devonport Market Hall, Granby Green, Devonport Park e St Aubyn's church). Nevertheless these planning initiatives were focused on Devonport, the Plymouth Plan and local regeneration tools (BIDs and partnerships) have excluded the area of Devonport until 2015.

In 2007, the CIC Real Ideas Organisation (RIO, www.realideas.org) settled in Devonport and took over responsibilities for the listed Devonport Guildhall. For the renewal of the latter, RIO received £1.75 million from the Big Lottery and NDC programme. RIO decided to situate the social enterprise in Devonport for two main reasons: *heritage* and *people*. «Devonport has a very exciting history, global story and a range of disused heritage buildings; these could be the key to a more prosperous future. [Regarding] people, Devonport has a strong sense of community and identity related to its history as a once separate town. This identity could be used to bring people together and create a sense of movement» (Interview #1, 2016).

In 2010, after restoration and safeguarding works, the Devonport Guildhall reopened as Cultural Social Enterprise. The project responded to the *triple bottom line*: people, environment, and profit. The Head of Enterprise and Regeneration explained «when considering any piece of work, we will ask the questions 'will it help people?', 'will it help the environment?' and 'will it make enough money to at least cover its costs?'.

[...] The reuse of the building essentially recycled the derelict building, improved the physical environment and clustered a number of organisations together, therefore creating a positive environmental outcome. Finally we felt that the building would work as business and pay for itself» (Interview #1, 2016).

To carry on these objectives, RIO had to take part to the planning process where the administrator (the City Council) and the funder (Big Lottery) need to find a compromise to manage and valorise this historical built heritage. Due to austerity policies, RIO had to be financially independent and generate revenue through social activities. Besides working as incubator, RIO wanted to attract more and more people to Devonport, thus giving Devonport's community an opportunity to restart and rethink its places. The attempt was not to attract people with a higher purchase power, but to make people more sensitive to the built heritage of Devonport. In the words of RIO's Head of Enterprise and Regeneration «the area needs a broader offer, hence there are a number of what we call vector buildings i.e. key derelict heritage building that needed a solution and would recreate purpose, direction and momentum for the community. We needed a range of offer to attract people to the Devonport» (Interview #4, 2016).

It could be argued that RIO's operators have identified their process as culture-led regeneration process [30]. They have assumed cultural participation, built heritage restoration and reuse as key assets to transform the urban landscape and its negative reputation. Having recognised the potential of that site and its existing heritage, RIO decided to restore and reopen another historical building, the Devonport Column, through a bottom-up public consultation process. RIO organised six different events - from City Council to families meeting - during which a questionnaire has been submitted to 155 participants. Results highlighted the key role of the cultural heritage as identities' driver, persisting negative perception about living and visiting Devonport, as well as the lack of shared awareness about potentialities of the area. The community's requirements helped RIO in establishing objectives and expected results: to restore and reopen the Column, to implement a sustainable conservation programme through civic participation and shared knowledge about Devonport's heritage, tracing top-down regeneration process carrying community desiderata. In 2013, the Column was opened, overcoming great difficulties in terms of management of built heritage, interaction with the Heritage Lottery Fund, and financial sustainability [31].

Especially, RIO aimed at improving public space, stimulate social engagement, creating regular job, and promoting sustainable tourism. The main requisite was to enter in the network of urban actors and intertwine new relationships among them to pursue these objectives. During this project, RIO interacted with several actors, including Devonport Regeneration Community Partnership, Plymouth City Council, Plymouth University, real estate developers, local churches, and various third sector infrastructure organisations. The scheduling of activities started shortly after and included architecture internships at the column construction site, university lectures about RIO's role in Devonport's regeneration process, voluntary activities, student competitions for part of the Column project, and heritage walking; the actors involved were educational organisations, universities and prison inmates. In 2013–2014, 1,169 students were involved, the number of tourists increased by 30%, the group "Friends of Devonport Column" was established and several linked projects were implemented – from internal extension works to a £5 million-project of the near Market Hall.

4 Ongoing Conclusions: Social Enterprise Engaged in Urban Planning and Regeneration

In the intricate *urbanisation of neoliberal austerity,* urban community responds with a series of alternative initiatives, from the stronger direct action to the softer public meetings. The study of Tore Sager highlighted a set of opportunities and alternatives behaviours that are challenging this complex economic and political scenario. Specifically, the figure of the *intermediate activist planner* offers a chance to work on power relations, which strongly affect the planning process. In this field, third sector organisations, in particular the social enterprise has been observed as actor participating at the planning and regeneration process. The hybrid nature of the social enterprise, which span social and entrepreneurial boundaries, could be an opportunity to link or mediate conflictual actors, to actively engage the local community, and to change urban and social priorities. In particular, the social enterprise aims at long-term and sustainable strategies and needs to be connected both with institutions and with citizens integrating top-down and bottom-up processes.

Author investigated these themes within the RIO organisation in which engagement and activities are: creating new jobs, sustaining emergent enterprises, promoting community capacity building and creating familiarity with its activities. A strong linkage with places of Devonport accompanied these conventional traits of RIO. Evidence from the fieldwork suggests that its location in the most representative building of Devonport, the Devonport Guildhall, has activated a new synergy between RIO's activities, built heritage and planning process. RIO's activities have achieved important results for Devonport community, for the context and for the whole city of Plymouth. The most important is that the reuse of Devonport Guildhall and Column has reinforced the ongoing regeneration process and has shed new light on completed projects (Devonport Park, Granby Green, and St. Aubyn's) increasing the identifiability and changing the reputation of Devonport within the city of Plymouth. This synergy has increased their value and started new economic activities in Devonport. Among these ones, the implementation of the Market Hall is the biggest outcome achieved in terms of funds and cultural resource: a grade II-listed building that will be transformed into new space for digital, arts, enterprise and visitors. The Head of Enterprise and Regeneration argued: «Market Hall is the next big step, capturing the community's heritage of innovation, creating jobs and providing key support for our schools» (http:// themarkethall.co.uk). The NDC promised £5.000.000 for its renewal but the extent of this intervention made the process very uncertain and slow. The success of the Devonport Column project gave to RIO the opportunity to meet actors involved in the Market Hall project, the Plymouth City Council, the Devonport Based Regeneration Agency and the University of Plymouth, to schedule operations and avoid misunderstandings. In brief, RIO became the leader manager for this partnership and in 2013 announced the Creative Digital Hub plans for Devonport Market Hall, having avoided loss of time and money by integrating different actors' perspectives and needs.

Devonport Community and RIO also obtained good results for: the northern area of the Column, which was transferred to RIO from the Council with an additional funding of £600,000; the Ker Street Social Club, in which RIO would create a Community

Benefit Society and has received funds for £800.000, and new facilities and wholesale operations for the Column Bakehouse (loan of £80,000) and Devonport Gallery.

RIO's activities have an impact on business increasing, on its activities, on context and city, in terms of new occupation, attraction of new people, of investments, better conditions of liveability and security of public spaces, new visibility as tourist destination. These results enable author to say that RIO's activities have been relevant to make the planning and regeneration process more fluid and effective, and that more projects were activated and completed thank to the mediation role of RIO between different institutions and citizens. This experience led to think about the necessary interaction between top-down and bottom-up planning approaches. These latter could interact only if each hybridise parts of itself. Top-down approaches, derived from current market-oriented urban policies, are potentially positive because they bring investments into action (in the case of RIO, the Big Lottery Fund). Often, however, they are not linked to the local context and could risk to be financially unsustainable and to be rejected by local community as foreign bodies. In this sense, top-down approaches must include new figures, which make the planning process more neutral in terms of its political and economic dimension and more rooted in the context and linked to people. On the other side, bottom-up approaches are well linked to community, encourage people to participate, but they do not have the financial dimension that allows them to create new jobs and manage new urban projects. In the UK, the social enterprise has become widespread in order to facilitate welfare reform and to discipline the third sector, and is officially declared as one of the organisations that has to make up for decentralisation of the state by producing goods and services for empowering communities. With these premises, social enterprises cannot be reduced to a production of goods and services because, as a hybrid organisation, it could fit into the common ground between top-down and bottom-up actions, thereby practicing a real hybridisation between social and economic sector very useful in terms of planning and decision-making processes. Social enterprise could stimulate, collect, diffuse community demands, and bring them to a higher level of the political and economic debate, by being a proactive actor within the planning process. In this sense, the community could be actually included in the decision-making process. Social entrepreneur could mediate conflicts and misunderstandings between political, regulatory and private actors acting on power relationships.

It takes on the entrepreneurial risk that allows it to change context conditions in terms of new occupation, built environment, new attractiveness and new relationships. It works for catering to a wider demand as compared to not-for-profit organisation, and for diversifying the offer. Besides, in the UK scenario, social enterprises could be politically neutral by being socially inclusive and financially independent, thus satisfying both left- and right-wings sensitivities, and make the decision process faster because some obstacles are avoided.

Taking into account these characteristics, the social entrepreneur could be considered an actor affecting planning process because she impacts on power relationships, it makes them broader, more neutral and enhance them by creating bridges between distant and/or conflicting parties, thus nourishing a more constructive dialogue about urban issues. Conversely, the state, privates, not-for-profit organisations and cooperatives are not able to create social outcomes developed by social enterprises, since only the latter works with a balanced equilibrium between *value capturing* and *value creation*.

Could this aspect position the social enterprise in a new market made of investments with a social return that could implement the planning discourse?

Acknowledgment. Within the research scheme coordinated by G. Esposito De Vita "Place-based strategies and policies for local development" at the IRISS, Stefania Ragozino developed the fieldwork during its Short Term Mobility Program at the Plymouth University granted by CNR (November–December 2014). Thanks to the RIO Head of Enterprise and Regeneration, Edward Whitelaw, for his precious collaboration during the research.

References

1. Peck, J.: Austerity urbanism: American cities under extreme economy. City **16**, 626–655 (2012)
2. Mayer, M.: First world urban activism. Beyond austerity urbanism and creative city politics, City **17**, 5–19 (2013)
3. Novy, J., Colomb, C.: Struggling for the right to the (creative) city in Berlin and Hamburg: new urban social movements, new "spaces of hope"? Int. J. Urban Reg. Res. **37**, 1816–1838 (2013)
4. Sager, T.: Activist planning: a response to the woes of neo-liberalism? Eur. Plan. Stud. **4313**, 1–19 (2016)
5. Allmendinger, P., Haughton, G.: The evolution and trajectories of english spatial governance: "neoliberal" episodes in planning. Plan. Pract. Res. **28**, 6–26 (2013)
6. Albrechts, L.: Planning and power: towards an emancipatory planning approach. Environ. Plan. C. **21**, 905–924 (2003)
7. Bailey, N.: The role, organisation and contribution of community enterprise to urban regeneration policy in the UK. Prog. Plann. **77**, 1 35 (2012)
8. Evans, M.: Who is for community participation? Who is community participation for? Exploring the well-being potential for involvement in regeneration. Educ. Knowl. Econ. **2**, 163–173 (2008)
9. Dacin, P.A., Dacin, M.T., Matear, M.: Social entrepreneurship: why we don't need a new theory and how we move forward from here. Acad. Manag. Perspect. **24**, 37–57 (2010)
10. Zahra, S.A., Gedajlovic, E., Neubaum, D.O., Shulman, J.M.: A typology of social entrepreneurs: Motives, search processes and ethical challenges. J. Bus. Ventur. **24**, 519–532 (2009)
11. Aziz, H.A., El Ebrashi, R.: A Business Model Design Process for Social Enterprises: The Critical Role of the Environment. World Acad. Sci. Eng. Technol. Int. J. Soc. Behav. Educ. Econ. Bus. Ind. Eng. **10**, 1536–1542 (2016)
12. Wagenaar, H., Healey, P., Laino, G., Healey, P., Vigar, G., Riutort Isern, S., Honeck, T., Beunderman, J., van der Heijden, J., Wagenaar, H.: The transformative potential of civic enterprise. Plan. Theory Pract. **16**, 557–585 (2015)
13. Davoudi, S., Madanipour, A.: Commentary. Localism and neo-liberal governmentality, Town Plan. Rev. **84**, 551–562 (2013)
14. Andrade, A.D.: Interpretive research aiming at theory building: adopting and adapting the case study design. Qual. Rep. **14**, 42–60 (2009)
15. Kunzmann, K.R., Koll-Schretzenmayr, M.: A planning journey across europe in the year 2015. disP – Plan. Rev. **51**, 86–90 (2015)

16. Tewdwr-Jones, M.: Mark Tewdwr-Jones -United Kingdom. disP-Plan. Rev. **51**, 84–85 (2015). https://doi.org/10.1080/02513625.2015.1038080
17. Murtagh, B., McFerran, K.: Adaptive utilitarianism, social enterprises and urban regeneration. Environ. Plan. C Gov. Pol. **33**, 1585–1599 (2015)
18. Community Interest Company Regulator: CIC Regulator: Chapter 1 Introduction (2016)
19. Allinson, G., Braidford, P., Houston, M., Robinson, F., Stone, I.: Business Support for Social Enterprises: Findings From a Longitudinal Study (2011)
20. Amin, A.: Extraordinarily ordinary: working in the social economy. Soc. Enterp. J. **5**, 30–49 (2009)
21. Santos, F.M.: A positive theory of social entrepreneurship. J. Bus. Ethics **111**, 335–351 (2012)
22. Doherty, B., Haugh, H., Lyon, F.: Social enterprises as hybrid organizations: a review and research agenda. Int. J. Manag. Rev. **16**, 417–436 (2014)
23. Pache, A.-C., Santos, F.: Inside the hybrid organization: Selective coupling as a response to competing institutional logics. Acad. Manag. J. **56**, 972–1001 (2013)
24. Defourny, J., Nyssens, M.: Defining social enterprise. In: Defourny, J., Nyssens, M. (eds.) Social Enterprise: At the Crossroads of Market, Public Policies and Civil Society, p. 352. Routledge, London (2007)
25. Borzaga, C., Defourny, J. (eds.): The Emergence of Social Enterprise. Routledge, London (2001)
26. Nicholls, A.: The institutionalization of social investment: the interplay of investment logics and investor rationalities. J. Soc. Entrep. **1**, 70–100 (2010)
27. Fearn, H.: Social enterprise city status - What it really means (2013). https://www.theguardian.com/. Accessed 15 Sept 2017
28. Plymouth Britain's Ocean City, Plymouth's Book of Wonder. Over 100 amazing facts (2014)
29. Atkinson, R.: Spatial planning, urban policy and the search for integration: the example of a medium-sized city. In: Cerreta, M., Concilio, G., Monno, V. (eds.) Making Strategies in Spatial Planning, pp. 101–122. Springer, Berlino (2010)
30. Ferilli, G., Sacco, P.L., Tavano Blessi, G., Forbici, S.: Power to the people: when culture works as a social catalyst in urban regeneration processes (and when it does not). Eur. Plann. Stud. **24**, 1–18 (2016)
31. Real Ideas Organisation, Report of the RIO's Public Consultation for the Devonport Column project (2014)

PPPs Palatability to Complete Unfinished Public Works in Italy

Manuela Rebaudengo$^{(\boxtimes)}$ ⓘ, Giuseppe Innocente ⓘ,
and Angelica Crisafulli ⓘ

Politecnico di Torino, 10125 Turin, Italy
manuela.rebaudengo@polito.it

Abstract. The New Code of Contracts and the new Programming Ministerial Decree state that all unfinished public works must be included by the contracting authorities in their three-year Programme to complete, even downsizing or transforming them by changing their destination, provided that they do not remain unfinished. While the request certainly seems to be acceptable, having committed (for now) public money in an unsuccessful way, one might wonder how credible it is to complete it. First of all for the lack of funds, which for most of them was the reason that interrupted the implementation process; then for the consistency between needs and interventions, for the condition of abandonment and for the period of construction. Can public-private partnerships, even in the new forms provided for by the Code, be a (at least partial) solution to the problem?

The paper presents the first results of an ongoing search aiming to define the national situation after the last census in July 2017, to identify the categories of works that are largely "not completed" on national areas and classifies them in hot/cold/ones, and to measure, with a first wide mesh sieve, the suitability and applicability of the PPP instrument.

Keywords: Unfinished public works · Economic and financial viability
PPPs suitability · Completing tool

1 The Census of Unfinished Works in Italy

Resuming previous studies [1–3], the work that was started but it's now not completed or is not usable by the community is defined as *unfinished*. The legislator, who has investigated the phenomenon, has established that this can be caused by for at least one of the following causes: (a) lack of funds; (b) technical reasons; (c) supervening new technical regulations or laws; (d) the bankruptcy of a contractor; (e) lack of interest at the completion by the Contracting Authority [4]. The MD no. 42/2013, aiming to regulate a national census-register of unfinished public works, provides that (art. 3, c.1) by March 31 of each year should be drawn up by each contracting authority where they list the works that are subsequently published by June 30 of that year.

The first census was done in 2013 and the last is that of July 2017. Overall, comparing the number of works surveyed and their amounts, we can describe a national situation synthetically represented below (Fig. 1).

© Springer International Publishing AG, part of Springer Nature 2019
F. Calabrò et al. (Eds.): ISHT 2018, SIST 100, pp. 635–642, 2019.
https://doi.org/10.1007/978-3-319-92099-3_71

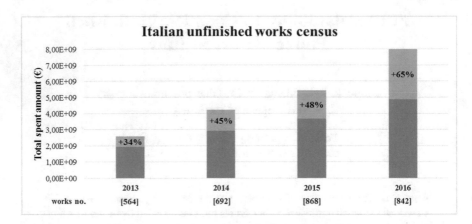

Fig. 1. The unfinished works: national analysis from 2013 (numbers and amounts; processing from a database created by the authors and updated annually with data from regional census)

With the exception of the unfinished works under ministry jurisdiction, at the end of 2016 the number of works is about 700, and the total amount has achieved about 2,500 million euro. Focusing only on the last year, we can say that the situation has improved: the number of works is smaller than in 2015 (about −15%) and the amounts are also more limited (about −11%) [5] (Fig. 2).

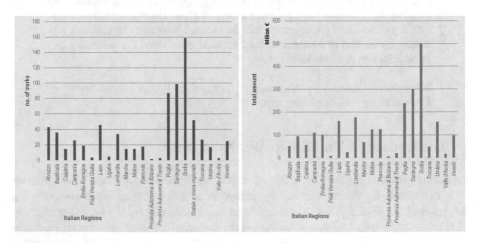

Fig. 2. The unfinished works (2017): regional distribution (numbers and amounts; processing by authors from regional census data [6])

2 What Kind of Unfinished Works? The Regional Framework

At national level, in 40% of cases, the construction works are interrupted after the contractual deadline for completion; in 50% of cases works are interrupted within the contractually foreseen deadline for completion, since the conditions for their restart are not met; in the last 10% of cases, the construction work is completed, but has not been formally tested within the prescribed time limit because the work does not meet all the requirements set out in the specifications and the relevant executive project, as verified during the test operations. There are few works whose completion (in terms of work progress percentage) is very close: on average the works have been started with 40% of the total amounts and therefore require a financial coverage of about 60% (Fig. 3). However, even if the figure seems credible, it should be pointed out that in 7% of cases the cross-referencing of information (% progress and missing costs) identifies critical points linked to the values declared in the census and then used by the authors to construct the database. If we refer to the possibility that the work is/will be usable, as the law indicates, this is foreseeable in more than 22% of cases. And this result is certainly not so encouraging… (Fig. 4).

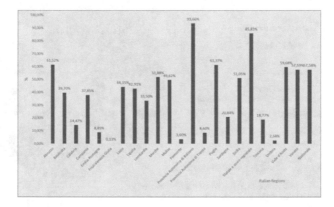

Fig. 3. Percentage of needed amount (regional distribution; processing by authors from regional census data [6])

In terms of work categories (as provided by law [7]), the national framework of unfinished is that shown in Fig. 5: the main category is social facilities (12%), followed by housing, sports and entertainment and roads, all of which account for about 11%.

Since the financial commitment required is considerable, given the inertia of completion over the five-year period (less than 10% per year), the strategy for completion should not only be mandatory (by law it was stipulated that unfinished works must be completed [8–11]) but should also include an in-depth study of possible funding channels.

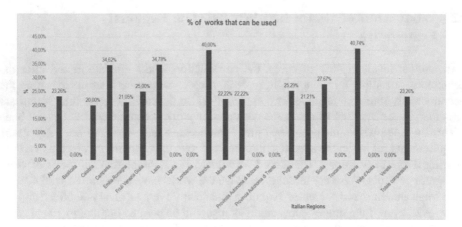

Fig. 4. Percentage of works that can be used (regional distribution processing by authors from regional census data [6])

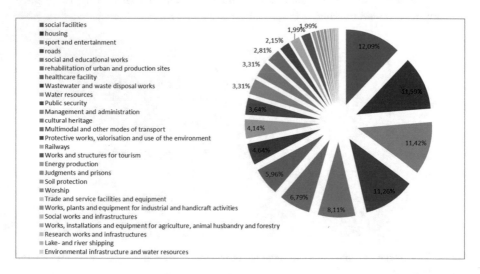

Fig. 5. Decreasing order of work categories that can be used (regional distribution; classification of authors according to categories of works in Table 2 of Annex 1, M.D. 24th October 2014 [7] of collected data from regional census)

Although it is possible that new funding channels may be created to complete the unfinished works [1]. First of all, we can imagine involving partners and private capital; however, this requires a specific assessment of the private palatability of the public work or the related service. Aiming to frame national leaning towards private public-private partnership on a national scale, to understand how much can be used for the completion of unfinished works, we need to introduce some general terms.

3   PPPS in Italy

Public-private partnerships (PPPs) generally refer to long-term forms of cooperation between public and private sectors in order to carry out public tasks, such as the financing, construction, renovation, operation or maintenance of an infrastructure or the provision of a service.

A recent study indicates [12] that in the last five years contracting authorities continue to apply PPPs procedures with the same leaning and that the main categories of works, to which PPP calls for tenders correspond, are sports, social and public housing, urban furniture and energy. Figures below show the dynamics of the PPPs market (Fig. 6) and the national distribution of tenders by size class of contracting authorities (Fig. 7).

Fig. 6. PPPs market in Italy: percentage of numbers and amount of PPPs procurements on public procurements [12]

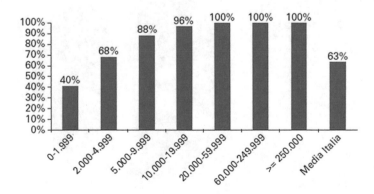

Fig. 7. PPPs market in Italy: percentage of contracting authorities which have used PPPs at least once per population size [12]

4 Completing Unfinished with PPPs

The law provides that the re-programming of an unfinished work may concern:

(A) its completion; (B) its reuse with alternative solutions; (C) its possible down-sizing; (D) its sale or sale as a remuneration for other public works construction process; (E) its demolition.

By current rules, it is clear the intention to promote their completion, but it seems very difficult to imagine that the causes that have led to incompleteness, primarily the lack of funds, are less successful. With the intention to investigate if it is possible that private capital could finance these public works, a new classification is appropriate.

We can distinguish between two large typology of PPP to measure in preliminary way the palatability of the private sector:

1. the so called "Opere Calde", projects able to generate income by revenue from user. Public Administration encourages and promotes conditions to enable the realization of the PPP project; Private Subject have to pay the public infrastructure costs during the entire life of the grant and receives as compensation the investment revenues. (for example *Project Financing*, *Concessions* and *Project Bond*).
2. the so called "Opere Fredde", projects in which the private provides direct services to the public administration. The Private Subject that builds and manage the work, gains its remuneration from payments made by the Public Administration considered as a public infrastructure user (for example *Lease of Public Property* and *Availability Agreement*).

Using this common definition, the unfinished contained in the database, already classified in terms of work categories, have been reclassified also by PPP typology.

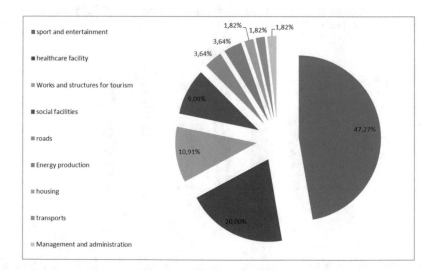

Fig. 8. "Opere calde": percentage per work categories (national distribution; classification of authors according to categories of works in Table 2 of Annex 1, M.D. 24th October 2014 [7] of collected data from regional census)

The result is not surprising: 8% of total works can be defined as "opere calde" (they may produce operating income that enables it to remunerate the initial investment); 79% of total works represents the so called "opere fredde" (their operating income do not provide a return on the initial investment); 12% can be both "cold" and "hot" and does not ensure economic and financial sustainability.

The total amount of investments shows interesting issues: the "opere calde" are few but large in economic values (about 40% of total amount); the "opere fredde" are more than two-thirds with economic values of 50% of total amount (smaller amounts). Observing only the "opere calde", the national framework is shown below (Fig. 8).

5   Conclusions

The recent initiatives of spontaneous reuse [1, 2, 11, 13–15] of spaces, buildings and places are very positive, which enhances the culture of common goods, sharing and promoting initiatives from the bottom. Unfortunately in Italy the abandonment landscapes are so widespread that it is difficult to imagine a reuse for everything: a careful study of the potentialities and vocations of the places will sometimes address, to the maintenance, sometimes to the re-naturalization [2]. In order to limit inertia upon completion, a preliminary investigation phase could be launched to verify the private attractiveness and therefore the contribution of private resources in the "reactivation" of currently blocked public works.

The ongoing research here summarized, shows first results of phenomenon characterization over last 5 years and proposes an initial approach to completion, taking advantage of the PPPs procedures provided by the law.

Few cases are currently expected to include public-private involvement, although the overall value required for completion is close to 40%. The next step, currently in progress, provides for a more detailed verification, for some regional realities in northern Italy and individual case studies, of the private appetite for the highlighted "opere calde".

Then, the attention will be given (second additional step) to the "opere tiepide" (12% of total works number; 10% of total works amount), in order to better characterize them and then identify first considerations on risk and profitability.

Lastly, for the "opere fredde" (about 80% of the unfinished works), the objective is to identify possible reconversions with a change of use, once again to attract private capital, in the plausible hypothesis that the public money available for completion is really limited. If verified the lack of interest of the private sector, the impossibility of conversion or the low value to be expected for sale, it should be considered the option of its demolition, not as a negative intervention or waste of resources but as the closure of a procedure that has not produced good results.

References

1. Prizzon, F., Rebaudengo, M.: Unfinished Public Works: A National Heritage to Develop? LaborEst, 11 (2015)
2. Prizzon, F., Rebaudengo, M.: Completing unfinished works in Italy: procedures, findings, synergies. In: 4th International Multidisciplinary Scientific Conference on Social Sciences and Arts SGEM 2017, Conference Proceedings, 24–30 August 2017, Book 1, vol. 3, pp. 115–122 (2017)
3. Rebaudengo, M., Prizzon F.: Assessing the investments sustainability after the new code on public contracts. In: Computational Science and Its Applications - ICCSA 2017, 17th International Conference, Trieste, Italy, 3–6 July 2017, Proceedings, Part VI. Lecture Notes in Computer Science vol. 10409, pp. 473–484. Springer (2017)
4. MD no. 42/2013, Ordinary Supplement to GURI no. 96 of 24th April 2013
5. MIT website, http://www.mit.gov.it/comunicazione/news/opere-incompiute-sistema-informa tivo-di-monitoraggio-delle-opere-incompiute. Accessed 21 May 2017
6. MIT, website - Sistema Informativo Monitoraggio Opere Incompiute https://www.serviziocontrattipubblici.it/simoi.html. Accessed 21 May 2017
7. MD 24th October 2014-Annex 1, Table 2; Ordinary Supplement to GURI no. 283 of 5th December 2014
8. Guiducci A., Ruffini P.: Opere incompiute e lavori realizzabili nel programma triennale dei lavori pubblici, Enti Locali&PA, 16 October 2017. http://www.quotidianoentilocali.ilsole24ore.com/art/fisco-e-contabilita/2017-10-13/opere-incompiute-e-lavori-realizzabili-programma-triennale-lavori-pubblici-173051.php?uuid=AEcp3wmC. Accessed 21 May 2017
9. Mammarella, P.: Opere pubbliche, la programmazione punta sulle incompiute, Edilportale, 25 September 2017. http://www.edilportale.com/news/2017/09/lavori-pubblici/operepubbliche-la-programmazione-punta-sulle-incompiute_59664_11.html. Accessed 21 May 2017
10. Muccioli, S.: Decreto del MIT sul programma degli acquisti e programmazione dei lavori pubblici, Appalti e contratti, 26 May 2017. http://www.appaltiecontratti.it/2017/05/26/decreto-mit-programma-acquisti-lavori-pubblici/. Accessed 21 May 2017
11. Tacconi, G.: Lavori pubblici, stop alle opere incompiute in Italia: il piano. Ingegneri.info, 5 October 2017. http://www.ingegneri.info/news/infrastrutture-e-trasporti/lavori-pubblici-stop-alle-opere-incompiute-in-italia-il-piano/. Accessed 21 May 2017
12. IFEL, Fondazione ANCI, La dimensione comunale del partenariato pubblico privato, pp. 59–230 (2017)
13. Fabian, L., Munarin, S. (a cura di): Re-Cycle Italy, Atlante, LetteraVentidue Edizioni, Siracusa (2017)
14. Fusco Girard, L.: Multidimensional evaluation processes to manage creative, resilient and sustainable city. Aestimum 59, 123–133 (2011)
15. Németh, J., Langhorst, J.: Rethinking urban transformation: temporary uses for vacant land. Cities 40, 143–150 (2014)

Airbnb Revenue Generation in the Urban Context: An Analysis of Renting Patterns and Dynamics

Irene Rubino$^{(\boxtimes)}$ ⓘ and Cristina Coscia ⓘ

Polytechnic of Turin, 10125 Turin, Italy
irene.rubino@polito.it

Abstract. Cities are multifunctional entities that attract a large variety of differently motivated travelers, and new approaches allowing to evaluate the economic impacts generated by tourists in the age of ICT and sharing economy are now definitely needed. More particularly, the spread of a digitally-enabled peer-to-peer accommodation system such as Airbnb is transforming the hospitality domain: if its economic consequences may affect cities and territories at large, the first stakeholders involved in this new economic dynamic are hosts renting their under-used properties to guests. In this framework, this article aims at providing an overview on Airbnb revenue generation patterns, focusing on Turin (Italy) as a case study. Firstly, the article provides insights about the performance of different types of accommodations (i.e. entire homes/apartments and private rooms), suggesting that hosts may adopt different strategies not only to get the most from their properties and appeal different targets of potential guests, but also depending on the physical attributes of their properties. Then, the article advocates for the implementation of methodological approaches combining quantitative and qualitative perspectives, as to better contextualize data-analysis and deepen the interpretation of this innovative economic phenomenon.

Keywords: Sharing economy · Tourism management · Economic value
Peer-to-peer accommodation systems

1 New Economic Dynamics in Attractive Cities: The Spread of Peer-to-Peer Accommodation Systems

Cities and metropolitan areas are now multifunctional entities that are characterized by density and diversity, whether of cultures, buildings, functions or facilities [1]. The complex and plural nature of these systems, combined with the development of transport infrastructures and the ever growing mobility of people, have progressively transformed cities into places that can be attractive and accessible for a large variety of differently motivated travelers [2]. Since it has been argued that most of urban facilities are used both by local citizens and non-residents city-users such as tourists [3], the evaluation and measurement of the economic impact generated by this latter category of people may be difficult [4]. An approach that has been traditionally used as a first step to quantify the people temporarily visiting the city for more than one day and then

© Springer International Publishing AG, part of Springer Nature 2019
F. Calabrò et al. (Eds.): ISHT 2018, SIST 100, pp. 643–651, 2019.
https://doi.org/10.1007/978-3-319-92099-3_72

estimate economic outcomes is the monitoring of the statistical data set about overnight stays in hotels and other official accommodation facilities. However, this scenario has been recently transformed by the spread of sharing economy web platforms that connect people willing to rent their properties and users seeking for short-term accommodations, such as *HomeAway*, *Waytostay*, *Tripping* and the leading firm *Airbnb*, which in 2017 has reached more than 3,000,000 listings through 191 countries and about 65,000 cities. Peer-to-peer accommodation systems have been defined as a disruptive innovation [5], and Airbnb (www.airbnb.com) is now expanding its market, conquering not only money-savers but new targets of end-users [6]. If the role played by Airbnb in terms of competition with hotels is still under investigation and may vary from context to context (see, for instance, the different results and research approaches reported in [7, 8]), the exponential growth of accommodations and reservations taking place through the platform makes the analysis of Airbnb supply and demand an essential step to better understand the new economic dynamics enabled by ICT. In fact, coherently with a smart communities framework in which citizens play an active role in shaping the current socio-economic scenario [9–11], peer-to-peer accommodation systems are affecting both the supply and demand side of the hospitality sector – since on the one hand they are providing more accommodation alternatives for guests and on the other one they are allowing private citizens to undertake new economic activities- and the implementation of appropriate methods of analysis is now needed.

2 Evaluating the Impacts of Airbnb: Stakeholders, Revenue Generation and Research Questions

The introduction of Airbnb in the hospitality market has affected a variety of stake-holders, including travelers, hosts, residents living in specific neighborhoods and local economic actors (e.g. hotels and other subjects working in the accommodation industries, retailers, food & wine entrepreneurs and so on). With regard to hosts, benefits can stem for instance by their interaction with guests and consequently be defined as personal and relational; however, since hosts charge guests for the use of their accommodation, an important goal for hosts is putting into value their under-used real estate properties and consequently gain revenues. More particularly, hosts may adopt different renting strategies depending on: (a) the type of accommodation they are able to provide (e.g. single room, entire apartment, etc.); (b) the availability of their accommodation through time (e.g. weekends only or selected months); (c) management conditions (e.g. availability of time and/or people to be dedicated to rental-related activities); (d) local regulations (e.g. taxation rules and limits concerning the number of days a property can be rent without a regular contract); (e) location and characteristics of the neighborhood, such as proximity to public transports; (f) demand trends; (g) hosts' revenue objectives.

With regard to the type of accommodation, it is worth-noting that across many countries entire homes/apartments represent 68.5% of the listings, whereas private rooms and shared rooms account to 29.8% and 1.7%, respectively [12]. These figures are particularly significant especially if we consider that in 2012 the percentage of Airbnb entire homes/apartments was definitely lower (57%), whereas private rooms represented 41% of the listings [12]. As underlined by some authors, short-term rentals are becoming in some cities not only an additional source of revenue for households but a deliberate real estate investment, which can in some cases subtract houses from the market and have negative social effects on potential residents [13, 14]. For instance, a recent study carried out by Schäfer and Braun [14] for the city of Berlin has pointed out that competition exists mainly for small apartments, characterized by one or two rooms and mainly located in areas convenient to the major points of attraction of the city. A high concentration of Airbnb accommodations in the city centre or nearby areas has been found by other authors too [15–17], and some of them have underlined a certain degree of inequality both in the distribution of revenues among hosts and between centre and periphery [15]. However, the analysis of the renting strategies adopted by hosts and of the revenues generated has not been fully explored yet. This paper aims both at identifying rental patterns and at estimating the economic impact generated for their hosts by single Airbnb listings, focusing on Turin (Italy) as a case study. More particularly, the research questions that lead the research are the following: which are the most common types of accommodations on offer on the short-term rental market, and which are their physical characteristics? Which are the most frequent rental strategies adopted by hosts, e.g. in terms of number of days made available for rent? Which are annual revenue patterns? Do occupancy rates suggest that accommodations have reasonably met hosts' presumed expectations?

3 The Revenues Generated by Airbnb for Hosts: The Case of Turin (Italy)

3.1 Turin: Characteristics, Tourism and Airbnb Growth

Turin is a city located in the North-West of Italy, spreading on a surface of around 130 km^2 and counting in 2017 a population of about 890,000 inhabitants, excluded its metropolitan area. After hosting the Winter Olympic Games in 2006, the city has registered a significant growth in the number of its tourists: in 2015 arrivals and presences amounted to 1,231,102 and 3,454,869 respectively, with a +48.5% and a +31.7% growth with respect to 2006 [18]. As other Italian and worldwide cities, Turin has recently been interested by an expansion of the Airbnb market: recent reports show that whereas in 2012 the number of listings was around 400, in August 2016 they exceeded 2,600 units, with an ever growing trend [19]. In 2016 listings were mainly entire homes/apartments (64.9%), followed by private rooms (32.5%) and shared rooms (2.5%); 73.8% were reported as being available for rent for more than 6 months and 42.1% as published by hosts managing more than one accommodation [19].

3.2 Research Methods

In order to describe Turin's Airbnb market dynamics and estimate the economic value generated by the properties made available for short-term rentals, we analyzed how listings performed over a period of one year, on the basis of the raw data provided by Airdna (www.airdna.co), a company that systematically collects data about accommodations listed on the Airbnb website. More precisely, we included in the research the listings that, beyond being present on the Airbnb website at the time of the study (i.e. November 2017), were created by the host no later than November 2016, as to analyze the performance over a period of 12 full months. We thus excluded the listings that, despite being active at the time of the research, were created by a host during the last 12 months. We also excluded both the listings that presented a booking calendar that was not updated by the host during the period under study ($n = 138$) - which might be interpreted as listings belonging to hosts who suspended their hospitality activity in the last year - and the ones that missed essential data ($n = 242$). Then, after the filtering, we carried out our analyses on a total of 1,967 listings. To get an overview of the phenomenon, we analyzed the set of data applying traditional descriptive statistics methods, focusing on the following metrics: *Type of listings*: it refers to the type of accommodation on offer (i.e. entire home/apartment, private room, shared room); *Average daily rate:* the average price per day for each listing; *Number of bookings:* the number of bookings registered for a listing, in the last 12 months (LTM); *Reservation days:* it refers to the number of reserved days in the LTM; *Available days:* the number of days that were available for rent in the LTM, but that were not actually booked by guests (i.e. days in which accommodations were vacant); *Rental days:* the number of days a listing was put on the market in the LTM (it is calculated as the sum of reservation days plus the number of available days, and we introduced this metric to identify hosts' rental behavior); *Occupancy rate:* it is calculated dividing the number of reservation days by the number of rental days; *Annual revenue:* it is calculated considering the number and type of days booked, the price per time unit (e.g. day or week) and cleaning fees (where applicable). For some metrics we deemed appropriate not only to calculate mean, median, standard deviation, minimum and maximum values, but also frequencies.

3.3   Results and Discussion

Coherently with current worldwide trends and previous studies, the most part of the accommodations analyzed in our research were entire homes/apartments (71.9%), followed by private rooms (26.6%) and shared rooms (1.5%). These results seem thus to indicate that in the last year the most part of the Airbnb accommodations in Turin were not actually shared by hosts when they were in their property, and that some rental patterns might have been driven either by investment strategies or hosts' mobility habits and needs. Table 1 shows that, on average, entire homes/apartments registered not only greater annual revenues, but also highest values in terms of bookings, booked days and rental days. Standard deviation (SD) values for annual revenues suggest that some listings generated peaks, whereas others generated limited added value for their owners.

Given that entire homes/apartments (EH) and private rooms (PR) are the most frequent type of accommodation, we then focused only on these two types.

Table 1. Airbnb accommodations in the last 12 months (LTM): descriptive statistics (authors' own elaboration on Airdna data)

		M	Mdn	Min	Max	SD
Annual revenue (euros)	Entire home/apt	6665	5427	46	37997	5777.44
	Private room	3333	2179	19	33820	3757.34
	Shared room	1458	778	50	6298	1528.51
Bookings	Entire home/apt	31	21	0	185	30.42
	Private room	29	15	0	206	35.83
	Shared room	19	7	1	81	23.75
Daily rate (euros)	Entire home/apt	69	60	17	552	38.16
	Private room	41	33	15	480	28.24
	Shared room	23	18	10	106	17.64
Reservation days	Entire home/apt	106	88	1	355	82.13
	Private room	91	64	1	311	80.66
	Shared room	78	49	1	270	83.73
Available days	Entire home/apt	98	93	0	296	63.68
	Private room	95	88	0	300	63.97
	Shared room	64	59	0	175	42.23
Rental days	Entire home/apt	204	217	2	365	110.77
	Private room	186	187	1	364	104.55
	Shared room	142	107	25	348	101.30
Occupancy rate	Entire home/apt	0.488	0.489	0.032	1.0	0.248
	Private room	0.446	0.417	0.032	1.0	0.268
	Shared room	0.485	0.513	0.032	1.0	0.301

Graph 1 shows that, with respect to bookings, the trends registered for both PR and EH are similar. The same can be said for reservation days (1–90 days: EH = 51%, PR = 60%; 91–180 days: EH – 27%, PR = 24%; 181–270 days: EH = 19%, PR = 13%; 271–365 days: EH = 3%, PR = 3%). The analysis of rental days shows that especially EH were put on the market for very long periods (1–90 days: EH = 22%, PR = 25%; 91–180 days: EH = 20% PR = 23%; 181–270 days: EH = 21%, PR = 25%; 271–365 days: EH = 37%, PR = 27%), suggesting that also in Turin peer-to-peer accommodation systems may subtract at least some properties from the regular rental market, as reported for other European cities [13, 14]. Occupancy rates show that EH may satisfy hosts' expectations better than PR: in fact, highest occupancy rates were registered especially for the former type of listing (0–0.25: EH = 22%, PR = 31%; 0.26–0.50: EH = 31%, PR = 29%; 0.51–0.75: EH = 31%, PR = 24%; 0.76–1: EH = 16%; PR = 16%). However, given that many listings registered occupancy rates lower than 0.50, it can be stated that some accommodations under-performed with respect to presumed hosts' expectations.

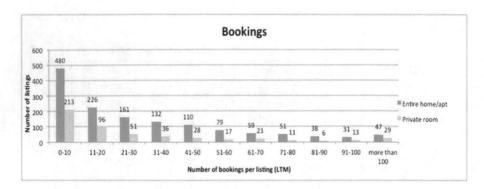

Graph 1. Number of listings registering a certain number of bookings (LTM) (authors' own elaboration)

Graph 2 shows that annual revenues present different patterns according to the type of accommodation, and that EH frequently register either limited revenues or significant ones, confirming patterns of income inequality [15].

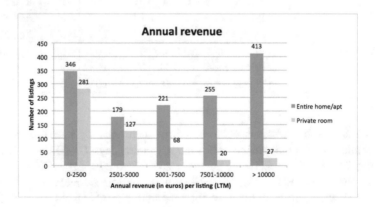

Graph 2. Annual revenue patterns (authors' own elaboration).

Graph 3 shows the cumulative distribution of annual revenues in relation to the number of bookings. When further investigating revenue inequality, we found that 15.3% of hosts published more than one listing, and that 35% of the accommodations analyzed was managed by multi-listing hosts. Even though 4 hosts manage between 11 and 21 EH each, the most frequent multi-listing pattern is represented by hosts managing 2 or 3 listings (65.6% and 23.5% respectively). Interestingly, multi-listing does not necessarily mean multi-property: in fact, the integration of data analysis with the direct and personal exploration of the photos published by hosts allowed to highlight that in some cases sub-portions of apartments with several bedrooms were promoted through different listings (e.g. two or more private bedrooms, one private bedroom and an entire home/apartment, and so on). This might thus represent a specific strategy adopted by hosts to get higher revenues from large apartments that otherwise could fall out of the market and expand the opportunities to appeal different types of guests (e.g. families, groups, singles and couples).

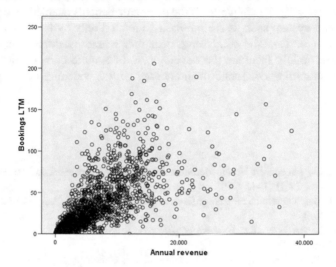

Graph 3. Scatter plot of annual revenues (in euros) and bookings (authors' own elaboration).

4   Conclusions, Limits and Future Steps of Research

This brief article has provided a preliminary overview of the renting patterns and dynamics observed for Airbnb accommodations in Turin (Italy), suggesting that the economic value generated by the properties made available for short-term rentals may vary. More precisely, EH generally seem both the most frequent and best performing type of accommodation, even though phenomena of income inequality exist. Further analyses could focus on the revenues generated for multi-listing hosts and on the economic benefits stemming from the promotion of a property via multiple listings. Then, an essential step of future research will be represented by the adoption of a spatial perspective, as to identify the areas of the city that are most affected by Airbnb supply and demand; this would allow both to analyze to what extent the location variable is affected by proximity to transport facilities, services and cultural hubs, and to expand urban-related research questions - e.g. are Airbnb accommodations mainly located in gentrified neighborhoods and/or in central areas where re-use policies are ongoing? Is the Airbnb phenomenon affecting neighborhoods far from city attractions but well served by public transports? -. The adoption of a spatial perspective would encourage not only to make wider considerations on the economic value generated by Airbnb - e.g. at the neighborhood and city scale- but also to compare current Airbnb dynamics with the regular rental market and real estate trends in given areas. In fact, as anticipated in the introductory paragraphs, the use of real estate units as temporary accommodations for short-stays may affect both rental and real estate market dynamics, with possible consequences on real estate values too. Other interesting steps of research could be represented by the conduction of field studies aiming at collecting information about guests' consumption patterns (e.g. in terms of time, goods and services enjoyed in the areas where the Airbnb accommodations are located), as to better investigate whether in given areas the recipients of the economic benefits enabled by a peer-to-peer accommodation system such as Airbnb are represented only by hosts or also by other economic players. Overall, combining quantitative and qualitative approaches of research would finally facilitate the development of a more comprehensive interpretation of this emerging economic phenomenon, with a widening of the perspectives adopted.

References

1. Ashworth, G., Page, S.J.: Urban tourism research: Recent progress and current paradoxes. Tour. Manage. **32**(1), 1–15 (2011)
2. Law, C.: Urban Tourism: The Visitor Economy and the Growth of Large Cities. Continuum, London (2002)
3. Pearce, D.G.: An integrative framework for urban tourism research. Ann. Tourism Res. **28** (4), 926–946 (2001)
4. De Filippi, F., Coscia, C., Boella, G., Antonini, A., Calafiore, A., Cantini, A., Guido, R., Salaroglio, C., Sanasi, L., Schifanella, C.: MiraMap: a we-government tool for smart peripheries in smart cities. IEEE ACCESS **4**, 3824–3843 (2016)
5. Guttentag, D.: Airbnb: disruptive innovation and the rise of an informal tourism accommodation sector. Curr. Issues Tourism **18**(12), 1192–1217 (2015)
6. Guttentag, D.: Airbnb: why tourists choose it and how they use it. https://www.dg-rsearch. com/Papers/Summary%20doc%20-%20Airbnb.pdf. Accessed 20 Nov 2017
7. Zervas, G., Proserpio, D., Byers, J.W.: The rise of the sharing economy: estimating the impact of Airbnb on the hotel industry. Boston U. School of Management research paper No. 2013-16 (2016). https://papers.ssrn.com/sol3/papers.cfm?abstract_id=2366898. Accessed 4 Nov 2017
8. Varma, A., Jukic, N., Pestek, A., Shultz, C.J., Nestorov, S.: Airbnb: Exciting innovation or passing fad? Tourism Manage. Perspect. **20**, 228–237 (2016)
9. Calzada, I., Cobo, C.: Unplugging: deconstructing the smart city. J. Urban Technol. **22**(1), 23–43 (2015)
10. De Filippi, F., Coscia, C., Cocina, G.: Piattaforme collaborative per progetti di innovazione sociale. Il caso MiraMap a Torino. Techne **14**, 218–225 (2017)
11. De Filippi, F., Coscia, C., Guido, R.: How technologies can enhance open policy making and citizen-responsive urban planning: MiraMap - a governing tool for the Mirafiori Sud district in Turin. Int. J. E-plann. Res. **6**(1), 23–42 (2017)
12. Ke, Q.: Sharing means renting? An entire-marketplace analysis of Airbnb. In: Proceedings of the 2017 ACM on Web Science Conference, pp. 131–139. ACM, New York (2017)
13. Guttentag, D.A., Smith, S.L.: Assessing Airbnb as a disruptive innovation relative to hotels: substitution and comparative performance expectations. Int. J. Hospitality Manage. **64**, 1–10 (2017)
14. Schäfer, P., Braun, N.: Misuse through short-term rentals on the Berlin housing market. Int. J. Hous. Markets Anal. **9**(2), 287–311 (2016)

15. Picascia, S., Romano, A., Teobaldi, M.: The airification of cities: making sense of the impact of peer to peer short term letting on urban functions and economy. In: Proceedings of the Annual Congress of the Association of European Schools of Planning, Lisbon (2017)
16. Gutiérrez, J., García-Palomares, J.C., Romanillos, G., Salas-Olmedo, M.H.: The eruption of Airbnb in tourist cities: comparing spatial patterns of hotels and peer-to-peer accommodation in Barcelona. Tour. Manage. **62**, 278–291 (2017)
17. Quattrone, G., Proserpio, D., Quercia, D., Capra, L., Musolesi, M.: Who benefits from the "Sharing" economy of Airbnb? In: Proceedings of the 25th International Conference on World Wide Web, pp. 1385–1394. ACM Press (2016)
18. Regione Piemonte, Flussi turistici in Piemonte 2015. Report (2016). http://www.regione. piemonte.it/pinforma/images/DOCUMENTI/flussi_turistici_2015.pdf. Accessed 4 Nov 2017
19. Federalberghi: Sommerso turistico e affitti brevi. Report (2016). http://intranet.federalberghi. it:8000/pubblicazioni/Pub/Sommerso%20turistico%20e%20affitti%20brevi/sommerso% 20turistico%20ed%20affitti%20brevi.pdf. Accessed 4 Nov 2017

Multidimensional Poverty Measures: Lessons from the Application of the MPI in Italy

Andrea Billi and Mia Scotti[✉]

Sapienza Università di Roma, 00185 Roma, Italy
{andrea.billi,mia.scotti}@uniroma1.it

Abstract. Several interesting methodologies are currently applied to assess multidimensional poverty around the world. Among these methods, the Multidimensional Poverty Index (MPI), developed from the Oxford Poverty & Human Development Initiative in 2010, has particular strengths in capturing multidimensional poverty described by simultaneous deprivations. The method has been widely applied to developing countries whilst its application to developed countries is still very limited. This paper, following the approach of Alkire Foster [1] and the work of Nicolai Suppa [2] applied to Germany, offers a first comparison for Italy, between the ISTAT monetary measure [3] of poverty and an ad hoc multidimensional poverty index (MPI).

Keywords: Multidimensional poverty index · Local policies planning
Multicriteria analysis

1 Poverty: Monetary Versus Multidimensional Approach

Policy actions against poverty are strictly related to how poverty is defined and to the theoretical framework underneath it. A deep knowledge of the effect of the normative implications and the theoretical framework might have on assessments and results is then essential to target public policies and to implement actions at a local level. In the multidimensional analysis, several aspects of human daily life are relevant to determine people's well-being: they are as complex as the heterogeneity of human beings themselves. What matters are the opportunities people have and the status of wellbeing they would like to achieve.

Political freedoms, economic infrastructures, social opportunities, transparency guarantees, security against crime and psychological violence are instrumental conditions to allow people enjoying a long, healthy and satisfying life [4, 5]. It follows that a development policy *"should aim to expand people's capabilities"* and *"a person's capability to achieve functionings that he or she has reason to value provides a general approach to the evaluation of social arrangements"* [6].

Soon then, three general issues have to be considered in policies planning to reduce poverty and thus to foster local development. Income is one of the means to escape from poverty but considered alone is not enough to give an exhaustive picture of what poverty is. Poverty is a multidimensional concept determined by different aspects of human daily life [7–9]. It may concern, with a different magnitude, less developed as well as developed countries. The main difference among monetary and

© Springer International Publishing AG, part of Springer Nature 2019
F. Calabrò et al. (Eds.): ISHT 2018, SIST 100, pp. 652–660, 2019.
https://doi.org/10.1007/978-3-319-92099-3_73

multidimensional poverty measures consist in focusing on what really matters for people, what it is really achieved and not merely what can be eventually be achieved through economic resources. Multidimensional poverty assessment, based on joint dataset analysis, enables policies makers to better target policies and actions. Indeed, the method allows to identify the population poorest subjects, to understand what is determining poverty across the different aspects of life, to identify possible correlation among dimensions, to capture the incidence and the intensity of poverty in the country, to decompose poverty analysis at subgroup level, to reduce errors and better target policies at micro level [1–10].

Microdata analysis, as proposed in the Alkire Foster method, enables to better identify the breath of poverty among groups accounting for several deprivations occurring at the same time [1–6].

2   The MPI: How It Works and Why It Can Be Helpful

The Alkire Foster method (AF) defines poverty as a condition of multiple deprivations. It is a counting approach where deprivations suffered by individuals at the same time are summed to identify or not a condition of poverty, If person i wellbeing, with $i = 1....n$, is constituted by j dimension with j accounting for d dimension, $j = 1....d$, deprivation of each dimension j will account in determining each person i poverty level. The matrix Y contains the available data, it is of size NxD, and describes for each individual the achievement in each dimension deemed relevant. The row vector z, with $z_d > 0$, describes the deprivation cut-offs, e.g., the achievements necessary for not being considered as deprived in the selected dimensions [1].

The Alkire Foster method is a way to assess multidimensional poverty. The method relies on the capability approach framework. Poverty is defined as capabilities deprivation and is assessed through the functionings achieved. Functionings constitute people wellbeing and thus its deprivation figures as poverty.

The basic concept of the AF is the dual cut-off level [11]. A cut-off is a threshold, a settled value that divides the population in two sets. In the AF, it is used in two moments: deprivation identification and poverty identification. The deprivation cut-off is a value under which a person is considered deprived for a certain dimension. The poverty cut-off is the value (number of dimensions) under which a person is identified as poor.

The application of the Alkire Foster class of measures brings several advantages; first, it is a way to assess poverty or wellbeing as a complex life status determined by a condition of simultaneously coexistent deprivations. Secondly it gives us the possibility to understand why poverty changes overtime. This because micro data joint analysis clearly allows to monitor how a particular deprivation is changed and so to identify which dimension is changing the poverty population profile. This property may be particularly relevant in local policy planning [1–10].

2.1 The Method – Identification, Aggregation and Weights

Among the axiomatic approaches, the application of the Alkire Foster method and the construction of an MPI index requires some essential steps [12] that we are going to scrutinize in detail:

- Establish the aims of the index. The following analysis has the objective to build an M_o index of multidimensional poverty in Italy for 2015. The intent is to obtain a multidimensional index able to capture joint deprivations among the population. Five key dimensions are considered. As part of the AF class of measure, M_o respects some of the properties as symmetry, normalization, poverty and deprivation focus, scale invariance, replication invariance, weak monotonicity and weak re-arrangement, dimensional monotonicity, subgroup decomposability, robustness to monotonic transformation of data [1–7]. As the measure has been built based on variables that aim to be proxies of critical value aspects of life, the measure itself aims to be an expression of un-freedoms that affect the reference population. For each dimension, outcomes indicators have been chosen as proxy of realized functioning or deprivations, so poverty is thus investigated in the space of the Human Development approach [5].
- Select the unit of analysis. The reference population of the investigation is the Italian population. The unit of identification are people living in Italy. Dataset is obtained by a statistical sample of about 20.000 households and 50.000 people that represents by its composition Italy as a country. Information is collected each year under the Aspects of Daily Life (ADF) investigation prompted by the Italian National Institute of Statistics (ISTAT). The ADF investigation collects information through assessment questionnaires directed to resident population.
- Identify the dimensions considered relevant to the analysis and the dimensional cut-offs. Five dimensions have been chosen to investigate on multidimensional poverty in Italy. In this empirical work, the dimensions judged critically relevant in describing poverty are schooling, health, economic security, housing and social relationships. Some basic principles guided this choice: first of all, the human development framework, relying particularly on Nussbaum works [12, 13], and the Global MPI approach. Secondly, the need to select relatively few dimensions, easy to understand, to be shared and to replicate. Most of the selected dimensions are included in recognized global and local measures of poverty and wellbeing as for example the MPI, BES, the Quality of life index, HDI, and others. The selection of dimensions made by this work to be considered as variables and weights is certainly open to criticism or modification. Here are briefly summarized:
- Health: the choice to look for a health dimension is probably the easiest to justify, it refers to every human and life itself. Being healthy or not is what this dimension aims to capture. Being healthy is multidimensional itself and asks for being physically and mentally healthy. Health conditions influence other capabilities as people' ability to relate with the exterior world and be part of the society.
- Schooling: education affects people capabilities to read and write as to express, senses, to imagine, think and reason, it affects being able to define the concept of "good" and to engage in critical reflection about the planning of one's life [13].

Education affects several aspects of human daily life that are universally recognized as essential.

- Economic security: the dimension wants to capture opportunities people have to do intrinsically important things or those that they judge so. Employment has been included in most of the multidimensional measures. Furthermore, economic resources have been and are even today the base of most common poverty measures adopted worldwide. It is recognized the essential role that economic resources have to realize several functioning issues people may value.
- Housing: Having a proper house may affect people health, social relationship, and life in general. Social relationships: Several activities related to the social space may be considered relevant in human life. Among those, the present work wants to focus on essential ones: the relationship with our family and friends. This choice is made to reduce possible critics around this dimension that could be judged as particularly subjective. The general aim of the dimension is to capture the affiliation and emotions capabilities human may enjoy and benefits.

Following are reported each dimension variables and cut-off adopted. The first principle guiding the indicators selection has been the urgency to focus on outcomes indicators able to synthetize functioning achieved by people. Thus their first aim is to measure deprivations in doing and being that individuals have reasons to value [5, 14, 15].

The health dimension is investigated through three variables: *The body mass index for people aged 18 and over*[1], *Being afflicted by chronic diseases or health problems of long duration, Perception on own health status*. A person is deprived in the health dimension if it scores one out of three on the variables chosen: if the individual suffers from obesity, if he suffers from at least one chronic disease or if he perceives his health condition as very bad.

Schooling. To cover the education dimension and capture its related functioning, the variable *achieved level of study* has been chosen. Deprivation in this dimension occurs if the compulsory qualification or diploma is missed.

The economic security dimension is measured through the variable *having sufficient resources to cover family needs in last twelve months*. The variable captures a subjective perception on available economic resources. As in some works the variables including subjective perception has been excluded, the present analysis retains such type of indicator able to capture the intrinsic value of security that can be different among people. Responds may answer as follow: more than sufficient, fair, inadequate or extremely inadequate.

Housing: four variables have been chosen to capture people functioning to enjoy a proper house, a place where find shelter and secure rest. These are: *respondents report their house is in bad conditions, the house is not connected to the sewerage system, the house does not have the heating system, problem in the water furniture in last twelve months*. A person is deprived if his house conditions totalize a score higher than 0.4 among the dimensions where the heating system and the sewer weight 0.4 and the others 0.1.

---

[1] A BMI that corresponds to obesity brings deprivation in the sub index.

Social relationship dimension is measured in this work through two indicators: *respondents report the satisfaction level for their relationships with friends and with their family in the last 12 months*. Two out of two problems must be reported for being deprived (dimensional cut-off). A person is deprived if for both indicators report a bad or extremely bad relationship.

2.2   Results: Aggregate Measure, Contributions and Subpopulation

The aim of this section is to synthetize some of the results obtained in the MPI construction and analysis. About 14% of the Italian population resulted multidimensionally poor in 2015, with an average share of deprivations among the poor population of about 62% (mean number of the five deprivations suffered by the poor part of the entire population). This means that 14% of Italians result concurrently deprived in three out of the five dimensions simultaneously. The set is distributed with a 23,13% of not deprived people, the 34,13% deprived in at least one dimension, the 28.49% deprived in two dimensions, 12,29%, 1,27% and 0,02% respectively in three, four and five dimensions. Poverty line cut-off determines the incidence of poverty in the population. It would then possible to divide the population in three segments as follows: at risk ofpoverty (28,49%), poor (12,54%) and extremely poor (1,54%). At National level, it is possible to identify what is determining poverty across the population: what matters more? Italian poor population result more deprived in the schooling dimension followed by economic security aspects, health, social relationship and housing.

The incidence of poverty among female population is higher than in male population (respectively about the 15,55% against the 12,45%) with slight differences in the incidence of each deprivations across the two groups. The order of importance of each deprivation between the two groups is stable so than poverty is determined in the same proportion by the five deprivations investigated. Men and women result equally deprived in the health dimension as in the economic security one, men more in schooling and less in social relationships. Some differences emerge comparing foreigners and Italian citizens. Poverty affect about 14% of the Italians respect to about 10% of foreigners living in Italy. Relevant inter deprivation differences between the two groups emerge in the housing and social relationship dimensions where foreigners are about two times more deprived than Italians. Italians results more deprived in the health dimension (29,34% against the 19,35%).

With regard to the age distribution of poverty among the population set, we observe the highest concentration of poverty among elderly people, with the 32,27% over of people over 75 years old afflicted by poverty. The class less affected by poverty results the 18–24 years old one with the 2,60% of people. Regarding the composition and the comparison of poverty among age groups, it would seem that the youngest population is more affected by social relationship and economic resource deprivations, less in schooling and health dimension. The poor in Italy are for 5% aged under 35 years old, 41% has among 35 and 65 years old and 54% are over 65 years old.

Two reverse and opposite trends are observable in what determines poverty among population. Deprivation in social relationship decreases at age growth as housing and economic insecurity. While health and schooling deprivations increase in the elderly. This highlight the vulnerability of elderly people also to the rapid changes in the globalized economy, such as issues relating to technology, the digital divide, financial literacy, etc....

The geographical analysis (see Figs. 1, 2 and 3) reveals that poverty incidence is lowest in Trentino Alto Adige Region where 6% of the population is affected by poverty. It is at the first place of the MPI, 0,038. Valle d'Aosta, Lombardy and Lazio are following. The worse performances are scored by Campania, Sardinia and Sicily, which are also the regions with a low performance in the implementation of EU structural Funds.

3   Poverty Measures and Policies in Italy: Which Level of Intervention for Each Dimension?

This section is focused on highlighting some of the results of comparing the monetary and multidimensional poverty in Italy[2] [16]. The first remarkable results of this investigation is that the incidence of poverty across the Italian population is following two reverse trends. Firstly the incidence of poverty across the population is considerably higher in the multidimensional measure in respect to monetary poverty. While in the ISTAT investigation poverty hits about 7% of the population, in the multidimensional analysis the incidence reaches about 14% of the population. Secondly, poverty in the multidimensional space is higher among elderly people respect to the youngest, while in the ISTAT official investigation, poverty higher incidence is measured among people under 35 years old. This result may be partially explained analysing the two poverty measure composition. Comparing how poverty is determined by each deprivation, it is evident that economic resources are one of first determinants of poverty across the Italian youngest population. As the ISTAT measure of poverty is calculated on individual expenditure, the result is consistent. So multidimensional and monetary poverty measures describe two different paths among the population in terms of age distribution of poverty.

Another reverse trend is observed between sexes. In the National measure male population results more affected with about 7,9% of the male population hit by poverty against 7% of women. In the multidimensional analysis, about 15% of women are affected by poverty in respect to male population (about 11%). The same pattern is confirmed by the AROPE index developed by EUROSTAT.

Common elements are evidenced in the geographical distribution. Both measures agreed that poverty incidence is higher in southern regions with Sicily, Campania, Sardinia and Calabria at the last ranking positions. Even if absolute poverty data are not available in respect to foreigners and to Italian citizenship poverty distribution, the analysis of aggregate absolute poverty in Italy highlights how foreigner households are much more affected by poverty. The analyses of the Italian MPI, reverse this prospective.

---

[2] The comparison is made with the official monetary measure offered by ISTAT through its open database considering both measures for the entire population excluded the under 18 years old. For more on this topics please refer to La povertà in Italia, ISTAT 2015 and to the official ISTAT website.

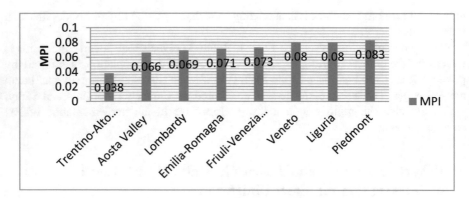

Fig. 1. MPI across northern, center and southern Italy. Multidimensional poverty index across northern Italy. An higher value of the index corresponds to an higher level of poverty in the reference region.

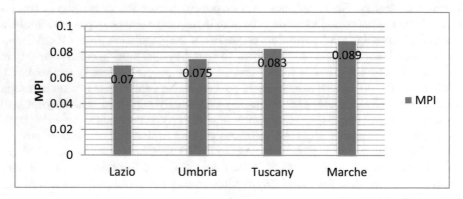

Fig. 2. Multidimensional poverty index across center Italy. An higher value of the index corresponds to an higher level of poverty in the reference region.

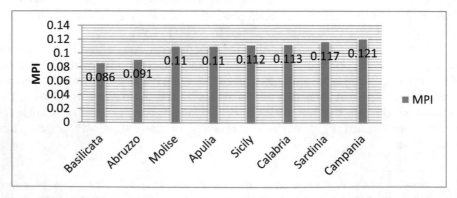

Fig. 3. Multidimensional poverty index across southern Italy. An higher value of the index corresponds to an higher level of poverty in the reference region.

Thanks to the results obtained by the work here briefly presented, it is possible to recommend target policies against poverty in Italy that will foster access to housing, implement work policies to improve economic security across younger population and foster aggregation and social exchange. At the same time it should give attention to rebalance women and men opportunities of access to economic resources. Results may matter in differentiating policies across regions. It is possible, indeed, thanks to the index construction methodology, to investigate particular poverty patterns across territories. This type of investigation may foster policy results increasing local development.

The analysis conducted at the Italian level highlights a specific consideration; there is not an unique proper and focused policy against multidimensional poverty in Italy. It is true that at a country/national level Italian government fails to plan a national policy against poverty intended as a multidimensional concept. The main initiatives and policy interventions are still strictly related to the income dimensions of poverty [17]. In Italy, even if a demonstrable growth in funding for poverty actions is demonstrable, these actions are still strongly anchored to income scarcity and material deprivations. From the comparison between the MPI, the censored head count ratio and the ISTAT index, some relevant aspects emerged: the three measures are only partially compliant in the poverty distribution among the population. The multidimensional analysis presented highlights a higher poverty level for Italy, more incident among the female population and in the elderly segment. About the geographical distribution of poverty, monetary and multidimensional poverty measures agreed that the northern Italy is the richer region. Health, schooling and economic security results the Italian three major drivers of poverty across the population.

In the end, there is still a work to do in Italy to enforce a multilevel governance and coordination of single policics and intervention to fight against poverty and foster local development.

References

1. Alkire, S., Foster, J.E., Seth, S., Santos, M.E., Roche, J.M., Ballon, P.: The Alkire-Foster counting methodology Oxford. In: Multidimensional Poverty Measurement and Analysis. Oxford University Press, Oxford (2015)
2. Suppa, N.: Towards a Multidimensional Poverty Index for Germany. OPHI Working Papers 98, University of Oxford (2016)
3. ISTAT: La misura della povertà assoluta, metodi e norme n.39, Istituto Nazionale di Statistica Via Cesare Balbo, 16. Roma (2009)
4. Sen, A.K.: Development as Freedom. Oxford University Press, Oxford (1999)
5. Alkire, S.: The Missing Dimensions of Poverty Data: An Introduction. Oxford University Press, Oxford (2007)
6. Sen, A.K.: Inequality Reexamined. Harvard University Press, New York (1995)
7. Bourguignon, F., Chakravarty, S.: The measurement of multidimensional poverty. J. Econ. Inequality 1, 25–49 (2003)
8. Sen, A.K.: Choice, Welfare and Measurement. Basil Blackwell, Oxford (1982)
9. UDR: Human Development Report 1990. Oxford University Press, Oxford, UK (1990)

10. Alkire S., Foster, J.: Counting and multidimensional poverty measurement. OPHI, working Paper 3. Oxford, UK (2009)
11. Alkire S.: Choosing Dimensions: The Capability Approach and Multidimensional Poverty. Chronic Poverty Research Centre Working Paper 88 (2007)
12. Sen, A.K., Nussbaum, M.: The Quality of Life. Clerendon Press, Oxford (1993)
13. Nussbaum, M.: Sex and Social Justice from Women and Human Development: The Capabilities Approach. Cambridge University Press, Cambridge (2000)
14. Burchi, F., De Muro, P.: Measuring human development in a high-income country: a conceptual framework for well-being indicators. Forum Soc. Econ. 45(2–3), 120–138 (2016)
15. Samman, E.: The Missing dimensions of poverty data. Oxford Dev. Stud. 35(4), 347–359 (2007)
16. ISTAT La povertà in Italia, statistiche report, Istituto Nazionale di statistica, Via Cesare Balbo, 16. Roma (2015)
17. Gori, C.: Caritas Italiana: Rapporto 2016 sulle politiche contro la povertà in Italia (2016)

The Local Provision of Public Services: Municipal Capitalism, PPP's Schemes and the Regulation Issues in Italy

Francesco Timpano[(⊠)] [iD]

Cattolica del Sacro Cuore University, 29122 Piacenza, Italy
francesco.timpano@unicatt.it

Abstract. The issue of the local provision of public services has becoming important in Italy due to several changes that have been implemented in the last years by the national government in the regulation of state owned companies and in the regulation of public services (or services of general economic interest). The regulatory role of national or local government and the increasing role played in some large companies managing important services have to be accurately discussed in order to reconcile efficiency of the companies with the need of properly represent the political orientations. Potential conflict of interests on one side but also the need to properly represent the interest of citizens, more than the interest of shareholders, are two issues that are discussed in the paper. A practical example of a small size Italian town is used in order to show the current situation in terms of regulations, planning of the service, pricing and management of the service in the most important sectors. Competition and public interest are not necessarily in conflict, but the whole setup must be reviewed.

Keywords: Local public services · Governance · Regulation

1 First Section

The public (vs private) provision of local public services is a relevant topic for a number of disciplines studying the different implications of the changing role played by those services in the local and territorial development dynamics.

Local public services are usually provided by locally-based firms, even when the company is actually external to the territory, determining relevant impact on local economies. In Europe traditionally local public services have been provided by public local companies or, in some cases, even directly by the local government apparatus. More recently, an increasing attention for efficiency has determined a change in regulation, aimed to promote a market orientation of the services' production obtained either by promoting the competition for the market (by implementing competition among different firms) or the competition in the market (liberalisation of the services).

In this transition, a number of local government-owned firms have been transformed in larger companies, specialised in local public services, operating on a larger scale, implementing agreements with private partners and operating according to

© Springer International Publishing AG, part of Springer Nature 2019
F. Calabrò et al. (Eds.): ISHT 2018, SIST 100, pp. 661–671, 2019.
https://doi.org/10.1007/978-3-319-92099-3_74

classical private firms' objectives, including in some cases those implied by the status of listed companies.

In this paper an analysis of the Italian case is proposed, starting from the evidence that in Italy is possible to find a number of different regimes in local public services management. An interesting issue concerns the implications of the so-called municipal capitalism, namely the situation where the local public services are provided by firms owned by local government, operating according the market oriented approach, and sometimes directly involved in the regulation activity. The implications of this situation, both from the normative than the positive point of view, are relevant and deserve a focus based on the practical experiences.

The topic is relevant also for the implications of the evaluation of the performance of companies involved in the provision of public services. In a recent work [1], an empirical analysis comparing the performance of public and private firms tries to disentangle the issue. In the literature there authors claiming for a superiority of private firms operating in public services provision with respect to public ones [2], some others claiming that the performance of companies owned by public government improves once they operate in a more competitive scenario [3]. What is not actually clear is whether the performance of the company depends on the ownership or on the level of competition of the market. In studying mixed companies (public-private partnerships) the efficiency of the company seems to improve when the private sectors' shares increases [4].

The issue is becoming even more crucial now in Italy, since national government has promoted a robust reform[1]. The reform is aimed to reduce the number of public-owned companies, to limit the number of locally owned companies and to promote increasing competition in the provision of public services at local level.

2 Municipal Capitalism: Latest Evidence in Italy

The phenomenon of the municipal capitalism [5] is the outcome of a process, founded in the push towards "market oriented" solutions in local services' provision, that attributes to the (central and local) government an increasing role in the regulation activity and a decreasing role in services management. An obstacle to a fully market-oriented organization of local public services is the evidence of existence of economies of scale in public services' production and then natural monopoly are sometimes preferable to a market provision of services [6]. An increasing attention to the impact of the behaviour of public owned companies is also related to the organizational implications and to the impact on public budget [7].

Most of public services are not anymore part of the income redistribution process through government budget, most of the time due to the decreasing resources available especially at local level but also to improve efficiency in the production of public

---

[1] The Italian national government has promoted a large-scale reform of the Public Administration on the basis of the delegation law n. 125 (7 August 2015). On this basis the Ministry Marianna Madia has introduced the Legislative decree n. 175, 2016, the so-called "Unique text for public owned companies".

services. Currently, the organization of the local public service has been often based on a funding mechanism fully paid by the customers (sometimes strictly related with the intensity of the utilization of the service) and not by the public budget. In most cases the economic risk of the service has been transferred from the public government to the company managing the service with the inflows of the service accruing directly to the company itself (the so-called net cost approach).

Nevertheless, different regimes have to be considered in these processes:

1. the first one is the fully privatization of the service. At local level no evidence of services provided by private companies on a market competition basis can be actually reported in Italy. At national level the market for telecommunications and partially the public transport have been characterized by a quasi-privatization approach;
2. the second regime is the competition for the market in the case of concession of public services. At local level, services for waste management, water distribution management, local public transports and natural gas local distribution have produced a setup where competition for the market is becoming the rule. In this case, the old local government-owned firms have been in some cases transformed in larger firms, operating with private partnerships and competing in different markets, and not only in the market of origin of the firm itself;
3. the third regime is the so-called in-house providing of the service, most of the time through a local government fully-owned company, operating only in the territory where the service is provided. Also in this case, the tendency has been towards the creation of a company, operating with a full political control but also formally independent from the local government.

The different regimes are a clear signal that at local level this process has been strongly characterized as a "reluctant privatization" [8]. At the same time, the role in regulation activity has increased for local levels of government. The idea was that the main role of local government should have focused on regulation of the services more than on management.

Nevertheless, in a number of cases the local government is at the same time regulator and player, most of the times in both cases along with other entities.

3   The Public Services Regulation in Italy and the Issue of the Optimal Territorial Level

In the cases of competition for the market, the Italian national/regional legislation has often pushed the creation of "optimal territorial level of regulation" (Ambito) greater than one single municipality (with some exceptions for larger cities). Most of the time the "optimal territorial level of regulation" is either provincial or regional and it is obviously characterized by territorial contiguity. The basis for the optimal territorial level is not the service, but the belonging of the cities to the same administrative level of government (regional or provincial) and no cases of interaction among territories at the border of different Regions have been experienced.

The evolution of Italian regulation at national level started in 2008 with the liberalisation of local public services with a substantial marginalisation of in-house providing and limitations to the so-called mixed firms[2] (art.23 bis Law 133/2008 - Government Berlusconi). In 2011 a national referendum eliminated the previous law with a strong "political" orientation towards the water management service, for which it was included the abolition of the returns on invested capital. In 2011 a new law reintroduced the same regulation of the Law 133/2008 excluding the water management service (art. 4 Law 48/2011). In the art. 9 Law 183/2011[3]. The principle of the strong preference for a competitive mechanism has been reinforced so that the direct concession of a public service could be applied only in the case of a mixed firm (where the private partner has been chosen through a competitive procedure) or in-house providing (less than 200 k€).

After that a relevant judgment of the Constitutional Court has abolished the previous situation in 2012 on the basis of the principle that the European alternatives concerning public services has to be fully available for any public entity operating in Italy and any limitation is not coherent with the orientation of European legislation. And then this judgment fully re-introduced the possibility of the choice among three "European" alternatives to deal with a public service:

- In-house providing (direct concession);
- Mixed firm: direct concession if the private partner has been chosen on the basis of a competition or by means of the so-called "tender for two objects"[4];
- Competition for the service.

More recently the evolution of the regulation in Italy has been mostly oriented to the reduction of the number of companies owned by public governments at local level and a new regime for public services starting with Law 124/2015. The up-coming legislation has introduced incentives to the dismission of the shares owned by local government in firms managing services[5] and has increased the restrictions of the sectors where it is allowed for local governments to own firms. Among the others only strategic companies can be kept by the local government and only one in each sector and the negative economic result of the company will have a direct impact on the budget of the local government. Moreover, the regulative role of municipalities will be gradually reduced by the increasing role given to national (and regional) regulatory agencies.

The expected side effect of the new regulations is an increase of the number of local services *de facto* not manageable anymore by the local company, that will probably be

---

[2] The mixed firms are entities managing public services, created on the basis of a partnership between the government and a private firm. The partnership is promoted by means of a tender.

[3] The budget law of the Governo Monti written in a situation of potential break-up of the Italian economy.

[4] Tender for two objects: "gara a doppio oggetto" where in the same tender the public body entitled with the concession asks to the market to provide offers for the service and for the partnership in a newco with itself.

[5] The more important incentive was that the revenues from shares' sales can be used freely without restrictions coming from the Internal Stability Pact.

managed by fully private company or public-owned company or even foreign company[6]. In the past, a standard history has been the merge of local companies that have been transformed in larger listed companies owned by local municipalities with more complex governance [9]. In all these cases along with the political influence an increasing potential conflict between management and ownership may systematically arise (especially after political changes) and moreover influenced by the presence of minority shareholders more profit oriented that the political side.

In Italy, the so-called "four sisters" are becoming increasingly important in a variety of services[7]. They are: Hera (owned by Bologna and other municipalities in Emilia Romagna), Acea (Rome), Iren (municipalities of Genova, Torino, Reggio Emilia, Parma and Piacenza) and A2A (municipalities of Milano and Brescia). All of them were local companies fully owned by the Municipality that, through a process of mergers and acquisitions, became larger companies owned by the Municipalities and also listed. They are managing local services in the cities of origin, but they have also expanded towards other smaller cities often "buying" the local companies.

In most cases, the presence of robust large companies like the "four sisters" has improved the quality of the service, ensuring investments and also a re-alignment of the prices to the need of the service abandoning a political based price. These companies are also often producing at local level important effects for the territories in terms of employment, demand for local firms and even innovation. This is happening in some sectors (water and waste management) more than in others (i.e. transport services). They are also actors able to limit the pressure coming from foreign firms looking for new markets. Finally, the value created by these companies sharply increased in this process. Some municipalities has started to sell at least part of their shares in order to face their budget difficulties or some others introduced, like in the case of Iren, a "golden share" for long lasting shareholders in order to protect the governance reducing the number of public-owned shares.

Nevertheless, the increasing number of companies external to the territory or the increasing size of the companies born from the merge of local companies has produced a number of side effects in the political debate. The weakened link between the company and the territory has produced increasing skepticism among local politicians and citizens. This is due to a governance too "far away" from the local interests, the perception of increasing profits for shareholders and decreasing quality for stakeholders and customers and the increasing prices are seen with suspect, even if in all these cases the pricing is determined by the regulatory local or national authorities.

Summing up, the national regulation pushes local services towards an increasing market-oriented approach regulated by national agencies. At local level, this trend determined an increasing role played by the municipalities (or local government) able to create larger companies operating at a larger territorial level (also in new areas with respect the origin of the company). This trend has partially limited the pressure coming

---

[6] This is the case for example of public transport where RATP (owned by the French central government) managed the service in some provinces of Tuscany and Emilia Romagna or ARRIVA (owned by the Deutsche Bahn, company owned by the German Federal Government).

[7] Namely electricity, natural gas, water and waste.

from foreign larger companies, most of them owned even by other national governments. An upcoming position against the larger size of the companies and the broadening distance between the very local political powers and the management of the services are seen with increasing suspect by local territories.

4   Local Public Services in an Italian Municipality: A Practical Case in Short, Piacenza

In order to make explicit the complexity of the situation in the Italian context, here is provided a table regarding the current situation of the Municipality of Piacenza. Piacenza[8] is a small municipality (103000 inhabitants) located in Emilia Romagna region, but close to Milan. It has a former agricultural based economy with a relevant industrial tradition in sectors like mechatronics, drilling technologies and agri-food and more recently interested by the development of the logistics sector due to the strategic position in Northern Italy. It is a high-medium income municipality with a relatively high quality of life, also due to the cultural traditions and a high level of public services. It shares with other small municipalities in the most developed areas of the country the difficulties from the economic crisis that influenced the economic environment and also the performance of local levels of government.

Water and Waste Management. In the last fifteen years has been interested by a deep transformation of the former local companies managing services in the water, waste and transport sectors. In the water and waste sectors, it has known the full range of opportunities offered by legislation and by the market. First, in the early years of the new millenium an already existing company (TESA) owned by the Municipality of Piacenza contributed to an improvement of the service at provincial level by aggregating all the forty-eight small municipalities of the province. Afterwards, it has been merged with two similar companies owned by the municipalities of the entire province of Reggio Emilia and a similar entity in Parma. The new company ENIA, born in 2005 has been listed and afterwards merged in 2010 to the Iride company owned by the municipality of Genova and Torino in the new listed company IREN, now operating in different services and also in new territories in Piemonte, Liguria and Emilia Romagna.

The planning activity is done by the Municipality of Piacenza through the ATERSIR agency born in 2011 on the basis of the regional regulation. ATERSIR operates at the optimal territorial level of the Region, where the service is organized through tenders that have to be launched both for the waste and for the water management activities, even if the plan of the service is done at provincial level. The pricing activity is based on the national regulation for the water service (obviously influenced by the local planning of the service) and on the tax level decided by the municipality, taking into account the provincial plan especially for the activity of waste disposal. The Municipality of Piacenza is one of the shareholder of Atersir and it has currently a seat in the Consiglio d'Ambito, the council that manages the agency. The decisional process in Atersir in influenced by collective political decisions.

---

[8] http://www.piacenzatheplace.it.

Transport. In the transport sector, a similar process has been developed in the last years. The local company owned by the Municipality and the Province of Piacenza has been merged in the larger SETA along with similar companies in Reggio Emilia and Modena. The merge by incorporation was led by the company in Modena, where the ownership was divided between the municipalities of the province and HERM, a company owned by RATP. In HERM, RATP was involved with TPER, another company managing transports in Bologna and other regional cities belonging to the Municipality of Bologna and the Regional Government of Emilia Romagna. RATP abandoned the company in 2014 and now TPER is the operative partner of the others municipalities of SETA, including the municipality of Piacenza.

The planning activity of the local transport sector is managed by Tempi Agenzia, owned by the Provincial Government of Piacenza and the Municipality itself. The territorial ambit is determined by the regional law and until now reflected the provincial level of Piacenza. In the future, the ambit will be added to the ambit of Parma. Also in this case a new tender has to be launched but there will probably be a renewed period of merging among local companies in Emilia Romagna, that may lead to a regional company owned by municipalities to deal with tenders in the Region.

Gas. In the gas distribution the municipality of Piacenza is historically an important place for the state-owned company (formerly Enelgas, now 2iRetegas) with a significant presence of smaller operators in the rest of the province. Also in this area a new tender has to be launched after the changes in the national legislation of the 2012.

In this field, the regulation is fully national, even if the tender of the service is organized locally at the optimal territorial level (in this case it is called, ATEM, minimum territorial ambit, ambito territoriale minimo) The new tenders that are about to be launched for the four services are necessarily influenced by the different levels of contestability of the markets. Actually, the level of investment in these sectors is relatively high and in these sectors there is a high level of "takeover capital" that has to be invested by potential rivals of the incumbent. This makes the contestability different according to the sector. The pricing regulation is now highly stable in the water and gas sector, strongly influenced by the political pressure in the waste and transport sector.

All this is summarized in the table below and it shows the complexity of local services in Italy in terms of regulatory regimes, current management situations and the future where tenders seem to be the rule. It is also clear that in such a situation it would be quite difficult for any municipality to go back to an in-house providing service, since the takeover capital prevents the municipalities to change regime. In local public services the history will strongly influence the future (see Table 1).

Table 1. Synoptic table of regulations of local public services: the case of Piacenza.

	Regulation - planning	Regulation - prices	Management of the service	Future
Water manag't	Regional and provincial with the active participation of the municipality	National authority Pricing now stable and more clearly regulated	**Currently, a listed company partly owned by municipalities (Piacenza < 2% but in pact with other municipalities)**	Tender for the new concession – moderately contestable
Waste manag't	Regional and provincial with the active participation of the municipality	National regulated tax with decision at municipal level – High political pressure for low taxes		Tender for the new concession – highly contestable
Transport	Regulatory framework provided at regional level – Service planning at provincial level	Regional objectives – Decision at municipal level High political pressure for low taxes	PPP owned by a private partner (at the beginning a foreign large company and now a regional large "public-owned" company) jointly with many municipalities in three territories	Tender for the new concession – highly contestable
Natural gas distribution	National regulation and local planning	National regulation – Highly stable	National company	Tender for the new concession – limited contestability

5   Policy Issues in the Future of Local Public Services in Italy

The current situation of local public services in Italy as depicted above, also with a practical case, allows to introduce relevant policy issues.

Regulation vs Management. In Italy this is a relevant issue: for different reasons there are local governments that play the role of regulator and the role of management of the service. This is implemented obviously through the action of managers/civil servants and by means of technicians and not necessarily directly by politicians, but the double role is often still existing. In theory, the public body should regulate and the management activity should be left to the market of private companies. Nevertheless, the companies producing local public services are often owned by public bodies (local or even national governments). This is not the case only in Italy, but also in other European countries. It is difficult to manage the transition to a fully market regime, with only private-owned shares. Somebody says that this would be also unfair. The implication of the transition would be the sale of public-owned companies to the private market and in the past this has led to inefficiencies and loss of value for the citizens. Moreover a public-owned company would be influenced by the preferences of the citizens, that are both shareholders and stakeholders, better than private-owned companies that are managed in the full interest of shareholders.

Nevertheless, the strict coincidence of regulation and management is a potential danger for the quality of competition but also for the quality of the service. Practically, a municipality plays a role in planning and in managing, but in the case of the municipality discussed in the previous paragraph this role is not played in autonomy, but it is strongly shared with other entities. Paradoxically, we can say that for all the municipalities that have been involved in mergers for the creation of larger companies, the risk of capture of the regulator is lower than in the cases where local companies are still operating.

Is Regulation Better If Assigned at National Level? One feature of the current process of the national legislation is moving the regulation at national level. This is the case for the water and gas sector and soon also for the waste activity. Actually what is named as regulation is a complex activity including different aspects. The first one is the planning activity of the service. The second one is the pricing decision. The third one is the activity connected with the issuing of the basis document of the tender. The fourth one is the control of the contract.

The planning activity and the control of the contract is still left at local level, but national or regional authorities implement the pricing mechanism and the normative regulatory activity. This might be a good tool in order to outweigh the power of local municipalities in the managing companies, but for those municipalities that are not owners of an operating company, it might be a problem if they can not influence the decision about the price.

Moreover, the tenders are in principle a robust tool for promoting competition. In order to produce optimal tenders, the level of competence of the regulatory agency has to be very high. There is a potential conflict between the local interest to have a strong influence in the planning of the tender and the competence level that might be reached by a local agency. A good compromise might be the organization of a technical agency at regional level, where the tender is implemented as a service for the territories. The local ambit (provincial or nearby) should implement the planning activity (that might be independent of the moment when the tender is actually launched, and actually this is the case) and should decide about the most important issues characterizing the tender, leaving to the technical body the practical production of the technical documents.

Municipal Capitalism. There is a last issue and it is about the efficiency of the municipal capitalism. There are different opinions about the need of a capitalism led by municipalities. The well known issues concerning the conflict of interest and the role played by politicians in important sectors of the life of the citizens are obviously to be always considered. But the question is the following one: does it make sense to leave to the private market the provision of public services? Is there any reason why the public provision of local public services is one of the options in the European laws? The reason is certainly connected to the evidence that the right level of local public services and the level of contribution of individual demand and public budget to the service must necessarily be set as a policy decision. The incentives of a public-owned company can be easily distorted by politics and this has to be considered. Larger companies, sometimes listed and owned by local governments but influenced by the market mechanism (i.e. the shares' market), could they be an ideal compromise to merge efficiency and preferences of the community?

One of the main problems coming from the Italian experience is certainly the reduction of the influence of politics on larger firms. This is so true that from the territories a complaint about the "distance" between the management of the companies and the preferences of the citizens is emerging. But another problem is the need of more freedom for those municipalities that want to opt out from a market oriented organisation of the service back to an in-house providing system should be allowed. The technical solution is not an easy task, but it should be implemented. Moreover, an important implication of the current situation is what should be when a service in a municipality is assigned after a tender to a company that is a competitor of the company (partially) owned by the Municipality. Is the ownership of the loser still strategic for the local government. Also in this case, an answer should be provided through an opting out solution that has not to penalize the citizens.

Mixed Companies as a Solution? In the Italian laws a tentative road towards mixed firms solutions has been included but not enough supported. Mixed firms are considered useful to reconcile political objectives with efficiency and productivity [10]. Actually, mixed firms are not easy to be implemented due to inadequate economic resources of Municipalities that have to contribute in equity to join the PPP, but also due to the average scarcity of competences among Municipalities in Italy in the field of shareholding. Besides, larger and specialised companies are not willing to make agreements with "political" bodies.

On the contrary, mixed companies contribute to calibrate the quality of the service, activating citizens' needs and to potentially reconcile economic efficiency with citizens' expectations. In the smaller cities, the public utilities are a crucial factor of development and competitiveness and an involvement of municipalities could be positive for creating value for the territory.

Compared to in-house providing the mixed companied would ensure the advantages of the incentives provided by the private actor with the orientation towards citizens of the local government. In Italy in-house providing would be limited by the financial constraint, especially in the start-up phase, and also it may imply a reduction in the control of the quality of the service, since no competition might reduce it.

6   Conclusions

Local public services' provision in Italy is a strategic issue also for the efficiency and the quality of the Italian economic environment. The Italian public finance crisis and the economic crisis has pushed the system towards market-based services, where tenders should be the way in which the competition for the market will be practically implemented. In this context, a major role is played by the tradition of the public services and by the recent changes. An important space for the municipal capitalism has been created during the nineties and in the last twenty years, also with the implicit objective to limit the foreign companies competition. Large public owned companies have been created in different sectors. The paper provides a practically guided investigation of the main issues coming from this situation. The setup of the regulation mechanism and the opportunity in keeping a municipal capitalism approach are the two topics that have been focused in the paper. A more distributed governance of the regulatory activity, favouring a setup that allows to take into account the positive political issues representing citizens' preferences with the incentives provided by the large (and sometimes listed) companies should be pushed in order to improve the quality of services that are crucial for local communities but also for the national economy.

References

1. Curci, N., Dapalo, D., Vadalà, E.: Municipal Socialism or Municipal Capitalism? The Performance of Local Public Enterprises in Italy. Banca d'Italia, Roma (2017). Quaderni di Economia e Finanza, Occasional Papers
2. Megginson, W., Netter, A.: From state to market: a survey of empirical studies on privatization. J. Econ. Lit. **39**(2), 321–389 (2001)
3. Martin, S., Parker, D.: The Impact of Privatisation: Ownership and Corporate Perfomance in the UK. Industrial Development Policy, Routledge (1997)
4. Bognetti, G., Robotti, L.: The provision of local public services through mixed enterprises: the Italian case. Ann. Publ. Cooper. Econ. **78**(3), 415–437 (2007)
5. Bianchi, P., Bortolotti, L., Pelizzola, C., Scarpa C.: Comuni Spa: Il capitalismo municipale in Italia, Il Mulino, Bologna (2009)
6. Bianco, M., Sestito P.: I servizi pubblici locali. Il Mulino, Bologna (2010)
7. Arcas, M., Bachiller, P.: Operating performance of privatized firms in Europe: organizational and environmental factors. In: International Journal of Public Administration, pp. 487–498. Taylor & Francis Online (2010)
8. Bortolotti, B., Faccio, M.: Reluctant privatization. Fondazione Eni Enrico Mattei Working Papers, p. 130, Milano (2004)
9. Elefanti, M., Cerrato, D.: La governance delle imprese pubbliche locali. In: Mele, R., Mussari, R., L'innovazione della governance e delle strategie nei settori delle public utilities, Il Mulino, Bologna (2009)
10. Boggio, M.: Municipal capitalism from state to mixed ownership in local public services provision. In: MPRA Paper, 46245 (2012)

A Paradigm Interpreting the City and the Analytic Network Process for the Management of Urban Transformations

Grazia Napoli[1](✉) iD, Salvatore Giuffrida[2] iD,
and Maria Rosa Trovato[2] iD

[1] University of Palermo, 90128 Palermo, Italy
grazia.napoli@unipa.it
[2] University of Catania, 95131 Catania, Italy

Abstract. When urban and environmental transformations occur in areas where the equilibrium between nature and culture is complex and fragile, public administrations could decide to induce private investments using several tools, such as financial contributions to those projects of refurbishment that better respect the purpose of improving the environmental quality and of preserving the local architecture. Multicriteria models may support public decision process regarding this issue, but it is essential to adopt a scientific paradigm that provides a major theoretical reference. This study proposes the development of a network model based on the scientific paradigm by Rizzo and the Analytic Network Process. The first one has been chosen because of its interpretation of the city as *autopoietic organization*, *dissipative structure* and *political-administrative system*, the second one because of its holistic representation of the decision problem in which the interactions between all the elements are made explicit. The network model has been applied to a case study that consists in ranking some alternative refurbishments of buildings in Favignana (Egadi islands, Italy) in order to grant public financial contributions.

Keywords: Analytic Network Process · Multicriteria analysis
Urban transformation · Decision aid

1 Introduction

The role of multicriteria decision models may be considered an action of supporting the evolutionary, adaptive, and cognitive process of reworking of the new data and knowledge acquired by the Decision Maker (DM), who operates in complex and adversarial social-territorial contexts [1, 2]. In fact, multicriteria analysis has the following several characteristics: it is able to represent the complex levels of an evaluation problem; it can involve many stakeholders in the evaluation process by using participation techniques that make clear all steps of the decision process and make explicit any social convergence or conflict; it clearly communicates the social value system and the judgments on which the decision process is based. This action, however, should be

© Springer International Publishing AG, part of Springer Nature 2019
F. Calabrò et al. (Eds.): ISHT 2018, SIST 100, pp. 672–680, 2019.
https://doi.org/10.1007/978-3-319-92099-3_75

founded on a scientific paradigm that provides an interpretative key to manage the extreme variety of multicriteria models. These models belong to different "families" of methods, each of them follows a proper logical-mathematical approach to elaborate data and to deal with specific decision problems (such as solving, ranking, sorting and designing) [3].

This study proposes to assume the interpretative paradigm by Rizzo [1, 4], which considers the city as *autopoietic organization, dissipative structure* and a *political-administrative system*, as the reference from which an Analytic Network Process - ANP- model is developed. The network is applied to a case study consisting in ranking some alternative refurbishments of buildings in Favignana (Egadi islands, Italy) in order to grant public subsidies to those projects that better preserve the local architecture and improve the environmental quality.

2 A Paradigm Interpreting the City and the ANP Model

The scientific paradigm conceived by Rizzo [4] interprets the city as an *autopoietic organization*, a *dissipative structure* and a *political-administrative system*. According to Prigogine's though [5], a *dissipative structure* is a structure that evolves itself creating order from disorder when it is distant from the equilibrium point and is subjected by perturbations. This theory, which was firstly developed with regard to chemical structures, is analogous to the description of *autopoietic structures* as structures that transform themselves though maintaining their own organization [6]. These concepts may be applied to the city because any urban transformations have to respect the principle of sustainability and minimize the consumption of energetic and natural resources, and because the city is able to maintain over time its organization and identity. Moreover a *political administrative system* provides development and competitiveness to the city when it is efficient and inclusive, e.g. reducing bureaucratic times and promoting social participation in the urban management.

The development of a multicriteria model for the application to a public decision process is more than a mere technical phase because it requests of being strictly referred to a scientific paradigm that provides a consensual linguistic dominion on which the decision are based [7]. This study has developed an Analytic Network Process model to solve decision problems regarding alternatives of urban transformation, according to the aim of taking into account the paradigm by Rizzo in the multicriteria analysis. In fact, nevertheless many different multicriteria models are available in the literature [8–13], the Analytic Network Process has the peculiarity of adopting a holistic approach that is able to best represent the complexity of the decision problem and the interrelations between all the elements of the network [14–18]. The ANP methodology is composed of several steps: the first one is to delineate the DM's objective identifying the alternatives, the clusters of nodes, as well as the feedback or dependence relationships between all of them. Afterward, the DM expresses pairwise comparison measurements according to a ratio scale. The ranking of the alternatives is obtained through the calculation of several supermatrixes [13, 14].

3   The Case Study: Refurbishment Projects of Urban Real Estate in Favignana Island

The network model has been developed for a case study in Favignana island. The "morphology" of Favignana is the outcome of the long-term relationship between natural environment and human actions; in particular the morphology of the cityscape is generated by the relationships between many elements, such as quarries, road system, private and public external spaces, and buildings [17]. That makes these natural/artificial contexts unique and generates the commitment of public administrations to implement sustainable development policies and measures in order to preserve them.

The decision problem of this case study consists of ranking some alternative refurbishments of real estate properties in the urban center of Favignana, in order to grant public financial contributions to those projects that allow maintaining the local architecture as well as increasing the market value or social use value of the buildings. The development of a multicriteria model depends on the need to take a choice on the basis not only of financial parameters such as cost of refurbishment, cost of energetic retrofit, or market prices [19–21], but also of qualitative parameters such as the environmental quality and the compliance with the "rules" of the local traditional architecture [22–24].

The alternatives are, as an example, the projects of refurbishment of five buildings, namely A, B, C, D, and E, which are located in an urban bock (Fig. 1). The projects have different characteristics with reference to: size and typology of the project - maintenance, alteration, partial demolition and rebuilding-, costs, technological installations, relationships between internal and external spaces or between building and public spaces, etc.

3.1   The Development of the ANP Model

The DM is formed by various experts, such as a public administration official, a local professional (architect), and a real estate developer. They interacted in a focus group for specifying the three interpretative keys of the city by Rizzo, formerly described, into a network model. During meetings, the model was discussed and reviewed until to reach a consensus on a network that consists of 6 clusters and 22 nodes. The 6 clusters are: "Urban real estate autopoiesis", "Consensual linguistic dominion", "Operative financial plasticity", "Decisional efficiency and social participation", "Non-entropy and energetic efficiency", and "Alternatives". The clusters are briefly described in Table 1, whereas Fig. 2 shows the structure of the network: the clusters, the nodes, and the relationships between the clusters.

Following the ANP procedure, for each cluster the DM expresses judgments on pairwise comparisons of elements -according to Saaty's fundamental scale [13] - in accordance with another element of the network to which they have a relationship. The pairwise comparisons follow the relationships -dependence, feedback, and loop- that were designed in the network by the DM. The judgments allow achieving the paired comparison matrixes and afterwards three supermatrixes are calculated: the unweighed

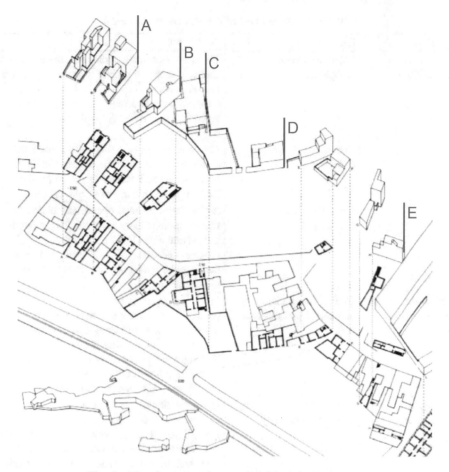

Fig. 1. The alternatives in an urban block in Favignana

one consists of N × N blocks Wj, where N is the number of the clusters, and n is the number of nodes of the cluster N-th, the weighed one that is calculated applying to the former matrix the eigenvector obtained from the cluster-level comparison, and finally the limit supermatrix by applying the formula (1).

$$\lim_{n \to \infty} (W)^{-1} \tag{1}$$

3.2   The Results of the Procedure

The implementation of the procedure, by using the SuperDecisions 2.8 software [25], gives as results: the priorities between the clusters, the priorities by clusters, and the ranking of the alternatives (Tables 2 and 3). The priorities between the clusters reveal that both the "Consensual linguistic dominion" and "Urban real estate autopoiesis" clusters have the greatest weights because the DM gives more importance to the

Table 1. Description of the clusters of the model

Criteria	Clusters	Description
Autopoietic organization	Urban real estate autopoiesis	The urban real estate autopoiesis describes the ability of a city to maintain its own structure and the local architecture through transformation and plus-valorization of the real estate properties by promoting investments
	Consensual linguistic dominion	The construction of a linguistic consensual dominion as a basis of the communication and of the social consensus is founded on the increase of the quality of public spaces
	Operative financial plasticity	A city has a financial operational plasticity when it is able: to employ effectively public and private capitals, to respect the programmed terms of realization, and to get the financial feasibility for the projects of transformation
Political administrative system	Decisional efficiency and social participation	In the city assumed as a political-administrative system, the increasing of the decisional efficiency plays an important role in terms of assuring the involvement of social groups, of reducing processing time, and of favoring the mixed public-private participation to urban transformations
Dissipative structures	Non-entropic and energetic efficiency	Energetic efficiency and supporting neg-entropic activities may be achieved through many actions, such as the refurbishing of existing buildings to reduce their deterioration and to improve their energetic efficiency

cityscape and the local architecture than to the minimization of financial costs or processing time.

The priority by clusters synthetizes the system of weights implicitly expressed by DM's judgments in order to apply the theoretical concept expressed by the clusters to the case study. For example, the DM judges the "Plus-valorization of the real estate" node as that element that best contributes to implement the "Urban real estate autopoiesis" because the increasing of prices corresponds to a semiotic translation -in monetary terms- of the significance that the social system gives to each category of capital goods and it also measures how much the investors believe in the future real estate market prices [26–28]. By contrast, all the nodes in the "Consensual linguistic dominion" cluster have similar weights as they have the same role in order to represent the characteristics of the local architecture; the low weight of the "Link building/quarry" node only depends on the fact that the quarry is almost nonexistent in the case study.

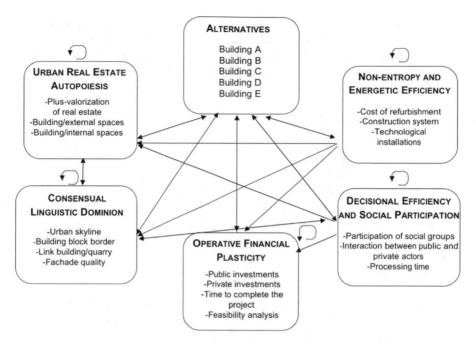

Fig. 2. The network model

Table 2. Priorities between and by clusters

Clusters	Nodes	Normalized	Limiting
Urban real estate autopoiesis 0.362	Plus-valorization real estate	0.79403	0.116899
	Building/external space	0.09098	0.013394
	Building/internal space	0.11500	0.016930
Consensual linguistic dominion 0.395	Urban skyline	0.34406	0.117568
	Building block border	0.26105	0.089202
	Link building/quarry	0.08505	0.029063
	Fachade quality	0.30985	0.105878
Decisional efficiency and social participation 0.046	Participation social groups	0.84845	0.071175
	Interaction private/public actors	0.03902	0.003273
	Processing time	0.11253	0.009440
Non-entropic and energetic efficiency 0.080	Cost of refurbishment	0.64471	0.021608
	Construction system	0.22276	0.007466
	Technological installation	0.13253	0.004442
Operative financial plasticity 0.117	Public investment	0.10319	0.006734
	Private investment	0.25003	0.016317
	Time to complete project	0.11941	0.007793
	Feasibility analysis	0.52737	0.034417

The final priority vector is coincident to a column of the limit supermatrix and provides the ranking of the alternatives. The buildings C and D are the best alternatives, whereas the others have very low scores, so it could be appropriate to suggest that the DM should set a threshold -minimum score- for obtaining public grants.

Table 3. Rankings of the alternatives

Alternatives	Ranking	Normals	Ideals
Building A	4	0.145465	0.466064
Building B	3	0.174619	0.559472
Building C	*1*	0.312114	1.000.000
Building D	*2*	0.253863	0.813364
Building E	5	0.113938	0.365052

4   Conclusions

The development of an ANP model on the basis of the paradigm by Rizzo, which considers the city an *autopoietic organization, dissipative structure,* and *political-administrative system,* and its application to a case study in Favignana have been aimed at supporting the public decision making process, especially when urban transformations are damaging the complex and fragile equilibrium between nature and culture, and public authorities are making attempts to preserve the local identity, architecture, and environment.

The model compares five alternatives of refurbishment of buildings and provides the ranking to establish the priority in the granting of public financial contributions in the case of limited public funds. The model could be improved by activating partici-patory process with a consistent number of stakeholders. The ranking may be a strong starting point to express the complexity of the social values and to support the public decision process for the implementing of a sustainable policy.

References

1. Rizzo, F.: Valore e valutazioni. La scienza dell'economia o l'economia della scienza. Franco Angeli, Milano (1999)
2. Maturana, H.R., Varela, F.J.: L'albero della conoscenza. Garzanti, Milano (1987)
3. Figueira, J., Greco, S., Ehrgott, M. (eds.): Multiple Criteria Decision Analysis. State of the Art Survey. Springer, New York (2005)
4. Rizzo, F.: Il capitale sociale della città. Valutazione, pianificazione e gestione. Franco Angeli, Milano (2002)
5. Prigogine, I., Stengers, I.: La nuova alleanza. Metamorfosi della scienza. Einaudi, Torino (1981)
6. Maturana, H.R., Varela, F.J.: Autopoiesi e cognizione. La realizzazione del vivente. Marsilio, Venezia (1985)

7. Luhmann, N.: Sistemi sociali. Fondamenti di una teoria generale. Il Mulino, Bologna (1990)
8. Bana e Costa, C.A., Vansnick, J-C.: Applications of the Macbeth approach in the framework of an additive aggregation model. J. Multi-Criteria Decis. Anal. **6**(2), 107–114 (1997)
9. Bouyssou, D., Marchant, T., Pirlot, M., Tsoukiàs, A., Vincke, P.: Evaluation and Decision Models: Stepping Stones for the Analyst. Springer, Berlin (2006)
10. Brans, J., Vincke, P.: A preference ranking organization method: the Promethee method. Manag. Sci. **31**, 647–656 (1985)
11. Greco, S., Matarazzo, B., Slowinski, R.: Rough sets theory for multicriteria decision analysis. Eur. J. Oper. Res. **129**, 1–47 (2001)
12. Roy, B., Bouyssou, D.: Aide multicritére à la décision: méthodes et case. Economica, Paris (1995)
13. Saaty, T.L.: The Analytic Hierarchy Process. McGraw Hill, New York (1980)
14. Saaty, T.L.: Theory and Applications of the Analytic Network Process. RWS Publications, Pittsburgh (2005)
15. Saaty, T.L., De Paola, P.: Rethinking design and urban planning for the cities of the future. Buildings **7**, 76 (2017)
16. Bottero, M., Ferretti, V.: An analytic network process-based approach for location problems: the case of a new waste incinerator plant in the province of Torino (Italy). J. Multi-Criteria Decis. Anal. **17**, 63–84 (2011)
17. Napoli, G., Schilleci, F.: An application of analytic network process in the planning process: the case of an urban transformation in Palermo (Italy). In: Murgante, B., et al. (eds.) Computational Science and Its Applications - ICCSA 2014. LNCS, vol. 8581, pp. 300–314. Springer, Cham (2014)
18. Napoli, G.: The value of the useless in the urban landscape of small islands. In: Piccinini, et al. (eds.) Proceedings of the 18th IPSAPA/ISPALEM International Scientific Conference: The Usefulness of the Useless in the Landscape Cultural Mosaic: Liveability, Typicality, Biodiversity, pp. 333–339 (2015)
19. Napoli, G., Gabrielli, L., Barbaro, S.: The efficiency of the incentives for the public buildings energy retrofit. The case of the Italian regions of the "objective convergence". Valori e valutazioni **18**, 25–39 (2017)
20. Bencardino, M., Nesticò, A.: Demographic changes and real estate values. A quantitative model for analyzing the urban-rural linkages. Sustainability **9**(4), 536 (2017)
21. Della Spina, L., Scrivo, R., Ventura, C., Viglianisi, A.: Urban renewal: negotiation procedures and evaluation models. In: Gervasi, O., et al. (eds.) Computational Science and Its Applications - ICCSA 2015. LNCS, vol. 9157, pp. 88–103. Springer, Cham (2015)
22. Nesticò, A., Sica, F.: The sustainability of urban renewal projects: a model for economic multi-criteria analysis. J. Property Investment Fin. **35**(4), 397–409 (2017)
23. Della Spina, L.: Integrated evaluation and multi-methodological approaches for the enhancement of the cultural landscape. In: Gervasi, O., et al. (eds.) Computational Science and Its Applications - ICCSA 2017. LNCS, vol. 10404, pp. 478–493. Springer, Cham (2017)
24. Calabrò, F.: Local communities and management of cultural heritage of the inner areas. An application of break-even analysis. In: Gervasi, O., et al. (eds.) Computational Science and Its Applications - ICCSA 2017. LNCS, vol. 10406, pp. 516–531. Springer, Cham (2017)
25. Superdecisions. https://www.superdecisions.com. Accessed 03 Dec 2017
26. Giuffrida, S., Ventura, V., Trovato, M.R., Napoli, G.: Axiology of the historical city and the cap rate. The case of the old town of Ragusa Superiore. Valori e Valutazioni **18**, 41–55 (2017)

27. Napoli, G., Valenti, A., Giuffrida, S.: The urban landscape and the real estate market. Structure and fragment of the axiological tessitura in a wide urban area of Palermo. In: Piccinini, L.C., et al. (eds.) Proceedings of the 19th IPSAPA/ISPALEM International Scientific Conference, Napoli, 2nd–3rd July 2015, pp. 67–78 (2016)
28. Nesticò, A., Galante, M.: An estimate model for the equalisation of real estate tax: a case study. Int. J. Bus. Intell. Data Min. **10**(1), 19–32 (2015)

A Fuzzy Multi-criteria Decision Model for the Regeneration of the Urban Peripheries

Marco Locurcio[1], Francesco Tajani[2(✉)], Pierluigi Morano[2],
and Carmelo Maria Torre[2]

[1] Sapienza University of Rome, 00197 Rome, Italy
[2] Polytechnic University of Bari, 70125 Bari, Italy
francescotajani@yahoo.it

Abstract. The regeneration of the urban peripheral areas is undoubtedly among the most complex issues with which the Public Administrations are currently facing. Different interests, often conflicting, coming from citizens, entrepreneurs and stakeholders, focus on these areas. In the present research a fuzzy multi-criteria decision model is proposed, as support of the Public Administration in the analysis of different scenarios, referred to different temporal moments. The application to four different urban peripheries located in the metropolitan area of Rome (Italy) highlights the potentialities of this model. The representation of the results through a radar diagram and a histogram makes them easily intelligible also by non-expert subjects.

Keywords: Urban redevelopment · Decision support models
Multi-criteria analysis · Scenario analysis · Fuzzy logic systems
Resource optimization

1 Introduction

The issue of the regeneration of the urban peripheries of the major European cities is particularly relevant and involves a multitude of multidisciplinary aspects: the refurbishment of the peripheral areas does not simply mean the revitalization of a multiplicity of buildings through new functions, but it presupposes an anthropological dimension of the urban initiative, associated with the needs of the local community and the growing necessity to create new places for the social development. The enhancement of public and private spaces in disuse would create a polycentric urban structure that, through a higher satisfaction of the public services, could favor a balanced economic growth, limiting the urban sprawl and reducing the soil sealing [1].

However, the current economic contingency and the excessive bureaucratization of the urban planning system have created a situation of immobility, that has diverted the interest of the private entrepreneurs. It should be added that, with the aim of promoting the peripheral regeneration initiatives through the involvement of the private investors, the frequent lack of necessary professional skills in the Public Administration (PA) precludes the possibility of addressing an effective public-private dialogue and establishing a negotiating platform in which both the parties are fully aware of the mutual conveniences.

© Springer International Publishing AG, part of Springer Nature 2019
F. Calabrò et al. (Eds.): ISHT 2018, SIST 100, pp. 681–690, 2019.
https://doi.org/10.1007/978-3-319-92099-3_76

2 Aims

With reference to the regeneration of the urban peripheries, the aim of this research is to provide a methodological support for the definition of a public-private negotiation platform, that allows to balance the public needs of collective services - especially in the peripheral areas - and the financial conveniences of the private entrepreneur. The proposed model is developed starting from the Urban Quality Protocol for the city of Rome (Italy), elaborated by the Abandoned Urban Areas Association (AUDIS), which constitutes an instrument for supporting the urban regeneration initiatives [2].

The Protocol borrows the logic of the participatory programming, aimed at taking into account the instances of the PA, the local community and the private entrepreneurs. The Protocol consists of 96 parameters, set out in a matrix, that represent 9 aspects of the urban quality (urban planning, architectural, public spaces, social, economic, environmental, energetic, cultural and landscape), to be considered in the phase of elaboration of an urban redevelopment initiative.

However, the Protocol is almost exclusively characterized by verbal indications, that can be hardly translated into objective criteria to support the decisions, especially in complex scenarios where the PA deals with different projectual alternatives for the same area, and, at the same time, the priority among different areas has to be identified. In order to overcome this limit, in the present research a multi-criteria fuzzy approach implemented to the Protocol is proposed [3–5]. The model allows to translate the different verbal indications into numerical scores and to obtain a synthetic indicator for each option: in this way, it constitutes a support for the decision maker in the final choice of the projectual solution to be realized and in the order processing phase of the redevelopment initiatives on different areas [6, 7].

The work is structured as follows. In the third section, the model is described with particular attention to the graphical tools as support of the analysis and the basic assumptions of the model are exposed. In the fourth section, the model is implemented to a case study and the results of the application are illustrated. Finally, the conclusions of the work are discussed.

3 Fuzzy Multi-criteria Decision Model

For the implementation of the Fuzzy Model, an algorithm has been created in Excel environment, that, through a series of steps [8], allows to combine the input values - "importance of the parameters" and "capacity of the solution in analysis to pursue the parameters" - in order to determine the output value that summarizes the scenarios described as follows. The shape of the membership function is assumed to be a triangular type: it is simple to be analytically treated but capable to offer a good approximation for the representation of the phenomenon. The value range, both for the input and the output variables, is (0–10). Having established the fuzzy rules with the support of the decision maker, the fuzzification phases of the data of the input variables- inference, composition and defuzzification - have been carried out. The operator chosen for the defuzzification is the Center of Gravity (COG), that ensures a high "sensitivity" of the model through the contribution of all the output variables [9].

The Fuzzy Multi-criteria Decision Model is aimed at evaluating, through three synthetic numerical indicators, three scenarios for different - real or hypothetical - moments associated with the urban regeneration project. The analyzed scenarios are thus named:

- "SF" – *current state*. It describes how much the current state of the project is in line with the original project and to what extent the project has been realized; in this way, the quality and the quantity of the project is analyzed;
- "CP" – *project completion*. It represents how much the hypothetical completion of the redevelopment initiative according to the indications of the original project corresponds to the new priorities identified for the area in analysis; in this way, the unfinished share of the original project is carefully evaluated by taking into account the new objectives identified by the PA;
- "IP" – *projectual hypothesis*. It indicates how much the new projectual hypothesis that the proposer is submitting to the PA, satisfies the new goals defined by the PA.

In order to apply the model to a concrete case study, a panel of public and private experts has been defined. This panel has identified thirty parameters of the Protocol, considered to be the most significant, that are listed and described respectively in columns 2 and 3 in Table 1. Based on these parameters, four redevelopment projects have been analyzed, that concern four peripheral areas, located in the city of Rome. Columns from 4 to 15 of Table 1 list the output values related to these parameters for various scenarios: the parameters have been grouped according to four "macro-qualities" listed in column 1 of Table 1, so that synthetic analyzes can be performed. In order to analyze the results of the models according to the different qualities considered, the "average quality" (R_S) has been obtained for each of the three scenarios $S = SF, CP, IP$, by averaging the output values assumed by the parameters. In columns 3, 4 and 5 of Table 2 the average value for each application case listed in the first two columns is represented.

Finally, the distribution of scores obtained for each parameter in the various scenarios has been analyzed, through a radar diagram and a histogram.

In the radar diagram three different scenarios, which correspond to three projectual hypotheses (SF, CP and IP), are represented by three polygons; the more the polygon vertices will tend outward, maximizing the numerical scores associated with each vertex, the more the solution will be close to the optimum project and vice versa. The standard behavior requires that the three polygons have a concentric trend, i.e. for each parameter the following inequality occurs:

$$R_{IP} > R_{CP} > R_{SF} \tag{1}$$

In practice, this means that the projectual hypothesis (IP scenario) is ameliorative - taking into account the current goals - compared to the project completion (CP scenario), which in turn is still a positive progress compared to the current state (SF scenario). If the inequality is not respected, i.e. the intersection of the three polygons is graphically observed, it is necessary to verify which of the possible combinations in terms of weights and judgments has generated this phenomenon and why. This check is

Table 1. Parameters and output variable values

Q	N	Description	Case A			Case B			Case C			Case D		
			R_{SF}	R_{CP}	R_{IP}	R_{SF}	R_{CP}	R_{IP}	R_{SF}	R_{CP}	R_{IP}	R_{SF}	R_{CP}	R_{IP}
URBAN PLANNING (U)	1	Morphological structure	6.1	7.0	8.8	7.0	6.4	6.6	7.5	6.4	7.5	7.0	8.8	10
	2	Conformation of the public spaces	5.0	6.6	7.5	5.0	6.5	8.8	7.1	6.4	7.5	5.0	6.1	7.5
	3	Relationship between the built environment and the public spaces	3.0	4.5	5.2	4.5	6.5	8.8	6.1	6.6	7.5	2.4	5.0	7.5
	4	Functional distribution	5.4	4.0	6.4	7.1	7.0	9.4	7.5	7.1	8.8	5.4	8.8	8.8
	5	Continuity of the urban fabric	5.0	7.5	9.4	6.4	7.0	7.0	6.1	5.0	7.0	7.0	4.1	7.5
	6	Continuity of the green spaces	4.1	7.5	0.0	6.5	7.1	9.4	4.6	6.1	7.5	5.4	8.8	8.8
	7	Infrastructure	3.5	5.5	7.1	5.0	7.0	8.8	7.5	8.8	8.8	5.0	7.1	9.4
	8	Parking distribution	6.5	6.4	7.1	8.8	8.8	8.8	7.5	8.8	0.0	7.5	0.0	6.4
ARCHITECTURAL-PUBLIC SPACES (A)	1	Selection criteria of the design team	6.1	5.0	7.1	6.4	5.9	7.1	7.0	6.5	8.8	7.0	0.0	7.5
	2	Flexibility of buildings to different functions over time	5.0	5.0	7.5	0.0	0.0	8.8	5.2	5.9	9.4	3.0	5.0	8.8
	3	Environmental sustainability	7.1	4.6	7.1	7.5	7.1	9.4	7.5	7.1	8.8	3.8	5.9	7.1
	4	Design of the public spaces	7.5	7.1	7.1	3.5	5.9	8.8	3.5	6.5	8.8	5.4	7.5	8.8
	5	Relationship with the entire city	5.0	7.5	8.8	4.0	7.1	8.8	5.5	7.1	8.8	5.8	0.0	9.4
	6	Accessibility	5.9	6.4	7.1	3.5	5.0	8.8	5.0	5.9	7.1	4.5	7.1	8.8
	7	Flexibility of theuses	4.1	4.6	5.5	0.8	6.5	8.8	6.6	7.1	8.8	2.9	6.1	7.5
SOCIAL–ECONOMIC(S)	1	Accessibility to the housing supply	8.8	5.5	7.1	9.4	5.9	8.8	8.8	6.5	8.8	7.1	0.0	0.0
	2	Tipology of the housing supply	8.8	6.6	7.5	7.5	6.1	7.5	7.5	7.0	9.4	8.8	0.0	0.0
	3	Presence of facilities for specific typologies of citizens	5.0	7.0	8.8	8.8	5.9	8.8	6.6	7.1	8.8	7.0	7.5	8.8
	4	Offer of jobs	3.0	7.1	7.1	8.8	6.4	7.5	8.8	8.8	0.0	7.1	7.5	9.4
	5	Public and private educational and cultural services	5.9	9.4	9.4	1.2	5.0	8.8	6.6	7.5	8.8	5.4	7.5	8.8
	6	Shops at neighborhood scale	5.0	6.6	7.5	0.0	0.0	8.8	8.8	10	0.0	5.0	6.4	7.5
	7	Provision of neighborhood services	5.9	7.5	8.8	0.0	0.0	7.5	7.5	8.8	9.4	5.9	6.1	7.5
	8	Shops at urban scale	5.9	6.4	7.1	10	0.0	0.0	10	0.0	0.0	10	0.0	0.0
	9	Economic sustainability	5.9	8.8	8.8	7.1	7.0	8.8	8.8	8.8	8.8	5.4	8.8	0.0
ENVIRONMENTAL–ENERGETIC (E)	1	Management of urban solid waste	3.4	3.4	5.8	9.4	7.1	9.4	6.6	0.0	0.0	0.0	0.0	6.4
	2	Vehicular traffic and environmental pollution	3.4	4.5	5.8	9.4	7.1	8.8	7.5	7.1	8.8	7.1	0.0	8.8
	3	Environmental and energetic sustainability	4.0	4.9	4.9	9.4	7.5	8.8	6.4	7.5	8.8	0.0	0.0	8.8
	4	Overall interest in the energy efficiency	4.5	5.9	7.1	7.5	7.1	8.8	5.9	0.0	8.8	5.8	0.0	7.5
	5	Landscape analysis	7.1	6.6	7.5	10	0.0	0.0	0.0	6.4	7.5	5.0	7.5	8.8
	6	Landscape integration	6.1	7.1	8.8	4.5	7.1	8.8	3.0	6.1	7.5	7.5	7.0	9.4

Table 2. Average quality scores for each scenario and respective percentage increases

Case	Periphery	Scenario			Increases		
		SF	CP	IP	$\Delta 1$	$\Delta 2$	$\Delta 3$
A	Ponte di Nona	5.3	6.1	7.1	14%	32%	15%
B	Europarco	6.0	5.6	8.0	−7%	32%	42%
C	Parco Leonardo	6.4	6.3	7.1	−1%	11%	13%
D	Porta di Roma	5.3	4.5	7.4	−16%	38%	65%

carried out through the analysis of the histogram that details the urban quality improvement in the three different scenarios, namely:

$$\Delta 1 = R_{CP} - R_{SF} \tag{2}$$

$$\Delta 2 = R_{IP} - R_{SF} \tag{3}$$

$$\Delta 3 = R_{IP} - R_{CP} \tag{4}$$

Synthetic results for these increments are summarized in columns 6, 7 and 8 of Table 2. The macro-behaviors and their meanings for incremental variations can be:

- $\Delta 1 \cong 0 \Rightarrow R_{CP} \cong R_{SF}$: this occurs when the project completion according to the indications of the original project, taking into account the current priorities, does not involve any significant changes compared to the current state.
- $\Delta 1 > 0 \Rightarrow R_{CP} > R_{SF}$: this occurs when the project completion is in line with the current goals that remain consistent with the initial objectives.
- $\Delta 1 < 0 \Rightarrow R_{CP} < R_{SF}$: this occurs when the original project is completed ($R_{CP} = 0$), or if the project completion is not consistent with the current goals, that are different from the initial objectives.
- $\Delta 2 \cong 0 \Rightarrow R_{IP} \cong R_{SF}$: this occurs when the new projectual hypothesis does not involve any significant improvements compared to the current state.
- $\Delta 2 > 0 \Rightarrow R_{IP} > R_{SF}$: this occurs when the new projectual hypothesis is ameliorative compared to the current state, or the current state does not provide anything about the parameter in analysis ($R_{SF} = 0$). This is the generally verified situation.
- $\Delta 2 < 0 \Rightarrow R_{IP} < R_{SF}$: this occurs when the new projectual hypothesis is pejorative compared to the current state, or - that is very frequent - when the original priority is currently of little or no relevance ($R_{IP} = 0$).
- $\Delta 3 \cong 0 \Rightarrow R_{IP} \cong R_{CP}$: this occurs when the new projectual hypothesis does not involve improvements compared to the project completion. In practice, this situation implies that the original objectives and the initial project are confirmed.
- $\Delta 3 > 0 \Rightarrow R_{IP} > R_{CP}$: this occurs when the new projectual hypothesis is ameliorative compared to the project completion.
- $\Delta 3 < 0 \Rightarrow R_{IP} < R_{CP}$: this occurs when the new projectual hypothesis is pejorative compared to the project completion.

In addition to the analysis of the individual incremental variations, it is very useful to study the possible combinations of the value variations and the related indications:

- $(\Delta 2 > 0 \wedge \Delta 3 > 0) \Rightarrow IP$. The new projectual hypothesis (IP scenario) associated with the parameters for which both of these inequalities are satisfied is the most incisive.
- $(\Delta 1 \gg 0 \wedge \Delta 3 \leq 0) \Rightarrow CP$. The project completion (CP scenario) associated with the parameters for which both of these inequalities are satisfied is consistent with the current goals and therefore the new projectual hypothesis does not imply any significant improvements.
- $(\Delta 1 < 0 \wedge \Delta 2 \leq 0) \Rightarrow SF$. The original project associated with the parameters for which both of these inequalities are satisfied is totally realized and is consistent with

the current goals (SF scenario) and therefore the new projectual hypothesis results poorly ameliorative.

4 Case Study

4.1 The Areas Studied

The model has been applied to four peripheral areas located in the city of Rome (Italy), described as follows:

- Case A: the "Porta di Roma" area is located in the north of the historic center, and it extends for about 330 ha. As a result of the change of the intended use in 1998, almost the half of the surface was devoted to an impressive construction project for about 10,000 new inhabitants. The project provided for a functional mix that was modified over time, through an increase in residences at the expense of neighborhood services.
- Case B: the "Ponte di Nona" area is located in the east and it extends for almost 160 ha. In 1994, a project of almost 480,000 m^2 of gross floor surface for about 16,000 new inhabitants was provided. Currently, about 70% of the projected residences were built, whereas the impressive road system provided has not yet been completed; the shopping center "Roma Est", that was not provided in the original project, represents the only existing polarity.
- Case C: the "Europarco" area is located in the south. The prevision of this project was to equip the city of Rome of a European business district. The project provided for the construction of 800,000 m^3 on about 63 ha. Started in 2005, the project has been characterized by a deferral of the completion; so far, only 50% of the original project has been realized.
- Case D: the "Parco Leonardo" area is located in the south-west of the city, and it extents for 160 ha. The original master plan included a neighborhood characterized by all services, shopping center, cinemas and 4,000 residential units. Currently, the cumbersome presence of the shopping center and the discontinuous pedestrian paths do not allow for the appropriate enjoyment of the neighborhood.

4.2 Results

The results of the application in the four different areas and the identification of some strategies are described as follows.

Case A: The Continuity of the Initial Project as "Zero" Strategy. The radar diagram of Fig. 1-A shows that it is not possible to identify a predominant quality that characterizes the original identity of the project (SF scenario). The new projectual hypothesis (IP scenario), although it is generally ameliorative compared to the other two scenarios, is not effective in the energetic and environmental quality improvement. The histogram of Fig. 2-A shows that the quality increments between the various scenarios are rather modest. The new projectual hypothesis (IP scenario) is only effective in the integration between the technical and projectual solutions in order to pursue the environmental sustainability goal ($\Delta 2 > 0 \land \Delta 3 > 0$). The average quality increases from one scenario to another (Table 2) are rather limited so as to be advisable to facilitate the project completion according to the original project rather than a new projectual hypothesis, that would probably result in a further financial burden on the PA.

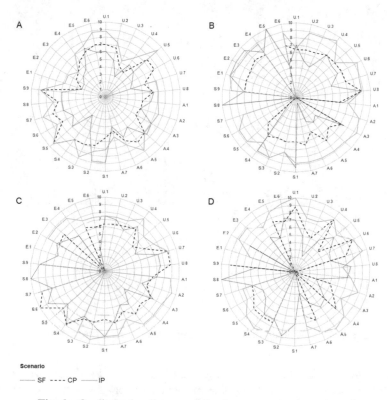

Fig. 1. Quality radar diagram of the three scenarios investigated

Case B: A New Proposal of Public Spaces. The radar diagram of Fig. 1-B outlines that the project identity of the "Europarco" (SF scenario) points to the search for the environmental, energetic and urban qualities. The histogram in Fig. 2-B shows that it is necessary to provide for a higher functional mix, as the relative parameters are not considered in the initial processing phase ($\Delta 2 > 0 \land \Delta 3 > 0$), whereas they are very important according with the current needs (IP scenario). The comparative analysis of the three scenarios (Table 2) highlights that, overall, the environmental and energetic qualities of the "Europarco" is significant, reaching a high score already in the current state (SF scenario), whereas the realization of the public spaces and public services is incomplete and poorly designed.

Case C: The Search for a Self-sustaining Neighborhood. The radar diagram of Fig. 1-C shows that the original project (SF scenario) is focused on the search for social

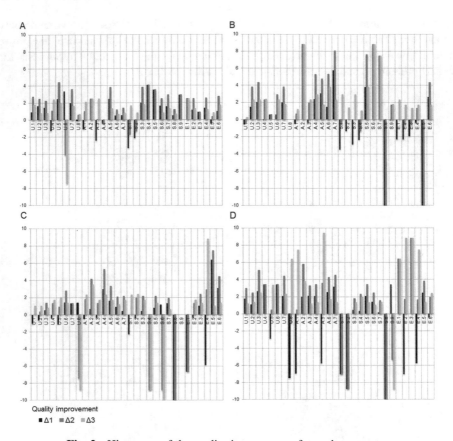

Fig. 2. Histogram of the quality improvement for each parameter

and economic qualities. In fact, the project is characterized by the realization of economically accessible housing units. The histogram of Fig. 2-C outlines that the new
projectual hypothesis (IP scenario) is particularly ameliorative ($\Delta2 > 0 \wedge \Delta3 > 0$)
compared to the other scenarios for certain aspects concerning the architectural,
energetic and environmental qualities and the quality of the public spaces. The comparative analysis of the three scenarios (Table 2) confirms that the initial project is
characterized by a high average quality, as the public and private services were
developed simultaneously to the residences; the completion of the redevelopment
initiative according to the indications of the original project is consistent with the new
objectives.

Case D: The Energy Efficiency Goal for the Neighborhood. The radar graph in
Fig. 1-D shows that the originally planned project identity (SF scenario) aims at the
search for social and economic qualities. In fact, the project provides a heterogeneous
housing supply, consisting of residential units of varying sizes and for a medium-high
buyer segment. The presence of the shopping center and urban-scale services significantly contributes to a high level of the social and economic qualities. The histogram in

Fig. 2-D shows that the new projectual hypothesis (IP scenario) is particularly effective with regard to the environmental and energetic quality parameters ($\Delta 2 > 0 \wedge \Delta 3 > 0$), given the lack or the low consideration of these qualities in the initial projectual phase. The comparative analysis of the three scenarios (Table 2) highlights that, through the new projectual hypothesis (IP scenario), an equally distributed result among all the qualities is obtained, except for the social and economic qualities, that have been well-represented in the current state (SF scenario).

5   Conclusions

The model developed in this research allows to reduce the uncertainty in the elaboration of complex urban redevelopment initiatives, using a fuzzy multi-criteria approach for the hierarchical fragmentation of the project studied in the range of the considerable aspects.

The rationalization through the fuzzy logic of the evaluation model allows *(i)* to improve the transparency and the effectiveness of the urban quality assessment procedure, *(ii)* to attribute differentiated weights to the regeneration targets, *(iii)* to mitigate the uncertainties and the linguistic ambiguities of the qualitative assessments, *(iv)* to evaluate the possible scenarios, *(v)* to compare different projectual solutions. In particular, the fuzzy approach allows to avoid the creation of a *"black box"* in which the logical linkage between the input and output elements of the decision-making problem is lost. Indeed, the method of a graphical synthesis representation through the radar diagram and the histogram increases the clarity of the model, leaving unchanged the link between the input value, the output value and the final synthetic value.

The model is characterized by a flexible structure that fits the different projects in analysis, thanks to the numerous parameters set up in the Urban Quality Protocol and to the ability to easily adapt the membership functions to the specific decision maker mechanism. Additionally, the application to the presented cases shows that the model is useful not only for comparing different projects related to the same area, but also for the possibility to carry out three different analyzes on three scenarios. This peculiarity allows to evaluate those areas for which the urban regeneration initiative is particularly effective, resulting in a significant improvement compared to the current state and where it is therefore more profitable to invest public resources [10, 11].

References

1. Torre, C.M., Morano, P., Tajani, F.: Saving soil for sustainable land use. Sustainability **9**(3), 1–35 (2017)
2. AUDIS: Il Protocollo della qualità di Roma Capitale. Definire e valutare la qualità dei progetti urbani complessi, Rome (2012)
3. Chen, S.J., Hwang, C.L.: Fuzzy multiple attribute decision making methods. In: Fuzzy Multiple Attribute Decision Making. Lecture Notes in Economics and Mathematical Systems, vol. 375, pp. 289–486. Springer, Heidelberg (1992)

4. Bagnoli, C., Smith, H.: The theory of fuzzy logic and its application to real estate valuation. J. Real Estate Res. **16**(2), 169–200 (1998)
5. Morano, P., Locurcio, M., Tajani, F., Guarini, M.R.: Fuzzy logic and coherence control in multi-criteria evaluation of urban redevelopment projects. Int. J. Bus. Intell. Data Min. **10**(1), 73–93 (2015)
6. Casas, G.L., Scorza, F.: Sustainable planning: a methodological toolkit. In: Apduhan, B.O., et al. (eds.) ICCSA 2016. LNCS, vol. 9786, pp. 627–635. Springer, Heidelberg (2016)
7. Del Giudice, V., De Paola, P.: Geoadditive models for property market. In: CEABM 2014, Applied Mechanics and Materials, vol. 584–586, pp. 2505–2509. Trans Tech Publications, China (2014)
8. Guarini, M.R.: Self-renovation in Rome: Ex Ante, in Itinere and Ex Post Evaluation. In: Murgante, B. et al. (eds.): ICCSA 2017. LNCS, vol. 9789, pp. 204–218. Springer, Heidelberg (2017)
9. Bai, Y., Wang, D.: Fundamentals of Fuzzy Logic Control - Fuzzy Sets, Fuzzy Rules and defuzzifications. In: Advanced Fuzzy Logic Technologies in Industrial Applications, pp. 17–36. Springer, London (2006)
10. Calabrò, F.: Local Communities and management of cultural heritage of the inner areas. an application of break-even analysis. In: Gervasi O., et al. (eds.) Computational Science and Its Applications - ICCSA 2017. Lecture Notes in Computer Science, vol 10406. Springer, Cham (2017)
11. Amato, F., Pontrandolfi, P., Murgante, B.: Using spatiotemporal analysis in urban sprawl assessment and prediction. In: International Conference on Computational Science and Its Applications. LNCS, vol. 8580, pp. 758–773. Springer, Heidelberg (2014)

The Use of Fuzzy Cognitive Maps for Evaluating the Reuse Project of Military Barracks in Northern Italy

Marta Bottero[✉], Giulia Datola, and Roberto Monaco

Politecnico di Torino, 10129 Turin, Italy
marta.bottero@polito.it

Abstract. Cities are complex systems and their changing are continuous. The evolution strictly depends on the relationships among the different aspects which compose the same structure, such as social, economic, political, environmental, historical aspects. The consideration of the complexity, the heterogeneity and reciprocal influences of these different elements becomes fundamental in urban regeneration projects that are characterized by many interconnected elements. Starting from a real project in the city of Pinerolo (Italy), the paper aims to investigate the potentiality of Fuzzy Cognitive Maps (FCMs) to represent the complexity of urban transformation processes, paying particular attention to the possibility of analyzing different scenarios simulation.

Keywords: Decision making-process · Urban regeneration
Dynamic behavior · Scenario analysis · Evaluation · Causal relationship

1 Introduction

Cities are complex systems that are described by different interconnected elements. Taking into consideration this complexity, it is of particular importance to provide the decision makers with integrated evaluation tools able to consider the multiplicity of objectives and values when dealing with urban regeneration processes and to include the opinions and the needs of the different stakeholders involved. The objective of this paper is to investigate an integrated evaluation approach [1] based on the method of Fuzzy Cognitive Maps (FCMs) [2], that represents an innovative tool for representing complex systems by networks and weighted interconnections and for supporting decision making processes.

2 Methodological Background

The technique of Fuzzy Cognitive Maps (FCMs) [2, 3] represents a natural extension of cognitive maps by embedding to them the use of Fuzzy Logic. FCMs have been introduced by Kosko [2], who suggested their use to those knowledge domains that involve a high degree of uncertainty.

© Springer International Publishing AG, part of Springer Nature 2019
F. Calabrò et al. (Eds.): ISHT 2018, SIST 100, pp. 691–699, 2019.
https://doi.org/10.1007/978-3-319-92099-3_77

FCMs are used to represent how complex systems work, by an aggregate network of concepts and weighted interconnections (Fig. 1). For their graphical representation, FCMs are also used to note experts' different knowledge about the behavior of the same system. The technique of FCMs is often employed to reveal the dynamic behavior of the system, describing how the system could evolve in time through causal relationships [4]. For this reason, this approach is considered as a useful tool in the context of scenario planning and decision making, because it can be used for the evaluation of alternatives by applying a complementary analysis [12].

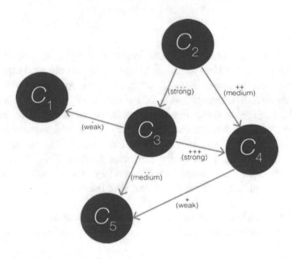

Fig. 1. FCM graphical representation (Source: author processing).

2.1 Properties

FCMs have two different representations:

- Graphical representation (Fig. 1), which include concepts, that represent the variables that compose the system, and directed and weighted edges, that symbolize the cause–effect relationship between concepts;
- Mathematical representation, that is made of state vector, which denotes the initial state, C_i concepts, that represent the variables that compose the system, and adjacency matrix $E = \{e_{ij}\}$ where e_{ij} is the weight of the direct edge C_iC_j. The values that are assigned to each relationship are included in the range $[-1; 1]$. The value 0 means that between the general concepts C_i and C_j there is not any causal relationship.

The mathematical representation is the responsible of FCMs ability to reveal the dynamic behavior of the system which they represent, since the simulation is the result of the iteration process between the state vector A and the adjacency matrix E.

The iteration process is performed by the following formula:

$$A_i^{(t+1)} = f\left(A_i^{(t)} \cdot E\right) \tag{1}$$

where $A_i^{(t+1)}$ is the value of concept C_i at moment $t + 1$, $A_i^{(t)}$ is the value of concept C_i at moment t. The function f is a threshold function [4, 5], used to normalize the values assumed by concepts, at each iteration, in the range $[-1; 1]$. According to Kok [4], in theory the procedure should be repeated at least $2 \times n$ (where n is the total number of concepts) times in order to identify indirect effects. In practice, the behavior of the system can emerge after 20–30 iterations.

3 Application

The case study considered in the present paper is related to requalification project of Bochard Caserma located in the city Pinerolo (Nothern Italy). This project arises to respond to a real need, and to give a solution to the big urban void in the center of the city, which was created after the demission of the military barracks [13–15]. Different alternative scenarios for the requalification of the site have been defined and evaluated by means of FCMs [6].

3.1 Structuring of the Decision Problem and the Identification of the Concepts

The first step for the application of the FCMs method refers to the identification of the concepts to be included in the model. For this purpose, an integrated framework based on SWOT analysis [6] has been proposed in the present application that aims at setting the problem and at highlighting the key elements. Table 1 describes the concepts that we identified to obtain a set of measurable attributes for the evaluation of the alternatives.

3.2 Alternative Transformation Projects

In this experimentation the integrated approach has been applied for the evaluation of five different regeneration projects, related to the restoration of the military barracks. Table 2 provides a synthesis of the main characteristics of the five alternatives [6] which have been evaluated according to the concepts highlighted in Sect. 3.1. In order to move on to the application of FMCs, the subsequent step consists in translating the performance of each alternative to the corresponding state vector, through a standardization operation. In particular, for the standardization of the performance of the different alternatives, the following formula has been used:

$$a_i = (x_i - x_{min}) / (x_{max} - x_{min}) \tag{2}$$

where x_i is the value of the initial performance of an alternative over a concept, x_{min} and x_{max} being their minimum and maximum value.   It is important to mention that

Table 1. Concepts description (Source: authors elaboration from [6])

C_1 Cultural functions	Inclusion within the project area of functions, such as classrooms, music schools or other activities
C_2 Mixité index	Index that describes the functional mix of the area
C_3 Smart policy attention	Presence of project elements oriented towards sustainability (e.g., use of renewable energy sources)
C_4 Green area	Total permeable surfaces in the project
C_5 Management cost	Operation and management costs of the structure
C_6 Opening hours	Opening hours of the structure
C_7 Total investment	Total cost for the requalification intervention
C_8 New jobs	Number of new jobs created by the project
C_9 Profitability	Income generated by the project
C_{10} Architectural quality	Quality of architectural project for the intervention
C_{11} Space flexibility	Possibility of space flexibility depending on the type of activities being carried out
C_{12} Accessibility	Facility to reach the area with several means of transport, public transport, pedestrian and cyclo-pedestrian paths

Table 2. Alternative projects description (Source: authors elaboration from [6])

Alternative projects	Description
Productive scenario	This strategy is based on the creation of shops for the sale of local products, co-working spaces, and craft workshops
Cultural scenario	This strategy is mainly devoted to the construction of spaces for co-working, classrooms and library
Artistic scenario	This strategy is oriented towards the construction of spaces for theatre, music school, artistic ateliers, exhibit rooms and concert hall
Productive scenario 2.0	This strategy considers the demolition of some parts of the military barracks while the restored part is dedicated to shops for the sale of local products, co-working spaces, and craft workshops
Cultural scenario 2.0	This strategy considers the demolition of some parts of the military barracks, while the restored part is devoted to co-working spaces, classrooms and library

the formula (2) was used for both the criteria that have to be minimized and for those that have to be maximized. This because in the application of the FCMs, the different concepts are interrelated to each other through the causal relations, that can result in increasing or decreasing the influenced concepts. Following the aforementioned procedure, the state vectors have been calculated (Table 3).

Table 3. State vector of the five alternatives (Source: author processing)

Concepts	Productive scenario	Cultural scenario	Artistic scenario	Productive 2.0	Cultural 2.0
C_1 Cultural functions	0	0,5	1	0,3	0.3
C_2 Mixité index	0,2	1	0	0,3	0,8
C_3 Smart policy attention	0,5	0,5	0	1	0,5
C_4 Green area	0,4	0	0	1	1
C_5 Management cost	0,5	0,5	1	0	0
C_6 Opening hours	0	1	0	0,5	0,5
C_7 Total investment	0,5	0,5	1	0	0
C_8 New jobs	0,8	1	0,7	0,1	0
C_9 Profitability	1	0,7	0	0,3	0
C_{10} Architectural quality	0	0,5	0	1	0,5
C_{11} Space flexibility	0	1	0,3	0,3	0,7
C_{12} Accessibility	0,5	1	0,5	0,5	1

3.3   FCMs and the Identification of the Relationship

When the different state vectors are defined, is it possible to move to real FCM application, drawing the FCMs and identifying the causal relationships among the concepts [7–10]. Figure 2 illustrates the FCMs drafted by a panel of three different experts in economic evaluation, architecture and urban design.

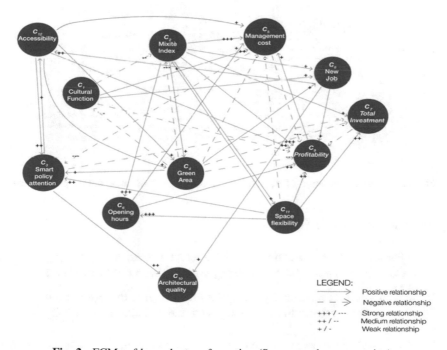

Fig. 2. FCMs of barracks transformation (Source: author processing).

4   Exploring Dynamic Scenarios

The subsequent step in the FCMs application involves the exploration of dynamic scenarios, obtained by the combination of the initial state vector and the adjacency matrix, with the formula (1), as explained in Sect. 2.1. Figure 3 details the graphs representing the output of the change of the state vector for the five alternative scenarios; more in details, the x axis represents the number of the iterations of the model while the y axis represents the value of the different considered concepts. It is important to recall that in the present application the state vector is represented by the initial configuration of the alternative transformation project for the area under examination. The dynamic behaviors have been obtained through the support of Excel software.

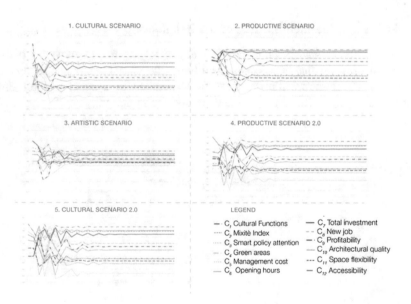

Fig. 3.   Scenario simulation of the five alternative projects (Source: author processing).

5   Discussion of the Results

The equilibrium and the evolution of the different scenarios can be explored and evaluated to find the preferable one, making use of different criteria related to:

1. Number of iteration required by the system for reaching the equilibrium;
2. Stability of the system: the frequency and the amplitude of the iteration are measured;
3. Final values of the concepts at the equilibrium considering the alternative systems.

According to the aforementioned criteria, the scenarios that can be considered most stable for number of iterations and amplitude of oscillations are the strategies named

"Cultural Scenario" and "Productive Scenario". In fact, these scenarios present smaller oscillations with respect to the other three options and at the same time they are characterized by higher values in the final equilibrium configuration [16].

Given the similarity of the scenario's development, it was necessary to further examine their evolution, monitoring the final values of their concepts. In particular, the final values of the two aforementioned scenarios have been considered and visualized in a radar diagram where the axis represent the concepts used for the evaluation and the data points are related to the final values of the two projects as resulting from the FCM application (Fig. 4). In this case it is possible to highlight that the preferable scenario is the strategy "Productive Scenario". In fact, this scenario shows an evolution that is more consistent with the overall objective of the evaluation, presenting more attention in smart policy, less management costs, a greater number of new jobs created, a greater profitability, a better architectonic quality and a better accessibility with respect to the "Cultural Scenario".

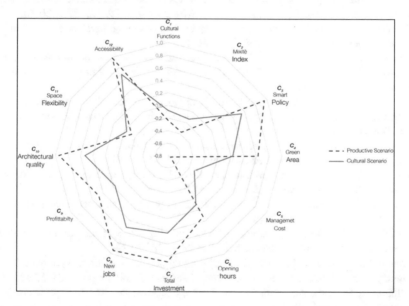

Fig. 4. Radar graph showing the final values of the concepts for "Cultural Scenario" and "Productive Scenario" projects (Source: author processing).

6 Conclusions

This paper represents one of the first applications of FCMs technique for the evaluation of urban regeneration operations [11]. This research reveals the versatility of the FCMs approach, that proved to be efficient in addressing the complexity of the decision problem under examination and in investigating semi-qualitative scenarios for modeling urban regeneration [17–19]. However, as a modelling tool, FCMs have not been exploited to their full capacity in urban and territorial planning. Therefore, despite the global coherence of the obtained results, further experimentations of real-world case studies will be necessary in order to validate the proposed approach.

Acknowledgments. The authors wish to thank Antonio De Rossi for providing the opportunity of the experimentation presented in the study, A special thanks goes also to Stefano Damiano and Georgui Boyanov Djarov for the data used in the research.

References

1. Bottero, M.: A multi-methodological approach for assessing sustainability of urban projects. Manag. Environ. Qual. Int. J. **26**(1), 138–154 (2015)
2. Kosko, B.: Fuzzy cognitive maps. Int. J. Man Mach. Stud. **24**, 65–75 (1986)
3. Axelrod, R.: Structure of Decision: The Cognitive Maps of Political Elites. Princeton University Press, Princeton (1976)
4. Kok, K.: The potential of Fuzzy Cognitive Maps for semi-quantitative scenario development, with an example from Brazil. Glob. Environ. Change **19**, 122–133 (2008)
5. Tsadiras, A.K.: Comparing the inference capabilities of binary, trivalent and sigmoid fuzzy cognitive maps. Inf. Sci. **178**, 3880–3894 (2008)
6. Damiano S., Boyanov Djarov, G.: Scenari di riuso per la caserma Bochard a Pinerolo. Master thesis, Politecnico di Torino (supervisors: professors Antonio De Rossi, Marta Bottero, Francesca Governa) (2017)
7. Özesmi, U., Özesmi, S.: A participatory approach to ecosystem conservation: fuzzy cognitive maps and stakeholder group analysis in Uluabat Lake Turkey. Environ. Manag. **31** (4), 518–531 (2003)
8. Ozemi, U., Ozemi, S.L.: Ecological models based on people knowledge: a multy-step fuzzy cognitive mapping approach. Ecol. Model. **176**, 43–64 (2004)
9. Jetter, A., Schweinfort, W.: Building scenarios with fuzzy cognitive maps: an exploratory study of solar energy. Futures **43**, 52–66 (2010)
10. Salmeron, J.L., Vidal, R., Mena, A.: Ranking fuzzy cognitive map based scenarios with TOPSIS. Expert Syst. Appl. **39**, 2443–2450 (2012)
11. Bottero, M., Mondini, G., Oppio, A.: Decision support systems for evaluating urban regeneration. Procedia Soc. Behav. Sci. **223**, 923–928 (2016)
12. Bottero, M., Datola, G., Monaco, R.: Exploring the resilience of urban systems using fuzzy cognitive maps. Lecture Notes in Computer Science, vol. 10406, pp. 338–353 (2017)
13. Caffaro, R., Barbero, A., Fenoglio, M., Drago, M.: Pinerolo, passato e presente. Borgaro Torinese, Canale Arte (2009)
14. Coscia, C., Fregonara, E., Rolando, D.: Project management, briefing and territorial planning, the case of military properties disposal. Territorio **73**, 135–144 (2015)
15. Pasqui, G.: Il Master Plan per le aree militari di Piacenza. Processo, attori e forme di conoscenza (The Master Plan for the military areas of Piacenza. Process, actors and forms of knowledge). Territorio **62**, 58–63 (2012)
16. Olazabal, M., Pascual, U.: Use of fuzzy cognitive maps to study urban resilience and transformation. Environ. Innov. Soc. Transitions **18**, 18–40 (2016)
17. Mondini, M.: Integrated assessment for the management of new social challenges. Valori e Valutazioni **17**, 15–17 (2016)

18. Morano, P.: Un modello multicriterio "fuzzy" per la valutazione degli interventi di riqualificazione urbana. Aestimum **39**, 81–122 (2000)
19. Morano, P., Locurcio, M., Tajani, F., Guarini, M.: Fuzzy logic and coherence control in multicriteria evaluation of urban redevelopment projects. Int. J. Bus. Intell. Data Min. **10**, 73–93 (2015)

Municipal Emergency Planning. The Strategic Planning Model and the Contribution of Evaluation Tools

Alessandro Rugolo[✉], Angela Viglianisi, and Claudio Zavaglia

Mediterranea University, 89100 Reggio di Calabria, Italy
alessandro.rugolo@unirc.it

Abstract. The numerous natural disasters that have devastated the Italian territory have ignited a strong political debate on the need to secure the Italian territory, which has now become fragile due to poor management characterized by decades of building abuses and poor choices of urban development.

The disastrous phenomena, at the same time, have highlighted the ineffectiveness of the current municipal emergency plans, which did not favor the management of the interventions by the rescuers because of poor risk assessment and planning in the emergency areas ex ante.

In part, the issue was tackled by the competent authorities at the highest levels, which have allocated considerable economic resources throughout the national territory for the implementation of risk prevention and mitigation measures. However, this is not enough if the planning of interventions on the territory is not structured according to a strategic approach that takes into account both the real conditions of vulnerability of the territory and the priorities of the mitigation measures required for the effectiveness of the rescue.

A new model of emergency planning, equipped with appropriate assessment tools, can therefore be a valuable measure both to ensure the effectiveness of emergency aid in the case of disasters and to address correct governance of the territory.

Keywords: Strategic planning · Emergency planning · Indicators
Multidimensional evaluation · Governance

1 The Theme of Emergency and the Task of Planning

The term "Civil Protection" refers to all the structures and activities put in place by the state to "*protect the integrity of life, property, settlements and the environment from damage or the danger of damage resulting from disasters and other calamitous events aimed at overcoming the emergency*" [1].

This is the result of the joint work of the authors. Although scientific responsibility is equally attributable, the abstract and the paragraphs nn. 3, 4 were written by Alessandro Rugolo; the paragraph nn. 2, was written by Angela Viglianisi; the paragraph nn. 1, was written by Claudio Zavaglia.

© Springer International Publishing AG, part of Springer Nature 2019
F. Calabrò et al. (Eds.): ISHT 2018, SIST 100, pp. 700–709, 2019.
https://doi.org/10.1007/978-3-319-92099-3_78

Civil Protection (in Italy, as well as in other EU countries) is not, therefore, an institution, but a public function to which all the components of the state contribute: from the municipalities up to the central administration through the National Department at the Presidency of the Council of Ministers, passing through the various levels of the public administration. A fundamental role is also assigned to citizens involved in voluntary associations active in the area [2].

The basic authority on the territory in case of emergency consists of the municipalities. To the mayors of the municipalities, pursuant to art. 15 of Law 225/1992, is entrusted the leading role in all civil-protection activities (prevention, rescue, and recovery from the emergency), and they have the duty to maintain an operational structure that is able to assist him in the preventive phases and organizational measures of the municipal civil-protection system, as well as in the operational phases aimed at recovery from the emergency.

The tool that must be provided by the mayors to deal with emergencies is represented by the *Municipal Emergency Plan*, governed by the Town Planning Law n. 19 of 16 April 2002 which, pursuant to art. 24, c. 1, letter g, is considered as an implementation plan of the Municipal Structural Plan [3].

The Municipal Emergency Plan defines all the operational procedures for intervention to deal with any calamity expected in the territory of the municipality. The plan, therefore, is the tool that enables the authorities to arrange and coordinate rescue interventions to protect the population and property in an area at risk.

In addition, the plan must identify in the municipal territory the *areas of assistance* (areas in charge of the rescue marshalling of rescue crew and equipment), the *waiting areas* (places of reception for the population in the first phase of the event), and *areas of shelter* (places where the first housing settlements will be installed or the structures to lodge the affected population) in numbers commensurate with the population at risk. It must also identify the *strategic viability* (ideally, both traffic- and risk free), which will enable rescuers to quickly and safely reach the affected areas and strategic buildings. Finally, it must indicate the *strategic structures* that must resist the calamitous event and be functional to support the relief efforts [4].

The identification of structures, areas, and emergency roads, however, poses a fundamental question for the preparation of the plan and for its effectiveness, that is to say, the preparatory role of the organization for the infrastructures dedicated to the emergency with respect to the organization of the operational phases of the emergency in which the personnel and means of civil protection will be engaged.

According to the pragmatic approach, the effectiveness of the emergency plan, as for any planning tool, can only be assessed downstream of the empirical process.

In the case of emergency planning, the effectiveness can be ascertained only and exclusively following the calamitous event, that is, when it will be possible to ascertain that every scenario previously and presumably hypothesized will have occurred in reality, and when each structure and every human resource in charge to the rescue will have actually been able to work for the resolution of the emergency and to carry out tasks efficiently.

The recent experiences of the earthquakes of L'Aquila (2009) and of Amatrice (2016) have, however, highlighted a harsh reality [5]:

1. Many emergency facilities that had been indicated in the emergency plans collapsed;
2. The emergency-aid machinery activated immediately after the emergency encountered physical and operational obstacles that have effectively prevented them from reaching the access areas;
3. The actions to restore conditions have not even been activated yet, and the state of emergency continues to exist years later.

From an analysis made of the various contingency plans elaborated at the national level, among the main causes of failure is, indeed, the confusion that is generated between the programmatic component of the plan as compared to the operational one, the latter often unidentifiable because it had been incorporated into the first, or vice versa. Basically, distinctions between the temporal phases of the plan, the previous one, and the one after the calamitous event are not clear.

2   The Limits of Traditional Planning Applied to the Topic of Emergency

The traditional planning approach has historically demonstrated its main element of criticality in the deterministic vision of future transformations, be they urban-territorial, economic, social, etc. According to this principle, the originator of the plan assumed the role of anticipating future phenomena without having the necessary tools to pre-figure and evaluate credible, future development scenarios, as well as to update the plan project to the continuous transformation of the real contexts [6].

The "static" of the plan, therefore, became the cause of that form of anachronism that was generated between the phase of drafting the plan and the process of continuous change to which any form of activity is subject, whether natural or anthropogenic [7].

The "*dynamism*" of the plan, or the ability to be able to adapt promptly to the continuous changes in the contexts, therefore, is a fundamental requirement for the *effectiveness* of the plan itself [8]. And if this applies to urban planning, it becomes even more important for *emergency planning*, where the need to guarantee the maximum speed of interventions and to preserve the citizens' safety cannot in any way be compromised by the unexpected (unforeseeable) resulting from poor risk assessment and/or the uncertainty of unprepared rescue forces to face critical situations in unknown contexts.

Not only that but another main limitation of the old planning method was that which prefigured the transformation of the territory neglecting the *priorities of objectives* and the specific *times* of realization of the interventions, as well as the *resources* (economic, infrastructural, human, etc.) necessary to enable the execution [6].

The unavailability, now consolidated for some time, of the requirements for the substantial resources necessary for the implementation of interventions greatly extends the time of implementation of the plan and actually makes it impossible to implement the provisions of the plan in parallel and consistently with the physical and socioe-conomic context, which in the meantime can undergo—and certainly will undergo—

autonomous and independent evolution. These have arisen at times from normal metamorphosis of socioeconomic processes, sometimes from local and parochial pressures that have little to do with a strategic vision of development, unfortunately.

The result is that the deferred implementation over time of the interventions, previously foreseen in the plan, may no longer be compatible with the territorial context which, in the meantime, has undergone changes according to various arguments and needs.

3 The Strategic Planning Approach to the Municipal Emergency Plan

When we talk about emergency planning, which aims to give concrete solutions when critical situations occur (catastrophic events due to earthquakes, floods, tsunamis, etc.), the main objective to be pursued can only be to enhance the *effectiveness of the rescue activities* following the event.

For the above reasons, to ensure the effectiveness of rescue operations by rescuers, it is clear that we must free the overall plan of the emergency plan (which is the result of a long-term general strategic vision that requires a lot of preparation time long), from the phase of management of emergency operations (which depend both on the organization of human resources and equipment available, and on the speed of relief for the victims).

The first factor on which it is necessary to intervene, therefore, is the clear distinction between the *planning phase* and the *operational phase* of the rescue. It is necessary to distinguish clearly the part of rescue management following the disaster, which mainly affects the resources (human and instrumental) and that in any case also involves training and pre-event information, from the part of the entire infrastructure for the emergency, which is capable of carrying out the function of the rescue services following the event.

In other words, it is necessary to distinguish the *planning of the emergency* in which the infrastructural structure is defined, consisting of emergency areas, shelter facilities, COC offices, etc., and the *management of the actual emergency* that involves the activation of emergency assistance of the Municipality, the Prefecture, the Department of Civil Protection Regional/National, Voluntary Associations, Law Enforcement, etc.

The first concerns technical/urban planning activities and is inevitably linked to public financial resources that are granted from time to time for the safety of structures (schools, municipal buildings, etc.) and/or natural and anthropized areas (landslides, embankments of watercourses, etc.) and, consequently, very long periods for the construction of the works.

The second concerns the organizational capacity of the authorities and of the human resources in general who will be involved in the rescue phases and who must be ready to face the emergency. The correct functioning of this last aspect, that is the management of the emergency, above all for matters concerning the municipal aid system, can be put in crisis by the change of the personnel in charge of the emergency who are usually replaced by the succession of administrations municipal policies.

Consequently, the principle of plan flexibility is of particular importance, or the ability of the plan to constantly adapt dynamically with changes in the context conditions, without altering its reliability over time [9]. This means that, independently of the long-term strategies set in the emergency plan, the persons in charge of the rescue functions can easily be replaced without compromising either the effectiveness of the operational organization during the emergency phase or the strategic vision of the emergency plan of long period. The latter, as mentioned, will require much more time to be implemented, especially as a function of public funding available, and is the result of a process of synthesis resulting from an in-depth analysis of the situation in the territory, available resources, and the detected morphological or infrastructural problems.

Based on these considerations, recalling the synthesis result of the long and articulated debate dealt with in the XXI National Congress of the National Institute of Urban Planning held in Bologna in 1995, it seems appropriate to apply to the new emergency planning the model of strategic planning, which separates the structural and strategic component (long-term planning and program plan) from the operational component (executive plan of limited duration).

Following the strategic planning approach, in drafting the *Municipal Emergency Plan (PEC)*, the *Structural Emergency Plan (PES)* should be distinguished, which will deal with the preparation of infrastructures for emergencies, from the *Operational Emergency Plan (PEO)* that will address issues related to the management of emergencies, in *ex ante, in itinere,* and *ex post* (Fig. 1).

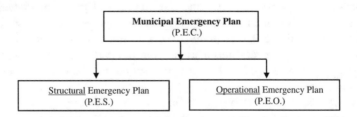

Fig. 1. Breakdown of the Municipal Emergency Plan

3.1   Structural Emergency Plan

In the indications of 31 March 2015 published by the Department of National Civil Protection emergency areas are defined as those *"places intended for civil protection activities and must be identified in advance in emergency planning"*.

The identification of the various types of emergency areas in the municipal territory is the result, by force of things, of a complex planning process (municipal and/or intercommunal). Through this process, depending on the specific morphological, hydrogeological, seismic, anthropological characteristics, infrastructures, etc. of the place, all those infrastructures intended for the rescue of citizens must be preventively and appropriately located even when it is not known where the emergency conditions will be manifest as a result of any calamitous event.

To establish the correct identification of the emergency areas, it is necessary to evaluate, in the *ex-ante* phase, the possible *future scenarios of the event* and the consequent qualification and quantification of the *risks* (seismic, hydrogeological, forest fire, etc.). The areas (and the road network), which are least exposed to risks, determine the plan choices.

It is therefore an activity of proper planning of the territory that can only be entrusted to the Structural Emergency Plan, as it represents a long-term planning of the territory, and it does not depend on small variations in the context.

In the proposed model, the planning process will take place according to the following logical-consequential phases:

1. Identification of the areas at risk (exposure) that must be protected. While in ordinary urban planning it is necessary to foresee the development of the whole municipal area, both urban and extra-urban, in the case of the emergency plan it is necessary to pay attention only to vulnerable areas that can suffer significant damage and therefore must be protected. It will therefore be necessary to identify parameters for defining vulnerability indices.
2. Processing of risk scenarios. For each of the risks (seismic, hydrogeological, etc.), an event scenario will be developed with the relevant effects induced on the portions of the territory considered (vulnerability). All the various scenarios are then accumulated until the map of all potentially vulnerable areas is produced.
3. Location of emergency areas. These portions of territory are the ones that, as indicated by computer simulations would be the safest and so where various emergency areas will be located.
4. Tracking of strategic viability. Once the emergency areas have been identified, the network of strategic viability will be set up to guarantee both the interconnection of the individual areas and accessibility from outside by the responders.
5. Internal evaluation of the effectiveness of the plan. This last section is useful to verify the effective functionality of the plan, or if there are any problems that can or cannot be overcome and the priority with which they will have to be resolved.

3.2   Operational Emergency Plan

P.E.O. is the document that illustrates the operation of the entire organizational machine, at the municipal or inter-communal level, consisting of human resources, means and equipment that must be activated in the context of civil protection emergencies. Basically it is about defining:

1. WHO … specifically assumes the various roles, based on the hierarchical levels of responsibility,
2. DOES WHAT … that is what are the tasks assigned to it,
3. WITH WHAT MEANS … that is what is the endowment of the means and tools assigned to it,
4. WHEN … or in which situations and in how much time must be performed the functions assigned,
5. WITH WHAT RESOURCES … this last point is fundamental to understand the feasibility (in both financial and human-resources term) of the actions necessary to

guarantee the safety threshold (in *ex-ante* phase with respect to the event), and emergency (*in itinere* and *ex-post*).

It is therefore a question of defining the operational organization of the emergency, which may be changing in the short term as it is completely independent of the planning choices entrusted to the PES. However, the operative effectiveness of the relief depends not only on the organizational quality of the civil-protection forces, but also on correct evaluations during the preparation of the plan.

4   Forecasting, Prevention and Emergency Planning. The Contribution of the Evaluation Discipline

The construction of risk scenarios is the result of a complex intellectual exercise aimed at the interpretation and simulation of real phenomena. This process expresses the peculiar and distinctive character of the evaluation: the *forecast* [10].

Whenever humans propose to solve problems, in reality they are trying to anticipate future events. In this sense, the evaluation is for humans the instrument with which they can fight time, or rather the uncertainty that is inherent in the unpredictable character of the future [11].

With regard to strategic emergency planning, the discipline of *economic evaluation of plans, programs and projects* can make a significant contribution through the development of two strands of scientific research, different from each other but both complementary to the pursuit of the principles of *efficiency* and *effectiveness* of the plan.

The first strand concerns the study of *evaluation indicators* necessary, in fact, for the elaboration of risk scenarios that, in order to be represented, require a very complex process of elaboration and synthesis of information (Fig. 2).

Commonly the term indicator identifies a tool able to provide information in synthetic form, through various representations (numbers, graphs, thematic maps, etc.), of a more complex phenomenon with a broader meaning, i.e., an instrument capable of making visible a trend or a phenomenon that is not immediately perceptible. This means that "an indicator has a more extensive meaning than simply measurable" [12].

The purpose of the indicators consists in "*quantifying*" (not necessarily in numerical form) the information, so that its meaning is more comprehensible and evident. It also means "*simplifying*" information relating to more complex phenomena, encouraging communication and comparison. These two concepts highlight a further fundamental characteristic of indicators that is *communicability*.

Moreover, the study of appropriate indicators of assessment, able to represent in a synthetic and intuitive very complex phenomena, such as, e.g., the effects of devastating meteorological events on a more-or-less vulnerable territory, are an indispensable ingredient of the correct and simple drafting of the Municipal Emergency Plan, a fundamental document for the protection of citizens and significant economic assets but whose drafting, unfortunately, due to the financial problems of the municipalities, is often entrusted to individual professionals who cannot possess specific skills in the

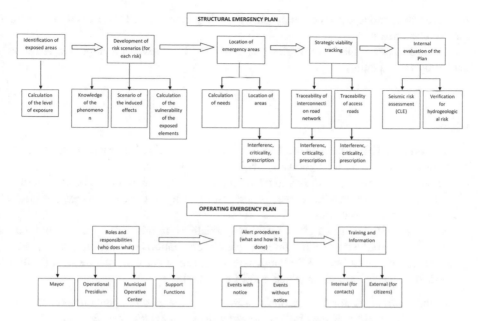

Fig. 2. Logical model of the contents of the Municipal Emergency Plan

various disciplinary sectors that are necessary to represent extremely complex phenomena.

Therefore, the indicators make it possible to rationalize the study of natural and anthropic phenomena by simulating the possible risk scenarios in the territory and, through them, to represent the level of territorial vulnerability and risk in qualitative and quantitative terms.

The estimate of vulnerability and risk is preliminary to the second contribution of the evaluation, which concerns the issue of prevention and protection of the territory.

The calamitous events, which more and more often affect the territory, testify to how it has become ever more urgent to reverse the trend in the management of the territory: if, in fact, landslides and floods are natural phenomena, the careless choices in land use, the excessive urbanization of some areas, illegal farming, deforestation of the slopes, and the alteration of the natural dynamics of the rivers amplify the risk, exposing citizens, goods and communities to serious danger.

Over the past few years, the competent authorities at national and regional level have promoted important actions to secure the territory (e.g., the Fund for the prevention of seismic risk, referred to in Article 11 of Law No. 77 of 24 June 2009 [13]; *Call for Safe Schools* of the Calabria Region of 2016 [14]). However, the selection criteria for the selection and financing of priority interventions do not always derive from a strategic vision of the whole linked to the governance of the territory.

This is particularly evident in many experiences of emergency plans, in which many measures to mitigate the vulnerability of strategic structures have been carried out, but these have proved useless because the strategic structures were located in areas not accessible during the emergency phase, or they were carried out in non-vulnerable territorial contexts.

In the literature, the level of risk (R) generated by a given phenomenon is represented by the equation:

$$R = P x V x E \qquad (1)$$

where

- the *exposure* (E), expresses the number of units (or "value") of each of the elements at risk (e.g., human lives, houses) present in a given area;
- the *vulnerability* (V) of the elements (people or things) that are exposed to certain dangers and represents the greater or lesser propensity of the elements exposed to be damaged by a given event;
- the *danger* (P) agent in a given territory (landslides, floods, earthquakes, sinkholes, liquefaction, industries with a major accident, emanations of harmful gases, coastal erosion, etc.) describes the probability that a given event will manifest itself with a certain intensity in a given place and in a given time.

If the *Danger* factor is obviously intrinsic to the natural phenomenon and is, therefore, independent of the choices of plan, the *Vulnerability* and *Exposure* are two variables on which it is possible to intervene to mitigate the Risk. Therefore, in particular, public action should concentrate its efforts on the most vulnerable areas, after reaching, however, a measured correlation with the *quantity* and *quality* of the exposed factors (human lives, material resources, landscape and environmental goods, services) present in a specific area.

Moreover, this balance should be governed according to the *Sustainability* criteria (environmental, social and economic) [15] which is to be used to optimize public spending with respect to the various sectoral policies that usually act due to parallel plans without integrated and coordinated logic [16].

It is clear, therefore, how the theme comprises a very complex field where several qualitative-quantitative variables operate simultaneously and that must necessarily be dealt with according to a multidimensional approach [17–22].

References

1. Legge 24 febbraio del 1992, n. 225 - Istituzione del Servizio nazionale di protezione civile
2. Protezione civile calabria. http://www.protezionecivilecalabria.it/index.php/it/chi-siamo/il-sistema-di-protezione-civile. Accessed 21 May 2017
3. Legge Regionale 16 aprile 2002, n. 19 - Norme per la tutela, governo ed uso del territorio—Legge Urbanistica della Calabria. (BUR n. 7 del 16 aprile 2002, supplemento straordinario n. 3)
4. Decreto del Capo Dipartimento del 2 febbraio 2015: indicazioni alle Componenti e alle Strutture operative del Servizio Nazionale per l'aggiornamento delle pianificazioni di emergenza ai fini dell'evacuazione cautelativa della popolazione della zona rossa dell'area vesuviana. 2 febbraio (2015)

5. Ioannilli, M. (a cura di): Linee Guida per la pianificazione comunale o intercomunale di emergenza di Protezione Civile, Regione Lazio-Assessorato Infrastrutture, Politiche abitative e Ambiente-Protezione Civile (2014)
6. Biancamano, P.F.: Pianificazione strategica e strutturale, integrazione e nuovi orizzonti. https://paolofrancobiancamano.wordpress.com. Accessed 21 May 2017
7. Della Spina, L., Lorè, I., Scrivo, R., Viglianisi, A.: An integrated assessment approach as a decision support system for urban planning and urban regeneration policies. Buildings **7**, 85 (2017). https://doi.org/10.3390/buildings7040085
8. Boscolo, E.: Beni comuni e consumo di suolo. Alla ricerca di una disciplina legislativa. In: Politiche urbanistiche e gestione del territorio: Tra esigenze del mercato e coesione sociale, Urbani, P. (a cura di), Giappichelli, G., Editore (2016)
9. Cappuccitti, A.: Le diverse "velocità" del Piano urbanistico comunale e il Piano strutturale. http://www.inu.it/wp-content/uploads/astengo/download/corsi/Corso_Piani_Strutturali_maggio 2008/Cappuccitti.pdf. Accessed 21 May 2017
10. Friedman, G.: The Next 100 Years: A Forecast for the 21st Century. Black Inc. (2010)
11. Roscelli, R.: Misurare nell'incertezza. Celid, Torino (2005)
12. Van der Grift, B., Van Dael J.G.F.: Un/Ece Task Force on Monitoring and Assessment (1999)
13. Protezione civile. http://www.protezionecivile.gov.it/jcms/it/piano_nazionale_art_11.wp. Accessed 21 May 2017
14. Calabriaeuropa. http://calabriaeuropa.regione.calabria.it/website/view/news/190/prevenzione-del rischio-sismico.html. Accessed 21 May 2017
15. Commissione mondiale per l'ambiente e lo sviluppo, Rapporto Brundtland. Agenda globale per il cambiamento (1987)
16. Florio, M.: La valutazione degli investimenti pubblici. FrancoAngeli, Milano (2002)
17. Fusco Girard, L. (a cura di), Nijkamp, P., Voogd, H.: Studi urbani e regionali. Conservazione e sviluppo: le valutazioni nella pianificazione fisica, Milano, Franco Angeli (1989)
18. Della Spina, L.: Integrated evaluation and multi-methodological approaches for the enhancement of the cultural landscape. In: Gervasi, O. et al. (eds.) Computational Science and Its Applications-ICCSA 2017. Lecture Notes in Computer Science, vol. 10404. Springer, Cham (2017). https://doi.org/10.1007/978-3-319-62392-4_35
19. Calabrò, F., Cassalia, G.: Territorial cohesion: evaluating the urban-rural linkage through the lens of public investments. In: Bisello, A., Vettorato, D., Laconte, P., Costa, S. (eds.): Smart and Sustainable Planning for Cities and Regions. Results of SSPCR 2017. Green Energy and Technology. Springer (2018). ISSN: 1865-3537. https://doi.org/10.1007/978-3-319-75774-2_39
20. Calabrò, F.: Local communities and management of cultural heritage of the inner areas. An application of break-even analysis. In: Gervasi O. et al. (eds.) Computational Science and Its Applications, ICCSA 2017. Lecture Notes in Computer Science, vol. 10406. Springer, Cham (2017). https://doi.org/10.1007/978-3-319-62398-6_37
21. Calabrò, F., Della Spina, L.: The public-private partnerships in buildings regeneration: a model appraisal of the benefits and for land value capture. Adv. Mater. Res. **931–932**, 555–559 (2014). https://doi.org/10.4028/www.scientific.net/AMR.931-932.555
22. Della Spina, L.: The integrated evaluation as a driving tool for cultural-heritage enhancement strategies. In: Bisello, A., Vettorato, D., Laconte, P., Costa S. (eds.): Smart and Sustainable Planning for Cities and Regions. Results of SSPCR 2017. Green Energy and Technology, Springer (2018). ISSN: 1865-3537. https://doi.org/10.1007/978-3-319-75774-2_40

Author Index

© Springer International Publishing AG, part of Springer Nature 2019
F. Calabrò et al. (Eds.): ISHT 2018, SIST 100, pp. 711–713, 2019.
https://doi.org/10.1007/978-3-319-92099-3

Printed in the United States
By Bookmasters